VIRAL INFECTIONS AND GLOBAL CHANGE

VIRAL INFECTIONS AND GLOBAL CHANGE

EDITED BY

Sunit K. Singh

WILEY Blackwell

Cover Design: Wiley
Cover Images: Vitruvian man, © iStockphoto.com/Mads Abildgaard; virus image, © iStockphoto.com/Baris Simsek; mosquito, © iStockphoto.com/Antagain; globe, © iStockphoto.com/Anton Balazh

Published by John Wiley & Sons, Inc., Hoboken, New Jersey.

Published simultaneously in Canada.

Library of Congress Cataloging-in-Publication data is available.

ISBN: 978-1-118-29787-2

Printed in Singapore.

10 9 8 7 6 5 4 3 2 1

Dedicated to my Parents

CONTENTS

FOREWORD

Tropical viruses, especially vector-borne and zoonotic viruses, contribute a huge although often hidden proportion of the world's burden of human infectious diseases. Yet despite their importance, these viruses are often severely neglected by the scientific community and global health policy makers. For instance, in the recently released Global Burden of Disease 2010 Study (Murray et al., 2012), only two vector-borne viral diseases (i.e., yellow fever and dengue) and one zoonotic viral infection (i.e., rabies) are listed by name. Together, dengue, yellow fever, and rabies cause roughly 40,000 deaths annually (Lozano et al., 2012), but the actual number of deaths and disability resulting from vector-borne and zoonotic viruses is undoubtedly much greater if we also consider other flaviviruses such as Japanese encephalitis and West Nile virus, as well as important alpha viruses such as Ross River, Chikungunya, and O'nyong-nyong virus, among others.

Dr. Sunit Singh's edited volume on these and related viral infections are a welcome addition to the biomedical literature. He and his colleagues have focused on the tropical viral infections that seldom get adequate attention relative to HIV/AIDS, hepatitis, and influenza, yet these are infections that in many respects may be almost as important. His book emphasizes unique aspects of these viral diseases and their etiologies including the important influence of climate change, as well as nonintuitive elements including socioeconomics, human behavior, travel, and animal and human migrations. The book really gets to the key elements of what helps to promote the emergence of vector-borne and zoonotic viruses and then perpetuate them in an endemic area.

While the viruses and their diseases emphasized in "Viral Infections and Global Change" are typically thought of as the major purview of low- and middle-income countries, we are seeing increasingly their emergence in wealthy countries. Important examples include Chikungunya in southern Europe, dengue in Singapore and coastal Brazil, and most recently a potentially serious dengue threat to the southern United States. Dengue type 2 has emerged in Texas, while dengue type 1 is now in Florida, setting up a possible scenario for dengue hemorrhagic fever one day appearing on the Gulf Coast of the United States (Hotez and Ryan, 2010). These viruses will receive increasing attention in the coming decade and in this sense, the current volume is ahead of the curve!

I wish to congratulate Dr. Singh and his colleagues for this important and timely volume on vector-borne and zoonotic viral infections. It will benefit both medical and graduate students interested in medical virology, as well as senior investigators working in this important but often neglected area!

REFERENCES

1. Murray CJL, Vos T, Lozano R, et al. 2012. Disability-adjusted life years (DALYs) for 291 diseases and injuries in 21 regions, 1990–2010: a systematic analysis for the Global Burden of Disease Study 2010. *Lancet* 380: 2197–2223.

2. Lozano R, Naghavi M, Foreman K, et al. 2012. Global and regional mortality from 235 causes of death for 20 age groups in 1990 and 2010: a systematic analysis for the Global Burden of Disease Study 2010. *Lancet* 380: 2095–2128.
3. Hotez PJ, Ryan ET. 2010. Gulf Coast must remain vigilant for new health threat. *Houston Chronicle*, July 17, 2010.

PETER HOTEZ, MD, PHD, FASTMH, FAAP
Dean, National School of Tropical Medicine
Professor, Pediatrics and Molecular Virology & Microbiology
Head, Section of Pediatric Tropical Medicine
Baylor College of Medicine, Houston, TX, USA
Texas Children's Hospital Endowed Chair of Tropical Pediatrics, Houston, TX, USA
Director, Sabin Vaccine Institute Texas Children's Hospital Center for Vaccine Development, Houston, TX, USA
President, Sabin Vaccine Institute, Washington, DC, USA
Baker Institute Fellow in Disease and Poverty, Rice University, Houston, TX, USA
Co-Editor-in-Chief, PLoS Neglected Tropical Diseases

PREFACE

Global changes have important health-related consequences. The emergence and reemergence of viral infections depend on the interaction between global changes and viruses and/or their reservoirs. Increased morbidity and mortality from emerging and reemerging viral outbreaks affect the growth of nations adversely in terms of economy and medical costs. Global changes may affect the health of human beings through the direct impact of changing environment, which may alter the geographic distribution and/or transmission dynamics of viruses. It is important to recognize complex factors that play a role in the distribution and transmission of viruses. Climate change as well as other factors that contribute to the emergence and reemergence of viral outbreaks such as social and demographic factors, geographical variables, global trade and transportation, land-use patterns, animal and human migration, and public health interventions affect the transmission and geographical distribution of viral infections. Population growth, changes in human behavior, livestock and agricultural farming practices, disturb the ecosystem and increase the risk of outbreak of viral infections.

It is important to study the effect of global changes on viral infections, including the ecological and environmental factors and influences of extreme weather fluctuations on viral outbreaks. The ease of international travel facilitates further the spread of viral infections globally. Systematic increases in mean temperature and precipitation, resulting in greater humidity, have facilitated the spread of many vector-borne viral diseases. Most vector-borne viral infections exhibit a distinct seasonal pattern, which suggests their relation with changes in weather conditions. Rainfall, temperature, and other climatic variables affect both the vectors and the viruses in several ways. High temperature can affect (increase or decrease) vector survival rate, depending on the vector, its behavior, ecology, and many other factors. However, viral infections that may be transmitted to humans from animals (zoonoses) continue to circulate in nature. Zoonotic viruses spread from animals to humans by broadening their host range, which increases their speed of transmission. The growing need to feed an increasing population has led to the adoption of intensive farming practices, which has led to close contact between farmworkers and animals. Animals contained in small areas/cages enable viruses to infect large numbers of animals at a single site, which might result into the generation of virulent forms through mutation or recombination processes.

In order to understand the mechanisms of the spread of viruses and strategies to deal with such outbreaks, we must understand the whole ecosystem in which diverse species such as humans, bats, and livestock coexist.

This book has been divided into two parts. Part I focuses on the general aspects of viral infections and global change. Part II deals with specific viral infections and their interrelationships with global change. This book is primarily targeted toward virologists, environmentalists, ecologists, health-care workers, clinicians, microbiologists, and students and research scholars of veterinary medicine, human medicine, or biology wishing to have an overview of the role of global changes and viral infections. I hope that this book will serve as a useful resource for all those who are interested in the field of viral infections. I am honored to have had a large panel of international experts as chapter contributors, whose detailed knowledge has greatly enriched this book.

We have so far conducted our studies related to wildlife, animal, and human viral infections separately. We should focus on a "one-health" approach to understand the complex interactions of the system (humans, wild animals, livestock production, and the environment) as a whole. The influence of global changes on the emergence and reemergence of vector- and non-vector-borne viral infections, the challenges associated with disease surveillance strategies, early detection of viral outbreaks, and international policy implications on human health are all major areas of investigation for infectious disease experts in the future.

Sunit K. Singh

CONTRIBUTORS

Assaf Anyamba Universities Space Research Association and Biospheric Science Laboratory, NASA Goddard Space Flight Center, Greenbelt, MD, USA

Steven M. Babin Johns Hopkins University Applied Physics Laboratory, Laurel, MD, USA

Luisa Barzon Department of Molecular Medicine, University of Padova, Padova, Italy

Microbiology and Virology Unit, Padova University Hospital, Padova, Italy

Peng Bi Discipline of Public Health, The University of Adelaide, Adelaide, Australia

Satesh Bidaisee Department of Public Health and Preventive Medicine, School of Medicine, St. George's University, Grenada, West Indies

Marshall E. Bloom Laboratory of Virology, Division of Intramural Research, National Institute of Allergy and Infectious Diseases, National Institutes of Health, Rocky Mountain Laboratories, Hamilton, MT, USA

Angela M. Bosco-Lauth Colorado State University, Fort Collins, CO, USA

Jérôme Bouquet UMR 1161 Virology, ANSES, Laboratoire de Santé Animale, Maisons-Alfort, France

UMR 1161 Virology, INRA, 94706 Maisons-Alfort, France

UMR 1161 Virology, Ecole Nationale Vétérinaire d'Alfort, Maisons-Alfort, France

Richard A. Bowen Colorado State University, Fort Collins, CO, USA

Aaron C. Brault Division of Vector-Borne Diseases, Centers for Disease Control and Prevention, Fort Collins, CO, USA

Anna L. Buczak Johns Hopkins University Applied Physics Laboratory, Laurel, MD, USA

Kathryn M. Carbone Division of Intramural Research, NIDCR/NIH, Bethesda, MD, USA

Carlos Castillo-Salgado Department of Epidemiology, Bloomberg School of Public Health, Johns Hopkins University, Baltimore, MD, USA

Bruno Chomel Department of Population Health and Reproduction, School of Veterinary Medicine, University of California, Davis, CA, USA

James Christopher Clegg Les Mandinaux, Le Grand Madieu, France

Richard Coker Communicable Diseases Policy Research Group (CDPRG), Department of Global Health and Development, London School of Hygiene and Tropical Medicine, London, UK

Faculty of Public Health, Mahidol University, Bangkok, Thailand

Saw Swee Hock School of Public Health, National University of Singapore, Singapore

Daniel Cowley Enteric Virus Group, Murdoch Childrens Research Institute, Royal Children's Hospital, Parkville, VIC, Australia

Jonathan F. Day Florida Medical Entomology Laboratory, Department of Entomology and Nematology, University of Florida, Vero Beach, FL, USA

Esteban Domingo Centro de Biología Molecular "Severo Ochoa" (CSIC-UAM), Consejo Superior de Investigaciones Científicas (CSIC), Campus de Cantoblanco, Madrid, Spain

Centro de Investigación Biomédica en Red de Enfermedades Hepáticas y Digestivas (CIBERehd), Barcelon, Spain

Celeste Donato Enteric Virus Group, Murdoch Childrens Research Institute, Royal Children's Hospital, Parkville, VIC, Australia

Department of Microbiology, La Trobe University, Bundoora, VIC, Australia

Agustín Estrada-Peña Department of Parasitology, Faculty of Veterinary Medicine, Miguel Servet, Zaragoza, Spain

Mari Paz Sánchez-Seco Fariñas Laboratory of Arbovirus and Imported Viral Diseases, Virology Department, National Center of Microbiology, Institute of Health "Carlos III", Ctra Pozuelo-Majadahonda, Madrid, Spain

Brian H. Feighner Johns Hopkins University Applied Physics Laboratory, Laurel, MD, USA

Kimberly Fornace Veterinary Epidemiology and Public Health Group, Royal Veterinary College, Hatfield, Hertfordshire, UK

Pieter L.A. Fraaij Department of Viroscience, Erasmus Medical Center, Rotterdam, The Netherlands

Brian D. Gushulak Migration Health Consultants, Inc., Qualicum Beach, BC, Canada

Alana Hansen Discipline of Public Health, The University of Adelaide, Adelaide, Australia

David O. Harley National Centre for Epidemiology and Population Health, The Australian National University, Canberra, ACT, Australia

Zdenek Hubálek Institute of Vertebrate Biology, v.v.i., Academy of Sciences of the Czech Republic, and Masaryk University, Faculty of Science, Department of Experimental Biology, Brno, Czech Republic

Melissa K. Jones Department of Molecular Genetics and Microbiology, University of Florida College of Medicine, Gainesville, FL, USA

Stephanie M. Karst Department of Molecular Genetics and Microbiology, University of Florida College of Medicine, Gainesville, FL, USA

Christopher Kilonzo Department of Population Health and Reproduction, School of Veterinary Medicine, University of California, Davis, CA, USA

Carl D. Kirkwood Enteric Virus Group, Murdoch Childrens Research Institute, Royal Children's Hospital, Parkville, VIC, Australia

Department of Microbiology, La Trobe University, Bundoora, VIC, Australia

Phillip T. Koshute Johns Hopkins University Applied Physics Laboratory, Laurel, MD, USA

A. **Desiree LaBeaud** Children's Hospital Oakland Research Institute, Center for Immunobiology and Vaccine Development, Oakland, CA, USA

Sheri H. Lewis Johns Hopkins University Applied Physics Laboratory, Laurel, MD, USA

Marco Liverani Department of Global Health and Development, London School of Hygiene and Tropical Medicine, London, UK

Calum N.L. Macpherson Department of Microbiology, School of Medicine, St George's University, Grenada, West Indies

Windward Islands Research and Education Foundation, Grenada, West Indies

Cheryl Cox Macpherson Bioethics Department, School of Medicine, St. George's University, Grenada, West Indies

Douglas W. MacPherson Migration Health Consultants, Inc., Qualicum Beach, BC, Canada

Verónica Martín Centro de Investigación en Sanidad Animal, Instituto Nacional de Investigación Agraria y Alimentaria, Valdeolmos, Madrid, Spain

Jonathan D. Mayer Department of Epidemiology, Geography, and Global Health, University of Washington, Washington, DC, USA

Kristin L. McNally Laboratory of Virology, Division of Intramural Research, National Institute of Allergy and Infectious Diseases, National Institutes of Health, Rocky Mountain Laboratories, Hamilton, MT, USA

Fleur M. Moesker Department of Viroscience, Erasmus Medical Center, Rotterdam, The Netherlands

Shigeru Morikawa Department of veterinary science, National Institute of Infectious Diseases, Tokyo, Japan

Stephen A. Morse Division of Foodborne, Waterborne, and Environmental Diseases, National Center for Emerging and Zoonotic Infectious Diseases, Centers for Disease Control and Prevention, Atlanta, GA, USA

Jessica Nagel U.S. Geological Survey, Patuxent Wildlife Research Center, Beltsville, MD, USA

Albert D.M.E. Osterhaus Department of Viroscience, Erasmus Medical Center, Rotterdam, The Netherlands

Monia Pacenti Microbiology and Virology Unit, Padova University Hospital, Padova, Italy

Sarah Paige University of Wisconsin, Madison, WI, USA

Giorgio Palù Department of Molecular Medicine, University of Padova, Padova, Italy

Microbiology and Virology Unit, Padova University Hospital, Padova, Italy

Nicole Pavio UMR 1161 Virology, ANSES, Laboratoire de Santé Animale, Maisons-Alfort, France

UMR 1161 Virology, INRA, Maisons-Alfort, France

UMR 1161 Virology, Ecole Nationale Vétérinaire d'Alfort, Maisons-Alfort, France

Celia Perales Centro de Biología Molecular "Severo Ochoa" (CSIC-UAM), Consejo Superior de Investigaciones Científicas (CSIC), Campus de Cantoblanco, Madrid, Spain

Centro de Investigación Biomédica en Red de Enfermedades Hepáticas y Digestivas (CIBERehd), Barcelon, Spain

Ann M. Powers Centers for Disease Control and Prevention, Fort Collins, CO, USA

Diann J. Prosser U.S. Geological Survey, Patuxent Wildlife Research Center, Beltsville, MD, USA

William K. Reisen Center for Vectorborne Diseases and Department of Pathology, Microbiology and Immunology, School of Veterinary Medicine, University of California, Davis, CA, USA

Leslie A. Reperant Department of Viroscience, Erasmus Medical Centre, Rotterdam, The Netherlands

Ivo Rudolf Institute of Vertebrate Biology, v.v.i., Academy of Sciences of the Czech Republic, and Masaryk University, Faculty of Science, Department of Experimental Biology, Brno, Czech Republic

Jonathan Rushton Veterinary Epidemiology and Public Health Group, Royal Veterinary College, Hatfield, Hertfordshire, UK

Linda J. Saif Food Animal Health Research Program, The Ohio State University, Wooster, OH, USA

Masayuki Saijo Department of Virology I, National Institute of Infectious Diseases, Tokyo, Japan

Juan Carlos Saiz Departamento de Biotecnología, INIA, Ctra. Coruña, Madrid, Spain

Noemí Sevilla Centro de Investigación en Sanidad Animal, Instituto Nacional de Investigación Agraria y Alimentaria, Valdeolmos, Madrid, Spain

Sunit K. Singh Laboratory of Neurovirology and Inflammation Biology, Centre for Cellular and Molecular Biology (CCMB), Hyderabad, India

Francisco Sobrino Centro de Biología Molecular "Severo Ochoa" (CSIC-UAM), Consejo Superior de Investigaciones Científicas (CSIC), Campus de Cantoblanco, Madrid, Spain

Laura Squarzon Department of Molecular Medicine, University of Padova, Padova, Italy

Thomas J. Stopka Department of Public Health and Community Medicine, Tufts University School of Medicine, Boston, MA, USA

Laura J. Sutherland Case Western Reserve University, Cleveland, Ohio, USA

Walter J. Tabachnick Florida Medical Entomology Laboratory, Department of Entomology and Nematology, University of Florida, Vero Beach, FL, USA

John Y. Takekawa U.S. Geological Survey, Western Ecological Research Center, Vallejo, CA, USA

Juan Carlos de la Torre IMM-6 The Scripps Research Institute, La Jolla, CA, USA

Ana Vazquez Laboratory of Arbovirus and Imported Viral Diseases, Virology Department, National Center of Microbiology, Institute of Health "Carlos III", Ctra Pozuelo-Majadahonda, Madrid, Spain

Anastasia N. Vlasova Food Animal Health Research Program, The Ohio State University, Wooster, OH, USA

William E. Walton Department of Entomology and the Center for Disease Vector Research, University of California, Riverside, CA, USA

Angela Weber Division of Foodborne, Waterborne, and Environmental Diseases, National Center for Emerging and Zoonotic Infectious Diseases, Centers for Disease Control and Prevention, Atlanta, GA, USA

Craig R. Williams Sansom Institute for Health Research, University of South Australia, Adelaide, SA, Australia

National Centre for Epidemiology and Population Health, The Australian National University, Canberra, ACT, Australia

Tomoki Yoshikawa Department of Virology I, National Institute of Infectious Diseases, Tokyo, Japan

Ying Zhang Discipline of Public Health, The University of Adelaide, Adelaide, Australia

School of Public Health, China Studies Centre, University of Sydney, Sydney, Australia

Shu Zhu Department of Molecular Genetics and Microbiology, University of Florida College of Medicine, Gainesville, FL, USA

ABOUT THE EDITOR

Dr. Sunit K. Singh completed his bachelor's degree program from GB Pant University of Agriculture and Technology, Pantnagar, India, and his master's degree program from the CIFE, Mumbai, India. He then joined the Department of Pediatric Rheumatology, Immunology, and Infectious Diseases, Children's Hospital, University of Wuerzburg, Wuerzburg, Germany, as a biologist. He completed his PhD degree from the University of Wuerzburg in the area of molecular infection biology.

Dr. Singh completed his postdoctoral training in the Department of Internal Medicine, Yale University, School of Medicine, New Haven, CT, United States, and the Department of Neurology, University of California Davis Medical Center, Sacramento, CA, United States, in the areas of vector-borne infectious diseases and neuroinflammation, respectively. He also worked as a visiting scientist in the Department of Pathology, Albert Einstein College of Medicine, New York, United States, the Department of Microbiology, College of Veterinary Medicine, Chonbuk National University, Republic of Korea, the Department of Arbovirology, Institute of Parasitology, Ceske Budejovice, Czech Republic, and the Department of Genetics and Laboratory Medicine, University of Geneva, Switzerland. Presently, he serves as a scientist and leads a research group in the area of neurovirology and inflammation biology at the prestigious Centre for Cellular and Molecular Biology, Hyderabad, India. His main areas of research interest are neurovirology and immunology. Dr. Singh has several awards to his credit, including the Skinner Memorial Award, Travel Grant Award, NIH-Fogarty Fellowship, and Young Scientist Award. He is also associated with several international journals of repute as associate editor and editorial board member.

I

GENERAL ASPECTS

1

CLIMATE CHANGE AND VECTOR-BORNE VIRAL DISEASES

Ying Zhang

Discipline of Public Health, The University of Adelaide, Adelaide, Australia
School of Public Health, China Studies Centre, University of Sydney, Sydney, Australia

Alana Hansen

Discipline of Public Health, The University of Adelaide, Adelaide, Australia

Peng Bi

Discipline of Public Health, The University of Adelaide, Adelaide, Australia

TABLE OF CONTENTS

1.1 Introduction 4
1.2 Epidemiology of VVD 4
1.2.1 What are VVD? 4
1.2.2 Temporal–spatial distribution of VVD around the world 5
1.2.3 Factors that affect the transmission 5
1.3 Association between climatic variables and emerging VVD 6
1.3.1 Dengue fever 6
1.3.2 Yellow fever 9
1.3.3 Viral encephalitis: Japanese encephalitis, Murray Valley encephalitis, and West Nile encephalitis 9
1.3.4 Ross River fever and Barmah Forest fever 10
1.3.5 Chikungunya fever 11

Viral Infections and Global Change, First Edition. Edited by Sunit K. Singh.

1.1 INTRODUCTION

There is an increasing threat of infectious disease due to globalization and climate change. It is now widely accepted in the scientific community that the Earth's climate system is changing, which has brought great public health challenges around the world. The impact of climate change on the transmission of infectious diseases, particularly on vector-borne diseases (VVD), has been examined in both developing and developed countries (Lafferty, 2009; Weaver and Reisen, 2010). In particular, emerging and reemerging VVD are expected to increase due to climate change and variability (Zell et al., 2008). For the first time, in the last decades, several VVD, such as West Nile virus infection, chikungunya virus infection, and viral hemorrhagic fevers, have been spreading geographically and recorded in areas outside their original ranges. However, the lack of knowledge and effective, safe vaccines and diagnosis for some VVD makes it challenging to prevent and reduce the burden of disease associated with the changing environment.

This chapter aims to present a synthetic view of the health impact of climate change and variability on VVD in order to assist evidence-based decision and policy making for disease prevention and control. We have divided this chapter into three parts. The first part will discuss the epidemiological background of VVD, including the scope; the temporal–spatial distribution around the world; a summary of the factors that affect their transmission, including the causal relationship between climatic factors and VVD; and its prevention and control. The second part will focus on the association between climatic variables and some emerging VVD, such as dengue fever (DF), Rift Valley fever (RVF), viral encephalitis, yellow fever (YF), and others. Implications and recommendations are provided in the third part of this chapter to inform decision and policy making for a range of stakeholders, for example, public health practitioners, doctors, and other health service providers at medical clinics, governments, researcher centers, and local communities.

1.2 EPIDEMIOLOGY OF VVD

1.2.1 What are VVD?

Vector-borne viral diseases are virus infections transmitted via vectors. To clarify the scope of the discussion in this chapter, we adopted epidemiological and biological definitions of *vector* and *virus* to select eligible VVD for discussion. Vectors are organisms that transmit infections from one host to another (Last, 2001). Vectors of human disease are typically arthropods (e.g., species of mosquitoes and ticks) that are able to transmit pathogens. A virus is a small infectious agent, consisting of nucleic acid in a protein coat, that can replicate only inside the living cells of organisms with a wide diversity of shapes and sizes (Koonin et al., 2006). There are more than 500 "vector-borne viruses" and about 100 of them are of veterinary and/or human importance, which can cause major epidemics (Moormann, 2012). Some selected emerging VVD have been listed in Table 1.1. Some other important infectious diseases that are sensitive to climate variations, for example, malaria (not viral) and hemorrhagic fever with renal syndrome (not vector borne), are not included in our discussion.

TABLE 1.1 Summary of Selected Emerging VVD and Their Vectors

Disease	Type of Virus	Vector	Main Reservoir
DF	*Flaviviridae*	*Aedes* mosquitoes	None (humans are the only hosts)
YF	*Flaviviridae*	*Ae. aegypti* and others, *Haemagogus*	Monkeys
JE/West Nile fever	*Flaviviridae*	*Aedes* sp., *Culex* sp.	Birds, pigs
Chikungunya fever	*Togaviridae*	*Aedes* mosquitoes	Monkeys
Ross River fever	*Togaviridae*	*Aedes* and other mosquitoes	Kangaroos, wallabies
RVF	*Bunyaviridae*	*Aedes* sp., *Culex* sp.	Sheep, cattle
OHF	*Flaviviridae*	Ticks (*Dermacentor*)	Field mouse
CCHF	*Bunyaviridae*	Ticks (*Hyalomma*)	Birds, crows, cows, ostriches

Source: Modified from WHO (2012a, b).

1.2.2 Temporal–spatial distribution of VVD around the world

Most VVD are restricted to the tropics and are often seen in temperate regions only as imported diseases, because of the required living environment for certain arthropod vectors like the *Anopheles* or *Aedes* mosquitoes. The majority of the mortality and morbidity burden of VVD occurs in Africa, South America, South Asia, and the Pacific Islands (WHO, 2004). However, geographical expanding of VVD has been reported around the world recently. This includes the emergence of West Nile virus in the Americas and Japanese encephalitis (JE) in Australasia, the spread of dengue, and the reemergence of YF virus in South America (Mackenzie and Williams, 2009; Mackenzie et al., 2004).

Due to the development of effective public health preventions and control measures targeted for VVD during the last century, many VVD, particularly mosquito-borne diseases, were controlled in many areas. However, over the last 20 years, some VVD, such as DF and West Nile virus infections, have reemerged in some areas, for example, Asia and the Americas (DeCarlo et al., 2011; Phillips, 2008; Rezza, 2012). Potential invasion of non-zoonotic VVD (only affect animals, not human beings) is of concern. For example, the world has recently witnessed the emergence and spread of a tick-borne VVD, that is, the outbreak of bluetongue, which currently affects sheep, goats, and cattle (Institute of Medicine of the National Academies, 2008).

1.2.3 Factors that affect the transmission

The epidemiology of VVD is influenced by the probability of contact between the vectors, the human population, and, for many viruses, the amplifying hosts, whether birds (most arboviral encephalitis), monkeys (YF virus), or rodents (hemorrhagic fever), which serve as reservoirs for the viruses. Like other infectious diseases, the transmission of VVD is influenced by social, economic, and environmental factors (Figure 1.1).

It is well established that climate is an important determinant of the spatial and temporal distribution of vectors and viruses (Bezirtzoglou et al., 2011; Slenning, 2010). The interplay of climate, vector, and host significantly influences the transmission of VVD (Sellers, 1980). Climate conditions affect the transmission of VVD mainly in three ways: altering the distribution of vector species and their reproductive cycles; influencing the reproduction of the virus within the vector organism, known as the external incubation period (EIP); and affecting human behaviors and activity that may increase the chance of contact with infected vectors (Zhang et al., 2008).

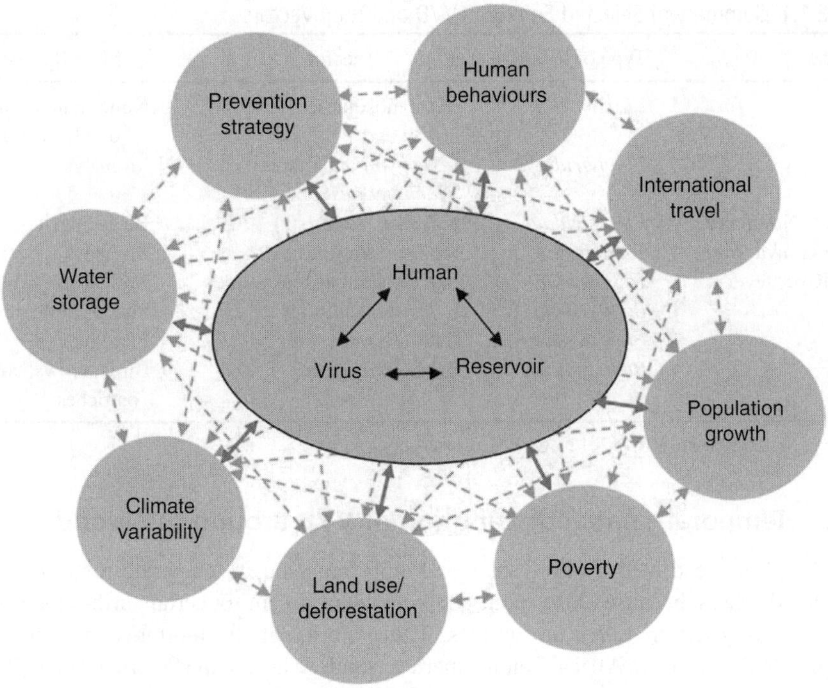

Figure 1.1. The VVD episystem showing interactions with influencing factors. For color detail, please see color plate section.

1.3 ASSOCIATION BETWEEN CLIMATIC VARIABLES AND EMERGING VVD

Some emerging and reemerging VVD have been selected for discussion in this chapter with a focus on their epidemiology and the association with climatic variability and climate change. These are DF, YF, several types of viral encephalitis, Ross River fever, Barmah Forest virus (BFV) disease, chikungunya fever, RVF, Omsk hemorrhagic fever (OHF), and Crimean–Congo hemorrhagic fever (CCHF). Findings from both historical data analyses and projective modelings indicate an increasing number of cases and expanding epidemic areas with projected climate change scenarios.

1.3.1 Dengue fever

Dengue is the most common arboviral infection in the world (Rezza, 2012). The disease, caused by the four dengue virus serotypes, ranges from asymptomatic infection, undifferentiated fever, and DF to severe dengue hemorrhagic fever (DHF) with or without shock. Symptoms may include fever, chills, and joint pain. It can be diagnosed by laboratory testing for virus isolation, viral antigen detection, or specific antibodies (serology). During the last 25 years, there have been increasing reports of dengue infection with unusual manifestations (Pancharoen et al., 2002). Great efforts are being made to understand the pathogenesis of this disease in order to develop a safe and effective dengue vaccine.

Dengue is transmitted by several species of mosquitoes within the genus *Aedes*, principally *Aedes aegypti*. *Ae. aegypti* has adapted well to urban environmental conditions

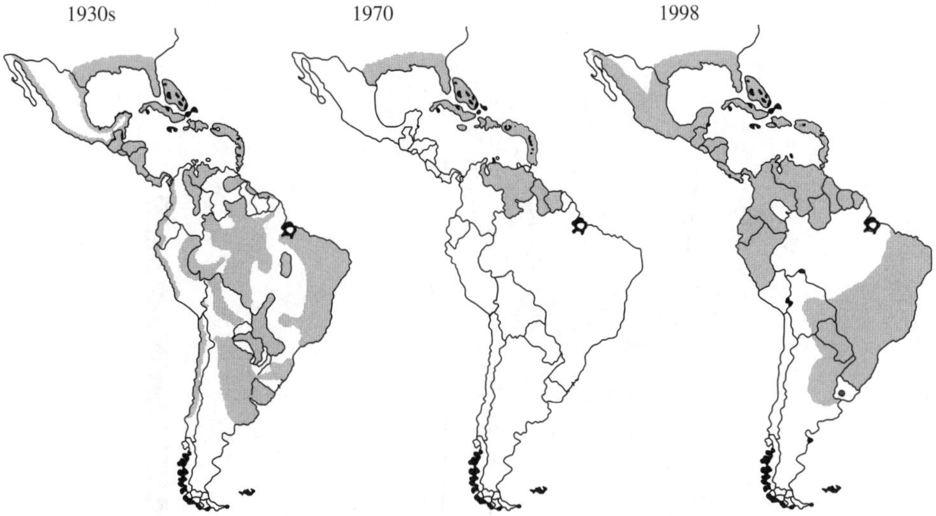

Figure 1.2. Reemergence of *Ae. aegypti* and dengue in the Americas. For color detail, please see color plate section. Source: Image from CDC.

such as poor housing, overcrowding, and inadequate sanitation, indicating the persistence of this species in regions with lower socioeconomic status, due to the close association between *Ae. aegypti*, humans, and the environment (Jansen and Beebe, 2010). The resilient vector may be a reason for the observed reemerging of dengue around the world (Bangs et al., 2007; Phillips, 2008; Rezza, 2012). Figure 1.2 shows the reemerging of dengue in the Americas following a successful hemispheric eradication campaign during the 1950s and 1960s (CDC).

The association between climatic variables and DF has been documented worldwide, indicating a positive relationship between notified cases and increasing temperature (Banu et al., 2011; Johansson et al., 2009; Patz et al., 1998; Russell et al., 2009; Vezzani and Carbajo, 2008). Climate change, in particular a warming climate, along with globalization and international traveling, may broaden the transmission range for *Ae. aegypti*. Accordingly, a slight increase in temperatures could result in epidemics of dengue in the world. But the vector population may develop independently from rainfall (Pontes et al., 2000), which could be due to the characteristics of the vector, *Ae. aegypti*, in the urban environment. Using logistic regression analysis, Hales et al. (2002) found that the annual vapor pressure (humidity) was the most important indicator of DF outbreak globally. This study was the only one to point out the very important effect of vapor pressure on dengue transmission, which indicated that the incidence of DF for the people living in humid areas could be 30% higher than people living in areas with less humidity (odds ratio 1.3). Recently, a series of papers studying the association between climatic factors and dengue have been published, which suggest a nonstationary influence (not with a single trend or a stable pattern) of climatic situation on dengue epidemics in Thailand (Nagao et al., 2012).

Projective modelings also provide evidence of potential expanding of DF to nonendemic areas. Globally, it is suggested that climate change could increase the number of people living in areas of higher dengue risk, from 1.5 billion in 1990 to about 50–60% of the global population by 2085 (Hales et al., 2002). It is estimated that there might be a dengue threat for southern parts of Australia where there have been no previous outbreaks, which may lead to 1.6 million people living in northern Australia at risk of dengue infection by 2050, due to the southwest expanding of suitable conditions for the transmission (Russell et al., 2005) (Figure 1.3).

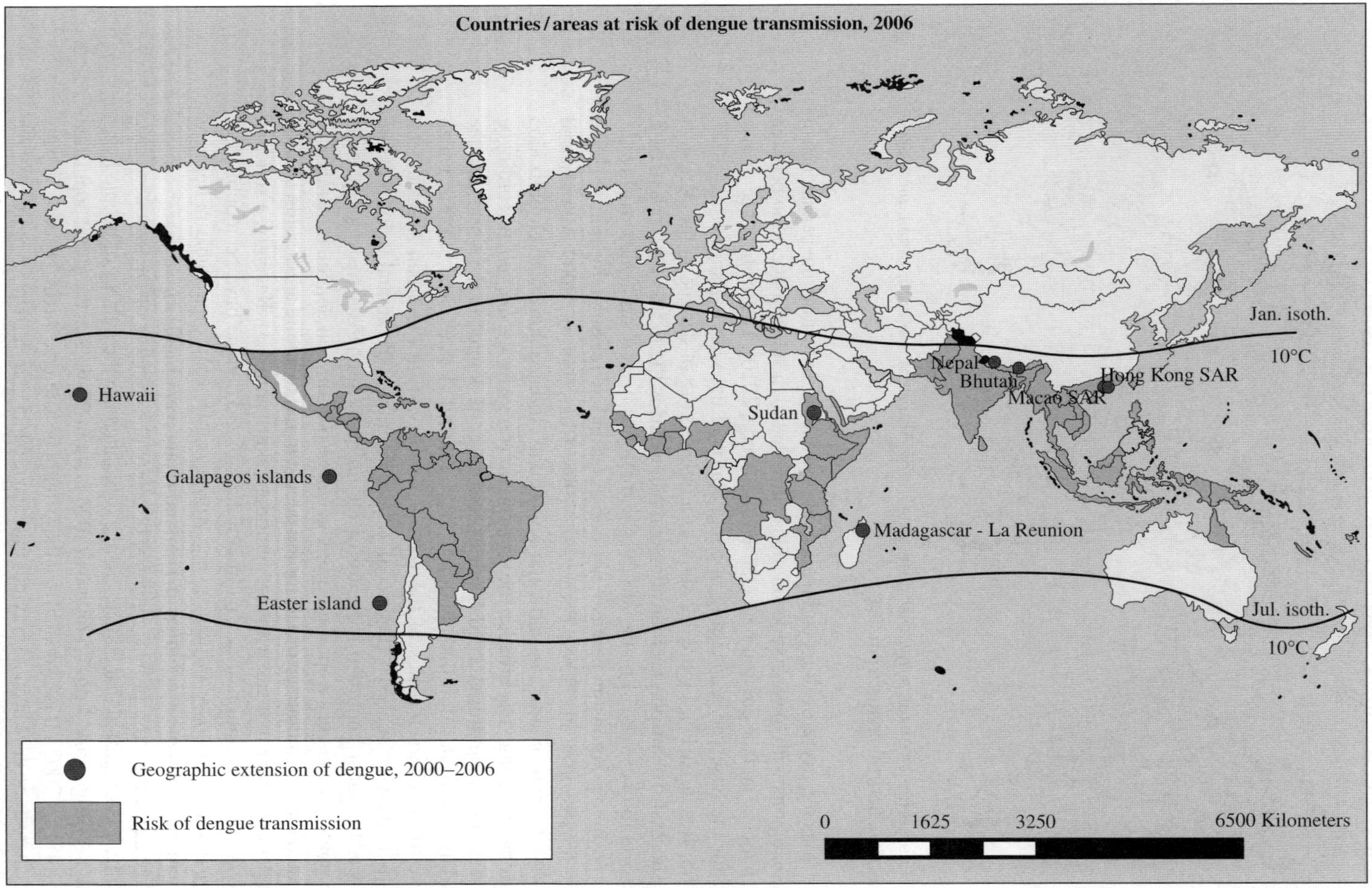

Figure 1.3. Areas at risk of dengue transmission. Source: Image from WHO. For color detail, please see color plate section.

1.3.2 Yellow fever

Yellow fever is a viral disease transmitted by infected mosquitoes from, principally, *Aedes* and *Haemagogus* mosquito species. Yellow fever varies in severity with some mild symptoms as fever, headache, chills, and back pain. Severe patients may have yellow coloring of the skin and kidney and liver function failure (Bell, 2007). Although safe and effective vaccination is available, there are an estimated 200 000 cases of YF, causing 30 000 deaths worldwide each year. When epidemics occur in unvaccinated populations, case-fatality rates may exceed 50% (WHO, 2011). The true number of cases is estimated to be 10–250 times what is now being reported due to underreporting.

Yellow fever is endemic in tropical and subtropical regions in African and Central/South American countries. The World Health Organization (WHO) has highlighted a greater risk of international spread of YF than in previous years (WHO, 2011). Historically, YF was once common in the United States but is no longer present mainly due to quarantine and mosquito control and improvements in living standards. However, the virus has the potential to spread to Asia and the Pacific, and the United States and Europe could expect a dramatic increase in imported YF cases for the same reasons as those applying to DF (Gubler et al., 2001). The tremendous growth in international travel as well as increasing temperature and rainfall will increase the risk of importation of YF in the United States (Monath and Cetron, 2002). The invasive YF mosquito species in Europe raised the concern of the potential risk of YF (Scholte et al., 2010). Although the disease has never been reported in Asia, the region is at risk due to the presence of the conditions required for transmission.

The number of YF cases has increased over the past two decades due to declining population immunity to infection, deforestation, urbanization, population movements, and climate change. Climate change will likely further exacerbate inequality in global health due to the potential to exacerbate endemic YF in developing countries, for example, India, where there is poor environmental sanitation, malnutrition, and a shortage of drinking water (Bush et al., 2011). Increasing temperature and rainfall were associated with the outbreak of YF in Brazil in 2000 (Vasconcelos et al., 2001). Concern regarding the reemergence of YF has risen in Brazil because of the large susceptible human population; high prevalence of vectors and primary hosts (monkeys); favorable climate conditions, especially increased rainfall; emergence of a new genetic lineage; and circulation of people and/or monkeys infected with the virus (Vasconcelos, 2010).

1.3.3 Viral encephalitis: Japanese encephalitis, Murray Valley encephalitis, and West Nile encephalitis

Like dengue virus, JE virus (endemic throughout Asia and the Pacific), Murray Valley encephalitis virus (mainly in northern Australia), and West Nile virus (spread in Africa, Europe, the Middle East, west and central Asia, Oceania, and, most recently, North America) are of the family *Flaviviridae* (Endy and Nisalak, 2002). Encephalitis (inflammation of the brain) is the most severe neurological symptom of these virus infections, although many may be asymptomatic. The viral encephalitis diseases are transmitted by mosquitoes, but the infected mosquito species vary according to geographical area, for examples, *Culex pipiens* in East United States and *Culex tritaeniorhynchus* in Asia. Birds are the main reservoir of these viruses. In recent years, these diseases have extended beyond their traditionally recognized boundaries. For example, outbreaks of JE occurred in the Torres Strait and north Queensland (Australia) in 1995 and 1998, and West

Nile virus epidemics recently occurred in America and Europe (Monaco et al., 2011; Reisen et al., 2008; van den Hurk et al., 2010).

The threats posed by the importation of exotic arboviruses, the introduction of exotic mosquitoes and reservoirs, and the potential geographic expansion of key local vectors are of great concern to public health. Climate change and anthropogenic influences provide additional uncertainty regarding the future health burden of viral encephalitis. Studies on JE indicate that increasing rainfall and temperature may bring more cases and high humidity may reduce the health burden in India (Murty et al., 2010). In China, it is suggested that the transmission of JE in both rural and urban areas may be affected by temperatures, rainfall, humidity, and air pressure with a threshold temperature of 21.0–25.2 °C and a lagged effect of 1–2 months (Bi et al., 2003, 2007). Incorporating variables for mosquito density, seasonal factors, and density of pigs (which are reservoir for the virus) can be of assistance in forecasting JE epidemics in Taiwan (Hsu et al., 2008).

There are very limited studies on the impact of climate variation on Murray Valley encephalitis virus in Australia, necessitating the need for analysis to prevent future potential risks relevant to climate change. Retrospective analysis of the epidemics of Murray Valley encephalitis in southern Australia during 1951 and 1974 indicated that mosquito longevity, extrinsic incubation period, and duration of the feeding cycle were the most important variables predisposing rapid amplification (Kay et al., 1987). Evidence also suggests that an early warning system for Murray Valley encephalitis could be effective when considering patterns of the Southern Oscillation, a synthetic indicator of climate change (Nicholls, 1986).

The relationship between climate variables and West Nile infection is not well understood. However, recent outbreaks with severe cases of West Nile encephalitis in America and European countries have led to more studies in this field. In New York, specific favorable weather conditions, for example, wet winter, warm and wet spring conditions, and dry summer, are associated with the increased local prevalence of West Nile virus among *Culex* mosquitoes (Shaman et al., 2011). Studies in other U.S. states have suggested that temperature and low precipitation alone are strong predictors of *Culex* vector population growth and more effective early warning systems can be achieved by including climate variables (Deichmeister and Telang, 2011; Liu et al., 2009; Ruiz et al., 2010). In 2009, an expansion of West Nile virus into the Canadian province of British Columbia was detected, and analysis shows that the establishment and amplification of West Nile virus in this region was likely facilitated by above-average nightly temperatures and a rapid accumulation of hot days in late summer (Roth et al., 2010). Expert opinions are consistent that climate change is predicted to increase the risk of incursion of vectors for West Nile virus infections in Canada (Gale et al., 2010). Analysis of the reemerging of West Nile infection in Russia in 2007 revealed the climate conditions favorable for the epidemic in mild winters and hot summers (Platonov et al., 2008). In addition, great concern of the potential risk of West Nile encephalitis for other countries has risen due to the changing climate and environment that are conducive to the vectors and reservoirs (van den Hurk et al., 2010).

1.3.4 Ross River fever and Barmah Forest fever

Ross River fever, caused by infection with Ross River virus (RRV), is the most common mosquito-borne disease in Australia. There have been more than 41 000 notified cases during the last decade with the most serious situation occurring in 1996. The RRV causes a flu-like illness with joint pains, rash, and fever in approximately 30% of infected people 3–11 days after being infected. The vertebrate reservoir hosts of RRV could include marsupials, placental mammals and birds, kangaroos, horses, and rats. There are over 40 species of mosquito vectors, with *Aedes vigilax*, *Aedes camptorhynchus* (saltmarsh along coastline), and *Culex annulirostris* (inland) being the most important (McMichael et al., 2003). Peak incidence of the disease is through the summer and autumn months, when the mosquito vectors are most abundant. Most affected people are middle aged and there seems to be no gender difference.

Barmah Forest virus and RRV demonstrate many similarities in disease symptoms and seasonal distribution (Flexman et al., 1998). They are both characterized by arthralgia, arthritis, and myalgia, often accompanied by fever and rash. Arthritis is more common and more prominent in RRV infection, and rash is more common and florid with BFV infection. These symptoms may continue for at least 6 months in up to 50% of patients with RRV, but in only about 10% of patients with BFV. Both diseases can be confirmed by serological tests. However, BFV infection has not yet been as intensively studied as RRV (Jacups et al., 2008a, b).

Studies suggest that climate variables, for example, temperature, rainfall, level of river flow, and sea level, are related to the transmission of Ross River fever, but associations vary in different regions due to different vector species and ecological situations (Bi et al., 2009; Tong et al., 2002, 2005, 2008). It has been suggested that climate variability might be a contributor to the spatial change of the disease in Queensland, Australia, over the period from 1985 to 1996 (Tong et al., 2001). The response of RRV to climate variability between coastline and inland regions is also different (Tong and Hu, 2002). A recent study in Queensland confirmed the various associations, depending on different environmental conditions (Gatton et al., 2005). Kelly-Hope et al. found that the environmental risks of RRV outbreaks varied among different regions throughout Australia and that the Southern Oscillation Index (SOI) could be a predictor only for the southeast temperate region (Kelly-Hope et al., 2004). Projections of RRV epidemics from regional weather data were conducted in different areas in Australia, demonstrating high accuracy of early warning models that combined data on local patterns of climate change and mosquitoes (Woodruff et al., 2002). In Western Australia, predicted climatic changes, especially rising sea level and greater rainfall and flooding, might significantly increase RRV activity (Lindsay et al., 1996). In Darwin, a model that included rainfall, minimum temperature, and three mosquito species was proven to have the best accuracy of disease prediction that could explain 63.5% model fit (Jacups et al., 2008a, b).

Projected various ecological scenarios could result in the transmission of RRV infection being different across Australia. Recent outbreaks of RRV in the state of Victoria suggest an increasing risk in southern Australia due to increasing global temperatures. Therefore, it would be useful to generate a complete map of the vulnerability of RRV infection by systematic ecological studies across the whole of Australia, noting the relationship between environmental factors and RRV, which will lead to a better understanding of RRV transmission and control strategies.

1.3.5 Chikungunya fever

Similar to RRV and BFV, chikungunya virus is an alphavirus, first identified in an outbreak in southern Tanzania in 1952. Characterized by sudden onset of fever and severe joint pain, symptoms are very similar to those of dengue but, unlike dengue, there is no hemorrhagic or shock syndrome form (Kamath et al., 2006). There is currently no specific treatment and no vaccine available for this disease. Both *Ae. aegypti* and *Aedes albopictus* mosquitoes have been implicated in large outbreaks of chikungunya. Whereas *Ae. aegypti* is confined within the tropics and subtropics, *Ae. albopictus* spreads in temperature areas. *Ae. albopictus* is generally considered to have a low vectorial capacity because of its lack of host specificity. Nevertheless, *Ae. albopictus* is abundant in rural and urban areas due to the diversity of the habitats (Delatte et al., 2008).

Chikungunya fever cases once occurred only in Africa, Asia, and the Indian subcontinent. However, there has recently been emergence of chikungunya globally, with thousands of people affected in Singapore, Malaysia, Thailand, and the Republic of the Congo (Kelvin, 2011; Pulmanausahakul et al., 2011). The major outbreak in 2005 in India resulted in a large number of imported cases in 2006 in Europe, and local transmission was first identified in Italy in 2007 (WHO, 2008).

In addition to the result of viral genetic mutations leading to the adaptation of new vector hosts and insecticide resistance, the rapid global spread may also be attributed to climate change. The relationship between climate change and reemerging of chikungunya has been reviewed in India, indicating potential risk of increased disease burden associated with changed patterns of temperature and rainfall (Dhiman et al., 2010). Using geographic information system (GIS) techniques, the investigation of the 2008/2009 incidence of chikungunya suggested that the direction of the outbreaks moved from south to north with a median speed of 7.5 km per week and the number of cases increased after 6 weeks of increasing cumulative rainfall with variation of average daily temperatures (23.7–30.7 °C) (Ditsuwan et al., 2011). Climate change that shortens the extrinsic incubation period was also considered as one of the drivers for an outbreak in Italy in 2007, the first large outbreak documented in a temperate climate country (Bezirtzoglou et al., 2011; Poletti et al., 2011). Hence, there is an increasing risk of tropical VVD in temperate climate countries, as a consequence of globalization and a changing climate.

1.3.6 Rift Valley fever

Rift Valley fever is caused by Rift Valley virus affecting humans and a wide range of animals. Infected patients usually suffer from only a mild illness with fever, headache, myalgia, and liver abnormalities. Outbreaks of RVF have occurred across Africa. Since the outbreak in Kenya/Tanzania in 1997–1998 when the Rift Valley virus claimed over 400 deaths, outbreaks in 2000 in Yemen and Saudi Arabia, 2007 in Kenya/Tanzania/Somalia, and 2008 in Sudan/Madagascar and South Africa have caused nearly 1000 deaths (WHO, 2010). The outbreak in 2000 marked the first reported occurrence of the disease outside the African continent and raised concerns that it could extend to other parts of Asia and Europe (Gale et al., 2010). With *Aedes* and *Culex* mosquitoes as major vectors, RVF virus has the potential to infect a wide range of vectors, for example, ticks and flies, unlike the majority of arboviruses that tend to be adapted to a narrow range of vectors (Pepin et al., 2010). Different species of vectors can play different roles in sustaining the transmission of the virus.

Although research on the association between climate change and RVF is limited, it is likely that the geography of RVF and the vectors will be altered by climate change and extreme weather events that will create the necessary conditions for RVF to expand its geographical range northward and cross the Mediterranean and Arabian seas (Martin et al., 2008). Consistent findings indicate that increasing rainfall, especially heavy rainfall events, is closely related to the outbreaks of RVF in Africa, Saudi Arabia, and Yemen (Andriamandimby et al., 2010; Hightower et al., 2012). A study in Senegal using remote-sensing techniques that simply relied upon rainfall distribution provided a new approach to enhance early warning systems for RVF based on both natural and anthropogenic climatic and environmental changes (Tourre et al., 2009). These findings will assist the development of forecasting models and early warning systems for RVF using climate forecasting data so as to better predict and respond to RVF outbreaks in Africa and other regions.

1.3.7 Omsk hemorrhagic fever and Crimean–Congo hemorrhagic fever

Omsk hemorrhagic fever (OHF) and Crimean–Congo hemorrhagic fever (CCHF) are tick-borne viral diseases. Omsk hemorrhagic fever virus is a member of the family *Bunyaviridae*, while CCHF virus is of the family *Flaviviridae* (Bajpai and Nadkar, 2011; Ruzek et al., 2010). They have symptoms of viral hemorrhagic fever but CCHF has a shorter incubation period (1–3 days) after tick bite than OHF (3–7 days). Specific symptoms for viral hemorrhagic fever vary, but initial signs often include marked fever, fatigue, dizziness, muscle aches, loss of strength, and exhaustion. Severe cases often show symptoms of bleeding under the skin, in internal organs, or from body orifices, for example, the mouth, eyes, or ears.

Omsk hemorrhagic fever has only been notified in Russia so far. Since the first description of OHF in the 1940s in Siberia, the clinical course, pathology, and epidemiology of the disease, as well as the ecology of the virus, vectors, and natural hosts (rodents), have been studied extensively, although English studies are scarce (Ruzek et al., 2010). Given the lack of a specific treatment or vaccine against OHF virus, elimination of wild rodents is a basic approach to reduce the disease burden. The morbidity from OHF has two seasonal peaks (autumn and winter, being the hunting seasons of muskrats) that correlate with activity of muskrats. There were outbreaks of OHF in the early 1990s after remarkably decreased incidence in 1970s and severe and fatal cases in 1998. As a family of tick-borne flaviviral disease, the transmission of OHF virus, like other vector-borne agents, is affected by a range of factors, including changes in climate and ecology. Climate conditions that favor the activities of muskrats and OHF ticks have not been studied. The reemergence of OHF in Russia is, nevertheless, an example of a human disease that emerged owing to human-mediated disturbance of an ecological function. Therefore, further research is necessary for a better understanding of the climatic factors that may contribute to the reemergence of OHF in Russia and other regions.

Crimean–Congo hemorrhagic fever is a severe disease in humans, with a high mortality rate. Treatment can be with antiviral agents such as ribavirin. However, review of ribavirin for patients with CCHF found that data are inadequate to support its efficacy in CCHF (Ascioglu et al., 2011). Crimean–Congo hemorrhagic fever is now endemic in many countries in Africa, Asia, and Europe and continues to emerge. The geographical distribution of the CCHF virus is widespread around the world (Leblebicioglu, 2010). Since the first disease was described in the Crimea in 1944 and later in 1956 in the Congo, the disease incidence has dropped significantly due to increasing living standards. However, reports of sporadic cases and outbreaks have recently increased significantly since 2000. In Turkey, the annual number of reported CCHF cases increased dramatically from 17 in 2002 to 1315 cases in 2008 and 1318 in 2009 (Yilmaz et al., 2009). In 2010/2011, outbreaks with severe and fatal cases were reported in Pakistan and India.

There is a strong need to better understand the underlying reasons of the reemergence of CCHF in different regions. Potential reasons for the emergence or reemergence of CCHF include climate changes, which may have a significant impact on the reproduction rate of the vector (*Hyalomma* ticks), as well as anthropogenic factors (Maltezou and Papa, 2010). A possible picture of the interaction among the potential factors for the emergence of

CCHF may include climate and other environmental change, migratory birds, increase in tick density, livestock movement, and increase in host animals (Leblebicioglu, 2010). Nevertheless, the association between climate change and CCHF, the virus, and its reservoir is not clear at all. Tick-borne disease has already been labeled as a "clear and present danger in Europe," and the probability of CCHF extending to other countries around the Mediterranean basin suggests that veterinarian and human surveillance should be enhanced (Chinikar et al., 2010; Heyman et al., 2010).

1.4 INVASION OF NONZOONOTIC VVD TO HUMANS

There are some nonzoonotic VVD, such as bluetongue, currently only affecting animals (e.g., sheep, goats, cattle) but not human beings. Recent emergence of bluetongue, an arthropod-borne viral disease of both domestic and wild ruminants in Europe, has raised concerns for the increasing risk to livestock and potential threats to humans. Bluetongue is widely distributed in Australia, the United States, Africa, the Middle East, Asia, and Europe, and the virulence of different strains of the virus varies considerably across countries (Maclachlan, 2011). However, disease invasion, that is, the emergence of a pathogen into a new host species, may happen in the future because of increased contact between humans and wildlife and potential unexpected impacts of a changing environment/climate on genetic mutations of the serotypes of these viruses (Daszak et al., 2001). Reasons for the spread are complex and the role changes in climate patterns and landscapes play is unclear, although increasing temperature and other meteorological variables may extend existing spatial and vector transmission models (Maclachlan, 2010). Surveillance for the vectors and vulnerable animals should be strengthened to ensure closer monitoring and better prediction of these diseases. This may avoid any unexpected transmission to humans within a changing world (Tabachnick, 2010).

1.5 IMPLICATIONS AND RECOMMENDATIONS FOR PREVENTION AND CONTROL

Despite a century of success in prevention and control efforts, infectious diseases remain a major challenge to public health around the world. Effective detection and control of VVD requires an integrated system that can provide effective early warning of an epidemic activity (Britch et al., 2008). It should also be built on a comprehensive surveillance system including diagnosis for VVD. In recent years, landscape epidemiology has used satellite remote sensing and GIS to develop predictive modelings that allow for incorporation with spatial and temporal climatic patterns that may influence the intensity of a vector-borne disease and predict risk conditions associated with future potential epidemics (Pinzon et al., 2005).

Some efforts to control VVD focus on one aspect of the episystem, for example, vaccination for certain viruses. Fortunately, effective and safe vaccination is already available or will hopefully be available in the next 5 years for some important VVD, for example, dengue and JE, tick-borne encephalitis, and YF (WHO, 2012a, b). Regardless of this, there has still been an observed increase or expansion of some of those diseases, for example, JE in Australasia. The vertebrate host and/or reservoir may also be the target for control measures. Reduction of vectors, such as control for mosquitoes and flies, has been proven to be a cost-effective public health method to reduce the risk of vector-borne disease in some areas (Tomerini et al., 2011). By recognizing the complex role of vector ecology that affects the transmission of vector-borne diseases, the WHO is calling for integrated vector management with cross-sectoral collaboration (WHO, 2012a, b).

It is clear that factors affecting the transmission of VVD, including the vectors, the viruses, and human beings, are influenced by a range of factors including climate variations and changes. Therefore, integrated approaches with considerations of the changing environment should be adopted to allow for climate-based health monitoring systems to respond to the emergence and reemergence of VVD around the world.

Some recommendations have been articulated in the following text:

- *Development of comprehensive surveillance strategies for new vectors and viruses, incorporating the ecoclimatic conditions associated with VVD*
  Given that climate variables, for example, maximum and minimum temperatures, rainfall, humidity, and El Niño events, are drivers for emerging and reemerging VVD, it is expected that early warning systems that include local meteorological variables would be useful in predicting future risks of VVD. In addition, the various types of vectors and complicated transmission routes from the viruses to the infection of human beings and integrated surveillance systems that monitor the activities of the virus inside the major vectors and host animals are key for VVD prevention and control, particularly when the associations between climate change and some VVD are not fully understood.
- *Modeling and risk assessments for future vulnerability*
  With the increasing concern of the projected global climate change and extreme weather events, the key questions for infectious disease prevention and control are "what is the vulnerability map?" for example, "who will be the most affected?" and "where would be the most affected regions?" Suitable time-series statistical methods (e.g., the seasonal autoregressive integrated moving average model) could be used to quantify the relationship between weather variables and the VVD because these models have intrinsic functions that can be used to effectively control for the autocorrelations, seasonal variations, any long-term trend, and lagged effects, to make the results of environmental health risk assessments more valid and reliable (Bi et al., 2001). In addition, the burden of disease attributed to future climate change or variations can be estimated to inform evidence-based decision and policy making.
- *Collective endeavor for the development of laboratory techniques for virus diagnosis and vaccines*
  Accurate and early diagnosis of VVD is the strong basis of an effective disease surveillance system. Due to the huge amount of money needed to develop new effective and safe vaccines and laboratory technologies, governments and other stakeholders need to endeavor to allocate enough research and development funding to fill the gaps in current clinical and epidemiological research. This will help to ensure adequate preparedness for future risks of VVD.
- *Identification of adaptive needs of society in different geographical areas so as to plan public health preparedness*
  Better adaptation to projected climate change and extreme weather events should be taken, with consideration of the local climate and geographic situation. For example, people living in temperate low-lying locations need be warned about possible risks of VVD that may not have been a public health concern previously and informed about adaptation strategies (e.g., using insect repellent and avoiding intensive

contact with certain wild animals) when unusual or extreme weather events are projected. The importance of public health education and health promotion in the community should be fully recognized in order to have better preparedness for public health emergencies.

- *Necessity of international and intersectoral cooperation*
  The spread of infectious diseases today is faster than ever in our "global village." When we humans are enjoying the increasing freedom of international travel, the viruses, vectors, and host animals also take advantage of this, making VVD prevention and control very challenging. Collaborations are often required beyond the health sector (e.g., environment sector, export/import economic sector, agriculture sector) and across borders of countries. Coordination by global leaders at high levels of government could certainly assist the WHO in preventing occurrences of VVD pandemics.

In summary, recognizing the complexity of the transmission of VVD, climate change may play a role in the emerging and reemerging of VVD around the world. Integrated prevention and control measures of the VVD should consider global climate change and local weather patterns to minimize projected adverse impacts on human health and society.

REFERENCES

Andriamandimby, S. F., A. E. Randrianarivo-Solofoniaina, E. M. Jeanmaire, et al. (2010). "Rift Valley fever during rainy seasons, Madagascar, 2008 and 2009." *Emerg Infect Dis* **16**(6): 963–970.

Ascioglu, S., H. Leblebicioglu, H. Vahaboglu, et al. (2011). "Ribavirin for patients with Crimean-Congo haemorrhagic fever: a systematic review and meta-analysis." *J Antimicrob Chemother* **66**(6): 1215–1222.

Bajpai, S. and M. Y. Nadkar (2011). "Crimean Congo hemorrhagic fever: requires vigilance and not panic." *J Assoc Physicians India* **59**: 164–167.

Bangs, M. J., R. Pudiantari, and Y. R. Gionar (2007). "Persistence of dengue virus RNA in dried *Aedes aegypti* (Diptera: Culicidae) exposed to natural tropical conditions." *J Med Entomol* **44**(1): 163–167.

Banu, S., W. Hu, C. Hurst, et al. (2011). "Dengue transmission in the Asia-Pacific region: impact of climate change and socio-environmental factors." *Trop Med Int Health* **16**(5): 598–607.

Bell, M. (2007). Viral hemorrhagic fevers. *Cecil Medicine. 23rd ed.* L. Goldman and D. Ausiello. Philadelphia, PA: Saunders Elsevier.

Bezirtzoglou, C., K. Dekas, and Charvalos, E. (2011). "Climate changes, environment and infection: facts, scenarios and growing awareness from the public health community within Europe." *Anaerobe* **17**(6): 337–340.

Bi, P., K. Donald, J. Hobbs, et al. (2001). "Climate variability and the dengue outbreak in Townsville, Queensland, 1992–93." *Environ Health* **1**: 54–60.

Bi, P., S. Tong, K. Donald, et al. (2003). "Climate variability and transmission of Japanese encephalitis in eastern China." *Vector Borne Zoonotic Dis* **3**(3): 111–115.

Bi, P., Y. Zhang, and K. A. Parton (2007). "Weather variables and Japanese encephalitis in the metropolitan area of Jinan city, China." *J Infect* **55**(6): 551–556.

Bi, P., J. E. Hiller, A. S. Cameron, et al. (2009). "Climate variability and Ross River virus infections in Riverland, South Australia, 1992–2004." *Epidemiol Infect* **137**(10): 1486–1493.

Britch, S. C., K. J. Linthicum, A. Anyamba, et al. (2008). "Satellite vegetation index data as a tool to forecast population dynamics of medically important mosquitoes at military installations in the continental United States." *Mil Med* **173**(7): 677–683.

Bush, K. F., G. Luber, S. R. Kotha, et al. (2011). "Impacts of climate change on public health in India: future research directions." *Environ Health Perspect* **119**(6): 765–770.

Chinikar, S., S. M. Ghiasi, R. Hewson, et al. (2010). "Crimean-Congo hemorrhagic fever in Iran and neighboring countries." *J Clin Virol* **47**(2): 110–114.

Daszak, P., A. A. Cunningham, A. D. Hyatt et al. (2001). "Anthropogenic environmental change and the emergence of infectious diseases in wildlife." *Acta Tropica* **78**: 103–116.

DeCarlo, C. H., A. B. Clark, K. J. McGowan, et al. (2011). "Factors associated with the risk of West Nile virus among crows in New York State." *Zoonoses Public Health* **58**(4): 270–275.

Deichmeister, J. M. and A. Telang (2011). "Abundance of West Nile virus mosquito vectors in relation to climate and landscape variables." *J Vector Ecol* **36**(1): 75–85.

Delatte, H., J. S. Dehecq, J. Thiria, et al. (2008). "Geographic distribution and developmental sites of Aedes albopictus (Diptera: Culicidae) during a Chikungunya epidemic event." *Vector Borne Zoonotic Dis* **8**(1): 25–34.

Dhiman, R. C., S. Pahwa, G. P. Dhillon, et al. (2010). "Climate change and threat of vector-borne diseases in India: are we prepared?" *Parasitol Res* **106**(4): 763–773.

Ditsuwan, T., T. Liabsuetrakul, V. Chongsuvivatwong, et al. (2011). "Assessing the spreading patterns of dengue infection and chikungunya fever outbreaks in lower southern Thailand using a geographic information system." *Ann Epidemiol* **21**(4): 253–261.

Endy, T. P. and A. Nisalak (2002). "Japanese encephalitis virus: ecology and epidemiology." *Curr Top Microbiol Immunol* **267**: 11–48.

Flexman, J. P., D. W. Smith, J. S. Mackenzie, et al. (1998). "A comparison of the diseases caused by Ross River virus and Barmah Forest virus." *Med J Aust* **169**(3): 159–163.

Gale, P., A. Brouwer, V. Ramnial, et al. (2010). "Assessing the impact of climate change on vector-borne viruses in the EU through the elicitation of expert opinion." *Epidemiol Infect* **138**(2): 214–225.

Gatton, M. L., B. H. Kay, and P. A. Ryan (2005). "Environmental predictors of Ross River virus disease outbreaks in Queensland, Australia." *Am J Trop Med Hyg* **72**(6): 792–799.

Gubler, D. J., P. Reiter, K. L. Ebi, et al. (2001). "Climate variability and change in the United States: potential impacts on vector- and rodent-borne diseases." *Environ Health Perspect* **109**(Suppl. 2): 223–233.

Hales, S., N. de Wet, J. Maindonald, et al. (2002). "Potential effect of population and climate changes on global distribution of dengue fever: an empirical model." *Lancet* **360**(9336): 830–834.

Heyman, P., C. Cochez, A. Hofhuis, et al. (2010). "A clear and present danger: tick-borne diseases in Europe." *Expert Rev Anti Infect Ther* **8**(1): 33–50.

Hightower, A., C. Kinkade, P. M. Nguku, et al. (2012). "Relationship of climate, geography, and geology to the incidence of Rift Valley fever in Kenya during the 2006–2007 outbreak." *Am J Trop Med Hyg* **86**(2): 373–380.

Hsu, S. M., A. M. Yen, and T. H. Chen (2008). "The impact of climate on Japanese encephalitis." *Epidemiol Infect* **136**(7): 980–987.

Institute of Medicine of the National Academies (2008). *Vector-Borne Diseases: Understanding the Environmental, Human Health, and Ecological Connections*. Washington, DC: The National Academies Press.

Jacups, S. P., P. I. Whelan, and B. J. Currie (2008a). "Ross River virus and Barmah Forest virus infections: a review of history, ecology, and predictive models, with implications for tropical northern Australia." *Vector Borne Zoonotic Dis* **8**(2): 283–297.

Jacups, S. P., P. I. Whelan, P. G. Markey, et al. (2008b). "Predictive indicators for Ross River virus infection in the Darwin area of tropical northern Australia, using long-term mosquito trapping data." *Trop Med Int Health* **13**(7): 943–952.

Jansen, C. C. and N. W. Beebe (2010). "The dengue vector *Aedes aegypti*: what comes next." *Microbes Infect* **12**(4): 272–279.

Johansson, M. A., D. A. Cummings, and G. E. Glass (2009). "Multiyear climate variability and dengue—El Nino southern oscillation, weather, and dengue incidence in Puerto Rico, Mexico, and Thailand: a longitudinal data analysis." *PLoS Med* **6**(11): e1000168.

Kamath, S., A. K. Das, and F.S. Parikh (2006). "Chikungunya." *J Assoc Physicians India* **54**: 725–726.

Kay, B. H., A. J. Saul, and A. McCullagh (1987). "A mathematical model for the rural amplification of Murray Valley encephalitis virus in southern Australia." *Am J Epidemiol* **125**(4): 690–705.

Kelly-Hope, L. A., D. M. Purdie, B. H. Kay (2004). "El Nino Southern Oscillation and Ross River virus outbreaks in Australia." *Vector Borne Zoonotic Dis* **4**(3): 210–213.

Kelvin, A. A. (2011). "Outbreak of Chikungunya in the Republic of Congo and the global picture." *J Infect Dev Ctries* **5**(6): 441–444.

Koonin, E. V., T. G. Senkevich, and V. V. Dolja (2006). "The ancient Virus World and evolution of cells." *Biol Direct* **1**: 29.

Lafferty, K. D. (2009). "The ecology of climate change and infectious diseases." *Ecology* **90**(4): 888–900.

Last, J., Ed. (2001). *A Dictionary of Epidemiology*. New York: Oxford University Press.

Leblebicioglu, H. (2010). "Crimean-Congo haemorrhagic fever in Eurasia." *Int J Antimicrob Agents* **36**(Suppl. 1): S43–S46.

Lindsay, M., N. Oliveira, E. Jasinska, et al. (1996). "An outbreak of Ross River virus disease in Southwestern Australia." *Emerg Infect Dis* **2**(2): 117–120.

Liu, A., V. Lee, D. Galusha, et al. (2009). "Risk factors for human infection with West Nile Virus in Connecticut: a multi-year analysis." *Int J Health Geogr* **8**: 67.

Mackenzie, J. S. and D. T. Williams (2009). "The zoonotic flaviviruses of southern, south-eastern and eastern Asia, and Australasia: the potential for emergent viruses." *Zoonoses Public Health* **56**(6–7): 338–356.

Mackenzie, J., D. Gubler, and L. R. Petersen (2004). "Emerging flaviviruses: the spread and resurgence of Japanese encephalitis, West Nile and dengue viruses." *Nat Med* **10**: S98–S109.

Maclachlan, N. J. (2010). "Global implications of the recent emergence of bluetongue virus in Europe." *Vet Clin North Am Food Anim Pract* **26**(1): 163–171.

Maclachlan, N. J. (2011). "Bluetongue: history, global epidemiology, and pathogenesis." *Prev Vet Med* **102**(2): 107–111.

Maltezou, H. C. and A. Papa (2010). "Crimean-Congo hemorrhagic fever: risk for emergence of new endemic foci in Europe?" *Travel Med Infect Dis* **8**(3): 139–143.

Martin, V., V. Chevalier, P. Ceccato, et al. (2008). "The impact of climate change on the epidemiology and control of Rift Valley fever." *Rev Sci Tech* **27**(2): 413–426.

McMichael, A., R. Woodruff, P. Whetton, et al., Eds. (2003). *Human Health and Climate Change in Oceania: A Risk Assessment 2002*. Canberra: Commonwealth of Australia.

Monaco, F., G. Savini, P. Calistri, et al. (2011). "2009 West Nile disease epidemic in Italy: first evidence of overwintering in Western Europe?" *Res Vet Sci* **91**(2): 321–326.

Monath, T. P. and M. S. Cetron (2002). "Prevention of yellow fever in persons traveling to the tropics." *Clin Infect Dis* **34**(10): 1369–1378.

Moormann, R. (2012). CVI research programme emerging vector borne viral diseases. Accessed Feb 2012. Lelystad, The Netherlands: Central Veterinary Institute.

Murty, U. S., M. S. Rao, and N. Arunachalam (2010). "The effects of climatic factors on the distribution and abundance of Japanese encephalitis vectors in Kurnool district of Andhra Pradesh, India." *J Vector Borne Dis* **47**(1): 26–32.

Nagao, Y., A. Tawatsin, S. Thammapalo, et al. (2012). "Geographical gradient of mean age of dengue haemorrhagic fever patients in northern Thailand." *Epidemiol Infect* **140**(3): 479–490.

Nicholls, N. (1986). "A method for predicting Murray Valley encephalitis in southeast Australia using the Southern Oscillation." *Aust J Exp Biol Med Sci* **64 (Pt 6)**: 587–594.

Pancharoen, C., W. Kulwichit, T. Tantawichien, et al. (2002). “Dengue infection: a global concern.” *J Med Assoc Thai* **85**(Suppl. 1): S25–S33.

Patz, J. A., W. J. Martens, D. A. Focks, et al. (1998). “Dengue fever epidemic potential as projected by general circulation models of global climate change.” *Environ Health Perspect* **106**(3): 147–153.

Pepin, M., M. Bouloy, B. H. Bird, et al. (2010). “Rift Valley fever virus(Bunyaviridae: Phlebovirus): an update on pathogenesis, molecular epidemiology, vectors, diagnostics and prevention.” *Vet Res* **41**(6): 61.

Phillips, M. L. (2008). “Dengue reborn: widespread resurgence of a resilient vector.” *Environ Health Perspect* **116**(9): A382–A388.

Pinzon, E., J. M. Wilson, and C. J. Tucker (2005). “Climate-based health monitoring systems for eco-climatic conditions associated with infectious diseases.” *Bull Soc Pathol Exot* **98**(3): 239–243.

Platonov, A. E., M. V. Fedorova, L. S. Karan, et al. (2008). “Epidemiology of West Nile infection in Volgograd, Russia, in relation to climate change and mosquito (Diptera: Culicidae) bionomics.” *Parasitol Res* **103**(Suppl. 1): S45–S53.

Poletti, P., G. Messeri, M. Ajelli, et al. (2011). “Transmission potential of chikungunya virus and control measures: the case of Italy.” *PLoS One* **6**(5): e18860.

Pontes, R., J. Freeman, J. W. Oliveira-Lima, et al. (2000). “Vector densities that potentiate dengue outbreaks in a Brazilian city.” *Am J Trop Med Hyg* **62**: 378–383.

Pulmanausahakul, R., S. Roytrakul, P. Auewarakul, et al. (2011). “Chikungunya in Southeast Asia: understanding the emergence and finding solutions.” *Int J Infect Dis* **15**(10): e671–e676.

Reisen, W. K., H. D. Lothrop, S. S. Wheeler, et al. (2008). “Persistent West Nile virus transmission and the apparent displacement St. Louis encephalitis virus in southeastern California, 2003–2006.” *J Med Entomol* **45**(3): 494–508.

Rezza, G. (2012). “Aedes albopictus and the reemergence of Dengue.” *BMC Public Health* **12**(1): 72.

Roth, D., B. Henry, S. Mak, et al. (2010). “West Nile virus range expansion into British Columbia.” *Emerg Infect Dis* **16**(8): 1251–1258.

Ruiz, M. O., L. F. Chaves, G. L. Hamer, et al. (2010). “Local impact of temperature and precipitation on West Nile virus infection in Culex species mosquitoes in northeast Illinois, USA.” *Parasit Vectors* **3**(1): 19.

Russell, R. C., C. R. Williams, R. W. Sutherst, et al. (2005). “Aedes (Stegomyia) albopictus—a dengue threat for southern Australia?” *Commun Dis Intell* **29**(3): 296–298.

Russell, R. C., B. J. Currie, M. D. Lindsay, et al. (2009). “Dengue and climate change in Australia: predictions for the future should incorporate knowledge from the past.” *Med J Aust* **190**(5): 265–268.

Ruzek, D., V. V. Yakimenko, L. S. Karan, et al. (2010). “Omsk haemorrhagic fever.” *Lancet* **376**(9758): 2104–2113.

Scholte, E., W. Den Hartog, M. Dik, et al. (2010). “Introduction and control of three invasive mosquito species in the Netherlands, July–October 2010.” *Euro Surveill* **15**(45): 19710.

Sellers, R. F. (1980). “Weather, host and vector—their interplay in the spread of insect-borne animal virus diseases.” *J Hyg (Lond)* **85**(1): 65–102.

Shaman, J., K. Harding, and S. R. Campbell (2011). “Meteorological and hydrological influences on the spatial and temporal prevalence of West Nile virus in Culex mosquitoes, Suffolk County, New York.” *J Med Entomol* **48**(4): 867–875.

Slenning, B. D. (2010). “Global climate change and implications for disease emergence.” *Vet Pathol* **47**(1): 28–33.

Tabachnick, W. J. (2010). “Challenges in predicting climate and environmental effects on vector-borne disease episystems in a changing world.” *J Exp Biol* **213**: 946–954.

Tomerini, D. M., P. E. Dale, and N. Sipe (2011). “Does mosquito control have an effect on mosquito-borne disease? The case of Ross River virus disease and mosquito management in Queensland, Australia.” *J Am Mosq Control Assoc* **27**(1): 39–44.

Tong, S. and W. Hu (2002). "Different responses of Ross River virus to climate variability between coastline and inland cities in Queensland, Australia." *Occup Environ Med* **59**(11): 739–744.

Tong, S., P. Bi, and McMichael, A. (2001). "Geographic variation of notified Ross River virus infections in Queensland, Australia, 1985–1996." *Am J Trop Med Hyg* **65**(3): 171–176.

Tong, S., P. Bi, K. Donald, et al. (2002). "Climate variability and Ross River virus transmission." *J Epidemiol Community Health* **56**(8): 617–621.

Tong, S., W. Hu, N. Nicholls, et al. (2005). "Climatic, high tide and vector variables and the transmission of Ross River virus." *Intern Med J* **35**(11): 677–680.

Tong, S., P. Dale, N. Nicholls, et al. (2008). "Climate variability, social and environmental factors, and ross river virus transmission: research development and future research needs." *Environ Health Perspect* **116**(12): 1591–1597.

Tourre, Y. M., J. P. Lacaux, C. Vignolles, et al. (2009). "Climate impacts on environmental risks evaluated from space: a conceptual approach to the case of Rift Valley Fever in Senegal." *Glob Health Action* **2**: 10.3402/gha.v2i0.2053.

van den Hurk, A. F., S. B. Craig, S. M. Tulsiani, et al. (2010). "Emerging tropical diseases in Australia. Part 4. Mosquitoborne diseases." *Ann Trop Med Parasitol* **104**(8): 623–640.

Vasconcelos, P. F. (2010). "Yellow fever in Brazil: thoughts and hypotheses on the emergence in previously free areas." *Rev Saude Publica* **44**(6): 1144–1149.

Vasconcelos, P. F., Z. G. Costa, E. S. Travassos Da Rosa, et al. (2001). "Epidemic of jungle yellow fever in Brazil, 2000: implications of climatic alterations in disease spread." *J Med Virol* **65**(3): 598–604.

Vezzani, D. and A. E. Carbajo (2008). "*Aedes aegypti*, Aedes albopictus, and dengue in Argentina: current knowledge and future directions." *Mem Inst Oswaldo Cruz* **103**(1): 66–74.

Weaver, S. C. and W. K. Reisen (2010). "Present and future arboviral threats." *Antiviral Res* **85**(2): 328–345.

WHO (2004). World Health Report.

WHO (2008). Chikungunya. Available at: http://www.who.int/mediacentre/factsheets/fs327/en/. Accessed July 6, 2013.

WHO (2010). Rift Valley fever. Available at http://www.who.int/mediacentre/factsheets/fs207/en/. Accessed July 6, 2013.

WHO (2011). Yellow Fever Fact Sheet. Available at: http://www.who.int/mediacentre/factsheets/fs100/en/. Accessed July 6, 2013.

WHO (2012a). Integrated vector management (IVM). Available at: http://www.who.int/heli/risks/vectors/vector/en/index.html. Accessed July 6, 2013.

WHO (2012b). Vector-Borne Viral Infections. Available at: http://www.who.int/vaccine_research/diseases/vector/en/index.html. Accessed July 6, 2013.

Woodruff, R. E., C. S. Guest, M. G. Garner, et al. (2002). "Predicting Ross River virus epidemics from regional weather data." *Epidemiology* **13**(4): 384–393.

Yilmaz, G. R., T. Buzgan, H. Irmak, et al. (2009). "The epidemiology of Crimean-Congo hemorrhagic fever in Turkey, 2002–2007." *Int J Infect Dis* **13**(3): 380–386.

Zell, R., A. Krumbholz, and P. Wutzler (2008). "Impact of global warming on viral diseases: what is the evidence?" *Curr Opin Biotechnol* **19**: 652–660.

Zhang, Y., P. Bi, and J. E. Hiller (2008). "Climate change and the transmission of vector-borne diseases: a review." *Asia Pac J Public Health* **20**(1): 64–76.

2

IMPACT OF CLIMATE CHANGE ON VECTOR-BORNE ARBOVIRAL EPISYSTEMS

Walter J. Tabachnick and Jonathan F. Day

Florida Medical Entomology Laboratory, Department of Entomology and Nematology, University of Florida, Vero Beach, FL, USA

TABLE OF CONTENTS

Viral Infections and Global Change, First Edition. Edited by Sunit K. Singh.

2.1 INTRODUCTION

Vector-borne diseases cause great suffering throughout the world. Table 2.1 provides recent estimates for the number of human cases for several of the most important vector-borne diseases. Among the various vector-borne pathogens, the arthropod-borne viruses (arboviruses) play a prominent role in human and animal morbidity and mortality. There are more than 500 known arboviruses representing diverse taxonomic groups. Most of the arboviruses are contained in five virus families, for example, Togaviridae, Flaviviridae, Bunyaviridae, Reoviridae, and Rhabdoviridae (Karabatsos, 1985). Many of the arboviruses cause a variety of disease symptoms in a great diversity of vertebrate hosts (Table 2.2).

One of the great challenges of the twenty-first century will be the control of vector-borne diseases and the reduction of the impact and suffering caused by these diseases. Many studies and reviews have provided dire predictions for the future effects of global climate change on vector-borne disease systems. Some scientists have concluded that global warming and the resulting climate change will undoubtedly cause an increase in vector-borne diseases throughout the world. A number of scientific publications illustrate this view, for example, Patz and Olson (2008), Epstein (2000, 2007), IPCC (2001, 2007),

TABLE 2.1. Estimated Annual Worldwide Impact of Selected Vector-Borne Diseases

Disease	Population at Risk (Billion)	Cases (Million)	Deaths
Malaria	3.3	247	881 000
Leishmaniasis	0.35	12–15	60 000
Filariasis	1.3	120	300
Chikungunya	2.5	1.64 (2005–2007)	3500–15 000
Dengue	2.5	50–100	20 000
Japanese encephalitis	3	0.035–0.05	15 000
Rift Valley fever	0.78	<0.01	
Yellow fever	0.9	0.2	30 000

TABLE 2.2. Selected Arboviruses That Cause Disease in Humans and Other Animals

Arbovirus	Hosts	Vectors
African horse sickness	Equids, dogs	Culicoides, biting midges
Bluetongue	Sheep, goats, cattle	Mosquitoes
Chikungunya	Humans	Mosquitoes
Eastern equine encephalitis	Horses, birds, humans	Mosquitoes
Dengue	Swine, humans	Mosquitoes
Japanese encephalitis	Rodents, humans	Mosquitoes
La Crosse encephalitis	Birds, humans	Mosquitoes
Rift valley	Sheep, cattle, humans	Mosquitoes
St. Louis encephalitis	Small mammals, humans	Mosquitoes
Venezuelan encephalitis	Birds, humans, horses	Mosquitoes
Vesicular stomatitis	Horses, cattle	Biting midges, blackflies
West Nile	Birds, humans	Mosquitoes
Western equine encephalitis	Horses, birds, humans	Mosquitoes
Yellow fever	Primates, humans	Mosquitoes

Hales et al. (2003), Githeko et al. (2000), Rogers and Randolph (2000), Martens (1999), Patz et al. (1998), and Jetten and Focks (1997). A considerable number of scientists have called attention to the complexity of vector-borne diseases and have noted that this complexity must be understood in order to predict the future consequences of global climate change. Experts have suggested that more information is needed concerning the complex ecology of vector-borne diseases to allow the capability to predict long-term trends in vector-borne disease systems with any assurance. This view is illustrated in the following selected references: Thai and Anders (2011), Randolph (2010), Randolph and Rogers (2010), Slenning (2010), Stresman (2010), Tabachnick (2010), Elliott (2009), Lafferty (2009), Rosenthal (2009), Russell (1998, 2009), Russell et al. (2009), Randolph (2009), Gould and Higgs (2009), Gubler (2002, 2008), Gage et al. (2008), Zell et al. (2008), Reiter et al. (2004), Sutherst (2004), Gubler et al. (2001), and Reiter (2001).

Predicting the future impact of climate change on vector-borne disease is a daunting challenge in large part because the vector-borne disease system consists of complex interacting components that are influenced by a variety of factors. Tabachnick (2004, 2010) used the term *episystem* to encompass the diverse components that comprise and influence vector-borne disease systems including the vertebrate host, the arthropod vector, the pathogen, and all of the environmental factors that affect each of these biological components and their collective interactions. Different vector-borne episystems may share qualities with one another, but likely there is great variation between them due to variation in the diversity of the individual components that comprise the episystems. There are likely different episystems for different pathogens, vectors, and hosts. Episystems may differ due to the spatial and temporal conditions that drive individual transmission cycles.

There are many examples illustrating the effects of environmental conditions on the different components of episystems. In this chapter we explore the potential for understanding the effects of climate change on vector-borne disease episystems. We differentiate climate effects from weather effects though they are interrelated. Here we define weather as the day-to-day, hour-to-hour changes in atmospheric conditions, while climate is the average of weather over time (Henson, 2007). There is ample evidence for the influence of weather conditions on vector-borne disease transmission cycles. For example, many studies have shown that an increase in temperature can modify a vector's competence for a particular pathogen (Hardy et al., 1983; Tabachnick, 1994). Temperature also influences the population growth and vector–host–pathogen interactions as well as the behaviors of many arboviral vectors. A second prominent weather factor, rainfall, can affect vector reproduction, abundance, age structure, and behavior. Rainfall also influences vector–host–pathogen interactions. Although it is well known that the environment plays a prominent role in episystems and that projections of worldwide climatic change will undoubtedly influence these episystems, there is little information available to predict precisely how episystems will respond to future global climatic changes (Tabachnick, 2010).

We know that the environment including weather is a prominent feature controlling episystems. We also know that climate has changed throughout the Earth's history and will continue to change into the future as a result of natural cycles and human environmental modification including those that affect global warming. Then why is it so difficult to predict with any assurance what the specific influences of anticipated climate change might be on episystems in general?

Vector-borne pathogens, like those listed in Table 2.2, are transmitted by a variety of arthropods including biting midges, blackflies, fleas, kissing bugs, mosquitoes, sandflies, ticks, and tsetse flies. Here we use selected insect-borne arbovirus episystems to illustrate the possible influence of climatic and environmental change on these systems and as an example of the complexity of predicting future trends and changes. Tabachnick (2010) illustrated the challenges to understanding the influence of climate and the environment on the episystems represented by bluetongue viruses (BTV) in Europe and West Nile virus (WNV) in North America. Here we expand the discussion of these episystems and also explore the North American dengue episystem with attention to recent mosquito-borne dengue transmission in Florida. Our focus is on selected arbovirus episystems; however, the discussion and conclusions are likely valid for a variety of vector-borne episystems. We conclude with some ideas about predicting the influence of climate change on arbovirus episystems in the future.

2.2 THE COMPLEX FACTORS INFLUENCING MOSQUITO-BORNE ARBOVIRUS EPISYSTEMS

All vector-borne disease cycles are comprised of dynamic interactions between the arthropod vector, the specific pathogen, and the host organism. For mosquito-borne arboviruses, this consists of the complex interactions between the mosquito, the virus, and vertebrate hosts including amplification and dead-end hosts. There are interactions between the mosquito and virus that include the growth of the virus in the mosquito, its replication in tissues that allow the mosquito to become infected, and subsequent transmission of the virus to a vertebrate host. This series of events is encompassed in the concept of vector competence. Another important series of interactions occurs between the mosquito vector and the vertebrate host. Vector behavior that facilitates these interactions includes contact and blood feeding on a vertebrate host and virus transfer from the infective vector to the susceptible vertebrate host. Interactions between the vertebrate host and pathogen include viral replication in the host with the possibility of neurological involvement in some hosts. Neurological involvement in infected vertebrate hosts is not however a prerequisite for continued cycling of virus between infected hosts and susceptible mosquito vectors because continued virus transmission depends on there being sufficient virus in the host circulatory system to infect blood-feeding mosquitoes.

Figure 2.1 is a diagram modified from Tabachnick (2010) illustrating the complex suite of environmental factors that influence mosquito-borne disease episystems. Factors such as poverty, environmental and cultural conditions, land and water use practices, human behavior, human and animal population size and growth, and human travel and commerce can influence episystems. These factors may change interactions between vectors and hosts by increasing vertebrate exposure to vectors, changing regional vector–host interactions, and reducing the ability of physicians and veterinarians to treat infected people and animals. Each of the individual environmental factors listed in Figure 2.1 impacts other environmental factors in unknown and complex ways. For example, changes in human population size likely influence poverty and land and water use patterns. Changes in human demographics directly affect urbanization. Each of these changes has a direct effect on mosquito-borne disease episystems. Increased poverty, changes in land and water use patterns that impact mosquito abundance and population age structure, and increased urbanization that expose humans to domestic vector mosquitoes like *Aedes aegypti* all increase the risk of the transmission of mosquito-borne arboviruses such as chikungunya, dengue, and yellow fever.

Addressing the interactions of so many environmental influences becomes more daunting with the realization that many of these environmental factors may be directly or indirectly influenced by weather and climate. Changes in weather and climate may affect a

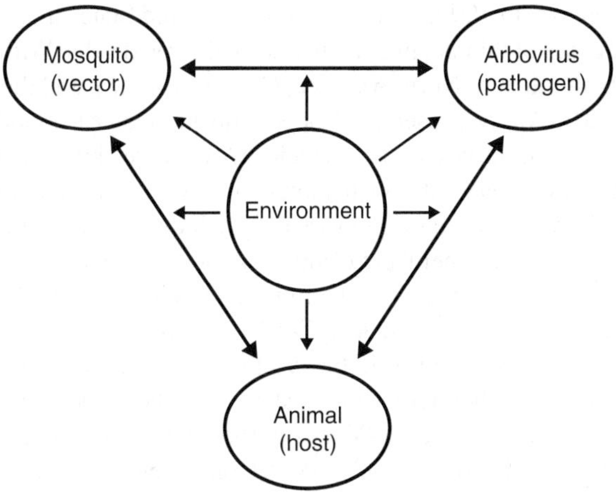

Figure 2.1. The interactions between the environment and components of the vector-borne disease episystem.

variety of environmental factors. Some changes may increase vector-borne disease; some may decrease disease transmission risk, while some may be neutral. Further, the specific climatic influence will likely be dependent on the spatial and temporal parameters for particular disease episystems. Any predictions based on a single climatic or biological factor, or subset of factors, must be made with extreme caution.

The challenge to predict future effects of climate change on episystems is difficult because of our poor understanding of the factors that play prominent roles in specific episystems. It is a daunting challenge to predict future climate effects on vector-borne disease episystems. How well can we currently assess influences of a changing climate on emerging or reemerging episystems? Do we know enough about vector-borne disease episystems to evaluate the current effects of climate change on these systems? Here we will focus on three vector-borne disease episystems: WNV in North America, dengue virus (DENV) in Florida, and the European BTV episystem. Similar reviews of malaria, Rift Valley fever virus, and Ross River virus have provided mixed assessments of climate change on the transmission of these important diseases. For example, evidence for a measurable impact of global warming on malaria transmission is mixed. Some believe that climate change has had a substantial impact on malaria transmission, while others have questioned this by providing equally likely alternative explanations that do not involve climate change and global warming (Chaves and Koenraadt, 2010; Reiter et al., 2004). Several studies have shown that Rift Valley fever virus transmission can be predicted by monitoring rainfall patterns (Anyamba et al., 2009). However, there is no direct evidence that long-term climate change has resulted in any specific vector-borne disease outbreak or that a change in these episystems is the direct result of changing climate.

2.3 WEST NILE VIRUS

West Nile virus (genus *Flavivirus*, family Flaviviridae) is transmitted by a number of mosquito species, and resulting infections can cause disease in humans. West Nile virus had historically been confined to the Old World and Asia where it periodically caused human disease outbreaks. In 1999, WNV entered the United States through New York City where it caused 62 cases of clinical human disease and three deaths. Within a decade WNV had spread throughout much of North America, Central America, and the Caribbean Basin. Between 1999 and 2011 WNV caused c. 31 000 clinical human cases and c. 1300 deaths in the United States (CDC: http://www.cdc.gov/ncidod/dvbid/westnile/surv&control_archive.htm). Prior to the entry of WNV into North America, it was known that the primary vectors in other regions of the world were mosquitoes that included members of the *Culex pipiens* complex. The North American members of this complex, *Culex pipiens pipiens* and *Culex pipiens quinquefasciatus*, have proven to be important WNV vectors. However, two other *Culex* species have also been prominent in WNV transmission in North America. *Culex tarsalis* is the primary vector of WNV in the western United States and *Culex nigripalpus* is a WNV vector in the southeastern United States. Neither species had ever previously been exposed to the virus, but once WNV was introduced to their geographic ranges, both species proved to be a highly competent WNV vector. Essentially, the *Cx. tarsalis* and *Cx. nigripalpus* WNV episystems have emerged de novo during the past 10 years.

2.3.1 Influence of climate on the North American WNV episystem

There is ample evidence for the importance of weather in influencing the North American WNV episystem. We know that increasing temperature directly influences vector competence for WNV and shortens its extrinsic incubation period. These temperature-related changes are tempered by complex interactions between the effect of temperature and other environmental factors such as those influencing virus doses and the mosquito age (Richards et al., 2007). The *Cx. nigripalpus* WNV episystem in Florida is heavily influenced by weather, particularly rainfall (Shaman et al., 2005) and winter freezes (Day and Shaman, 2009). In more temperate habitats, such as eastern Colorado, winter snow accumulation and spring melt patterns drive the WNV episystem (Shaman et al., 2010). Precipitation is important to arboviral episystems in that it drives vector reproductive cycles and the overall age structure of vector populations. The important North American vectors are *Culex* mosquitoes (*Culex quinquefasciatus*, *Cx. pipiens*, *Cx. tarsalis*, and *Cx. nigripalpus*) that rely on newly flooded habitats for oviposition or rainfall conditions that saturate the habitats surrounding their daytime resting sites and allow the nocturnal dispersal of host-seeking females. Because of these oviposition preferences, rainfall and drought cycles drive mosquito blood feeding, oviposition, infection status, and population age structure (Shaman et al., 2002, 2003, 2004). The presence or absence of freshwater oviposition sites that are favored by WNV vector mosquitoes is what determines vector abundance and vector population age structure. Years with extreme drought and years with excessive rainfall may act against the efficient cycling of WNV or SLEV and result in low viral amplification and transmission. By contrast, during years where rainfall and drought cycle is in predictable patterns, viral amplification is extremely efficient resulting in large numbers of infected and infective vector mosquitoes (Day and Curtis, 1993, 1994, 1999).

Although weather in the form of increased rainfall may figure prominently in the Florida WNV episystem, there is no evidence that the entry of WNV into Florida was due to any climatic change recorded during the past two decades. This is also true for the *Cx. tarsalis* WNV episystem in the western United States. Rather than attributing the rapid spread of WNV throughout the United States to climate change, it is more likely that the climatic conditions conducive for WNV amplification and transmission and the presence of *Culex* species that were highly competent WNV vectors were already in place to support the rapid spread of WNV once it entered North America. Although weather, such as daily

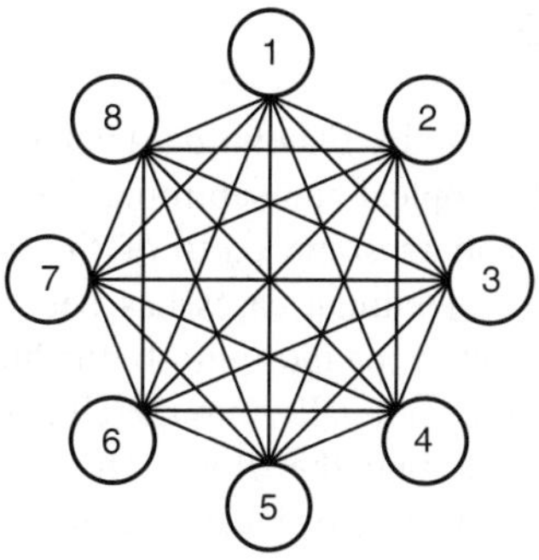

Figure 2.2. Examples of selected environmental factors and their interactions with one another that influence the vector-borne disease episystem (adapted from Tabachnick, 2010): 1, climate; 2, poverty; 3, human population growth; 4, land use and forestation; 5, international commerce and travel; 6, water storage and irrigation; 7, human behavior and prevention strategies; and 8, government and political stability.

rainfall or temperature, is important in WNV epidemiology, a change in climate was not a major factor in its introduction and subsequent spread through North America.

2.3.2 Effects of future changes in climate on the North American WNV episystem

Can we predict future climatic changes that may alter WNV transmission patterns in Florida or elsewhere in North America? To do this with accuracy requires a complete understanding of the factors that influence WNV episystems in different regions of North America. We know that aspects of weather are important components of the WNV episystem. However, quantifying individual episystem components and measuring and understanding their relationships and interactions make the prediction of transmission risk difficult (Figure 2.2). For example, how would the changes in rainfall patterns that might accompany global warming alter the North American WNV episystem? Would the current rainfall–drought cycles in Florida change in a way that makes the *Cx. nigripalpus*–bird amplification cycles more efficient resulting in an increased number of human epidemics? Would changes in annual rainfall–drought cycles and other environmental factors in Florida result in a decrease in WNV amplification efficiency with a decrease in the number of human epidemics? How might the current water storage and irrigation practices in Florida change under conditions of increasing drought, and how would these changes alter *Cx. nigripalpus* abundance, age structure, vector competence, and behavior? Could changing long-term climatic patterns alter the interaction of *Cx. nigripalpus* with predators, vertebrate hosts, or other vector mosquitoes, thus reducing the WNV vector competence of *Cx. nigripalpus*? Climate change might alter WNV episystems elsewhere in North America. The winter of 2011–2012 was unusually warm throughout all of North America and particularly in the Midwestern states. What might the influence of this change in weather be on *Cx. tarsalis* populations? Could the warm winter influence the abundance, behavior, or vector competence of future mosquito populations? Is a rise in *Cx. tarsalis* or *Cx. pipiens* transmitted WNV in the future? Raising these possibilities as examples reinforces the challenges and the difficulty of predicting the effects of long-term climatic changes even at a local level. These are complex issues and there is little hope that accurate long-term predictions can be made regarding the impact of climate change on vector-borne disease episystems.

2.4 DENGUE IN FLORIDA

Dengue virus (genus *Flavivirus*, family Flaviviridae) is the cause of dengue infection and disease throughout much of the tropical and subtropical world. It is generally transmitted by the mosquito *Ae. aegypti* and in some parts of the world by the mosquito *Aedes albopictus*. The annual number of human cases has been estimated at 50–100 million, making dengue the most prevalent of the mosquito-borne human arboviruses.

The DENV has been imported annually into the United States by travelers returning from dengue endemic regions around the globe. However, for nearly 60 years there have been only a few documented cases of locally acquired mosquito transmitted DENV in North America. These have been largely confined to a small area on the United States–Mexico border in southwest Texas. Even though Florida has extensive *Ae. aegypti* and *Ae. albopictus* populations and annually reports dengue-infected travelers returning to the state, there has been no record of autochthonous dengue transmission in Florida since 1954 (Day, 2008).

In 2009, a focal outbreak of dengue transmitted by *Ae. aegypti* that resulted in 22 human cases was reported in Key West, Florida, which has the only tropical climate in the Continental United States. A serosurvey of the Key West human population provided data that indicated approximately 1000 Key West residents had been infected with DENV serotype 1 (DENV-1) during 2009 (Radke et al., 2012). Dengue transmission subsided during the 2009/2010 dry season. However, in the early summer of 2010, DENV transmission resumed resulting in 66 human cases in a population of ca. 20 000. The virus circulating in 2010 was again DENV-1. The incidence of c. 330 cases/100 000 population was among the highest in the world in 2010 (i.e., Martin et al. 2010). Considering the estimated 1000 residents infected in 2009 resulting in 22 clinical cases, it is likely that an additional 3000 Key West residents were infected in 2010. The 2009–2010 Key West DENV transmission resulted in approximately 20% of the Key West residents being infected with DENV-1. Further, in 2010 a human dengue case (DENV-2) that was clearly the result of local transmission was reported in Dade County, Florida. Two additional locally acquired dengue cases (DENV-3) were reported from Broward County. It became clear in 2010 that the Florida DENV episystem was again becoming active. In 2011 and 2012, Key West did not report any human dengue cases despite the presence of *Ae. aegypti* at levels similar to 2009 and 2010. However, during the summer of 2011, locally acquired human dengue cases were reported in Hillsborough County (one case), Palm Beach County (two cases), Dade County (two cases), and Martin County (one case), clearly indicating that the Florida dengue episystem was changing.

2.4.1 Did climate change play a role in the reemergence of dengue in Florida?

Why is dengue transmission now occurring in Florida after an absence of nearly 60 years? How did DENV-1 arrive in Key West and result in an epidemic? Why did transmission disappear in 2011 and 2012? Why have sporadic cases occurred in widespread areas of southeast Florida? Is this the result of climate change? Will there be other dengue cases in Key West in the future?

We will likely never know the answer to some of these questions. There is no evidence that Key West experienced unusual changes in either temperature or rainfall patterns during 2008, 2009, 2010, or 2011 (NCDC, 2011). Temperatures in Key West are typical of the tropics and did not fluctuate significantly from the normal 50 year means during 2009–2011. Rainfall patterns were typical each year with no observable differences between years with dengue and those without dengue transmission. Therefore, it is unlikely the meteorological conditions played a primary role in the recent reintroduction of dengue into the Florida DENV episystem. It is likely that DENV arrived in Key West in an infected traveler, perhaps someone traveling on one of the many cruise ships that arrive in Key West from elsewhere in the Caribbean basin where dengue has been epidemic during recent years. The reasons why DENV transmission disappeared in Key West during 2011 and 2012 are unclear. Perhaps the virus failed to survive the 2010–2012 dry season when there were greatly reduced *Ae. aegypti* populations throughout the city. On the other hand, perhaps the Key West residents previously infected with DENV-1 (20% of the Key West population) represented the high-risk humans for dengue transmission, and once infected, those high-risk individuals were removed from the dengue episystem in essence shutting down the transmission cycle. Unfortunately, the entry of another DENV serotype into Key West will result in a high risk for a more severe and devastating dengue outbreak (dengue hemorrhagic fever or dengue shock syndrome) in the 20% of the Key West population already infected with DENV-1. However, though weather conditions have been thought to have played a role in elevated dengue transmission in other parts of the world (Johansson et al., 2009), there is currently no evidence that meteorological conditions were responsible for the reintroduction of DENV into the Florida dengue episystem.

2.5 BLUETONGUE

Bluetongue virus (genus *Orbivirus*, family Reoviridae) is the cause of bluetongue disease that primarily occurs in ruminants such as sheep, goats, and cattle. There are 24 different serotypes of BTV, all of which are transmitted by biting midges in the genus *Culicoides*. Not all species of *Culicoides* serve as vectors of BTV and different *Culicoides* species serve as primary BTV vectors in different regions of the world (Tabachnick, 2004). Bluetongue disease can cause significant morbidity and mortality in sheep, but the virus only rarely has been known to cause clinical signs in cattle. The economic consequences of BTV on livestock industries around the world can be severe due to livestock morbidity and mortality and also as the result of restrictions on the movement of livestock from bluetongue endemic to bluetongue-free areas.

2.5.1 Influence of climate on the European bluetongue episystem

The BTV were historically confined to Africa, Asia, the New World, and Australia. Although there had been a few sporadic outbreaks in Europe, BTV did not become established there until 1998 when it was reported for the first time in Italian sheep populations. During subsequent years nine different BTV serotypes entered Europe, where all had previously been absent up to that time. Millions of ruminants were infected and a variety of indigenous *Culicoides* species proved to be efficient BTV vectors. Bluetongue virus spread throughout Europe and entered the United Kingdom for the first time in 2007. The European BTV episystem represents another episystem that has emerged during the last decade. In total, this new European episystem resulted in millions of BTV-infected sheep, cattle, goats, and wild ruminants (including deer and elk). There were substantial losses in sheep populations, and in 2007 it was found that BTV-8 was the cause of clinical symptoms in substantial numbers of cattle that, up to that time, had not previously been reported with BTV infections. In an effort to control BTV in livestock, several countries initiated a BTV vaccination program using modified live BTV vaccines that proved effective and reduced the number of bluetongue cases and greatly reduced the risk of BTV transmission. In 2010, as a result of this campaign, the United Kingdom was declared BTV-free.

2.5.2 The role of climate change in the European BTV episystem

The recent emergence of BTV in Europe is perhaps the best example of the possible influence of weather, and perhaps climate change, on a vector-borne disease episystem (Rogers and Randolph, 2006). A variety of studies have described a link between changes in European climate during the first decade of the twenty-first century and a possible link to the introduction and spread of BTV (Guis et al., 2011; Purse et al., 2006). The European BTV episystem emerged coincidently with a significant warming trend in the region. Seven of the 10 warmest European winters reported during the previous 50 years occurred between 1998 and 2007 with the winter of 2006–2007 being the warmest on record (Wilson and Mellor, 2008).

It is likely that the climate change (specifically increasing temperature) in Europe during this period had an influence on BTV epidemiology. Guis et al. (2011) provide a convincing model that correlates the climate variables of temperature and rainfall with many aspects of Europe's BTV epidemiology. They further use their model to predict that under similar conditions and under the same model assumptions, the risk of bluetongue in susceptible ruminants will rise in Europe due to anticipated climate change and cause an estimated 10% and 30% increase in Northwest and Southwest Europe, respectively, by the year 2050 (Guis et al., 2011). Tabachnick (2010) called attention to the possibility that other factors, such as those depicted in Figure 2.1, may have played a role in the introduction and subsequent spread of BTV in Europe. It is clear that climate change (increasing mean winter temperature) likely played a prominent role in the spread of BTV throughout Europe, though it is less clear precisely what the role of climate might have been in influencing other episystem drivers like those shown in Figure 2.1. In addition, it is possible that other environmental drivers may have been more prominent in the BTV European episystem than climate change. Factors including changes in animal management practices, increases in susceptible wildlife populations, increases in exotic animal theme parks, and implementation of BTV vaccination programs (Tabachnick, 2010) may have played a prominent role in the introduction and spread of BTV in Europe, though these may have been of minor importance compared with climate change (Guis et al., 2011). However, despite the concordance of climate-driven models with the BTV episystem in Europe during the past 10 years, Guis et al. (2011) recognized that their model does not suggest climate as the only driver of the BTV episystem. They recognize that other BTV episystem drivers that supplant the effects of present-day climate change might arise in the future. Further, Guis et al. (2011) emphasize that it is essential to disentangle the interactions between known episystem drivers in order to apply their model to the European BTV episystem.

Most experts agree that using models to forecast long-range predictions about the influence of climatic change in any episystem is fraught with uncertainty and must be approached with great caution (Tabachnick, 2010). Despite explicit statements about caution concerning the use of predictive models to forecast episystem dynamics, there continue to be well-publicized examples attaching more certainty than is warranted to the outcomes of these models. For example, predictions based on models have resulted in media assertions that "bluetongue will increase with climate change" (Environmental Research Web http://environmentalresearchweb.org/cws/article/news/47167) or "bluetongue set to rise with climate change" (Planet Earth online http://planetearth.nerc.ac.uk/news/story.aspx?id=1024).

Figures 2.1 and 2.2 summarize the complexity of vector-borne disease episystems that cast doubt on the accuracy of such statements. Some models make predictions about BTV transmission 50 years into the future. Consider the analogy of someone using U.S. stock market data from 1961 to predict stock market returns in 2011. The environment that influences stocks in 2011 did not even exist in 1961. The same is true of BTV episystems. The conditions in the future, for example, the factors in Figures 2.1 and 2.2, do not yet exist and are difficult to imagine.

2.6 CONCLUSIONS

Weather is a major environmental driver influencing vector-borne disease episystems. A variety of environmental factors are influenced by weather and climate, but these environmental conditions are also influenced by other biotic and abiotic factors and relationships, all of which affect vector-borne disease episystems. For example, factors involving the suite of ecological and physiological conditions that influence vectors, hosts, and pathogens, as well as socioeconomic and human behavioral factors, are influenced by weather and climate and interact with one another and affect disease episystems. As a result, predicting weather, climate, climate change, and environmental effects on specific disease system becomes problematic because of the following:

A. We do not know all of the primary drivers of specific vector-borne disease episystems.
B. We do not understand how the primary drivers of these episystems interact with one another to influence vector-borne disease transmission.
C. We do not know the primary mechanisms that affect vector competence and how these are influenced by the primary biotic and abiotic drivers.
D. There may be different controlling mechanisms in different vector, pathogen, and vertebrate host populations that interact with the environment and with the genomes of the biological components of each episystem.

The effects of weather and climate must be studied in the context of the complex interacting factors that influence vector-borne disease episystems. This will likely have to be done on a microscale involving small, localized regions since the complex factors, including variation within vector species, will likely show differences over different spatial and temporal scales. The challenge is great, but it is naïve to expect that general findings about climate change can be extrapolated broadly when there is so much biological diversity in different episystems throughout the world. Humankind must prepare for the consequences of anticipated climate and environmental changes on vector-borne disease episystems. Improving our understanding of complex episystems is essential. There is no doubt that vector-borne disease transmission patterns will change in response to changing environmental conditions, including changes in weather and climate. Undoubtedly some episystems will prosper, exposing more humans and animals to greater risk of disease transmission.

Humans also influence disease risk through human activities that change the environment. Attempts toward climatic stability may be impossible to attain and maintain, might be undesirable, and may have unpredictable consequences. Other drivers of vector-borne diseases might have greater influence over vector-borne disease episystems than weather and climate and may be more easily modified through human intervention. Improving social infrastructure, public health, and socioeconomic status by reducing poverty throughout the world would likely have a greater immediate impact on reducing vector-borne diseases than efforts to maintain a static climate. Reducing global warming is clearly desirable for a variety of reasons. However, even reducing global warming would not necessarily decrease the risk of vector-borne disease transmission nor reduce future changes in vector-borne disease episystems. Many factors are at play. It is essential for humankind to address all of the factors that may influence vector-borne disease transmission risk to effectively reduce the impact of these episystems on our world.

ACKNOWLEDGEMENT

We thank Mr. James Newman for his assistance with the figures. Mr. Timothy G. Hope's assistance with preparing the manuscript was greatly appreciated.

REFERENCES

Anyamba, A., Chretien, J-P., Small, J., et al., 2009. Prediction of a Rift Valley fever outbreak. *Proc. Natl. Acad. Sci. U.S.A.* **106**, 955–959.

Chaves, L.F., Koenraadt, C.J., 2010. Climate change and highland malaria: fresh air for a hot debate. *Q. Rev. Biol.* **85**(1), 27–55.

Day, J.F., 2008. A brief history of dengue virus in Florida. *Florida J. Environ. Health* 201(Winter 2008), 7–10.

Day, J.F., Curtis, G.A., 1993. Annual emergence patterns of *Culex nigripalpus* females before, during and after a widespread St. Louis encephalitis epidemic in south Florida. *J. Am. Mosq. Control Assoc.* **9**, 249–255.

Day, J.F., Curtis, G.A., 1994. When it rains, they soar—and that makes *Culex nigripalpus* a dangerous mosquito. *Am. Entomol.* **40**, 162–167.

Day, J.F., Curtis, G.A., 1999. *Culex nigripalpus* (Diptera: Culicidae) blood feeding and oviposition before, during and after a widespread St. Louis encephalitis epidemic in Florida. *J. Med. Entomol.* **36**, 176–181.

Day, J.F., Shaman, J., 2009. Severe winter freezes enhance St. Louis encephalitis virus amplification and epidemic transmission in peninsular Florida. *J. Med. Entomol.* **46**, 1498–1506.

Elliott, R.M., 2009. Bunyaviruses and climate change. *Clin. Microbiol. Infect.* **15**, 510–517.

Epstein, P.R., 2000. Is global warming harmful to health? *Sci. Am.* **283**, 50–57.

Epstein, P.R., 2007. Chikungunya fever resurgence and global warming. *Am. J. Trop. Med. Hyg.* **76**(3), 403–404.

Gage, K.L., Burkot, T.R., Eisen, R.J., et al., 2008. Climate and vector-borne diseases. *Am. J. Prev. Med.* **35**(5), 436–450.

Githeko, A.K., Lindsay, S.W., Confalonieri, U.E., et al., 2000. Climate change and vector-borne disease: a regional analysis. *Bull. World Health Org.* **78**(9), 1136–1147.

Gould, E.A., Higgs, S., 2009. Impact of climate change and other factors on emerging arbovirus diseases. *R. Soc. Trop. Med. Hyg.* **103**, 109–121.

Gubler, D.J., 2002. The global emergence/resurgence of arboviral diseases as public health problems. *Arch. Med. Res.* **33**, 330–342.

Gubler, D.J., 2008. The global threat of emergent/reemergent diseases, in: *Vector-borne Diseases: Understanding the Environmental, Human health and Ecological Connections*, Institute of Medicine. Washington, DC: The National Academies Press, pp. 43–64.

Gubler, D.J., Reiter, P., Kristie, L.E., et al., 2001. Climate variability and change in the United States: potential impacts on vector- and rodent-borne diseases. *Environ. Health Perspect.* **109**(Suppl. 2), 223–233.

Guis, H., Caminade, C., Calvete, C., et al., 2011. Modeling the effects of past and future climate on the risk of bluetongue emergence in Europe. *J. R. Soc. Interface* **9**(67), 339–350.

Hales, S., de Wet, N., Macdonald, J., et al., 2003. Potential effect of population and climate changes on global distribution of dengue fever: an empirical model. *Lancet* **360**, 830–834.

Hardy, J.L., Houk, E.J., Kramer, L.D., et al., 1983. Intrinsic factors affecting vector competence of mosquitoes for arboviruses. *Ann. Rev. Entomol.* **28**, 229–262.

Henson, R., 2007. *The Rough Guide to Weather*. London: Rough Guides Ltd., 432 pp.

Intergovernmental Panel on Climate Change (IPCC), 2001. Third Assessment Report, Cambridge University Press, Cambridge, UK.

Intergovernmental Panel on Climate Change (IPCC), 2007. Fourth Assessment Report, Cambridge University Press, Cambridge, UK.

Jetten, T.H., Focks, D.S., 1997. Potential changes in the distribution of dengue transmission under climate warming. *Am. J. Trop. Med. Hyg.* **57**, 285–297.

Johansson, M.A., Dominici, F., Glass, G.E., 2009. Local and global effects of climate on dengue transmission in Puerto Rico. *PLoS Negl. Trop. Dis.* **3**(2), e382.

Karabatsos, N., (ed.), 1985. *International Catalogue of Arboviruses Including Certain Other Viruses of Vertebrates*, 3rd. ed. San Antonio, TX: American Society of Tropical Medicine and Hygiene, 1147 pp.

Lafferty, K.D., 2009. The ecology of climate change and infectious diseases. *Ecology* **90**(4), 888–900.

Martens, P., 1999. How will climate change affect human health. *Am. Sci.* **87**(6), 534–541.

Martin, J.L.S., Brathwaite, O., Zambrano, B., et. al., 2010. The epidemiology of dengue in the Americas over the last three decades: a worrisome reality. *Am. J. Trop. Med.* Hyg. **82**, 128–135.

National Climate Data Center (NCDC), 2011. NOAA Satellite and Information Service: http://www.ncdc.noaa.gov/oa/ncdc.html. Accessed July 6, 2013.

Patz, J.A., Olson, S.H., 2008. Climate change and health: global to local influences on disease risk, in: *Vector-borne Diseases: Understanding the Environmental, Human health and Ecological Connections*, Institute of Medicine. Washington, DC: The National Academies Press, pp. 88–103.

Patz, J.A., Martens, W.J., Focks, D.A., et al., 1998. Dengue fever epidemic potential as projected by general circulation models of global climate change. *Environ. Health Perspect.* **106**(3), 147–153.

Purse, B.V., Mellor, P.S., Rogers, D.J., et al., 2006. Climate change and the recent emergence of bluetongue in Europe. *Nat. Rev.* **3**, 171–182.

Radke, E.G., Gregory, C.J., Kintziger, K.W., et al., 2012. Dengue outbreak in Key West, Florida, USA, 2009. *Emerg. Infect. Dis.* **18**(1), 135–137.

Randolph, S.E., 2009. Perspectives on climate change impacts on infectious diseases. *Ecology* **90**(4), 927–931.

Randolph, S.E., 2010. To what extent has climate change contributed to the recent epidemiology of tick-borne diseases? *Vet. Parasitol.* **167**, 92–94.

Randolph, S.E., Rogers, D.J., 2010. The arrival, establishment and spread of exotic diseases. *Nat. Rev. Microbiol.* **8**, 361–371.

Reiter, P., 2001. Climate change and mosquito-borne disease. *Environ. Health Perspect.* **109**(Suppl. 1), 141–161.

Reiter, P., Thomas, C.J., Atkinson, P.M., et al., 2004. Global warming and malaria: a call for accuracy. *Lancet* **4**, 323–324.

Richards, S.L., Mores, C.N., Lord, C.C., et al., 2007. Impact of extrinsic incubation temperature and virus exposure on vector competence of *Culex quinquefasciatus* Say (Diptera: Culicidae) for West Nile virus. *Vector-borne Zoon. Dis.* **7**, 629–636.

Rogers, D.J., Randolph, S.E., 2000. The global spread of malaria in a future, warmer world. *Science* **289**, 1763–1766.

Rogers, D.J., Randolph, S.E., 2006. Climate change and vector-borne diseases. *Adv. Parasitol.* **62**, 345–381.

Rosenthal, J., 2009. Climate change and the geographic distributions of infectious diseases. *Ecohealth* **6**, 489–495.

Russell, R.C., 1998. Mosquito-borne arboviruses in Australia: the current scene and implications of climate change for human health. *Int. J. Parasitol.* **28**, 955–969.

Russell, R.C., 2009. Mosquito-borne disease and climate change in Australia: time for a reality check. *Aust. J. Entomol.* **48**, 1–7.

Russell, R.C., Currie, B.J., Lindsay, M.D., et al., 2009. Dengue and climate change in Australia: predictions for the future should incorporate knowledge from the past. *Med. J. Aust.* **190**(5), 265–268.

Shaman, J., Day, J.F., Stieglitz, M., 2002. Drought-induced amplification of St. Louis encephalitis virus in Florida. *Emerg. Infect. Dis.* **8**, 575–580.

Shaman, J., Day, J.F., Stieglitz, M., 2003. St. Louis encephalitis virus in wild birds during the 1990 south Florida epidemic: the importance of drought, wetting conditions, and the emergence of *Culex nigripalpus* (Diptera: Culicidae) to arboviral amplification and transmission. *J. Med. Entomol.* **40**, 547–554.

Shaman, J., Day, J.F., Stieglitz, M., 2004. The spatial–temporal distribution of drought, wetting, and human cases of St. Louis encephalitis in south-central Florida. *Am. J. Trop. Med. Hyg.* **71**, 251–261.

Shaman, J., Day, J.F., Stieglitz, M., 2005. Drought-induced amplification and epidemic transmission of West Nile virus in south Florida. *J. Med. Entomol.* **42**, 134–141.

Shaman, J., Day, J.F., Komar, N., 2010. Hydrologic conditions describe West Nile virus risk in Colorado. *Int. J. Med. Environ. Res. Public Health* **7**, 494–508.

Slenning, B.D., 2010. Global climate change and implications for disease emergence. *Vet. Pathol.* **47**, 28–33.

Stresman, G.H., 2010. Beyond temperature and precipitation: ecological risk factors that modify malaria transmission. *Acta Trop.* **116**, 167–172.

Sutherst, R.W., 2004. Global change and human vulnerability to vector-borne diseases. *Clin. Microbiol. Rev.* **17**(1), 136–173.

Tabachnick, W.J., 1994. The role of genetics in understanding insect vector competence for arboviruses. *Adv. Dis. Vector Res.* **10**, 93–108.

Tabachnick, W.J., 2004. *Culicoides* and the global epidemiology of bluetongue virus. *Vet. Ital.* **40**(4), 145–150.

Tabachnick, W.J., 2010. Challenges in predicting climate and environmental effects on vector-borne disease episystems in a changing world. *J. Exp. Biol.* **213**, 946–954.

Thai, K.T., Anders, K.L., 2011. The role of climate variability and change in the transmission dynamics and geographic distribution of dengue. *Exp. Biol. Med.* **236**(8), 944–954.

Wilson, A., Mellor, P.S., 2008. Bluetongue in Europe: vectors, epidemiology and climate change. *Parasitol. Res.* **103**(Suppl. 1), s69–s77.

Zell, R., Krumbholz, A., Wutzler, P., 2008. Impact of global warming on viral diseases: what is the evidence? *Curr. Opin. Biotechnol.* **19**, 652–660.

3

INFLUENCE OF CLIMATE CHANGE ON MOSQUITO DEVELOPMENT AND BLOOD-FEEDING PATTERNS

William E. Walton

Department of Entomology and the Center for Disease Vector Research, University of California, Riverside, CA, USA

William K. Reisen

Center for Vectorborne Diseases and Department of Pathology, Microbiology and Immunology, School of Veterinary Medicine, University of California, Davis, CA, USA

TABLE OF CONTENTS

Viral Infections and Global Change, First Edition. Edited by Sunit K. Singh.

3.1. INTRODUCTION

Human activities are rapidly causing unprecedented changes in global climate that will impact mosquito vectors and the viruses they transmit. Atmospheric carbon dioxide (CO_2) levels are currently higher than anytime during the last 100 000 years (Siegenthaler et al., 2005). Concomitant warming has elevated global mean temperature above levels found during the past 1200 years (Esper et al., 2002). The mean global temperature is predicted to increase between 1.5 and 4.5 °C during the next century (IPCC, 2007), but these changes will not occur homogeneously across latitude. Climate warming at the equator is proceeding slower than at high latitudes in the Nearctic and Palearctic, and summer temperatures are warming comparatively slower than are winter temperatures (IPCC, 2001, 2007). Winter temperatures in the Arctic (Alaska, western Canada, and eastern Russia) have increased by 2–4 °C (4–7 °F) over the last half century (ACIA, 2004). Diel temperature patterns are changing as well, with nighttime temperatures warming comparatively more than daytime temperatures (IPCC, 2001).

Changes of global temperature also will affect the distribution of precipitation. Northern high latitudes will be wetter, whereas Africa and Asia will be drier than current conditions (IPCC, 2007). Despite considerable spatial variation in rainfall across the land masses, the planet is drier today than it was at any time since 1900. However, storms on a warmer planet are predicted to be more intense and variable, leading to increased flooding (IPCC, 2007).

Changes associated with continued warming (IPCC, 2001) that are likely to affect vector-borne viral diseases include (i) the expansion of biogeographic ranges of vectors and arboviruses (Altizer and Pederson, 2008), (ii) an earlier arrival of spring and later arrival of winter that will extend the annual activity period of mosquitoes and viruses in regions where cold weather currently limits the activity of vectors, (iii) changes in the distribution of plants and animals that may influence the distribution of the vertebrate arbovirus reservoirs that differ in their potential to infect vector mosquitoes, (iv) changes in reservoir and vector population structure in response to abiotic stresses, and (v) changes in precipitation leading to heavy rain and snowfalls that will cause flooding in some regions but extended droughts in other regions. Besides the effect of higher mean global temperature and changing precipitation patterns on individual species, the rate at which warming is occurring is predicted to have a dramatic impact on biodiversity (Parmesan, 2006). Biodiversity loss and habitat destruction can increase the spread of nonindigenous vectors and pathogens (Pongsiri et al., 2009). Extensive circumglobal urbanization has compounded this process by further reducing diversity and creating a suite of peridomestic niches that are readily exploited by commensal species.

This chapter addresses the responses of the mosquito vectors of viral pathogens to increased temperature, changing patterns of precipitation, and increased carbon dioxide concentration associated with global climate change. We focus on immature development and related life history changes and on the blood-feeding patterns that influence the demography and vectorial capacity of mosquito populations. We discuss briefly the impact of climate change on mosquito-borne viral infections but refer the reader to other chapters that discuss viral pathogens in greater detail. We conclude that while the predicted changes in abiotic factors associated with climate warming are likely to increase the size and distribution of vector populations and enhance the transmission rates of viral pathogens, the epidemiologies of diseases caused by viral pathogens will not be independent of the changes in human population size and land use, especially the storage and use of water across the landscape.

3.2. MOSQUITO DEVELOPMENT

3.2.1. Temperature

Mosquitoes are poikilotherms and are intimately linked to aquatic environments for immature development. Because the temperature of a mosquito fluctuates with that of the environment, metabolic rate, the time between hatching and emergence, incubation period of viral pathogens, and other biological processes of mosquitoes are related to ambient temperature. Temperature, nutrition, and larval density are the principal extrinsic factors that affect rates of growth and development of the subadult stages in the mosquito life cycle (Clements, 1992). In nature, these and other (e.g., salinity, water depth) factors can interact to influence developmental rates (Clements, 1992; Padmanabha et al., 2011; Reisen et al., 1989). Although the relationship between development time and these extrinsic factors can be complex in nature, results from experiments carried out under controlled laboratory conditions indicate that the effects of environmental temperature on immature mosquito development differ among mosquito species (Pritchard and Mutch, 1985) and among populations of the same species (Bradshaw and Holzapfel, 2010) living at different latitudes.

Adult mosquitoes can behaviorally thermoregulate by selecting different resting and activity space. *Culex tarsalis*, for example, escapes the heat of the day by resting in refugia, such as vegetation or rodent burrows, and egresses after dusk when temperatures are lower and humidity generally higher (Meyer et al., 1990). During extremely hot and dry conditions, the timing of the egress and host-seeking activities may be delayed until after temperatures have reached optimal conditions (Reisen et al., 1997b). In contrast, the immature stages are subject to the conditions of their aquatic habitat, selected by the ovipositing female. Although the diel means may be similar, in general, small and shallow surface pools fluctuate more widely than do large and deep pools, thereby subjecting the immatures to a wider range of temperature. Warming scenarios may increase the duration of exposure of larval stages to temperatures approaching their thermal maximum.

The hypothetical thermal sensitivity of a fitness-related trait can be portrayed by a performance curve (Figure 3.1a). Fitness-related activities include the optimal temperature (T_{opt}) at which performance is maximized. The breadth of the performance curve indicates the extent of thermal specialization, and the limits of fitness-related activities are defined by the variables T_{min} and T_{max}. The asymmetry of the performance curve is related to the changes in enzyme structure and function with temperature. Enzymatic reactions are slowed at cold temperatures, and the precipitous decline in performance at high temperatures is caused by changes in the shape of proteins, which either slow or inhibit metabolism (Gilchrist and Folk, 2008). Assuming that fitness is proportional to the level of performance, the summation of the amount of time at each temperature multiplied by the performance at that temperature is an estimate of the lifetime fitness of an individual (Gilchrist, 1995, 2000). Enzyme kinetics-based models have been used to predict the rate of development of vectors, extrinsic cycles of viral pathogens, and temperature-related blood digestion (Focks et al., 1993) and have been extended to climate change scenarios (Martens et al., 1997).

Climate change will influence the distribution of temperatures during the lifetime of an individual and can impose selection on the shape and position of the thermal performance curve. Gilchrist and Folk (2008) suggested that if thermal preference remains unchanged and the area under the performance curve is constant, then a shift in the performance curve along the temperature axis (Figure 3.lb) or an evolutionary shift of either thermal limit (Figures. 3.1c and 3.ld) in response to a warmer thermally stressful environment will generally lower rates of population growth and reduce fitness in populations undergoing

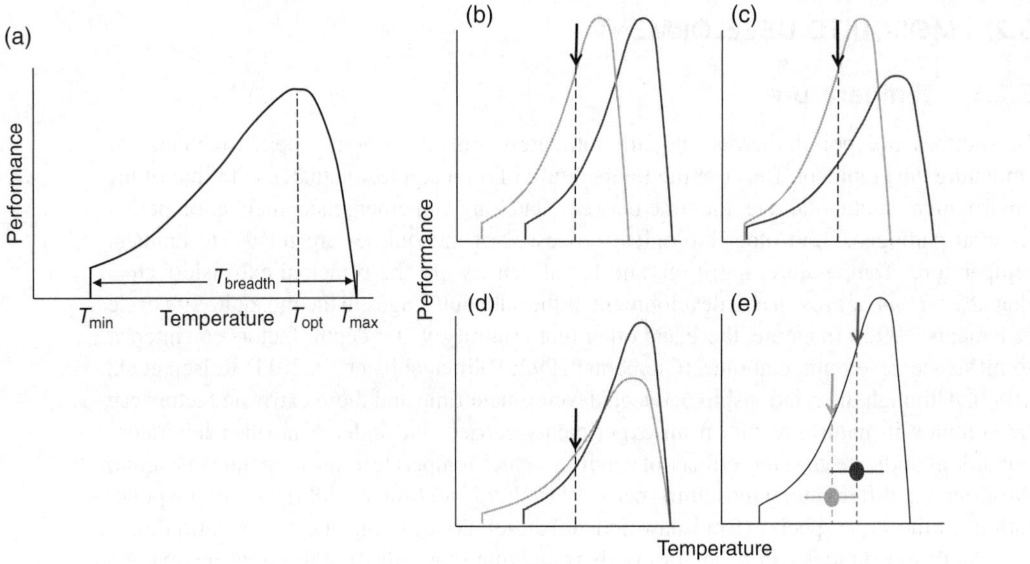

Figure 3.1. Hypothetical performance curve. A hypothetical performance curve illustrating the relationship between a performance-based measure of fitness and environmental temperature **(a)** and scenarios for performance curve or thermal preference evolution in response to climate warming **(b–e)** (redrawn from Gilchrist and Folk, 2008). The optimal temperature (T_{opt}), maximum limit of performance (T_{max}), minimum limit of performance (T_{min}), and degree of temperature specialization as indicated by the performance breadth ($T_{breadth}$) are depicted. For **(b)** through **(d)**, a constant area under the performance curve constrains performance curve evolution. Current conditions are shown in blue and future conditions under global warming are shown in red. The black arrows indicate thermal preference prior to selection. Selection by climate warming generally decreases performance at the preference point. **(b)** The performance curve shifts horizontally with climate warming. **(c)** Evolution of the maximum thermal performance limit and the minimum thermal limit is constrained. **(d)** Evolution of the minimum thermal limit and the maximum thermal limit is constrained. **(e)** The performance curve does not change with global warming, but the temperature preference changes (from the blue arrow to the red arrow). The mean and variation in the temperature regimes of two hypothetical climate states are shown under the curve. A warming climate increases the risk of thermal damage as the population resides in an environment that is closer to the collapse in performance at high environmental temperature. For color detail, please see color plate section.

directional selection resulting from climate change. Relaxing the assumption of a constant area under the performance curve can also result in lower fitness of a population under thermal stress imposed by a warmer climate. An increase in mortality and reproductive failure under thermally stressful conditions exert selective pressure for increased tolerance of the stress; the effects on local demographics will be greatest at the ends of species ranges (Gilchrist and Folk, 2008).

Alternatively, thermal preference can change in response to a warming climate. If the relationship of the performance curve with temperature is unchanged but thermal preference shifts, then a shift in thermal preference to a warmer temperature will situate the population nearer to the temperatures that cause metabolic collapse on the right side of the performance curve (Figure 3.1e). Gilchrist and Folk (2008) suggest that an increase in thermal preference without a change in the performance curve is unlikely to be a successful long-term response to climate change. Probably least likely is a scenario of climatic matching in which climate change does not cause an evolutionary shift in both the performance curve and the thermal preference.

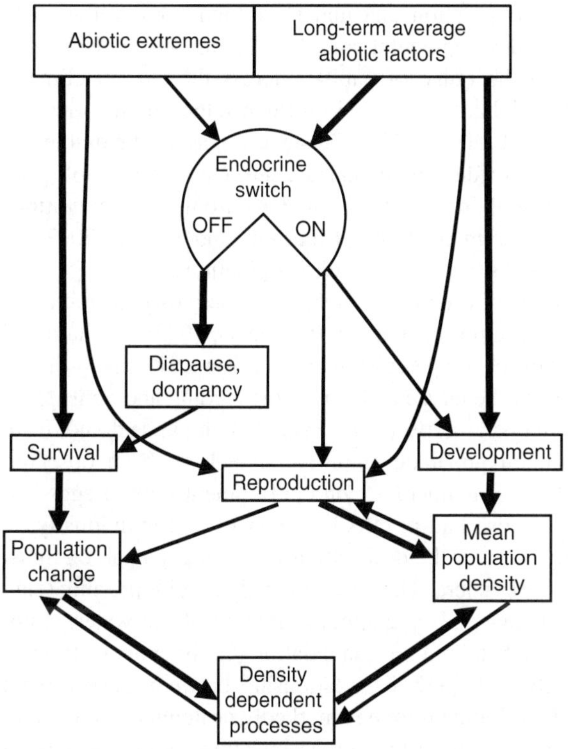

Figure 3.2. The main effects of abiotic extremes (weather, microclimate) and long-term averages of abiotic factors (climate) on a mosquito population. Modified from Varley et al. (1973). The demographic consequences of climate change will influence mean population density, whereas vagaries of abiotic factors associated with local weather affect survival and consequently population change. Physiological adaptation to climate change will be manifest primarily through changes in the response of the endocrine system to temperature-compensated responses to day-length cues used to time major events of mosquito life histories (see Bradshaw and Holzapfel, 2010).

The demographic consequences of short-term variation in climate or weather versus long-term changes in average climatic conditions are likely to be very different. Varley et al. (1973) distinguished between the effects of long-term average conditions at a location (climate) and the extremes of abiotic factors at a smaller temporal scale (e.g., daily weather) on insect populations. Climate zones defined by temperature and the distribution of precipitation can be represented as isotherms or rainfall isopleths, which are influenced by latitude and altitude. The average abiotic conditions have a large influence on the mean population density of a species, whereas the extremes of weather influence survival and population variation (Figire 3.2). Varley et al. (1973) argued further that weather can influence insect physiology in four fundamental ways by modifying (i) the activity of the endocrine system, (ii) survival, (iii) development, and (iv) reproduction. Variation in weather can directly affect survival; however, the effects of microclimate variability on reproduction and development are modulated through endocrine responses to environmental conditions (Figure 3.2).

Endocrine-mediated changes in life histories made in response to proximate cues used to predict environmental conditions facilitate species persistence at a particular latitude. Responses to photoperiods that cue processes such as reproduction, initiation and cessation of development, and migration may be comparatively more important than temperature responses for long-term persistence under conditions of climate change (Bradshaw and Holzapfel, 2006, 2010). However, in *Culex* mosquitoes reproductive diapause requires cool temperature re-enforcement of short photoperiods during the 4th instar to induce diapause (Eldridge, 1968; Reisen, 1986; Reisen et al., 1986), and females exposed to only short photoperiod will not enter reproductive arrest. Therefore, under warming conditions, southern populations may lose their ability to enter diapause.

The genetic structure of a vector population that underlies phenotypic variation in thermal tolerance, bioenergetic, and developmental and behavioral mechanisms that facilitate survival and reproduction in the range of habitats across the current climatic gradient will determine the extent that global warming will influence the life histories and biogeography of insects (Sweeney et al., 1992). Yet, during the periods favorable for development, temperature-related effects on developmental rate, survival, and physiological tradeoffs that influence reproduction will interact with other environmental conditions (e.g., nutritional conditions and trophic interactions; Blaustein and Chase, 2007) to determine the population growth rate and fitness of mosquito populations.

The relationship between development time for each stage of the mosquito life cycle and temperature is typically described as a power function (Clements, 1992; Pritchard and Mutch, 1985). The potential interaction of temperature with nutrition and density on the development time of eggs is less than for the larvae and pupae, the feeding and nonfeeding stages of the aquatic portion of the life cycle, respectively. Even though pupae do not feed, carry-over effects of larval nutrition on pupal development time have been observed (Pritchard and Mutch, 1985). The effect of temperature on development time of eggs laid directly on the water surface, such as those laid in rafts by *Culex* or laid individually by *Anopheles*, is determined more easily than for the desiccation-resistant eggs laid on moist substrates by *Aedes* and other floodwater genera. The latter typically require priming (e.g., a prescribed period of desiccation; changes in oxygen concentration of the water, atmospheric pressure, and salinity) to initiate hatching after inundation (Gerberg et al., 1994).

Well-fed larval mosquitoes exhibit a hyperbolic relationship between development time and temperature (Figure 3.3a). Development time is markedly prolonged at cold temperatures above freezing and then declines rapidly with increasing temperature. As ambient temperature approaches the upper thermal limit for survival (>30–40 °C for most mosquitoes), development time increases slightly (Figure 3.3a). The findings of studies on the relationship between temperature and rate of mosquito growth and development indicate that (i) growth and development occur within a temperature range defined by a lower developmental threshold (the developmental zero) and an upper lethal temperature; (ii) within the temperature range favorable for development, rate of growth and development is positively correlated with temperature; and (iii) the temperature ranges favorable for growth and shape of the growth curve differ among mosquito species (Clements, 1992).

Over the range of temperature favorable for development (i.e., the central region of the hyperbola), the product of developmental period and temperature is often assumed to be constant (but see Pritchard and Mutch, 1985):

$$t(T-c)=k \quad (3.1)$$

where t is the development time (days or hours), T is the mean ambient temperature (°C), c is the estimated developmental zero temperature (°C), and k is the thermal constant (degree-days (DD)). The minimum value of the development time versus temperature curve represents a measure of the optimal temperature for growth. Survival, reproduction, and other important life history characteristics often attain maxima at, or very near, this mean environmental temperature (Vannote and Sweeney, 1980). This is the classical DD method used to quantify the thermal requirements for development of insects (Vannote and Sweeney, 1980; Varley et al., 1973) and corresponds to T_{opt} in the models of Gilchrist and Folk (2008).

An alternative to using the duration of development is to plot the developmental rate versus the rearing temperature, again corrected for the developmental zero. Developmental

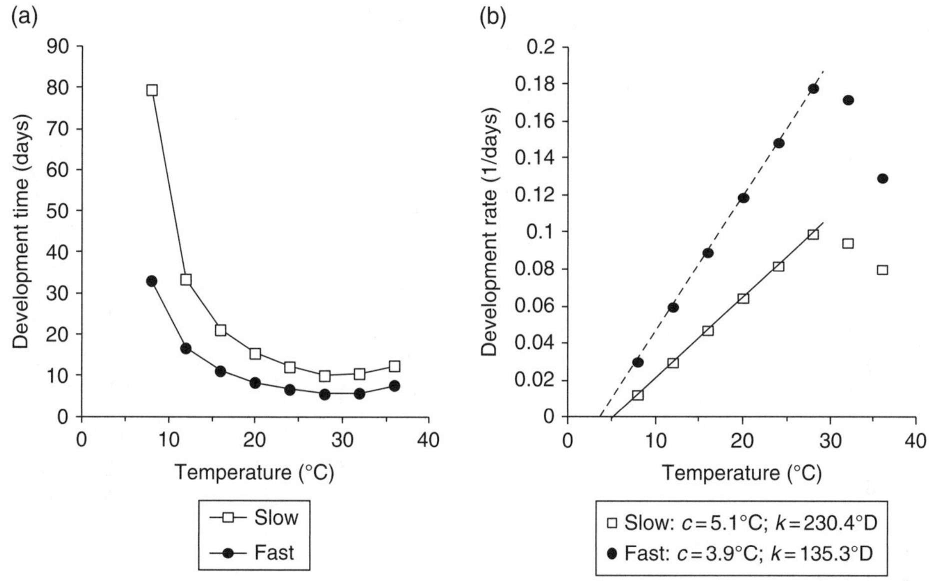

Figure 3.3. The relationship between development time (**a**) and developmental rate (**b**) for two hypothetical mosquitoes. One species develops rapidly across the range of environmental temperatures, and the second species develops comparatively slowly. The faster developing species exhibits a lower developmental zero (*c*) and a lower thermal requirement to complete immature development (*k*) than does the slower developing species.

velocity is defined as the reciprocal of the developmental duration (t) and is positively correlated with temperature:

$$\frac{1}{t} = \frac{T - c}{k} \tag{3.2}$$

This relationship has been modeled as a sigmoid curve across a broad range of environmental temperatures because developmental rate declines under stressful high temperatures and is low under cold conditions near the developmental zero (Clements, 1992). The central region of the curve relating developmental velocity with temperature is effectively linear (Figure 3.3b), and the slope ($1/k$) indicates thermal adaptation across the range of favorable thermal environments. The developmental zero is estimated by extrapolation from the linear region of the curve to the temperature axis. However, c may be higher than the actual developmental threshold because the larvae of some mosquito species may grow at low temperatures but do not complete development (Clements, 1992).

Differences in the developmental processes between mosquitoes currently found in the tropics versus species occurring in the temperate zones provide an indication of the expected changes of the thermal limits on growth and temperature-related growth rates as the climate continues to warm. Among mosquitoes, the trend for a decreasing slope and decreasing intercept for the relationship between the rate of development and temperature in species that range further north (Figure 3.4a) is viewed as indicating that metabolic adjustments have occurred in relation to the temperature regimes in environments colonized since the last period of glaciation (Pritchard and Mutch, 1985). This finding contrasts

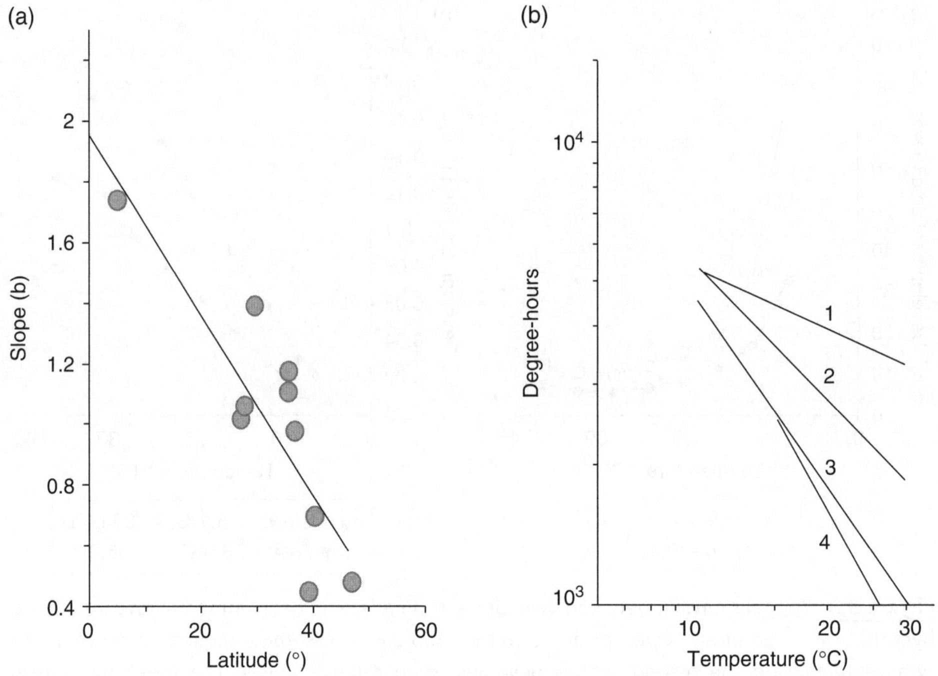

Figure 3.4. The relationship between the slope for developmental rate and temperature. The relationship between the slope for developmental rate with temperature for mosquitoes living at different latitudes (**a**) and the thermal requirement for development of four mosquito species (**b**) (redrawn from Pritchard and Mutch, 1985). The slope of the log transformation of the relationship $D=aT^b$ is the dependent variable in panel (**a**). D is development time. T is temperature (°C, uncorrected for developmental zero). The numbers in panel (**b**) correspond to *Ae. sticticus* from 50N (1), *Aedes vexans* from 40N (2), *Anopheles quadrimaculatus* from 32.5N (3), and *Toxorhynchites brevipalpis* from 7S (4). For color detail, please see color plate section.

those for the aquatic immature stages of some other insects, such as dragonflies, which appear to have been very conservative in the evolution of the temperature–developmental rate relationship.

Adaptation to cool environments with short growing seasons is manifest in species residing in the high-latitude temperate zone by a gain in relative growth efficiency: that is, a requirement of fewer heat units to complete development at low temperatures. The annual heat budget for lotic systems decreases linearly with latitude from about 6500 DD (relative to 0 °C) at 31 N to 1500 DD at 50 N and then nonlinearly at even higher latitudes (first- to fifth-order streams and rivers; Sweeney et al., 1992). Mosquitoes are not found in flowing water, but the general trend of decreasing thermal units with latitude is biologically relevant. A lower absolute slope value for the relationship of development time with temperature may be interpreted as showing adaptation to a wider range of temperatures. Pritchard and Mutch (1985) posited that temperate-zone mosquito species that belong to families that have mainly tropical distributions are expected to have temperature–growth rate relationships more similar to those of their tropical relatives than to sympatric species that belong to predominantly temperate-zone families.

Climate warming is expected to favor vector populations that have phenotypes that are comparatively more tropical for the relationship between development and temperature. A warmer temperature regime is expected (i) to favor individuals that develop comparatively quickly under warmer conditions, (ii) to shift T_{min} to a higher temperature, and (iii) to increase the slope of the relationship between the rate of immature development and temperature. As the climate warms, the developmental rate of mosquitoes is expected to increase directly with warming until the upper lethal limit of development is approached. Tewksbury et al. (2008) speculated that tropical ectothermic vertebrates were at greater risk of extinction than temperate species because animals living in tropical environments exist across a narrower range of environmental temperatures and climate change would push populations close to the precipitous decline in performance at high temperatures. It is unknown whether this is also true for mosquitoes. The extent that thermal tolerances and developmental rates will change as climate warms is currently unknown.

The geographic distributions of most mosquito vectors are bounded by temperature and are expected to move poleward and increase in altitude, as conditions warm. A 1 °C rise in temperature will correspond to an extension equivalent to 90 km of latitude and 150 m of altitude in the range of a vector (Peters, 1993). A 4 °C rise in temperature will change the thermal regime of a midlatitude aquatic habitat to correspond to a thermal regime that is currently 680 km toward the equator (Sweeney et al., 1992). For example, *Aedes albopictus* is bounded by the mean 0 °C isotherm in the coldest month of the year in the northern hemisphere. Even if no evolution in its performance profile occurs, the geographic distribution will change with climate warming. Potential changes in performance profiles will influence demographics and distribution of vectors: rapid evolutionary responses to climate change may decrease performance at higher temperatures or cause shifts in thermal preference that alter geographical distribution, with little change in vectorial capacity and vector competence. By altering vector population size, biting rates, and daily survival, the net effect of larger vector populations and increased biting rates at warmer temperatures on vectorial capacity might be offset by a decrease in daily survival.

However, warm temperatures during larval development may decrease the vector competence of *Culex* females for per os infection with alphaviruses, but not flaviviruses. Repeated sampling of both foothill (Hardy et al., 1990) and desert (Reisen et al., 1996) populations showed a 2- to 3-order magnitude decrease in the infectious dose required to infect 50% of *Cx. tarsalis* females with Western equine encephalomyelitis virus (WEEV), but not St. Louis encephalitis virus (SLEV), during midsummer. Because these females were collected as immatures, it was presumed that temperature during larval development reduced adult female size at emergence and susceptibility of the midgut to infection. However, the mechanism for this temporal association was not resolved by detailed subsequent laboratory experiments that held genotype constant (Hardy and Reeves, 1990).

The direct relationship between developmental rate and temperature will cause the number of generations per annum of multivoltine mosquitoes to increase in warmer climates. Short-term warming cycles of climate (i.e., El Niño southern oscillation (ENSO) events; Heft and Walton, 2008) can increase the potential number of generations per year by 40–50% above the number of generations under long-term average conditions for *Culex* mosquitoes. For example, the strong El Niño cycle that occurred between 1997 and 1999 increased the number of generations of *Cx. tarsalis* from the long-term average of 5.1 ± 0.7 per year to 7.0 ± 1.1 per year at four urban locations in the greater metropolitan Los Angeles area during the warmest year of the cycle. In the warmest of the four locations (Irvine vs. Long Beach, Port Hueneme, and Santa Monica) within the LA basin, this comparatively strong ENSO cycle increased the potential number of generations to 8.4 per annum (assuming 287.4 DD were required to complete immature development and egg laying activity did not occur at temperatures less than 12 °C). An increase in the mean population size of vectors is likely to occur in warmer climates; however, the negative impact of a warmer climate on adult survival and the effect of warmer climates on blood-feeding patterns will ultimately determine the change in virus transmission.

3.2.2. Precipitation

Changing precipitation patterns will affect the abundance and distribution of developmental sites and alter survival of adult vectors. The increased unpredictability and severity of storms on a warmer planet are likely to reduce the survival of adult arbovirus vectors independently of vector density. The impact of changing precipitation patterns on mosquito vectors will be influenced additionally by local evaporation rates (saturation deficit), soil type, slope of the terrain, and hydrological factors such as the proximity to water courses that flood. Changing rainfall patterns will have lesser effects on the populations and virus epidemiology for vectors associated with domestic habitats than for vector populations residing outside the peridomestic zone. In contrast in peridomestic urban environments, high runoff from impervious surfaces actually serves as an important mortality factor for species developing in municipal drainage systems (Metzger, 2004; Su et al., 2003).

Mosquito populations generally increase with increasing precipitation. Flooding associated with snowmelt and rainfall provides developmental sites for mosquitoes found in permanent wetlands or semipermanent lentic habitats (Wegbreit and Reisen, 1994). Rainfall also increases the abundance of container-dwelling mosquitoes (Vezzani et al., 2004) and floodwater species that utilize intermittently flooded areas for developmental sites. A low saturation deficit (high humidity) will prolong vector survival but may lower survival through increased susceptibility to fungal and bacterial pathogens (Mellor, 2004). The net effect on vector abundance from the interplay between the reduction of immature vector density from flushing and drowning associated with precipitation and the enhancement of vector populations caused by greater availability of developmental sites will differ spatially among habitats. Even comparatively short-term changes in weather (e.g., ENSO) can have a significant impact of vector abundance and species composition (Heft and Walton, 2008) and arthropod-borne diseases (Maelzer et al., 1999; Poveda et al., 2001).

A high saturation deficit (dehydration) will decrease survival of adult vectors and reduce the availability of developmental sites. Dry conditions may increase blood feeding to compensate for water loss (Mellor, 2004). Changes in the temperature regime and distribution of rainfall have the potential to alter the biogeographic distribution and timing of seasonal events in life histories of mosquitoes (Bradshaw and Holzapfel, 2010), as well as increase the population sizes and number of generations per annum.

3.2.3. Effects of elevated CO_2 concentration

Elevated CO_2 concentrations can decrease larval mosquito growth and survival by altering the chemical properties of water in mosquito developmental sites and reducing the quantity and quality of food resources for larvae (i.e., decreasing decomposition rates of organic matter and reducing the quality of detritus derived from higher plants). Elevated atmospheric CO_2 concentrations are predicted to increase diffusion of CO_2 into aquatic ecosystems, which will lower the pH of poorly buffered aquatic habitats. The metabolism of heterotrophic consumers causes most freshwater habitats to be net sources of CO_2 (Wetzel, 2001). Eutrophic waters that commonly support mosquito and other virus vectors (i.e., Ceratopogonidae) are unlikely to be greatly affected by elevated levels of atmospheric CO_2.

Elevated CO_2 levels cause higher lignin concentrations in plant tissues and increase the ratio of carbon to nitrogen (Cotrufo and Ineson, 1996; Cotrufo et al., 1994, 1999; Frederiksen et al., 2001; Trumble and Butler, 2009). Such changes in plant composition are predicted to slow degradation of leaf litter. Tuchman et al. (2003) found that the high lignin concentrations associated with elevated CO_2 slowed litter decomposition and decreased bacterial productivity, lowering potential food resources for larval *Aedes triseriatus*, *Ae. albopictus*, *Aedes aegypti*, and *Armigeres subalbatus*.

Alto et al. (2005) found that the effects of decomposition of organic matter on pH and dissolved oxygen concentration outweighed the effects of elevated CO_2. A doubling of atmospheric CO_2 concentration did not significantly change water quality in small containers. Furthermore, the degradation rate of oak leaf litter was not affected by elevated atmospheric CO_2. The decomposition of litter of plants grown under elevated CO_2 did not differ from that of plants grown at ambient CO_2 levels.

In general, the negative effects of elevated CO_2 on container-dwelling mosquito vectors were either nonsignificant or limited to mostly small differences in development time for a few mosquito species (*Ae. triseriatus*, *Ae. aegypti*, and *Ar. subalbatus*; Strand et al., 1999; Tuchman et al., 2003) or survivorship of *Ae. albopictus* (Alto et al., 2005; Tuchman et al., 2003).

3.2.4. Photoperiodic cues

The fitness of a vector will be determined by its ability to take advantage of favorable periods for growth and development as well as the abilities to predict and avoid extinction during periods unfavorable for population growth. Reducing development time is important to maximize the number of generations that occur during favorable periods; yet, tradeoffs of reproduction and survival (but see Bradshaw and Holzapfel, 1996) also are important traits influencing the net reproductive rate and size of the vector population. Equally important, especially in temperate and polar regions, are abilities to deal with unfavorable periods for growth and reproduction. The ability to predict reliably the onset of unfavorable abiotic (Bradshaw and Holzapfel, 2010) or biotic (Hairston and Walton, 1986) conditions is an important component of fitness. Animals typically predict the change in seasons to escape in time or space through dormancy or migration by one of three mechanisms: photoperiod (the duration of light in a light/dark cycle), photoperiodism (the ability to use the length of day or night to regulate seasonal behavior or physiology), or a circannual clock (an internal, self-sustained clock with a period of oscillation of approximately a year) (Bradshaw and Holzapfel, 2010). Short-lived animals with a short period of reproductive maturation, such as mosquitoes, rely on absolute day length to interpret time of year (Bradshaw and Holzapfel, 2010).

As climate warms, the day length that organisms use to cue physiological changes necessary to begin development, to mature gonads, or to migrate poleward in spring and then to cease reproduction, to enter dormancy for overwintering, or to migrate toward the equator in the autumn will decrease. Favorable periods for rapid development will occur earlier in spring and later in autumn than what happens presently. The altered timing of seasonal events with climate change will impose selection on the interpretation of light and its hormonal integration (Figure 3.2). Individual phenotypic plasticity will initially facilitate appropriate responses to cues that change with increasing climate warming. As conditions continue to change, compensation in the timing of major life cycle events by individual phenotypic plasticity will be eventually exceeded and evolution by natural selection will take place (Bradshaw and Holzapfel, 2010). Genetic shifts toward an increased number of generations, a later entry into diapause, and shorter, more southern critical photoperiods in insects are likely (Bradshaw and Holzapfel, 2001, Gomi et al., 2007).

Although a genetic shift in thermal tolerance or thermal optima is predicted to occur in *Drosophila* as climate changes (Gilchrist and Folk, 2008) and the latitudinal changes for the relationship of immature development with temperature for mosquitoes (Pritchard and Mutch, 1985) presumably indicate the potential for evolutionary changes in response to climate change, the principal target of selection by recent changes in climate is the timing of seasonal events (Bradshaw and Holzapfel, 2010). As climate warmed during the last 30 years of the twentieth century, populations of the pitcher plant mosquito, *Wyeomyia smithii*, have evolved a shorter critical photoperiod, with the greatest effect at higher latitudes. Populations from high latitudes have a longer critical photoperiod triggering entry into diapause than do lower-latitude populations. In the late 1990s, high-latitude populations waited to enter diapause 9 days later than during the 1970s (Bradshaw and Holzapfel, 2001).

The responses of mosquitoes to climate change and their impact on disease transmission are not likely to be due to changes in thermal tolerance but to adaptation to changing seasons at high latitudes. The importance of photoperiodism has been shown by the loss of fitness of more than 80% when incorrect day-length information is perceived; in contrast, fitness in the temperate zones improves in ectotherms experiencing warmer thermal conditions (Bradshaw et al., 2004).

3.3. BLOOD-FEEDING PATTERNS

3.3.1. Temperature

Blood feeding by arthropods is critical for nutrient acquisition for egg production but also provides the primary mechanism for pathogen acquisition and distribution by vector species. The behavioral and physiological processes associated with blood feeding determine the frequency of host–vector–pathogen interaction and therefore essentially which hosts are infected and how frequently with which pathogens. Because mosquitoes are poikilothermic, blood-feeding frequency typically increases as a function of warming temperature, which reduces the time required for the gonotrophic cycle (GC) of blood digestion, egg development and oviposition, and refeeding. This increase in the feeding rate frequently is paralleled by increased population growth. Conversely, daily survivorship tends to decrease as a function of increasing temperature. However, the extrinsic incubation period (EIP) required for pathogen infection, growth, dissemination, and transmission tends to decrease dramatically with temperature so that vector-borne pathogen transmission usually is most efficient under warm conditions.

The interplay among the temperature and mosquito biology may be exemplified by research on *Cx. tarsalis*, an important vector of arboviruses throughout the western United States (Reisen and Reeves, 1990). In the laboratory, eggs take ca. 1.5–2.5 days for hatching (Miura et al., 1978), after which larval development progresses as a direct function of temperature both in the laboratory (Reisen, 1995; Reisen et al., 1984) and the field (Reisen et al., 1989, 1997a). If the median time to pupation is converted to a daily rate and plotted as a function of temperature, a linear regression can be used to estimate physiological time for development in DD (Figure 3.5). However, larval survival was maximal at ca. 26–30 °C, with low survival at

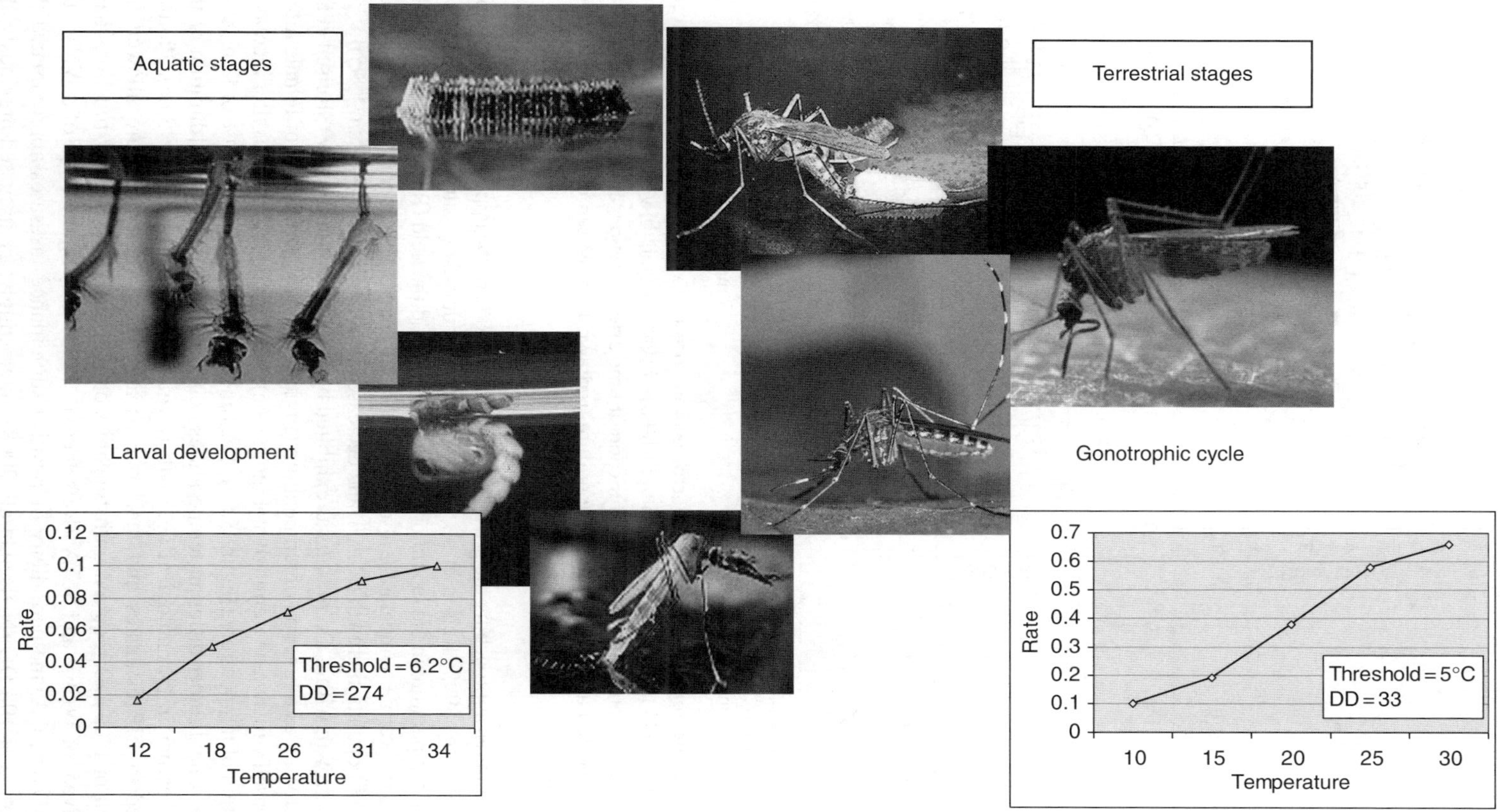

Figure 3.5. Life history of *Cx. tarsalis* showing the aquatic and terrestrial life stages. Graphs show the median rate of immature development from larval eclosion to emergence [left] and of the GC [right] plotted as a function of temperature.

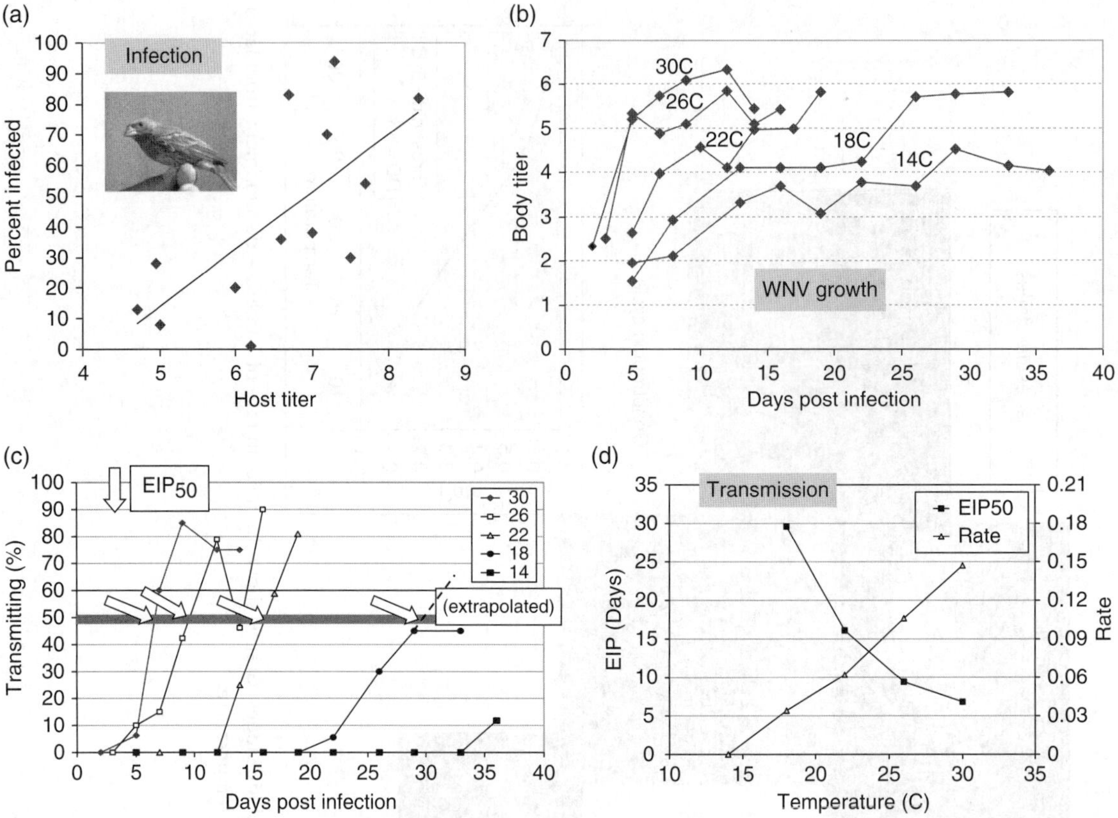

Figure 3.6. Effects of host titer and temperature on the vector competence of *Cx. tarsalis* for WNV. (**a**) Percent infected as a function of host viremia in $\log_{10}$ plague forming units [PFU] of WNV per ml, (**b**) virus growth in PFU per mosquito as a function of days, (**c**) percent transmission as a function of time after infection, and time to 50% transmission and (**d**) rate of the EIP plotted as a function of temperature (redrawn from Reisen et al., 2006). For color detail, please see color plate section.

cool temperatures approaching the growth minimum of 6.2 °C or the upper survival asymptote of 34 °C (Reisen, 1995; Reisen et al., 1984). Similarly adult daily survival decreased as a function of temperature in the laboratory (Reisen, 1995) and the field (Reeves et al., 1994). In the laboratory, warmer temperature also decreased adult female size at emergence, increased estimates of population growth (R_0, r_m), and decreased generation times (Reisen, 1995).

Viral growth within *Cx. tarsalis* as exhibited by West Nile virus (WNV) (Reisen et al., 2006) increased as a function of temperature (Figure 3.6). The percentage of female infection increased with avian viremia titer (Figure 3.6a), but after infection, virus titer increased as a function of temperature (Figure 3.6b). The time from infection until 50% of females in each cohort held under different temperatures could potentially transmit [duration of the EIP] was earlier in chronological age with warmer temperatures (Figure 3.6c). Plotting these median EIP estimates and their rate as a function of temperature allowed the estimation of total DD required until transmission (Figure 3.6d). Similar functions have been derived for WEEV and SLEV (Reisen et al., 1993). During the onset of a WNV outbreak in Davis, CA, the periodicity of positive surveillance measures approximated the accumulation of 108 DD (Nielsen et al., 2008). Interestingly, as discussed previously, the onset of this outbreak seemed related to increasing nocturnal temperatures.

3.3 BLOOD-FEEDING PATTERNS

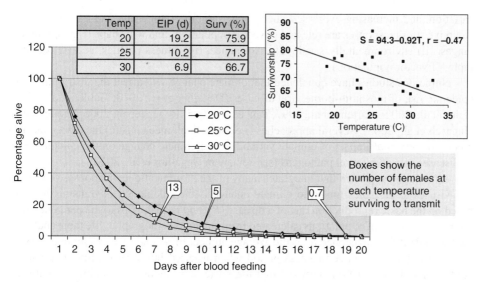

Figure 3.7. Effects of temperature on a number of infected females alive to transmit at the end of the EIP. Percentage of females surviving each day plotted as a function of temperature and the percentage alive at the end of the EIP at three temperatures. For color detail, please see color plate section.

Percent daily survivorship estimated by mark–release–recapture experiments using curvilinear regression (Nelson et al., 1978) decreased as a function of mean temperature at release (Reeves et al., 1994). A linear regression model fit to these data indicated that daily survivorship decreased about 1% for every degree Celsius increase in temperature (Figure 3.7). Using this regression function, the percentage of females alive on each day was plotted as a function of days for three representative temperatures along with the duration of the EIP estimated from the DD model (Figure 3.6). These data showed clearly that warmer temperatures potentially allowed more females to complete their WNV EIP earlier in life, thereby enabling rapid pathogen amplification (Figure 3.7).

3.3.2. Humidity

The amount of moisture the air can hold increases as a function of temperature. For many insects including disease vectors, survival at warm temperatures is significantly improved if humidity is increased. When humidity is low, some subtropical *Culex* such as *Culex nigripalpus* actually suspend flight and host-seeking activity (Day and Curtis, 1989). If the EIP of SLEV or WNV is completed within these periods of arrested flight, intermittent rainfall raises humidity and releases flight activity and provides oviposition sites leading to synchronized refeeding and transmission events (Day et al., 1990). Transmission also is facilitated by the concentration of suitable avian hosts at hammocks where the mosquitoes find suitable vegetative harborage (Day, 2001).

3.3.3. Cumulative impact on mosquito-borne viral infections

The evolutionary responses of mosquitoes to environmental change are not yet well known but are expected to alter abundance and geographical distributions (Bradshaw and Holzapfel, 2010; Gilchrist and Folk, 2008). Environmental stress can alter patterns of adaptive evolution by increasing recombination and mutation rates, maintaining genetic variation, and increasing expressed phenotypic variation (Hoffmann and Parsons, 1991). Adaptive stress responses are relevant to vector populations and will shape (i) geographical ranges, (ii) environmentally sensitive performance, (iii) fitness profiles, and (iv) demography of vector populations.

Numerous studies have concluded that climate change will affect the distribution and epidemic risk of mosquito-borne viral diseases (Degallier et al., 2010; Gubler et al., 2001; Hales et al., 2002; Martens et al., 1997; Patz et al., 2005; Rogers et al., 2006; Shope, 1992; Unnasch et al., 2005). Several approaches (i.e., statistical, mechanistic) have been used to model the potential impact of climate variability and climate change on the epidemic risk of diseases caused by viral pathogens (see review in Degallier et al., 2010).

A modeling approach that integrates many of the aforementioned responses of vectors and viruses to variables linked to climate change is the calculation of transmission risk based on the basic reproduction rate of a disease (R_0) and is analogous to the per generation multiplication rate of populations calculated from standard demographic/life table analyses. Macdonald (1957) envisioned R_0 as the average number of secondary cases produced by an index case during its infectiousness period when inserted into a susceptible population. Anderson and May (1991) surmised that the basic reproduction rate coincides with the threshold that breaks the stability of the disease-free steady state for a disease that involves only one host and one vector. The entomological aspects of the Macdonald model were summarized into a formula for vectorial capacity (Garrett-Jones, 1964) modified slightly for arboviruses (Reisen, 1989) as follows:

$$C = ma^2 P^n V / -\log_e P \tag{3.3}$$

where C is the number of cases per case per day, ma is the host biting rate, a is the ratio of the host selection index (HI) and the duration of the GC in days, P is the probability of daily survival, n is the duration of the EIP (days), and V is vector competence (Garrett-Jones, 1964).

Parameters such as biting rate (ma), the duration of the GC, EIP (n), and daily survivorship (P) are all influenced strongly by temperature and to some degree humidity. Parameters such as vector competence or the probability of infection of an uninfected vector by a viremic host are assumed to be either constant or independent of climate (Degallier et al., 2010). Recently, the relationship between EIP and GC has been shown to function as an inverse of R_0 and provide a useful indication of risk for WNV transmission (Hartley et al., 2012).

Climate change will likely increase the abundance of mosquito vectors and the transmission of arthropod-borne human viral pathogens such as dengue virus, encephalitides, and other viral pathogens. In a warmer global climate, mosquitoes will develop faster, population sizes and the number of generations will increase during an expanded annual period of activity, and biogeographic ranges will expand. Transmission rates of viral pathogens also will likely increase under warmer conditions because the EIP of viruses will decrease and mosquitoes that emerge smaller when they develop in warmer water must blood feed more frequently (Juliano and Lounibos, 2005; Patz et al., 2003). Whereas marked changes in the biogeographic distributions of mosquito vectors have not yet been found, some insect pests have moved poleward by 180–185 miles during the increase of about 2 °C in the past 25 years (green stinkbug: *Acrosternum hilare*, Parmesan, 2006; *Dendroctonus ponderosae*, Logan and Powell, 2001). While some northern hemisphere species have moved poleward, they have declined at the southern end of the previous range (Parmesan, 2006).

There is some evidence for the evolution of thermal traits in *Drosophila* following introduction and range expansion into new habitats (Gilchrist and Folk, 2008); however, evidence of comparable changes in the thermal physiology of vector mosquitoes is currently lacking (Bradshaw and Holzapfel, 2010). Biotic responses in mosquitoes to changes in abiotic conditions of climate fall into two principal categories: the changes of species' ranges and altered timing of seasonal phenologies (Bradshaw and Holzapfel, 2010). We need a better understanding of how performance profiles associated with stress responses to changes in climate variables impact mosquitoes and vector competence.

Degallier et al. (2010) found that the global distribution of dengue fever corresponded well with the epidemic risk of dengue transmission, except in a few places where uncertainties in climate data (China and Arabian Peninsula) or ongoing vector control activities (southeastern United States) reduced dengue incidence in regions where dengue epidemics were predicted. Moreover, transmission risk changed seasonally, expanding poleward during warmer seasons. Lastly, the prevalence of the disease was not exclusively correlated with climate conditions; the density of the human population interacted with climate risk. At comparatively low climate risk of dengue epidemics, dengue prevalence in high-density human populations is comparatively higher than at low human population density.

The relative epidemic potential of dengue virus is predicted to increase with increasing temperature up to 40 °C; above this temperature increased vector biting rates and the accelerated development of the parasite cannot compensate for the decrease of survival of the mosquito vector (Martens et al., 1997). The increase in transmission potential of dengue virus in tropical and subtropical countries varies between 31% and 47% under different climate change scenarios (Marten et al., 1997). Expansion of epidemic area through the changing of biogeographic distributions of mosquito vectors and of epidemiological and entomological factors with climate change puts human populations in non- or low-endemic areas at greatest risk. These populations currently reside at the periphery of current epidemic zones, at altitudes above and at latitudes higher than the sites currently prone to epidemics. The lack of naturally acquired immunity in naïve human populations exacerbates the intensity of arbovirus transmission and infection as the epidemic area expands with climate warming. Martens et al. (1997) concluded that the prevalence of dengue infection will change little in current highly endemic areas as climate changes.

The sensitivity of vector-borne disease to changes in climate is not independent of socioeconomic development, local environmental conditions, human behavior and immunity, and the effectiveness of control measures (Degallier et al., 2010; Institute of Medicine, 2003; Martens et al., 1997). The incidence of mosquito-borne diseases is largely determined by public health capacity and socioeconomic factors such as affluence and lifestyle (Brunkard et al., 2007; Harrigan et al., 2010; Reiter, 2001). An increasing unpredictability of water supply coupled with the demands for potable water of an ever-increasing human population will likely increase standing water across the landscape. The fiscal realities of a global economy that will likely include between 8 and 10 billion people by 2050 (Brown, 2010) will necessitate the use of decentralized facilities for the treatment and reclamation of wastewater. Such landscape features have the potential to increase developmental sites for arbovirus vectors and, subsequently, the prevalence of vector-borne disease, especially in the tropics (Johnson et al., 2010; Walton, 2012). Other societal priorities may take precedence to even the most basic water reclamation infrastructure. The combination of ever warmer climate and widespread enriched, untreated, or minimally treated wastewaters, as well as comparatively small water storage containers in urban environments, has potentially significant implications for the abundance of mosquito vectors and the prevalence of diseases caused by viral pathogens.

REFERENCES

ACIA (Arctic Climate Impact Assessment). 2004. *Impacts of a Warming Arctic*. Cambridge: Cambridge University Press.

Altizer, S. and Pederson, A. B. 2008. Host–pathogen evolution, biodiversity, and disease: risks for natural populations. In *Conservation Biology: Evolution in Action*, S. P. Carroll and C. W. Fox (eds.), pp. 259–277. New York: Oxford University Press.

Alto, B. W., Yanoviak, S. P., Lounibos, L. P. et al. 2005. Effects of elevated atmospheric CO_2 on water chemistry and mosquito (Diptera: Culicidae) growth under competitive conditions in container habitats. *Florida Entomologist* **88**:372–382.

Anderson, R. M. and May, R. M. 1991. *Infectious Diseases of Humans: Dynamics and Control*. New York: Oxford University Press.

Blaustein, L. and Chase, J. M. 2007. Interactions between mosquito larvae and species that share the same trophic level. *Annual Review of Entomology* **52**:489–507.

Bradshaw, W. E. and Holzapfel, C. M. 1996. Genetic constraints to life-history evolution in the pitcher-plant mosquito, *Wyeomyia smithii*. *Evolution* **50**:1176–1181.

Bradshaw, W. E. and Holzapfel, C. M. 2001. Genetic shift in photoperiodic response correlated with global warming. *Proceedings of the National Academy of Sciences of the United States of America* **98**:14509–14511.

Bradshaw, W. E. and Holzapfel, C. M. 2006. Evolutionary response to rapid climate change. *Science* **312**:1477–1478.

Bradshaw, W. E. and Holzapfel, C. M. 2010. Light, time, and the physiology of biotic response to rapid climate change in animals. *Annual Review of Physiology* **72**:147–166.

Bradshaw, W. E., Zani, P. A. and Holzapfel, C. M. 2004. Adaptation to temperature climates. *Evolution* **58**:1748–1762.

Brown, L. R. 2010. *World on the Edge*. New York: Earth Policy Institute, W. W. Norton and Company.

Brunkard, J. M., Robles López, J. L., Ramirez, J., et al. 2007. Dengue fever seroprevalence and risk factors, Texas–Mexico border, 2004. *Emerging Infectious Diseases* **13**:1477–1483.

Clements, A. N. 1992. *The Biology of Mosquitoes, Volume 1, Development, Nutrition, and Reproduction*. New York: Chapman and Hall.

Cotrufo, M. F. and Ineson, P. 1996. Elevated CO_2 reduces field decomposition rates of *Betula pendula* (Roth.) leaf litter. *Oecologia* **106**:525–530.

Cotrufo, M. F., Ineson, P. and Rowland, A. P. 1994. Decomposition of tree leaf litters grown under elevated CO_2: effect of litter quality. *Plant and Soil* **163**:121–130.

Cotrufo, M. F., Raschi, A., Lanini, M., et al. 1999. Decomposition and nutrient dynamics of *Quercus pubescens* leaf litter in a naturally enriched CO_2 Mediterranean ecosystem. *Functional Ecology* **13**:343–351.

Day, J. F. 2001. Predicting St. Louis Encephalitis Virus epidemics: lessons from recent, and not so recent, outbreaks. *Annual Review of Entomology* **46**:111–138.

Day, J. F. and Curtis, G. A. 1989. Influence of rainfall on *Culex nigripalpus* (Diptera: Culicidae) blood-feeding behavior in Indian River County, Florida. *Annals of the Entomological Society of America* **82**:32–37.

Day, J. F., Curtis, G. A. and Edman, J. D. 1990. Rainfall-directed oviposition behavior of *Culex nigripalpus* (Diptera: Culicidae) and its influence on St. Louis encephalitis virus transmission in Indian River County, Florida. *Journal of Medical Entomology* **27**:43–50.

Degallier, N., Favier, C., Menkes, C., et al. 2010. Toward an early warning system for dengue prevention: modeling climate impact on dengue transmission. *Climatic Change* **98**:581–592.

Eldridge, B. F. 1968. The effect of temperature and photoperiod of blood feeding and ovarian development in mosquitoes of the *Culex pipiens* complex. *American Journal of Tropical Medicine and Hygiene* **17**:133–140.

Epstein, P. R., Diaz, H. F., Elias, S., et al. 1998. Biological and physical signs of climate change: focus on mosquito-borne diseases. *Bulletin of the American Meteorological Society* **79**:409–417.

Esper, J., Cook, E. R. and Scheweingruber, F. H. 2002. Low-frequency signals in long tree-ring chronologies for reconstructing past temperature variability. *Science* **295**:2250–2253.

Focks, D.A., Haile, E., Daniels, E., et al. 1993. Dynamic life table model for *Aedes aegypti* (Diptera: Culicidae): analysis of the literature and model development *Journal of Medical Entomology* **30**:1003–1017.

Frederiksen, H. B., Rønn, R. and Christensen, S. 2001. Effect of elevated atmospheric CO_2 and vegetation type on microbiota associated with decomposing straw. *Global Change Biology* **7**:313–321.

Garrett-Jones, C. 1964. Prognosis for interruption of malaria transmission through assessment of the mosquito's vectorial capacity. *Nature* **204**:1173–1175.

Gerberg, E. J., Barnard, D. R. and Ward, R. A. 1994. *Manual for Mosquito Rearing and Experimental Techniques*. American Mosquito Control Association, Bull. No. 5 (revised), Lake Charles, LA: American Mosquito Control Association, Inc.

Gilchrist, G. W. 1995. Specialists and generalists in changing environments. 1. Fitness landscapes of thermal sensitivity. *American Naturalist* **146**:252–270.

Gilchrist, G. W. 2000. The evolution of thermal sensitivity in changing environments. In *Cell and Molecular Responses to Stress. Vol. 1. Environmental Stressors and Gene Responses*, K. B. Storey and J. M. Storey (eds.), pp. 55–70. Amsterdam: Elsevier Science.

Gilchrist, G. W. and Folk, D. G. 2008. Evolutionary responses to environmental change. In *Conservation Biology: Evolution in Action*, S. P. Carroll and C. W. Fox (eds.), pp. 164–180. New York: Oxford Univ. Press.

Gomi, T., Nagasaka, M., Fukkuda, T., et al. 2007. Shifting of the life cycle and life-history traits of the fall webworm in relation to climate change. *Entomologia Experimentalis et Applicata* **125**:179–184.

Gubler, D. J., Reiter, P., Ebi, K. L., et al. 2001. Climate variability and change in the United States: potential impacts on vector- and rodent-borne diseases. *Environmental Health Perspectives* **109** (Suppl. 2):223–233.

Hairston, N. G. Jr. and Walton, W. E. 1986. Rapid evolution of a life history trait. *Proceedings of the National Academy of Sciences of the United States of America* **83**:4831–4833.

Hales, S., Wet, N., Maindonald, J. et al. 2002. Potential effect of population and climate changes on global distribution of dengue fever: an empirical model. *Lancet* **360**:830–834.

Hardy, J. L., Meyer, R. P., Presser, S. B., et al. 1990. Temporal variations in the susceptibility of a semiisolated population of *Culex tarsalis* to peroral infection with western equine encephalomyelitis and St. Louis encephalitis viruses. *American Journal of Tropical Medicine and Hygiene* **42**:500–511.

Hardy, J. L. and Reeves, W. C. 1990. Experimental studies on infection in vectors In *Epidemiology and control of mosquito-borne arboviruses in California, 1943–1987. W. C.* Reeves (ed.), pp. 145–250. Sacramento: California Mosquito and Vector Control Association.

Harrigan, R. J., Thomassen, H. A., Buermann, W., et al. 2010. Economic conditions predict prevalence of West Nile virus. *PLoS ONE* **5**(11): e15437.

Hartley, D. M., Barker, C. M., Le Menac'h, A., et al. 2012. The effects of temperature on the emergence and seasonality of West Nile virus in California. *American Journal of Tropical Medicine and Hygiene* **86**:884–894.

Heft, D. E. and Walton, W. E. 2008. Effects of El Niño - Southern Oscillation (ENSO) cycle on mosquito populations in southern California. *Journal of Vector Ecology* **33**:17–29.

Hoffmann, A. A. and Parsons, P. A. 1991. *Evolutionary Genetics and Environmental Stress*. Oxford: Oxford University Press.

Institute of Medicine. 2003. *Emerging Infections, Microbial Threats to Health in the United States*. Washington, D.C.: The National Academies Press.

International Panel on Climate Change (IPCC). 2001. *Climate Change 2001: The Scientific Basis. Contribution of Working Group I to the Third Assessment Report of the Intergovernmental Panel on Climate Change*. Geneva: IPCC Secr.

International Panel on Climate Change (IPCC). 2007. *Climate Change 2007: The Physical Basis. Contribution of Working Group I to the Fourth Assessment of the Intergovernmental Panel on Climate Change*. Geneva: IPCC Secr.

Johnson, P. T. J., Townsend, A. R., Cleveland, C. C., et al. 2010. Linking environmental nutrient enrichment and disease emergence in humans and wildlife. *Ecological Applications* **20**:16–29.

Juliano, S. A. and Lounibos, L. P. 2005. Ecology of invasive mosquitoes: effects on resident species and on human health. *Ecology Letters* **8**:558–574.

Logan, J. A. and Powell, J. A. 2001. Ghost forests, global warming and the mountain pine beetle. *American Entomologist* **47**:160–173.

Macdonald, G. 1957. *The Epidemiology and Control of Malaria*. London: Oxford University Press.

Maelzer, D., Hales, S., Weinstein, P., et al. 1999. El Niño and arboviral disease prediction. *Environmental Health Perspectives* **107**:817–818.

Martens, W. J. M., Jetten, T. H., and Focks, D. A. 1997. Sensitivity of malaria, schistosomiasis and dengue to global warming. *Climatic Change* **35**:145–156.

Mellor, P. S. 2004. Environmental influences on arbovirus infections and vectors. In *Microbe-vector Interactions in Vector-borne Diseases*, S. H. Gillespie, G. L. Smith and A. Osbourn (eds.), pp. 181–197. Cambridge: Society for General Microbiology Symposium 63, Cambridge University Press.

Metzger, M. A. 2004. *Managing Mosquitoes in Stormwater Treatment Devices*. Oakland: Division of Agriculture and Natural Resources, Publ. No. 8125.

Meyer, R. P., Hardy, J. L., and Reisen, W. K. 1990. Diel changes in adult mosquito microhabitat temperatures and their relationship to the extrinsic incubation of arboviruses in mosquitoes in Kern County, California, USA. *Journal of Medical Entomology* **27**:607–614.

Miura, T., Takahashi, R. M., Reed, D. E., et al. 1978. An empirical method for predicting the hatching data of *Culex tarsalis* eggs in early spring in Fresno County, California. *Proceedings of the California Mosquito and Vector Control Association* **46**:47–49.

Nelson, R. L., Milby, M. M., Reeves, W. C., et al. 1978. Estimates of survival, population size, and emergence of *Culex tarsalis* at an isolated site. *Annals of the Entomological Society of America* **71**:801–808.

Nielsen, C. F., Armijos, M. V., Wheeler, S., et al. 2008. Risk factors associated with human infection during the 2006 West Nile virus outbreak in Davis, a residential community in Northern California. *American Journal of Tropical Medicine and Hygiene* **78**:53–62.

Padmanabha, H., Bolker, B., Lord, C. C., et al. 2011. Food availability alters the effects of larval temperature on *Aedes aegypti* growth. *Journal of Medical Entomology* **48**:974–984.

Parmesan, C. 2006. Ecological and evolutionary responses to recent climate change. *Annual Review of Ecology, Evolution and Systematics* **37**:637–669.

Patz, J. A., Githeko, A. K., McCarthy, P., et al. 2003. Climate change and infectious diseases. In *Climate Change and Human Health: Risks and Responses*, A. J. Michael, D. H. Campbell-Lendrum, and C. F. Corvalan (eds.), pp. 103–132. Geneva: World Health Organization.

Patz, J. A., Campbell-Lendrum, D., Holloway, T. et al. 2005. Impact of regional climate change on human health. *Nature* **438**:310–317.

Peters, R. L. 1993. Conservation of biological diversity in the face of climate change. In: *Global Warming and Biological Diversity*, R. L. Peters and T. E. Lovejoy (eds.), pp. 15–30. New Haven: Yale Univ. Press.

Pongsiri, M., Roman, J., Ezenwa, V. O., et al. 2009. Biodiversity loss affects global disease ecology. *Bioscience* **59**:945–954.

Poveda, G., Rojas, W., Quiñones, M. L., et al. 2001. Coupling between annual and ENSO timescales in the malaria-climate association in Colombia. *Environmental Health Perspectives* **109**: 489–493.

Pritchard, G. and Mutch, R. A. 1985. Temperature, development rates and origins of mosquitoes. In *Ecology of Mosquitoes: Proceedings of a Workshop*, L. P. Lounibos, J. R. Rey and J. H. Frank (eds.), pp. 237–249. Vero Beach: Florida Medical Entomology Laboratory.

Reeves, W. C., Hardy, J. L., Reisen, W. K. et al. 1994. Potential effect of global warming on mosquito-borne arboviruses. *Journal of Medical Entomology* **31**:323–332.

Reisen, W. K. 1986. Overwintering studies on *Culex tarsalis* (Diptera: Culicidae) in Kern County, California: life stages sensitive to diapause induction cues. *Annals of the Entomological Society of America* **79**:674–676.

Reisen, W. K. 1989. Estimation of vectorial capacity: relationship to disease transmission by malaria and arbovirus vectors. *Bulletin of the Society of Vector Ecology* **14**:39–40.

Reisen, W. K. 1995. Effect of temperature on *Culex tarsalis* (Diptera: Culicidae) from the Coachella and San Joaquin Valleys of California. *Journal of Medical Entomology* **32**:636–645.

Reisen, W. K. and Reeves, W. C. 1990. Bionomics and ecology of *Culex tarsalis* and other potential mosquito vector species. In: *Epidemiology and Control of Mosquito-Borne Arboviruses in California, 1943–1987*. W. C. Reeves (ed.), pp. 254–329. Sacramento: California Mosquito and Vector Control Association.

Reisen, W. K., Milby, M. M., and Bock, M. E. 1984. The effects of immature stress on selected events in the life history of *Culex tarsalis*. *Mosquito News* **44**:385–395.

Reisen, W. K., Meyer, R. P. and Milby, M. M. 1986. Overwintering studies on *Culex tarsalis* (Diptera: Culicidae) in Kern County, California: survival and the experimental induction and termination of diapause. *Annals of the Entomological Society of America* **79**:664–673.

Reisen, W. K., Meyer, R. P., Shield, J., et al. 1989. Population ecology of preimaginal *Culex tarsalis* (Diptera: Culicidae) in Kern County, California, USA. *Journal of Medical Entomology* **26**:10–22.

Reisen, W. K., Meyer, R. P., Presser, S. B., et al. 1993. Effect of temperature on the transmission of western equine encephalomyelitis and St. Louis encephalitis viruses by *Culex tarsalis* (Diptera: Culicidae). *Journal of Medical Entomology* **30**:151–160.

Reisen, W. K., Hardy, J. L., Presser, S. B. et al. 1996. Seasonal variation in the vector competence of *Culex tarsalis* (Diptera: Culicidae) from the Coachella Valley of California for western equine encephalomyelitis and St. Louis encephalitis viruses. *Journal of Medical Entomology* **33**:433–437.

Reisen, W. K., Hardy, J. L. and Presser, S. B. 1997a. Effects of water quality on the vector competence of *Culex tarsalis* (Diptera: Culicidae) for western equine encephalomyelitis (Togaviridae) and St. Louis encephalitis (Flaviviridae) viruses. *Journal of Medical Entomology* **34**:631–643.

Reisen, W. K., Lothrop, H. D. and Meyer, R. P. 1997b. Time of host-seeking by *Culex tarsalis* (Diptera: Culicidae) in California. *Journal of Medical Entomology* **34**:430–437.

Reisen, W. K., Fang, Y. and Martinez, V. M. 2006. Effects of temperature on the transmission of West Nile virus by *Culex tarsalis* (Diptera: Culicidae). *Journal of Medical Entomology* **43**:309–317.

Reiter, P. 2001. Climate change and mosquito-borne disease. *Environmental Health Perspectives* **109** (Suppl. 1):141–161.

Rogers, D. J., Wilson, A. J., Hay, S. I., et al. 2006. The global distribution of yellow fever and dengue. *Advances in Parasitology* **62**:181–220.

Shope, R. E. 1992. Impacts of global climate change on human health: spread of infectious disease. In *Global Climate Change: Implications, Challenges and Mitigation Measures*, S. K. Majumdar, L. S. Kalkstein, B. Yarnal, E. W. Miller and L. M. Rosenfeld (eds.), pp. 363–370. Easton: The Pennsylvania Academy of Science.

Siegenthaler, U., Stocker, T. F., Monnin, E., et al. 2005. Stable carbon cycle–climate relationship during the late Pleistocene. *Science* **310**:1313–17.

Strand, M., Herms, D. A., Ayers, M. P., et al. 1999. Effects of atmospheric CO_2, light availability and tree species on the quality of leaf detritus as a resource for treehole mosquitoes. *Oikos* **84**:277–283.

Su, T., Webb, J. P., Meyer, R. P., et al. 2003. Spatial and temporal distribution of mosquitoes in underground storm drain systems in Orange County, California. *Journal of Vector Ecology* **28**:79–89.

Sweeney, B. W., Jackson, J. K., Newbold, J. D., et al. 1992. Climate change and the life histories and biogeography of aquatic insects in eastern North America. In *Global Change and Freshwater Ecosystems*, P. Firth and S. G. Fisher (eds.), pp. 143–176. New York: Springer-Verlag.

Tewksbury, J. J., Huey, R. B., and Deutsch, C. A. 2008. Putting the heat on tropical animals. *Science* **320**:1296–1297.

Trumble, J. T. and Butler, C. D. 2009. Climate change will exacerbate California's insect pest problems. *California Agriculture* **63**(2):73–78.

Tuchman, N. C., Wahtera, K. A., Wetzel, R. G., et al. 2003. Nutritional quality of leaf detritus altered by elevated atmospheric CO_2: effects on development of mosquito larvae. *Freshwater Biology* **48**:1432–1439.

Unnasch, R. S., Sprenger, T., Katholi, C. R., et al. 2005. A dynamic transmission model of eastern equine encephalitis virus. *Ecological Modelling* **192**:425–440.

Vannote, R. L. and Sweeney, B. W. 1980. Geographic analysis of thermal equilibria: a conceptual model for evaluating the effect of natural and modified thermal regimes on aquatic insect communities. *American Naturalist* **115**:667–695.

Varley, C. G., Gradwell, G. R. and Hassell, M. P. 1973. *Insect Population Ecology: An Analytical Approach*. Berkeley: Blackwell Scientific Publications, University of California Press.

Vezzani, D., Velazquez, S.-M. and Schweigmann, N. 2004. Seasonal pattern of abundance of *Aedes aegypti* (Diptera: Culicidae) in Buenos Aires city, Argentina. *Memorias do Instituto Oswaldo Cruz* **99**:351–355.

Walton, W. E. 2012. Design and management of free water surface constructed wetlands to minimize mosquito production. *Wetlands Ecology and Management* **20**:173–195.

Wegbreit, J. and Reisen, W. K. 1994. Relationships among weather, mosquito abundance, and encephalitis virus activity in California: Kern County 1990–98. *Journal of the American Mosquito Control Association* **16**:22–27.

Wetzel, R. G. 2001. *Limnology: Lake and River Ecosystems*. 3rd ed. San Diego: Academic Press.

ENVIRONMENTAL PERTURBATIONS THAT INFLUENCE ARBOVIRAL HOST RANGE: INSIGHTS INTO EMERGENCE MECHANISMS

Aaron C. Brault

Division of Vector-Borne Diseases, Centers for Disease Control and Prevention, Fort Collins, CO, USA

William K. Reisen

Center for Vectorborne Diseases, Department of Pathology, Microbiology and Immunology, School of Veterinary Medicine, University of California, Davis, CA, USA

TABLE OF CONTENTS

4.1	Introduction	57
4.2	The changing environment	59
4.3	Deforestation and the epizootic emergence of venezuelan equine encephalitis virus	62
4.4	Rice, mosquitoes, pigs, and Japanese encephalitis virus	63
4.5	*Culex pipiens* complex, House sparrows, urbanization, and West Nile virus	66
4.6	Urbanization, global trade, and the reemergence of Chikungunya virus	70
4.7	Conclusions	71
	References	71

4.1 INTRODUCTION

This chapter attempts to describe how environmental perturbations provide emergence opportunities for highly adaptable viral pathogens, capable of rapidly evolving and exploiting unique niches generated during ecological change. No viruses exemplify this capacity better than RNA viruses, especially arthropod-borne viruses or arboviruses for which biological propagation requires an arthropod vector. The dependency of these viruses on the biology of their poikilothermic vectors brings into prominence environmental factors including temperature and rainfall (Reisen et al., 2008a). Herein, we provide several examples where arboviruses have emerged as a direct result of deforestation, changes in agricultural practices, urbanization, and increases in temperature.

As with all RNA viruses, arboviruses contain inherently error-prone RNA-dependent RNA polymerases that lack proofreading capacity (Domingo, 1997) and may generate approximately 1 mutation per 10 000 nucleotides replicated. As most of these viruses contain genomes of approximately 10^4 nucleotides, this translates into one mutation per genome replicated (Weaver, 1995). With population sizes of 10^{10} viral particles per ml of sera during acute infection in some hosts, the presence of multiple mutations at numerous genetic positions in a single infected animal is not uncommon. Although most mutations have either a neutral or detrimental fitness effect on the virus and are either neutrally or negatively selected, respectively, a small percentage impart a fitness benefit to the virus in a given environment (Weaver, 1995). It is these adaptations on which this chapter is focused, with fitness defined as the replicative (and/or transmissibility) success of the virus in a changing environment (Domingo et al., 1996). The changing landscape in which these viruses circulate among a complex suite of arthropod vectors and vertebrate hosts overlaid by climatic and other environmental change results in a dynamic system in which the virus must mutate to keep pace with the genetics of vector and host as well as the environment in which they exist.

Multiple factors dictate the complex interactions of arboviruses with their replicative environment. Changing human demography has resulted in a global trend toward urbanization and deforestation that moves humans into close proximity with each other as well as adapted enzootic arboviral transmission cycles (Brault et al., 2004b), initiating novel transmission cycles with altered disease spectrums. In addition to the movement of humans and goods around the world, increased world commerce has facilitated the movement of vectors and arboviruses to new locations in which autochthonous transmission cycles have been established (Maclachlan, 2010). Urbanization without adequate municipal utilities necessitates water storage practices optimal for the invasion of container-breeding mosquitoes such as *Aedes albopictus*, thereby providing endemic viruses the opportunity for adaptation to new mosquito vectors such as those identified with the emergence of chikungunya virus (CHIKV) from East Africa to the Indian Ocean islands and Indian subcontinent (Tsetsarkin et al., 2007). Finally, climate change can effect vector distributions and dynamics by altering the duration of maturation, fecundity, and longevity and by modulating extrinsic incubation that is dependent upon the rate of replication of the virus (Dobson, 2009; Reisen et al., 2010). The interactions between virus and vector/vertebrate host for which these changing landscapes dictate relative fitness may be modulated by an almost limitless array of constantly changing factors, and it is not our intent to document all the ways that these modulations have resulted in arboviral emergence or the manner in which this could occur. Rather, we have selected examples that represent some of these processes for which scientific evidence has documented the ecological perturbation and the resulting viral change to exploit that niche with subsequent epidemiological effects. We have included examples of arboviral emergence from typical rural or enzootic habitats to modified peridomestic patterns of transmission associated with novel disease presentation (Venezuelan equine encephalitis virus (VEEV)), agroecosystems (Japanese encephalitis virus (JEV)), urbanized transmission (West Nile virus (WNV)), and anthroponotic cycles (CHIKV).

4.2 THE CHANGING ENVIRONMENT

The world's human population has tripled over the previous 70 years to more than seven billion, and this increasing trend shows no immediate signs of decline, although one United Nations' projection includes a possible decrease over the next 70 years (Figure 4.1). There is marked spatial disparity, with the most dramatic increases recorded in Asia, but the highest growth rates realized in Africa. To support this rapid increase in abundance, humans have concurrently exploited the earth's resources, resulting in marked and perhaps irreversible changes to our planet's ecosystems. Consistent with population growth has been the expanding use of energy (O'Neill et al., 2010) that has resulted in increasing atmospheric concentrations of CO_2 and global temperatures (Figure 4.2 and Figure 4.3a) that generally have tracked population growth (Figure 4.1)

Although the actual mechanisms linking these temporal associations are not clear and there have been similar warming trends in the earth's geological past, there can be no doubt that the earth's human population is increasing rapidly and that global temperatures have increased over the past 50 years. Numerous models project continued increase, but these range from 2 to 5 °C. Global warming has not been uniform (Figure 4.3b), with the greatest increases seen at northern latitudes. Warm tempera-

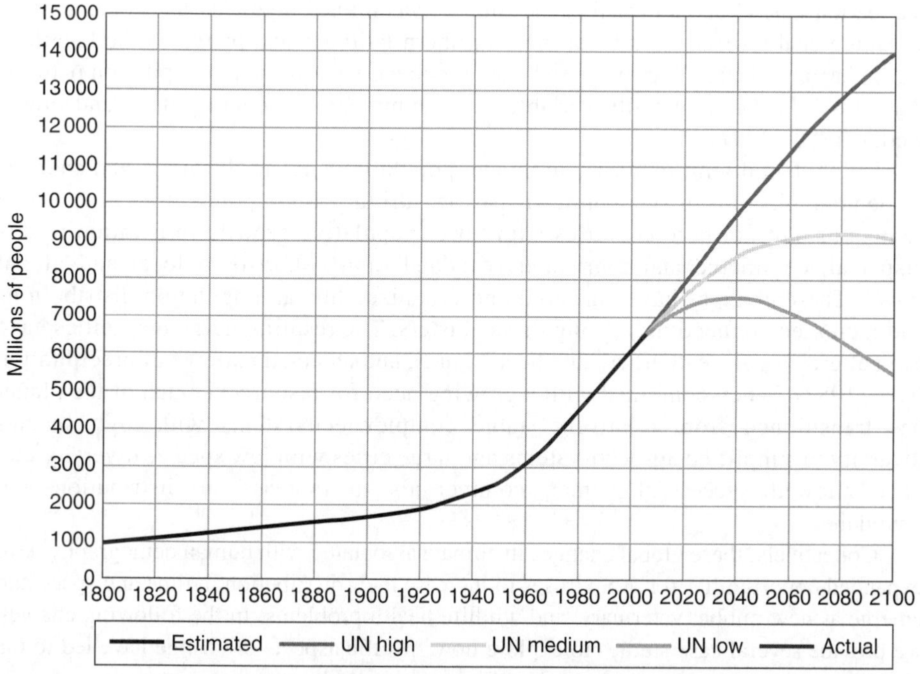

Figure 4.1. Global human population size. From: United Nations, Department of Economic and Social Affairs, Population Division (2011). World Population Prospects: The 2010 Revision, CD-ROM Edition. For color detail, please see color plate section.

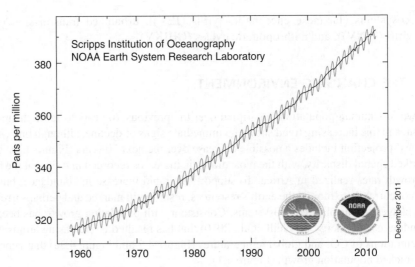

Figure 4.2. Change in atmospheric CO_2 concentrations measured at Mauna Loa, Hawaii (from P. Tans (www.esrl.noaa.gov/gmd/ccgg/trends/) and R. Keeling (scrippsco2.ucsd.edu/); accessed 1 Jan 2012). For color detail, please see color plate section.

tures are critical for effective arbovirus transmission, because vectors digest blood faster and thereby feed more frequently increasing the rate of host–vector contact, and the virus grows faster within the vectors decreasing the length of the extrinsic incubation period and shortening the time between viral ingestion and transmission. Warming enables transmission at more northern latitudes and increases the length of transmission seasons. Warming has been associated with altered precipitation patterns that also have been spatially variable, with some areas becoming drier and others wetter (IPPC, 2007).

Coincidental with increases in human population size has been marked changes in demography due to economic disparities driving rural populations into urban centers or into northern countries with lower population growth. In addition to this dispersal, commerce and tourism have added rapid, short-term local and global travel. These changes have clustered and urbanized human population distributions and increased connectivity among these clusters. The resulting large megacities have even altered local weather, created heat islands, and changed patterns of precipitation (Oke, 1982). When combined with a growing need for resources, much of the planet has transitioned from relatively stable complex ecosystems with high species diversity to simplified agroecosystems and large cities with low species diversity that have allowed successful human commensals to increase in distribution and abundance.

Collectively, these global changes in climate associated with human demography have expanded opportunities for arboviruses to increase their distribution in time and space and emerge as new public, veterinary, and wildlife health problems. In the following chapter, we provide several case studies to explore how specific aspects of change have led to the evolution and emergence of arboviral problems on multiple continents.

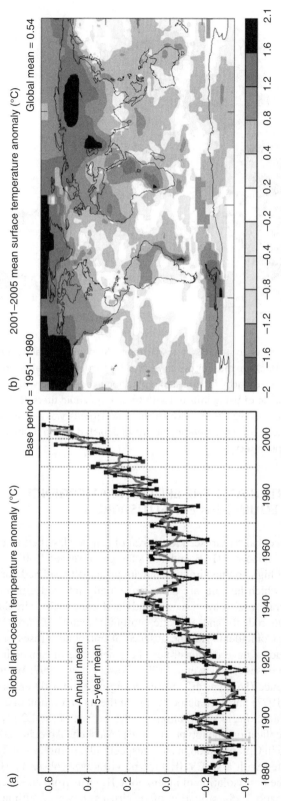

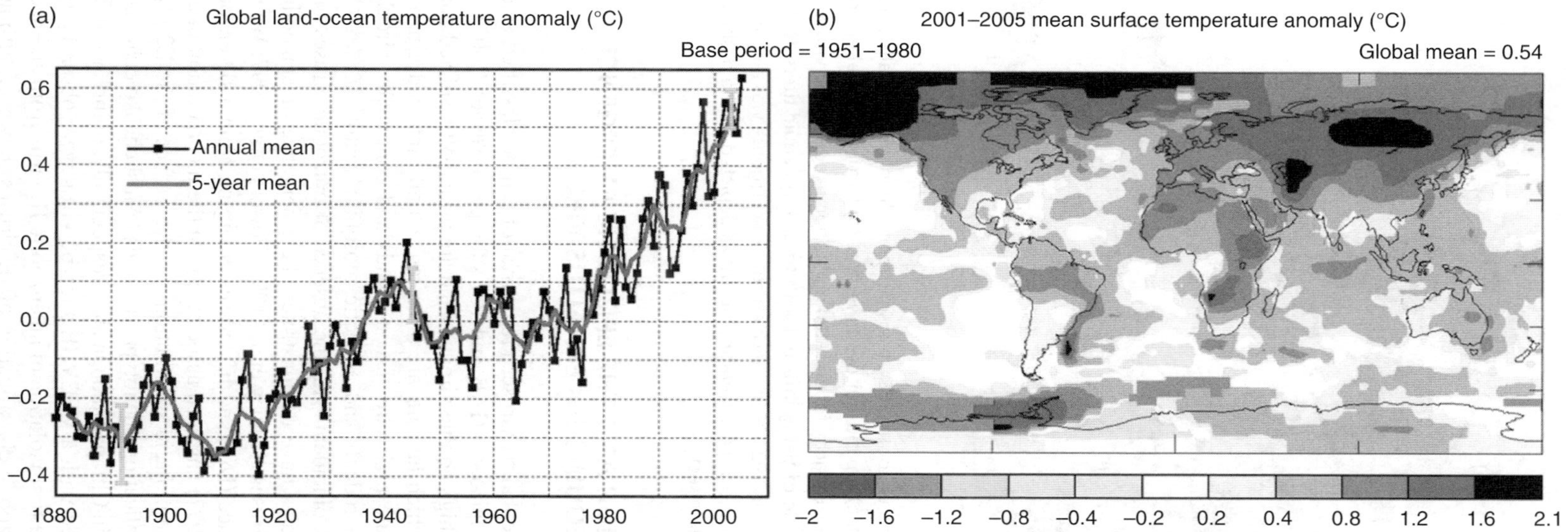

Figure 4.3. Surface temperature anomalies relative to 1951–1980 from surface air measurements at meteorological stations and ship and satellite sea surface temperature (SST) measurements: **(a)** global annual mean anomalies and **(b)** temperature anomaly for the first half decade of the twenty-first century (Hansen et al., 2006). For color detail, please see color plate section.

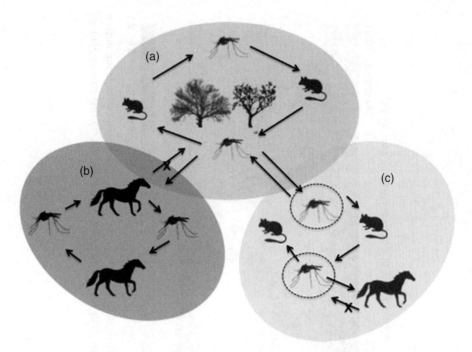

Figure 4.4. (a) Depicits an enzootic VEEV transmission cycle in which small rodents serve as reservoir hosts and *Culex (Mel.)* spp mosquitoes as vectors. (b) Depicts classical epizootic emergence in which viruses acquire the capacity for elevated replication in equids and viruses are transmitted by epizootic vectors capable of being infected with these high equid titers. (c) Depicts the novel epizootic emergence event in 1993/96 in which epizootic emergence was mediated by adaptation to an epizootic mosquito vector (*Ae. taeniorhynchus*) as depicted by the dashed circle. For color detail, please see color plate section.

4.3 DEFORESTATION AND THE EPIZOOTIC EMERGENCE OF VENEZUELAN EQUINE ENCEPHALITIS VIRUS

Venezuelan equine encephalitis viruses (*Togaviridae*) can be classified as either enzootic (ID, IE, IF, II–VI) or epizootic (IAB and IC) subtypes (Weaver, 2001); however, only epizootic viruses elicit high equine viremias and are principally isolated during epizootics of equine disease, in which humans are often tangentially infected. Enzootic viruses, in contrast, do not generate high viremias in equids and continually circulate in sylvatic habitats between small rodents and enzootic mosquito vectors [*Culex* (*Melanoconion*) spp.] (Weaver, 2001). Coincident with the absence of epizootic viruses in interepidemic periods, numerous studies have demonstrated the emergence of epizootic viruses from enzootic predecessor viruses (Kinney et al., 1992; Powers et al., 1997; Wang et al., 2001). This epizootic emergence (IAB and IC viruses emerging from a ID enzootic progenitor virus) (Powers et al., 1997) has been associated with the incorporation of a select number of positively charged amino acid substitutions on the surface of the E2 glycoprotein that encode an increased replication phenotype in equids (Anishchenko et al., 2006).

In 1993 and 1996 in Chiapas and Oaxaca, respectively, outbreaks of equine encephalitic disease associated with a VEEV IE subtype virus were observed (Oberste et al., 1998). This marked the first isolation of an enzootic IE virus from an encephalitic equid and questioned the enzootic classification of IE viruses; however, although generating low viremias consistent with enzootic VEEV strains, experimental inoculation studies indicated encephalitis in a small proportion of exposed horses (Gonzalez-Salazar et al., 2003). Additional viral genetic studies indicated that a single E2 mutation was present in the Chiapas/Oaxacan IE strain that was associated with increased oral infectivity of a salt marsh epizootic mosquito vector, *Aedes taeniorhynchus* (Figure 4.4).

This mutation was not found in IE enzootic strains that exhibited lower infectivity for this epizootic mosquito species (Brault et al., 2004b). The areas of Chiapas and Oaxaca in which the epizootic VEEV IE subtypes emerged previously had been heavily forested; however, by 1993 and 1996, much of the primary forest had been replaced with mango plantations. The loss of old-growth forest habitat for the IE enzootic mosquito vector, *Culex (Melanoconion) taeniopus*, potentially resulted in a selective adaptation for a highly abundant mosquito vector that breeds in the salt marshes and mangrove swamps prevalent in the area at the time of the 1993 and 1996 epizootics.

4.4 RICE, MOSQUITOES, PIGS, AND JAPANESE ENCEPHALITIS VIRUS

As described earlier, the earth's human population has been increasing markedly. Perhaps no place else on earth has seen the dramatic increase in total numbers as Asia and with it increases in both the acreage and intensity of rice farming. Growth in China over the past 500 years has been representative of the region (Figure 4.5). As reviewed recently (Erlanger et al., 2009), in the past 50 years, the population of Asia has more than doubled from 1.7 to 3.5 billion people, of which 3 billion reside in areas at risk from JEV infection. Not only has the number of people increased, but the per capita caloric consumption has increased from 1900 to 2600 calories per day.

Concurrently to feed this growing population, rice production has more than doubled in the past 50 years from 226 to 529 million tons per year by increasing acreage and intensifying production per acre. Efficiency and the number of crops per year vary markedly among rice-producing countries largely due to land area, population density, economic development,

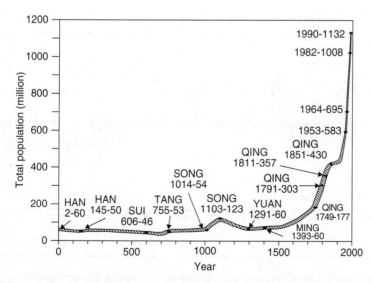

Figure 4.5. Growth of population in China (http://afe.easia.columbia.edu/special/china_1950_population.htm).

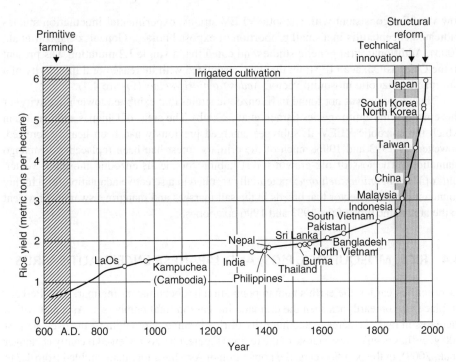

Figure 4.6. Increase in rice production efficiency in Japan. Individual Asian countries are placed on the chart according to when Japan was producing rice at the level at which they are producing rice today.

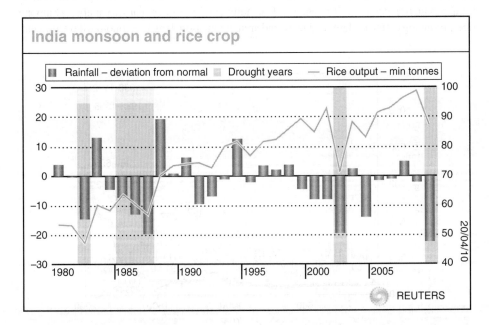

Figure 4.7. Annual rainfall anomaly and rice production in India. *Source*: Government of India. For color detail, please see color plate section.

and rice strains grown (Figure 4.6 and Figure 4.7). The change has created vast rice monocultures with intermittent standing water suitable for the production of mosquitoes and wading birds including ardeids that feed on the insects and other aquatic fauna. The use of chemical fertilizers has increased the yield of both rice and mosquitoes (Sunish et al., 2003; Victor and Reuben, 2000). Overall, the population dynamics of adult *Culex tritaeniorhynchus*, the most important and widespread vector of JEV, has been linked directly in time and space to rice production and the number of crops produced per year (Reisen et al., 1976).

Pork provides much of the meat throughout this rice-growing region, with pigs typically raised in small numbers by individual families living near their paddy. This brings a suitable blood meal host into proximity with mosquito production sites. The evolution of this expanding rural agroecosystem, interspersed with small villages and, with population growth, small cities exploiting higher elevations less prone to flooding, has created a mosaic of vector production and host availability perfect for the transmission of JEV.

Japanese encephalitis virus is the type *Flavivirus* for the JEV serocomplex that includes WNV discussed later. In humans and equines, infection can lead to neuroinvasive disease and death, whereas infection in porcine hosts produces an elevated viremia without overt disease, although fetal death and abortion are common (Burke and Leake, 1988). JEV apparently evolved within the Indonesian Archipelago where all five lineages co-circulate with relatively low levels of clinical disease (Solomon et al., 2003). Human neuroinvasive disease cases were first recorded in Japan in the late 1800s when this temperate rice-producing area was invaded by lineage I of the virus (Burke and Leake, 1988; van den Hurk et al., 2009). Later in-depth studies showed that JEV was maintained and amplified at ardeid rookeries near Tokyo (Buescher et al., 1959) and that peridomestic pigs (Scherer et al., 1959) were an important amplifying host that provided a link between the rural avian cycle and the human population (Burke and Leake, 1988). The primary mosquito vectors, *Cx. tritaeniorhynchus* and other members of the *Culex vishnui* complex, feed frequently on large mammals as well as birds (Iha, 1971; Mitchell et al., 1973; Reisen and Boreham, 1979), thereby bridging the virus from pigs to the human population. Tracking expanding rice and pig culture, outbreaks of JEV were documented in northeast (Korea, China, and Taiwan), southeast (Vietnam and Thailand), and most recently central (Nepal, India, and Pakistan) and south central (Sri Lanka) Asia (Weaver and Reisen, 2010). Transmission patterns were altered progressively as JEV moved into south and western Asia where pig production declined due to changes in religion. Here, water birds were frequently infected and presumably served as amplification hosts (Boyle et al., 1983; Khan and Banerjee, 1980; Pandit and Work, 1957), although pockets of pig production have remained important (Impoinvil et al., 2011).

Although spatially defined by the distribution of rice culture, the intensity of transmission may be dictated by climate variation that determines water availability for rice and therefore mosquito production and temperature that drives transmission dynamics. This is especially true for drier regions such as India where rice production is affected strongly by the strength of the annual monsoon rainfall (Figure 4.7). Conversely, in temperate climates with highly developed irrigation systems, outbreaks may be more closely related to temperature and the intensity of virus amplification (Mogi, 1983). Climate change altering rainfall patterns and warming northern latitudes is expected to have a marked impact on JEV transmission dynamics and may have contributed to the recent invasion of Kathmandu Valley in Nepal by JEV (Partridge et al., 2007).

In recent years the occurrence of human infections in NE Asia has been greatly reduced by control programs that have vaccinated school-age children and coalesced hog production into centralized facilities away from the peridomestic environment. Although these programs have been generally successful (Arai et al., 2008; Elias et al., 2009; Erlanger et al., 2009), enzootic transmission remains and is annually detected by surveillance programs near major cities. As regional temperatures continue to warm and regional successful control programs wane, there remains a high risk of resurgence of JEV in these heavily populated areas.

4.5 *CULEX PIPIENS* COMPLEX, HOUSE SPARROWS, URBANIZATION, AND WEST NILE VIRUS

A variety of anthropogenic impacts and climate variation contributed to the successful invasion of North America (NA) by WNV. The primary vectors, members of the *Culex pipiens* complex, most likely arose within the Ethiopian region and were dispersed circumglobally throughout tropical and temperate latitudes by human movements (Farajollahi et al., 2011). In NA, this complex includes *Culex quinquefasciatus* in the south and *Cx. pipiens* with form molestus in the north and considerable introgression within intervening latitudes (Barr, 1957, 1967; Harbach, 2011). Urban peridomestic containers, drainage systems, and human and domestic animal waste management (or mismanagement) systems have created highly eutrophic and productive larval habitats that have led to large almost unmanageable *Cx. pipiens* populations within and around most cities. In addition, peridomestic landscaping has provided ample sugar sources and resting habitats for adult *Cx. pipiens* and created multiple niches filled by several commensal avian species. This avifauna includes the ubiquitous House sparrow (*Passer domesticus*) (Zimmerman, 2009) and pigeon (*Columba livia*) (Biechman, 2007) that were intentionally introduced into NA as well as native species such as American robins (*Turdus migratorius*) and House finches (*Carpodacus mexicanus*). Collectively, these and other avian species provide abundant blood meal sources for urban mosquitoes. In addition, several species in the family *Corvidae* also exploit urban centers and waste disposal sites as food sources (Reisen et al., 2006a); however, their utilization as frequent mosquito blood meal sources seems uncertain (Kilpatrick, 2011). Historically, these conditions in NA may have contributed to the expansion of St. Louis encephalitis virus (SLEV) at the end of the nineteenth century (Baillie et al., 2008) and set the stage for the invasion of WNV.

West Nile virus (a Flavivirus closely related to SLEV) was discovered in rural Uganda in 1937 (Hayes, 2001) but was not recognized as a significant health problem until an outbreak of febrile illness was investigated in the Nile delta of Egypt (Taylor et al., 1956). Later genetic studies (Brault et al., 2007) showed that this outbreak strain in lineage I contained a proline substitution at the NS3-249 position that imparted elevated viremia and mortality in crows and was associated with outbreaks of human neuroinvasive disease in Russia and human and avian disease in Israel. In 1999, a similar strain of lineage I WNV containing the same NS3 substitution invaded New York City (Lanciotti et al., 1999), where it led to a focal outbreak of human neuroinvasive disease, mortality in zoo birds, an extensive die-off of American crows, and several cases of equine neuroinvasive disease. *Cx. pipiens* was incriminated as the primary vector mosquito and House sparrows the primary maintenance host (Komar et al., 2001, 2003). Later, mosquito blood meal identification studies would incriminate the American robin as an important host throughout eastern NA, where it was fed upon frequently by *Cx. pipiens* (Apperson et al., 2002, 2004). The hallmark of the initial and subsequent outbreaks of WNV in NA has been large die-offs of corvids, especially American crows (Eidson, 2001; Eidson et al., 2001a,b,c). The invading virus with the NS3 mutation produced extraordinarily elevated viremias during acute infection in American crows (Brault et al., 2004a; Komar et al., 2003) that were able to effectively drive WNV into moderately susceptible urban *Cx. pipiens* populations (Turell et al., 2000). The importance of corvids as a critical amplifying host became evident as the virus invaded

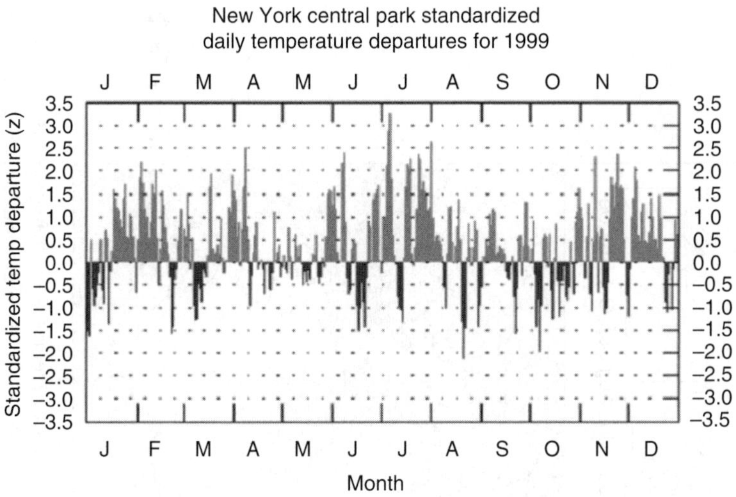

Figure 4.8. Departures from 50-year average temperature in Central Park, New York City, during 1999 (data from http://www.climatestations.com/new-york-city/). For color detail, please see color plate section.

California causing large outbreaks in Los Angeles and Sacramento associated with large American crow die-offs, but failing to produce outbreaks in areas such as Palm Springs where few corvids are found (Reisen et al., 2006a).

The introduction and amplification of a tropical virus into temperate New York City during the summer of 1999 was undoubtedly enabled by anomalously warm temperatures that averaged up to 3.5 °C above normal throughout much of that summer (Figure 4.8), making it the hottest summer in New York City history. Above normal temperatures extended the duration of the mosquito season, most likely increased *Cx. pipiens* abundance, and facilitated WNV transmission by reducing the length of the vector's gonotrophic cycle, increasing the frequency of vector–host contact, and reducing the duration of the extrinsic incubation period of the virus (Hartley et al., 2012; Reisen et al., 2006b).

Unexpectedly, infected *Cx. pipiens* females were collected during the winter months (Nasci et al., 2001), and this tropical virus successfully survived the cold NE U.S. winter. As *Cx. pipiens* typically enter reproductive diapause during winter, these mosquitoes were presumed to be infected by vertical transmission (Anderson and Main, 2006; Anderson et al., 2008); however, later studies showed that parous *Cx. pipiens* females also overwinter in the New York area (Andreadis et al., 2010). During the subsequent 5-year period, WNV rapidly invaded the New World, extending its distribution west to the Pacific Coast, north into Canada, and south to Argentina (Kramer et al., 2008), with major outbreaks in Chicago, New Orleans, Colorado, and then Los Angeles. As the virus invaded the west, *Culex tarsalis* emerged as an important rural vector, especially in the Canadian prairies above the northern distribution of the *Cx. pipiens* complex. Here, outbreaks centered in Saskatchewan during 2003 and 2007 were associated with exceptionally warm temperatures that exceeded 11 °C above 50-year normals and extended the duration of the transmission season from 3 to 5 months (Figure 4.9). Similar trends were noted in the northern U.S. prairie states (Reisen et al., 2006b) that typically have had the highest incidence of infection in the United States, especially during La Niña periods. Above normal temperatures at these northern latitudes were characteristic of a global warming trend expected to extend the northern distribution of mosquitoes and viruses as well as the duration of the transmission

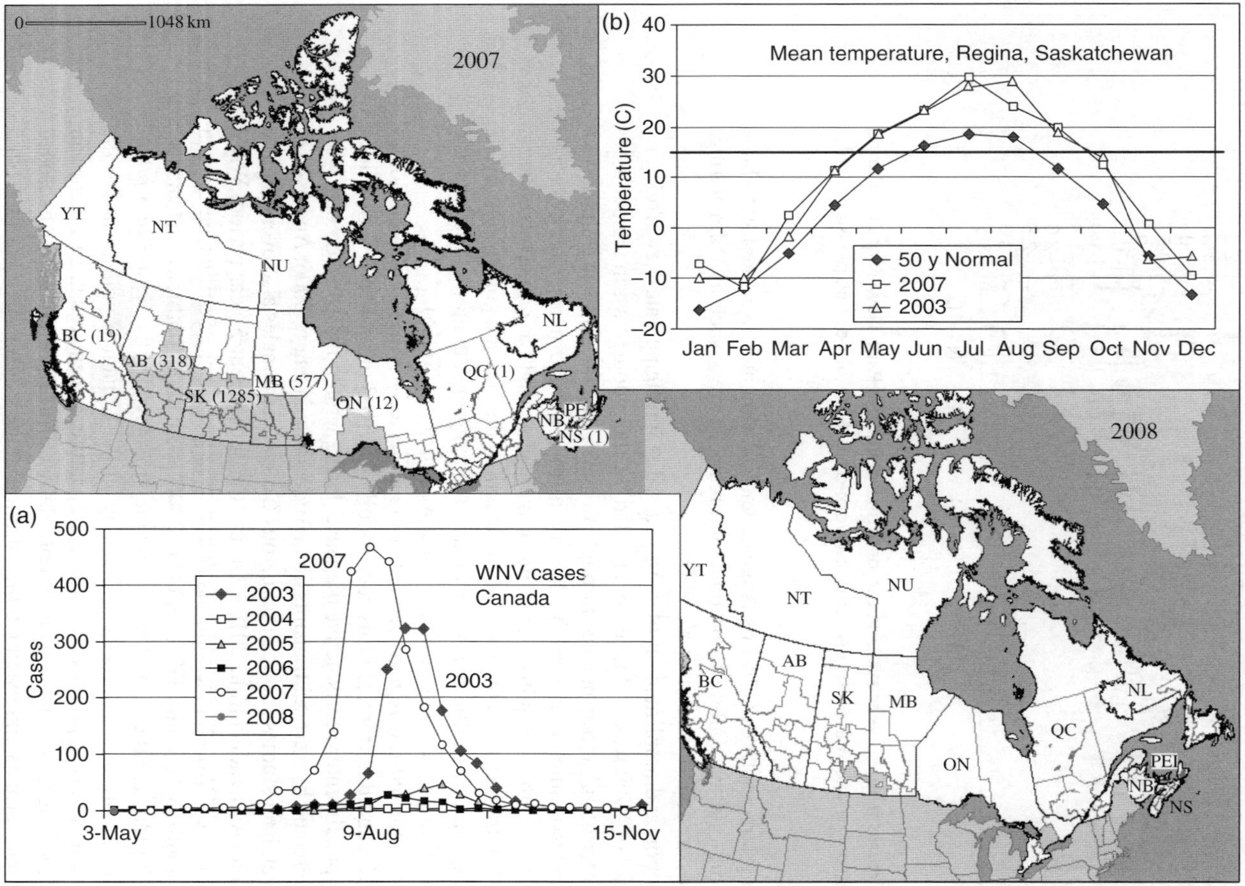

Figure 4.9. WNV in Canada: (**a**) number of cases per week during 2003–2008 period and (**b**) average mean monthly temperature in Regina, Saskatchewan, compared to 50-year normals. Bar shows the minimal threshold for WNV replication in *Cx. tarsalis*. Also shown are the contrasting distributions of human cases during 2007 and 2008 (data redrawn from Health Canada, http://www.phac-aspc.gc.ca/wnv-vwn/ and http://www.climate.weatheroffice.gc.ca).

season. Outbreaks during exceptionally warm summers were facilitated by virus transmission dynamics within the vector mosquitoes (Reisen et al., 2006b) in concert with the acquisition by the virus of an envelope gene mutation that enabled rapid transmission by *Culex* mosquitoes at warm temperatures (Kilpatrick et al., 2008; Moudy et al., 2007). This new WN02 strain retained the NS3 mutation of the founding NY99 strain and soon displaced the NY99 strain throughout NA (Brault, 2009), leaving a strain highly virulent in some avian hosts and more effectively transmitted by *Culex* vectors.

Repeated outbreaks as the virus moved across NA revealed several patterns related to anthropogenic change. WNV amplified to outbreak levels most effectively in highly modified urban environments where ecosystem complexity was simplified to several competent avian host species (Loss et al., 2009) and *Cx. pipiens* mosquitoes. A repeated 3-year pattern of virus activity was observed, with quiet invasion during year 1, an explosive outbreak during year 2, and then rapid subsidence during year 3 (Hayes et al., 2005). Subsidence was related, in part, to elevated herd immunity in peridomestic passerine maintenance hosts (Kwan et al., 2010) and depopulation of amplifying corvid hosts (Ladeau et al., 2007; Wheeler et al., 2008, 2009). Avian population recovery was based on the replacement rates of both species groups and required from 3 to 4 years in urban Los Angeles where House finches were the principal avian host (Kwan et al., 2012; Molaei et al., 2010; Thiemann et al., 2012).

The economic crisis in the United States driven by the collapse of the housing market during the WNV epidemic produced unexpected consequences. Marked increases in home foreclosures produced a suburban landscape dotted with unmaintained swimming pools, Jacuzzis, and ornamental ponds, habitats readily exploited by both the *Cx. pipiens* complex and *Cx. tarsalis* (Reisen et al., 2008b). Home foreclosures that reduced property tax revenues to support control programs concurrently produced a landscape of productive mosquito sources difficult to detect and control and associated with increases in human cases in some cities.

A major problem for the persistence of the North American encephalitides including WNV is surviving the overwintering period when transmission is interrupted by cold temperatures. At southern latitudes, moderate temperatures allow *Cx. quinquefasciatus* to survive winter in a quiescent state (Eldridge, 1968) that may be periodically interrupted by periods of warm temperatures allowing intermittent transmission (Tesh et al., 2004). With climate change these interruptions are anticipated to be of longer duration and gradually move northward. In California, diapause termination and associated blood feeding by both *Cx. tarsalis* and the *Cx. pipiens* complex commenced as early as the first week in January, although nullipars were collected through early March indicating a gradual response to increasing temperature or reduced feeding success during cool evenings by the overwintering cohort (Reisen et al., 2010).

In summary, anthropogenic and associated climate change set the stage for the invasion the New World by WNV. Unintentional and intentional transport of vector and host species that efficiently exploited urban environments brought components of the transmission cycle to the peridomestic environment. Urban heat islands exacerbated the situation by providing above normal temperatures during the year of invasion and summers that followed and enabled rapid western expansion. Viral fitness concurrently was enhanced by mutations that facilitated virulence in key amplification hosts and transmission by the primary *Culex* vectors. In combination, these factors have resulted in the addition of a novel and now permanent public, veterinary, and wildlife health problem to the New World.

4.6 URBANIZATION, GLOBAL TRADE, AND THE REEMERGENCE OF CHIKUNGUNYA VIRUS

Chikungunya virus diverged into two principle lineages over 500 years ago (Volk et al., 2010). In West Africa, CHIKV circulates in relatively silent enzootic cycles between non-human primates and sylvatic *Aedes* mosquitoes, with occasional epidemics emerging within adjacent rural villages where *Aedes aegypti* transmits the virus between viremic humans (Powers and Logue, 2007). In contrast, in arid eastern Africa, water storage within rural villages and recently major cities has led to the urbanization of CHIKV within an anthroponotic cycle involving humans and *Ae. aegypti*. During the European colonial period, extensive travel between East Africa and the Indian subcontinent first introduced the urban vector, *Ae. aegypti*, and then within the past 100 years (Volk et al., 2010) the original CHIKV strain that spread rapidly throughout Southeast Asia. From the 1950s through the 1970s, CHIKV became recognized as a cause of febrile epidemics in Asia, although its health significance was probably under-recognized as dengue viruses predominated in the same regions (Sudeep and Parashar, 2008). Interestingly, this original invading strain remained genetically conserved (Powers et al., 2000), was confined to urban areas and *Ae. aegypti*, and did not exploit other *Stegomyia* mosquitoes such as *Ae. albopictus* and Asian primates, even though *Ae. albopictus* was a more competent vector than *Ae. aegypti* (Turell et al., 1992). After 20 years of repeated urban outbreaks, CHIKV essentially disappeared as an Asian health problem only to reemerge 30 years later.

Over the past 20 years, *Ae. albopictus* has been dipsersed circumglobally from Asia largely by expanded global trade, most notably used tires (Craven et al., 1988) and agriculture products such as "lucky bamboo" (Linthicum et al., 2003; Madon et al., 2002). In 2004–2005, a CHIKV outbreak occurred in Kenya with an African genotype that subsequently spread in 2005 to the neighboring Indian Ocean islands of Comoros, La Réunion, Mauritius, and the Seychelles and in 2006 to Mayotte and Madagascar, areas with established *Ae. albopictus* populations. Extremely high infection rates with tens of thousands of human clinical cases of arthralgia were reported on these islands. Drought conditions and the storage of water in cisterns were associated with the increases in *Ae. albopictus* and the emergence CHIKV in East Africa (Chretien et al., 2007). Additionally, many of these islands are of volcanic origin and have little to no natural freshwater sources, necessitating water cisterns and other storage containers, creating mosquito breeding sources and sufficient vector populations for the epidemic transmission of CHIKV. Subsequently, the virus was moved by infected travelers and perhaps mosquitoes along historical routes into the Indian subcontinent and Malaysia where the new strain has largely displaced existing Asian CHIKV genotypes (Charrel et al., 2007), resulting in millions of symptomatic human infections in India alone. The emergence of the East-Central African genotype of CHIKV in Kenya and then the Indian Ocean islands and Asia has been largely associated with a single amino substitution in the E1 surface glycoprotein that increased the infectivity CHIKV for *Ae. albopictus* but not *Ae. aegypti* (Tsetsarkin et al., 2007). The emergence of this novel CHIKV genotype with the capacity to efficiently exploit the invading *Ae. albopictus* vector has significantly increased the areas at risk for introduction of this virus. Traveler-associated viremic CHIKV cases have been reported in numerous countries and autochthonous transmission of CHIKV was identified in Italy in 2007 (Bordi et al., 2008). Increased temperatures in northern Europe have been associated with the expansion of *Ae. albopictus* into new geographic locales (Charrel et al., 2007). This fact, coupled with the high competency of this mosquito vector for the newly emerged CHIKV strain (Tsetsarkin et al., 2007), portend a likely emergence of CHIKV outbreaks in areas not previously associated with the disease agent. Increased evolutionary rates have been documented in the new Asian CHIKV strains that circulate in urban cycles as opposed to enzootic transmission cycles (Volk et al., 2010), and this will likely increase the capacity of CHIKV to exploit novel urban niches and further establish itself as a peridomestic anthroponotic agent.

4.7 CONCLUSIONS

This chapter presented several examples highlighting arboviral emergence due to environmental change fostered by altered human demographics, agricultural practices, and land use. Increased temperatures have had and will continue to have a profound effect on the reproductive rate of arthropod vectors, shortening virus extrinsic incubation periods and expanding both vector and virus geographic ranges. Factors that increase the movement of arboviral agents such as human travel and commerce provide new opportunities for these highly plastic viruses to adapt to novel transmission cycles. Movement of viral agents and vectors overlaid with environmental change creates selective pressures for the emergence of genetic variants with altered pathologic capacity. The emergence of VEE subtype IE viruses highlights the flexibility of RNA viruses to adapt to changing ecological environments and the resulting consequences that such adaptation can have on the emergence of new epidemiological and disease patterns. As there is no indication of global stabilization of human populations, climate, transport of vectors and viruses, land use, or the multitude of other factors that dictate the replicative environment of arboviruses, it will be very likely that the continued emergence of these viruses will be fostered by these disruptions.

REFERENCES

Anderson, J.F., Main, A.J., 2006. Importance of vertical and horizontal transmission of West Nile virus by Culex pipiens in the Northeastern United States. *J Infect Dis* **194**, 1577–1579.

Anderson, J.F., Main, A.J., Delroux, K., et al., 2008. Extrinsic incubation periods for horizontal and vertical transmission of West Nile virus by Culex pipiens pipiens (Diptera: Culicidae). *J Med Entomol* **45**, 445–451.

Andreadis, T.G., Armstrong, P.M., Bajwa, W.I., 2010. Studies on hibernating populations of Culex pipiens from a West Nile virus endemic focus in New York City: parity rates and isolation of West Nile virus. *J Am Mosq Control Assoc* **26**, 257–264.

Anishchenko, M., Bowen, R.A., Paessler, S., et al., 2006. Venezuelan encephalitis emergence mediated by a phylogenetically predicted viral mutation. *Proc Natl Acad Sci USA* **103**, 4994–4999.

Apperson, C.S., Harrison, B.A., Unnasch, T.R., et al., 2002. Host-feeding habits of Culex and other mosquitoes (Diptera: Culicidae) in the Borough of Queens in New York City, with characters and techniques for identification of Culex mosquitoes. *J Med Entomol* **39**, 777–785.

Apperson, C.S., Hassan, H.K., Harrison, B.A., et al., 2004. Host feeding patterns of established and potential mosquito vectors of West Nile virus in the eastern United States. *Vector Borne Zoonotic Dis* **4**, 71–82.

Arai, S., Matsunaga, Y., Takasaki, T., et al., Vaccine Preventable Diseases Surveillance Program of Japan, 2008. Japanese encephalitis: surveillance and elimination effort in Japan from 1982 to 2004. *Jpn J Infect Dis* **61**, 333–338.

Baillie, G.J., Kolokotronis, S.O., Waltari, E., et al., 2008. Phylogenetic and evolutionary analyses of St. Louis encephalitis virus genomes. *Mol Phylogenet Evol* **47**, 717–728.

Barr, A.R., 1957. The distribution of *Culex p. pipiens* and *C.P. quinquefasciatus* in North America. *Am J Trop Med Hyg* **6**, 153–165.

Barr, A.R., 1967. Occurrence and distribution of the *Culex pipiens* complex. *Bull World Health Organ* **37**, 293–296.

Biechman, A., 2007. *Pigeons – the Fascinating Saga of the World's Most Revered and Reviled Bird. St Lucia, Queensland, Australia.* University of Queensland Press.

Bordi, L., Carletti, F., Castilletti, C., et al., 2008. Presence of the A226V mutation in autochthonous and imported Italian chikungunya virus strains. *Clin Infect Dis* **47**, 428–429.

Boyle, D.B., Dickerman, R.W., Marshall, I.D., 1983. Primary viraemia responses of herons to experimental infection with Murray Valley encephalitis, Kunjin and Japanese encephalitis viruses. *Aust J Exp Biol Med Sci* **61** (Pt 6), 655–664.

Brault, A.C., 2009. Changing patterns of West Nile virus transmission: altered vector competence and host susceptibility. *Vet Res* **40**, 43.

Brault, A.C., Langevin, S.A., Bowen, R.A., et al., 2004a. Differential virulence of West Nile strains for American crows. *Emerg Infect Dis* **10**, 2161–2168.

Brault, A.C., Powers, A.M., Ortiz, D., et al., 2004b. Venezuelan equine encephalitis emergence: enhanced vector infection from a single amino acid substitution in the envelope glycoprotein. *Proc Natl Acad Sci USA* **101**, 11344–11349.

Brault, A.C., Huang, C.Y., Langevin, S.A., et al., 2007. A single positively selected West Nile viral mutation confers increased virogenesis in American crows. *Nat Genet* **39**, 1162–1166.

Buescher, E.L., Scherer, W.F., Mc, C.H., et al., 1959. Ecologic studies of Japanese encephalitis virus in Japan. IV. Avian infection. *Am J Trop Med Hyg* **8**, 678–688.

Burke, D., Leake, C., 1988. Japanese encephalitis. In: Monath, T.P. (Ed.), *The Arboviruses: Epidemiology and Ecology*. CRC Press, Boca Raton, FL, pp. 63–92.

Charrel, R.N., de Lamballerie, X., Raoult, D., 2007. Chikungunya outbreaks – the globalization of vectorborne diseases. *N Engl J Med* **356**, 769–771.

Chretien, J.P., Anyamba, A., Bedno, S.A., et al., 2007. Drought-associated chikungunya emergence along coastal East Africa. *Am J Trop Med Hyg* **76**, 405–407.

Craven, R.B., Eliason, D.A., Francy, D.B., et al., et al., 1988. Importation of *Aedes albopictus* and other exotic mosquito species into the United States in used tires from Asia. *J Am Mosq Control Assoc* **4**, 138–142.

Dobson, A., 2009. Climate variability, global change, immunity, and the dynamics of infectious diseases. *Ecology* **90**, 920–927.

Domingo, E., 1997. Rapid evolution of viral RNA genomes. *J Nutr* **127**, 958S–961S.

Domingo, E., Escarmis, C., Sevilla, N., et al., 1996. Basic concepts in RNA virus evolution. *FASEB J* **10**, 859–864.

Eidson, M., 2001. "Neon needles" in a haystack: the advantages of passive surveillance for West Nile virus. *Ann N Y Acad Sci* **951**, 38–53.

Eidson, M., Komar, N., Sorhage, F., et al., 2001a. Crow deaths as a sentinel surveillance system for West Nile virus in the northeastern United States, 1999. *Emerg Infect Dis* **7**, 615–620.

Eidson, M., Kramer, L., Stone, W., et al., 2001b. Dead bird surveillance as an early warning system for West Nile virus. *Emerg Infect Dis* **7**, 631–635.

Eidson, M., Miller, J., Kramer, L., et al., 2001c. Dead crow densities and human cases of West Nile virus, New York State, 2000. *Emerg Infect Dis* **7**, 662–664.

Eldridge, B.F., 1968. The effect of temperature and photoperiod on blood-feeding and ovarian development in mosquitoes of the Culex pipiens complex. *Am J Trop Med Hyg* **17**, 133–140.

Elias, C., Okwo-Bele, J.M., Fischer, M., 2009. A strategic plan for Japanese encephalitis control by 2015. *Lancet Infect Dis* **9**, 7.

Erlanger, T.E., Weiss, S., Keiser, J., et al., 2009. Past, present, and future of Japanese encephalitis. *Emerg Infect Dis* **15**, 1–7.

Farajollahi, A., Fonseca, D.M., Kramer, L.D., et al., 2011. "Bird biting" mosquitoes and human disease: a review of the role of *Culex pipiens* complex mosquitoes in epidemiology. *Infect Genet Evol* **11**, 1577–1585.

Gonzalez-Salazar, D., Estrada-Franco, J.G., Carrara, A.S., et al., 2003. Equine amplification and virulence of subtype IE Venezuelan equine encephalitis viruses isolated during the 1993 and 1996 Mexican epizootics. *Emerg Infect Dis* **9**, 161–168.

Hansen, J., Sato, M., Ruedy, R., et al., 2006. Global temperature change. *Proc Natl Acad Sci USA* **103**, 14288–14293.

Harbach, R.E., 2011. Classification within the cosmopolitan genus Culex (Diptera: Culicidae): the foundation for molecular systematics and phylogenetic research. *Acta Trop* **120**, 1–14.

Hartley, D., Barker, C., Le Menac'h, A., et al., 2012. The effects of temperature on the emergence and seasonality of West Nile virus in California. *Am J Trop Med Hyg* **86**, 884–894.

Hayes, C.G., 2001. West Nile virus: Uganda, 1937, to New York City, 1999. *Ann N Y Acad Sci* **951**, 25–37.

Hayes, E.B., Komar, N., Nasci, R.S., et al., 2005. Epidemiology and transmission dynamics of West Nile virus disease. *Emerg Infect Dis* **11**, 1167–1173.

Iha, S., 1971. Feeding preference and seasonal distribution of mosquitoes in relation to the epidemiology of Japanese encephalitis in Okinawa Main Island. *Trop Med* **12**.

Impoinvil, D.E., Solomon, T., Schluter, W.W., et al., 2011. The spatial heterogeneity between Japanese encephalitis incidence distribution and environmental variables in Nepal. *PLoS One* **6**, e22192.

IPPC, 2007. Assessment of observed changes and responses in natural and managed systems. In: *Climate Change 2007: Impacts, Adaptation and Vulnerability, Contribution of Working Group II to the Fourth Assessment Report of the Intergovernmental Panel on Climate Change*. Cambridge University Press.

Khan, F.U., Banerjee, K., 1980. Mosquito collection in heronries and antibodies to Japanese encephalitis virus in birds in Asansol-Dhanbad region. *Indian J Med Res* **71**, 1–5.

Kilpatrick, A.M., 2011. Globalization, land use, and the invasion of West Nile virus. *Science* **334**, 323–327.

Kilpatrick, A.M., Meola, M.A., Moudy, R.M., et al., 2008. Temperature, viral genetics, and the transmission of West Nile virus by Culex pipiens mosquitoes. *PLoS Pathog* **4**, e1000092.

Kinney, R.M., Tsuchiya, K.R., Sneider, J.M., et al., 1992. Molecular evidence for the origin of the widespread Venezuelan equine encephalitis epizootic of 1969–1972. *J Gen Virol* **73**, 3301–3305.

Komar, N., Panella, N.A., Burns, J.E., et al., 2001. Serologic evidence for West Nile virus infection in birds in the New York City vicinity during an outbreak in 1999. *Emerg Infect Dis* **7**, 621–625.

Komar, N., Langevin, S., Hinten, S., et al., 2003. Experimental infection of North American birds with the New York 1999 strain of West Nile virus. *Emerg Infect Dis* **9**, 311–322.

Kramer, L.D., Styer, L.M., Ebel, G.D., 2008. A global perspective on the epidemiology of West Nile Virus. *Annu Rev Entomol* **53**, 61–81.

Kwan, J.L., Kluh, S., Madon, M.B., et al., 2010. West nile virus emergence and persistence in Los Angeles, California, 2003–2008. *Am J Trop Med Hyg* **83**, 400–412.

Kwan, J., Kluh, S., Reisen, W., 2012. Antecedent avian immunity limits tangential transmission of West Nile virus to humans. *PLoS One* **7**, e34127.

Ladeau, S.L., Kilpatrick, A.M., Marra, P.P., 2007. West Nile virus emergence and large-scale declines of North American bird populations. *Nature* **447**, 710–713.

Lanciotti, R.S., Roehrig, J.T., Deubel, V., et al., 1999. Origin of the West Nile virus responsible for an outbreak of encephalitis in the northeastern United States. *Science* **286**, 2333–2337.

Linthicum, K.J., Kramer, V.L., Madon, M.B., et al., 2003. Introduction and potential establishment of *Aedes albopictus* in California in 2001. *J Am Mosq Control Assoc* **19**, 301–308.

Loss, S.R., Hamer, G.L., Walker, E.D., et al., 2009. Avian host community structure and prevalence of West Nile virus in Chicago, Illinois. *Oecologia* **159**, 415–424.

Maclachlan, N.J., 2010. Global implications of the recent emergence of bluetongue virus in Europe. *Vet Clin North Am Food Anim Pract* **26**, 163–171.

Madon, M.B., Mulla, M.S., Shaw, M.W., et al., 2002. Introduction of *Aedes albopictus* (Skuse) in southern California and potential for its establishment. *J Vector Ecol* **27**, 149–154.

Mitchell, C.J., Chen, P.S., Boreham, P.F., 1973. Host-feeding patterns and behaviour of 4 Culex species in an endemic area of Japanese encephalitis. *Bull World Health Organ* **49**, 293–299.

Mogi, M., 1983. Relationship between number of human Japanese encephalitis cases and summer meteorological conditions in Nagasaki, Japan. *Am J Trop Med Hyg* **32**, 170–174.

Molaei, G., Cummings, R.F., Su, T., et al., 2010. Vector–host interactions governing epidemiology of West Nile Virus in Southern California. *Am J Trop Med Hyg* **83**, 1269–1282.

Moudy, R.M., Meola, M.A., Morin, L.L., et al., 2007. A newly emergent genotype of West Nile Virus is transmitted earlier and more efficiently by *Culex mosquitoes*. *Am J Trop Med Hyg* **77**, 365–370.

Nasci, R.S., Savage, H.M., White, D.J., et al., 2001. West Nile virus in overwintering *Culex mosquitoes*, New York City, 2000. *Emerg Infect Dis* **7**, 742–744.

O'Neill, B.C., Dalton, M., Fuchs, R., et al., 2010. Global demographic trends and future carbon emissions. *Proc Natl Acad Sci USA* **107**, 17521–17526.

Oberste, M.S., Fraire, M., Navarro, R., et al., 1998. Association of Venezuelan equine encephalitis virus subtype IE with two equine epizootics in Mexico. *Am J Trop Med Hyg* **59**, 100–107.

Oke, T., 1982. The energetic basis of the urban heat island. *Quart J Roy Meterol Soc* **108**, 1–24.

Pandit, C., Work, T., 1957. Japanese B encephalitis in India. *Proc 9th Pacific Sci Conf* **17**, 56–59.

Partridge, J., Ghimire, P., Sedai, T., et al., 2007. Endemic Japanese encephalitis in the Kathmandu valley, Nepal. *Am J Trop Med Hyg* **77**, 1146–1149.

Powers, A.M., Logue, C.H., 2007. Changing patterns of chikungunya virus: re-emergence of a zoonotic arbovirus. *J Gen Virol* **88**, 2363–2377.

Powers, A.M., Oberste, M.S., Brault, A.C., et al., 1997. Repeated emergence of epidemic/epizootic Venezuelan equine encephalitis from a single genotype of enzootic subtype ID virus. *J Virol* **71**, 6697–6705.

Powers, A.M., Brault, A.C., Tesh, R.B., et al., 2000. Re-emergence of Chikungunya and O'nyong-nyong viruses: evidence for distinct geographical lineages and distant evolutionary relationships. *J Gen Virol* **81** Pt 2, 471–479.

Reisen, W.K., Boreham, P.F., 1979. Host selection patterns of some Pakistan mosquitoes. *Am J Trop Med Hyg* **28**, 408–421.

Reisen, W.K., Aslamkhan, M., Basio, R.G., 1976. The effects of climatic patterns and agricultural practices on the population dynamics of Culex tritaeniorhynchus in Asia. *Southeast Asian J Trop Med Public Health* **7**, 61–71.

Reisen, W.K., Barker, C.M., Carney, R., et al., 2006a. Role of corvids in epidemiology of west Nile virus in southern California. *J Med Entomol* **43**, 356–367.

Reisen, W.K., Fang, Y., Martinez, V.M., 2006b. Effects of temperature on the transmission of West Nile virus by *Culex tarsalis* (Diptera: Culicidae). *J Med Entomol* **43**, 309–317.

Reisen, W.K., Cayan, D., Tyree, M., et al., 2008a. Impact of climate variation on mosquito abundance in California. *J Vector Ecol* **33**, 89–98.

Reisen, W.K., Takahashi, R.M., Carroll, B.D., et al., 2008b. Delinquent mortgages, neglected swimming pools, and West Nile virus, California. *Emerg Infect Dis* **14**, 1747–1749.

Reisen, W.K., Thiemann, T., Barker, C.M., et al., 2010. Effects of warm winter temperature on the abundance and gonotrophic activity of Culex (Diptera: Culicidae) in California. *J Med Entomol* **47**, 230–237.

Scherer, W.F., Moyer, J.T., Izumi, T., et al., 1959. Ecologic studies of Japanese encephalitis virus in Japan. VI. Swine infection. *Am J Trop Med Hyg* **8**, 698–706.

Solomon, T., Ni, H., Beasley, D.W., et al., 2003. Origin and evolution of Japanese encephalitis virus in southeast Asia. *J Virol* **77**, 3091–3098.

Sudeep, A., Parashar, D., 2008. Chikungunya: an overview. *J Biosci* **33**, 443–449.

Sunish, I.P., Rajendran, R., Reuben, R., 2003. The role of urea in the oviposition behaviour of Japanese encephalitis vectors in rice fields of South India. *Mem Inst Oswaldo Cruz* **98**, 789–791.

Taylor, R.M., Work, T.H., Hurlbut, H.S., et al., 1956. A study of the ecology of West Nile virus in Egypt. *Am J Trop Med Hyg* **5**, 579–620.

Tesh, R.B., Parsons, R., Siirin, M., et al., 2004. Year-round West Nile virus activity, Gulf Coast region, Texas and Louisiana. *Emerg Infect Dis* **10**, 1649–1652.

Thiemann, T., Lemenager, D.A., Kluh, S., et al., 2012. Spatial variation in the host feeding patterns of *Culex tarsalis* and the *Culex pipiens* complex in California (Diptera: Culicidae). *J Med Entomol* **49**, 903–916.

Tsetsarkin, K.A., Vanlandingham, D.L., McGee, C.E., et al., 2007. A single mutation in chikungunya virus affects vector specificity and epidemic potential. *PLoS Pathog* **3**, e201.

Turell, M.J., Beaman, J.R., Tammariello, R.F., 1992. Susceptibility of selected strains of *Aedes aegypti* and *Aedes albopictus* (Diptera: Culicidae) to chikungunya virus. *J Med Entomol* **29**, 49–53.

Turell, M.J., O'Guinn, M., Oliver, J., 2000. Potential for New York mosquitoes to transmit West Nile virus. *Am J Trop Med Hyg* **62**, 413–414.

van den Hurk, A.F., Ritchie, S.A., Mackenzie, J.S., 2009. Ecology and geographical expansion of Japanese encephalitis virus. *Annu Rev Entomol* **54**, 17–35.

Victor, T.J., Reuben, R., 2000. Effects of organic and inorganic fertilisers on mosquito populations in rice fields of southern India. *Med Vet Entomol* **14**, 361–368.

Volk, S.M., Chen, R., Tsetsarkin, K.A., et al., 2010. Genome scale phylogenetic analyses of chikungunya virus reveal independent emergences of recent epidemics and varying evolutionary rates. *J Virol* **84**, 6497–6504.

Wang, E., Bowen, R.A., Medina, G., et al., 2001. Virulence and viremia characteristics of 1992 epizootic subtype IC Venezuelan equine encephalitis viruses and closely related enzootic subtype ID strains. *Am J Trop Med Hyg* **65**, 64–69.

Weaver, S.C., 1995. Evolution of alphaviruses. In: Gibbs, A.J., Calisher, C.H., Garcia-Arenal, F. (Eds.), *Molecular Basis of Virus Evolution*. Cambridge University Press, Cambridge, pp. 501–530.

Weaver, S.C., 2001. Venezuelan equine encephalitis. In: Service, M.W. (Ed.), *The Encyclopedia of Arthropod-transmitted Infections*. CAB International, Wallingford, UK, pp. 539–548.

Weaver, S.C., Reisen, W.K., 2010. Present and future arboviral threats. *Antiviral Res* **85**, 328–345.

Wheeler, S., Barker, C., Carroll, B., et al., 2008. Impact of West Nile virus on California birds. *Proc. Mosq. Vector Control Assoc. Calif* 76.

Wheeler, S.S., Barker, C.M., Fang, Y., et al., 2009. Differential impact of West Nile virus on California birds. *Condor* **111**, 1–20.

Zimmerman, E., 2009. House sparrow history. Sialis. http://www.sialis.org/hosphistory.htm

Shiraki, W. K., Hoshi, J. I., Ishino, T. et al. 1999. Ultrastructure of Japanese encephalitis virus in [illegible] biting [illegible]. Am J Trop Med Hyg 8: 698–706.

Solomon, T., Ni, H., Beasley, D.W., et al. 2003. Origin and evolution of Japanese encephalitis virus in southeast Asia. J Virol 77: 3091–3098.

Sudeep, A., Parashar, D. 2008. Chikungunya: an overview. J Biosci 33: 443–449.

Suom, [illegible], Reuben, R., [illegible] 1982. The role of [illegible] in the oviposition behavior of Culex [illegible] in the fields of South India. Ind J Med Res 75: 77.

Turell, M.J., O'Guinn, M., Oliver, J. 2000. Potential for New York mosquitoes to transmit West Nile virus. Am J Trop Med Hyg 62: 413–414.

[illegible] R. [illegible] Holmes, [illegible] Smith, [illegible] et al. 2004. Continental West Nile virus activity. [illegible] Emerg Infect Dis 10: 1040–1042.

[illegible] [illegible] D.A., Kuhn, [illegible] 2012. Spatial variation in the breeding patterns of Culex [illegible] and the Culex [illegible] complex in [illegible]. Culicidae [illegible] J Med Entomol 49: [illegible].

Tsetsarkin, K.A., Vanlandingham, D.L., McGee, C.E., et al. 2007. A single mutation in chikungunya virus affects vector specificity and epidemic potential. PLoS Pathog 3: e201.

Turell, M.J., Beaman, J.R., Tammariello, R.F. 1992. Susceptibility of selected strains of Culex pipiens and Aedes albopictus to chikungunya virus. J Med Entomol 29: 49–53.

Turell, M.J., O'Guinn, M., Oliver, J. 2000. Potential for New York mosquitoes to transmit West Nile virus. Am J Trop Med Hyg 62: 413–414.

van den Hurk, A.F., Ritchie, S.A., Mackenzie, J.S. 2009. Ecology and geographical expansion of Japanese encephalitis virus. Annu Rev Entomol 54: 17–35.

Victor, T.J., Reuben, R. 2000. Effects of organic and inorganic fertilisers on mosquito populations in the rice fields of southern India. Med Vet Entomol 14: 361–368.

Volk, S.M., Chen, R., Tsetsarkin, K.A., et al. 2010. Genome-scale phylogenetic analyses of chikungunya virus reveal independent emergences of recent epidemics and various evolutionary rates. J Virol 84: 6497–6504.

Wang, H., Fu, S., [illegible] et al. 2008. Mosquito [illegible] Japanese encephalitis [illegible] of the [illegible] and [illegible] virus closely related to [illegible]. [illegible]

Weaver, S.C. 2006. Evolutionary influences in arboviral disease. In: [illegible] Quasispecies: Concept and Implications for Virology. Cambridge University Press, Cambridge, pp. 411–438.

Weaver, S.C. 2001. Venezuelan equine encephalitis. In: Service, M.W., ed. The Encyclopedia of Arthropod-transmitted Infections. CABI International, Wallingford, UK, pp. 539–548.

Weaver, S.C., Reisen, W.K. 2010. Present and future arboviral threats. Antiviral Res 85: 328–345.

[illegible] 2001. [illegible] Conn [illegible]

[illegible], L.S., [illegible] 2001 [illegible] of West Nile virus, California, [illegible]

Zimmerman, [illegible] Culex [illegible] of the [illegible] [illegible]

5

THE SOCIO-ECOLOGY OF VIRAL ZOONOTIC TRANSFER

Jonathan D. Mayer

Departments of Epidemiology, Geography, and Global Health, University of Washington, Seattle, WA, USA

Sarah Paige

University of Wisconsin, Madison, WI, USA

TABLE OF CONTENTS

Viral Infections and Global Change, First Edition. Edited by Sunit K. Singh.

5.1 INTRODUCTION

With the unexpected emergence of SARS, better understanding of complexes producing influenza, and the emergence of hemorrhagic fevers such as Lassa fever and Ebola fever, the public suddenly became aware of the fact that transfer of pathogens between animals and humans is a threat to human health. Recently, two books written for the interested public have concentrated specifically on the species transfer of pathogens and especially of viruses. Of course this awareness already existed in epidemiology, ecology, virology, and many other scientific and social scientific disciplines. High-throughput sequencing technologies, better understanding of the evolution of virulence, and advances in molecular and genetic epidemiology have begun to clarify questions such as the origin of HIV-1 and HIV-2. Meanwhile, social scientific understanding of species transfer and the underlying conditions that produce species transfer has lagged behind. In this chapter, we establish a conceptual model for understanding how culture, society, agriculture, and human behavior contribute to the transfer of zoonotic viruses to humans. The basic question of this chapter is: what broader social and ecological factors establish the necessary and sufficient context for zoonotic transfer? We outline the analytical framework, use specific examples to highlight the social and ecological bases of viral transfer, and discuss the implications of zoonotic transfer for surveillance and control systems and for public health institutions.

Much of the understanding surrounding this issue comes from the fact that the majority of emerging human pathogens come from animals. The scientific understanding of new and emerging infections has burgeoned since the publication of the Institute of Medicine's (IOM) foundational report that highlighted the threat of emerging infections for public health and predicted that new and emerging infections would, if anything, increase in importance for human health (Lederberg et al., 1992). While generalizations concerning the roles of demographic and land-use change, changes in technology, and the other "factors of emergence" specified in the IOM abound, empirical studies specifically linking zoonotic disease to these factors lag behind our understanding of the molecular mechanisms of zoonotic pathogenesis. Indeed, the report noted that "[t]he significance of zoonoses in the emergence of human infections cannot be overstated" (Lederberg et al., 1992). Human activities as basic as nutrition can be high risk. As Nathan Wolfe concludes, hunting, probably the earliest source of human nutrition, continues to be a human activity with a high risk of infection and disease from zoonotic pathogens (Wolfe et al., 2005).

Extensive biomedical research demonstrates that people and animals can and do share many pathogens. Zoonotic infectious agents are estimated to cause between 60% and 70% of emerging and reemerging diseases in humans (Bengis et al., 2004; Jones et al., 2008).

5.2 HISTORICAL PERSPECTIVE

The roots of species transfer between nonhuman species and hominids have probably been present since humans developed as a species. Overlaps in human and animal habitats and the spaces in which multiple species travel provided the opportunity for species transfer. Archeologic and paleontologic data suggest the overlap between hominid and animal pathogens (Richard and Fiennes, 1979). It is hard to conceive of any period in hominid history where humans (or their ancestors) and animals did not exist in ecological complexes that facilitated species transfer of pathogens. Virchow, Koch, and Pasteur were aware of the fact that humans and animals shared either pathogens or diseases, and the contemporary call for "one medicine"—the unification of veterinary and human medical understanding—can be attributed as far back as these founders of microbiology (Atlas, 2012).

5.3 HUMAN–ANIMAL INTERFACE

Very little research has been done that explores the drivers of the events that place humans and animals in such contact that enables a viral zoonoses spillover. Instead, the human–animal interface is typically described as part of a contact-tracing scenario. Little is known about the socio-ecological history of interaction, the implications of changing human behavior, or the socio-ecological drivers of significant or mundane human–animal contact. This is changing as more scholars embrace partnerships across disciplines in order to understand the reasons behind human–animal contact in order to prevent risky contact in the future.

The zoonotic viral infection we are perhaps most familiar with is HIV. However, other examples include Ebola, Nipah virus, Hanta virus, many subtypes of influenza, and West Nile virus. The emergence of zoonotic viral infections may occur as viruses jump from an animal reservoir directly to humans, may move through a series of animal hosts before spilling over into humans, may cross into humans and mutate, or any combination of these events. The opportunity for novel viruses to "jump" from animals to people can occur in all environments, that is, from domestic animals to their urban owners (Cleaveland et al., 2001) or from wildlife to bushmeat butchers (Wolfe et al., 2005).

5.4 SURVEILLANCE

Extensive human and animal surveillance methods have been developed and implemented across the globe to track and respond to the emergence of novel zoonotic viruses. Currently, surveillance efforts are focused on either humans or animals (Drewe et al., 2012). Human surveillance activities primarily include field-based research and health department surveillance efforts. For example, Wolfe et al. (2004) used field-based molecular epidemiologic methods to detect and identify simian retroviruses in West African hunters. In the United States, public health departments may practice passive surveillance by responding to an outbreak. Through outbreak investigation and contact tracing, they may uncover an animal source (Proctor et al., 1998).

Animals are also used in zoonotic disease surveillance. Field-based methods in animal surveillance rely heavily on biological sample collection. For example, through the collection and analysis of blood samples, novel retroviruses have been identified in primates in western Uganda, with possible implications for human health (Lauck et al., 2011).

Despite the current structure of zoonotic disease surveillance in humans and animals, many scholars and institutions recommend restructuring emerging zoonotic disease surveillance initiatives. Through integrating human and animal surveillance programs, results could more quickly inform prevention and control policies and practice (Drewe et al., 2012; Gubernot et al., 2008).

In addition to molecular epidemiology of emerging zoonotic viruses, social scientists study the suite of human and ecological factors that enable viral movement from animals to people. However, research into the social drivers of emerging zoonoses in humans is extremely limited, especially compared to what is known about the molecular epidemiology of cross-species pathogen exchange. This remains true despite multiple calls for deeper transdisciplinarity between the biological and social sciences to identify, describe, and prevent events that facilitate zoonotic disease emergence (Parkes et al., 2005; Weiss and McMichael, 2004). Here, we strive to bridge the gap by examining the human and ecological drivers of zoonotic virus disease emergence. We present models or scenarios of human–animal interactions that facilitate emerging zoonoses and then suggest the applicability of these scenarios in exercises envisioning zoonotic risk under different climate change futures.

This chapter focuses on the socio-ecology of zoonotic *viral* infections. However, we want to emphasize that the processes described are applicable to not only the emergence of new viruses but also pathways of emergence for other infectious zoonotic agents, such as bacteria and parasites.

We use the term "socio-ecology" to reflect the intersection of human social, economic, cultural, political, and behavioral factors with the surrounding physical environment. We also assert that such a relationship is not unidirectional. Instead, as people enact changes to their environment, they also are responsive and adaptive to their environment. This interplay is fundamental to the study of climate change as well as in the study of emerging viral zoonoses.

"Socio-ecology" is a term currently in use by many but without a singular definition. Here we refer to the actions taken by people on an environment to extract specific ecological services as well as the broader social drivers that of human behavior. The theoretical foundation of socio-ecology is based on disease ecology and its assertion that disease and well-being are influenced by the interaction of a population, its habitat, and behaviors. Socio-ecology expands the scope of the disease ecology framework by probing more deeply into the social and structural drivers of "behavior." Socio-ecology is therefore is a holistic and multidimensional framework to guide the study of emerging zoonotic viruses.

In this chapter we offer in-depth examples of the relationship between socio-ecology with zoonotic disease emergence and climate change. With each example we describe the interplay of select social and ecological factors that enable the zoonotic disease spillover event. Ample opportunity exists for further research into the socio-ecology of all zoonotic disease emergence. What we have provided are suggestions or indications of viewing such emergence through the socio-ecological framework to gain greater clarity on how to identify and prevent future zoonotic disease emergence.

5.5 DEFORESTATION AND FRAGMENTATION

Landscape changes that result in deforestation simultaneously impact climate change and the emergence of viral zoonoses. Healthy forests sequester atmospheric carbon on a global scale and stabilize temperature and rainfall patterns on a local scale (Patz and Olson, 2006). Large-scale development projects that result in extreme landscape changes, like mining and timber extraction in Central Africa, contribute to climate change through deforestation and facilitate the emergence of zoonotic disease (Daszak et al., 2000; Wolfe et al., 2005, 2007). Deforestation facilitates contact between people and novel zoonotic viruses. For example, large-scale natural resource projects draw laborers and their families to extraction sites, resulting in increased population density and corresponding use of the forest itself. As people work and live in these degraded ecological systems, they may come into contact with novel zoonoses through direct contact such as hunting, butchering, and consumption of wildlife (Bowen-Jones and Pendry, 1999; Wolfe et al., 2005) or through indirect contact including ecological overlap of human livelihood spaces with wildlife habitat. Selective deforestation, whereby humans extract high-value timber, increases spatial fragmentation, thereby increasing the variety of zoonotic pathogens that can infect hunters and humans engaged in other activities (Wolfe et al., 2005).

While large commercial extraction projects influence sweeping landscape changes, swidden agriculture also contributes to climate change (through landscape fragmentation) and zoonotic disease emergence. Forest fragmentation refers to islands or pockets of forest that persist and are primarily utilized as natural resources for subsistence farmers. In western Uganda, Kibale National Park (Kibale) and the landscape immediately surrounding it is a classic example of a fragmented landscape.

Kibale is a 795 km^2 forested park near the foothills of the Rwenzori Mountains. A complex and interconnected history of resource extraction, development, national and regional conflicts, poverty, and migration has created a mosaic of habitats surrounding Kibale. The landscape includes household settlements, trading centers, gardens, banana plantations, tea fields, eucalyptus plots, pasture, wetlands, and forest fragments (Hartter, 2009; Hartter and Southworth, 2009). Primates still live in many of the small forest fragments (Goldberg et al., 2008).

Because of their small size, proximity to humans, and "permeable" borders, these forest fragments are the locations of intense and frequent interaction and contact between people and primates (Goldberg et al., 2008). "Contact" between people and primates in these systems refers to ecological overlap, where human activity spaces overlap with primate habitat as well as to direct contact, when people may hunt crop-raiding primates or prepare primate carcasses for their dogs. Ecological overlap has resulted in increased rates of *Giardia duodenalis* and *Cryptosporidium* spp. in primates living in fragments compared to primates of undisturbed forests, indicating a likely anthropogenic influence of infection (Salzer et al., 2007). This relationship has been subjected to molecular work as well, where *Escherichia coli* has shown more genetic similarity between people, livestock, and fragment-dwelling primates than with forest-dwelling primates (Goldberg et al., 2008). Several studies in the United States trace the links between fragmentation and Lyme borreliosis, caused by *Borrelia burgdorferi*.

The critical moments of contact, both direct and indirect, that facilitate disease exchange are made possible by the socio-ecology underlying the system. Forest fragments support the subsistence livelihood of local residents while simultaneously serving as primate habitats. Human activities that occur within fragments include firewood and forest product collection, water collection, and charcoal production. Daily livelihood activities are enough to put people at risk of exposure to the next primate pathogen (and, for the sake of conservation, vice versa). This relationship enhances pathogen exchange between people and primates (Goldberg et al., 2008; Salzer et al., 2007) and reinforces the role of socio-ecology as a critical factor in zoonotic viral disease.

5.6 URBANIZATION

For the first time in history, the world's urban population now exceeds its rural population. The UN estimates that this happened in 2007, although there is no way to confirm this because of variable quality in demographic data and differences in the definition of "urban" between countries and agencies. The urban population in 2011, according to the UN, was nearly about 5.7 billion, out of a global population of approximately 7 billion (UN, 2012). Moreover, virtually all of the net population growth globally in the next 50 years is expected to be in urban areas. The relative and absolute increases in urban population are a result of in-migration as well as natural increase (births–deaths). Increasing urbanization has many public health consequences, ranging from increased levels of air and water pollution to greater ease of transmission of infection and particularly of respiratory infections. It is beyond the scope of this chapter to describe the demographics of increased levels of urbanization, but briefly salient features for the topic at hand include mixing of land uses, the nature of human–animal contact within the context of great population density, and the transformation of animal husbandry in these dense settlements.

Much of the global urban growth will be in dense slums in developing countries. The conditions in such settlements are ideal for rapid transmission of infectious diseases, and people typically will be in close contact with livestock and poultry—in many urban slums, livestock and poultry can even be in a housing unit. The potential for viral transfer through vectors, fomites, oral–fecal, respiratory, and direct contact is great. There is a dearth of research documenting this or establishing transmission dynamics for zoonoses in such environments.

In the past decade, there has been increasing attention given to the role of "wet markets" in scientific and popular literature.

5.7 EXAMPLES

5.7.1 Nipah virus

Bats have been implicated in emerging viral infections including Ebola, Nipah, Hendra, and SARS. Nipah virus illustrates the role of bats in new and emerging infectious disease. A severe and sometimes fatal series of encephalitis cases in Malaysia received attention during and after its occurrence in 1998 and 1999. The infection was associated with some overlapping symptoms in pigs. Epidemiologic investigation indicated that the majority of symptomatic humans had contact with sick pigs. While these cases were first attributed to Japanese encephalitis virus (JEV)—endemic in that part of the world—there were some notable discrepancies in the symptoms reported in this outbreak and typical JEV cases. Thus, further investigation of the pathogen resulted in the isolation of a new pathogen that has come to be known as Nipah virus, named after the settlement in Malaysia where the specimen came from (Chua et al., 2000).

Antibodies to the newly isolated pathogen were found in some species of bats—notably, *Pteropus vampyrus* and *Pteropus hypomelanus*. The bats themselves did not appear to be symptomatic. The putative ecology of Nipah virus is probably due to commercial pig farming in endemic areas on land that also had, or was near, fruit trees. The fruit appears to be an attractive food for bats, and the bats would drop partially eaten fruit into pens containing the commercial pigs. The pigs are packed into tight spaces, facilitating easy transmission of the virus via the respiratory route. Human contact with the pigs allowed transfer into the human population.

Thus, Nipah virus outbreaks appear to be linked to the presence of humans, bats, and pigs in areas with fruit trees. Thus, the density of pig farming and the frequency of human contact with the commercially important pigs are important, if not essential, to Nipah outbreaks. Regional demand for pig products, facilitated by ready transportation from the farm, explains the initial outbreaks of Nipah. Economic, agricultural, and transportation considerations were responsible for these outbreaks.

5.7.2 Hendra

Like Nipah virus, fruit bats (*Pteropus* spp.) are the critical reservoir for Hendra virus. Hendra virus also emerged in the Pacific Rim but in Australia in 1994 as a fatal respiratory disease (Ksiazek et al., 2011). The virus is named "Hendra" as it was initially identified in a suburb of Brisbane, named Hendra. Hendra virus is a quintessential example of the role that urbanization plays in the emergence and spread of novel zoonoses. Breeding, raising, training, and showing a horse require space as well as an economic base, which makes the suburbs of a wealthy city an ideal location for such a venture. However, as the suburbs encroach upon rural areas, spatial overlap between indigenous wildlife, domestic animals, and humans occurs.

The social drivers of economics, entertainment, and industry underpin such an ecological context and ultimately drive the spillover events of virus from bats to horses and then on to humans. Fruit bats roost in trees around equine clinics and horse farms. Horses are exposed to the virus through excreta, and fall ill. As owners and handlers seek care for their ill animals, veterinarians and their assistants are exposed to secretions and tissues from infected animals and have fallen ill and even perished as a result.

5.7.3 Influenza

Influenza is usually cited as one of the prototype diseases involving transfer of virus from animal species to humans. Typically, this involves some complex of humans, avian species, and swine. Much attention has been devoted to these complexes in Southeastern China, as well as in the North America, and, to a lesser extent, elsewhere.

A complete review of these interactions is beyond the scope of this chapter. There have been excellent analyses that analyze the importance for human health and disease of influenza strains and variants that affect nonhumans.

"Avian flu," which is the popular name for a strain of H5N1 influenza, caused significant panic in Hong Kong with its appearance in 1997. Shortly thereafter, a massive die-off of local poultry, particularly in markets, and the link was apparent to the public. This was confirmed by careful epidemiologic investigation. Case–control studies indicated that most individuals who were severely ill either had handled birds or poultry or were in their close proximity. Public health authorities thereby began a massive and rapid culling of poultry in Hong Kong. Moreover, imports of poultry from Guangdong Province, China, were banned.

Though it had been apparent before to some, the scientific understanding and the public health measures that were necessary in this outbreak served as the genesis of the "one-health" movement that has argued that it is impossible to impose an arbitrary separation between pathogens affecting humans and pathogens affecting animal species (Peiris et al., 2009). This is not only because of the obvious avian–human link but also because of the links between culture, agriculture, economy, and transportation that were crucial in explaining and understanding the H5N1 outbreak in Hong Kong and surrounding areas. There was great demand for poultry for food in Hong Kong and surrounding areas. Much of the poultry was produced in areas surrounding Hong Kong on commercialized poultry farms with highly dense conditions conducive to rapid spread of respiratory pathogens. This was in response to demand for poultry as an economic commodity. There was plenty of labor in poultry-producing areas, and there was a great need for jobs on the part of laborers. Thus, poultry and labor came together, servicing the economic demand for Hong Kong. But that is not the end of the story. Transportation systems, built because of the need to move humans and commodities between Hong Kong and surrounding provinces, were crucial in explaining this outbreak. Finally, the conventional way of marketing poultry once transported to Hong Kong was in "wet markets," where live birds were sold to individual consumers. This brought millions of individuals into close contact with infected birds, sold in markets, and produced in dense agricultural conditions. For a detailed analysis of the social, economic, and financial conditions that explain this epidemic, the reader is referred to the work of Robert Wallace (Wallace, 2009).

5.8 CONCLUSION

However, it is important to keep in mind that new infectious diseases are emerging all the time, so what is important to glean from this chapter is not the complete model or system, but that those critical moments of human–animal contact are underpinned by a complex social and ecological scenario. It is the theoretical foundation that is transportable across landscape, time, and scale. The full extent of the ecology behind the emergence of any viral zoonoses is not completely understood. Instead we relate knowledge through rough scales of the human, animal, and environment to make our point.

REFERENCES

Atlas, R. M. (2012). "One health one health: its origins and future." *Curr Top Microbiol Immunol*, Springer-Verlag, Berlin, Heidelberg. DOI:10.1007/82_2012_223.

Bengis, R. G., F. A. Leighton, J. R. Fischer, et al. (2004). "The role of wildlife in emerging and re-emerging zoonoses." *Rev Sci Tech* **23**(2): 497–511.

Bowen-Jones, E. and S. Pendry (1999). "The threat to primates and other mammals from the bushmeat trade in Africa, and how this threat could be diminished." *Oryx* **33**: 233–246.

Chua, K. B., W. J. Bellini, P. A. Rota, et al. (2000). "Nipah virus: a recently emergent deadly paramyxovirus." *Science* **288**(5470): 1432–1435.

Cleaveland, S., M. K. Laurenson, L. H. Taylor, et al. (2001). "Diseases of humans and their domestic mammals: pathogen characteristics, host range and the risk of emergence." *Philos Trans R Soc Lond B—Biol Sci* **356**(1411): 991–999.

Daszak, P., A. A. Cunningham, A. D. Hyatt, et al. (2000). "Emerging infectious diseases of wildlife—threats to biodiversity and human health." *Science* **287**: 443–449.

Drewe, J. A., L. J. Hoinville, A. J. Cook, et al. (2012). "Evaluation of animal and public health surveillance systems: a systematic review." *Epidemiol Infect* **140**(4): 575–590.

Goldberg, T. L., T. R. Gillespie, I. B. Rwego, et al. (2008). "Forest fragmentation as cause of bacterial transmission among nonhuman primates, humans, and livestock, Uganda." *Emerg Infect Dis* **14**(9): 1375–1382.

Gubernot, D. M., B. L. Boyer, M. S. Moses (2008). "Animals as early detectors of bioevents: veterinary tools and a framework for animal–human integrated zoonotic disease surveillance." *Public Health Rep* **123**(3): 300–315.

Hartter, J. (2009). "Attitudes of rural communities toward wetlands and forest fragments around Kibale National Park, Uganda." *Human Dimensions of Wildlife* **14**: 433–447.

Hartter, J. and J. Southworth (2009). "Dwindling resources and fragmentation of landscapes around parks: wetlands and forest patches around Kibale National Park, Uganda." *Landsc Ecol* **24**: 643–656.

Jones, K. E., N. G. Patel, M. A. Levy, et al. (2008). "Global trends in emerging infectious diseases." *Nature* **451**(7181): 990–993.

Ksiazek, T. G., P. A. Rota, P. E. Rollin (2011). "A review of Nipah and Hendra viruses with an historical aside." *Virus Res* **162**(1–2): 173–183.

Lauck, M., D. Hyeroba, A. Tumukunde, et al. (2011). "Novel, divergent simian hemorrhagic fever viruses in a wild Ugandan red colobus monkey discovered using direct pyrosequencing." *PLoS One* **6**(4): e19056.

Lederberg, J., Shope, R. E., Oakes, S. C., Jr. (eds.) (1992). *Emerging Infections: Microbial Threats to Health in the United States*. Institute Of Medicine, National Academy Press: Washington, DC.

Parkes, M. W., L. Bienen, J. Breilh, et al. (2005). "All hands on deck: transdisciplinary approaches to emerging infectious disease." *EcoHealth* **4**: 258–272.

Patz, J. A. and S. H. Olson (2006). "Climate change and health: global to local influences on disease risk." *Ann Trop Med Parasitol* **100**(5–6): 535–549.

Peiris, J. S., L. L. Poon, Y. Guan (2009). "Emergence of a novel swine-origin influenza A virus (S-OIV) H1N1 virus in humans." *J Clin Virol* **45**(3): 169–173.

Proctor, M. E., K. A. Blair, J. P. Davis (1998). "Surveillance data for waterborne illness detection: an assessment following a massive waterborne outbreak of Cryptosporidium infection." *Epidemiol Infect* **120**(1): 43–54.

Richard, N. T-W-Fiennes. (1979). *Zoonoses and the Origins and Ecology of Human Disease*. London, Academic Press.

Salzer, J. S., I. B. Rwego, T. L. Goldberg, et al. (2007). "*Giardia* sp. and *Cryptosporidium* sp. infections in primates in fragmented and undisturbed forest in western Uganda." *J Parasitol* **93**(2): 439–440.

UN (2012). World Urbanization Prospects, The 2011 Revision: Highlights.

Wallace, R. G. (2009). "Breeding influenza: the political virology of offshore farming." *Antipode* **41**(5): 916–951.

Weiss, R. A. and A. J. McMichael (2004). "Social and environmental risk factors in the emergence of infectious diseases." *Nat Med* **10**(12 Suppl): S70–S76.

Wolfe, N. D., W. M. Switzer, J. K. Carr, et al. (2004). Naturally acquired simian retrovirus infections in central African hunters. *Lancet* **363**(9413): 932–937

Wolfe, N. D., P. Daszak, A. M. Kilpatrick, et al. (2005). "Bushmeat hunting, deforestation, and prediction of zoonoses emergence." *Emerg Infect Dis* **11**(12): 1822–1827.

Wolfe, N. D., C. P. Dunavan, J. Diamond (2007). "Origins of major human infectious diseases." *Nature* **447**(7142): 279–283.

6

HUMAN BEHAVIOR AND THE EPIDEMIOLOGY OF VIRAL ZOONOSES

Satesh Bidaisee

Department of Public Health and Preventive Medicine, School of Medicine, St. George's University, Grenada, West Indies

Cheryl Cox Macpherson

Bioethics Department, School of Medicine, St. George's University, Grenada, West Indies

Calum N.L. Macpherson

Department of Microbiology, School of Medicine, St. George's University, Grenada, West Indies
Windward Islands Research and Education Foundation, Grenada, West Indies

TABLE OF CONTENTS

Viral Infections and Global Change, First Edition. Edited by Sunit K. Singh.

6.1 INTRODUCTION

In the eighteenth century in Europe, smallpox affected all levels of society, led to 400 000 deaths annually, and blinded one-third of its survivors (Barquet and Domingo, 1997). Toward the end of the eighteenth century, the renowned physician Edward Jenner noted that milkmaids did not suffer from smallpox and correctly attributed this to the cross protection afforded by their exposure to cowpox through their occupation as milkmaids. A poem coined at the time read "Where are you going, my pretty maid?" "I'm going a milking sir." "What is your fortune, my pretty maid?" "My face is my fortune sir." The observation that an occupation that required close contact between humans and animals led to protection from a virus obtained from animals was significant and gave rise to the term zooprophylaxis (Nelson, 1974). Jenner subsequently used this finding to inoculate humans with material from cows (vacca—Latin for cow). The procedure of inoculating the product of cows into humans was termed vaccination. This discovery revolutionized our means of controlling an infectious disease and also represented an example of how human behavior could affect viral infections from vertebrate animal origins.

Today, we have a clearer understanding of the heterogeneity of susceptibility to infection, and knowledge of the routes of transmission and the epidemiology of the viral zoonoses is being obtained through technological advances in molecular, genetic, immunological, and newer and more powerful epidemiological tools. Most zoonotic infections that are recognized today are viral in origin and are either emerging or reemerging

(Venkatesan et al., 2010). Of the 1415 microbial diseases affecting humans, 61% are zoonotic, and among emerging infectious diseases, 75% are zoonotic with viruses from wildlife being one of the major sources of infection (Taylor et al., 2001). In the past 20 years, a new virus has emerged almost every year including hemorrhagic fevers and arboviruses transmitted by vectors and also shown to be transmitted through direct contact with animals (Peters and Khan, 2002; Wilke and Hass, 1999).

Viral zoonoses are transmitted between different wild and domestic vertebrate host species and humans in a variety of ways including ingestion (Enterovirus, Hepatovirus, Rotavirus, Astrovirus, Norovirus, Reovirus, Coxsackie virus, and hepatitis A and E), inhalation (severe acute respiratory syndrome (SARS), avian influenza (AI) H5N1, influenza type A H1N1, Hantavirus, Hendra virus (HeV), Nipah virus (NIV), Menangle virus), contact (Marburg, Ebola, Crimean–Congo hemorrhagic fever, yellow fever, Rift Valley fever (RVF), Lassa virus, foot-and-mouth disease, monkeypox, La Crosse (LAC) encephalitis, Japanese encephalitis (JE), NIV, Cercopithecine herpesvirus I, vesicular stomatitis virus, sandfly fever virus, Toscana virus, Chagres virus, Punta Toro virus, lymphocytic choriomeningitis (LCM) virus, and rabies virus), and inoculation by a variety of diurnal (West Nile virus (WNV), yellow fever, Chikungunya, tick-borne encephalitis (TBE), tick fever, eastern equine encephalitis (EEE), western equine encephalitis, LAC virus) and nocturnal (rabies virus) vector species. Each of these different modes of transmission of viral zoonoses to humans is influenced substantially by human behavior.

This chapter will explore the role of human behavior, in the widest context, in the epidemiology of viral zoonoses. It will examine how human behavior has contributed and continues to contribute to the emergence and reemergence of zoonotic viruses through the convergence of complex but interrelated factors. These include numeric and spatial changes in human and wild and domestic animal populations and concomitant changes to the earth's climate, environment, land-use patterns, and water resources.

6.2 SOCIETAL CHANGES AND THE EPIDEMIOLOGY OF VIRAL ZOONOSES

6.2.1 The human–animal relationship

Human–animal relationships contribute to the epidemiology of viral zoonoses. Culture was once widely perceived as being determined by geographical environment (Galvin et al., 2001). Cultural practices evolved from nomadic hunters and gatherers (some societies of which still exist) to the domestication and farming of many animal and plant species. The domestication of dogs, cats, cattle, goats, horses, and sheep occurred between 13 000 and 2500 BC (Diamond, 2002). Domestic animals were selectively bred for specific qualities and now differ genetically from their wild ancestors. The most suitable animals for domestication are those that naturally live in groups and allow human dominance, and animal domestication is strongly linked with the development of human civilization (Rose and Lauder, 1996).

The evolution of cultural and commercial practices associated with animal domestication, production, transportation, and use has resulted in closer association between humans and animals and greater risk of cross-species transmission of viral and other infections. Animal domestication has had a profound effect on the population distribution and density of all animal species. Wild animal populations were once able to adapt to their selective environments, but human-induced environmental changes have altered their ability to adapt and confronted animal populations with greater disturbances (Vitousek et al., 1997). Deforestation and other forms of environmental destruction harm both wild and domestic animal populations and have caused the extinction of some species. The exposure to, and transmission of, viral and other zoonotic infections is facilitated by steadily increasing human and animal population densities and increasingly close relationships between them. Human population growth and the associated demand for more land and water for shelter, agriculture, and hygiene accentuate these conditions.

6.2.2 Migration and population movements

The world's population now exceeds 7 billion and is continuously moving within and between countries. In 2009, it was estimated that more than 125 million people moved from rural to urban areas (UN, 1999). This human migration has occurred at such a rate that in 2010 for the first time in history more people lived in urban than rural areas (UN, 1999). Less than 3% of the world's population lived in urban areas less than 200 years ago. In 1950, New York and London were the only megacities (>10 million people), and by 2015 it is estimated that there will be more than 38 megacities including 23 in Asia (UN, 2010). This concentrates huge numbers of people and their domestic pets. One consequence of this concentration is the increasing number of dog bites, which may transmit rabies (Bengis et al., 2004). Worldwide it is estimated that 30 000–60 000 people die of rabies every year (Knobel et al., 2005). The number of people receiving postexposure treatment, mostly after dog bites, is annually approximately 3.5 million (Bögel and Meslin, 1990; Bögel and Motschwiller, 1986). Provision of adequate sanitation and safe drinking water to these shifting populations poses an enormous logistical engineering problem, and waterborne transmission of viral zoonoses is facilitated by lack of sanitation and clean drinking water. Over 1 billion people lack access to safe drinking water, and 2.4 billion lack access to adequate sanitation (UN, 2003). In the tropics, increased vector densities will also manifest and provide transmission opportunities for vector-borne diseases. The latter are also facilitated through climactic changes. Better understanding of how specific environmental and behavioral factors influence the virulence, transmission, and epidemiology of viral zoonoses is needed to guide prevention and control. Measures to improve sanitation reduce the risk of viral zoonotic exposure and transmission and include treatment and disposal of human waste and improved personal hygiene and food safety including the handling and processing of animals and animal products (Venkatesan et al., 2010).

The International Business Times (IBT) in 2011 reported that the International Air Transport Association (IATA) forecasted that there will be 3.3 billion air travelers by 2014, up from 2.5 billion in 2009 (IBT, 2011) with more than a million people in the air at any one time, and no two cities are today more than 24 h travel by air apart (IATA, 2007). The current and projected increase in air travel will continue to increase the risk of global spread of viral zoonoses in the future. Additionally, air travel facilitates the rapid transmission of viruses within their short incubation periods in the time taken to travel around the world. Viruses are impossible to detect during the incubation phase.

Globalization encourages the production, transportation, and consumption of wild and domestic animals and animal products. These are legally imported and exported for exhibition, scientific education, research, conservation, companionship, and traditional "medicinal" products and as food (Marano et al., 2007). Such movement and trade of animals and animal products also facilitate the transmission of viral zoonoses, and this became evident in 2003 when monkeypox was transmitted to humans after a shipment of African Gambian giant rats was housed with prairie dogs (Marano et al., 2007).

The magnitude of global transportation of animals is staggering: 37 808 179 live amphibians, birds, mammals, and reptiles were legally imported into the United States from 163 countries between 2000 and 2004 (Jenkins et al., 2007). The illegal trade is likely to be much larger. In addition to the human-mediated transportation of animals, there are huge natural migrations of animals and, perhaps more significantly for the spread of viral zoonoses, birds. It is estimated that more than 5 billion birds migrate across North America each year (US Fish and Wildlife Service, 2010). The migration of birds has been implicated in the transmission of viral zoonoses including AI and WNV (Rao et al., 2009).

Emerging viral zoonoses like SARS and AI are consequences of human encroachment into wildlife habitat and increases in global animal trade and human international travel. Routes of transmission of viral agents to animals and humans are changing due to altering environments and increasingly close human–animal relationships; this is evident in the transmission of hantavirus by human contact with animal excreta (Marr and Calisher, 2003).

Interactions between the environment and human demographics and behavior are complex and influence the emergence and reemergence of zoonotic viruses in ways that are not always apparent. Ongoing discoveries of new zoonoses suggest that the viruses known today comprise only a fraction of those that exist (Murphy, 1998). RNA viruses are particularly capable of adapting rapidly to changing environmental conditions and are among the most prominent emerging pathogens (Ludwig et al., 2003).

Some viruses have increasing importance and may cause flu-like symptoms (Alkhurma virus infection, influenza A), respiratory symptoms (SARS), lesions in the fingers and hands (pox), hemorrhagic fever (Marburg, Ebola, and Hantavirus), or encephalitides (herpesvirus complex) (Manojkumar and Mrudula, 2006).

6.2.3 Climate change and vectors

Climate change is occurring because greenhouse gas emissions are accumulating in earth's atmosphere faster than they can dissipate. It manifests as long-term increases in air and sea temperatures, rising sea levels, and increasingly frequent and severe weather including heat waves, storms, and floods. These impacts will last for decades or longer and are occurring globally with unique impacts in different geographical regions. Leading health organizations are responding to the diverse health impacts by implementing surveillance, targeted interventions, and educational programs including webinars for health professionals (Centers for Disease Control (CDC), 2012; Health Canada, 2012; National Health Service (NHS), 2012; WHO, 2012). These impacts include the emergence, reemergence, and burden of vector-borne, communicable, non-communicable, and mental illnesses (Haines et al., 2007; Jarvis et al., 2011; McMichael et al., 2007). Some of the health impacts are consequences of disruptions in the salinity of freshwater, coastal estuary and marine life, and, notably, the habitats of marine and migratory birds. Children in rural populations are particularly vulnerable to such health impacts and warrant special attention (Macpherson, 2010).

These environmental changes will also facilitate the growth and expansion of arthropod populations and thereby increase human and animal exposure to arthropod vectors. Increases in arthropod populations that share human and animal space have contributed to vector-borne viral zoonoses such as RVF, equine encephalitis, and JE (Kruse et al., 2004). The paradox is that the benefits of human behaviors including travel and industrial agricultural practices raise the risk of human and animal exposure to viral zoonoses and other health impacts of climate change. Ethical perspectives on the associated risks and harms would help to inform which means of mitigating some of the problems will be most sustainable and effective.

6.3 VIRAL ZOONOSES AND HUMAN SOCIETAL VALUES

Behavioral sciences are central to reducing the impacts of anthropogenic factors that contribute to environmental changes and the emergence and reemergence of viral zoonoses. The need to integrate insights about human behavior, biological methods of prevention and control, and technological advances requires transdisciplinary collaborations.

The agent, host, and environmental ecologic model for disease causation proposes that human psychology and social behavior exert causal influences on humans' ontogenic experiences with infectious diseases (Thornhill et al., 2010). Societal practices, characterized by increasingly close human–animal relationships, increase the potential for exposure to viral zoonoses. The environmental ecologic model for disease causation implies that viral zoonoses are a result or consequence of interactions that occur between humans and the environment. The presence of viral zoonoses in the ecology of human health may reflect cultural value systems, political ideologies, and/or individual and societal practices. Value systems typically encourage adherence to existing traditions and norms, limit the inclination to deviate from these norms, and may do so in ways that bear on exposure to and transmission of viral zoonoses (Nettle, 2005).

6.3.1 Individual and collective responsibility

Collective value systems typically integrate ethnocentric attitudes, adherence to existing traditions, and behavioral conformity, while individualist value systems encourage innovation, tolerance of idiosyncratic behavior, and neophilia (Gelfand et al., 2004). The extent to which a population is affiliated with collectivism or individualism varies with cultural values and socioeconomic and political structures and corresponds to variations in disease prevalence and zoonotic burdens (Thornhill et al., 2009). Close proximity to a zoonotic animal reservoir increases risk of exposure, and gregarious and extraverted individuals may be at greatest risk in conditions in which a disease is highly prevalent (Hamrick et al., 2002). The connections between exposure and individual dispositions seem similar across many cultures although variations exist (McCrae, 2002).

Public health goals can improve both individual and collective health but are primarily associated with the principle of utility that seeks to balance the benefits, harms, and risks of any action or intervention. Conversely, the traditional focus and goal of medicine is each individual patient with emphasis on their individualism and autonomy. This tends to obscure complex interactions between individuals, the natural environment, and animals. That individual behaviors develop through socialization and acculturation during childhood (Atran and Norenzayan, 2004), derive from value systems, and are expressed through societal structures including governments (Conway et al., 2006) bears on what interventions will best modify behaviors and reduce exposure to and transmission of viral and other zoonoses.

Health is affected, inter alia, by behavior, environment, and policies. Waste disposal can pollute the environment and introduce rotavirus and hepatitis A. The growing extent of animal domestication and commercialization alters disease patterns by reducing the boundaries between humans and animals and exposing populations to new diseases. Individuals, nations, and health systems all share some moral responsibility for maintaining health, and their behaviors as consumers, citizens, and policy-makers have environmental consequences that bear on health and exposure to viral zoonoses. While the extent to which individuals bear responsibility for the harms of climate change is publicly debated (Fahlquist, 2008), individuals in most cultures perceive themselves as having a moral responsibility to their children and descendants (Richardson, 1999). To what extent does seeking personal gain from legal or illegal trade in animals or animal products violate this responsibility? Answers to this and similar ethical questions depend at least in part on how well informed those who engage in this behavior are about the impacts of this trade on health and climate change and also on what alternative behaviors and livelihoods are available to them. Governments and industries therefore share a moral responsibility to provide adequate information and ensure that alternative livelihoods and means of waste disposal, for example, are widely accessible.

The policies of governments and institutions directly and/or indirectly affect the health impacts of climate change including the emergence and reemergence of viral zoonoses. Such policies, like public health policies, can be used to alter human behaviors and promote healthy and environmentally friendly lifestyles and consumption patterns (Frumkin and McMichael, 2008; Williams, 2008). Individual behaviors involved in fossil fuel harvesting, deforestation, confined livestock operations, wildlife markets, and others are often sanctioned by governments and consumers although they compromise ecosystem stability and increase the potential emergence and movement of viral zoonotic pathogens. SARS, for example, was transmitted from a corona-like virus in civet cats in Southeast Asia to humans in densely populated wildlife market operations. Socioeconomic and cultural factors encourage individuals to participate in wildlife markets, and governments and industries in rich and poor nations share the moral responsibility for conditions that facilitate such behaviors.

The short incubation period of viral infections and the severity of their clinical outcomes indicate that educational programs can be effective in altering behavior to reduce exposure. Emerging and reemerging viral zoonoses may have enormous economic consequences and can easily induce panic. Those responsible for the delivery of educational messages must be aware of this potential and responsibly ensure that appropriate information is transmitted. In contrast to viral zoonoses, parasitic infections have long incubation periods, are difficult to correlate with clinical outcomes, and are more challenging to control through sustained behavioral changes (Macpherson, 2005).

Solutions to the emergence and reemergence of viral zoonoses require transdisciplinary collaborations. These may focus on the application of technological advances and mathematical modeling to prevention and control; analyses of geographic, cultural, and socioeconomic conditions that drive individual and collective behaviors associated with exposure; and determinations about the responsibilities of industries and nations to health and health promotion. The responsible design of programs to reduce or prevent exposure through behavior change involves ethical analyses of probable risks, harms, and benefits, some of which are discussed further.

6.4 HUMAN BEHAVIOR AND THE EPIDEMIOLOGY OF VECTOR-BORNE VIRAL ZOONOSES

At the start of the twentieth century, epidemic vector-borne diseases were among the most important global health problems. Yellow fever and dengue fever caused explosive epidemics. Subsequently, effective prevention and control measures accelerated in the post-World War II years with the advent of new insecticides, drugs, and vaccines (Gubler, 1998). By the 1960s, the majority of vector-borne viruses had been effectively controlled.

The 1970s ushered in a 25-year period characterized by decreasing resources for infectious diseases. Coincident with this period of complacency global climate change evolved resulting in the reemergence of vector-borne viral zoonoses (Gubler, 1989). The once limited geographical and host ranges of many vector-borne viral zoonoses are expanding, spurred largely by anthropogenic factors. Epidemics of dengue and other formerly contained vector-borne viruses are on the rise in the developing world. In recent decades, the United States witnessed the introduction of WNV in New York City and its subsequent spread to other areas in the country.

Significant variations in temperature and rainfall affect vector-borne viral transmission (NCBI, 2008). The global warming effect of climate change is linked with outbreaks of a variety of arthropod-borne diseases. In the case of RVF, this association was sufficiently strong to allow scientists to develop risk maps that successfully predicted a major outbreak in Africa in 2006–2007. Vector-borne pathogens are particularly sensitive to climate variations because these influence vector survival and reproduction and bear on biting and feeding patterns, pathogen incubation and replication, and the efficiency of pathogen transmission among multiple hosts (NCBI, 2008).

Of the vector-borne diseases, arboviruses have become the most important causes of reemergent epidemic disease. In 2007, there were few places on earth where there was no risk of infection of viral disease transmitted by mosquitoes (Gubler, 1996). The more important reemergent epidemic arboviral diseases include members of the Togaviridae, Flaviviridae, and Bunyaviridae.

6.4.1 Yellow fever (urban yellow fever, sylvatic or jungle yellow fever)

Yellow fever is endemic in tropical Africa between 15° north latitude and 15° south latitude and in the northern and eastern parts of South America, and epidemics occur in parts of Central America (Varma, 1989). The main vector *Aedes aegypti* originated in Africa and was carried to the New World with the slave trade in the sixteenth century. Epidemics have occurred in the coastal areas of South America and the United States as far north as New York. Other outbreaks occurred in Ethiopia in 1960–1962 and in 1986–1987 in West Africa (Varma, 1989). All the urban epidemics in the New World were transmitted by *A. aegypti*. Urban yellow fever is now controlled by vaccination, but the risk associated with rural yellow fever in enzootic areas of Africa and South America results in human cases.

6.4.2 WNV

WNV, a flavivirus, is a virus that originated in Africa. The virus was enzootic throughout Africa, West and Central Asia, the Middle East, and the Mediterranean. Since WNV was first isolated in 1937, it caused only occasional epidemics, and the illness in humans, horses, and birds was either asymptomatic or mild neurological diseases (Marfin and Gubler, 2001). In 1999, there was widespread transmission throughout New York City with rapid movement of the disease westward across the United States to the west coast. The spread of WNV continued northward into Canada and south into Mexico and Central America and the Caribbean. In 2002, WNV caused the largest epidemic of meningoencephalitis in U.S. history with approximately 3000 cases and 284 deaths (Campbell et al., 2002).

Migrating birds have played a role in the spread of WNV in the western hemisphere (Owen et al., 2006). The westward movement of WNV across the North American continent maps the migratory patterns of birds in North America. The virus has been isolated from 62 species of mosquitoes, 317 species of birds, and more than 30 non-avian vertebrates that all contribute to the successful enzootic and epizootic WNV transmission (CDC, 2007).

6.4.3 TBE

TBE is caused by two closely related viruses of the family Flaviviridae (Peterson and Gubler, 2006). The eastern subtype of the TBE viruses is transmitted by *Ixodes persulcatus* and causes spring–summer encephalitis that occurs from Eastern Europe to China (Peterson and Gubler, 2006). The Western subtype is transmitted by *Ixodes ricinus* and causes Central European encephalitis that occurs from Scandinavia in the north to Greece and Serbia and Montenegro in the south. The Eastern subtype has a higher mortality (5–20%) when compared with less than 2% for the Western subtype. The two subtypes are maintained in natural cycles involving a variety of mammals and ticks. Human exposure occurs during work and recreational activities during spring and summer months in temperate zones and during fall and winter months in the Mediterranean that coincide with the most active period of the vectors.

6.4.4 Encephalitides

Zoonotic viral encephalitis is caused by a number of arboviruses belonging to the family Flaviviridae, Togaviridae, Bunyaviridae, and Reoviridae (Peterson and Gubler, 2006).

6.4.5 LAC encephalitis

LAC virus is the most pathogenic member of the California encephalitis serogroup. LAC is maintained in a cycle involving *Aedes triseriatus* mosquitoes and a number of mammalian hosts including chipmunks, tree squirrels, and foxes (McJunkin et al., 2001).

6.4.6 JE

JE is the most important global cause of arboviral encephalitis with approximately 45 000 cases reported annually. JE is spread throughout Asia and has also been detected in Australia and the Pacific region. JE virus is maintained in a natural enzootic cycle involving *Culex* spp. mosquitoes and birds that live close to water (Peterson and Gubler, 2006).

6.4.7 Saint Louis encephalitis (SLE)

Saint Louis encephalitis (SLE) virus is a Flaviviridae that is prevalent throughout the western hemisphere from Canada in the north to Argentina in the south (Peterson and Gubler, 2006). The infection is maintained between wild birds and *Culex* spp. of mosquitoes.

6.4.8 EEE

EEE virus is a Togaviridae and widely distributed throughout North, Central, and South America and the Caribbean (Cupp et al., 2003). Human infection is sporadic and occurs as small outbreaks mainly in the Atlantic and Gulf coast regions of the United States. In North America, wild birds and *Culiseta melanura*, a swamp-dwelling mosquito, maintain the virus.

6.4.9 Venezuelan equine encephalitis (VEE)

Venezuelan equine encephalitis (VEE) like EEE is a Togaviridae of several subtypes and antigenic variants. VEE is distributed from Florida to South America and is transmitted by several different genera of mosquitoes (Peterson and Gubler, 2006).

Other zoonotic arboviruses of note include the Colorado tick fever and Chikungunya. The Colorado tick fever virus is a Reoviridae that is transmitted to humans in the United States and Canada by the wood tick, *Dermacentor andersoni* (Peterson and Gubler, 2006). Chikungunya virus is a Togaviridae found in Africa and Asia and is transmitted by *Aedes* spp. mosquitoes (Laras et al., 2005). After a 20-year absence of the disease, a reemergence occurred in Indonesia from 2001 to 2005 (Laras et al., 2005).

There are numerous vector-borne viral zoonoses that cause different diseases in humans throughout the world. What is most significant is the dramatic reemergence of some vector-borne viral zoonoses over the last two decades and the expansion of the geographical vector–host ranges. The increased observations of vector-borne disease are closely linked to the global demographic, economic, and societal changes that have occurred during the last two decades. These include the unprecedented rates of population growth, primarily in the cities of the developing world, which facilitated the transmission of vector-borne viruses. Population growth is a major driver of environmental change in rural areas too. Anthropogenic behaviors including deforestation and diversion of natural water courses for agricultural use and animal husbandry increase the population and alter the distribution of vectors and their proximity to humans and the zoonotic viral agents they transmit.

6.5 HUMAN BEHAVIOR AND THE EPIDEMIOLOGY OF RESPIRATORY VIRAL ZOONOSES

From the emergence of HeV and Menangle virus in Australia to the global pandemics of SARS and the influenzas (both H5N1 and H1N1), there has been a surge of zoonotic respiratory virus outbreaks in the last two decades. These outbreaks are due to a true emergence of new pathogens. Globalization of travel and trade, changes in agricultural practices, and climate change are some of the drivers responsible for the emergence of novel respiratory viruses affecting humans (Wang, 2011). Respiratory zoonotic diseases rank among the potentially most devastating infectious diseases in humans. The pandemic influenza outbreak of 1918/1919 was caused by an AI virus, which was directly transmitted to humans without further reassortment (Mettenleiter, 2006). The 1918/1919 pandemic resulted in an estimated 50 million deaths. The highly pathogenic AI (HPAI) of subtype H5N1 is also currently affecting the poultry industry of the world and has the capacity to infect humans by close contact with infected wild birds or poultry. Human deaths due to HPAI H1N1 continue to be recorded in 8 countries in Asia and Africa (Mettenleiter, 2006). The appearance of influenza type A H1N1 also created a pandemic in 2009 and had a direct impact by causing human morbidity and mortality as well as indirect effect through restrictions in travel and trade among most countries of the world. Animals are the source of these zoonotic viral infections, which infect humans via the respiratory tract and cause either respiratory symptoms or infection of other organs. The identity of the animal reservoirs especially wildlife as a reservoir for potentially zoonotic infectious agents remains arcane. This lack of knowledge makes it necessary to bring together different disciplines that are essential to understanding and subsequently to control infections in humans.

6.5.1 HeV

HeV is a Paramyxoviridae transmitted to people through close contact with infected horses or their body fluid. Infection in people is known to cause respiratory and neurological disease and death in horses (WHO, 2009a,b). HeV was first recognized in 1994 during an outbreak of acute respiratory disease among 21 horses in Australia. Two people were infected and one died. Since then, there have been another 10 outbreaks all in Australia, and three of the outbreaks involved human cases. With the travel of horses for racing, HeV has the potential for significant geographical spread.

6.5.2 NIV

NIV is an emerging zoonotic virus that causes severe illness characterized by inflammation of the brain (encephalitis) or respiratory disease. NIV was first recognized in 1999 during an outbreak among pig farmers in Malaysia, and since that time, there have been several other outbreaks in Asia (WHO, 2009a,b). During the initial outbreaks in Malaysia and Singapore, human infections resulted from direct contact with infected pigs or their contaminated tissues. Transmission is thought to have occurred via respiratory droplets, contact with throat or nasal secretions from the pigs, or contact with the tissue of a sick animal. In the Bangladesh and Indian outbreaks, consumption of fruits or fruit products contaminated with urine or saliva from infected fruit bats was the most likely source of infection. Human infections range from asymptomatic to influenza-like symptoms of fever, headaches, myalgia, vomiting, and sore throat. Some people also experience atypical pneumonia and severe respiratory problems including acute respiratory distress (WHO, 2009a,b).

6.5.3 SARS

The SARS corona-like virus was responsible for the first serious and widespread zoonotic disease outbreak of the twenty-first century. The great global impact on health, travel, and the economy was intensified by the delay in identifying the causative agent of the disease (Wang, 2011). From November 2002, a disease termed "atypical pneumonia" was reportedly spreading in Southern China. The virus spread from Asia to North America in a few days and affected patients and health-care workers alike. Incredibly, the outbreak was brought under control within 3 months of the identification of the caudated agent and its mode of transmission through appropriate behavioral change and public health measures (Wang, 2011).

6.5.4 Influenza H1N1

The 2009 H1N1 influenza pandemic was the first global pandemic since 1968 (Condon and Sinha, 2010). The 2009 H1N1 influenza outbreak emerged in Mexico and then spread further into North America and to most countries worldwide. By July 2001, the WHO declared a pandemic. Two hundred and twelve countries and more than 15 000 laboratory confirmed deaths comprised the pandemic. The H1N1 virus evolved from the avian origin 1918 H1N1 pandemic virus, which was thought to have entered human and swine populations (Jhung et al., 2011).

6.5.5 Influenza H5N1

HPAI H5N1 has been reported in domestic poultry, wildlife, and human populations since 1996. Risk of infection is associated with direct contact with infected birds. The mode of H5N1 spread from Asia to Europe, Africa, and the Far East is unclear; risk factors such as legal and illegal domestic poultry and exotic bird trade and migratory bird movements have been documented (Yee et al., 2009). Measures used to control disease such as culling, stamping out, cleaning and disinfecting, and vaccination have not been successful in eradicating H5N1 in Asia but have been effective in Europe.

6.5.6 Travel and respiratory viral zoonoses

The movement of people has facilitated the passage of pathogens. Among the more notable examples are the great plagues that swept into Europe from Asia during the middle ages, the importation of smallpox to the Americas by European explorers, and the reverse movement of syphilis into Europe in those same returning explorers. Today's increasing numbers, frequency, and speed of travel have enhanced the opportunities for disease spread (Condon and Sinha, 2010).

The two forms of influenza that have recently emerged have been distributed largely due to the translocation of the disease through travel. H5N1, which was first observed on a limited scale in Hong Kong in 1997, emerged in Vietnam in 2003. Since 2003, there have been human cases in 15 countries. No human-to-human transmission has been reported, and the movement of the virus between countries is associated with goods carried by conveyances and animals (Condon and Sinha, 2010). In contrast, the rapid spread of the 2009 influenza type A H1N1 virus was due to infected travelers. Travelers who were either symptomatic or in the incubation stage support the spread (WHO, 2011a,b). An analysis of air traffic patterns found a very strong correlation between the volume of air travel from Mexico to a country and the likelihood H1N1 was subsequently identified in that country.

The SARS epidemic of 2003 is another example of a rapid worldwide spread of a virus through role of travel. A case study in February 2003 was a professor from Southern China who had been treating patients with an unrecognized respiratory illness. He traveled to Hong Kong while he was ill to attend a family wedding. His infection spread to 10 other travelers in his hotel who then boarded airplanes to other parts of Asia, North America, and Europe setting off a global pandemic of SARS that resulted in 8098 cases and 774 deaths in 29 countries (Condon and Sinha, 2010).

Respiratory viral zoonoses are an increasing challenge to human health, and this burden is occurring in parallel with the increase in global travel. The inextricable link between human behavior and viral zoonoses is once again exemplified by respiratory viral infections. Based on current and projected trends of global travel, respiratory viral zoonoses are likely to continue to increase.

6.6 HUMAN BEHAVIOR AND THE EPIDEMIOLOGY OF WATERBORNE VIRAL ZOONOSES

The nature of viruses includes an absolute requirement for living cells in which to replicate, and they cannot be found free living in the environment. The waterborne route for viral transmission is common typically as enteric viruses enter waterways via sewage.

Contamination of water by human practices can also occur through land runoff, agricultural practices such as manure sludge, and domestic and industrial use of water (Pardio Sedas, 2007). In addition to human practice-induced contamination, anthropogenic climate change is measurably affecting ecosystems. These human outcomes may have impacts on ecosystem balances, potentially leading to new diseases associated with environmental changes such as extreme temperature variation and violent weather events (Walther et al., 2002). The broad categories of weather conditions are associated with several diseases including Enteroviruses and Norwalk and Norwalk-like viruses (Rose et al., 2001). Understanding the links between infectious diseases and climate is difficult given the multivariate nature of climate change and nonlinear thresholds in both disease and climate processes. Studies report a statistically significant association between excess rainfall and waterborne viral disease outbreaks (Curriero et al., 2011). Excessive rainfall results in surface water contamination, which, in highly dense urban populations, affects many communities in developing countries (Harvell et al., 2002). The decrease in salinity that accompanies climate change promotes viral pathogens in water as extreme Ph inactivates viral pathogens (Hunter, 2003). Temperature that is increasing with climate change serves to promote wet weather to support viral pathogen persistence in the environment (Hunter, 2003).

Climate change through warming, abnormal rainfall, and changes in salinity are examples of factors driving the global emerging resurgence and redistribution of infectious diseases. The effects also seem to have a disproportionate impact on regions of the world where there is a low socioeconomic status, which is often accompanied by political instability.

6.6.1 Waterborne zoonotic viruses

Enterovirus and Hepatovirus are members of the Picornaviridae family of RNA viruses that are enterically transmitted. Enteroviruses are able to persist in the environment and have been associated with infection from eating soft fruits, vegetables, and shellfish. Hepatitis A virus is most commonly associated with inadequate water supplies and poor hygiene. In hepatitis E, outbreaks frequently follow heavy rains and flooding (Carter, 2005). Rotaviruses belong to the family Reoviridae and account for more than 800 000 deaths each year in young children from underdeveloped countries. The virus is spread readily via the fecal–oral route (Carter, 2005). Noroviruses are part of the family Caliciviridae provisionally described as "Norwalk-like viruses." Swimming areas and contaminated drinking water as well as food are likely sources of the infection. Astrovirus similarly is associated with swimming areas (Huffman et al., 2003). Adenoviruses that cause gastroenteritis in humans are also found in fecal polluted water (Vasickova et al., 2005).

Among the human enteroviruses, Coxsackie viruses are known to have a possible animal carriage including swine (Cotruvo et al., 2004). There are many animal rotaviruses, and both bovine and porcine rotaviruses have been detected in drinking water (Gratacap-Cavallier et al., 2000). In areas where humans and cows live in close proximity, it seems plausible that fecal contaminated water could play a role in the disease transmission between bovine species and humans. The porcine digestive tract is considered similar to that of humans. Hepatitis E virus in swine is similar to human hepatitis E virus, and people are at risk of infection by contact with swine viruses through contaminated water. There are clearly human and animal viruses that are excreted into waterways, and there is evidence that humans in contact with contaminated water can be infected. The enteric viruses are stable in water and in the environment and can move during rainfall and flood. Large numbers of persons can also be affected because of the common source, use, and exposure to contaminated *water*.

6.6.2 Epidemiology of waterborne viral zoonoses

Waterborne viral zoonoses are increasing globally in terms of the number of agents, outbreaks, and human cases. While developing countries are disproportionately affected by contaminated water due to poor sanitation standards and hygiene practices, developed countries are also affected. To estimate the global illness and deaths caused by rotavirus disease, studies from 1986 to 2000 on deaths caused by diarrhea from rotaviral infections was conducted by the World Bank (Parashar et al., 2003). Each year, rotavirus accounted for approximately 111 million episodes of gastroenteritis, 25 million clinic visits, 2 million hospitalizations, and a median of 440 000 deaths in children less than 5 years of age (Parashar et al., 2003). Also, by the age of 5 years, nearly every child will have an episode of rotavirus gastroenteritis (Parashar et al., 2003). This enormous incidence associated with rotavirus disease underscores the global burden of the virus. The significant increase of waterborne viral zoonoses comes at a time when mortality rates from diarrhea have in fact declined over the last two decades. This is attributed to improvements in hygiene and sanitation. The proportional increase in waterborne viral zoonoses such as rotaviruses probably reflects climatic factors including increased rainfall, flooding, and water temperature and Ph. These together with human behaviors and population growth and increasing urban population density promote the spread of waterborne zoonotic viral pathogens.

6.6.3 Prevention and control

Prevention and control of waterborne viruses will include measures for preventing fecal contamination of water, treatment of water, and adequate disposal of waste as well as promoting the improvement of sanitation standards and hygiene practices among humans. In this section, the specific prevention and control of waterborne zoonotic viruses will focus on the zoonotic agents in animal reservoirs. There are generic procedures that can be applied to prevent the establishment of zoonotic pathogens in domestic animal reservoirs. These include establishing groups of "clean" animals, keeping animals free from infection, and providing clean housing, food, and water for animal management (Cotruvo et al., 2004). Other tactics such as selective breeding for disease resistance in animals, competitive exclusion, and active and passive immunization may also be helpful in controlling specific waterborne viral zoonotic diseases in animal reservoirs. Control in wild animal populations presents an even greater challenge (Cotruvo et al., 2004).

The increasingly affluent populations in many developing countries necessitate a commensurate increase in animal production. This trend is likely to continue (Cotruvo et al., 2004). This trend needs to be balanced by concerns about the effects of industrial animal production on the environment and on human health. Proper disposal of animal waste, including from abattoirs, has become a significant problem around the world as this plays a role in the transmission of waterborne zoonotic viruses.

International coordination of national animal health programs for more than 164 countries is carried out through the World Organization for Animal Health (OIE). Control of zoonoses is considered to be of socioeconomic and/or public health significance and it is conducted through national disease surveillance; control or eradication of specific diseases using methods such as quarantine, test and slaughter, and vaccination; and animal importation control systems based on assessment of the risks of importation from specific countries and diagnostic testing of animals before export or upon entry (Cotruvo et al., 2004).

6.7 HUMAN BEHAVIOR AND THE EPIDEMIOLOGY OF WILDLIFE-ASSOCIATED VIRAL ZOONOSES

The complex interactions between humans and the environment that result in viral zoonoses include large numbers of wild animal reservoirs distributed throughout the world. Every landscape and habitat supports a variety of vertebrate species, and each of the species harbors an array of zoonotic pathogens including viruses. Viral zoonoses that originate in wild animals have become increasingly important throughout the world in recent decades. Wildlife-originated viral zoonoses have substantial impacts on human health and agricultural production and also invite commentary on topics of wildlife-based economies and wildlife conservation. The emergence of these pathogens as significant health issues is associated with a range of causal factors, most of them limited to the sharp and exponential rise of global human activity. The expanding human and animal populations encroach upon habitats that were previously exclusively wildlife areas, shrinking wildlife habitats and increasing interaction between wildlife, domestic animals, and humans, and the enormous international trade in wildlife species facilitates the transmission of viral zoonoses. Examples given previously include SARS, influenza respiratory viruses, and West Nile vector-borne viruses. Additional viral agents include rabies and the various hemorrhagic fevers including Ebola and Marburg viruses (Bengis et al., 2004). Occupational risk factors for the transmission of wildlife zoonoses include any prolonged stay in wildlife areas and hunting and the practice of preparing and eating "bushmeat."

6.7.1 Case study: bushmeat hunting in Cameroon

One example of the increasing risk of viral zoonotic transmission is the bushmeat trade in Cameroon. The large and growing urban demand for bushmeat is facilitated by the construction of roads into the forests of eastern Cameroon by logging companies. These roads facilitate the extraction of timber and increase access for bushmeat hunters. The roads have also led to the fragmentation of the forest in conjunction with the opening up of logging concessions in the East Province (Wolfe et al., 2005). The bushmeat market plays an important dietary role among poor and wealthy households, so everyone involved in the bushmeat trade including hunters, butchers, and consumers is at risk for zoonotic transmission due to bites, cuts, and other exposures to fluids or tissue. Ebola, Marburg, and monkeypox have emerged as human pathogens from the hunting of bushmeat (Wolfe et al., 2005). Habitat fragmentation and destruction can lead to the loss of some vertebrate species and may result in increased abundance of highly competent reservoirs and increase the risk for transmission of viral zoonoses to humans (LoGiudice et al., 2003). The global emergence of viral zoonoses from wildlife accounts for a significant proportion of the observed emerging infectious diseases.

6.7.2 Epidemiology of viral zoonoses from wildlife

The lymphocytic choriomeningitis (LCMV), Lassa Fever (Lassa virus) and Argentine (Junin virus), Bolivian (Machupo virus), Venezuelan (Guanarito), and Brazilian (Sabia virus) hemorrhagic fevers are all viral infections that are transmitted through direct contact with wild rats and mice. There is a worldwide distribution of diseases that manifest clinically with aseptic meningitis and influenza symptoms. Crimean–Congo hemorrhagic fever is a tick-borne disease characterized by fever and hemorrhages. This viral disease is found in sub-Saharan Africa, Eastern Europe, Russia, the Middle East, and Western China (Hoch et al., 1995). Ebola and Marburg viruses are hemorrhagic fever viruses that are transmitted by direct contact with infected nonhuman primates (Feldmann and Klenk, 1996). Ebola virus is distributed in the humid rain forests in Central and Western Africa, whereas Marburg virus is located in Central and Eastern Africa (Feldmann and Klenk, 1996). Marburg virus was first recognized in 1967 through the importation of infected monkeys from Uganda to Germany and Yugoslavia, which resulted in laboratory exposure to humans. RVF is another viral zoonosis that can be mosquito borne or spread through aerosols of sick sheep and goats (Lacy and Smego, 1996). RVF is found in Kenya, sub-Saharan Africa, Egypt, and Saudi Arabia (Lacy and Smego, 1996).

There are many other examples of wildlife-associated viral zoonoses. What is common to all is that viral zoonoses from wildlife are inevitably linked to areas of the world where there is encroachment of humans into wildlife areas. The increasing proximity between humans and wildlife facilitates transmission. The forest of Central Africa and the sub-Saharan region are hotspots for viral zoonoses.

The international trade in wildlife also poses health implications from imported wildlife. For example, in 2003, monkeypox was introduced to the United States when a shipment of African Gambian giant rats was sold to dealers, one of whom housed the rats with prairie dogs that led to infections in humans (Marano et al., 2007).

Viral infections with zoonotic potential are a significant challenge to human health. The new epidemiological patterns of infectious diseases that emerge from wildlife suggest that measures to control these diseases are complicated. As far as wildlife is concerned, difficulties can arise in controlling the population dynamics whether it is at the level of the animal reservoir, the pathogen, or the susceptible host. International organizations, such the WHO and OIE, has listed the reportable diseases of human and animal species. Among the list are several zoonotic viral agents known to occur in wildlife including AI and rabies (Artois, 2003). An additional OIE list includes wildlife species that have the potential to cause problems by serving as reservoirs of viral zoonoses (Artois, 2003). New emerging diseases constantly require the list of diseases and implicated wildlife species to be updated. Wildlife viral zoonoses management is based on the knowledge of the infectious state in wildlife. This knowledge is used to develop surveillance for the various diseases to collect structured data. There is no universal system for surveillance, but some reporting systems are based on specific zoonotic viral diseases of economic importance and mainly oriented to domestic animals. Specialized diagnosis departments for wildlife exist in Sweden, Germany, Switzerland, and France (Smith and Harris, 1991). Several laboratories at research institutes, notably the Pasteur Institute, are dealing with wildlife disease surveillance and diagnosis (Lamarque and Artois, 1997). There is clearly a great deal of interest in wildlife disease surveillance in Europe. The dedicated centers and institutes for zoonotic disease control are situated outside of the main hotspots.

In managing viral zoonoses of wildlife origin, a clear distinction should be made between wildlife control and pathogen or disease control. The approach to controlling a viral zoonotic pathogen in wildlife is either the eradication of the pathogen or preventing transmission to domestic animals and humans. The rationale for the need of a control program is the threat to human health. Strategies targeting the reservoir, the pathogen, or the transmission are rarely addressed effectively. In practice, control programs are frequently designed to fit the perception of what is feasible in the short term and aim to decrease the level of disease impact. The need for management programs for wildlife-sourced viral zoonoses is critical. There is an argument for nature to be allowed to take its course and achieve a balance (Gilmour and Munro, 1991). The anthropogenic environmental degradation that is occurring does not allow for the natural course to take place. Techniques that have been applied include oral vaccination through the use of bait placement in wildlife habitats. Immunizations against rabies through oro-mucosal application using oral vaccines in baits on wild foxes in Europe have been successful (Wandeler, 1991). Lethal control is also a strategy using culling to control foxes by providing bounties to nonprofessionals who turned in evidence of fox kill. This strategy has been unsuccessful (Macdonald et al., 1981).

Prevention and control of human diseases require a multidisciplinary approach to understand the mechanisms of transmission of pathogens and determine predictive indicators of potential reemergence of pathogens. It is clear that human behavior can reduce the human-to-wildlife interface but may also increase the risk of pathogen transmission.

6.8 THE ROLE OF HUMAN BEHAVIOR IN THE CONTROL OF VIRAL ZOONOSES

Viral zoonoses are affected at the interface between humans and animals by behaviors including those associated with globalization; population growth and shifts; urbanization; trade in animals and animal products; agricultural technologies; increasingly close human–animal relationships; and changes in ecosystems, vector and reservoir etiology, land use, and patterns of hunting and consumption of wildlife.

Significant successes in the control of viruses have been made with the eradication of smallpox and near eradication of polio in humans and rinderpest in cattle. These successes were due inter alia to the use of effective vaccines and well-coordinated global programs. Control of other emerging and reemerging viruses has been limited (Daszak et al., 2001). Viral zoonoses of significant public health impact can be managed with innovative approaches incorporating virology, epidemiology, and other disciplines. The detection of early warning signs may require syndromic surveillance and participatory epidemiological approaches that involve livestock, wildlife, humans, and the environment (WHO, 2010). The incorporation of indicators of ecosystem health into models and simulations that predict emerging viral zoonoses could help map the diversity of viral pathogens and their connections to and interactions with the environment, identify early stages of infection, and mitigate transmission through behavior change. Social network tools could be used to engage the public and facilitate real-time reporting. Rapid, efficient, simple, and cost-effective diagnostics for isolation and identification of specific pathogens that could be used in the field for surveillance and screening would be helpful. The development of diagnostics for pathogens that are not yet known will require microarray chips for field polymerase chain reactions and early characterization of pathogen groups and subgroups.

6.8.1 Communication

Data sharing is essential to managing viral zoonoses, and making technical data available to all stakeholders will enhance risk analyses and better inform interventions (WHO, 2010). Challenges to transparency of methods and interpretations, and to open access to databases, include technical and political associated with ownership. Virologists and epidemiologists must collaborate with specialists in ethics, information technology, and communication, among other disciplines, in order to overcome such challenges.

Commercial marketing programs have changed behaviors involving patterns of consumption. Public health campaigns can similarly change behaviors that bear on exposure to and transmission of the health impacts of climate change (Frumkin and McMichael, 2008). The design and implementation of such campaigns for the control of viral zoonoses require greater understanding of sociological, cultural, economic, anthropological, and ethical aspects of behavior and risk perception. Incentives to encourage transdisciplinary collaborations are needed. Collaborative efforts to control the HPAI H5N1 and A H1N1 in 2009 demonstrate the value of managing public perceptions, and that transparency between partners and stakeholders is important.

Other disciplines that may have significant collaborative roles in controlling the emergence and reemergence of viral zoonoses include the biological, chemical, and environmental sciences; the behavioral and social sciences; veterinary and human medicine; and international law. Cooperation across sectors and geographical borders is critical for effective surveillance, diagnoses, and interventions including those that involve communication and public education. Control may require government commitments to international cooperation in minimizing import and export of vectors and diseases and to improving public health (Filder, 2003). International Health Regulations (IHR) aim "to ensure the maximum protection against the international spread of disease with minimum interference with world traffic" and require governments to notify other nations about outbreaks of specific diseases and to maintain capabilities to control disease exit and entry (WHO, 1983). International law dealing with infectious diseases also helps control the movement of disease, and various treaties and conventions like the General Agreement on Tariffs and Trade specify legal duties for any given infectious disease (Fidler, 2001; Goodman, 1971).

Vertical strategies addressing environmental degradation, air quality, and water safety are particularly relevant to viral zoonoses. Nongovernmental organizations (NGOs) have important roles in strategies that involve emerging and reemerging infectious diseases (Charnovitz, 1997). Community-based participatory programs can improve access to medications, vaccinations, sanitation, and education on preventing infectious diseases (Taylor, 2002). NGO successes in such efforts attract private and public funding and support (WHO, 2011a,b). An indirect but important strategy is to incorporate health concerns and socioeconomic determinants of health into policies and into marketing strategies aimed at motivating behavior change (Frumkin and McMichael, 2008).

REFERENCES

Artois, M., 2003. Wildlife infectious disease control in Europe. *J. Mt. Ecol.* 7(Suppl.): 89–97.

Atran, S. and Norenzayan, A., 2004. Religion's evolutionary landscape: counter intuition, commitment, compassion, communion. *Behav. Brain Sci.* **27**: 713–770.

Barquet, N. and Domingo, P., 1997. Smallpox: the triumph over the most terrible of the ministers of death. *Ann. Intern. Med.* **127**(8 Pt 1): 635–642.

Bengis, R.G., Leighton, L.A., Fischer, J.R., et al., 2004. The role of wildlife in emerging and re-emerging in zoonoses. *Rev. Sci. Tech.* **23**: 497–511.

Bögel, K. and Motschwiller, E., 1986. Incidence of rabies and post-exposure treatment in developing countries. *Bull. World Health Organ.* **64**: 883–887.

Bögel, K. and Meslin, F., 1990. Economics of human and canine rabies elimination: guidelines for programme orientation. *Bull. World Health Organ.* **68**: 281–291.

Campbell, G.L., Marfin, A.A., Lanciotti, R.S., et al., 2002. West Nile virus. *Lancet Infect. Dis.* **2**(9): 519–529.

Carter, M. J., 2005. Enterically infecting viruses: pathogenicity, transmission and significance for food and waterborne infection. *J. Appl. Microbiol.* **98**(6): 1354–1380.

CDC, 2007. West Nile virus maps and data. Accessed on April 15, 2012, from http://www.cdc.gov/ncidod/dvbid/westnile/surv&control.htm

CDC, 2012. Climate and health program. Accessed on July 12, 2012, from http://www.cdc.gov/climatechange/

Charnovitz, S., 1997. Two centuries of participation: NGOs and international governance. *Michigan J. Int. Law* **18**: 183–286.

Condon, B.J. and Sinha, T., 2010. The effectiveness of pandemic preparations: legal lessons from the 2009 influenza epidemic. *Florida J. Int. Law* **22**(1): 1–30.

Conway, L.G., III, Sexton, S.M., and Tweed, R.G., 2006. Collectivism and governmentally initiated restrictions: a cross-sectional and longitudinal analysis across nations and within a nation. *J. Cross-Cultural Psychol.* **37**: 20–41.

Cotruvo, J.A., Dufour, A., Rees, G., et al., 2004. Waterborne Zoonoses. London: IWA Publishing.

Cupp, E.W., Klingler, K., and Hassan, H.K., 2003. Transmission of eastern equine encephalomyelitis virus in central Alabama. *Am. J. Trop. Med. Hyg.* **68**: 495.

Curriero, F.C., Patz, J.A., Rose, J.B., et al., 2011. The association between extreme precipitation and waterborne disease outbreaks in the United States, 1948–1994. *Am. J. Public Health* **91**: 1194–1199.

Daszak, P., Cunningham, A.A., and Hyatt, A.D., 2001. Anthropogenic environmental change and the emergence of infectious diseases in wildlife. *Acta Trop.* **78**: 103–116.

Diamond, J., 2002. Evolution, consequences, and future of plant and animal domestication. *Nature*, August 8 418(6898): 700-707.

Fahlquist, J.N., 2008. Moral responsibility for environmental problems—individual or institutional? *J. Agric. Environ. Econ.* 22(2): 109–124

Feldmann, H. and Klenk, H.D., 1996. Marburg and Ebola viruses. *Adv. Virus Res.* **47**: 1–52.

Fidler, D.P., 2001. International law and global infectious disease control. Commission on Macroeconomics and Health Working Paper No. WG2: 18 (2001). Accessed on April 16, 2012, from http://www.cmhealth.org/cmh_papers&reports.htm#Working%20Group%202

Filder, D.P., 2003. Emerging Trends in International Law Concerning Global Infectious Disease Control. Perspective, Vol 9, No. 3, March 2003. Accessed on April 16, 2012, from http://wwwnc.cdc.gov/eid/article/9/3/02-0336_article.htm

Frumkin, H. and McMichael, A.J., 2008. Climate change and public health: thinking, communicating, acting. *Am. J. Prev. Med.* **35**: 403–410.

Galvin, K.A., Randall, B.B., Smith, N.M., et al., 2001. Impacts of climate variability on East African pastoralists: linking social science and remote sensing. *Clim. Res.* **19**: 161–172.

Gelfand, M.J., Bhawuk, D.P.S., Nishii, L.H., et al., 2004. Individualism and collectivism. In R.J. House, P.J. Hanges, M. Javidan, (Eds.), Culture, Leadership, and Organizations: The GLOBE Study of 62 Societies (pp. 437–512). Thousand Oaks, CA: Sage Publications.

Gilmour, J.S. and Munro, R., 1991. Wildlife disease: management or masterly inactivity? *J. Nat. Hist.* **25**: 537–541.

Goodman, N.M., 1971. International Health Organizations and Their Work, 2nd ed. London: Churchill Livingstone.

Gratacap-Cavallier, B., Genoulaz, O., Brengel-Pesce, K., et al., 2000. Detection of human and animal rotavirus sequences in drinking water. *Appl. Environ. Microbiol.* **66**(6): 2690–2692.

Gubler, D.J., 1989. *Aedes aegypti* and *Aedes aegypti*-borne disease control in the 1990s: top down or bottom up. *Am. J. Trop. Med. Hyg.* **40**(6): 571–578.

Gubler, D.J., 1996. The global resurgence of arboviral diseases. *Trans. R. Soc. Trop. Med. Hyg.* **90**(5): 449–451.

Gubler, D.J., 1998. Resurgent vector-borne diseases as a global health problem. *Emerg. Infect. Dis.* **4**(3): 442–450.

Haines, A., Smith, K.R., Anderson, D., et al., 2007. Policies for accelerating access to clean energy, improving health, advancing development, and mitigating climate change. *Lancet* **370**: 1264–1281.

Hamrick, N., Cohen, S., and Rodriguez, M.S., 2002. Being popular can be healthy or unhealthy: stress, social network diversity, and incidence of upper respiratory infection. *Health Psychol.* **21**: 294–298.

Harvell, C.D., Mitchell, C.E., Ward, J.R., et al., 2002. Climate warming and disease risk for terrestrial and marine biota. *Science* **296**: 2158–2162.

Health Canada, 2012. Climate change and health. Accessed on July 12, 2012, from http://hc-sc.gc.ca/ewh-semt/climat/index-eng.php

Hoch, S.P.F., Khan, J.K., Rehman, S., et al., 1995. Crimean-congo hemorrhagic fever treated with oral ribavirin. *Lancet* **346**: 472–475.

Huffman, D.E., Nelson, K.L., and Rose, J.B., 2003. Calicivirus—an emerging contaminant in water: state of the art. *Environ. Eng. Sci.* **20**(5): 503–515.

Hunter, P.R., 2003. Climate change and waterborne and vector-borne diseases. *J. Appl. Microbiol.* **94**: 37S–46S.

IATA, 2007. Press Release, 24th October 2007. International Air Transport Association (IATA) website. Accessed on July 5, 2012, from http://www.iata.org/pressroom/pr/Pages/2007-24-10-01.aspx

IBT, 2011. Global airline profits likely to drop in 2011; but number of passengers to surge by 800-mln in 2014. Accessed on July 10, 2012, from http://www.ibtimes.com/articles/112134/20110214/airline-profits-passengers.htm

Jarvis, L., Montgomery, H., Morisetti, N., et al., 2011. Climate change, ill health, and conflict. *BMJ* **342**: d1819 (published 5 April 2011).

Jenkins, P.T., Genovese, K., and Ruffler, H., 2007. Broken Screens: The Regulation of Live Animal Importation in the United States. Washington, DC. Accessed on June 4, 2013, from http:// Defenders of Wildlife

Jhung, M.A., Swerdlow, D., Olsen, S.J., et al., 2011. Epidemiology of 2009 pandemic influenza A (H1N1) in the United States. *Clin. Infect. Dis.* **52**(Suppl. 1): S13–S26.

Knobel, D.L., Cleaveland, S., Coleman, P.G., et al., 2005. Re-evaluating the burden of rabies in Africa and Asia. *Bull. World Health Organ.* **83**: 360–368.

Kruse, H., Kirkemo, A.M., and Handeland, K., 2004. Wildlife as a source of zoonotic infections. *Emerg. Infect. Dis.* **10**: 2067–2072.

Lacy, M.D. and Smego, R.A., 1996. Viral hemorrhagic fevers. *Adv. Pediatr. Infect. Dis.* **12**: 21–53.

Lamarque, F. and Artois, M., 1997. Surveillance of wildlife diseases in France: the SAGIR network. *Epidemiol. Sante Anim.* (31–32): 07.b.31.

Laras, K., Sukri, N.C., Larasati, R.P., et al., 2005. Tracking the re-emergence of epidemic chikungunya virus in Indonesia. *Trans. R. Soc. Trop. Med. Hyg.* **99**: 128.

LoGiudice, K., Ostfeld, R.S., Schmidt, K.A., et al., 2003. The ecology of infectious disease: effects of host diversity and community composition on Lyme disease risk. *Proc. Natl. Acad. Sci. U.S.A.* **100**: 567–571.

Ludwig, B., Kraus, F.B., Allwinn, R., et al., 2003. Viral zoonoses—a threat under control. *Intervirology* **46**: 71–78.

Macdonald, D.W., Bunce, R.G.H., and Bacon, P.J., 1981. Fox populations, habitat characterization and rabies control. *J. Biogeogr.* **8**: 145–151.

Macpherson, C.N.L., 2005. Human behavior and epidemiology of parasitic zoonoses. *Int. J. Parasitol.* **35**: 1319–1331.

Macpherson, C.C., 2010. Public health, bioethics and policy: protecting our health and environment. *Int. Public Health J.* **2**: 535–540.

Manojkumar, R. and Mrudula, V., 2006. Emerging viral diseases of zoonotic importance-review. *Int. J. Trop. Med.* **1**: 162–166.

Marano, N., Arguin, P.M., and Pappaioanou, M., 2007. Impact of globalization and animal trade on infectious disease epidemiology. *Emerg. Infect. Dis.* **13**(December (12)): 1807–1809.

Marfin, A.A. and Gubler, D.J., 2001. West Nile encephalitis: an emerging disease in the United States. *Clin. Infect. Dis.* **33**(10): 1713–1719.

Marr, J.S. and Calisher, C.H., 2003. Alexander the Great and West Nile virus encephalitis. *Emerg. Infect. Dis.* **9**: 1599–1603.

McCrae, R.R., 2002. NEO-PI-R data from 36 cultures: further intercultural comparisons. In R.R. McCrae and J. Allik (Eds.), The Five-Factor Model of Personality Across Cultures (pp. 105–126). New York: Kluwer Academic/Plenum.

McJunkin, J.E., de los Reyes, E.C., Irazuzta, J.E., et al., 2001. La Crosse encephalitis in children. *N. Engl. J. Med.* **344**: 801.

McMichael, A.J., Powles, J.W., Butler, C.D., et al., 2007. Food, livestock production, energy, climate change, and health. *Lancet* **370**: 1253–1263.

Mettenleiter, T.C., 2006. Zoonotic Respiratory Diseases Caused by Viruses. Symposium 1: Respiratory Zoonoses. Accessed on April 19, 2012, from http://www.the-vcrs.org/ 2006meetingupdates/Symposium-I-Res-Zoonoses.pdf

Murphy, F.A., 1998. Emerging zoonoses. *Emerg. Infect. Dis.* **4**: 429–435.

National Center for Biotechnology Information (NCBI), 2008. U.S. National Library of Medicine. Vector-Borne Diseases: Understanding the Environmental, Human Health, and Ecological Connections, Workshop Summary. Copyright © 2008, National Academy of Sciences.

Nelson, G.S., 1974. Zooprophylaxis with special reference to schistosomiasis and filariasis. In E.J.L. Soulsby (Eds.), Parasitic Zoonoses Clinical and Experimental Studies (pp. 273–285). Academic Press, London.

Nettle, D., 2005. An evolutionary approach to the extraversion continuum. *Evol. Human Behav.* **26**: 363–373.

NHS, 2012. Confederation on Climate Change and Sustainability. Accessed on July 12, 2012, from http://www.nhsconfed.org/Training/climate-change/Pages/Climate_Change.aspx

Owen, J., Moore, F., and Panell, N., 2006. Migrating birds as dispersal vehicles for West Nile virus. *EcoHealth* **3**(2): 79–85.

Parashar, U.D., Hummelman, E.G., Breese, J.S., et al., 2003. Global Illness and Deaths Caused by Rotavirus Disease in Children. Accessed on April 19, 2012, from www.cdc.gov.eid/article/9/5/02-0562 article.htm

Pardio Sedas, V.T., 2007. Influence of environmental factors on the presence of *Vibrio cholerae* in the marine environment: a climate link. *J. Infect. Dev. Countries* **1**(3): 224–241.

Peters, C.J. and Khan, A.S., 2002. Hanta virus pulmonary syndrome the new American haemorrhagic fever. *Clin. Infect. Dis.* **34**: 1224–1231.

Peterson, L.R. and Gubler, D.J., 2006. Viral zoonoses. In Dale D.C. and Federman D.D. (eds.), ACP Medicine. New York: BC Decker.

Rao, J.R., Millar, B.C., and Moore, J.E., 2009. Avian influenza, migratory birds and emerging zoonoses: unusual viral RNA, enteropathogens and Cryptosporidium in poultry litter. *Biosci. Hypotheses* **2**(6): 363–369.

Richardson, H.S., 1999. Institutionally divided moral responsibility. In E.F. Paul, F.D. Miller, and J. Paul (Eds.), Responsibility. Cambridge: Cambridge University Press.

Rose, M.R. and Lauder, G.V. (Eds.), 1996. Adaptation. New York: Academic Press.

Rose, J.B., Epsetin, P.R., Lipp, E.K., et al., 2001. Climate variability and change in the Unites States: potential impacts on water and foodborne diseases caused by microbiological agents. *Environ. Health Perspect.* **109**(Suppl. 2): 211–222.

Smith, G.C. and Harris, S., 1991. Rabies in urban foxes (*Vulpes vulpes*) in Britain: the use of a spatial stochastic simulation model to examine the pattern of spread and evaluate the efficacy of different control regimes. *Philos. Trans. R. Soc. Lond. B* **334**: 459–479.

Taylor, A.L., 2002. Global governance, international health law and WHO: looking towards the future. *Bull. World Health Organ.* **80**: 975–980.

Taylor, L.H., Latham, S.M., and Woolhouse, M.E., 2001. Risk factors for human disease emergence. *Philos. Trans. R. Soc. Lond. B, Biol. Sci.* **356**: 983–989.

Thornhill, R., Fincher, C.L., and Aran, D., 2009. Parasites, democratization and the liberalization of values across contemporary countries. *Biol. Rev.* **84**: 113–131.

Thornhill, R., Fincher, C.L., Murray, D.R., et al., 2010. Zoonotic and non-zoonotic diseases in relation to human personality and societal values: support for the parasite-stress model. *Evol. Psychol.* **8**(2): 151–168.

United Nations (UN), 1999. Population Issues: Migration and Urbanization. Accessed on July 5, 2012, from http://www.unfpa.org/6billion/populationissues/migration.htm

UN, 2003. 1st United Nations World Water Development Report 'Water for People, Water for Life'. Accessed on April 22, 2012, from http://waterwiki.net/index.php/1st_United_Nations_World_Water_Development_Report_'Water_for_People,_Water_for_Life'

UN, 2010. Population Division, Concise Report on the World Population Situation in 2010. p. 26.

US Fish and Wildlife Service, 2010. Bird Migration. Accessed on July 12, 2012, from http://www.fws.gov/wheeler/observation/birdmigration.html

Varma, M.G.R., 1989. Geographical distribution of arthropod-borne diseases and their principal vectors. Unpublished document WHO/VBC/89.967. Geneva: World Health Organization.

Vasickova, P., Dvorska, L., Lorencova, A., et al., 2005. Viruses as a cause of foodborne diseases: a review of the literature. *Vet. Med.* **50**(3): 89–104.

Venkatesan, G., Balamurugan, V., Gandhale, P.N., et al., 2010. Viral zoonosis: a comprehensive review. *Asian J. Anim. Vet. Adv.* **5**: 77–92.

Vitousek, P.M., Mooney, H.A., Lubchenco, J., et al., 1997. Human domination of Earth's ecosystems. *Science* **277**: 494–499.

Walther, G.R., Post, E., Convey, P., et al., 2002. Ecological responses to recent climate change. *Nature* **416**: 389–395.

Wandeler, A.I., 1991. Oral immunization of wildlife. In Baer G.M. (Ed.), The Natural History of Rabies, 2nd ed. (pp. 485–503). Boca Raton: CRC Press.

Wang, L., 2011. Discovering novel zoonotic viruses. *New South Wales Public Health Bull.* **22**(July (5–6)): 113–117.

World Health Organization (WHO), 1983. International Health Regulations, 3rd ed. Geneva: World Health Organization.

WHO, 2009a. Hendra Virus Fact Sheet, July 2009. Accessed on April 23, 2012, from http://www.who.int/mediacentre/factsheets/fs329/en/index.html

WHO, 2009b. Nipah Virus Fact Sheet, July, 2009. Accessed on April 23, 2012, from http://www.who.int/mediacentre/factsheets/fs262/en/

WHO, 2010. Influenza and Other Emerging Zoonotic Diseases at the Human–Animal Interface. FAO/OIE/WHO Joint Scientific Consultation, 27–29 April, 2010, Verona.

WHO, 2011a. Pandemic Influenza A (H1N1) Donor Report. 1st March 2011.

WHO, 2011b. What is the Green Light Committee? Accessed on April 20, 2012, from http://www.who.int/tb/publications/en/

WHO, 2012. Health Topics: Climate change. Accessed on July 12, 2012, from http://www.who.int/topics/climate/en/

Wilke, I.G. and Haas, L., 1999. Emerging of new viral zoonoses. *Dtsch. Tierarztl. Wochenschr.* **106**: 332–338.

Williams, G., 2008. Responsibility as a virtue. *Ethical Theory Moral Pract.* **11**(4): 455–470.

Wolfe, N.D., Daszak, P., Kilpatrick, A.M., et al., 2005. Bushmeat hunting, deforestation, and prediction of zoonotic disease emergence. *Emerg. Infect. Dis.* **11**(December (2)) 1822–1827.

Yee, K.S., Carpenter, T.E., and Cardona, C.J., 2009. Epidemiology of H5N1 avian influenza. *Comp. Immunol. Microbiol. Infect. Dis.* 32(July (4)): 325–340.

WHO (2010) Health Topics: Dengue. Online. Accessed on July 1, 2012. http://www.who.int/topics/dengue/en/

Wilde, L.G. and Hass, L. (1996) Encyclopedia of Infectious Diseases. Oxford University Press, Oxford.

Williams, R. (2008) Emerging infectious disease. Environ Health Perspect, 116 (11): 466–470.

Wolfe, N.D., Daszak, P., Kilpatrick, A.M. et al. (2005) Bushmeat hunting, deforestation, and prediction of zoonotic disease emergence. Emerg Infect Dis, 11 (12): 1822–1827.

Yan, X., Carpenter, T.E., and Cardona, C.J. (2009) Epidemiology of infectious disease. Emerg Infect Dis, 15: 587–590.

7

GLOBAL TRAVEL, TRADE, AND THE SPREAD OF VIRAL INFECTIONS

Brian D. Gushulak and Douglas W. MacPherson

Migration Health Consultants, Inc., Qualicum Beach, BC, Canada

TABLE OF CONTENTS

Viral Infections and Global Change, First Edition. Edited by Sunit K. Singh.

7.1 INTRODUCTION

Throughout all of recorded human history and evident through earlier time periods, travel and mobility have been fundamental elements of human activity. The mobility of individuals and populations has taken humanity to the furthest reaches of the planet and, beginning in the twentieth century, beyond the limits of earth itself. When people move, they take many elements of their environment with them. Some of those elements such as their genetic and biological makeup are physical; others reflect the social and economic background and status of the travelers. Each and all of the various aspects of travel and mobility can influence and affect the incidence, prevalence, and spread of disease and illness present in those traveling themselves or localized in the geographic regions or areas that the travelers pass through or settle.

Together the interface between human mobility and the presence of disease-causing organisms creates a fluid and evolving pattern of disease presentation and epidemiology. The basic principles and outcomes of this interaction have been present since humans set out on their first journey. Virus infecting humans can accompany human movements and people may move to areas where new or previously unknown viruses capable of causing infection exist. At the same time human mobility is often accompanied by the movement of animals and plants, which themselves may harbor or transmit viruses within and between species. The mobility-associated ebb and flow of the distribution of viruses across the planet has been an ongoing component of history (Oldstone, 2010).

Seasonal, periodic, and occasionally pandemic distributions of endemic viruses have always been affected by human mobility and travel. The dynamics of seasonal human influenza provide perhaps one of the best understood examples in both the historical and modern context (Potter, 2001). Human mobility and its consequences are also implicated in the spread and transmission of newly emerging or previously unrecognized viruses. Historical examples include the spread of smallpox to the Americas associated with European colonization (Fenner et al., 1998). A more modern example is provided by the SARS outbreak in 2003 (Wilder-Smith, 2006). While many of these relationships have remained constant over time, recent developments in the nature, dynamics, and mode of travel have affected and changed several of the interactions between human mobility and viral infections (Jones et al., 2008).

The human mobility-related effects and influences on global viral epidemiology are not limited simply to the process of travel alone. The expansion of human populations has resulted in the peopling of regions and locations where materials, foodstuffs, and goods differed between locations. Barter, exchange, and commerce ensured that disparities in local access can be met through the transport and trade of goods and economic resources. The process of moving goods and materials across and between geographic areas and human communities can also affect the extent and distribution of viral infections. Viruses may be physically transported with the material or the trade and commercial activity may allow or facilitate the movement of vectors or other methods of transmitting viral infections.

Both of these activities, human mobility and travel and trade and commerce, have exerted effects on the distribution and spread of viral infections throughout history. Future developments and the continuing evolution of travel and trade in an increasingly globalized and interconnected world will continue to impact the spread and extent of viral infections. Climate change and the further evolution of the planetary environment can be expected to both modify the existing processes and perhaps introduce new influences on the interactions between viruses and mobile human populations. This chapter will describe how human travel, population mobility, and trade influence and affect the extent and spread of viruses and then review some of the possible implications of sustained climatic change on these processes.

As there are thousands of viruses on the planet (Fauquet et al., 2005), detailed descriptions of how travel and mobility affect their distribution at the individual level are beyond the scope of this chapter. General principles supported by specific examples will be used to describe the relationships and outcomes. This chapter is primarily focused on the processes through which human population mobility and trade can influence the spread of viruses. Detailed information on specific viruses can be found in other chapters in this volume.

7.2 BASIC PRINCIPLES

The movement and spread of viral infections can take place in one of two ways: extension or expression.

7.2.1 Extension

The virus itself can physically be moved to new locations where infection was historically limited or absent. A modern example would be the introduction of West Nile virus into North America in the late 1990s (Huhn et al., 2003).

7.2.2 Expression

Competent hosts, vectors, or methods of transmission may appear in areas where the virus is present but was limited in its capacity to infect susceptible hosts. An example would be increased incidence of endemic viral diseases such as dengue virus or eastern equine encephalitis following the introduction of *Aedes albopictus* mosquitoes (Gratz, 2004).

The net result of the two processes is the new presentation or appearance of viral infections in locations where they were previously nonexistent or occurred in a limited fashion. Both of these processes are long-term, recurring activities that are influenced by both natural and man-made events. Travel, population mobility, and trade can be associated with both processes. Viruses can accompany the movement or travel of humans, animals, goods, products, or items, extending the spread of infection. At the same time, population mobility and trade can create situations where susceptible populations, vectors, or new methods of transmission are presented to existing reservoirs of viruses.

7.3 AN OVERVIEW OF POPULATION MOBILITY

Travel and movement have been a part of the human existence since man evolved on the earth. Growing populations required resources that exceeded local capacities and created a need for new territories, the acquisition of materials and economic resources often required travel, greed and conflict have often been associated with the acquisition of new lands, and the human desire for knowledge and exploration has caused people to travel for no other reason than to find out what was "over the mountain" or "beyond the sea." The movement of the original human population from its origin in Africa across the globe has been an integral and important component of evolution and development (Mellars, 2006).

The process of travel and mobility continues to reflect the human condition. People continue to seek new and better destinations to live, work, and provide for their families.

Others are forced to move as a consequence of natural, environmental, social, political, or military reasons. Still more journey temporarily for employment, business, pleasure, or the extension of civil or military actions (IOM, 2011).

In an attempt to provide a generally applicable approach that encompasses and compensates for all aspects of modern human travel, we will use the term population mobility to describe patterns of current travel and migration (Gushulak and MacPherson, 2006). Under the umbrella of population mobility, it is possible to consider human travel and movement in terms of the processes involved. These processes can be related to specific elements including the movement of pathogens and diseases or to the impact of external forces such as globalization or changing climatic conditions. Within this contextual framework, modern population mobility can be considered as a composite of specific travel-related elements that include duration, direction, disparities, and diversities. In turn each of those elements can be considered against the characteristics and demographics of the affected populations themselves and finally considered in the light of the geographic locations in which the travel takes place.

The direction of human mobility may be one way, as observed in the traditional immigration/emigration paradigm. Alternatively it may be temporary with a sojourn at a new destination followed by a return home, a pattern that is seen in both migrant labor and recreational travel. In addition, in today's age of dual or multiple citizenship, the process can be circular with repetitive journeys between locations. As will be described in further detail, the biological and epidemiological outcomes, including the spread of microbes, can be related to the direction of the journey.

In terms of disparities, depending on location and situation, the travel may involve the crossing of multiple levels and differences in geographic, biological, social, economic, medical, and health environments. In biological context, humans may travel across and between differences in the incidence and prevalence of infections and biological and ecological niches in vector and host distributions. Travel also traverses boundaries and differences in terms of disparities and variability in health-care services, public health, and disease control practices and capacities, all of which can influence the incidence and outcomes of viral infections. In addition mobile populations frequently move between a variety of inspection, control, quarantine, customs, and importation rules and practices. The crossing of these disparate and diverse parameters ultimately can have implications for health and disease at both individual and population levels (Gushulak and MacPherson, 2004). Those disparities ultimately determine and affect efficacy of programs and practices designed to limit or mitigate the spread of microbes and pathogens and secondarily affect the spread of viruses and their vectors.

Understanding travel population mobility in this context is important when considering the forces influencing the spread of viruses. In today's world as many of the forces, influences and aspects of modern travel and mobility are much different from those of only few decades previously. Many of the approaches towards the movement of people, health, and international disease control have been challenged or weakened by the evolution of globalization and modern population mobility. The impact of a changing climate can be expected to introduce even further pressures and influences on the interaction between mobility and the spread of viruses (Kuhn et al., 2004).

7.4 THE DYNAMICS OF MODERN POPULATION MOBILITY

As the global population has reached levels never before encountered, it is practically a tautology to note that the current level of population mobility is historically unprecedented.

TABLE 7.1. Examples of Viruses Potentially Spread by Travel and Trade

Agent	Transmission
Chikungunya[a]	Insect vector
Dengue[b]	Insect vector
Influenza (human)[c]	Human to human
Influenza (zoonotic)[c]	Inter-/intraspecies
Hemorrhagic fever viruses (Ebola, Marburg, Lassa)	Host contact/human to human
Monkeypox	Host contact/human to human
West Nile virus[d]	Insect vector
Yellow fever	Insect vector
Crimean Congo hemorrhagic fever	Insect vector
Foot and mouth	Direct contact/fomites
Infectious salmon anemia	Unknown
Rift Valley fever	Insect vector

[a] Simon et al. (2008).
[b] Gubler DJ (2002).
[c] Russell et al. (2008).
[d] Lanciotti et al. (1999).

The total human planetary population was approximately one billion people in 1850. By 2011 it had increased sevenfold and is still rising (United Nations, Department of Economic and Social Affairs, Population Division, 2011). Many of those people are mobile and travel. Table 7.1 describes some of the magnitude of permanent and temporary human mobility. However, current flows of people across and between global regions and areas are a product of more than the total global population. Globalization; global integration of economics, commerce, and trade; and the availability, ease, and low cost of high-speed travel provide new and novel influences and forces affecting human mobility. The result is a planet marked by population flows occurring at rates and time durations for which there is no parallel historical experience. The 214 million international migrants present today, for example, if considered as a nation would comprise the fifth largest county in the world.

7.5 HUMAN POPULATION MOBILITY AND THE SPREAD OF VIRUSES

Travel is important in the context of the spread of virus because humans may act as both the host and vector for numerous viruses (Wilson, 2007). Therefore, individuals with clinical viral infections or those who harbor latent viruses will carry the organisms with them when they travel. Should such individuals arrive or travel to locations where suitable hosts and/or vectors exist, imported infection can result. Imported infections create the potential for secondary transmission to humans or animals, and may result in the establishment of a new focus of disease.

A secondary aspect of the introduction of viral infections into new geographic areas or environments is that the organism may arrive as either an acute or a chronic infection. The former is demonstrated by increased incidence rates of disease following the arrival of the microbe and frequently represents the development of acute infection in unexposed or unprotected hosts. The second pathway involves the migration of large populations with chronic or latent infections to regions where prevalence is low. Over time and with sustained migration flows, the prevalence of chronic infections can increase. Depending on the nature of the organism, secondary transmission to unexposed hosts may increase the incidence of the disease, or the prevalence may simply increase due to the arrival of those infected earlier.

Historical examples of the spread of acute infections are provided by spread of smallpox (Fenner et al., 1998), yellow fever (Vainio and Cutts, 1998), and measles (Black et al., 1971) following the routes of colonization and trade (Ray, 1976). More recent examples have documented the global spread of seasonal influenza following the end of the First World War (Cox and Subbarao, 2000) and the extension of West Nile virus infection into North America beginning in the late 1990s (Hayes et al., 2005). Examples where travel and migration have affected the prevalence of chronic viral infections are represented by the development of the global human immunodeficiency virus pandemic (De Cock et al., 2011) and the migration-associated increase in hepatitis B and C infection in low-incidence nations (Mitchell et al., 2011).

The modern aspects of human mobility and travel influence and affect the global spread of viruses on multiple levels. The growing numbers numbers of people on the move mathematically increase the likelihood that more infected individuals will be traveling than in comparison with previous historical periods. In the absence of increased preventive or mitigating strategies, the greater biological burden of these viruses in mobile human populations increases the chance of travel-related spread viruses increases in relation to the proportion of infected travelers. Recent examples can be demonstrated in gastrointestinal viral infections in ocean-going tourist vessels that now carry several thousand individuals (Verhoef et al., 2008). The spread of chronic viral diseases can likewise be related to growth in population mobility. It is estimated that more than 350 million people are chronically infected with hepatitis B, for example (Shepard et al., 2006). Many of those infected are mobile and populations migrating from areas of high prevalence to lower prevalence areas are influencing both the epidemiology and biological burden of the disease in those low-prevalence locations (Marschall et al., 2008).

Secondly, the diversity and expansion of modern travel and migration patterns has introduced different pathways of travel and destinations, for both temporary travelers and longer-term or permanent migrants. Some of these journeys expose viral hosts and vectors to viral threats in new geographic areas as is observed with travel to areas of yellow fever or dengue virus endemic locations (Field et al., 2010). Other journeys can involve the return of travelers with infections such as chikungunya (Chen and Wilson, 2010) or yellow fever (Robertson et al., 1996) to locations where the infection is currently rare but where the potential for secondary or sustained transmission exists. A related aspect involves the growth in volume of travel to and from isolated or rural locations that historically have been remote or visited by relatively few. Increased travel to those rarely visited locations can increase the number of individuals exposed to the risk of previously geographically restricted zoonotic infections. Population mobility that increases such exposure through travel, migration, or commercial activity is globally important as the majority of modern emerging infectious diseases are zoonotic (Jones et al., 2008).

Finally, the speed of modern travel has increased the risk of viral spread. This is the relationship that has become increasingly important as the speed of travel has increased over the past 150 years (Cliff and Haggett, 2004). Before the development of the steam engine, human mobility was limited to the speed of animal-drawn or wind-powered conveyances. Circumnavigation of the globe in the 1850s, for example, required almost a full year. Long-distance travelers with viral infections present at the time of the beginning of their journey had either survived the disease and were no longer infectious or succumbed to its effects before its completion. This reduced the importation and transport of geographically isolated viral infections and prevented their introduction into health care treatment or control programs at the destination.

Modern travel volumes present unprecedented opportunities for the global distribution of pathogens (Tatem et al., 2006). During the 1957 influenza pandemic, the majority of international travel still took place through sea travel. It took approximately 6 months for the virus responsible for that pandemic to spread globally. Air travel has markedly reduced these dispersion times. In 2003 and 2009 a newly recognized respiratory disease and a novel strain of influenza, respectively, spread to many nations across the world in a matter of a few weeks. A study examining the role of air travel in the global spread of influenza noted that in 2 months of 2008, 2.35 million travelers boarded an aircraft in Mexico and flew to over one thousand cities in 164 nations (Khan et al., 2009).

The use of increasing rapid travel has effectively reduced the barriers to viral extension historically posed by distance. Beginning with the use of steam power and extending even further with the development of air travel, it is now possible to cover great distances well within the periods of incubation or clinical illness for many infections. This allows for the potential transmission of communicable disease across previous historical barriers and supports the rapid and wide dissemination of newly emerging or reemerging viral infections. An example of the former includes Japanese encephalitis virus (JEV) imported into North America (CDC, 2011a), while the SARS represents an example of the rapid spread of a newly emerging disease (Goubar et al., 2009).

The importance of the speed of travel on the spread of viral infections is not a new phenomenon. The shift from sail to steam produced observable impact on the geographic spread of human virus over a century ago. As exemplified by measles imported to Fiji, shorter travel times meant that a disease that previously only affected passengers and crew and that was self-limited by the duration of the voyage became a post-arrival risk (see Figure 7.1). As travel time decreases, the risk of incubating or active disease in those arriving from endemic areas increases. This relationship also challenges border and frontier medical screening and disease control practices, many of which originated in the days of wind-powered long-distance travel.

7.6 THE BIOLOGICAL ASPECTS OF POPULATION MOBILITY AND THE SPREAD OF VIRUSES

The human response to viral exposure and possible infection can be affected and governed by several biological and genetic factors. At the genetic level, cellular structural and biochemical factors under host genetic control can affect aspects of infection such as viral attachment to cell components, entry into host cells, and viral reproduction. At the same time many host defense mechanisms both cellular and humeral are genetically controlled (Hughes, 2002). Other genetic factors can influence the severity of viral disease once acquired or the development of serious adverse outcomes (Stephens, 2010).

Repeated exposure to pathogenic organisms can apply selective genetic pressures at population level, which can, over time, be associated with differences at biological and

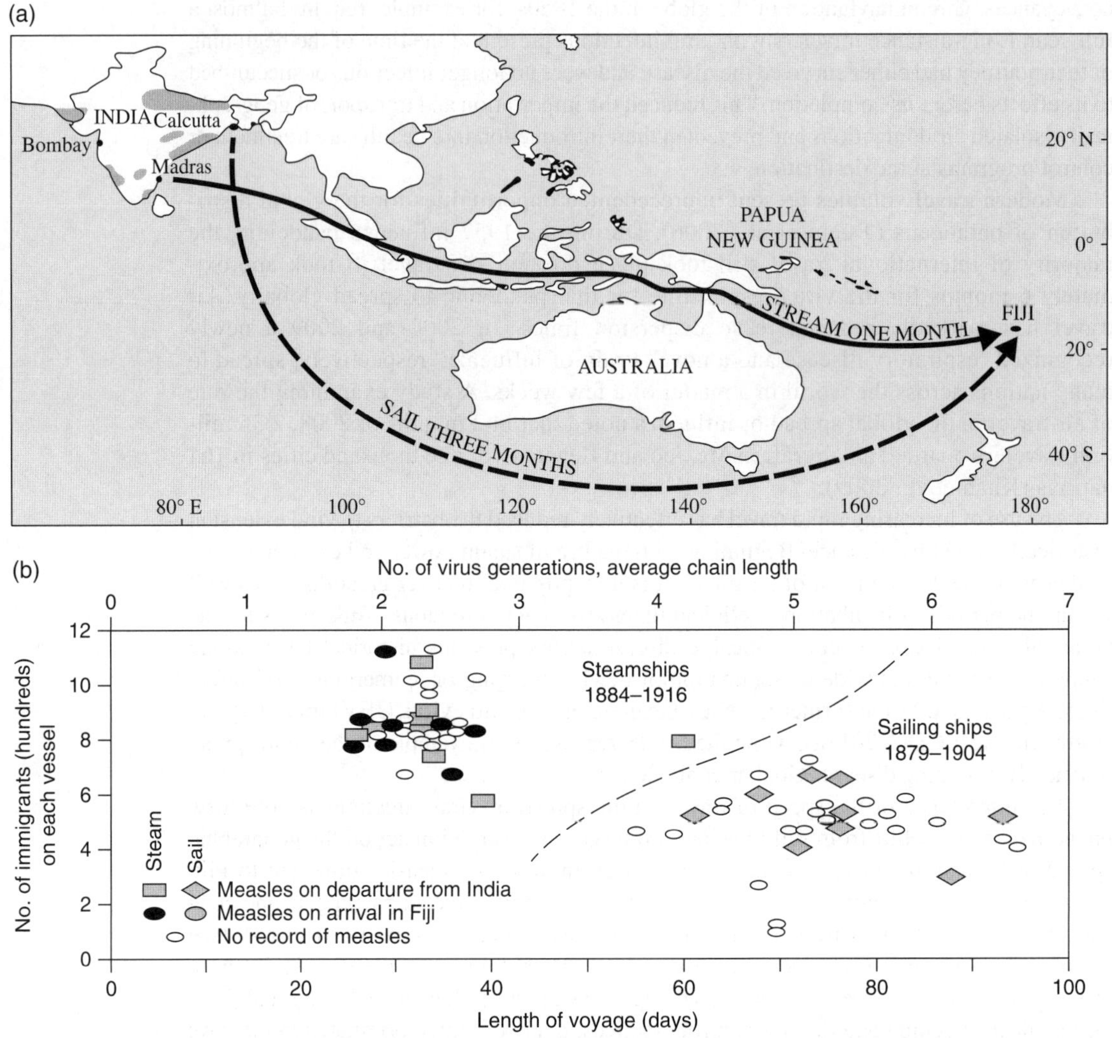

Figure 7.1. Measles outbreaks associated with two modes of international voyages (Source: Cliff A, Haggett P. (2004) Time, travel and infection. Br. Med. Bull. 69, page 94).

genetic level in the response to viral infections (Fumagalli et al., 2010). Thus, at a population level, mobile populations will bring elements of their genetic characteristics that can influence aspects of viral infections (Frodsham, 2005). Historical examples are provided by the differential yellow fever mortality noted between Europeans (immunologically naïve) and local African residents (endemically exposed) during the fifteenth-century period of European colonization (Watts, 2001). More modern examples are represented by North American travelers, who may have reduced immunity to measles through low levels of vaccination and low herd immunity due to interrupted domestic transmission visiting destinations in Europe, Asia, and Africa where the incidence of the disease is higher (CDC, Measles update).

At the biological level, previous exposure to viruses can have important and significant impact on subsequent reexposure or infection. Long-lasting and protective immunity can develop following viral infections or vaccination. That biological "history" of viral exposure will accompany mobile populations who travel across and between different viral exposure and prevalence environments. Together the genetic and biological determinants of response to viral exposure and infection will influence both individual and population outcomes following movement or travel to destinations where the microbial environment may be different. Understanding the nature and importance of these determinants is an important component of the current management of the health aspects of population mobility. Examples include pretravel vaccination and practices for reducing vector contact for tourists, migrant workers, military forces, and humanitarian workers.

7.7 THE DEMOGRAPHIC ASPECTS OF POPULATION MOBILITY AND THE SPREAD OF VIRUSES

7.7.1 Elements related to the volume of travel

As noted earlier, the unprecedented number of people living on the planet coupled with the capacity and widespread availability of high-speed travel allows for the rapid extension and circulation of viruses. New or novel viruses, particularly those that have high pathogenic potential, have always posed global risks to susceptible populations as demonstrated by the 1918 influenza pandemic. However, the speed and volume of modern travel have reduced the time between recognition of an event and its international and subsequent global extension. While it is theoretically possible to contain isolated outbreaks if antivirals or vaccines can be delivered before those beyond the immediate area are exposed, this is practically very difficult to accomplish. As witnessed in 2003 with SARS and in 2009 with pandemic H1N1 (pH1N1) influenza, population mobility has made even initially localized outbreaks of new or novel infections now practically immediate global threats. Interrupting or delaying the spread of infections without paralyzing global travel trade and commerce is very challenging.

As such, modern patterns of human travel and mobility will provide opportunities for the rapid extension of some viral infections. At the same time, the numbers of people traveling and the speed with which they move exceed the capacities of traditional border control practices. Consequently, current and future activities will tend to be focused on surveillance, early recognition, and mitigation of these events. The important role played by human mobility in the international spread of viral infections is reflected in all modern control and mitigation strategies. Some of those strategies are beginning to consider the future implications of climate change on the patterns of population mobility.

7.7.2 Elements related to disparities in health practices

Viral disease prevention and mitigation programs and practices frequently differ between nations and regions. Those differences may result from disparities in local risk and prevalence of the organisms, different health system priorities, variations in social and public health capacities, and unequal levels of economic development. These variabilities can manifest themselves in different patterns of immunization, exposure to infection, and access to treatment between geographic locations and populations. The movement of people between or across these differences may be associated with the spread of viral infections.

In some cases viruses are spread as a consequence of differences in immunization levels. In areas of very low incidence or absence of transmission of a viral infection, routine immunization rates and coverage may decrease with a diminished perception of risk of the disease. At the same time low rates of domestic disease reduce the effects of herd immunity. Individuals and groups who arrive from regions where the disease remains endemic may reintroduce the infection into the domestic population. Additionally, short-term travelers who have incomplete or inadequate immunization may acquire viral infections during travel and generate transmission on return. Recent examples have included outbreaks of measles arriving with travelers to North America (CDC, 2011b) and Europe and the transmission of hepatitis A in children of migrants returning to low-incidence environments after visiting friends and relatives in areas endemic for the disease (Suijkerbuijk et al., 2009).

7.7.3 Situations where population mobility and travel can affect the spread of viruses

7.7.3.1 ***Migration.*** Modern global migration (individuals residing outside of their country of birth) is very diverse and the 214 million international migrants originate in locations and nations that cross both the development spectrum and many viral disease prevalence levels. Viral diseases may accompany migrants on their journeys and migration can affect the epidemiology of viral infections. In the case of acute infections such as influenza, measles, mumps, and varicella, the movement of viruses with migrants is frequently lost in the pattern of diseases arriving with the much larger number of international travelers. However, some specific situations occur where migration-specific impacts can be observed. These usually involve infections with long latency periods or situations where prevalence differences between origin and destination are high.

The movement of large numbers of migrants from global regions of high endemicity for hepatitis B virus (HBV) to regions of low prevalence is affecting the national epidemiological patterns of the disease. In nations where large proportions of new immigrants have arrived from HBV endemic regions, the foreign born represent significant populations burdened with chronic disease.

Migration does not have to be international to be associated with the spread of viral infections. Rural–urban migration or the internal displacement of populations within national borders can also impact on the distribution of viruses. For example, in locations where sylvatic yellow fever is endemic, large numbers of rural residents migrating to urban areas can increase the risk of urban yellow fever outbreaks in the presence of competent arthropod vectors of the disease.

7.7.3.2 ***Visiting Friends and Relatives.*** Globalization and ease and availability of travel now offer many migrants the opportunity to visit their places of origin more often than previous waves of immigrants were able to. These populations can be reexposed to virus exposures not present at their new home. Known Visiting Friends and Relatives (VFR) travelers represent a community at risk of acquiring disease during travel (Angell and Cetron, 2005). When accompanied by children who were born after the family immigrated, these return visits can expose the children to risks that exceed their parents, who may retain immunity acquired before migration (Hendel-Paterson and Swanson, 2011). Increasing numbers of travel-acquired infections are to be reported in VFR travelers.

7.7.3.2.1 SHORT-TERM TRAVEL. As noted in Table 7.2, over one billion international journeys are made annually. The majority of this travel takes place between locations where prevalence and nature of virus epidemiology is similar at origin and destination. However,

TABLE 7.2. A Snapshot of Current Population Mobility

Population	Number (000 s)	Source
International migrants (2010)	214 000	UN[a]
Refugees (2010)	10 550	UNHCR[b]
Internationally displaced populations (2010)	25 200	UNCHR
Migrant workers (2010)	105 000	ILO[c]
International tourists (2010)	980 000	WTO[d]

[a] United Nations, Department of Economic and Social Affairs, Population Division (2011).
[b] UNHCR (2011).
[c] ILO (2010).
[d] World Tourism Organization (2012).

the high volume of international travel means that large numbers of people travel across gaps in the prevalence of virus infection or the presence of viral vectors (Shi et al., 2010). The importance of travel in the spread of viral infections has been an issue of increasing global public health concern, heightened by the 2003 SARS event (Anderson et al., 2004b), highly pathogenic avian influenza (Ferguson Neil et al., 2004), and the 2009 experience with pH1N1 influenza. As a result of this awareness, all contingency and mitigation strategies designed to limit or control the spread of infectious diseases of international public health importance include elements focused on the risks posed by modern travel patterns (Bell, 2004; Katz, 2009; WHO, 2007).

In addition to viral infections of global public health importance, short-term travelers periodically carry a variety of viral infections acquired in association with their journeys. Many of these infections are common such as viral gastrointestinal and upper respiratory tract infections and represent limited risk to the individual or those around them (O'Brien et al., 2006). Less frequently, other more serious viruses may be observed in travelers. Not surprisingly these infections reflect the viral epidemiology present during the journey. Uncommon or unusual viral infections acquired by short-term travelers can be associated with serious illness such as tick-borne encephalitis (Reusken et al., 2011) or rabies. A few infections of this type may also represent local or regional public health risks if there is the possibility of secondary transmission after travel. Because these serious events are low in occurrence and may not be recognized through regular national public health surveillance systems, integrated multinational surveillance networks that are focusing on travelers have been developed during the past two decades.

Religious pilgrimages and events represent some of the largest movements of humanity in terms of numbers of participants. The Hajj, for example, brings millions of people together from all regions of the globe. The Ardh Kumbh (also known as Kumbh Mela) in India involves more than 60 million people every 12 years (BBC, 2007). These events may amplify the risks for disease importation, transmission among the pilgrims, and international spread on their return (Ahmed et al., 2006; Memish et al., 2009). Not all of the risks of viral spread associated with pilgrimages are related to human infections. Some of these events such as Eid al-Fitr are associated with the concomitant transportation of animals and livestock. In the Arabian Peninsula and Horn of Africa, the movement of animals and livestock associated with religious events is a concern for the potential spread of Rift Valley fever, foot-and-mouth disease (FMD), and camelpox (FAO, 2007).

7.7.4 Humanitarian and complex emergencies

Complex humanitarian emergencies and situations can result from natural or man-made forces and events. They can often be accompanied by the displacement and movement and travel of large numbers of people, goods, and animals. As such, these events provide several potential opportunities for the movement of viral infections. Climate change is anticipated to affect both the frequency and extent of future complex emergencies (Frumkin et al., 2008). In some of these events, the displacement may be short term as can be seen with earthquakes, tsunamis, or severe weather events. However, in other events, both natural such as large volcanic eruptions and sustained famine and man-made in the case of emergencies resulting from human conflict, the population displacement may be long term or even permanent. Depending on the situation these displacements may result in the movement of hosts and vectors of viral diseases.

Viruses may also be spread via the journey of humanitarian and relief workers to areas of natural or man-made calamity and disaster. Depending on the size of the event and the numbers of people involved, hundreds and sometimes thousands of humanitarian workers can be involved in relief and rehabilitation activities. Relief workers traveling to assist in these humanitarian emergencies and deal with their consequences may be exposed to viral infections not prevalent or occurring at their normal place of residence. Recent examples have included dengue virus infection in missionary travelers to Haiti who participated in visits as short as 1 week in late 2010 (Sharp et al., 2012). In immunologically naive travelers, these situations can result in significant rates of infection. For example, in the study noted earlier, infection with dengue virus was observed in 25% of those participating in the mission.

Should these humanitarian workers originate in locations where competent vectors for the travel-acquired viral infection reside, it is possible that they could introduce or reintroduce the infection on their return. It is also possible that humanitarian or relief workers can import or transport viral infections from their place of origin to the site of the calamity or event.

7.7.5 Social and economic aspects of population mobility

The movement of individuals and communities is only a part of the process of population mobility. The human journey is accompanied by the associated transport of a variety of goods, material, and items as well as social, cultural, economic, and mercantile practices. Short-term travelers may only bring their immediate personal effects and some food and nutritional items for the journey, but those who are mobile for the longer-term and permanent migrants can bring many elements of their original residence with them. Once established at their new destination, many mobile populations import or acquire items and aspects of their previous environments.

Items that can accompany travelers and mobile populations can include personal goods, animals and pets, foodstuffs and nutritional items, medicines and pharmaceuticals, personal and manufactured goods, as well as personal and recreational items. Each of those social and economic activities has the potential to influence or affect the introduction and distribution of microbes. For example, rabies-free areas have experienced the introduction of rabid animals, which have accompanied travelers arriving from endemic locations (Gautret et al., 2011), while insect vectors of diseases may be transported in luggage or on the person of the traveler (Heath and Hardwick, 2011).

Many mobile populations bring social and cultural aspects of the lives with them when they travel. Some of those activities can involve the potential for the extension of virus. An example is provided by the personal importation of the meat of wild African animals known as bushmeat. The movement of bushmeat is unregulated and not well quantified but is estimated to be extensive. As primates are often the source of bushmeat, viruses capable of human infection may be transmitted through this source. Simian immunodeficiency virus (SIV), simian T-lymphotropic virus (STLV), and simian foamy virus (SFV) have been isolated from bushmeat (Smith et al., 2012).

7.7.6 An overview of trade

Trade and commerce have accompanied the migration of humanity across the globe. The recognition that illness and disease could accompany the movement of goods as well as the movement of people was an integral component of early quarantine practice. The development of international standards to limit the spread of diseases was integrally associated with managing the risks posed by goods and commerce. In a manner analogous to the processes described earlier for human travel and mobility, modern trade, economic, and commercial activity may result in the direct spread of viruses themselves or nonhuman hosts and vectors across and between epidemiological and geographic boundaries.

Also similar to the role played by human travel and population mobility, the recent influences and changes in trade produced by an integrated, globalized, ever more rapidly moving world have new implications for the spread of viruses. The volume of goods and material being traded or shipped has increased and evolved in relation to regional population growth, economic wealth, and globalized manufacturing. Total global merchandise trade has increased more than threefold during the past two decades (see Figure 7.2).

For example, the global production of pork increased more than threefold between 1961 (24.7 million tonnes) and 2002 (86.6 million tonnes), and this increased production has been matched by a nearly 10% annual growth in the trade of pig meat since 1992 (FAO, 2011). As noted earlier, increased volumes of material increase the risk of viral movement and can compromise inspection and quarantine practices by making the inspection of all items impractical or unreasonably expensive.

In a manner similar to the dynamics of modern population mobility, the increase in growth has been influenced by globalization. Expanding trade volumes have been accompanied by an expansion in the diversity and distribution of source and destinations involved in international trade. Several trade patterns for commodities are markedly different than they were only a few decades ago as witnessed by the economic growth of Asia and parts of Africa during the past 30 years (see Figure 7.3). These new patterns may increase the risk of introduction of viruses, hosts, or vectors in manners that do not follow historical patterns. As demonstrated by the international spread of the mosquito *A. albopictus* through the evolution of the international trade in used tires (Enserink, 2008), new patterns of trade may have significant and unanticipated outcomes in terms of viral extension.

The speed of transportation measured in terms of direct shipping times and the impact of containerized shipment where loading and unloading of cargo are much quicker has reduced the time between origin and destination. At the same time the centralization of processing, manufacturing, and shipping has resulted in fewer more integrated enterprises handling ever greater volumes of material. The centralized processing of goods, such as foodstuffs, in situations of contamination or product infection can result in the exposure of large numbers of individuals increasing the risks of infection far beyond historical episodes where a greater number of smaller processing centers were in operation.

Not all aspects of trade and commerce that affect virus distribution involve the direct movement of the viruses or vectors themselves. Commercial activities that indirectly support trade, such as shifting patterns of agricultural production to more lucrative crops, can influence viral epidemiology. More intensive use of irrigated methods of rice production in South and Southeast Asia, for example, can affect mosquito populations and influence the incidence of JEV (Keiser et al., 2005). When these changes in rice irrigation

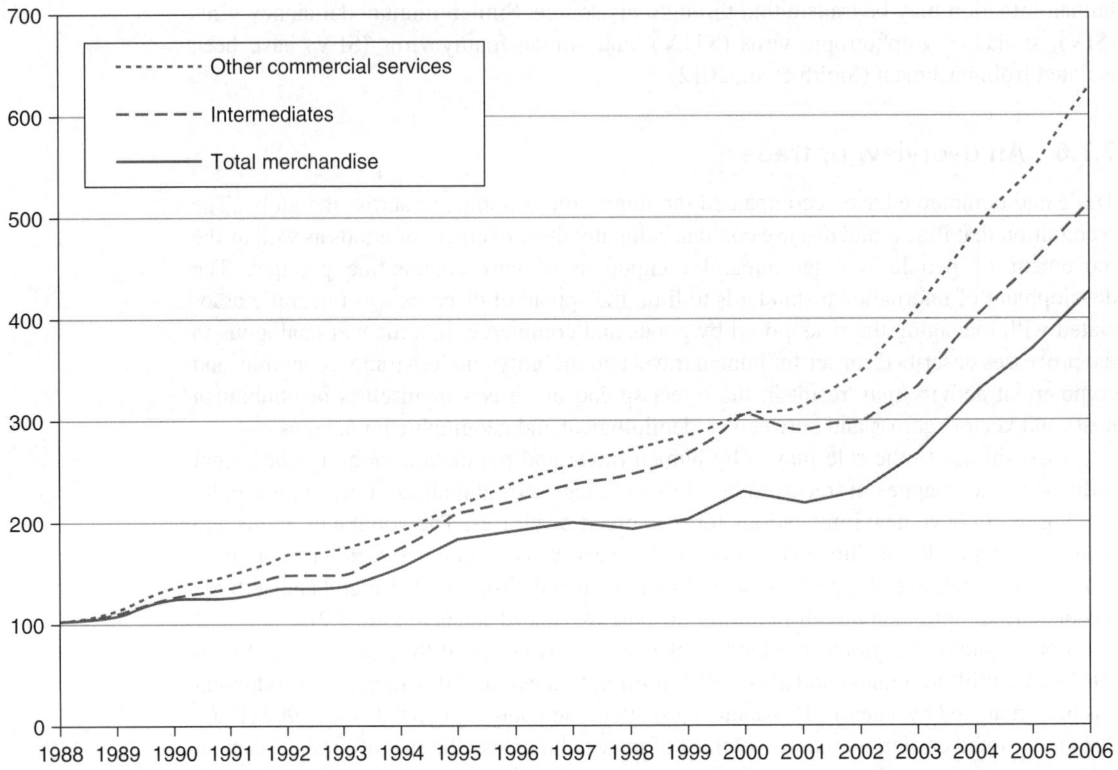

Figure 7.2. Trends in world trade of total merchandise, intermediate goods, and other commercial services, 1998–2006 (Source: WTO World Trade Report 2008: Trade in a globalizing world. Available from URL: http://www.wto.org/english/res_e/booksp_e/anrep_e/world_trade_report08_e.pdf. Accessed February 12, 2012).

are also associated with increased pork production, the virus, which normally infects birds, may be transmitted to pigs, further increasing the risk to humans (Erlanger et al., 2009). A similar situation has been observed in the case of the newly emerging viral infection, Nipah virus. This virus, which can cause serious human infections, is normally found in fruit bats, but encroachment of the bats territory through the construction of pig farms in endemic areas increases the risks of human infection (Epstein et al., 2006).

Such examples are not limited to viruses infecting humans. Some modern production processes designed to increase production of food commodities can create environments that support viral expression. The growth of the aquaculture industry provides some examples. In these situations aquatic animals such as shrimp and fish may be raised in high-density environments sometimes using artificial feeds. Some viral disease may be associated with these endeavors (Walker and Winton, 2010). Infectious salmon anemia provides an example. This viral disease, which was first identified in 1984, is a major pathogen for Atlantic salmon grown in fish farms. First reported in Norway, outbreaks have occurred in Scotland, Eastern Canada, the United States, and Chile (The Center for Food Security and Public Health, 2010).

7.7.7 Trade and the spread of viruses

Trade and commerce can facilitate the spread of viral infections through the movement of either hosts or vectors. The risks of viral spread include viruses that infect humans, animals,

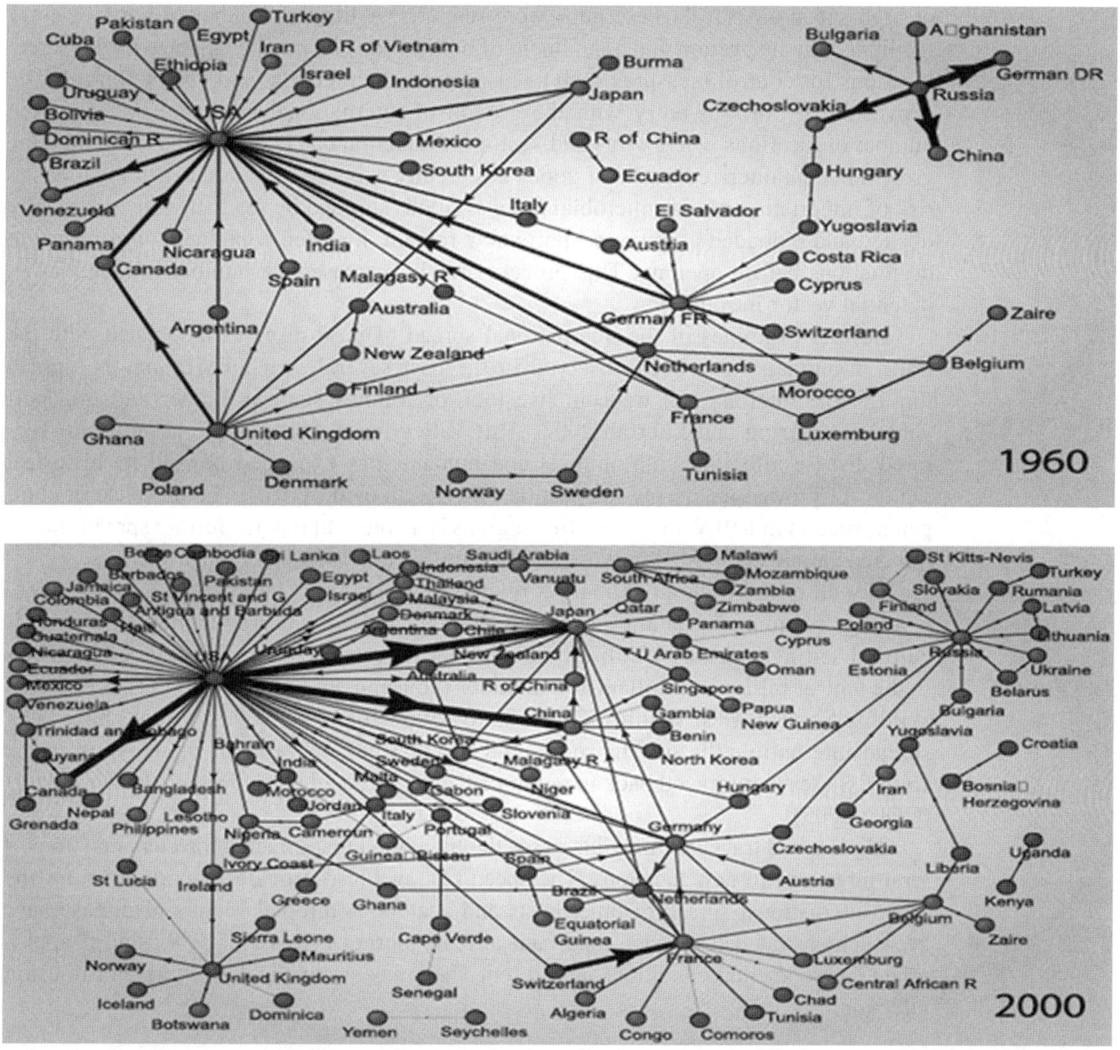

Figure 7.3. Evolution of modern trade patterns (Source: Patterns of dominant flows in the world trade web. Serrano MA, Bogunia M, Vespignani A J. Econ. Interac. Coor. 2007;2:111. Available from URL: http://arxiv.org/pdf/0704.1225v1.pdf. Accessed February 3, 2012). For color detail, please see color plate section.

and plants. Viruses may be transported through the trade of the plants and animals themselves or through seeds and agricultural products. As discussed in the case of virus extension through human travel, the spread of a virus through trade and commerce may not always initially be followed by adverse outcome. Newly arriving viruses may remain contained or localized by the absence of vectors or environmental conditions conducive to wider spread. Well-studied examples include a virus infecting citrus plants, Citrus tristeza virus, which is spread to citrus plants by an aphid insect vector (Moreno et al., 2008). The implications of the arrival may be mild and only become apparent after the subsequent arrival of the vector (Anderson et al., 2004a).

Historical travel and trade patterns usually involved the arrival of people and goods at and through a limited number of transportation hub locations. Before the development of high-speed air travel, these hubs were seaports. A limited number of ports of entry facilitated the inspection and assessment of both goods and people and provided discreet locations for control measures such as inspection, disinfection, and vector control. The expansion of ports of entry with the growth of international air travel increased the number of locations where virus and vector introduction can occur. At the same time the growth in containerized traffic of goods has greatly expanded the number of locations at risk of importation of new microbial threats. Goods shipped by sea that were historically opened and unloaded at major seaports now remain in sealed containers until arrival at destinations before opening. This process greatly increases the number of locations of potential vector introduction.

Several examples of the international spread of viral diseases associated with the trade in animals have been observed in the past several years. FMD may be spread through the international trade in livestock or fomites such as fodder and transport vehicles (Geering and Lubroth 2002). Rift Valley fever, for example, a mosquito-vectored disease affecting both animals and humans, has expanded beyond its historical regions of prevalence during the past decade (Pepin et al., 2010). The presence of competent vectors in Rift Valley fever-free regions is a potential risk for further spread should infected animals be traded to these areas (Turell et al., 2008).

Not all of the trade in animals is related to agriculture (Pavlin et al., 2009). There is a large volume of global trade in live animals and animal products outside of agricultural settings such as wildlife trade and pet industry. This trade is valued annually in the tens of billions of dollars and includes thousands of animal species (US Fish and Wildlife Service, U.S. Wildlife Trade: an overview for 1997–2003). Viruses have been spread internationally via this route, and in 2003 human cases of monkeypox in the United States were traced back to pet prairie dogs infected by imported African rodents (CDC, 2003).

High-speed travel not only increases the likelihood that infected travelers may import viral infections to new locations. The speed of travel also provides opportunities for the rapid international spread of other hosts and vectors. Arthropod vectors of disease have been demonstrated to survive long-haul aircraft journeys (Russell, 1989), and if appropriate environmental conditions are present, they may survive and proliferate or transmit infection.

7.8 POTENTIAL IMPACT OF CLIMATE CHANGE

In the context of human trade and travel, climate change may influence the spread of viruses through several means. Firstly, an evolving climate will affect and impact already existing patterns of population mobility and trade. Ever larger numbers of people may become part of the current flow of humanity on the planet. Those increasing numbers of mobile individuals will, as described earlier, bring with them their genetic, biological, behavioral, and social factors that affect viral distribution on the planet. They will also bring their livestock, pets, and patterns of trade and commerce with them. These impacts can be anticipated to be volumetric expansions of currently existing processes.

In terms of trade, climatic alterations and the consequential impacts on temperatures, moisture patterns, and weather events may affect the extension and survival of viruses and their vectors in plants and animals outside of historical ranges. Some viruses and vectors arriving through trade and commerce in climate-affected regions may have new opportunities for emergence or reemergence, while others present in areas of historical distribution may be negatively affected (Gubler et al., 2001).

In regard to human travel, an important way in which climate change will affect the global spread of viruses through travel and trade will be through the creation of new and novel patterns and directions of population flows. Estimates place the number of people displaced by climate change at levels of 200 (Norwegian Refugee Council, 2009) to 250 (UK Treasury, 2005) million people by the year 2050. At the same time warming may allow for increased settlement in regions of the world currently too cold to support large populations. As described earlier in this chapter, new patterns of population mobility have two effects on virus spread. Those leaving uninhabitable locations may take viruses to areas where infections could be established or reestablished. Alternatively, climate change-induced population movements could put new hosts and vectors into places where there is a preexisting viral presence.

The net result may be the same, increased rates of infections and the growth in consequential adverse outcome. In terms of control and mitigation, however, it is the latter process where climate change induces new patterns of mobility that can be expected to be the most challenging. This is because the resulting patterns and presentation of viral infection may be novel, unusual, and unanticipated. Existing surveillance, monitoring, and control mechanisms for important viral infections can simply be expanded to deal with growing volumes of travelers. New patterns of travel-associated infections and novel or unexpected situations may not be detected early in their presentation.

7.9 CONCLUSION

Travel and trade have long been associated with the international spread of viruses and viral infections of humans, animals, and plants. This relationship will continue in an ever more integrated and globalized world where the rapid movement of increasing numbers of people and continued growth of trade and commerce are integral components of modern society. Global climate change can be expected to influence and affect human travel, trade, and commerce and consequently impact the global distribution of virus, hosts, reservoirs, and vectors. The result of this complex interface between virus, travel, and trade will create situations and environments suitable for the extension and expression of recognized virus infections as well as increasing the likelihood of the emergence and recognition of new presentations.

For example, as this volume was being prepared, several nations and organizations in Europe were investigating and dealing with the emergence of a previously unrecognized *Orthobunyavirus* affecting domestic ruminant animals (Hoffmann et al., 2012). Named for the location where it was first isolated, Schmallenberg virus was first detected in November of 2011, and epidemiological studies suggest it is transmitted by insect vectors and vertically from mother to offspring (OIE: Schmallenberg Virus, 2012). The manner of its introduction into Europe is at this time undefined. It represents, however, a metaphoric example of the implications of the spread of viruses in a world of integrated travel and economic activity.

The threats posed by viral extension and expression into new areas and populations require continual and extensive epidemiological surveillance, systems to provide early recognition, and strategies, policies, and programs for mitigation and control. It is important that anticipating and planning for the impact of climate change becomes an integral component of those strategies and policies.

REFERENCES

Ahmed QA, Arabi YM, Memish ZA. Health risks at the Hajj. *Lancet.* 2006;**367**:1008–1015.

Anderson PK, Cunningham AA, Patel NG, et al. Emerging infectious diseases of plants: pathogen pollution, climate change and agrotechnology drivers. *Trends Ecol Evol.* 2004a;**19**:535–544.

Anderson RM, Fraser C, Ghani AC, et al. Epidemiology, transmission dynamics and control of SARS: the 2002–2003 epidemic. *Philos Trans R Soc Lond B—Biol Sci.* 2004b;359:1091–1105.

Angell SY, Cetron MS. Health disparities among travelers visiting friends and relatives abroad. *Ann Intern Med.* 2005;**142**:67–72.

BBC. Millions bathe at Hindu festival. *BBC News*, 2007 January 3. Available from URL: http://news.bbc.co.uk/2/hi/south_asia/6226895.stm. Accessed February 3, 2012.

Bell DM. Public health interventions and SARS spread, 2003. *Emerg Infect Dis.* 2004;**10**: 1900–1906.

Black FL, Hierholzer W, Woodall JP, et al. Intensified reactions to measles vaccine in unexposed populations of American Indians. *J Infect Dis.* 1971;**124**:306–317.

CDC. Measles update. Available from URL: http://wwwnc.cdc.gov/travel/notices/watch/measles. Accessed June 8, 2013.

CDC. Update: multistate outbreak of monkeypox—Illinois, Indiana, Kansas, Missouri, Ohio, and Wisconsin, 2003. Morb Mortal Wkly Rep. 2003;52:642–646.

CDC. Japanese encephalitis in two children—United States, 2010. Morb Mortal Wkly Rep. 2011a; 60:276–278.

CDC. Measles imported by returning U.S. travelers aged 6–23 months, 2001–2011. Morb Mortal Wkly Rep. 2011b;60:397–400.

Chen LH, Wilson ME. Dengue and chikungunya infections in travelers. *Curr Opin Infect Dis.* 2010;**23**:438–444.

Cliff A, Haggett P. Time, travel and infection. *Br Med Bull.* 2004;**69**:87–99.

Cox NJ, Subbarao K. Global epidemiology of influenza: past and present. *Annu Rev Med.* 2000; **51**:407–421.

De Cock KM, Jaffe HW, Curran JW. Reflections on 30 years of AIDS. Emerg Infect Dis. [serial on the Internet]. 2011 June. Available from URL: http://wwwnc.cdc.gov/eid/article/17/6/pdfs/10-0184.pdf. Accessed January 21, 2012.

Enserink M. A mosquito goes global. *Science.* 2008;**320**:864–866.

Epstein JH, Field HE, Luby S, et al. Nipah virus: impact, origins, and causes of emergence. *Curr Infect Dis Rep.* 2006;**8**:59–65.

Erlanger TE, Weiss S, Keiser J, et al. Past, present, and future of Japanese encephalitis. *Emerg Infect Dis.* 2009;**15**:1–7.

FAO. Rift Valley Fever could spread with movement of animals from East Africa. Empres Watch, October 10, 2007. Available from URL: ftp://ftp.fao.org/docrep/fao/011/aj215e/aj215e00.pdf. Accessed June 6, 2013.

FAO. Classical Swine Fever. Empres Transbound Anim Dis Bull. 2011;**39**:46–51. Available from URL: http://www.fao.org/docrep/015/i2530e/i2530e00.pdf. Accessed February 16, 2012.

Fauquet CM, Mayo MA, Maniloff J, et al. Virus taxonomy: classification and nomenclature of viruses. San Diego: Elsevier Academic; 2005.

Fenner F, Henderson DA, Arita I, et al. The history of smallpox and its spread around the world. In: Smallpox and its eradication. Geneva: WHO; 1998.

Ferguson NM, Fraser C, Donnelly CA, et al. Public health. Public health risk from the avian H5N1 influenza epidemic. *Science.* 2004;**304**:968–969.

Field V, Gautret P, Schlagenhauf P, et al. Travel and migration associated infectious diseases morbidity in Europe, 2008. *BMC Infect Dis.* 2010; **10**:330.

Frodsham AJ. Host genetics and the outcome of hepatitis B viral infection. *Transpl Immunol.* 2005;**14**:183–186.

Frumkin H, Hess J, Luber G, et al. Climate change: the public health response. *Am J Public Health.* 2008;**98**:435–445.

Fumagalli M, Pozzoli U, Cagliani R, et al. Genome-wide identification of susceptibility alleles for viral infections through a population genetics approach. *PLoS Genet.* 2010;**6**(2): e1000849.

Gautret P, Ribadeau-Dumas F, Parola P, et al. Risk for rabies importation from North Africa. Emerg Infect Dis [serial on the Internet]. December 2011. Available from URL: http://wwwnc.cdc.gov/eid/article/17/12/pdfs/11-0300.pdf. Accessed June 6, 2013.

Geering WA, Lubroth J. Preparation of Foot-and-Mouth Disease Contingency Plans. *FAO Animal Health Manual No. 16.* Food and Agriculture Organization of the United Nations. Rome, 2002. Available from URL: ftp://ftp.fao.org/docrep/fao/005/y4382e/y4382e00.pdf. Accessed June 6, 2013.

Goubar A, Bitar D, Cao WC, et al. An approach to estimate the number of SARS cases imported by international air travel. *Epidemiol Infect.* 2009;**137**:1019–1031.

Gratz N. Critical review of the vector status of *Aedes albopictus*. *Med Vet Entomol.* 2004;**18**: 215–227.

Gubler DJ. The global emergence/resurgence of arboviral diseases as public health problems. *Arch Med Res.* 2002;**33**:330–342.

Gubler DJ, Reiter P, Ebi KL, et al. Climate variability and change in the United States: potential impacts on vector- and rodent-borne diseases. *Environ Health Perspect.* 2001 May;**109** Suppl 2:223–233.

Gushulak BD, MacPherson DW. Population mobility and health: an overview of the relationships between movement and population health. *J Travel Med.* 2004;**11**:171–178.

Gushulak BD, MacPherson DW. Migration medicine & health: principles & practice. Hamilton: BC Decker; 2006.

Hayes EB, Komar N, Nasci RS, et al. Epidemiology and transmission dynamics of West Nile virus disease. *Emerg Infect Dis.* 2005;**11**:1167–1173.

Heath AC, Hardwick S. The role of humans in the importation of ticks to New Zealand: a threat to public health and biosecurity. *N Z Med J.* 2011 July 29;**124**(1339):67–82.

Hendel-Paterson B, Swanson SJ. Pediatric travelers visiting friends and relatives (VFR) abroad: illnesses, barriers and pre-travel recommendations. *Travel Med Infect Dis.* 2011;**9**:192–203.

Hoffmann B, Scheuch M, Höper D, et al. Novel orthobunyavirus in cattle, Europe, 2011. Emerg Infect Dis [serial on the Internet]. March 2012. Available from URL: http://wwwnc.cdc.gov/eid/article/18/3/pdfs/11-1905.pdf. Accessed June 6, 2013.

Hughes AL. Natural selection and the diversification of vertebrate immune effectors. *Immunol Rev.* 2002;**190**:161–168.

Huhn GD, Sejvar JJ, Montgomery SP, et al. West Nile virus in the United States: an update on an emerging infectious disease. *Am Fam Physician.* 2003;**68**:653–661.

ILO. International labour migration: a rights based approach. The Organization. Geneva, 2010.

IOM. World migration report 2011. Geneva: The Organization; 2011. Available from URL: http://publications.iom.int/bookstore/free/WMR2011_English.pdf. Accessed June 6, 2013.

Jones KE, Patel NG, Levy MA, et al. Global trends in emerging infectious diseases. *Nature.* 2008; **451**:990–993.

Katz R. Use of revised international health regulations during influenza A (H1N1) epidemic. *Emerg Infect Dis.* 2009;**15**:1165–1170.

Keiser J, Maltese MF, Erlanger TE, et al. Effect of irrigated rice agriculture on Japanese encephalitis, including challenges and opportunities for integrated vector management. *Acta Trop.* 2005; **95**:40–57.

Khan K, Arino J, Hu W, et al. Spread of a novel influenza A (H1N1) virus via global airline transportation. *N Engl J Med.* 2009;**361**:212–214.

Kuhn K, Campbell-Lendrum D, Haines A, et al. Using climate to predict infectious disease outbreaks: a review. Geneva: World Health Organization; 2004. Publication WHO/SDE/OEH/04.01. Available from URL: http://www.who.int/globalchange/publications/en/oeh0401.pdf. Accessed February 6, 2012.

Lanciotti RS, Roehrig JT, Deubel V, et al. Origin of the West Nile virus responsible for an outbreak of encephalitis in the northeastern United States. *Science* 1999;**286**:2333–2337. Available from URL: http://preview.ncbi.nlm.nih.gov/pubmed/10600742. Accessed June 6, 2013.

Marschall T, Kretzschmar M, Mangen MJ, et al. High impact of migration on the prevalence of chronic hepatitis B in the Netherlands. *Eur J Gastroenterol Hepatol.* 2008;**20**:1214–1225.

Mellars, P. Why did modern human populations disperse from Africa ca. 60,000 years ago? A new model. *Proc Natl Acad Sci USA.* 2006;**103**:9381–9386.

Memish ZA, McNabb SJ, Mahoney F, et al. Establishment of public health security in Saudi Arabia for the 2009 Hajj in response to pandemic influenza A H1N1. *Lancet.* 2009;**374**:1786–1791.

Mitchell T, Armstrong GL, Hu DJ, et al. The increasing burden of imported chronic hepatitis B—United States, 1974–2008. *PLoS One*. 2011;**6**(12):e27717.

Moreno P, Ambrós S, Albiach-Martí MR, et al. Citrus tristeza virus: a pathogen that changed the course of the citrus industry. *Mol Plant Pathol*. 2008;**9**:251–268.

Norwegian Refugee Council. Climate change: people displaced. 2009. The Organization. Available from URL: http://www.nrcfadder.no/arch/img.aspx?file_id=9904600. Accessed January 25, 2012.

O'Brien DP, Leder K, Matchett E, et al. Illness in returned travelers and immigrants/refugees: the 6-year experience of two Australian infectious diseases units. *J Travel Med.* 2006;**13**:145–152.

OIE: Schmallenberg Virus. Technical fact sheet. February 16, 2012. Available from URL: http://www.oie.int/fileadmin/Home/fr/Our_scientific_expertise/docs/pdf/A_Schmallenberg_virus.pdf. Accessed June 6, 2013.

Oldstone MBA. Viruses, plagues and history; past, present and future. New York: Oxford University Press; 2010.

Pavlin BI, Schloegel LM, Daszak P. Risk of importing zoonotic diseases through wildlife trade, United States. *Emerg Infect Dis.* 2009;**15**:1721–1726.

Pepin M, Bouloy M, Bird BH, et al. Rift Valley fever virus (Bunyaviridae: Phlebovirus): an update on pathogenesis, molecular epidemiology, vectors, diagnostics and prevention. *Vet Res.* 2010;**41**:61.

Potter CW. A history of influenza. *J Appl Microbiol.* 2001;**91**:572–579.

Ray AJ. Diffusion of diseases in the western interior of Canada, 1830–1850. *Geogr Rev.* 1976;**66**: 139–157.

Reusken C, Reimerink J, Verduin C, et al. Case report: tick-borne encephalitis in two Dutch travellers returning from Austria, Netherlands, July and August 2011. Euro Surveill. 2011;16(44):pii=20003. Available from URL: http://www.eurosurveillance.org/ViewArticle.aspx?ArticleId=20003. Accessed March 12, 2012.

Robertson SE, Hull BP, Tomori O, et al. Yellow fever: a decade of reemergence. *J Am Med Assoc.* 1996;**276**:1157–1162.

Russell RC. Transport of insects of public health importance on international aircraft. *Travel Med Int.* 1989;**7**:26–31.

Russell CA, Jones TC, Barr IG, et al. The global circulation of influenza A (H3N2) viruses. *Science* 2008;**320**:340–346.

Sharp TM, Pillai P, Hunsperger E, et al. A cluster of dengue cases in American missionaries returning from Haiti, 2010. *Am J Trop Med Hyg*. 2012;**86**:16–22.

Shepard CW, Simard EP, Finelli L, et al. Hepatitis B virus infection: epidemiology and vaccination. *Epidemiol Rev*. 2006;**28**:112–125.

Shi P, Keskinocak P, Swann JL, et al. The impact of mass gatherings and holiday traveling on the course of an influenza pandemic: a computational model. *BMC Public Health.* 2010 December 21;**10**:778.

Simon F, Savini H, Parola P. Chikungunya: a paradigm of emergence and globalization of vector-borne diseases. *Med Clin North Am.* 2008;**92**:1323–1343.

Smith KM, Anthony SJ, Switzer WM, et al. Zoonotic viruses associated with illegally imported wildlife products. *PLoS One.* 2012;**7**(1):e29505.

Stephens HA. HLA and other gene associations with dengue disease severity. *Curr Top Microbiol Immunol.* 2010;**338**:99–114.

Suijkerbuijk AW, Lindeboom R, van Steenbergen JE, et al. Effect of hepatitis A vaccination programs for migrant children on the incidence of hepatitis A in The Netherlands. *Eur J Public Health.* 2009;**19**:240–244.

Tatem AJ, Rogers, DJ, Hay SI. Global transport networks and infectious disease spread. *Adv Parasitol.* 2006;**62**:294–343.

The Center for Food Security and Public Health. Infectious salmon anemia. March 2010. Available from URL: http://www.cfsph.iastate.edu/Factsheets/pdfs/infectious_salmon_anemia.pdf. Accessed June 6, 2013.

Turell MJ, Dohm DJ, Mores CN, et al. Potential for North American mosquitoes to transmit Rift Valley fever virus. *J Am Mosq Control Assoc.* 2008;**24**:502–507.

UK Treasury. Stern review on the economics of climate change. 2005. Available from URL: http://webarchive.nationalarchives.gov.uk/±/http://www.hm-treasury.gov.uk/independent_reviews/stern_review_economics_climate_change/stern_review_report.cfm. Accessed January 25, 2012.

UNHCR. Global trends 2010. The Organization, Geneva. 2011. Available from URL: http://www.unhcr.org/4dfa11499.html. Accessed June 6, 2013.

United Nations, Department of Economic and Social Affairs, Population Division. World population 2010 (Wall Chart). ST/ESA/SER.A/307. 2011. Available from URL: http://esa.un.org/unpd/wpp/documentation/pdf/wpp2010_wallchart_text.pdf. Accessed January 20, 2012.

United Nations, Department of Economic and Social Affairs, Population Division. The age and sex of migrants 2011 (Wallchart). Available from URL: http://www.un.org/esa/population/publications/2011Migration_Chart/wallchart_2011.pdf. Accessed June 6, 2013.

United States Fish and Wildlife Service Office of Law Enforcement Intelligence Unit. US Wildlife Trade: An Overview for 1997 -2003, Washington, DC.

Vainio J, Cutts F. Yellow fever. Document WHO/EPI/GEN/98.11. Geneva: WHO; 1998. Available from URL: http://www.who.int/immunization_delivery/adc/yf/yellow_fever.pdf. Accessed June 6, 2013.

Verhoef L, Depoortere E, Boxman I, et al. Emergence of new norovirus variants on spring cruise ships and prediction of winter epidemics. *Emerg Infect Dis.* 2008;**14**:238–243.

Walker PJ, Winton JR. Emerging viral diseases of fish and shrimp. *Vet Res.* 2010;**41**:51.

Watts S. Yellow fever immunities in West Africa and the Americas in the age of slavery and beyond: a reappraisal. *J Soc Hist.* 2001;**34**:955–967.

WHO. The international health regulations 2005, 2nd ed. Geneva: World Health Organization; 2007. Available from URL: http://whqlibdoc.who.int/publications/2008/9789241580410_eng.pdf. Accessed February 5, 2012.

Wilder-Smith A. The severe acute respiratory syndrome: impact on travel and tourism. *Travel Med Infect Dis.* 2006;**4**:53–60.

Wilson ME. Population mobility and the geography of microbial threats. In: Apostolopoulos Y, Sonmez S, editors. Population Mobility and Infectious Diseases. New York: Springer; 2007.

World Tourism Organization. Press Release. International tourism to reach one billion in 2012. January 16, 2012. Madrid. Available from URL: http://media.unwto.org/en/press-release/2012-01-16/international-tourism-reach-one-billion-2012. Accessed June 6, 2013.

8

EFFECTS OF LAND-USE CHANGES AND AGRICULTURAL PRACTICES ON THE EMERGENCE AND REEMERGENCE OF HUMAN VIRAL DISEASES

Kimberly Fornace

Veterinary Epidemiology and Public Health Group, Royal Veterinary College, Hatfield, Hertfordshire, UK

Marco Liverani

Department of Global Health and Development, London School of Hygiene and Tropical Medicine, London, UK

Jonathan Rushton

Veterinary Epidemiology and Public Health Group, Royal Veterinary College, Hatfield, Hertfordshire, UK

Richard Coker

Communicable Diseases Policy Research Group (CDPRG), Department of Global Health and Development, London School of Hygiene and Tropical Medicine, London, UK
Faculty of Public Health, Mahidol University, Bangkok, Thailand
Saw Swee Hock School of Public Health, National University of Singapore, Singapore

TABLE OF CONTENTS

Viral Infections and Global Change, First Edition. Edited by Sunit K. Singh.

8.1 INTRODUCTION

Land use can be defined as the human management and modification of terrestrial surfaces, including changes to animal and plant populations, soil, surface waters, and topography (Turner et al., 2007). Current levels of land-use changes are unprecedented in history. While humans have transformed and managed natural landscapes since the beginning of agriculture 10 000 years ago, over the past decades changes have been particularly marked as a result of urban expansion, the construction of new roads and waterways, and extensive agricultural development. Today, between half and two-thirds of the earth's ice-free land surface has been transformed (Haberl et al., 2007), with 4.9 billion hectares or close to 38% of the total surface covered by agriculture alone (FAO, 2012).

These developments vary greatly across different regions of the world as land is transformed for agriculture or settled by human populations (Figure 8.1) (Ellis et al., 2010; Lambin et al., 2001). With the global population expected to reach nine billion by 2050, increasing levels of urbanization, higher levels of food consumption, and demand for natural resources and biomass-derived energy, the intensity of land use should be expected to rise as well (Haberl et al., 2007). Rapid transformations in land uses are likely to result in further changes to ecosystems and land fragmentation.

What are the implications of these changes for human health and the environment? On the positive impacts, agricultural development has contributed to socioeconomic growth and food security worldwide by providing cheaper and safer food. Improved agricultural productivity can additionally provide essential ecosystem services such as carbon sequestration, regulation of soil and water, and support of pollinating insects (Power, 2010).

However, concerns have been raised that land-use changes alter the ecological systems leading to environmental damage and contributing to novel hazards to animal and human health. Research has long documented the environmental consequences of land modification and agricultural development, including the depletion of soil nutrients, land degradation, contributions to climate change, disruption of fundamental ecosystems, and loss of biodiversity (Patz et al., 2000; Sherbinin, 2002). In the past few years, increasing attention has been directed to the effects on communicable diseases. Especially after the emergence of highly pathogenic diseases such as severe acute respiratory syndrome (SARS) and Nipah virus in areas undergoing rapid changes in land use, the link between agricultural practices and disease emergence has become more apparent (Patz et al., 2004; Weiss and McMichael, 2004; Wilcox and Gubler, 2005). Given the zoonotic nature of a majority of emerging diseases in humans (Woolhouse et al., 2005), there are concerns that changes in land use may facilitate the emergence of new ecological niches that microbes may exploit and the pathogen exchange between wild animals, livestock, and human populations. It is now

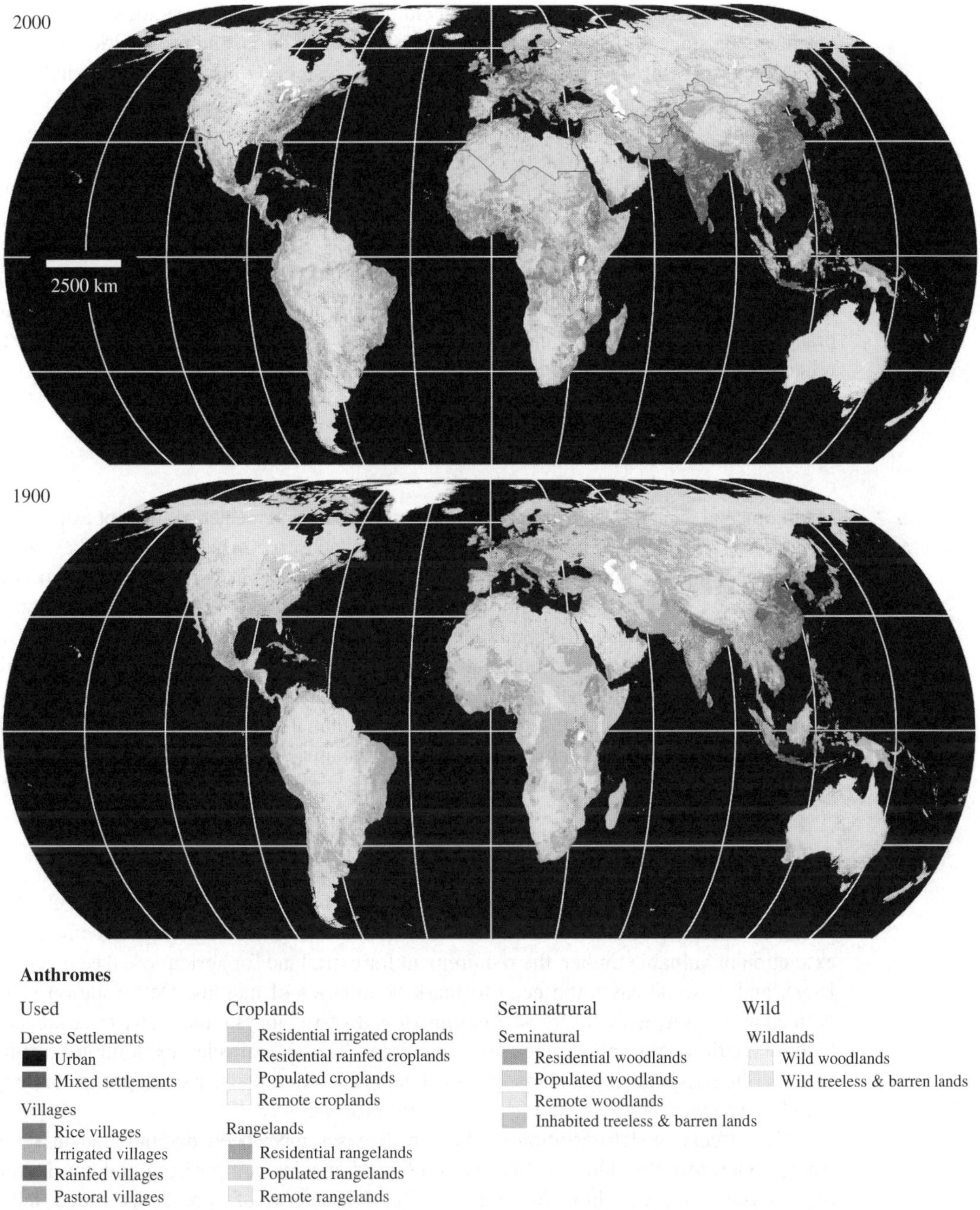

Figure 8.1. Anthropogenic biomes, 1900–2000, reproduced from Ellis et al. (2010). For color detail, please see color plate section.

clear that man-made interventions such as deforestation, urbanization, and agricultural expansion alter fundamental ecosystems and the geographical distribution of human populations and animal species, with mixed effects that may increase or reduce the risk of infectious disease emergence or reemergence. Land use influences in potentially complex ways microbial transmission dynamics, including the likelihood of pathogens progressing through each stage of disease emergence—exposure, infection, and propagation—exerting evolutionary pressures on pathogens, hosts, and ecosystems, while transmission in humans can be amplified by migration, trade, and societal changes.

Despite sustained research, however, the processes that link land-use change with disease emergence are still largely unknown. Disease emergence is not a simple causal effect; it is rather the result of complex multifactorial interactions, requiring pathogens to overcome numerous ecological and evolutionary barriers to switch hosts and establish in human populations. This chapter aims to shed some light on this complexity by reviewing recent scientific works at the interface between ecological and health sciences. Specifically, the chapter examines the role of changes to the environment and ecosystems; agricultural practices, including agricultural expansion and intensification of livestock production; and related demographic changes, including urbanization, trade, and migration.

8.2 ECOLOGICAL AND ENVIRONMENTAL CHANGES

Environmental changes, such as deforestation and habitat fragmentation, disrupt existing ecosystems and change the physical characteristics of landscape. This can affect the size and movement of human, animal, and pathogen populations and can increase the interface between different areas, bringing previously separated populations into contact and changing the population dynamics regulating community structures. Humans entering new environments can be exposed to new pathogens and may spread these pathogens to different communities. Landscape changes additionally modify the physical characteristics of the environment, determining the suitability of different areas as habitats for animal and insect populations and altering migration patterns and connectivity between different areas. These changes can transform the structure of ecological communities, affecting species richness, abundance, and composition.

8.2.1 Deforestation

Deforestation is one of the largest anthropogenic changes to land cover, with 2–3% of planetary forests lost each year (Patz et al., 2004). It is driven by various demands, the extraction of valuable timber, the requirement for extra land for agriculture (see discussion later), and in some cases the need to mark boundaries of nations. Deforestation can be defined as the conversion of forest areas into less diverse biosystems, such as pasture, cropland, plantations, transport routes, or urban areas. The uncontrolled exploitation of forest areas is often associated with large personal gains for individuals or small groups of people with large negative impacts and local disruption for many.

The effects of deforestation on human diseases have been documented in several studies. Deforestation alters habitat structure and species composition and can increase the interface between humans and wildlife. Disruption of forest cover and habitat fragmentation creates ecotones, transition areas between wild and man-made environments with increased contact between hosts, vectors, and pathogens (Lambin et al., 2010). The "edge effect" of these areas was identified as a driver of increased numbers of cases of human tick-borne encephalitis (TBE) cases in Latvia, where human activities and tick habitats overlapped (Vanwambeke et al., 2010). The emergence of yellow fever has also been associated with a forest ecotone. Yellow fever virus is maintained in monkey and mosquito species predominantly living in the deep forest canopy. Encroachment of human settlements and agricultural lands disrupted transmission cycles, infecting domestic *Aedes* mosquito species and exposing humans to the virus (Despommier et al., 2006). Comparable changes from jungle to sylvatic and domestic transmission cycles may have also contributed to the emergence of dengue and other arboviruses in the edges of the Amazon rainforest (Vasconcelos et al., 2001).

Human activity in forests, through logging or other practices associated with deforestation, can further contribute to zoonotic risk. While clear-cutting forests may reduce wildlife diversity, selective extraction of high-value timber species is more likely to sustain biodiversity and thus a wider pool of potential pathogens. Moreover, selective extraction can increase the contact rate between humans and pathogens in the wildlife, as workers venture into forest areas to select and collect wood (Wolfe, 2005). Initial human exposure to Marburg and Ebola viruses most likely occurred through contact with wildlife reservoirs while humans were active in forest environments (Cascio et al., 2011). Additionally, encroachment of agriculture and livestock into forest areas can contribute to disease spread by supporting vector or pathogen populations. Grazing cattle in forest areas in the Mysore State of India provided a source of blood meals for ticks carrying the previously undetected Kyasanur Forest disease virus. The associated increase of tick populations and human exposure led to the emergence of Kyasanur Forest disease in people living around forest areas (Simpson, 1978).

The construction of roads and waterways into forests for the transport of logs and other material can create further opportunities for disease spread into human communities. In some cases, new transport routes may stimulate an increase in deforestation and associated disease risk. Medeiros et al. found a dramatic increase in human hantaviral cases following the completion of a highway through the Brazilian Amazon. Traffic through the region provided new economic opportunities for the inhabitants of the area surrounding the highway, encouraging wood collection, agriculture, and other forest activities (Medeiros et al., 2010).

Deforestation has also been associated with risk resulting from bushmeat consumption; poor individuals living in proximity to wildlife are more likely to hunt and consume bushmeat as an alternative protein source (Brashares et al., 2011). With the identification of chimpanzees as the natural reservoir for HIV, there is strong evidence HIV originated from hunting and butchering nonhuman primates in central Africa (Keele et al., 2006). Bushmeat hunters and other people reporting contact with nonhuman primates have increased exposures to other simian viruses, such as simian foamy virus (Wolfe et al., 2004). Circulation of simian and human T-lymphotropic viruses (HTLV) between African hunters and nonhuman primate populations suggests bushmeat hunting could contribute to HTLV emergence (Wolfe et al., 2005).

8.2.2 Habitat fragmentation

Changes to the environment, whether through deforestation or other changes to land cover, alter the species abundance, distribution, and connectivity of habitats. Habitat fragmentation occurs when the habitat is spatially separated due to land changes, creating a mosaic pattern of different land types. Fragmentation can increase the risk of disease emergence by increasing the area of edges forming ecotones and increasing crowding and contact rates within the habitat. For example, increased density of birds in smaller forest fragments is correlated with higher tick infestation rates, likely due to the increased probability of ticks successfully finding a competent host (Ogrzewalska et al., 2011). However, habitat fragmentation may also decrease risk of disease emergence by isolating susceptible species from disease hosts (Vogeli et al., 2011). Geographical separation of habitats affects host migration patterns and use of land by different species.

The varied effects of habitat fragmentation on hantavirus distribution illustrate the complexity of ecological disturbances on disease dynamics. Phylogenetic studies indicate hantaviruses coevolved with their rodent hosts; geographical distribution of viral strains is determined primarily by the ecological niche and migration patterns of the primary rodent hosts (Wei et al., 2011). Changing land use affects the population size of rodent hosts in different ways depending on the biogeography of the host. In Central America, disturbed areas around the edges of natural habitats were correlated with an increased density of rodent hosts (Suzan et al., 2008). In contrast, land disturbances in North America altered rodent population structures, causing a high turnover, younger ages, and decreased serological evidence of exposure to the Sin Nombre hantavirus (Calisher et al., 2001). Other studies suggest hantavirus infection may not occur below a certain density of rodent hosts (Tersago et al., 2008).

Landscape and habitat fragmentation also affect the connectivity of different habitats. While increased Puumala hantavirus carriage was detected in European bank voles living in large connected habitats, geographically isolated and smaller habitats were associated with decreased hantavirus prevalence (Guivier et al., 2011). Overlapping spatial distribution of different rodent hosts plays a role in virus evolution through host switching and reassortment events. Related hantaviral strains were detected in different rodent species living in humid subtropical forests and temperate grasslands. Although these species live in very different ecological regions, anthropogenic changes to habitats led to transition areas used by previously geographically isolated species (Chu et al., 2006).

8.2.3 Structural changes in ecosystems

Structural changes, such as change in the numbers or behavior of different species, within ecosystems can produce further risk of disease emergence by changing species richness, abundance, and composition. For example, land use may affect the microclimatic variables, changing suitability of areas for vector habitats. Ross River virus emerged in Australia after clearing vegetation decreased transevaporation levels and increased water levels dissolved salt deposits leading to the formation of saline pools. The main vector of Ross River virus, the *Aedes camptorhynchus* mosquito, is salt tolerant and thrives in areas of dryland salinity. Wide-scale habitat changes led to conditions favoring this vector leading to an increase in human cases (Jardine et al., 2008).

In addition, anthropogenic land use and the modification of natural habitats negatively impact biodiversity. The effects of biodiversity on disease emergence and transmission are complex, altering the abundance, behavior, and condition of hosts, vectors, or pathogens (Keesing et al., 2010). In heterogeneous populations, the effect of biodiversity on disease dynamics is primarily determined by structural changes to ecological communities. Whether changing host community structure amplifies or reduces the risk of emergence depends on the disease competence of the new community structure and the transmission mode of the pathogen. Loss of species that are not disease hosts or are suboptimal disease hosts will not affect pathogen transmission rates. Additional species may provide sources of blood meals for vectors, increasing vector population and disease risk. Highly biologically diverse areas can increase the zoonotic pool of pathogens, acting as a source for further disease emergence events (Randolph and Dobson, 2012). If the amplifying host species declines as biodiversity decreases, there will be a reduction in disease risk. Conversely, if the amplifying host thrives as biodiversity decreases, disease risk will increase. Influences on different species and populations in multi-host and vector-borne disease systems may have opposing effects.

Protective effects of biodiversity were initially described for vector-borne parasitic diseases, such as malaria and Lyme disease. Higher densities of noncompetent hosts can either dilute disease effects by "wasting insect bites" or amplify disease risks by supporting populations. Biodiversity decreases risk when vectors preferentially feed on noncompetent hosts; increased species diversity of wild birds, the primary reservoirs of West Nile virus (WNV), was suggested to decrease mosquito infection risk by increasing numbers of suboptimal hosts. Lower virus amplification rates were found in areas with higher non-passerine wild bird species richness (Ezenwa et al., 2006). Further studies found the force of WNV infection, the number of infectious mosquitoes resulting from feeding on a host, depended on changes in host selection and reservoir competence rather than measures of avian diversity (Hamer et al., 2011). Presence of noncompetent hosts can alternatively amplify disease risk by increasing abundance of vector populations. For example, increased deer density is correlated with increased human tick-borne encephalitis virus (TBEV) cases; although deer cannot transmit the virus, increased deer abundance supports tick populations (Rizzoli et al., 2009). Changes in plant diversity can also alter the vector habitats, affecting insect population abundance and feeding behavior (Randolph and Dobson, 2012).

8.3 AGRICULTURAL CHANGE

Agriculture is the largest human use of land and one of the main drivers of ecological changes and environmental modification. Part of the deforestation described earlier relates to a demand for new agricultural land and is not solely about a drive to extract resources. Population growth and changing consumption patterns contribute to rising demand for agricultural products. Global demand for livestock products is predicted to increase from 6 to 23 kg per person per year by 2050, with demand per person more than doubling in sub-Saharan Africa (Thornton and Herrero, 2010). Global biofuel production is also growing rapidly and expected to quadruple in the next 15–20 years due to rising oil prices (Patz et al., 2008). Meeting this demand entails both expanding land use and increasing productivity levels through intensification. Agricultural intensification is characterized by selection for increasingly specialized breeds, mechanization, and use of technology and industrial management. Crop yields have been improved through wide-scale use of irrigation, fertilizers, and pesticides, while livestock industries have adopted higher stocking densities, the use of concentrated feed, and increased pharmaceutical product use. Intensification changes the relative importance and dynamics of diseases, and it can drive disease emergence. In addition by selecting for homogeneous host populations and increasing movement and trade, the impact of disease presence is increased. Biosecurity and husbandry practices affect contact rates with wild populations and management of waste and the environment. Disease control practices may decrease susceptibility to infection or alternatively select for new or more pathogenic infections.

8.3.1 Agricultural expansion

Spatial expansion of agriculture necessitates clearing and modifying lands for crop or livestock production. As discussed earlier, large-scale changes in land surface and deforestation may create favorable ecological conditions to disease emergence and transmission. In some cases, these changes can act in combination with technological interventions to produce high-risk environments. In Bolivia, for example, wide-scale deforestation to create croplands triggered an increase in the wild *Calomys* mouse population, which carried the Machupo virus and fed off agricultural waste. Applications of pesticides such as DDT killed cats, the main predators of mouse populations. As a result, the mouse population further expanded leading to an outbreak of Machupo hemorrhagic fever in humans exposed to mouse excretions. The virus killed close to 15% of the local population (McMichael, 2004; Simpson, 1978).

Expansion of crop production also involves new irrigation systems and construction of waterways, providing further opportunities for transmission of waterborne diseases. Contaminated irrigation water is the suspected source of norovirus, rotavirus, and hepatitis E viruses detected on strawberries grown for human consumption (Brassard et al., 2012). In addition, irrigation modifies surface waters and may create new breeding sites for arthropod vectors of viral diseases; research suggests that water management practices in rice paddies can significantly increase or reduce the number of mosquitoes carrying Japanese encephalitis virus (JEV) (Keiser et al., 2005). Other characteristics of rice paddies, such as the height of plants and dissolved oxygen and nitrogen levels, may also influence their suitability as mosquito-breeding sites (Sunish and Reuben, 2002). JEV is maintained in wild bird populations and can be further amplified by pigs; proximity of rice paddies to wild bird habitats and pig farming influences risk of mosquito viral carriage (Mackenzie and Williams, 2009). Of patients with confirmed cases of JE admitted to a hospital in Assam, India, over 78% reported working in rice cultivation and 55% reported close association with pigs (Phukan et al., 2004).

8.3.2 Intensification of livestock production

The growing intensification of livestock systems further influences the emergence and dynamics of zoonotic diseases. Animal production, particularly for pigs and poultry, has become increasingly intensified in both developed and developing countries. Large numbers of genetically homogeneous animals are kept at high densities in controlled environments under industrial management, frequently integrating all aspects of production from feed production to processing (FAO, 2006). Industrial farms may be easier to regulate, have improved levels of biosecurity, and expose fewer workers to disease risks. Also, industrial farms keep single species in isolation, reducing opportunities for viral recombination between species.

However, large populations of standardized animals can amplify risk of contact and pathogen transmission. Selection for productivity and high throughput also changes the population structure resulting in shorter lifespans and similar ages of farmed animals. Decreased diversity and close contact can facilitate establishment and amplification of pathogens, allowing rapid transmission through the population. Models of avian influenza in Thailand found that both backyard and industrial poultry contributed to outbreaks but industrial production systems were disproportionately more infectious (Walker et al., 2012). When industrial farms have contact with traditional backyard farms, industrial farms may amplify the effects of disease introductions. Additionally, metabolic stress from overcrowding and poor conditions on badly managed commercial farms can weaken animal immune systems, causing increased viral shedding and hyperinfection (Maillard and Sparagano, 2008).

Managing disease risks in industrialized farms necessitates biosecurity and disease control programs. For example, the swine industry has adopted regulations to limit proximity and contact between poultry and pig Concentrated Animal Feeding Operations (CAFOs), which would be difficult to enforce in traditional backyard farming systems (Gilchrist et al., 2007). Risk of epidemic disease outbreaks was speculated to be higher in traditional extensively managed farms due to limited biosecurity measures and frequent contact with wildlife. However, studies of highly pathogenic avian influenza (HPAI) have demonstrated the ability of pathogens to breach biosecurity measures of large commercial farms. In Thailand, for example, large industrial farms had higher relative risks of HPAI infection (Otte et al., 2007). In addition, effective surveillance and disease control programs can limit the spread of disease by isolating or culling infected animals or minimize the consequences of disease introduction, such as through vaccination and treatment. However, antimicrobial resistance due to indiscriminate or excessive agricultural use of antibiotics frequently leads to resistant strains of bacteria. Widespread antibiotic use can alter the host immunity by disrupting normal microbiota and symbiotic bacteria (Dethlefsen et al., 2007). Exposure to pesticides can also negatively affect immunity, increasing susceptibility to infection (Straube et al., 1999). In some cases, vaccination exerts selection pressure for more virulent pathogens, as seen with Marek's disease vaccine selecting for increasingly pathogenic strains of virus (Gimeno, 2008).

Intensified production systems can present additional risks of disease spread due to the need to manage the environment. Large intensive units require ventilation systems to regulate temperature and humidity. Viruses can survive in the dust generated and expelled by ventilation systems, as demonstrated by the airborne spread of influenza between geographically separated poultry units in Canada (Graham et al., 2008). Industrial farming also generates large amounts of animal waste in concentrated geographical areas. In the United States, over half of livestock farms produce more manure than can be feasibly spread on their land (Gollehon et al., 2001). Management of animal waste is poorly regulated in many countries, and its removal and disposal presents further opportunities for disease dissemination. Wastewater lagoons are commonly used to store swine wastes in the United States. Environmental contamination and pathogen spread can result from lagoons breaking or seeping into groundwater. Other methods of managing waste, such as through land application or spray fields, can distribute zoonotic viruses over a larger geographical area (Cole et al., 2000). Increase in zoonotic human hepatitis E cases has been linked to poor sanitation and waste management, as humans are most commonly infected through fecal–oral transmission and contamination of water supplies (Aggarwal and Naik, 2009).

Inputs into livestock systems, such as feed and water, present an additional route of transmission. Feeding livestock animal by-products can expose animals to new pathogens, as demonstrated by the exposure of cattle to contaminated meat and bone meal causing the bovine spongiform encephalopathy epidemic (Imran and Mahmood, 2011). Viruses can also be spread through animal by-products and may not always be activated by feed processing measures (Vinneras et al., 2012). In addition to introducing waterborne diseases, dependence of livestock on groundwater sources can increase risk of exposure to inorganic elements such as fluorine, bromine, arsenic, and lead. Exposure to different constituents can contaminate the food chain and negatively affect livestock susceptibility to disease (Meyer and Casey, 2012).

8.4 DEMOGRAPHIC CHANGES

While population growth and the need for resources drive changes in land use and expansion of agriculture, the availability of food and natural resources also supports population growth and increasing levels of urbanization. Expansion of urban environments can facilitate disease transmission by changing ecosystems, increasing population density, and changing human behavioral patterns. Large urban centers require the development of transport networks, facilitating the movement of people and goods from different areas of the world. Limited resources with high population densities can stress public health measures, further driving disease spread.

8.4.1 Urbanization

By 2050, over six billion people, close to 70% of the global population, are predicted to live in cities (FAO, 2012). Urban development in industrialized countries is usually correlated with improvements in socioeconomic status, and increased levels of urbanization are usually associated with increases in per capita income. In contrast, rapid urbanization in developing countries has caused an expansion of informal settlements leading to health disparities and creation of urban environments favoring disease transmission (Alirol et al., 2011). Of the 100 fastest-growing cities from 1950 to 2000, 75 were in Asia and Africa (Satterthwaite, 2007). In Mumbai, the most densely populated city, populations approach 30 000 people per square kilometer (World Bank, 2012). The overcrowding of people in urban spaces allows rapid transmission of communicable diseases. Human behavioral patterns also affect disease propagation, particularly for sexually transmitted diseases. Changing family structures and behavioral norms and reduced social capital can lead to unsafe sexual behavior and increase in addictions, including intravenous drug use. Analysis of data from national surveys in sub-Saharan Africa found that urban adolescents were more likely to have multiple sexual partners. On the other hand, the same analysis found that urban adolescents were more likely to use condoms due to increased access to sexual education and disease control programs (Doyle et al., 2012). Improvements in sanitation and disease control programs in urban settings can also have positive outcomes, as the reduction of hepatitis A cases in many low- and middle-income countries demonstrated (Tufenkeji, 2000).

In many other contexts, however, urban sanitation systems and infrastructure have struggled to keep up with the rapid population expansion, particularly in slums and low-income areas. The WHO estimates 600 million people in urban settings lack access to adequate sanitation systems and a further 167 million do not have access to safe drinking water (WHO, 2010). Raw sewage collected from cities in the United States, Ethiopia, and Spain was found to harbor over 230 known species of viruses. The high levels of viral diversity present opportunities for recombination and reassortment (Cantalupo et al., 2011). Poor sanitation and water systems increases transmission of waterborne diseases and may create new breeding sites for insect vectors.

The recent resurgence of dengue, a mosquito-borne flavivirus, exemplifies the complex relationships between urban development and infectious diseases. Dengue has reemerged as a major public health problem in urban and peri-urban areas of the tropical and subtropical countries. The spatial distribution of dengue has increased substantially, infecting 50 million people annually. Dengue is associated with urban areas where poor housing and lack of sanitation systems have increased the areas of vector-breeding sites. In the cities of Doula and Yaoundé in Cameroon, urban agricultural practices were shown to influence insecticide resistance levels in mosquitoes (Antonio-Nkondjio et al., 2011). Dengue outbreaks are frequently centered in large cities. For example, the epicenter of the dengue fever outbreak in Brazil was the densely populated city of Rio de Janeiro (Alirol et al., 2011). The level of urbanization was also shown to affect risk of dengue infection in Vientiane, Laos, with infection higher in recently developed urban areas around the periphery of the city center (Vallee et al., 2009). The high mobility of many urban dwellers contributes to the spread of pathogens between different areas of the city and between different cities. Large numbers of migrant workers from dengue-endemic areas have most likely led to the increase of dengue cases in the Middle East (Amarasinghe and Letson, 2012). Local movements and commuting can also introduce diseases into urban environments. Kubiak et al. modeled disease transmission between a small village and city, showing most communities are sufficiently connected to allow disease transmission between commuter towns and urban centers despite spatial segregation (Kubiak et al., 2010).

Urban expansion can also influence animal distribution, creating habitats favorable for some wildlife species. Wildlife such as skunks, foxes, raccoons, and feral dogs inhabit urban areas, scavenging off refuse. Rabies has become a major urban disease problem due to the adaption of mammalian hosts carrying the virus to city environments (Aguirre et al., 2000). Bat populations have also adapted to urban environments; Ebola and Lagos filoviruses were isolated from a bat captured in the urban area of Accra, Ghana, presenting a risk for infection of the human population (Hayman et al., 2010). Urbanization additionally leads to increases in rodent populations, presenting increased opportunities for transmission of rodent-borne diseases. For example, Lassa fever reemerged in West Africa following an explosion of the rodent populations, exacerbated by poor quality housing and hygiene practices allowing infected rodents to access food (Bonner et al., 2007).

Urbanization is also associated with increased levels of stress and pollution, potentially affecting the immunity of both animal reservoirs and people. Physiological responses to stress, such as production of the cortisol hormone, modulate the immune system and can increase susceptibility to disease. Wild birds were found to have higher stress indexes due to interspecific competition in urban environments (Bradley and Altizer, 2007). Exposure to pollutants can produce similar immunosuppressive effects. Ozone, one of the most abundant air pollutants in urban areas, induces oxidative stress, causing airway inflammation and increased susceptibility to respiratory diseases. High levels of ozone pollution were recorded during the emergence of H1N1 pandemic influenza in Mexico City in 2009, and ozone was subsequently shown to increase susceptibility to influenza viruses (Kesic et al., 2012). However, high levels of urban pollution may have opposing effects on some diseases by decreasing survival rates of disease vectors (Awolola et al., 2007).

8.5 LAND USE, DISEASE EMERGENCE, AND MULTIFACTORIAL CAUSATION

As we have seen, changes in land use have important effects on viral diseases, which can be observed throughout the process of disease emergence, from pathogen exposure to the stages of infection and transmission (Figure 8.2). Agricultural development leads to changes to environments and habitats, affecting both the size and contact rates of different populations. Practices such as deforestation and agriculture may create new pathways of pathogen exchange by keeping multiple species in close proximity, increasing opportunities for species jumps from wild and domestic animals to humans, as well as creating additional selection pressures for pathogen evolution. Land use can influence the biological characteristics of viral infections and therefore their chance to survive and spread. Effects of land use can also change the susceptibility of the individuals, such as through weakening immune systems by exposure to pollution or other pathogens. Finally, land uses can facilitate the movement of pathogen, vector, and hosts populations, increasing potential for disease propagation and transmission.

These changes in land use exert complex evolutionary pressures at different levels, from the pathogen and individual host to ecosystem levels. These pressures can both increase and decrease the risk of a disease passing through each stage of emergence and alter the risk of pathogen amplification and spread to new locations. Moreover, land

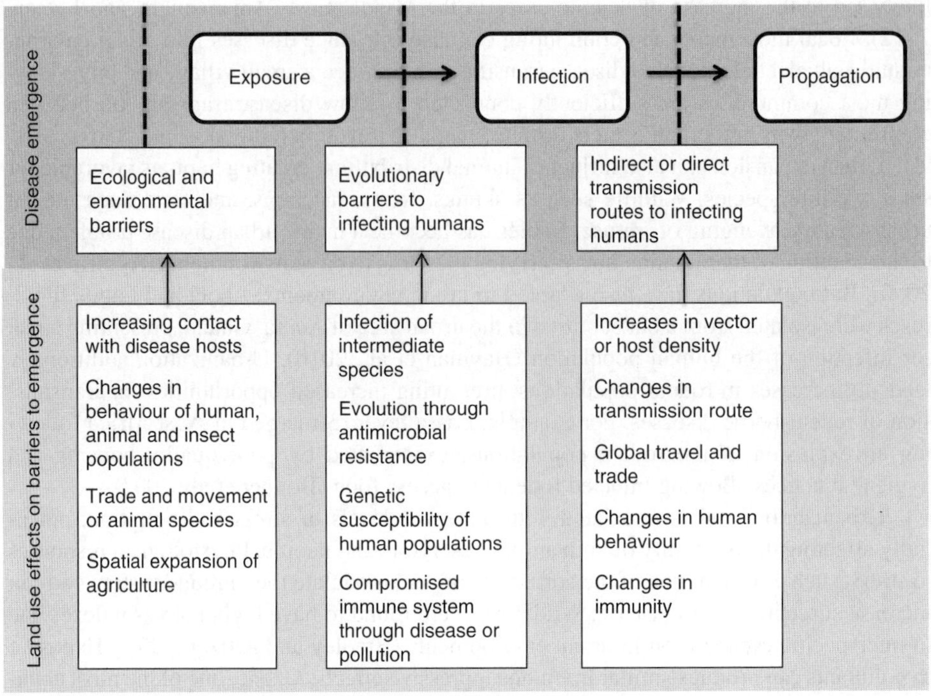

Figure 8.2. Effects of land-use changes on the process of disease emergence. For color detail, please see color plate section.

BOX 8.1 Multiple Effects of Land-Use Changes Causing Nipah Virus Emergence

The case of Nipah virus illustrates how different changes in land use and agricultural systems can interact to cause human disease. The first human outbreaks of Nipah virus, a virus harbored by fruit bats, occurred in Malaysia in late 1998 and 1999, causing over 100 human deaths and costing US$500 million (Field, 2009). The index case for these outbreaks was traced to a 30 000-unit intensive pig farm in the Malaysian forest, where pigs had contact with Nipah-infected fruit bats. The virus was probably passed to the pigs through urine and masticated pellets dropped when the bats fed in overhanging fruit trees. During the initial outbreaks, most human infections resulted from direct contacts with sick pigs or contaminated tissues.

Multiple biological, environmental, and social factors contributed to these complex pathways of disease transmission from wildlife to humans. Increased population growth and demand for food led to spatial expansion and intensification of agriculture, tripling the production of pigs and mangoes within peninsular Malaysia between 1970 and the late 1990s. Increases in pig and mango production were loosely correlated, with mango trees frequently planted in the proximity of commercial pig farms (Pulliam et al. 2012). Mangoes were also used as a food source by wild fruit bats, attracting bat populations to agricultural areas. Wide-scale deforestation and destruction of bat habitats in the region may have additionally contributed to the change in fruit bat migration patterns and their increasing use of lands surrounding mango trees [81]. Finally, high density of homogeneous and susceptible pig populations in intensive production farms facilitated viral persistence and amplification (Chua et al. 2002), while regional trade and pig movements further contributed to transmission to humans.

changes may act synergistically or independently on different disease systems, increasing risks of emergence for one pathogen while simultaneously decreasing risks for another pathogen. There is a need to account for the combined effects of multiple changes on the process of emergence (Box 8.1).

8.6 CONCLUSION

Anthropogenic changes in land use are driven by the need to extract resources and provide food, water, and housing. Rapid population growth combined with income increases also leads to changes in consumption patterns resulting in a rising demand for livestock products and resulting spatial expansion and intensification of agriculture. These positive aspects of land-use change need to be balanced against the environmental changes from both agricultural expansion and natural resource extraction, which if poorly regulated and managed disrupt ecosystems, affecting the composition, density, and behavior of human, animal, and insect populations. Moreover, as populations become increasingly urbanized, this expands trade and travel networks to supply food to cities and allow movement between different areas.

Given the complexity of ecological and evolutionary processes at work, research at the interface between land use, agricultural development, and infectious diseases needs to embrace a holistic approach, able to account for the multiple interactions contributing to produce risk environments in particular ecosystems.

Negative externalities generated from land uses, including disruption of essential ecosystem services and potential for disease emergence and reemergence, need to be valued and where appropriate regulated with a structured decision-making process. At the same time, the policy imperative to manage health hazards needs to be balanced against the many benefits of land uses and agricultural development. Assessing impacts of land-use decisions requires multidisciplinary approaches, evaluating effects on ecosystems, wild, and domestic populations and human health and well-being. Initiatives such as the Millennium Ecosystem Assessment have begun to quantify and value human health costs and benefits from ecosystems and land uses (MEA, 2005). Further research is needed to understand the complex set of drivers influencing each stage of disease emergence and how land planning and management can limit negative consequences of land use in ways appropriate to local contexts, taking into account the local environment, ecology, and socioeconomic development while conforming to international policy, standards and regulations, and surveillance systems.

REFERENCES

Aggarwal, R., Naik, S., 2009. Epidemiology of hepatitis E: current status. *J Gastroenterol Hepatol* **24**, 1484–1493.

Aguirre, A.A., Pearl, M.C., Patz, J., 2000. Urban expansion impacts on the health of ecosystems, wildlife and humans: panel contribution to the PERN Cyberseminar on Urban Spatial Expansion. Population Environment Research Network. Available at: http://www.populationenvironmentresearch.org/seminars112004.jsp. Accessed July 6, 2013.

Alirol, E., Getaz, L., Stoll, B., et al., 2011. Urbanisation and infectious diseases in a globalised world. *Lancet Infect Dis* **11**, 131–141.

Amarasinghe, A., Letson, G.W., 2012. Dengue in the Middle East: a neglected, emerging disease of importance. *Trans R Soc Trop Med Hyg* **106**, 1–2.

Antonio-Nkondjio, C., Fossog, B.T., Ndo, C., et al., 2011. Anopheles gambiae distribution and insecticide resistance in the cities of Douala and Yaounde (Cameroon): influence of urban agriculture and pollution. *Malar J* **10**, 154.

Awolola, T.S., Oduola, A.O., Obansa, J.B., et al., 2007. Anopheles gambiae s.s. breeding in polluted water bodies in urban Lagos, southwestern Nigeria. *J Vector Borne Dis* **44**, 241–244.

Bonner, P.C., Schmidt, W.P., Belmain, S.R., et al., 2007. Poor housing quality increases risk of rodent infestation and Lassa fever in refugee camps of Sierra Leone. *Am J Trop Med Hyg* **77**, 169–175.

Bradley, C.A., Altizer, S., 2007. Urbanization and the ecology of wildlife diseases. *Trends Ecol Evol* **22**, 95–102.

Brashares, J.S., Golden, C.D., Weinbaum, K.Z., et al., 2011. Economic and geographic drivers of wildlife consumption in rural Africa. *Proc Natl Acad Sci USA* **108**, 13931–13936.

Brassard, J., Gagne, M.J., Genereux, M., et al., 2012. Detection of human food-borne and zoonotic viruses on irrigated, field-grown strawberries. *Appl Environ Microbiol* **78**, 3763–3766.

Calisher, C.H., Mills, J.N., Sweeney, W.P., et al., 2001. Do unusual site-specific population dynamics of rodent reservoirs provide clues to the natural history of hantaviruses? *J Wildl Dis* **37**, 280–288.

Cantalupo, P.G., Calgua, B., Zhao, G., et al., 2011. Raw sewage harbors diverse viral populations. *mBio* **2**, 1–11.

Cascio, A., Bosilkovski, M., Rodriguez-Morales, A.J., et al., 2011. The socio-ecology of zoonotic infections. *Clin Microbiol Infect* **17**, 336–342.

Chu, Y.K., Milligan, B., Owen, R.D., et al., 2006. Phylogenetic and geographical relationships of hantavirus strains in eastern and western Paraguay. *Am J Trop Med Hyg* **75**, 1127–1134.

Chua, K.B., Chua, B.H., Wang, C.W., 2002. Anthropogenic deforestation, El Nino and the emergence of Nipah virus in Malaysia. *Malays J Pathol* **24**(1), 15–21.

Cole, D., Todd, L., Wing, S., 2000. Concentrated swine feeding operations and public health: a review of occupational and community health effects. *Environ Health Perspect* **108**, 685–699.

Despommier, D., Ellis, B.R., Wilcox, B.A., 2006. The role of ecotones in emerging infectious diseases. *EcoHealth* **3**, 281–289.

Dethlefsen, L., McFall-Ngai, M., Relman, D.A., 2007. An ecological and evolutionary perspective on human–microbe mutualism and disease. *Nature* **449**, 811–818.

Doyle, A.M., Mavedzenge, S.N., Plummer, M.L., et al., 2012. The sexual behaviour of adolescents in sub-Saharan Africa: patterns and trends from national surveys. *Trop Med Int Health* **17**(7), 796–807. Available at: http://www.ncbi.nlm.nih.gov/pubmed/22594660. Accessed July 6, 2013.

Ellis, E.C., Goldewijk, K.K., Siebert, S., et al., 2010. Anthropogenic transformation of the biomes, 1700 to 2000. *Global Ecol Biogeogr* **19**, 589–606.

Ezenwa, V.O., Godsey, M.S., King, R.J., et al., 2006. Avian diversity and West Nile virus: testing associations between biodiversity and infectious disease risk. *Proc Biol Sci* **273**, 109–117.

FAO, 2006. World agriculture: towards 2030/2050. Interim report. Global Perspective Studies Unit, Food and Agriculture Organization of the United Nations, Rome.

FAO, 2012. FAOSTAT. Food and Agriculture Organization of the United Nations. Rome Available at: http://faostat3.fao.org/home/index.html. Accessed July 6, 2013.

Field, H.E., 2009. Bats and emerging zoonoses: henipaviruses and SARS. *Zoonoses Public Health* **56**, 278–284.

Gilchrist, M.J., Greko, C., Wallinga, D.B., et al., 2007. The potential role of concentrated animal feeding operations in infectious disease epidemics and antibiotic resistance. *Environ Health Perspect* **115**, 313–316.

Gimeno, I.M., 2008. Marek's disease vaccines: a solution for today but a worry for tomorrow? *Vaccine* **26**(Suppl. 3), C31–C41.

Gollehon, N., Caswell, M., Ribaudo, M., et al., 2001. Confined Animal Production and Manure Nutrients. Agriculture Information Bulletin. US Department of Agriculture, Washington, DC.

Graham, J.P., Leibler, J.H., Price, L.B., et al., 2008. The animal–human interface and infectious disease in industrial food animal production: rethinking biosecurity and biocontainment. *Public Health Rep* **123**, 282–299.

Guivier, E., Galan, M., Chaval, Y., et al., 2011. Landscape genetics highlights the role of bank vole metapopulation dynamics in the epidemiology of Puumala hantavirus. *Mol Ecol* **20**, 3569–3583.

Haberl, H., Erb, K.H., Krausmann, F., et al., 2007. Quantifying and mapping the human appropriation of net primary production in earth's terrestrial ecosystems. *Proc Natl Acad Sci USA* **104**, 12942–12947.

Hamer, G.L., Chaves, L.F., Anderson, T.K., et al., 2011. Fine-scale variation in vector host use and force of infection drive localized patterns of West Nile virus transmission. *PloS One* **6**, e23767.

Hayman, D.T., Emmerich, P., Yu, M., et al., 2010. Long-term survival of an urban fruit bat seropositive for Ebola and Lagos bat viruses. *PloS One* **5**, e11978.

Imran, M., Mahmood, S., 2011. An overview of animal prion diseases. *Virol J* **8**, 493.

Jardine, A., Lindsay, M.D., Johansen, et al., 2008. Impact of dryland salinity on population dynamics of vector mosquitoes (Diptera: Culicidae) of Ross River virus in inland areas of southwestern Western Australia. *J Med Entomol* **45**, 1011–1022.

Keele, B.F., Van Heuverswyn, F., Li, Y., et al., 2006. Chimpanzee reservoirs of pandemic and nonpandemic HIV-1. *Science* **313**, 523–526.

Keesing, F., Belden, L.K., Daszak, P., et al., 2010. Impacts of biodiversity on the emergence and transmission of infectious diseases. *Nature* **468**, 647–652.

Keiser, J., Maltese, M.F., Erlanger, T.E., et al., 2005. Effect of irrigated rice agriculture on Japanese encephalitis, including challenges and opportunities for integrated vector management. *Acta Trop* **95**, 40–57.

Kesic, M.J., Meyer, M., Bauer, R., et al., 2012. Exposure to ozone modulates human airway protease/antiprotease balance contributing to increased influenza A infection. *PloS One* **7**, e35108.

Kubiak, R.J., Arinaminpathy, N., McLean, A.R., 2010. Insights into the evolution and emergence of a novel infectious disease. *PLoS Comput Biol* **6**, e1000947. Available at: http://www.ncbi.nlm.nih.gov/pubmed/20941384. Accessed July 6, 2013.

Lambin, E.F., Turner, B.L., Geist, H.J., et al., 2001. The causes of land-use and land-cover change: moving beyond the myths. *Global Environ Chang* **11**, 261–269.

Lambin, E.F., Tran, A., Vanwambeke, S.O., et al., 2010. Pathogenic landscapes: interactions between land, people, disease vectors, and their animal hosts. *Int J Health Geogr* **9**, 54.

Mackenzie, J.S., Williams, D.T., 2009. The zoonotic flaviviruses of southern, south-eastern and eastern Asia, and Australasia: the potential for emergent viruses. *Zoonoses Public Health* **56**, 338–356.

Maillard, J.C., Sparagano, O.A., 2008. Animal biodiversity and emerging diseases prediction and prevention. Introduction. *Ann N Y Acad Sci* **1149**, xvii–xix.

McMichael, A.J., 2004. Environmental and social influences on emerging infectious diseases: past, present and future. *Philos Trans R Soc Lond B—Biol Sci* **359**, 1049–1058.

MEA, 2005. Ecosystems and Human Well-Being, Millennium Ecosystem Assessment. World Health Organization, Geneva.

Medeiros, D.B., da Rosa, E.S., Marques, et al., 2010. Circulation of hantaviruses in the influence area of the Cuiaba-Santarem Highway. *Mem Inst Oswaldo Cruz* **105**, 665–671.

Meyer, J.A., Casey, N.H., 2012. Establishing risk assessment on water quality for livestock. *Anim Front* **2**, 44–49.

Ogrzewalska, M., Uezu, A., Jenkins, C.N., et al., 2011. Effect of forest fragmentation on tick infestations of birds and tick infection rates by rickettsia in the Atlantic forest of Brazil. *EcoHealth* **8**, 320–331.

Otte, J.M., Pfeiffer, D., Tiensin, T., et al., 2007. Highly pathogenic avian influenza risk, biosecurity and smallholder adversity. *Livest Res Rural Dev* **19**. Available at: http://www.lrrd.org/lrrd19/7/otte19102.htm. Accessed July 6, 2013.

Patz, J.A., Graczyk, T.K., Geller, N., et al., 2000. Effects of environmental change on emerging parasitic diseases. *Int J Parasitol* **30**, 1395–1405.

Patz, J.A., Daszak, P., Tabor, G.M., et al., 2004. Unhealthy landscapes: policy recommendations on land use change and infectious disease emergence. *Environ Health Perspect* **112**, 1092–1098.

Patz, J.A., Olson, S.H., Uejio, C.K., et al., 2008. Disease emergence from global climate and land use change. *Med Clin North Am* **92**, 1473–1491, xii.

Phukan, A.C., Borah, P.K., Mahanta, J., 2004. Japanese encephalitis in Assam, northeast India. *Southeast Asian J Trop Med Public Health* **35**, 618–622.

Power, A.G., 2010. Ecosystem services and agriculture: tradeoffs and synergies. *Philos Trans R Soc Lond B—Biol Sci* **365**, 2959–2971.

Pulliam, J.R.C., Epstein, J.H., Dushoff, J., et al., 2012. Agricultural intensification, priming for persistence and the emergence of Nipah virus: a lethal bat-borne zoonosis. J R Soc Interface 9, 89–101.

Randolph, S.E., Dobson, A.D., 2012. Pangloss revisited: a critique of the dilution effect and the biodiversity-buffers-disease paradigm. *Parasitology* **139**, 847–863.

Rizzoli, A., Hauffe, H.C., Tagliapietra, V., et al., 2009. Forest structure and roe deer abundance predict tick-borne encephalitis risk in Italy. *PloS One* **4**, e4336.

Satterthwaite, D., 2007. The transition to a predominantly urban world and its underpinnings. Human Settlements Discussion Paper Series: Theme: Urban Change- 4. International Institute for Environment and Development. London.

Sherbinin, A.D., 2002. A Guide to Land-Use and Land-Cover Change (LUCC). Columbia University, New York.

Simpson, D.I., 1978. Viral haemorrhagic fevers of man. *Bull World Health Organ* **56**, 819–832.

Straube, E., Straube, W., Kruger, E., et al., 1999. Disruption of male sex hormones with regard to pesticides: pathophysiological and regulatory aspects. *Toxicol Lett* **107**, 225–231.

Sunish, I.P., Reuben, R., 2002. Factors influencing the abundance of Japanese encephalitis vectors in ricefields in India—II. Biotic. *Med Vet Entomol* **16**, 1–9.

Suzan, G., Marce, E., Giermakowski, J.T., et al., 2008. The effect of habitat fragmentation and species diversity loss on hantavirus prevalence in Panama. *Ann N Y Acad Sci* **1149**, 80–83.

Tersago, K., Schreurs, A., Linard, C., et al., 2008. Population, environmental, and community effects on local bank vole (Myodes glareolus) Puumala virus infection in an area with low human incidence. *Vector Borne Zoonotic Dis* **8**, 235–244.

Thornton, P.K., Herrero, M., 2010. The inter-linkages between rapid growth in livestock production, climate change, and the impacts on water resources, land use and deforestation. Policy Research Working Paper. World Bank, Washington, DC.

Tufenkeji, H., 2000. Hepatitis A shifting epidemiology in the Middle East and Africa. *Vaccine* **18**(Suppl. 1), S65–S67.

Turner, II, B.L., Lambin, E.F., Reenberg, A., 2007. The emergence of land change science for global environmental change and sustainability. *Proc Natl Acad Sci USA* **104**, 20666–20671.

Vallee, J., Dubot-Peres, A., Ounaphom, P., et al., 2009. Spatial distribution and risk factors of dengue and Japanese encephalitis virus infection in urban settings: the case of Vientiane, Lao PDR. *Trop Med Int Health* **14**, 1134–1142.

Vanwambeke, S.O., Sumilo, D., Bormane, A., et al., 2010. Landscape predictors of tick-borne encephalitis in Latvia: land cover, land use, and land ownership. *Vector Borne Zoonotic Dis* **10**, 497–506.

Vasconcelos, P.F., Travassos da Rosa, A.P., Rodrigues, S.G., et al., 2001. Inadequate management of natural ecosystem in the Brazilian Amazon region results in the emergence and reemergence of arboviruses. *Cad Saude Publica* **17**(Suppl.), 155–164.

Vinneras, B., Samuelson, A., Emmoth, E., et al., 2012. Biosecurity aspects and pathogen inactivation in acidified high risk animal by-products. *J Environ Sci Health A Tox Hazard Subst Environ Eng* **47**, 1166–1172.

Vogeli, M., Lemus, J.A., Serrano, D., et al., 2011. An island paradigm on the mainland: host population fragmentation impairs the community of avian pathogens. *Proc Biol Sci* **278**, 2668–2676.

Walker, P., Cauchemez, S., Hartemink, N., et al., 2012. Outbreaks of H5N1 in poultry in Thailand: the relative role of poultry production types in sustaining transmission and the impact of active surveillance in control. *J R Soc Interface* **9**(73), 1836–1845. Available at http://www.ncbi.nlm.nih.gov/pubmed/22356818. Accessed July 6, 2013.

Wei, L., Qian, Q., Wang, Z.Q., et al., 2011. Using geographic information system-based ecologic niche models to forecast the risk of hantavirus infection in Shandong Province, China. *Am J Trop Med Hyg* **84**, 497–503.

Weiss, R.A., McMichael, A.J., 2004. Social and environmental risk factors in the emergence of infectious diseases. *Nat Med* **10**, S70–S76.

WHO, 2010. Progress on Sanitation and Drinking Water 2010 Update. World Health Organization, Geneva.

Wilcox, B.A., Gubler, D.J., 2005. Disease ecology and the global emergence of zoonotic pathogens. *Environ Health Prev Med* **10**, 263–272.

Wolfe, N.D., 2005. Bushmeat hunting, deforestation, and prediction of zoonotic disease emergence. *Emerg Infect Dis* **11**, 1822–1827.

Wolfe, N.D., Switzer, W.M., Carr, J.K., et al., 2004. Naturally acquired simian retrovirus infections in central African hunters. *Lancet* **363**, 932–937.

Wolfe, N.D., Heneine, W., Carr, J.K., et al., 2005. Emergence of unique primate T-lymphotropic viruses among central African bushmeat hunters. *Proc Natl Acad Sci USA* **102**, 7994–7999.

Woolhouse, M.E., Haydon, D.T., Antia, R., 2005. Emerging pathogens: the epidemiology and evolution of species jumps. *Trends Ecol Evol* **20**, 238–244.

World Bank, 2012. World Bank Data. World Bank, Washington, DC.

9

ANIMAL MIGRATION AND RISK OF SPREAD OF VIRAL INFECTIONS

Diann J. Prosser and Jessica Nagel

U.S. Geological Survey, Patuxent Wildlife Research Center, Beltsville, MD, USA

John Y. Takekawa

U.S. Geological Survey, Western Ecological Research Center, Vallejo, CA, USA

TABLE OF CONTENTS

Viral Infections and Global Change, First Edition. Edited by Sunit K. Singh.

9.1 INTRODUCTION

9.1.1 Animal migration and disease

Tens of thousands of species of birds, mammals, fish, herpetofauna, and insects undergo migrational movements each year (Aidley, 1981) (Figure 9.1). These long- and short-distance movements are stimulated by seasonal changes in food availability, temperature, precipitation, day length, or environmental conditions such as overcrowding (Dingle, 1996; Rankin and Burchsted, 1992). Distances moved over migration range from local (<1 km for salamanders) (Schmidt et al., 2007) to annual circumpolar movements of up to 70 000 km by the Arctic Tern, *Sterna paradisaea* (Egevang et al., 2010).

The extent to which animals can migrate with and disperse viruses over long distances depends on many factors related to host–pathogen relationships, the majority being understudied (Altizer et al., 2011). In recent decades, the majority of emerging infectious diseases that pose a significant risk to human health have originated from animals (known as zoonoses) (Jones et al., 2008). A significant fraction of these zoonoses in humans are derived from viruses (Taylor et al., 2001). Because these viral pathogens can have significant impacts on public health and economies worldwide (Meslin, 2008), monitoring the origin and spread of these diseases has become increasingly important. In addition, understanding how animals contribute to the long-distance dispersal of viral zoonoses can be difficult to assess; however, it is critical in controlling and preventing diseases for humans and wild species alike.

9.2 DOES ANIMAL MIGRATION INCREASE RISK OF VIRAL SPREAD?

There are several possible mechanisms by which animal migration could lead to long-distance dispersal of viral diseases: (i) a migratory animal may become infected prior to migration and spread virus along its migration route if it can withstand the rigors of migration while infected and if sufficient viremia (presence of virus in the bloodstream or tissues) is maintained; (ii) a latent infection may become active during or following migration, upon which viral material is shed; or (iii) an intermediary process may occur whereby a migratory animal directly transports infected arthropod vectors to new locations or indirectly by infecting resident vectors following a blood meal from an infected migrant. Migratory animals represent a mechanism for annual reintroduction of viruses to vectors along migration route. This is especially important in regions where transmission is interrupted by cooler seasons and where overwintering of the virus through vertical vector transmission (i.e., transmission from adult vectors to future generations of vector) does not occur.

The potential contribution of migration towards the spread of disease is as varied as the ecology of the pathogens themselves and their host populations (Table 9.1). Evidence in the literature indicates both the geographic expansion of diseases and the reduction of viral prevalence due to animal movements (Altizer et al., 2011). A general hypothesis supporting virus spread by animal migration is that migrant populations may be exposed to more pathogens than resident populations due to greater exposure to individuals outside the population and new habitats at stopover locations during migration. Alternatively, hypotheses of reduced viral spread and prevalence include concepts of "migratory escape" and "migratory culling" (Altizer et al., 2011; Loehle, 1995), whereby migratory populations escape high levels of pathogens by migrating

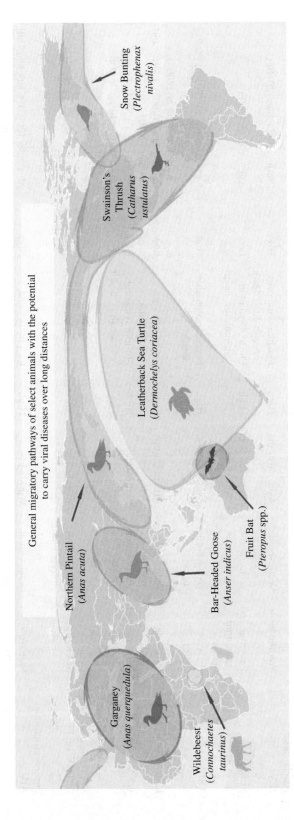

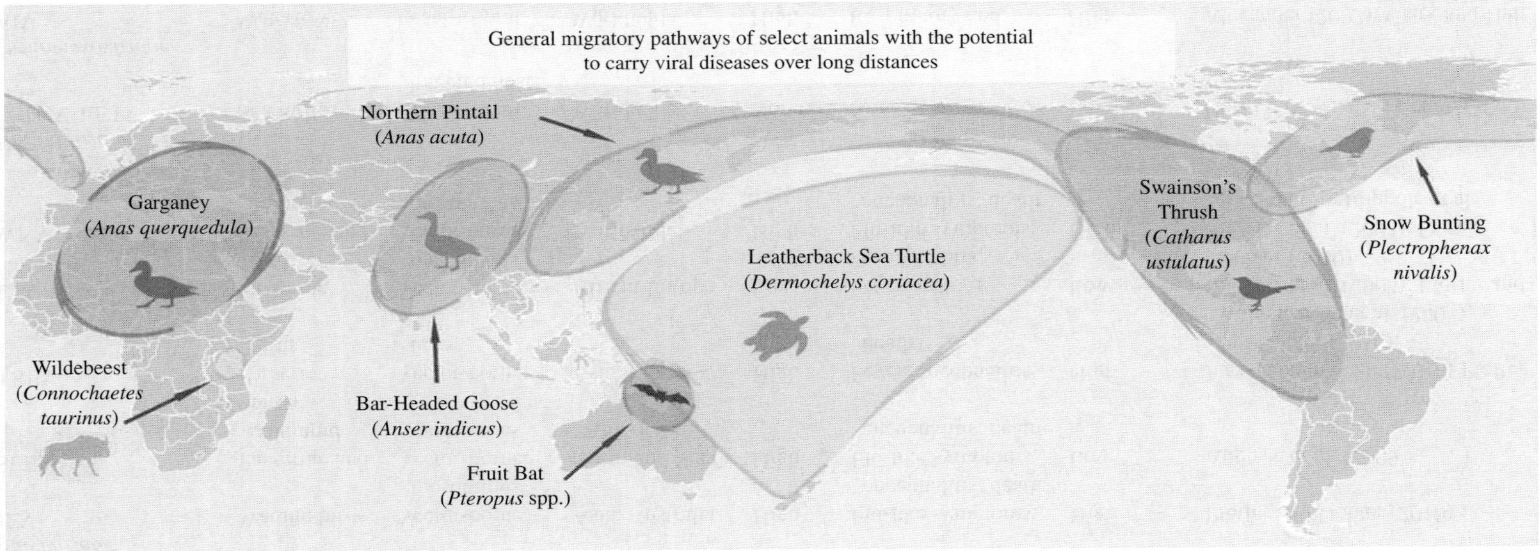

Figure 9.1. Examples of general migratory pathways for select animals that may contribute to the dispersal of viral diseases over long distances. Migratory routes for each species mentioned earlier were obtained and modified from the following sources: garganey (Gaidet et al., 2008), wildebeest (Serneels and Lambin, 2001), bar-headed goose (Takekawa et al., 2009), northern pintail (Miller et al., 2005; Yamaguchi et al., 2010), fruit bat (Breed et al., 2010), leatherback sea turtle (Benson et al., 2011), Swainson's thrush (Mack and Yong, 2000), and snow bunting (Lyngs, 2003). Pathways shown are examples of reported pathways from the literature but do not necessarily represent all known pathways for each species. For color detail, please see color plate section.

TABLE 9.1. Viruses with the Potential to Be Dispersed by Migratory Hosts. Mode of transmission, distribution in wildlife, risks to human health and agriculture and economy, potential effects on human health, and references relevant to migratory hosts are also included

Virus	Migratory Host	Mode of Transmission	Distribution in Wildlife	Risk to Human Health	Effects on human health	Risk to agriculture or economy	References
Coronaviridae							
SARS-like coronaviruses	Bats	Contact with infected host	Africa; Asia; Europe	High	Respiratory and flu-like symptoms; death	High	Balboni et al. (2012); Li et al. (2005)
Filoviridae							
Ebola virus	Pteropid bats	Contact with infected host	Africa	High	Fever and flu-like symptoms; hemorrhaging; death	High	Leroy et al. (2009)
Flaviviridae							
JEV	Wading birds	Vector-borne; mosquitoes	Asia; Australia	High	Flu-like symptoms; encephalitis; death	High	Pfeffer and Dobler (2010)
SLEV	Passerine and columbid birds	Vector-borne; mosquitoes	North and South America	High	Flu-like symptoms; encephalitis; death	Low	Auguste et al. (2009)
TBEV	Passerine birds	Vector-borne; ticks	Asia; Europe	High	Fever; encephalitis; death	High	Jaaskelainen et al. (2010); Pfeffer and Dobler (2010); Waldenstrom et al. (2007)
USUV	Passerine birds	Vector-borne; mosquitoes	Africa; Europe	Low	Fever; rash; encephalitis	Low	Buckley et al. (2003); Pfeffer and Dobler (2010)
WNV	Passerine and columbid birds	Vector-borne; mosquitoes	Worldwide	High	Flu-like symptoms; encephalitis; death	High	Buckley et al. (2003); Dusek et al. (2009); Rappole et al. (2000)
Herpesviridae							
FPTHV, LETV	Sea turtles	Contact with infected host	Worldwide	None	None	Low	Herbst (1994), Curry et al. (2000)
Orthomyxoviridae							
AIV	Waterfowl and passerine birds	Fecal-to-oral route; contact with infected host	Worldwide	High	Respiratory and flu-like symptoms; death	High	Alexander (2007); Clark and Hall (2006); Fouchier et al. (2007); Webby and Webster (2001)

Paramyxoviridae						
Distemper viruses	Mammals	Contact with infected host	Worldwide	None	None	Low	Hall et al. (2006); Harkonen et al. (2006)
Hendra virus	Pteropid bats	Contact with infected host	Australia	High	Respiratory and flu-like symptoms; encephalitis; death	High	Breed et al. (2006); Drexler et al. (2012); Plowright et al. (2011)
Nipah virus	Pteropid bats	Contact with infected host	Asia	High	Respiratory and flu-like symptoms; encephalitis; death	High	Breed et al. (2010); Calisher et al. (2006); Drexler et al. (2012)
NDV	Birds	Contact with infected host	Worldwide	Low	Conjunctivitis; flu-like symptoms	High	Hubalek (2004); Muller et al. (1999)
Picornaviridae						
FMD virus	Bovine mammals	Contact with infected host	Worldwide	Low	Fever; malaise; lesions	High	Thomson et al. (2003)
Rhabdoviridae						
IHNV	Salmonid fish	Contact with infected host	Asia; Europe; North America	None	None	Low	Amend (1975); Kurath et al. (2003)
Rabies	Bats	Contact with infected host	Africa; Asia; Europe; North and South America	High	Neurological and flu-like symptoms; death	Low	Calisher et al. (2006)
Togaviridae						
EEEV	Passerine and wading birds	Vector-borne; mosquitoes	North and South America	Low	Flu-like symptoms; encephalitis; death	Low	Hubalek (2004); Pfeffer and Dobler (2010)
SINV	Passerine birds	Vector-borne; mosquitoes	Africa; Asia; Australia; Europe	Low	Fever; rash; arthralgia	Low	Buckley et al. (2003); Kurkela et al. (2008); Sammels et al. (1999)
VEEV	Birds	Vector-borne; mosquitoes	North and South America	High	Flu-like symptoms; encephalitis; death	High	Hubalek (2004); Pfeffer and Dobler (2010)
WEEV	Passerine and columbid birds	Vector-borne; mosquitoes	North and South America	Low	Flu-like symptoms; encephalitis; death	Low	Hubalek (2004); Pfeffer and Dobler (2010)

away from diseased areas during peak pathogen seasons and infected individuals are culled from the population during the migration process.

Although quantifying the effects of migration on disease prevalence can be difficult to measure, multiple examples supporting these hypotheses exist. For example, the Delaware Bay along the east coast of the United States is a hotspot stopover location for more than a million shorebirds (Scolopacidae family) migrating from wintering sites in South America to their Arctic breeding grounds in late spring. High concentrations of breeding horseshoe crabs (*Limulus polyphemus*) attract shorebird species including red knots (*Calidris canutus*), ruddy turnstones (*Arenaria interpres*), sanderlings (*Calidris alba*), semipalmated sandpipers (*Calidris pusilla*), and dunlin (*Calidris alpina*) where these birds nearly double their body weight within a 2-week period in May by feeding on billions of horseshoe crab eggs (Mizrahi et al., 2009). Long-term monitoring of avian influenza viruses (AIVs) at this critical stopover location indicated the highest prevalence rates of AIVs globally (Krauss et al., 2010). Birds at this site had 17 times greater risk than all other surveillance sites worldwide (5.2% from 490 isolates/9474 samples at Delaware Bay vs. 0.3% from 49 isolates/15 848 samples worldwide), providing a high-risk region for viral exposure among long-distance migrating populations.

An example of phocine distemper virus (PDV) in harbor seals (*Phoca vitulina*) in Europe also shows how migratory animals may increase the spread of viral diseases. In two distinct epidemics in 1988 and 2002, outbreaks in harbor seals began on an island off the coast of Denmark and subsequently spread to new sedentary populations of harbor seals along the European coasts. Retrospective analysis of the timing and spatial movements of the disease indicated that a migratory species, the asymptomatic grey seal (*Halichoerus grypus*), was likely the vector that spread virus among the nonmigratory harbor seal populations (Harkonen et al., 2006). In total, more than 50 000 seals died between the two epidemics.

On the great plains of the open Serengeti in eastern Africa, wildebeest (*Connochaetes* spp.) have undergone long migrations following the rains across the seasons (Dingle, 1996). In the early twentieth century, rinderpest virus (also known as cattle plague) reached Africa through links with the cattle system in Italy, causing devastating effects on domestic stock and wildlife alike (Dobson, 1995). The wildebeest population was severely reduced from an estimated 2 million to fewer than 100 000. In this situation, an example of migratory culling could be considered if the prevalence of rinderpest in wildebeest was reduced due to the removal of infected individuals from the population; alternatively, migratory escape could be considered if healthy individuals were able to escape infection by moving to disease-free regions during their migration cycle. In 2011, following the loss of millions of domestic and wild ungulates, and a decades-long global battle against the virus, rinderpest became the first animal disease to be globally eradicated (Food and Agriculture Organization of the United States and World Organization for Animal Health, 2011).

Several studies have shown that viral infection can lead to compromised migration, thereby reducing the risk of long-distance virus transmission. For example, van Gils et al. (2007) have shown that wild migratory Bewick's swans (*Cygnus columbianus bewickii*) infected with low pathogenic avian influenza (LPAI) delayed their onset of migration by more than a month, traveled shorter distances, and exhibited reduced feeding rates in comparison to noninfected individuals. However, another study that examined resighting rates of greater white-fronted geese found that AIV infection was not related to resighting probability nor to maximum distance traveled (191 km) (Kleijn et al., 2010), showing that pathogenicity can vary greatly depending on the viral strain and host species.

In the following text, we outline multiple examples of viral diseases in animal populations and their mechanisms of viral spread. Although many species of insects, mammals, fish, and birds exhibit migratory behavior and have the potential to disperse diseases over long distances, the majority of studies available on viral zoonoses have focused on birds and bats, due to their highly migratory life histories.

9.3 EXAMPLES OF MIGRATORY ANIMALS AND SPREAD OF VIRAL DISEASE

9.3.1 Birds

Birds are known reservoirs for many viral zoonoses, including avian influenza and various vector-borne viruses, which can significantly impact human health. It has been estimated that approximately 50 billion of the world's 200–400 billion birds undertake migrations, traveling distances from tens to thousands of kilometers each year (Berthold, 2001). Viral diseases in animals seem to be best studied in the Aves group, and given their extensive annual migrations, this group shows high potential for risk of viral transmission to new areas. However, as outlined in the examples later, the conditions necessary for long-distance movement of viruses can be complex and rely on a number of factors including pathogenicity of the virus to individual host species, the effect of a compromised immune system on an individual's ability to migrate, and the general ecology of the host species.

9.3.1.1 Newcastle disease: Paramyxoviridae. Newcastle disease is a highly contagious disease found in domestic and wild avian species. Its causative agent is the Newcastle disease virus (NDV) avian paramyxovirus type 1 (APMV-1), which comprises many strains that can cause mild to lethal clinical symptoms in its hosts. NDV was first isolated from great cormorants (*Phalacrocorax carbo*) and European shags (*Phalacrocorax aristotelis*) in Scotland in 1897 (MacPherson, 1956) and since has been detected in more than 241 species of 27 taxonomic orders (Kaleta and Baldauf, 1988). In wild species, however, it has been most commonly found in cormorants (*Phalacrocorax* spp.). Risk to human health is generally low, but mild cases of conjunctivitis and influenza-type symptoms have been reported in humans exposed to infected birds, mainly poultry-processing workers, vaccinators, and laboratory personnel (Leighton and Heckert, 2007). The greatest threat to humans is economic loss to the poultry industry. NDV is listed as one of the World Organization for Animal Health's (OIE) notifiable diseases due to its high pathogenicity in poultry species and the past and potential future losses to this important economic industry (OIE, 2012b).

NDV is transmitted readily among susceptible hosts, whereby virus is shed in the feces, body fluids, eggs, and tissues (Leighton and Heckert, 2007). Cormorants shed virus for more than 20 days postinfection, and the virus can survive for long periods outside its host (Kuiken, 1999; MacPherson, 1956). The migratory patterns of the double-crested cormorants (*Phalacrocorax auritus*) in North America span the north–south extent of the United States, with breeding grounds in northern United States and Canada and wintering grounds along the Atlantic and Gulf Coasts (Dolbeer, 1991). The long duration of viral shedding in combination with subclinical infection points towards the potential for cormorants to spread NDV along their migratory corridor. In a study of migration patterns of bean and white-fronted geese (*Anser fabalis* and *Anser albifrons*, respectively) that had previously been infected with NDV, the authors suggest that these example species may also have the potential to transmit NDV over long distances (Muller et al., 1999).

9.3.1.2 Sindbis virus: Togaviridae. Sindbis virus (SINV) and other related viruses such as Ockelbo virus have the potential to be dispersed over long distances via migratory birds. SINV is a mosquito-borne alphavirus that has led to small-scale outbreaks of fever, rash, and arthralgia in humans in Europe, Africa, China, and Australia (Kurkela et al., 2008). The primary amplifying host reservoirs for SINV are birds, which do not typically exhibit clinical signs of infection. The virus is sustained by transmission cycles between mosquitoes (mainly genus *Culex*) and birds including grouse (*Tetraonidae*), thrushes (*Turdidae*), crows (*Corvidae*), and shrikes (*Laniidae*), among others. Despite its widespread prevalence in birds worldwide, recent introductions of SINV to resident birds in Finland (Kurkela et al., 2008), the United Kingdom (Buckley et al., 2003), and Australia (Sammels et al., 1999) have been linked to migratory birds. Phylogenetic analyses of SINV strains have also pointed to migratory birds as the likely mechanism of dispersal across the continent of Australia (Sammels et al., 1999) and between South Africa and northern Europe (Shirako et al., 1991). Although birds are the main reservoir for SINV, the virus has been isolated from mammals, amphibian, and tick species less commonly, indicating the potential of a reservoir base beyond birds (Thomas et al., 2007).

9.3.1.3 West Nile virus: Flaviviridae. West Nile virus (WNV) is a mosquito-borne flavivirus that can cause fever, flu-like symptoms, encephalitis (inflammation of the brain), and death in humans, other mammals, and birds. Birds are the primary amplifying reservoir of WNV, and migratory birds have long been suspected as the principal means of introduction of WNV to new regions (Rappole et al., 2000). WNV is widespread throughout the world and has exhibited an increase in the frequency and severity of outbreaks in recent decades. In 1999, WNV was reported for the first time in North America, where it caused massive die-offs in horses and both domestic and wild birds. It has been contracted by nearly two million humans since its introduction (Kilpatrick, 2011). At the time of this writing, the beginning of potentially the largest year for human cases in the United States is underway with hotspot areas in Texas and Louisiana (>1500 cases by mid-August 2012; CDC, 2012c). While it is unclear whether migratory birds led to the introduction of WNV to North America (Rappole and Hubálek, 2003; Reed et al., 2003), recent studies have shown that migratory passerine birds are capable of being dispersal vehicles for WNV (Dusek et al., 2009; Owen et al., 2006). Furthermore, now that WNV is established in North America, neotropical species migrating between North America, the Caribbean, and Central and South America present a potential mechanism for dispersal of WNV across these regions. A risk model assessment of potential pathways of WNV introduction to Barbados suggests that migratory birds pose a significant risk for introduction of the disease to that country (Douglas et al., 2007).

9.3.1.4 Encephalitis viruses: Flaviviridae and Togaviridae. In addition to WNV and SINV, there are numerous other mosquito-borne encephalitis viruses that are maintained in a bird–mosquito cycle and may be dispersed by migratory birds. These include flaviviruses belonging to the Japanese encephalitis virus (JEV) group and alphaviruses belonging to the equine encephalitis virus group. Although humans are generally considered incidental dead-end hosts of these viruses and often do not exhibit clinical symptoms, the widespread occurrence and ability of these viruses to cause severe infections in domestic animals or humans can have significant ramifications for human health and global economies (including health-care costs, surveillance, and vector control programs) (Table 9.1).

JEV is widespread throughout Asia, affecting 30–50 000 humans annually and causing flu-like symptoms, encephalitis, long-term neurological symptoms, and death in humans and commercially important animals such as horses (OIE, 2012a) and less commonly a wide range of host species including birds, canines, ruminants, bats, amphibians, and reptiles. The primary host–vector cycle involves wading birds such as the Asian cattle egret (*Bubulcus ibis coromandus*) or the black-crowned night heron (*Nycticorax nycticorax*) and mosquitoes (*Culex* sp.), all of which are associated with flooded agricultural fields in Southeast Asia. Additionally, pigs can develop sufficient viremia to serve as amplifying reservoirs for JEV. Short-distance migrations of wading birds from rural to urban areas may contribute to JEV dispersal (Pfeffer and Dobler, 2010). A 1995 outbreak in the Torres Strait off the coast of Australia was thought to have originated from infected bird movements between islands, which led to the establishment of mosquito–pig and mosquito–bird cycles on the islands (Mackenzie et al., 2004).

Another mosquito-borne virus in the JEV group that may be dispersed by migratory birds is St. Louis encephalitis virus (SLEV), which can cause fever, encephalitis, and death in humans. SLEV is endemic to both North and South America, affecting several hundred humans annually (Reisen, 2003). Although human cases in South America have been rare previously, recent outbreaks in Brazil (Mondini et al., 2007) and Argentina (Spinsanti et al., 2008) have brought attention to SLEV as a reemerging disease. Passerine (Passeriformes) and columbid (family Columbidae, pigeons and doves) birds appear to be the primary amplifying hosts (Hubalek, 2004; Reisen, 2003). A genetic analysis of SLEV strains revealed patterns consistent with migratory birds returning to their northern breeding grounds after acquiring infection in the wintering region of the Gulf of Mexico (Auguste et al., 2009).

Usutu virus (USUV) is another mosquito-borne virus in the JEV group that has gained attention in recent years. The virus originated in Africa, where it has caused mild fever and rash in humans. In 2001, USUV emerged in Austria, causing mass mortalities of blackbirds (*Turdus merula*) and great grey owls (*Strix nebulosa*) (Weissenbock et al., 2001) similar to the WNV outbreak in North America. Since that time, the virus has spread to Hungary, Germany, Italy, Switzerland, Czech Republic, and Poland, causing bird mortalities in those countries (Weissenbock et al., 2007). Migratory birds have been suggested as a possible means of USUV introduction to Central Europe (Becker et al., 2012; Pfeffer and Dobler, 2010). Buckley et al. (2003) concluded that USUV had been introduced to resident birds in the United Kingdom by migratory birds; however, there is no evidence to date of clinical USUV infection of either birds or humans in that country. In 2009, the first neuro-invasive human cases of USUV were reported in Italy, prompting concern for monitoring and surveillance of this emerging disease (Vazquez et al., 2011).

The equine encephalitis viruses, including eastern equine encephalitis virus (EEEV), western equine encephalitis virus (WEEV), and Venezuelan equine encephalitis virus (VEEV), represent another group of mosquito-borne viruses that can be carried over long distances by migratory birds. These diseases are endemic to North and South America, where they are known to cause fever, encephalitis, and death in humans and equines. Reports of clinical infections of EEEV and WEEV in humans are relatively rare (<100 cases annually and case fatality rates of 1–4%) (CDC, 2012a). In contrast, recent outbreaks of VEEV in South America have resulted in hundreds to as many as tens of thousands of human cases in each outbreak and case fatality rates as high as 30% in humans and greater than 50% in horses (Navarro et al., 2005; Weaver, 1998). WEEV and EEEV are primarily maintained in a bird–mosquito cycle; however, bats, amphibians, and reptiles may also serve as host reservoirs for EEEV (Pfeffer and Dobler, 2010). In addition to birds, rodents and equines appear to be important reservoir hosts for VEEV (Weaver, 1998). Although the role of migratory birds in contributing to recent outbreaks of these diseases is yet unclear, all of these viruses have been isolated previously from birds undergoing active migration (Hubalek, 2004; Pfeffer and Dobler, 2010).

Another vector-borne encephalitis virus is tick-borne encephalitis virus (TBEV), a flavivirus transmitted by *Ixodes* sp. tick vectors. TBEV is widespread throughout Europe and Asia, affecting 10–20 000 people per year, and can cause fever, nausea, headaches, meningitis, permanent damage to the central nervous system, and, in some cases, death (Nuttall and Labuda, 2005). The virus has been isolated from a range of host species including ruminants, canines, rodents, birds, horses, and humans. While it is not clear to what extent birds act as reservoirs for TBEV, Waldenstrom et al. (2007) reported that four passerine bird species migrating from western Russia to Sweden carried TBEV-infected ticks, illustrating the possible role of migratory birds in the geographic dispersal of TBEV. Similarly, a study of viral strains of TBEV found in ticks in Finland and Russia suggested that migratory birds may have been responsible for introduction of new TBEV strains to Finland (Jaaskelainen et al., 2010).

9.3.1.5 Avian influenza: Orthomyxoviridae. Few model systems exist within the young field of research on animal migration and disease dynamics, avian influenza being one of the more studied groups. AIV is an RNA virus of the family Orthomyxoviridae. There are five genera in this family, including Thogotovirus, Isavirus, and influenza types A, B, and C. The natural hosts of type A influenza are birds (hence the common name, avian influenza or bird flu), although mammals such as humans, horses, pigs, cats, and seals have also acquired infection from this virus. Type A influenza is the most commonly distributed of the group and can cause infections ranging from subclinical to highly lethal in its hosts. Wild waterbirds, particularly those within the orders Anseriformes (waterfowl) and Charadriiformes (shorebirds and gulls), are the natural reservoirs for the low pathogenic (nonlethal) form of avian influenza (LPAI) (Alexander, 2000; Clark and Hall, 2006; Muzaffar et al., 2006; Olsen et al., 2006; Stallknecht and Shane, 1988). LPAI viruses are replicated in the intestinal tract, shed through feces, and transmitted via the fecal–oral route. LPAI has a wide host range, with isolates from more than 110 species of wild birds from 26 families (Munster et al., 2007; Olsen et al., 2006). Surveillance efforts have shown distinct geographic and temporal variations in prevalence between migrating waterfowl and shorebirds, with waterfowl having high prevalence rates in the fall and shorebirds having high rates in the spring (Krauss et al., 2004; Slemons et al., 2003; Stallknecht and Shane, 1988). Surveillance studies in North America showed LPAI prevalence in ducks as high as 60% in late fall before the southward migration, followed by decreasing prevalence on the wintering grounds (0.4–2%) and the northward return spring migration (0.3%). The high prevalence of LPAI in ducks in the fall has been attributed to dense congregations of juvenile waterfowl (immunologically naive individuals of the year) at fall staging areas.

Although LPAIs are perpetuated by wild bird reservoirs, the highly pathogenic forms (HPAI) historically have been restricted to domestic systems such as large-scale chicken, duck, and turkey farming. The terminology "high" versus "low pathogenic avian influenza" is defined by the pathogenicity of an AIV in chickens. LPAI is defined as causing no symptoms to mild disease (nasal secretions, decreased egg production) in domestic chickens, and HPAI as causing severe mortality and rapid spread resulting in complete flock mortality within 48 h of exposure. The terms were designed for chickens due to the economic importance of the poultry industry and with the intent of developing a consistent measure of pathogenicity in a single host type (World Organization of Animal Health, 2005). These designations, however, do not predict pathogenicity in other types of hosts such as wild birds, humans, or even other types of poultry such as domestic ducks.

With the onset of HPAI subtype H5N1 (Asian lineage, hereafter referred to as H5N1), in 1997 and subsequently 2003 (Sims and Brown, 2008), have come major changes in historic patterns of HPAI. H5N1 is the first HPAI to (i) spill back from poultry to the wild bird community (Webster et al., 2007), (ii) affect such a wide diversity of host species (Cardona et al., 2009), (iii) replicate and spread efficiently via the respiratory tract (van Riel et al., 2006), and (iv) reproduce silently in large populations of domestic ducks (Sturm-Ramirez et al., 2005), each contributing to the virus's unique and deadly ability to persist and reemerge, which has occurred in more than 60 countries across Asia, Europe, and Africa over the past 16 years. The unique characteristics of this virus have also sparked confusion and debate among the global scientific community regarding the role wild migratory birds play in the spread of HPAI (Feare, 2010; Gauthier-Clerc et al., 2007; Yasue et al., 2006). The first occurrence of HPAI "spill-back" occurred on the remote plateau of western China in 2005, killing more than 6000 waterfowl, gulls, and cormorants at the famous breeding grounds of Qinghai Lake National Nature Reserve (Liu et al., 2005). In the months following this event, H5N1 spread beyond Asia and into Europe and Africa, with wild migratory birds being implicated upon each new outbreak event. However, at that time, neither were there data on the pathogenicity of H5N1 in the affected wild species nor knowledge of the movement ecology of the birds in these regions to support or refute the implications. In 2007 the United States Geological Survey (USGS) began a satellite telemetry program to study the migration movements of waterbird species from two H5N1 hotspot regions in China: Qinghai Lake in western China and Poyang Lake in southeastern China (Takekawa et al., 2010b). More than 100 waterbirds (mainly waterfowl) of primary species involved in outbreaks were marked between the two locations, including bar-headed geese (*Anser indicus*), ruddy shelducks (*Tadorna ferruginea*), Palla's gulls (*Larus ichthyaetus*), Eurasian wigeon (*Anas penelope*), northern pintail (*Anas acuta*), common teal (*Anas crecca*), falcated teal (*Anas falcata*), Baikal teal (*Anas formosa*), mallard (*Anas platyrhynchos*), garganey (*Anas querquedula*), and Chinese spotbill (*Anas poecilorhyncha*). Through this work new migratory connections were discovered: between Qinghai Lake and Mongolia to the north (Prosser et al., 2009) and Lhasa, Tibet, to the south (Prosser et al., 2011). Along the Central Asian Flyway (CAF), which encompasses north-central Siberia south through western China and India, analysis of wild bird movements in relation to poultry and H5N1 outbreaks indicated the potential for bar-headed geese and ruddy shelduck to serve as vectors of H5N1 transmission (Newman et al., 2012). On the Qinghai–Tibet Plateau, movements of bar-headed geese from wintering grounds surrounding Lhasa indicated spatial and temporal overlap with captive bar-headed geese and H5N1 outbreaks in poultry. In addition, migration along the 1200 km corridor between the Lhasa wintering grounds and Qinghai Lake was achieved in 5 days (Prosser et al., 2011), the approximate amount of time that experimentally infected bar-headed geese remained asymptomatic before onset of clinical symptoms of H5N1 infection (Brown et al., 2008), indicating a potential for movement of H5N1 via migration before onset of debilitating symptoms. In the East Asian Flyway (eastern Siberia south through eastern China to Australia), which supports large numbers of wild waterfowl as well as dense human and poultry populations, the telemetry results indicated a spatial but not temporal concordance between wild waterfowl and H5N1 outbreaks (Takekawa et al., 2010a) demonstrating the complexity in the potential for H5N1 to be transported by various species in different regions. A revelation from the telemetry studies was that using molecular phylogenies alone to explain movements of virus in relation to wild birds is not sufficient, but rather an understanding of the ecology of the host species is also necessary, including patterns of pathogenicity among species as well as their behaviors and usage of different habitats (Prosser et al., 2009; Takekawa et al., 2010a). The program expanded beyond China, in partnership with the United Nations Food and Agriculture Organization and many international agencies to study migration movements of more than 500 waterfowl from H5N1 hotspot regions across Asia, Europe, and Africa (Prosser et al., 2012).

9.3.2 Mammals

In comparison to birds, fewer mammal species are required to migrate due to the evolutionary adaptation of hibernation. Nonetheless, migration does exist for many mammal species with the more mobile marine and flying mammals having larger tendencies to migrate. Ungulates also commonly undergo migration, following either summer–winter cycles as in the case of North American caribou (*Rangifer tarandus*) or wet–dry seasons as in the case of the African wildebeest. In the following text, we describe migration and virus transmission for multiple mammal taxa (ungulates, seals, canines, raccoons, primates, etc.); however, virus transmission as related to migration is best studied in bats and therefore is a main focus of the following description.

Bats are known reservoirs of many viral zoonoses such as SARS-like coronaviruses (Balboni et al., 2012), henipaviruses (Breed et al., 2006, 2010; Drexler et al., 2012), and rhabdoviruses (Calisher et al., 2006) that can be transmitted across species to cause lethal infections in humans, livestock, and other wildlife. The main methods of transmission of these viruses to humans are through direct contact with infected bats or secondary hosts (e.g., livestock) that have acquired the disease from bats, inhalation of infectious particles from secretions or guano, and consuming infected bat meat (Calisher et al., 2006). The high degree of lethality in recent human outbreaks of these diseases has led to an increased demand for knowledge of bat movements and the potential for further dispersal of diseases.

While human encroachment on existing bat habitat has been suggested as a main driver behind the recent emergence of bat-related diseases (Daszak et al., 2006), migratory movements by bats may contribute to long-distance dispersal of these viruses. Approximately 3% of bat species exhibit migratory behavior (Fleming and Eby, 2003), but those that do migrate are capable of traveling distances of several thousand kilometers (Breed et al., 2010; Richter and Cumming, 2008). Using satellite telemetry, Breed et al. (2010) demonstrated that three fruit bat species known to carry Hendra and Nipah viruses exhibited long-distance movements (~100s–1000s of km) between Australia and Asia, thus illustrating the potential for movement of disease via migration.

9.3.2.1 Coronaviruses: Coronaviridae. In recent years, several viral zoonoses that have resulted in human epidemics have originated from mammals, with bats being of particular importance as reservoir hosts. From 2002 to 2003, an epidemic of a previously undiscovered coronavirus, severe acute respiratory syndrome-associated coronavirus (SARS-CoV), led to 8000 infections and 700 deaths in humans worldwide (CDC, 2012b). While early investigations into the SARS-CoV outbreak pointed to marketplace mammals such as the Himalayan palm civet (*Paguma larvata*) as the source, further studies suggested that those mammals may have been intermediate hosts and that the primary reservoir host was likely bats (Li et al., 2005). The etiological agent or source species of the outbreak is yet to be determined; however, since the outbreak, studies have shown that bats in Asia, Africa, and Europe are hosts to a wide variety of SARS-like coronaviruses that are genetically very similar to the strain isolated from humans (Balboni et al., 2012 and references therein).

9.3.2.2 Henipaviruses, distemper virus: Paramyxoviridae. Two lethal mammalian paramyxoviruses have emerged recently in Asia and Australia, and both of these viruses may have originated from migratory bats. Since the 1990s, several outbreaks of Nipah virus have occurred in Southeast Asia, where human infections of hundreds of people resulted in flu-like symptoms and encephalitis with a high degree of lethality (~30–70%) (WHO, 2012b). The primary reservoir hosts for Nipah virus are pteropid bats (Drexler et al., 2012), which undergo seasonal migration in search of fruit crops associated with agricultural fields. Outbreaks in Malaysia and Bangladesh appear to be associated with two different strains of Nipah virus since Malaysian outbreaks have been linked to pigs acting as intermediate hosts in the transmission of the virus to humans, and Bangladesh outbreaks did not involve a livestock host but were also capable of human-to-human transmission (Daszak et al., 2006). A related virus, Hendra virus, emerged in 1994 in Australia and is also linked to migratory pteropid bats. In the following decade, five subsequent outbreaks occurred, each typically involving horses as the intermediate host in the viral transmission to humans (Calisher et al., 2006). Although only a relatively small number of human and equine cases (<50 combined) have been reported, Hendra virus can cause fatal respiratory illness or encephalitis in both humans and equines (WHO, 2012a).

Canine distemper virus (CDV) is a paramyxovirus that is naturally hosted by domestic and feral dogs but also affects wild carnivorous animals. Transmitted via close contact with an infected host, CDV has caused massive die-offs in recent decades in previously unexposed wild carnivores such as African lions (*Panthera leo*), hyenas (*Crocuta crocuta*), foxes (family Canidae), and Caspian seals (*Phoca caspica*) (Deem et al., 2000). An epidemic of a related virus, PDV, has recently led to deaths of tens of thousands of harbor seals in Europe (Hall et al., 2006).

9.3.2.3 Rabies and lyssavirus: Rhabdoviridae. Rabies is a zoonotic disease caused by an RNA virus of the genus Lyssavirus, family Rhabdoviridae. The virus is one of the more widely distributed zoonotic diseases, having been reported in all mammal orders except dolphins and whales. The virus most commonly affects raccoons (*Procyon lotor*), dogs, bats, foxes, and skunks (family Mephitidae). Exposure occurs via infected saliva that enters the body through a bite or skin wound and eventually travels to the brain where it causes inflammation. The incubation period is approximately 3–10 weeks. Symptoms in wild animals include lethargy, loss of awareness and fear of humans, lack of coordination, aggression, paralysis, and death. In humans, if a preventive vaccine is administered within 24 h of exposure (including five doses over 28 days), rabies symptoms are prevented; however, if treatment is not administered before symptoms appear, the end result is death usually due to respiratory failure. Historically, transmission potential from animals to humans was highest in dogs, but since widespread vaccination programs have been put in place in many developed countries, bats and raccoons have become the major sources of infection for humans. Spread of rabies via wild animal migration is generally low as the main hosts are not long-distance migrants (except for the case of some bats); however, Roscoe et al. (1998) discovered that rabid raccoons travel linear distances six times that of non-rabid raccoons, increasing opportunity for transmission among susceptible populations.

Australian bat lyssavirus (ABLV) is a rare virus similar to the rabies virus that is found in pteropid and insectivorous bat species in Australia. It causes a fatal, rabies-like illness in humans and has contributed to two human deaths (Calisher et al., 2006).

9.3.2.4 Foot-and-mouth disease: Picornaviridae. Foot-and-mouth disease (FMD) virus is a highly contagious disease found in domesticated and wild ruminants and pigs. Symptoms include high fever followed by blistering in the mouth and feet, often causing lameness and possibly death. Most animals eventually recover, however, often with reduced growth and milk production. The economic consequences of FMD have caused catastrophic losses to the livestock industry—billions of dollars per outbreak—and have included the culling of millions of animals to prevent spread to new regions (Thomson et al., 2003). FMD has been reported across much of the globe, with some countries being disease-free for many years.

Cloven-hoofed wild animals such as feral pigs (*Sus scrofa*) and deer (family Cervidae) can potentially transmit the disease to domestic livestock and are susceptible to infection via domestic animals. Some wild ruminants have the potential to become silent carriers, allowing the virus to persist in the absence of obvious signs of disease without prohibiting migrational or local movements. In Africa, the buffalo (*Syncerus caffer*) is considered a main source of FMD virus for livestock (Vosloo et al., 2002), but other wild ruminants such as impala (*Aepyceros melampus*) may act as a vector for transmission between buffalo and cattle (Bastos et al., 2000; Hargreaves et al., 2004). Attempts to prevent disease spread by separating livestock from wildlife using fences have been successful but also have caused ecological disturbance and wildlife mortality (Morgan et al., 2006).

9.3.2.5 Ebola virus: Filoviridae. Human Ebola virus infection causes hemorrhagic fever and death within a few days. The most lethal strains cause up to 88% mortality and occur in Gabon, the Republic of Congo, and the Democratic Republic of Congo of central Africa. Epidemiological investigations show that outbreaks typically result from a variety of animal sources due to hunting and handling of dead animals such as gorillas, chimpanzees, or ungulates (Leroy et al., 2004). After testing more than a thousand small vertebrates after an Ebola outbreak between 2001 and 2003, evidence of asymptomatic infection by Ebola virus was found in three species of fruit bat, indicating that these animals may act as a reservoir for the virus (Leroy et al., 2005). A reemergence of Ebola in 2007 in the Democratic Republic of Congo led to further investigation and identified a massive annual fruit bat migration as the most likely cause (Leroy et al., 2009). Each year in April, thousands of bats arrive and use the islands of Ndongo and Koumulele as migratory stopover sites, refueling on fruit. While the bats take advantage of this important food resource, local hunters shot and killed several dozen roosting animals to sell or take home for consumption. Further examination of the initial human cases and how villagers handled the dead bats suggested strong evidence of the connection between the fruit bats and the Ebola outbreak.

9.3.3 Fish and herpetiles

Fish and herpetiles (amphibians and reptiles) vary greatly in their propensity to migrate. Migration in fish can range from daily movements of a few meters to annual movements over thousands of kilometers. These migration patterns occur vertically (at varying depths in the water column) and laterally (e.g., from saltwater to freshwater systems). One of the most well-known migrations of fish includes that of the anadromous salmon species (family Salmonidae) that are born in freshwater, migrate to the ocean to mature, and return upstream to their natal freshwater lakes to breed. Other highly migratory fish include deepwater pelagic species such as tuna (family Scombridae), swordfish (family Xiphiidae), and sharks (class Chondrichthyes). Migration in amphibians and reptiles varies greatly as well, with the majority exhibiting limited movements related to the breeding season. Sea turtles (superfamily Chelonioidea), however, are widely dispersed and migrate over long distances, often thousands of kilometers between seasons (Russell et al., 2005a).

The economic importance of aquaculture has prompted studies of viral infection in fish species since the mid-twentieth century (Crane and Hyatt, 2011); however, study of viruses in wild populations of fish as well as amphibians and reptiles remains very young fields with limited literature (Hendrix, 2005; Marschang, 2011). Recent studies have shown that these taxa may act as important secondary hosts for some viruses such as WNV and other arboviruses, providing a mechanism for virus to persist over winter (Marschang, 2011). Here we focus on a few examples of current viral threats to fish and herpetiles. The known risks to human health are low; however, we included the following examples due to conservation importance, such as endangered sea turtles, and commercial or recreational importance such as Pacific salmon.

9.3.3.1 Herpesvirus: Alloherpesviridae and Herpesviridae. Herpesviruses are double-stranded DNA viruses that can affect nearly all animal groups including humans, birds, mammals, fish, and herpetiles. The majority of herpesviruses found in fish and amphibians are of the family Alloherpesviridae and are distantly related to those of family Herpesviridae that affect reptiles, birds, and mammals (Hanson et al., 2011). Disease symptoms caused by viral strains of herpesviruses range from conjunctivitis and pneumonia to tumorlike growths on the skin's surface. In sea turtles, recent common herpesviruses include lung–eye–trachea virus (LETV) and fibropapilloma-associated turtle herpesvirus (FPTHV). LETV causes respiratory illness, conjunctivitis, and sometimes death. Some turtles may become chronically ill, while others die after several weeks of clinical infection. The virus is transmitted by direct contact or through contaminated sediments and seawater where it can remain infectious for a minimum of 5 days (Curry et al., 2000).

FPTHV causes tumor masses on both the dermal and epidermal skin layers of infected sea turtles (Herbst, 1994). In advanced cases, tumors can grow so large that they interfere with internal organ function and an individual's ability to see and feed, thereby increasing risk of death. Exposure to FPTHV is thought to occur as migrating turtles return to their natal breeding grounds (Work et al., 2004). Recent studies have shown that changing land-use patterns and pollution increase exposure of marine turtles to FPTHV. Van Houtan et al. (2010) studied nearly 4000 Hawaiian green sea turtle (*Chelonia mydas*) strandings over three decades and found elevated disease rates linked to high-nitrogen coastal runoff areas. These unnaturally nutrient-rich regions provide ideal conditions to support blooms of invasive species of macroalgae. The macroalgae sequester environmental N in the form of arginine, an amino acid. Arginine, when consumed by foraging turtles, can trigger activation of dormant FPTHV within the turtles. More than 90% of the strandings in this study were caused by FPTHV, highlighting the importance of human impacts on this endangered migratory species.

9.3.3.2 Infectious hematopoietic necrosis virus: Rhabdoviridae. Infectious hematopoietic necrosis virus (IHNV) is an RNA virus of the Rhabdoviridae family of economic importance in wild and commercial salmonid fish. The virus causes internal hemorrhaging and necrosis of the kidney, spleen, and liver, often leading to death particularly in young fish. Recovered fish may become asymptomatic carriers. IHNV was first detected in western North America but has also been reported in Europe. Transmission occurs via infected feces, external mucus, and sexual fluids (Amend, 1975). It can survive in water for a minimum of 30 days (Amend, 1970). Kurath et al. (2003) studied the phylogeny of IHNV isolates from North America and found that the oceanic migration ranges of the salmon species as well as aquaculture practices contributed to the evolution of this virus.

9.4 CLIMATE CHANGE EFFECTS ON ANIMAL MIGRATION AND VIRAL ZOONOSES

The body of scientific evidence that the earth is warming has been building over the past three decades including documented rising temperatures, rising sea levels, increased frequency and severity of storm events, and changes in precipitation patterns (IPCC, 2007). The hottest month in recorded history (July 2012) occurred at the time of this writing, 3.3°F above the twentieth-century average (NOAA, 2012). Simultaneously, the Arctic ice sheet has reached its lowest minimum area in recorded history, 70 000 km^2 below the 2007 record according to the National Snow and Ice Data Center (NSIDC, 2012). These landmark changes may forewarn of serious potential effects on ecosystem and human health, and there is a need for the global community of public health officials, scientists, veterinarians, epidemiologists, agriculturists, wildlife biologists, and the general public to be aware of these changes in order to work towards providing solutions against them (Slenning, 2010).

Although patterns in global warming are evident, climate change has not been occurring uniformly worldwide; for example, rates of rising temperatures have been higher towards the poles, and changes in precipitation patterns have resulted in some regions experiencing drought-like conditions, while others exhibit increases in severe storm and flooding events (IPCC, 2007). Because the effects of climate change vary regionally and migrants have the potential to travel through multiple regions, the response of migrants to climate change may be complex, variable, and difficult to predict (Gordo, 2007).

9.5 SHIFTS IN TIMING OF MIGRATION AND RANGE EXTENTS

A number of studies have reported evidence of changes in the timing of animal migrations in response to climate change. The majority indicate an advancement of spring migration, with few or inconclusive results for fall migration. Rates of climate warming have been higher in Europe than eastern North America (Hansen et al., 2006); however, relationships between warming temperatures and earlier spring arrival dates have been observed along both shores (Both et al., 2006; Root, 1988). An analysis of 46 years of passerine bird banding data in the northeastern United States, for example, showed that spring migration generally came earlier for all species while autumn migration was advanced for long-distance (neotropical) migrants and delayed in short-distance migrants (Van Buskirk et al., 2009). Year-to-year changes in timing were also correlated with local temperatures, indicating a fine-scale temporal relationship between migration and local climate.

It is well established that in recent decades the phenology, or timing of seasonal activities such as plant production, animal breeding, and migration, has advanced in response to warming temperatures associated with climate change (Saino et al., 2010; Walther et al., 2002). A meta-analysis of phenological response to climate change in more than 200 plant and animal species showed advanced phenology across most taxa studied, with amphibian rates being twice as fast as trees, birds, and butterflies (Parmesan, 2007). For migratory species, the timing of the arrival to breeding grounds, where there is often a narrow window of food availability, can significantly affect both fitness and reproductive success (Baker et al., 2004; Visser and Both, 2005). Thus, long-distance migrants may be particularly susceptible to climate change, especially where they have not advanced their arrival to breeding grounds to coincide with an earlier onset of food production (Both et al., 2010). In habitats with relatively short time scales of food production, failure to adjust to changes in timing may result in a trophic mismatch between timing of breeding and peak food availability, which can lead to reduced overall fitness and reproductive success as well as population declines in long-distance migrants. Such is the case of the pied flycatcher (*Ficedula hypoleuca*) whose population declined by 90% over two decades because food provisions peaked before egg hatching, leaving many nestlings to starve (Both et al., 2006).

Changes in the geographic distribution, migration routes, and range extents have also been observed in relation to climate change over the past decades. For example, a number of species have demonstrated shifts in distributions to higher latitudes and altitudes in response to warming (Walther et al., 2002), such as red and Arctic foxes (*Vulpes vulpes* and *Alopex lagopus*) in Canada, butterfly species in North America and Europe (200 km over 27 years), bird species in Europe (19 km over 20 years), alpine plants in the European Alps (1–4 m elevational shift per decade), and tropical lowland bird species in Costa Rica (from bottom slopes to higher mountain areas). A meta-analysis by Hickling et al. (2006) reported that 83% of 329 species across 16 taxa in Europe showed an average northward shift of 31–60 km. Climate change and warmer winters have also led to decreasing migration distances; for example, in northwestern Europe between 1932 and 2004, 12 of 24 avian species shortened their migration to wintering sites, with the greatest reduction by species from dry open areas versus wet open and forested areas (Visser et al., 2009). Godet et al. (2011) and Devictor et al. (2008) looked at community assemblages of shorebird species (Charadrii) from Europe and Africa and found a sharp increase in a community index that measures the average temperature across each species' range over the period of 1977–2009. They note a reassembly of the wintering shorebird species composition across a large spatiotemporal scale, which may have implications for interspecies niche relationships and related community dynamics.

9.6 COMBINED EFFECTS OF CLIMATE CHANGE, DISEASE, AND MIGRATION

Predicting the combined effects of climate change on migratory patterns of host species and epidemiology of viral pathogens is complex and not fully realistic. However, taking consideration of multiple factors that are most likely to change in response to climate change is one way for us to begin to understand the implications of these integral stressors (Figure 9.2). Some potential and observed key changes related to climate change that could affect viral transmissions include (i) changes of host population migratory behavior such as altered patterns and timing of migration and shifting of distribution ranges; (ii) changes to the molecular biology of viral pathogens that could affect survival rates in the environment, virulence rates, and transmission function among hosts; and (iii) changes to vector biology that could affect mechanisms such as feeding frequency, metabolic rates, larval maturation times, length of breeding seasons, and extent of vector range (Figure 9.2). Changes to farming practices and land-use patterns are additional indirect factors that may be influenced by climate change (Gale et al., 2009).

Some of the cumulative stressors listed earlier have already begun to be observed. For example, a recent model analysis of pteropid bats and Henipavirus emergence in Australia suggests that a reduction in migratory behavior may increase risk of virus transmission lethal to humans and horses (Plowright et al., 2011). In the rivers of western Canada, increasing water temperatures have resulted in high losses of sockeye salmon (*Oncorhynchus nerka*) migrating from the Pacific Ocean to upstream spawning grounds, a result of decreased oxygen delivery and increased development of viral infection (Miller et al., 2011). Off the coast of Norway, Atlantic salmon (*Salmo salar*)

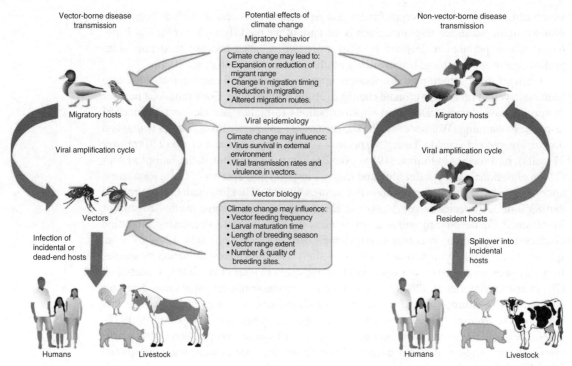

Figure 9.2. Potential effects of climate change on disease transmission pathways of vector-borne (left) and non-vector-borne (right) viruses associated with migratory animals. Symbols used in this figure were provided courtesy of the Integration and Application Network, University of Maryland Center for Environmental Science (ian.umces.edu/symbols/). For color detail, please see color plate section.

are staying at sea for longer periods and delaying maturation due to ocean warming and freshwater conditions (Otero et al., 2012).

Regarding changes to the molecular biology and survival rates of virus in the environment, we expect to see differences in the longevity of avian influenza strains in water or moist environments; however, the direction of these changes might not be clear or consistent. For example, a few studies that have explored relationships between water temperature and persistence of AIVs reported decreased persistence with increasing temperature (Stallknecht et al., 1990a, b; 2010); however, Gilbert et al. (2008) point out that several HPAI H5N1 epizootic outbreaks have occurred under consistently hot conditions (Indonesia, sub-Saharan Africa), emphasizing our lack of understanding of the direct influence of environmental factors such as climate change on the epidemiology of AIVs. In the vector-borne group of diseases, genetic mutation of a single amino acid substitution on VEEV has allowed the virus to adapt from its traditional forest-dwelling mosquito vector, *Culex taeniopus*, to the open habitat species, *Ochlerotatus taeniorhynchus*, illustrating an example of rapid expansion to new regions in response to changes in the biology of the virus (Brault et al., 2004).

Increases in temperature and changes in precipitation patterns can have direct and indirect effects on the biology of vector populations and virus prevalence. For example, both the larval and adult stages of mosquitoes are sensitive to temperature changes (Githeko et al., 2000). At higher temperatures, mosquito larvae tend to mature more quickly, leading to an increase in the number of offspring and number of susceptible vectors produced within a season (Rueda et al., 1990). Adult mosquitoes digest blood more quickly at higher temperatures, causing them to feed more regularly, which can lead to increased vector-to-host encounters. Milder winters can reduce mortality in mosquito and tick populations and extend the length of their breeding season, leading to increased disease activity. Alternatively, hotter, drier summers could also increase tick mortality, imparting a negative effect on disease transmission (Githeko et al., 2000). Indirectly, climate warming can lead to a change in habitat for vector species, such as increasing (or decreasing) the quality and total area of mosquito breeding sites through changes in precipitation (Gage et al., 2008) and expanding the latitudinal or altitudinal range for mosquito and tick species (Danielova et al., 2010; Reeves et al., 1994).

Recent studies of North American strains of WNV have shown that higher temperatures, such as those during abnormally warm summers, result in more efficient transmission (Reisen et al., 2006; Ruiz et al., 2010) and increased viral replication of WNV within mosquito vectors (Kilpatrick et al., 2008). However, transmission or prevalence of some vector-borne viral zoonoses will not necessarily increase with increasing temperatures. Reeves et al. (1994) suggest that while both WEEV and SLEV may expand northward in North America with increasing temperatures, WEEV may actually disappear in more southern regions as temperatures exceed its optimal temperature of 21 °C.

While publications regarding the effects of climate change on the spread of viral diseases are increasing, the ecological response is not uniform, and we have much to learn in this realm.

9.7 CONCLUSIONS AND FUTURE DIRECTIONS

Our world is changing at an unprecedented pace. The effects of a human dominated landscape go beyond climate change to include increased human population growth, alterations to land-use patterns, loss of natural habitats, intensification of agricultural and livestock production, and globalization of trade and human travel. With these pressures we observe changes in our natural communities and a resulting increase of emerging diseases in both wildlife and human populations (Jones et al., 2008). In response, we must think critically about how best to address and prepare for new emerging zoonotic threats, here with particular focus on relationships to wild animal migration. A major goal should be the integration of the fields of wildlife ecology and virology and epidemiology, which can be accomplished in multiple ways.

First, disease surveillance can be brought to existing monitoring networks such as long-term banding stations, hunter check stations (often for mammals and waterfowl), or recreational and commercial fisheries. Seventy percent of published studies on diseases in wildlife focus on two viruses: avian influenza and WNV (Fuller et al., 2012). Using existing frameworks that handle live or harvested wildlife can provide an opportunity to collect valuable baseline data on existing pathogens in these wild species.

Second, for emerging or ongoing threats such as in the case of HPAI H5N1, coupling the ecology of host species with virology and phylogenetics is key to understanding how migratory animals may contribute to virus transmission across large geographic extents (Newman et al., 2012; Prosser et al., 2009, 2011; Takekawa et al., 2010a). Advancements in tracking technology such as satellite telemetry, geolocators, stable isotopes, and genetic markers provide important tools for following animals over long distances and across political borders (Bridge et al., 2011). Understanding migratory connectivity, timing, and habitat use will provide critical baseline data for informing disease surveillance and prevention programs.

Integrating modeling and field data is an important approach to predicting, preparing for, and responding to emerging viral threats. Models can be as simple as geographic analyses that exhibit transmission potential between host species (Prosser, 2012) or between vectors and susceptible populations (Teadrow, 2009). Models can incorporate simulations under different anthropogenic and environmental conditions or under different control strategies (Clements and Pfeiffer, 2009; Ferguson et al., 2001). The availability of remote sensing data has advanced model applications in a geographic context (Orme-Zavaleta et al., 2006; Russell et al., 2005b) including a wide array of climatic and environmental predictors applicable to disease factors (Hay et al., 2000, 2006). The development of more powerful computer systems and open software allows for more complex models and larger simulation runs that can aid in the assessment of uncertainty analyses and risk modeling (Morgan and Henrion, 1990; Prosser, 2012). Combining epidemiological models with environmental and demographic predictors, along with the ecology of host species, allows for powerful spatial and temporal predictions of high-risk transmission areas (Prosser et al., 2013 In Press). However, comprehensive spatial data for wildlife distributions are generally lacking (Prosser et al., 2013 In Review), which is another area that could use further study (Morrison et al., 2006).

Efforts to bridge the fields of wildlife ecology and epidemiology are young, with many exciting opportunities for exploration and improvement of our current state of knowledge. Learning from existing disease models such as avian influenza and WNV is a good place to start, along with a proactive approach of increasing baseline data on existing pathogens in our migratory wildlife communities.

ACKNOWLEDGEMENTS

We would like to thank Lisa Vormwald for contributions to sections of this book chapter and Dr. R. Michael Erwin for useful comments on earlier drafts of this chapter.

REFERENCES

Aidley, D.J. (1981) “Animal migration”. Cambridge University Press.

Alexander, D.J. (2000) A review of avian influenza in different bird species. *Veterinary Microbiology* **74**(1–2), 3–13.

Alexander, D. J. (2007). An overview of the epidemiology of avian influenza. *Vaccine* **25**(30), 5637–5644.

Altizer, S., Bartel, R., and Han, B. A. (2011). Animal migration and infectious disease risk. *Science* **331**(6015), 296–302.

Amend, D. F. (1970). Control of infectious hematopoietic necrosis virus disease by elevating the water temperature. *Journal of the Fisheries Research Board of Canada* **27**(2), 265–270.

Amend, D. F. (1975). Detection and transmission of infectious hematopoietic necrosis virus in rainbow trout. *Journal of Wildlife Diseases* **11**, 471–478.

Auguste, A. J., Pybus, O. G., and Carrington, C. V. F. (2009). Evolution and dispersal of St. Louis encephalitis virus in the Americas. *Infection, Genetics and Evolution* **9**(4), 709–715.

Baker, A. J., Gonzalez, P. M., Piersma, T., et al. (2004). Rapid population decline in Red Knots: Fitness consequences of decreased refuelling rates and late arrival in Delaware Bay. *Proceedings of the Royal Society B: Biological Sciences* **271**(1541), 875–882.

Balboni, A. F., Battilani, M., and Prosperi, S. (2012). The SARS-like coronaviruses: The role of bats and evolutionary relationships with SARS coronavirus. *New Microbiologica* **35**(1), 1–16.

Bastos, A. D. S., Boshoff, C. I., Keet, D. F., et al. (2000). Natural transmission of foot-and-mouth disease virus between African buffalo (*Syncerus caffer*) and impala (*Aepyceros melampus*) in the Kruger National Park, South Africa. *Epidemiology and Infection* **124**(3), 591–598.

Becker, N., Jost, H., Ziegler, U., et al. (2012). Epizootic emergence of usutu virus in wild and captive birds in Germany. *PLoS One* **7**(2), e32604.

Benson, S. R., Eguchi, T., Foley, D. G., et al. (2011). Large-scale movements and high-use areas of western Pacific leatherback turtles, *Dermochelys coriacea. Ecosphere* **2**(7), art84.

Berthold, P. (2001). "Bird Migration: A General Survey." Oxford University Press, Oxford.

Both, C., Bouwhuis, S., Lessells, C. M., et al. (2006). Climate change and population declines in a long-distance migratory bird. *Nature* **441**(7089), 81–83.

Both, C., Van Turnhout, C. A. M., Bijlsma, R. G., et al. (2010). Avian population consequences of climate change are most severe for long-distance migrants in seasonal habitats. *Proceedings of the Royal Society B: Biological Sciences* **277**(1685), 1259–1266.

Brault, A. C., Powers, A. M., Ortiz, D., et al. (2004). Venezuelan equine encephalitis emergence: Enhanced vector infection from a single amino acid substitution in the envelope glycoprotein. *Proceedings of the National Academy of Sciences of the United States of America* **101**(31), 11344–11349.

Breed, A. C., Field, H. E., Epstein, J. H., et al. (2006). Emerging henipaviruses and flying foxes—Conservation and management perspectives. *Biological Conservation* **131**(2), 211–220.

Breed, A. C., Field, H. E., Smith, C. S., et al. (2010). Bats without borders: Long-distance movements and implications for disease risk management. *Ecohealth* **7**(2), 204–212.

Bridge, E. S., Thorup, K., Bowlin, M. S., et al. (2011). Technology on the move: Recent and forthcoming innovations for tracking migratory birds. *Bioscience* **61**(9), 689–698.

Brown, J. D., Stallknecht, D. E., and Swayne, D. E. (2008). Experimental infection of swans and geese with highly pathogenic avian influenza virus (H5N1) of Asian lineage. *Emerging Infectious Diseases* **14**(1), 136–142.

Buckley, A., Dawson, A., Moss, S. R., et al. (2003). Serological evidence of West Nile virus, Usutu virus and Sindbis virus infection of birds in the UK. *Journal of General Virology* **84**(10), 2807–2817.

Calisher, C. H., Childs, J. E., Field, H. E., et al. (2006). Bats: Important reservoir hosts of emerging viruses. *Clinical Microbiology Reviews* **19**(3), 531–545.

Cardona, C. J., Xing, Z., Sandrock, C. E., et al. (2009). Avian influenza in birds and mammals. *Comparative Immunology Microbiology and Infectious Diseases* **32**(4), 255–273.

Center for Disease Control CDC (2012a). Eastern equine encephalitis. http://www.cdc.gov/easterne-quineencephalitis/ (accessed September 8, 2012).

CDC (2012b). Severe Acute Respiratory Syndrome (SARS). http://www.cdc.gov/sars/index.html (accessed September 14, 2012).

CDC (2012c). West Nile Virus 2012. http://www.cdc.gov/ncidod/dvbid/westnile/index.htm (accessed on August 29, 2012).

Clark, L., and Hall, J. (2006). Avian influenza in wild birds: Status as reservoirs, and risks to humans and agriculture. *Ornithological Monographs* (60), 3–29.

Clements, A. C. A., and Pfeiffer, D. U. (2009). Emerging viral zoonoses: Frameworks for spatial and spatiotemporal risk assessment and resource planning. *Veterinary Journal* **182**(1), 21–30.

Crane, M., and Hyatt, A. (2011). Viruses of fish: An overview of significant pathogens. *Viruses* **3**, 2025–2046.

Curry, S. S., Brown, D. R., Gaskin, J. M., et al. (2000). Persistent infectivity of a disease-associated herpesvirus in green turtles after exposure to seawater. *Journal of Wildlife Diseases* **36**(4), 792–797.

Danielova, V., Schwarzova, L., Materna, J., et al. (2010). Integration of a tick-borne encephalitis virus and Borrelia burgdorferi sensu lato into mountain ecosystems, following a shift in the altitudinal limit of distribution of their vector, *Ixodes ricinus* (Krkonose mountains, Czech Republic). *Vector-Borne and Zoonotic Diseases* **10**(3), 223–230.

Daszak, P., Plowright, R. K., Epstein, J. H., et al. (2006). The emergence of Nipah and Hendra virus: Pathogen dynamics across a wildlife-livestock-human continuum. In: "Disease Ecology: Community Structure and Pathogen Dynamics" (S. K. Collinge, and C. Ray, Eds.), pp. 186–201. Oxford University Press, Oxford.

Deem, S. L., Spelman, L. H., Yates, R. A., et al. (2000). Canine distemper in terrestrial carnivores: A review. *Journal of Zoo and Wildlife Medicine* **31**(4), 441–451.

Devictor, V., Julliard, R., Couvet, D., et al. (2008) Birds are tracking climate warming, but not fast enough. Proceedings of the Royal Society B: *Biological Sciences* **275**(1652), 2743–2748.

Dingle, H. (1996). "Migration: The Biology of Life on the Move." Oxford University Press, New York.

Dobson, A. (1995). Ecology and epidemiology of Rinderpest virus in Serengeti and Ngorongoro conservation area. In: "Serengeti II: Dynamics, Management, and Conservation of an Ecosystem" (A. R. E. Sinclair, and P. Arcese, Eds.), pp. 485–505. University of Chicago Press.

Dolbeer, R. A. (1991). Migration patterns of double-crested cormorants east of the rocky mountains. *Journal of Field Ornithology* **62**(1), 83–93.

Douglas, K. O., Kilpatrick, A. M., Levett, P. N., et al. (2007). A quantitative risk assessment of West Nile virus introduction into Barbados. *West Indian Medical Journal* **56**(5), 394–397.

Drexler, J. F., Corman, V. M., Muller, M. A., et al. (2012). Bats host major mammalian paramyxoviruses. *Nature Communications* **3**, 796.

Dusek, R. J., McLean, R. G., Kramer, L. D., et al. (2009). Prevalence of West Nile Virus in migratory birds during spring and fall migration. *American Journal of Tropical Medicine and Hygiene* **81**(6), 1151–1158.

Egevang, C., Stenhouse, I. J., Phillips, R. A., et al. (2010). Tracking of Arctic terns *Sterna paradisaea* reveals longest animal migration. *Proceedings of the National Academy of Sciences of the United States of America* **107**(5), 2078–2081.

Feare, C. J. (2010). Role of wild birds in the spread of highly pathogenic avian influenza virus H5N1 and implications for global surveillance. *Avian Diseases* **54**(1), 201–212.

Ferguson, N. M., Donnelly, C. A., and Anderson, R. M. (2001). The foot-and-mouth epidemic in Great Britain: Pattern of spread and impact of interventions. *Science* **292**(5519), 1155–1160.

Fleming, T. H., and Eby, P. (2003). Ecology of bat migration. In: "Bat Ecology" (T. H. Kunz, and M. B. Fenton, Eds.), pp. 156–208. University of Chicago Press, Chicago.

Food and Agriculture Organization of the United States, and World Organization for Animal Health (2011). Joint FAO/OIE Committee on Global Rinderpest Eradication: Final Report, 26 pp.

Fouchier, R. A. M., Munster, V. J., Keawcharoen, J., et al. (2007). Virology of avian influenza in relation to wild birds. *Journal of Wildlife Diseases* **43**(3), S7–S14.

Fuller, T., Bensch, S., Müller, I., et al. (2012). The ecology of emerging infectious diseases in migratory birds: An assessment of the role of climate change and priorities for future research. *Ecohealth*, **9**(1), 80–88.

Gage, K. L., Burkot, T. R., Eisen, R. J., and Hayes, E. B. (2008). Climate and vectorborne diseases. *American Journal of Preventive Medicine* **35**(5), 436–450.

Gaidet, N., Newman, S. H., Hagemeijer, W., et al. (2008). Duck migration and past influenza A (H5N1) outbreak areas [letter]. *Emerging Infectious Diseases*. Available from http://wwwnc.cdc.gov/eid/article/14/7/07–1477.htm (accessed on May 29, 2013).

Gale, P., Drew, T., Phipps, L. P., et al. (2009). The effect of climate change on the occurrence and prevalence of livestock diseases in Great Britain: A review. *Journal of Applied Microbiology* **106**(5), 1409–1423.

Gauthier-Clerc, M., Lebarbenchon, C., and Thomas, F. (2007). Recent expansion of highly pathogenic avian influenza H5N1: A critical review. *Ibis* **149**(2), 202–214.

Gilbert, M., Slingenbergh, J., and Xiao, X. (2008). Climate change and avian influenza. *Revue Scientifique et Technique-Office International Des Epizooties* **27**(2), 459–466.

Githeko, A. K., Lindsay, S. W., Confalonieri, U. E., et al. (2000). Climate change and vector-borne diseases: A regional analysis. *Bulletin of the World Health Organization* **78**(9), 1136–1147.

Godet, L., Jaffre, M., and Devictor, V. (2011) Waders in winter: long-term changes of migratory bird assemblages facing climate change. *Biological Letters* **7**(5), 714–717.

Gordo, O. (2007). Why are bird migration dates shifting? A review of weather and climate effects on avian migratory phenology. *Climate Research* **35**(1–2), 37–58.

Hall, A. J., Jepson, P. D., Goodman, S. J., et al. (2006). Phocine distemper virus in the North and European Seas: Data and models, nature and nurture. *Biological Conservation* **131**(2), 221–229.

Hansen, J., Sato, M., Ruedy, R., et al. (2006). Global temperature change. *Proceedings of the National Academy of Sciences of the United States of America* **103**(39), 14288–14293.

Hanson, L., Dishon, A., and Kotler, M. (2011). Herpesviruses that infect fish. *Viruses* **3**, 2160–2191.

Hargreaves, S. K., Foggin, C. M., Anderson, E. C., et al. (2004). An investigation into the source and spread of foot and mouth disease virus from a wildlife conservancy in Zimbabwe. *Revue Scientifique et Technique-Office International Des Epizooties* **23**(3), 783–790.

Harkonen, T., Dietz, R., Reijnders, P., et al. (2006). The 1988 and 2002 phocine distemper virus epidemics in European harbour seals. *Diseases of Aquatic Organisms* **68**, 115–130.

Hay, S. I., Randolph, S. E., and Rogers, D. J. (2000). "Remote Sensing and Geographical Information Systems for Epidemiology." Academic Press, San Diego, CA.

Hay, S. I., Tatem, A. J., Graham, A. J., et al. (2006). Global environmental data for mapping infectious disease distribution. *Advances in Parasitology* **62**, 37–77.

Hendrix, S. S. (2005). Viral, bacterial, algal, and fungal diseases of fishes. In: "Wildlife Diseases: Landscape Epidemiology, Spatial Distribution and Utilization of Remote Sensing Technology" (S. K. Majumdar, J. E. Huffman, F. J. Brenner, and A. I. Panah, Eds.), pp. 144–150. The Pennsylvania Academy of Science, Easton, PA.

Herbst, L. H. (1994). Fibropapillomatosis of marine turtles. *Annual Review of Fish Diseases* **4**, 389–425.

Hickling, R., Roy, D. B., Hill, J. K., et al. (2006). The distributions of a wide range of taxonomic groups are expanding polewards. *Global Change Biology* **12**(3), 450–455.

Hubalek, Z. (2004). An annotated checklist of pathogenic microorganisms associated with migratory birds. *Journal of Wildlife Diseases* **40**(4), 639–659.

IPCC (2007). "Climate Change 2007: Synthesis Report. Contribution of Working Groups I, II and III to the Fourth Assessment Report of the Intergovernmental Panel on Climate Change" (R. K. Pachauri, and A. Reisinger, Eds.) IPCC, Geneva, CA, 104 pp.

Jaaskelainen, A. E., Sironen, T., Murueva, G. B., et al. (2010). Tick-borne encephalitis virus in ticks in Finland, Russian Karelia and Buryatia. *Journal of General Virology* **91**, 2706–2712.

Jones, K. E., Patel, N. G., Levy, M. A., et al. (2008). Global trends in emerging infectious diseases. *Nature* **451**(7181), 990–993.

Kaleta, E. F., and Baldauf, C. (1988). Newcastle disease in free-living and pet birds. In: "Newcastle Disease" (D. J. Alexander, Ed.), pp. 197–246. Kluwer Academic Publishers, Boston, MA.

Kilpatrick, A. M. (2011). Globalization, land use, and the invasion of West Nile virus. *Science* **334**(6054), 323–327.

Kilpatrick, A. M., Meola, M. A., Moudy, R. M., et al. (2008). Temperature, viral genetics, and the transmission of West Nile Virus by *Culex pipiens* mosquitoes. *PLoS Pathogens* **4**(6): e1000092.

Kleijn, D., Munster, V. J., Ebbinge, B. S., et al. (2010). Dynamics and ecological consequences of avian influenza virus infection in greater white-fronted geese in their winter staging areas. *Proceedings of the Royal Society B: Biological Sciences* **277**(1690), 2041–2048.

Krauss, S., Walker, D., Pryor, S.P., et al. (2004) Influenza A viruses of migrating wild aquatic birds in North America. *Vector-Borne and Zoonotic Diseases* **4**(3), 177–189.

Krauss, S., Stallknecht, D. E., Negovetich, N. J., et al. (2010). Coincident ruddy turnstone migration and horseshoe crab spawning creates an ecological 'hot spot' for influenza viruses. *Proceedings of the Royal Society B: Biological Sciences* **277**(1699), 3373–3379.

Kuiken, T. (1999). A review of Newcastle disease in cormorants. *Waterbirds* **22**, 333–347.

Kurath, G., Garver, K. A., Troyer, R. M., et al. (2003). Phylogeography of infectious haematopoietic necrosis virus in North America. *Journal of General Virology* **84**(4), 803–814.

Kurkela, S., Raetti, O., Huhtamo, E., et al. (2008). Sindbis virus infection in resident birds, migratory birds, and humans, Finland. *Emerging Infectious Diseases* **14**(1), 41–47.

Leighton, F. A., and Heckert, R. A. (2007). Newcastle disease and related avian paramyxoviruses. In: "Infectious Diseases of Wild Birds" (N. J. Thomas, D. B. Hunter, and C. T. Aktkinson, Eds.), pp. 3–16. Blackwell Publishing, Ames, IA.

Leroy, E. M., Rouquet, P., Formenty, P., et al. (2004). Multiple Ebola virus transmission events and rapid decline of central African wildlife. *Science* **303**(5656), 387–390.

Leroy, E. M., Kumulungui, B., Pourrut, X., et al. (2005). Fruit bats as reservoirs of Ebola virus. *Nature* **438**(7068), 575–576.

Leroy, E. M., Epelboin, A., Mondonge, V., et al. (2009). Human Ebola outbreak resulting from direct exposure to fruit bats in Luebo, Democratic Republic of Congo, 2007. *Vector-Borne and Zoonotic Diseases* **9**(6), 723–728.

Li, W. D., Shi, Z. L., Yu, M., et al. (2005). Bats are natural reservoirs of SARS-like coronaviruses. *Science* **310**(5748), 676–679.

Liu, J., Xiao, H., Lei, F., et al. (2005). Highly pathogenic H5N1 influenza virus infection in migratory birds. *Science* **309**(5738), 1206.

Loehle, C. (1995). Social barriers to pathogen transmission in wild animal populations. *Ecology* **76**(2), 326–335.

Lyngs, P. (2003). Migration and winter ranges of birds in Greenland. An analysis of ringing recoveries. *Dansk Ornitologisk Forenings Tidsskrift* **97**, 1–167.

Mack, D. E., and Yong, W. (2000). "Swainson's Thrush (*Catharus ustulatus*), The Birds of North America Online" (A. Poole, Ed.). Cornell Lab of Ornithology. Ithaca, NY. Retrieved from the Birds of North America Online: http://bna.birds.cornell.edu/bna/species/540 (accessed on May 29, 2013).

Mackenzie, J. S., Gubler, D. J., and Petersen, L. R. (2004). Emerging flaviviruses: The spread and resurgence of Japanese encephalitis, West Nile and dengue viruses. *Nature Medicine* **10**(12 Suppl): S98–S109.

MacPherson, L. W. (1956). Some observations on the epizootiology of Newcastle disease. *Canadian Journal of Comparative Medicine* **10**, 55–168.

Marschang, R. E. (2011). Viruses infecting reptiles. *Viruses* **3**(11), 2087–2126.

Meslin, F.-X. (2008). Public health impact of zoonoses and international approaches for their detection and containment. *Veterinaria Italiana* **44**(4), 583–590.

Miller, M. R., Takekawa, J. Y., Fleskes, J. P., et al. (2005). Spring migration of Northern Pintails from California's Central Valley wintering area tracked with satellite telemetry: Routes, timing, and destinations. *Canadian Journal of Zoology-Revue Canadienne De Zoologie* **83**(10), 1314–1332.

Miller, K. M., Li, S., Kaukinen, K. H., et al. (2011). Genomic signatures predict migration and spawning failure in wild Canadian salmon. *Science* **331**(6014), 214–217.

Mizrahi, D. S., Peters, K. A., Tanacredi, J. T., et al. (2009). "Relationships Between Sandpipers and Horseshoe Crab in Delaware Bay: A Synthesis Biology and Conservation of Horseshoe Crabs," pp. 65–87. Springer, New York.

Mondini, A., Cardeal, I. L. S., Lazaro, E., et al. (2007). Saint Louis encephalitis virus, Brazil. *Emerging Infectious Diseases* **13**(1), 176–178.

Morgan, M. G., and Henrion, M. (1990). "Uncertainty: A Guide to Dealing with Uncertainty in Quantitative Risk and Policy Analysis." Cambridge University Press, New York.

Morgan, E. R., Lundervold, M., Medley, G. F., et al. (2006). Assessing risks of disease transmission between wildlife and livestock: The Saiga antelope as a case study. *Biological Conservation* **131**(2), 244–254.

Morrison, M. L., Marcot, B. G., and Mannan, R. W. (2006). "Wildlife–Habitat Relationships: Concepts and Applications." Third Edition. University of Wisconsin Press, Madison, WI.

Muller, T., Hlinak, A., Muhle, R. U., et al. (1999). A descriptive analysis of the potential association between migration patterns of bean and white-fronted geese and the occurrence of newcastle disease outbreaks in domestic birds. *Avian Diseases* **43**(2), 315–319.

Munster, V.J., Baas, C., Lexmond, P., et al. (2007) Spatial, temporal, and species variation in prevalence of influenza A viruses in wild migratory birds. *PLoS Pathogens* **3**(5), 630–638.

Muzaffar, S.B., Ydenberg, R.C., and Jones IL (2006) Avian influenza: an ecological and evolutionary perspective for waterbird scientists. *Waterbirds* **29**(3), 243–257.

Navarro, J.-C., Medina, G., Vasquez, C., et al. (2005). Postepizootic persistence of Venezuelan equine encephalitis virus, Venezuela. *Emerging Infectious Diseases* **11**(12), 1907–1915.

Newman, S. H., Hill, N. J., Spragens, K. A., et al. (2012). Eco-virological approach for assessing the role of wild birds in the spread of avian influenza H5N1 along the Central Asian Flyway. *PLoS One* **7**(2), e30636.

NOAA (2012). State of the Climate Report. http://www.ncdc.noaa.gov/sotc/national/2012/7 (accessed August 30, 2012).

NSIDC (2012). Arctic Sea Ice Extent 2007 Record Low. http://nsidc.org/arcticseaicenews/ (accessed August 30, 2012).

Nuttall, P. A., and Labuda, M. (2005). Tick-borne encephalitis. In: "Tick-Borne Diseases of Humans" (J. L. Goodman, D. T. Dennis, and D. E. Sonenshine, Eds.), pp. 150–163. ASM Press, Washington, DC.

OIE (2012a). Japanese Encephalitis. http://www.oie.int/fileadmin/Home/eng/Animal_Health_in_the_World/docs/pdf/JAPANESE_ENCEPHALITIS_FINAL.pdf (accessed September 8, 2012).

OIE (2012b). OIE Listed Diseases. http://www.oie.int/animal-health-in-the-world/oie-listed-diseases-2012/ (accessed August 29, 2012).

Olsen, B., Munster, V.J., Wallensten, A., et al. (2006) Global patterns of influenza a virus in wild birds. *Science* **312**(5772), 384–388.

Orme-Zavaleta, J., Jorgensen, J., D'Ambrosio, B., et al. (2006). Discovering spatio-temporal models of the spread of West Nile virus. *Risk Analysis* **26**(2), 413–422.

Otero, J., Jensen, A. J., L'Abée-Lund, J. H., et al. (2012). Contemporary ocean warming and freshwater conditions are related to later sea age at maturity in Atlantic salmon spawning in Norwegian rivers. *Ecology and Evolution*, **2**(9): 2192–2203.

Owen, J., Moore, F., Panella, N., et al. (2006). Migrating birds as dispersal vehicles for West Nile virus. *Ecohealth* **3**(2), 79–85.

Parmesan, C. (2007). Influences of species, latitudes and methodologies on estimates of phenological response to global warming. *Global Change Biology* **13**(9), 1860–1872.

Pfeffer, M., and Dobler, G. (2010). Emergence of zoonotic arboviruses by animal trade and migration. *Parasites and Vectors* **3**(1), 35–50.

Plowright, R. K., Foley, P., Field, H. E., et al. (2011). Urban habituation, ecological connectivity and epidemic dampening: The emergence of Hendra virus from flying foxes (*Pteropus* spp.). *Proceedings of the Royal Society B: Biological Sciences* **278**, 3703–3712.

Prosser, D. J. (2012). Wild birds and emerging diseases: modeling avian influenza transmission risk between domestic and wild birds in China, Ph.D Dissertation. University of Maryland, College Park, MD.

Prosser, D. J., Takekawa, J. Y., Newman, S. H., et al. (2009). Satellite-marked waterfowl reveal migratory connection between H5N1 outbreak areas in China and Mongolia. *Ibis* **151**, 568–576.

Prosser, D. J., Cui, P., Takekawa, J. Y., et al. (2011). Wild bird migration across the Qinghai-Tibetan plateau: A transmission route for highly pathogenic H5N1. *PLoS One* **6**(3): e17622.

Prosser, D. J., Takekawa, J. Y., and Newman, S. (2012). USGS–UNFAO Partnership: Understanding Migratory Birds and Their Role in Highly Pathogenic Avian Influenza Transmission. https://www.pwrc.usgs.gov/resshow/prosser/USGS-FAOWildBirdAIProgram.pdf (accessed on May 29, 2013).

Prosser, D. J., Hungerford, L. L., Erwin, R. M., et al. (2013 In Press). Mapping high risk areas of disease transfer between domestic and wild birds: The case of highly pathogenic avian influenza in China. *Ecohealth, Spatial and Spatio-temporal Epidemiology* (*OR Ecological Modelling*).

Prosser, D. J., Lee, D., Ding, C., et al. (2013 In Review). Species distribution modeling in a region of high need and limited data: China's Anatidae waterfowl. *Frontiers in Ecology and the Environment* (*OR Ecological Modelling OR Biological Conservation*).

Rankin, M. A., and Burchsted, J. C. A. (1992). The cost of migration in insects. *Annual Review of Entomology* **37**, 533–559.

Rappole, J. H., and Hubálek, Z. (2003). Migratory birds and West Nile virus. *Journal of Applied Microbiology* **94**, 47–58.

Rappole, J. H., Derrickson, S. R., and Hubalek, Z. (2000). Migratory birds and spread of West Nile virus in the Western Hemisphere. *Emerging Infectious Diseases* **6**(4), 319–328.

Reed, K. D., Meece, J. K., Henkel, J. S., et al. (2003). Birds, migration and emerging zoonoses: West nile virus, lyme disease, influenza A and enteropathogens. *Clinical Medicine & Research* **1**(1), 5–12.

Reeves, W. C., Hardy, J. L., Reisen, W. K., et al. (1994). Potential effect of global warming on mosquito-borne arboviruses. *Journal of Medical Entomology* **31**(3), 323–332.

Reisen, W. K. (2003). Epidemiology of St. Louis encephalitis virus. In: "Advances in Virus Research," Vol. 61, pp. 139–183. Academic Press, New York.

Reisen, W. K., Fang, Y., and Martinez, V. M. (2006). Effects of temperature on the transmission of West Nile Virus by *Culex tarsalis* (Diptera: Culicidae). *Journal of Medical Entomology* **43**(2), 309–317.

Richter, H. V., and Cumming, G. S. (2008). First application of satellite telemetry to track African straw-coloured fruit bat migration. *Journal of Zoology* **275**(2), 172–176.

Root, T. (1988). "Atlas of Wintering North American Birds: An Analysis of Christmas Bird Count Data." University of Chicago Press, Chicago, IL.

Roscoe, D. E., Holste, W. C., Sorhage, F. E., et al. (1998). Efficacy of an oral vaccinia-rabies glycoprotein recombinant vaccine in controlling epidemic raccoon rabies in New Jersey. *Journal of Wildlife Diseases* **34**(4), 752–763.

Rueda, L. M., Patel, K. J., Axtell, R. C., et al. (1990). Temperature-dependent development and survival rates of *Culex quinquefasciatus* and *Aedes aegypti* (Diptera: Culicidae). *Journal of Medical Entomology* **27**(5), 892–898.

Ruiz, M., Chaves, L., Hamer, G., et al. (2010). Local impact of temperature and precipitation on West Nile virus infection in Culex species mosquitoes in northeast Illinois, USA. *Parasites & Vectors* **3**(1), 19.

Russell, A. P., Bauer, A. M., and Johnson, M. K. (2005a). Migration in amphibians and reptiles: An overview of patterns and orientation mechanisms in relation to life history strategies. In: "Migration of Organisms: Climate, Geography, Ecology" (A. M. Elewa, Ed.), pp. 151–184. Springer-Verlag, Berlin.

Russell, C. A., Smith, D. L., Childs, J. E., et al. (2005b). Predictive spatial dynamics and strategic planning for raccoon rabies emergence in Ohio. *PLoS Biology* **3**(3), 382–388.

Saino, N., Ambrosini, R., Rubolini, D., et al. (2011). Climate warming, ecological mismatch at arrival and population decline in migratory birds. *Proceedings of the Royal Society B: Biological Sciences*, **278**(1707), 835–842.

Sammels, L. M., Lindsay, M. D., Poidinger, M., et al. (1999). Geographic distribution and evolution of Sindbis virus in Australia. *Journal of General Virology* **80**(3), 739–748.

Schmidt, B., Schaub, M., and Steinfartz, S. (2007). Apparent survival of the salamander *Salamandra salamandra* is low because of high migratory activity. *Frontiers in Zoology* **4**(1), 19.

Serneels, S., and Lambin, E. F. (2001). Impact of land-use changes on the wildebeest migration in the northern part of the Serengeti–Mara ecosystem. *Journal of Biogeography* **28**(3), 391–407.

Shirako, Y., Niklasson, B., Dalrymple, J. M., et al. (1991). Structure of the Ockelbo virus genome and its relationship to other sindbis viruses. *Virology* **182**(2), 753–764.

Sims, L. D., and Brown, I. H. (2008). Multicontinental epidemic of H5N1 HPAI virus (1996–2007). In: "Avian Influenza" (D. E. Swayne, Ed.), pp. 251–286. Blackwell Publishing, Ames, IA.

Slemons, R.D., Hansen, W.R., Converse, K.A., et al (2003). Type A influenza virus surveillance in free-flying, nonmigratory ducks residing on the eastern shore of Maryland. *Avian Diseases* **47**, 1107–1110.

Slenning, B. D. (2010). Global climate change and implications for disease emergence. *Veterinary Pathology* **47**(1), 28–33.

Spinsanti, L. I., Diaz, L. A., Glatstein, N., et al. (2008). Human outbreak of St. Louis encephalitis detected in Argentina, 2005. *Journal of Clinical Virology* **42**(1), 27–33.

Stallknecht, D.E., and Shane, S.M. (1988) Host range of avian influenza-virus in free-living birds. *Veterinary Research Communications* **12**(2–3), 125–141.

Stallknecht, D. E., Kearney, M. T., Shane, S. M., et al. (1990a). Effects of pH, temperature, and salinity on persistence of avian influenza-viruses in water. *Avian Diseases* **34**(2), 412–418.

Stallknecht, D. E., Shane, S. M., Kearney, M. T., et al. (1990b). Persistence of avian influenza viruses in water. *Avian Diseases* **34**(2), 406–411.

Stallknecht, D. E., Goekjian, V. H., Wilcox, B. R., et al. (2010). Avian influenza virus in aquatic habitats: What do we need to learn? *Avian Diseases* **54**(1), 461–465.

Sturm-Ramirez, K. M., Hulse-Post, D. J., Govorkova, E. A., et al. (2005). Are ducks contributing to the endemicity of highly pathogenic H5N1 influenza virus in Asia? *Journal of Virology* **79**(17), 11269–11279.

Takekawa, J. Y., Heath, S. R., Douglas, D. C., et al. (2009). Geographic variation in bar-headed geese *Anser indicus*: Connectivity of wintering and breeding areas and migration across a broad front. *Wildfowl* **59**, 102–125.

Takekawa, J. Y., Newman, S., Xiao, X., et al. (2010a). Migration of waterfowl in the East Asian Flyway and spatial relationship to HPAI H5N1 outbreaks. *Avian Diseases* **54**, 466–476.

Takekawa, J. Y., Prosser, D. J., Newman, S., et al. (2010b). Victims and vectors: Highly pathogenic H5N1 and the ecology of water birds. *Avian Biology Research* **3**, 51–73.

Taylor, L. H., Latham, S. M., and Woolhouse, M. E. J. (2001). Risk factors for human disease emergence. *Philosophical Transactions of the Royal Society of London. Series B: Biological Sciences* **356**(1411), 983–989.

Teadrow, C. A. (2009). Ph.D. Dissertation. George Mason University, Fairfax, VA.

Thomas, N. J., Hunger, D. B., and Atkinson, C. T. (Eds.) (2007). "Infectious Diseases of Wild Birds." Blackwell Publishing, Oxford.

Thomson, G. R., Vosloo, W., and Bastos, A. D. S. (2003). Foot and mouth disease in wildlife. *Virus Research* **91**(1), 145–161.

Van Buskirk, J., Mulvihill, R. S., and Leberman, R. C. (2009). Variable shifts in spring and autumn migration phenology in North American songbirds associated with climate change. *Global Change Biology* **15**(3), 760–771.

van Gils, J. A., Munster, V. J., Radersma, R., et al. (2007). Hampered foraging and migratory performance in swans infected with low-pathogenic avian influenza A virus. *PLoS One* **2**(1), e184.

Van Houtan, K. S., Hargrove, S. K., and Balazs, G. H. (2010). Land use, macroalgae, and a tumor-forming disease in marine turtles. *PLoS One* **5**(9), e12900.

van Riel, D., Munster, V. J., de Wit, E., et al. (2006). H5N1 virus attachment to lower respiratory tract. *Science* **312**(5772), 399.

Vazquez, A., Jimenez-Clavero, M., Franco, L., et al. (2011). Usutu virus: Potential risk of human disease in Europe. *Euro Surveill* **16**(31).

Visser, M. E., and Both, C. (2005). Shifts in phenology due to global climate change: The need for a yardstick. *Proceedings of the Royal Society B: Biological Sciences* **272**(1581), 2561–2569.

Visser, M. E., Perdeck, A. C., Van Balen, J. H., et al. (2009). Climate change leads to decreasing bird migration distances. *Global Change Biology* **15**(8), 1859–1865.

Vosloo, W., Bastos, A. D. S., Sangare, O., et al. (2002). Review of the status and control of foot and mouth disease in sub-Saharan Africa. *Revue Scientifique et Technique De L Office International Des Epizooties* **21**(3), 437–449.

Waldenstrom, J., Lundkvist, A., Falk, K. I., et al. (2007). Migrating birds and tickborne encephalitis virus. *Emerging Infectious Diseases* **13**(8), 1215–1218.

Walther, G.-R., Post, E., Convey, P., et al. (2002). Ecological responses to recent climate change. *Nature* **416**(6879), 389–395.

Weaver, S. (1998). Recurrent emergence of Venezuelan equine encephalomyelitis. In: "Emerging Infections" (W. M. Shields, D. Armstrong, and J. M. Hughes, Eds.), pp. 27–42. American Society for Microbiology, Washington, DC.

Webby, R. J., and Webster, R. G. (2001). Emergence of influenza A viruses. *Philosophical Transactions of the Royal Society of London Series B: Biological Sciences* **356**(1416), 1817–1828.

Webster, R. G., Krauss, S., Hulse-Post, D., et al. (2007). Evolution of influenza a viruses in wild birds. *Journal of Wildlife Diseases* **43**(3), S1–S6.

Weissenbock, H., Kolodziejek, J., Url, A., et al. (2001). Emergence of Usutu virus, an African mosquito-borne flavivirus of the Japanese encephalitis virus group, central Europe. *Emerging Infectious Diseases* **8**(7), 652–656.

Weissenbock, H., Chvala-Mannsberger, S., Bayoni, T., et al. (2007). Emergence of Usutu virus in Central Europe: Diagnosis, surveillance and epizootiology. In: "Emerging Pests and Vector-borne Diseases in Europe 2007" (W. Takken, and B. G. J. Knols, Eds.), pp. 153–168. Wageningen Academic Publishers, The Netherlands.

WHO (2012a). Hendra virus. http://www.who.int/mediacentre/factsheets/fs329/en/index.html (accessed September 14, 2012).

WHO (2012b). Nipah virus. http://www.who.int/mediacentre/factsheets/fs262/en/ (accessed September 14, 2012).

Work, T. M., Balazs, G. H., Rameyer, R. A., et al. (2004). Retrospective pathology survey of green turtles *Chelonia mydas* with fibropapillomatosis in the Hawaiian Islands, 1993–2003. *Diseases of Aquatic Organisms* **62**, 163–176.

World Organization of Animal Health (2005). Manual of diagnostics tests and vaccines for terrestrial animals. In: "Avian Influenza" (World Organization of Animal Health, Ed.), Paris, France.

Yamaguchi, N., Hupp, J. W., Higuchi, H., et al. (2010). Satellite-tracking of northern pintail *Anas acuta* during outbreaks of the H5N1 virus in Japan: Implications for virus spread. *Ibis* **152**, 262–271.

Yasue, M., Feare, C. J., Bennun, L., et al. (2006). The epidemiology of H5N1 avian influenza in wild birds: Why we need better ecological data. *Bioscience* **56**(11), 923–929.

10

ILLEGAL ANIMAL AND (BUSH) MEAT TRADE ASSOCIATED RISK OF SPREAD OF VIRAL INFECTIONS

Christopher Kilonzo

Department of Population Health and Reproduction, School of Veterinary Medicine, University of California, Davis, CA, USA

Thomas J. Stopka

Department of Public Health and Community Medicine, Tufts University School of Medicine, Boston, MA, USA

Bruno Chomel

Department of Population Health and Reproduction, School of Veterinary Medicine, University of California, Davis, CA, USA

TABLE OF CONTENTS

Viral Infections and Global Change, First Edition. Edited by Sunit K. Singh.

10.1 INTRODUCTION

It is estimated that approximately 75% of the pathogens causing emerging infectious diseases today are zoonotic, with wildlife playing a major role in such emergence (Cunningham, 2005). Illegal wildlife trade is the second-largest black market worldwide, after narcotics (Toledo et al., 2012). During recent years, researchers have increasingly worked to assess the origins and worldwide trends in emerging infectious diseases (Jones et al., 2008; Wolfe et al., 2007). It has been suggested that changes in human ecology and behavior, including migration and increased population density (Murphy, 2008), social conflict and war, changes in personal behavior, and increased deforestation, have played an important role in the emergence of new infectious diseases (Weiss and McMichael, 2004). Climate change, agricultural practices, antibiotic resistance, and immunodeficiency attributed to the human immunodeficiency virus (HIV) pandemic (LeBreton et al., 2007), perhaps the most devastating modern-day emerging infectious disease of animal origin, have also influenced global trends in the emergence and reemergence of infectious diseases (Jones et al., 2008). Enhanced surveillance, particularly in developed countries, has led to an increased frequency and rapidity with which emerging diseases are detected in present time (Jones et al., 2008; Weiss and McMichael, 2004). We conducted a thorough review of the scientific literature in an effort to provide a detailed overview of current research and knowledge surrounding the bushmeat trade, illegal animal trade, and emerging zoonoses and to suggest future directions in research and policy. We consider HIV as a case study to emphasize the worldwide impact that zoonoses can have on public health and to suggest steps that can be taken to prevent and address future zoonotic pandemics.

10.2 SEARCH STRATEGY AND SELECTION CRITERIA

Literature for this review was obtained from searches of PubMed, CAB, and Google Scholar®. Search terms included "bushmeat," "bush meat," "bushmeat and zoonoses," "bushmeat hunting," "bushmeat trade," "bushmeat hunting and emerging zoonoses," "illegal trade and zoonoses," and "bushmeat hunting and infectious disease." References cited by relevant journal articles were also obtained for review. The reviewed literature was limited to English language articles and abstracts. We excluded from our study cases associated with consumption of game in developed countries that can also lead to viral zoonoses such as hepatitis E associated with consumption of wild boar and wild deer raw meat, as reported in Japan (Matsuda et al., 2003; Tei et al., 2003).

10.3 THE BUSHMEAT TRADE

Bushmeat, a moniker for the meat of wild animals, includes any type of terrestrial animal, from rodents to elephants. Duikers (*Philantomba monticola*), a type of small antelope; cane rats (*Thryonomys swinderianus*); brush-tailed porcupines (*Atherurus africanus*); and various nonhuman primates (NHP) are reported as the most commonly eaten bushmeat (Kümpel, 2005). Some researchers indicate that a number of marine animals, such as cetaceans, sea turtles, and sirenians ("marine bushmeat"), could also be included in a broad definition for bushmeat (Clapham and Van Waerebeek, 2007). During recent years, a growing number of emerging zoonotic diseases have been linked to bushmeat hunting, butchering, trade, and consumption (Wolfe et al., 2005b).

Bushmeat plays an important role in local economies, human sustenance, nutrition, international trade, culture, and the reduction of biodiversity (Brashares et al., 2004; Rosen and Smith, 2010; Chomel et al., 2007; Vliet et al., 2011). While it has been an important dietary staple in many locations internationally for generations, increased access to new and expanded hunting areas has been made possible by logging campaigns and development of related roadways (Kümpel, 2005; Patz et al., 2004; Wolfe et al., 2005b). The bushmeat trade is a multibillion dollar endeavor that provides a substantial boost to many local economies (Brashares et al., 2004). Consumers living in developed countries are willing to pay high prices for bushmeat, considered culturally as a highly valued product. It was estimated that bushmeat resale in specific markets in Paris, France, in June 2008 was ranging between 20 and 30 Euros (US$ 31–46) per kg compared to an average of 15 Euros (US$ 23) per kg for domestic meat sold in French supermarkets (Chaber et al., 2010). The development of luxury markets such as this one is of concern as it creates an even higher demand for endangered species (Chaber et al., 2010). More recently, retroviruses and herpesviruses were detected in NHP products illegally imported in the USA (Smith et al., 2012). On the Ivory Coast alone, the value of the bushmeat trade has been estimated at US$ 150 million (Fa et al., 2002). Wild meat harvests range from hundreds of thousands of tons per year in Latin America to more than three million tons annually in Central Africa (Fa et al., 2002; Kümpel, 2005; Wolfe et al., 2005b). Concomitant with increased bushmeat hunting, butchering, trade, and consumption, a growing number of emerging zoonotic diseases have led to concerning levels of morbidity and mortality among human and animal populations.

10.4 BUSHMEAT HUNTING AND EMERGING INFECTIOUS DISEASES

Increased human contact with wild animals through hunting, butchering, and consumption of bushmeat has led to cross species transmission of numerous infectious diseases, several of which have led to devastating outcomes across the globe (LeBreton et al., 2006). Transmission across species through bushmeat hunting and butchering has been connected to outbreaks of monkeypox virus (Khodakevich et al., 1988; LeBreton et al., 2006), Ebola virus (Amblard et al., 1997; Cunningham, 2005; Leroy et al., 2004), simian foamy viruses (Calattini et al., 2007; Wolfe et al., 2004b), human T-lymphotropic viruses (HTLV) (Wolfe et al., 2005a; Zheng et al., 2010), and even Lassa fever by consumption of peri-domestic rodents (Ter Meulen et al., 1996). Recent molecular research indicates that many variants, highly divergent strains, and subtypes of these and other zoonoses have crossed between NHP and humans (Calattini et al., 2009; Carr et al., 2010; Gao et al., 1998), sometimes on

TABLE 10.1. Zoonotic Viral Pathogens Transmissible Through Bushmeat

Virus	Type of bushmeat	Location	Reference
Ebola	Chimpanzee	Gabon, Africa	Amblard et al. (1997); Leroy et al. (2004)
Monkeypox	NHP, rodents, squirrels	Central, Western Africa	Khodakevich et al. (1988); Rimoin et al. (2010)
HIV-1/SIVcpz	Chimpanzees	Central/Eastern Africa	Carr et al. (2010); Gao et al. (1999)
HIV-2/SIVsm	Sooty mangabey (*C. atys*)	West Africa	Van Heuverswyn and Peeters (2007)
Simian foamy virus	NHP	Cameroon	Calattini et al. (2007); Wolfe et al. (2004a,b)
SARS	Bats (civets)	China	Poon et al. (2004)
HTLV1	NHP	Central Africa	Courgnaud et al. (2004); Wolfe et al. (2005a)
HTLV2	NHP	Central Africa	Peeters (2004)
HTLV3	NHP	Central Africa	Wolfe et al. (2005a); Zheng et al. (2010)
HTLV4	NHP	Central Africa	Wolfe et al. (2005a)
STLV	NHP	Central Africa	Sintasath et al. (2009)
Hepatitis E	Deer, wild boar	Japan	Matsuda et al. (2003); Tei et al. (2003)
Lassa	Rodents	Guinea	Ter Meulen et al. (1996)

NHP, nonhuman primates.

multiple occasions (Gao et al., 1999). Several outbreaks of Ebola virus in Western Africa have been traced back to index cases having butchered dead chimpanzees (Fa et al., 2002). Cross species transmission of the simian immunodeficiency virus (SIV) from NHP to humans is linked to bushmeat hunting and is widely believed to be the precursor to the HIV pandemic (Apetrei et al., 2005; Peeters, 2004; Takehisa et al., 2009). Extensive research on NHP suggests that African primates represent a very large reservoir for a number of lentiviruses and have the potential to infect other species (including humans) in their natural habitats (Hahn et al., 2000). It has been estimated that more than 20% of NHP hunted for food are infected with SIV (Peeters et al., 2002). HTLV, especially human T-lymphotropic virus type 1 (HTLV1) and human T-lymphotropic virus type 2 (HTLV2), are related to distinct lineages of simian T-lymphotropic viruses (STLV) 1 and 2 (Courgnaud et al., 2004; Sintasath et al., 2009; Wolfe et al., 2005a). A large diversity of HTLV- and STLV1-like viruses have been identified in Central African bushmeat hunters, suggesting that exposure to NHP contributed to the emergence of HTLV in humans (Wolfe et al., 2005a). In Gabon, simian foamy retroviruses (SFVs) were characterized in wild-borne NHP, and cross species transmission to humans could be detected (Mouinga-Ondémé et al., 2012). A total of 497 NHP samples composed of 286 blood and 211 tissue (bushmeat) samples were collected. Anti-SFV antibodies were detected in 31 (10.5%) of 286 plasma samples. Novel SFVs were detected in several *Cercopithecus* species. Of the 78 humans, mostly hunters, who had been bitten or scratched by NHP, 19 were SFV seropositive, with 15 cases confirmed by PCR. All but one were infected with ape SFV (Table 10.1).

While the majority of research focused on bushmeat hunting and emerging zoonotic diseases focuses on Africa, it is important to acknowledge the diverse contexts of zoonotic transmission of simian retroviruses in Asia (Jones-Engel et al., 2005). Economic growth and development of new infrastructure has transformed many regions of Asia. This is particularly evident in South and Southeast Asia, where some of the world's most densely packed human populations are situated in close proximity to some of the world's richest sources of biodiversity. It has been demonstrated that the international severe acute respiratory syndrome (SARS) outbreak in 2003 may have been triggered by interest in exotic meats (Peiris et al., 2004; Poon et al., 2004). Asian cultures have long histories of venerating NHP, often placing animals in direct contact with people. Such close contact also occurs during bushmeat hunting and consumption in some regions of Asia. Each of these diverse contexts has the potential to facilitate zoonotic transmission of simian foamy viruses (Jones-Engel et al., 2008) and perhaps a wider spectrum of zoonotic diseases in Asia.

Bushmeat hunting in Latin America, believed to be smaller in sheer harvest tonnage (Fa et al., 2002), is still quite prevalent among a number of indigenous populations, particularly in the Amazon region of South America, where the woolly monkey (*Lagothrix lagotricha*), spider monkey (*Ateles chamek*), and white-lipped peccary (*Tayassu pecari*), among others, are often hunted (Ohl-Schacherer et al., 2007). Such close contact between humans and NHP and other mammals may also contribute to transmission of zoonotic diseases in Latin America.

10.5 RISK FACTORS AND MODES OF TRANSMISSION

10.5.1 Human–nonhuman primate overlap

A number of ecological and behavioral factors contribute to increased risk of exposure to zoonoses through the bushmeat trade. Research in Africa indicates that primate hunters are susceptible to zoonotic infections such as naturally acquired simian retrovirus infections and that simian foamy virus infections were acquired from three distinct NHP lineages (Wolfe et al., 2004b), supporting evidence that emerging zoonotic diseases connected to wildlife and bushmeat are well established in their natural habitats and can easily adapt and jump from species to species.

Increased demand for bushmeat in urban centers and increased access to bushmeat due to improved access to prey via new logging roads have increased human exposure to retroviruses and other zoonotic disease agents (Chomel et al., 2007). Once isolated and hard to reach, several regions of the world are now easily accessible due to a growing network of rural roads that bisect natural habitats for wild animals. Habitat fragmentation caused by logging roads can impact the density of reservoir hosts, in some regions increasing the relative abundance of highly competent reservoirs, while fragmentation due to road building may also increase the overlap and interaction between human populations and reservoir hosts (Wolfe et al., 2005b). A recent study in Congo indicates that such overlap is particularly high among individuals who migrate to logging developments—increasing the population by 69% in five logging towns over the course of 6 years (Poulsen et al., 2009). In an effort to meet nutritional needs without stable resources, migrants to logging towns may put themselves at increased risk by hunting and consuming bushmeat at elevated rates, with an increase of 64% in logging towns' bushmeat supplies during the same time period (Poulsen et al., 2009). Similarly, war-torn areas are at increased risk of bushmeat demand and supply. In Congo, a fivefold increase of bushmeat sales was reported in war-torn areas, where family members of military officers who are heavily armed are well equipped for bushmeat hunting and are reported to sell the meat at local markets (De Merode and Cowlishaw, 2006).

10.5.2 Behavioral risks

Zoonotic disease connected to the bushmeat trade can be directly linked to three primary human risk behaviors: (i) hunting (medium risk), (ii) butchering (high risk), and (iii) consumption (low risk due to less contact with blood) of infected animals. Microbial transmission can occur during multiple points in the hunting and butchering process. During hunting activities, for instance, individuals are more susceptible to bites and scratches from infected primates and rodents, particularly if they have open wounds (LeBreton et al., 2006). Research in Cameroon with a general population sample and a sample of bushmeat hunters found that hunters who experienced serious bites, scratches, wounds, or other injuries from animals, especially gorillas and apes, were at highest risk for simian foamy virus infection (Calattini et al., 2007). Sick animals may be more easily hunted or captured, increasing the odds that transmission from animals to humans occurs. Twenty-one people were infected with Ebola, for instance, by one infected chimpanzee that was found dead in Africa (Leroy et al., 2004).

Exposure through such risk behaviors varies across different subpopulations. The factors most associated with hunting and butchering activities include gender and type of landscape surrounding villages (LeBreton et al., 2006). In Cameroon, men have been found to be more likely than expected to have hunted bushmeat, while women have been more likely than expected to have butchered bushmeat (Jones et al., 2008; Wolfe et al., 2004a). Investigators found that a higher percentage of participants had eaten wild game than had butchered wild game, a higher percentage had butchered wild game than had hunted it, and all three activities were more likely in forested areas than in other types of terrain (Wolfe et al., 2004a). Overall, men comprised 83% of participants reporting direct contact, and most reports of direct contact involved monkeys (Wolfe et al., 2004a). Of participants who reported direct contact with NHP blood and body fluids, 91% reported butchering primates, 73% reported hunting NHP, and 43% reported keeping an NHP as a pet, a behavior that may also place individuals at increased risk of zoonotic disease acquisition. In Congolese logging communities, in order to meet their nutritional sustenance needs, migrants are at increased risk of exposure compared to their native counterparts, as they were responsible for 72% of bushmeat supply (hunting) and 66% of bushmeat consumption (Poulsen et al., 2009).

10.6 CONSERVATION AND WILDLIFE SUSTAINABILITY

While habitat conservation and wildlife sustainability may not be foremost on the mind of disease prevention specialists, opportunities to engage members of diverse communities exist that could lead to mutually beneficial collaboration across such disciplines. Exploitation of bushmeat represents the largest threat to wildlife sustainability in some regions of the world (Kümpel et al., 2010; Ohl-Schacherer et al., 2007; Waite, 2007). In fact, investigators have asserted that the "ape bushmeat crisis" cannot last long when considering the relatively low population of chimpanzees remaining worldwide (~200 000 in 1999) and the thousands of kills that occur to fulfill bushmeat hunting needs annually (Weiss and Wrangham, 1999).

Bennett et al. (2007) discuss the occurrence of a "bushmeat crisis" as it relates to conservation. In some regions of South America, however, where relatively small indigenous communities rely on subsistence hunting of local wild species, it appears that such a crisis is less of a concern as bushmeat hunting can remain sustainable while human population sizes remain stable (Fa et al., 2002; Ohl-Schacherer et al., 2007). It has been recommended that, ultimately, policies regarding bushmeat hunting need to seek to attain human needs (i.e., health, sustenance, and nutrition) while having benign consequences on exploited species (Waite, 2007).

In one study, feral pig hunting in the Brazilian Pantanal was found to positively impact wildlife conservation efforts (Jean Desbiez et al., 2011). Feral pigs, an abundant and invasive species in the area, have led to a reduction of hunting pressure on native wildlife by effectively acting as a replacement species for the hunters. In addition to being culturally acceptable, the pigs provide a readily available source of meat and oil to the inhabitants.

Opportunities for conservation and disease prevention also exist across species and terrains. The relationship between supply and demand of bushmeat and fish resources has been explored with great interest in recent years in order to better understand how exploitation of one influences exploitation of the other (Rowcliffe et al., 2005). Strong evidence exists for a direct link of wild meat and fish consumption (Brashares et al., 2004). It is surmised that the primary anthropogenic threat to six sea turtle species, three sirenians, and an unknown number of cetaceans may be the bushmeat trade. This is particularly notable in South American and West African coastal countries (Clapham and Van Waerebeek, 2007).

Bushmeat hunting has even been discussed as a climate threat in that it diminishes dispersion of large tree seeds by diminishing populations of large animals and, in turn, decreases carbon storage in trees (Brodie and Gibbs, 2009). All of these issues remind us of the intertwined issues across research disciplines and across diverse communities.

10.7 CASE STUDY: THE ROLE OF THE BUSHMEAT TRADE IN THE EVOLUTION OF HIV

HIV, the cause of the acquired immunodeficiency syndrome (AIDS), ranks as one of the most devastating zoonotic infectious diseases that man has encountered in recent decades. By the year 2000, less than two decades since AIDS was first recognized, it had resulted in more than 16 million deaths internationally (Hahn et al., 2000). In 2006, 40 million individuals were infected with HIV-1 worldwide, and more than 50% were inhabitants of sub-Saharan Africa (LeBreton et al., 2007).

Humans were not the initial and natural hosts of HIV. The bushmeat trade and the "viral chatter" that is prevalent in its general vicinity have played an integral role in the evolution, transmission, and distribution of HIV/AIDS. It is now generally accepted that these viruses entered the human population as a result of zoonotic disease transmission during the latter half of the twentieth century (Hahn et al., 2000; Worobey et al., 2008), originating in Central Western Africa where multiple parental strains have been detected (Carr et al., 2010) and where bushmeat hunting, butchering, and consumption have been commonplace for generations.

The search for the origins of HIV has led researchers to learn that the common chimpanzee (*Pan troglodytes*) is the common reservoir for HIV-1 and has been the source of human infection on at least three occasions (Gao et al., 1999). The ancestral strains of HIV-1 groups M and N still persist in today's wild chimpanzee populations (*Pan troglodytes troglodytes*) in south Cameroon, and HIV-1 group O-related viruses have been identified in western gorillas (*Gorilla gorilla*), with chimpanzees most likely the original reservoir of this virus (Takehisa et al., 2009). HIV-2 is the result of at least eight distinct cross species transmissions of SIV from sooty mangabeys (*Cercocebus atys*) in West Africa (Hahn et al., 2000; Van Heuverswyn and Peeters, 2007).

Among HIV-1-infected persons surveyed across 17 rural villages in Cameroon, almost 80% reported butchering wild animals, 26% reported hunting them, and 96% reported eating wild animals (LeBreton et al., 2007), perhaps indicative of a long-standing overlap between humans and NHP. Initially, research indicated that hunted primates had a high prevalence of zoonoses such as SIV (Peeters et al., 2002), the virus that led to HIV. More recent research, however, indicates that SIV prevalence differs across species and within species depending on sampling locations. Further, low SIV prevalence has been detected within primate species that are most frequently hunted as well as a high genetic diversity across and within SIV lineages (Aghokeng et al., 2010).

In addition to the tremendous challenges that exist in detecting, treating, and curtailing HIV infection, this zoonotic disease also leads to conditions that foster the emergence and reemergence of a number of additional zoonotic diseases. It is believed that immunosuppression, due largely to HIV, could favor the process of adaptation and subsequent emergence of new zoonotic pathogens (LeBreton et al., 2007). It is surmised that the frequency of exposure to zoonoses such as SIV has increased during recent decades (Kalish et al., 2005), and such increases may continue into the future if bushmeat hunting and related activities are left unabated. Further, HIV-related immunosuppression likely enhances the risk for acquisition, adaptation, and emergence of zoonotic diseases infecting other animals that are frequently hunted such as monkeypox and hantaviruses in rodents and *Lyssavirus* in bats (Khodakevich et al., 1988; LeBreton et al., 2007). In fact, a major increase (20-fold) in the number of human cases of monkeypox has been reported for the last 30 years in the Democratic Republic of the Congo, living in forested areas being one of the significant risk factors, likely associated with hunting behaviors (Rimoin et al., 2010). Coinfection with HIV-1 and additional zoonoses, such as simian foamy virus, has led to increased concerns that infection with multiple zoonoses may further complicate public health matters in the near future (Switzer et al., 2008).

Frequent and widespread international travel has also contributed to rapid and thorough transmission of HIV/AIDS—bringing pandemics across continents and oceans over the course of weeks and months rather than years and decades. Other zoonotic diseases related to bushmeat hunting, as with HIV, can be carried to locations that are often far removed from the point of original cross species transmission during the months and years to come.

10.8 ILLEGAL TRADE OF DOMESTIC ANIMALS AND EXOTIC PETS

Illegal trade can also be a possible source of zoonotic human infection. Travelers returning to Europe from North Africa or other regions of the world where they adopted stray puppies have been exposed to rabies and have brought rabid animals in rabies-free areas (Brugère-Picoux and Chomel, 2009). Similarly, trade of exotic African rodents has been involved in an outbreak of monkeypox in prairie dogs and humans in the United States (Reed et al., 2004). Cases of tularemia were detected in people who had purchase wild-caught prairie dogs (Avashia et al., 2004). A highly pathogenic avian influenza (HPAI) A H5N1 virus from crested hawk eagles smuggled into Europe by air travel has been isolated and characterized (Van Borm et al., 2005); fortunately, screening of human and avian contacts indicated that no dissemination had occurred.

In Vietnam, 95% of Hanoian bird vendors appear unaware of trade regulations, and across Vietnam vendors buy birds sourced outside of their province (Edmunds et al., 2011). Approximately 25% of the species common to Vietnam's bird trade are known to be HPAI virus H5N1 susceptible. The anthropogenic movement of birds within the trade chain and the range of HPAI-susceptible species, often traded alongside poultry, increase the risk Vietnam's bird trade presents for the transmission of pathogens such as HPAI H5N1.

10.9 DISCUSSION AND FUTURE DIRECTIONS

A number of cultural, socioeconomic, and geographic dimensions are associated with bushmeat hunting and emerging zoonoses. Movement and entry of human populations into new locations that bridge NHP and human habitats, concomitant entry of additional monies, weaponry and local economies, interest in exotic and luxurious food items, and civil unrest are but a few of the dynamic issues that have influenced the bushmeat trade and related transmission of zoonotic diseases. Prevention and control efforts that incorporate culturally appropriate health education, conservation, and supply and demand-focused interventions are needed. Interdisciplinary collaboration, new technologies, and enhanced surveillance can provide essential tools to improve efforts to detect and decrease zoonoses tied to bushmeat hunting.

10.10 PREVENTION AND CONTROL: FROM SUPPLY AND DEMAND TO HEALTH EDUCATION TECHNIQUES

The reduction of bushmeat hunting and butchering has the potential to decrease the prevalence of zoonotic disease emergence (Wolfe et al., 2004b). Halting the exotic meat trade altogether would, on the surface, seem like a reasonable solution to prevent the spread of emerging infectious diseases (Weiss and McMichael, 2004). The hunting and butchering of bushmeat in many parts of the world, however, is part of daily food preparation and sustenance. It has been suggested that, instead, a risk reduction approach may be more realistic and effective (Pike et al., 2010). Conservation interventions that aim to reduce the commercial bushmeat trade need to account for likely shifts in individual spending that may ensue and the secondary effects on household economies, which can be highly dependent on bushmeat income (Coad et al., 2010).

Increased access to domestic animal meat, fish, and fresh foods has also been deemed worthy of consideration. Research indicates that consumers of meat in West/Central Africa make a much clearer distinction between fresh and frozen foods than between bushmeat, domestic meat, and fish. Fresh foods are strongly preferred over frozen foods and consumption of fresh foods increases with income (East et al., 2005). Consumption patterns are also related to tribe and nationality. "Bringing the trade in resilient species into the formal economy could provide the impetus needed to monitor and manage stocks effectively while improving protection for vulnerable species (East et al., 2005). This could make bushmeat amenable to the kinds of policy tools open to fisheries" (Rowcliffe et al., 2005). East et al. (2005) claim that controlled demand for bushmeat could be met from common, highly productive species that are relatively robust to exploitation. Improving the supply of underdeveloped commodities, particularly domestic livestock, could also offset demand for bushmeat. Finding the appropriate balance between what is sustainable in theory and what is manageable in practice remains as the primary challenge (Bennett et al., 2007). Some experts point to the transition from bushmeat to domestic meat in Asia as a possible glimpse into the future for Africa (East et al., 2005). In both Asia and Africa, people from forest communities prefer bushmeat to domestic meat. In Asia, as populations grew and forests diminished, people made the shift to domestic meat. Questions persist as to whether Africans will go in the same direction once wildlife and wild meat are gone (Kümpel, 2005). Poulsen et al. (2009) assert that, while "enforcement of hunting laws and promotion of alternative sources of protein may help curb the pressure on wildlife, the best strategy for biodiversity conservation may be to keep saw mills and the towns that develop around them out of forests."

Educational interventions that act to reduce contact with wild animal blood and body fluids through hunting and butchering have been recommended (LeBreton et al., 2006). Wolfe et al. (2004a) suggest that gender-based interventions are merited to decrease potential exposures to NHP. Education on the risks associated with NHP contact is essential in communities where the bushmeat trade is prevalent. Pike et al. have developed a "healthy hunter" educational program in an effort to reduce risk of zoonotic infection (2010). Some researchers, however, caution that similar targeted education campaigns that focus on a "butcher bushmeat with care" campaign may be perceived as approval of bushmeat hunting and consumption and may ultimately lead to increases in bushmeat-harvesting rates and a higher likelihood of interaction and transmission of zoonoses among humans and NHP (Monroe and Willcox, 2006).

Perceived risk among individuals involved with different aspects of the bushmeat trade may play an important role in public health education campaigns that aim to reduce exposure to zoonotic diseases associated with bushmeat. LeBreton et al. (2006) found that hunters and butchers who perceived personal risks were significantly less likely to butcher wild animals. Perception of risk, however, was not associated with hunting and eating bushmeat. It is unclear whether perceived risk is associated with fear from zoonotic disease or fear of going against spiritual or cultural practices tied to butchering, which frown upon contact with animal blood (Monroe and Willcox, 2006). Nevertheless, inclusion of health education messages that address and capitalize on perceived risk behaviors that may, in fact, be high-risk behaviors for zoonotic disease transmission could enhance the effectiveness of prevention campaigns (LeBreton et al., 2006).

Support for prevention, care, and treatment of HIV-1 infection is also strongly encouraged in order to diminish human risks associated with zoonotic disease infection and to prevent outbreaks of new viral pathogens among individuals involved with the bushmeat trade (LeBreton et al., 2007). If HIV incidence and prevalence decrease, the susceptibility of the population to other infectious diseases—zoonotic and non-zoonotic—can be significantly decreased.

10.11 NEW TECHNOLOGIES

Through the late 1990s, five lines of evidence were used to substantiate zoonotic transmission of many viral pathogens, such as lentiviruses: (i) similarities in viral genome organization, (ii) phylogenic relatedness, (iii) prevalence in the natural host, (iv) geographic coincidence, and (v) plausible routes of transmission (Gao et al., 1999). While these lines of evidence still play an integral role, guiding research to determine transmission of zoonotic diseases connected to the bushmeat trade, new technologies have facilitated significant leaps forward.

10.11.1 Laboratory tools

Enhanced laboratory techniques are keys to our understanding of detecting, treating, and controlling zoonoses connected to the bushmeat trade. Hahn et al. (2000) have suggested that priority should be given to the full molecular and biological characterization of all known SIV lineages while also searching for others. They have also recommended that phylogenic analyses be performed to characterize the evolutionary history of such viruses to determine how frequently they cross primate species barriers. Such analyses are essential from a public health perspective, to determine whether additional virus strains exist that are not currently detectable through existing screening tests and to avert outbreaks of other zoonotic disease epidemics.

10.11.2 Surveillance tools

Research that establishes connections between bushmeat and emerging zoonoses provides strong evidence for the need to maintain and bolster existing zoonotic disease surveillance systems. While an international surveillance system that can effectively monitor, trace, and predict zoonotic disease evolution and transmission does not currently exist, a number of efforts have been put forth during the past decade that offer examples of multidisciplinary collaborative work across international boundaries that are worth supporting and expanding in an effort to curtail the negative impacts and outcomes tied to bushmeat hunting and zoonotic diseases (Jones et al., 2008; Wolfe et al., 2007).

Mathematical modeling, computer modeling, spatial analysis, genomics, and satellite tracking can all play an important role in future efforts to monitor, trace, and predict zoonotic diseases. Enhanced research and surveillance of zoonotic infections connected to bushmeat hunting could have several benefits. Wolfe et al. (2005b) suggest that it may be possible to predict and prevent viral emergence by conducting longitudinal surveillance of exposed populations and by the creation of interventions to decrease risk factors such as primate hunting. Zheng et al. (2010) indicate that recent discovery of the human T-lymphotropic virus type 3 (HTLV3) highlights the importance of expanded surveillance to better understand the prevalence and public health impact of new retroviruses. Patz et al. (2004) emphasize the importance of studying the effects of landscape fragmentation on public health threats, highlighting three key research components: (i) collection of baseline data, including historical data where possible; (ii) health impact modeling; and (iii) development of decision-support tools. The use of geographic information systems (GIS) and spatial analyses in zoonosis surveillance has only recently begun to be explored on an international level. These tools offer tremendous potential in our better understanding of past, current, and future transmission routes and hot spots for the emergence of new zoonoses. While not specific to bushmeat-related zoonoses, Riolo (2010) offers a good example of zoonoses monitoring through web-based GIS. Since 2004, an online reporting system and central data repository has been used by European Union member states to submit, manage, and analyze zoonoses data. Cutler et al. assert that "human disease surveillance must be combined with enhanced longitudinal veterinary surveillance in food-producing animals and wildlife" (Cutler et al., 2010). A surveillance system that is implemented similarly across lower and upper income nations is of utmost importance if surveillance bias is to be avoided (Jones et al., 2008).

10.12 COLLABORATION: MULTIDISCIPLINARY ADVANCES AND NEXT STEPS

Collaborative approaches focus on the importance of interdisciplinary work among epidemiologists, wildlife biologists, veterinarians, microbiologists, and ecologists, among others (Daszak et al., 2007). Future efforts might be bolstered by inclusion of geographers, anthropologists, sociologists, climatologists, and clinicians—facilitating exploration of societal, cultural, environmental, biological, statistical, and epidemiological factors that may combine to provide the "perfect storm" that facilitates further evolution, transmission, and widespread distribution of future zoonotic pandemics tied to the bushmeat trade. Such collaborative efforts are being enhanced under the "One Medicine–One Health" concept (Daszak et al., 2007), and interventions related to bushmeat activities could benefit from joint programs, such as the campaign to combine vaccination of humans and animals against zoonotic agents, as described in Chad (Zinsstag et al., 2007).

Interdisciplinary collaboration between the veterinary and human public health fields is essential to the development, implementation, and sustainability of effective prevention and control efforts at the crossroads of bushmeat and emerging zoonotic disease. Collaboration across agencies and departments and a collective response to policy recommendations on land use change, habitat preservation, and infectious disease prevention could prove invaluable (Patz et al., 2004). Such collaboration is not always easily facilitated by existing governmental public health agencies (Murphy, 2008). As Woodford (2009) highlights, "the interests of a doctor of human medicine overlap with those of the veterinarian where zoonoses and anthropozoonoses are concerned. Many other disciplines can also benefit and data can be used to aid in the making and implementation of policy." Such intersections across disciplines are equally important when it comes to research and development of appropriate policy for bushmeat hunting and emerging zoonoses.

The Field Veterinary Program (FVP), a collaborative effort between FVP staff and a multidisciplinary team of scientists (e.g., acarologists, mycobacterium experts, biologists, and ecologists), was an early example of wildlife conservation medicine that could lead toward favorable outcomes in wildlife and human public health (Deem et al., 2000).

10.13 CONCLUSION

Increases in emerging and reemerging infectious diseases during recent decades have led to new challenges to human and veterinary public health. Bushmeat hunting, butchering, and consumption are associated with the cross species transmission of a number of zoonotic diseases, several of which have brought about devastating rates of morbidity and mortality internationally. Ongoing and increased support for development of multidisciplinary policies, prevention interventions, and research is needed to assess and address the latest developments and trends in emerging zoonoses associated with bushmeat hunting.

CONFLICTS OF INTEREST

The authors do not have any conflicts of interest to report.

REFERENCES

Aghokeng, A.F., Ayouba, A., Mpoudi-Ngole, E., et al., 2010. Extensive survey on the prevalence and genetic diversity of SIVs in primate bushmeat provides insights into risks for potential new cross-species transmissions. Infect. Genet. Evol. **10**, 386–396.

Amblard, J., Obiang, P., Edzang, S., et al., 1997. Identification of the Ebola virus in Gabon in 1994. Lancet **349**, 181–182.

Apetrei, C., Metzger, M.J., Richardson, D., et al., 2005. Detection and partial characterization of simian immunodeficiency virus SIVsm strains from bush meat samples from rural Sierra Leone. J. Virol. **79**(4), 2631–2636.

Avashia, S.B., Petersen, J.M., Lindley, C.M., et al., 2004. First reported prairie dog-to-human tularemia transmission, Texas, 2002. Emerg. Infect. Dis. **10**(3), 483–486.

Bennett, E.L., Blencowe, E., Brandon, K., et al., 2007. Hunting for consensus: reconciling bushmeat harvest, conservation, and development policy in West and Central Africa. Conserv. Biol. **21**, 884–887.

Brashares, J., Arcese, P., Sam, M., et al., 2004. Bushmeat hunting, wildlife declines, and fish supply in West Africa. Science **306**, 1180–1183.

Brodie, J., Gibbs, H., 2009. Bushmeat hunting as climate threat. Science **326**, 364–365.

Brugère-Picoux, J., Chomel, B., 2009. Importation of infectious diseases to Europe via animals and animal products: risks and pathways. Bull. Acad. Natl. Méd. **193**, 1805–1818.

Calattini, S., Betsem, E., Froment, A., et al., 2007. Simian foamy virus transmission from apes to humans, rural Cameroon. Emerg. Infect. Dis. **13**, 1314–1320.

Calattini, S., Betsem, E., Bassot, S., et al., 2009. New strain of human T lymphotropic virus (HTLV) type 3 in a Pygmy from Cameroon with peculiar HTLV serologic results. J. Infect. Dis. **199**, 561–564.

Carr, J., Wolfe, N.D., Torimiro, J., et al., 2010. HIV-1 recombinants with multiple parental strains in low-prevalence, remote regions of Cameroon: evolutionary relics? Retrovirology **7**, 39.

Chaber, A., Allebone-Webb, S., Lignereux, Y., et al., 2010. The scale of illegal meat importation from Africa to Europe via Paris. Conserv. Lett. **3**, 317–321.

Chomel, B.B., Belotto, A., Meslin, F.X., 2007. Wildlife, exotic pets, and emerging zoonoses. Emerg. Infect. Dis. **13**(1), 6–11.

Clapham, P., Van Waerebeek, K., 2007. Bushmeat and bycatch: the sum of the parts. Mol. Ecol. **16**, 2607–2609.

Coad, L., Abernethy, K., Balmford, A., et al., 2010. Distribution and use of income from bushmeat in a rural village, Central Gabon. Conserv. Biol. **24**(6), 1510–1518.

Courgnaud, V., Van Dooren, S., Liegeois, F., et al., 2004. Simian T-cell leukemia virus (STLV) infection in wild primate populations in Cameroon: evidence for dual STLV type 1 and type 3 infection in agile mangabeys (*Cercocebus agilis*). J. Virol. **78**, 4700–4709.

Cunningham, A., 2005. A walk on the wild side—emerging wildlife diseases. BMJ **331**, 1214–1215.

Cutler, S., Fooks, A., van der Poel, W., 2010. Public health threat of new, reemerging, and neglected zoonoses in the industrialized world. Emerg. Infect. Dis. **16**, 1–7.

Daszak, P., Epstein, J., Kilpatrick, A., et al., 2007. Collaborative research approaches to the role of wildlife in zoonotic disease emergence. Curr. Top. Microbiol. Immunol. **315**, 463–475.

Deem, S., Kilbourn, A., Wolfe, N., et al., 2000. Conservation medicine. Ann. N. Y. Acad. Sci. **916**, 370–377.

De Merode, E., Cowlishaw, G., 2006. Species protection, the changing informal economy, and the politics of access to the bushmeat trade in the Democratic Republic of Congo. Conserv. Biol. **20**, 1262–1271.

East, T., Kümpel, N.F., Milner-Gulland, E.J., et al., 2005. Determinants of urban bushmeat consumption in Río Muni, Equatorial Guinea. Biol. Conserv. **126**, 206–215.

Edmunds, K., Roberton, S.I., Few, R., et al., 2011. Investigating Vietnam's ornamental bird trade: implications for transmission of zoonoses. Ecohealth **8**, 63–75.

Fa, J., Peres, C., Meeuwig, J., 2002. Bushmeat exploitation in tropical forests: an intercontinental comparison. Conserv. Biol. **16**, 232–237.

Gao, F., Bailes, E., Robertson, D., et al., 1999. Origin of HIV-1 in the chimpanzee *Pan troglodytes troglodytes*. Nature **397**, 436–441.

Gao, F., Robertson, D.L., Carruthers, C.D., et al., 1998. An isolate of human immunodeficiency virus type 1 originally classified as subtype I represents a complex mosaic comprising three different group M subtypes (A, G, and I). J. Virol. **72**, 10234–10241.

Hahn, B., Shaw, G., De Cock, K., et al., 2000. AIDS as a zoonosis: scientific and public health implications. Science **287**, 607–614.

Jean Desbiez, A.L., Keuroghlian, A., Piovezan, U., et al., 2011. Invasive species and bushmeat hunting contributing to wildlife conservation: the case of feral pigs in a Neotropical wetland. Oryx **45**, 78–83.

Jones, K., Patel, N., Levy, M., et al., 2008. Global trends in emerging infectious diseases. Nature **451**, 990–993.

Jones-Engel, L., Engel, G.A., Schillaci, M.A., et al., 2005. Primate-to-human retroviral transmission in Asia. Emerg. Infect. Dis. **11**, 1028–1035.

Jones-Engel, L., May, C., Engel, G., et al., 2008. Diverse contexts of zoonotic transmission of simian foamy viruses in Asia. Emerg. Infect. Dis. **14**, 1200–1208.

Kalish, M.L., Wolfe, N.D., Ndongmo, C.B., et al., 2005. Central African hunters exposed to simian immunodeficiency virus. Emerg. Infect. Dis. **11**, 1928–1930.

Khodakevich, L., Jezek, Z., Messinger, D., 1988. Monkeypox virus: ecology and public health significance. Bull. World Health Org. **66**, 747–752.

Kümpel, N., 2005. The Bushmeat Trade. POSTnote 236. Parliamentary Office of Science and Technology: London.

Kümpel, N., Milner-Gulland, E., Cowlishaw, G., et al., 2010. Assessing Sustainability at Multiple Scales in a Rotational Bushmeat Hunting System. Conserv. Biol. **24**(3), 861–871.

LeBreton, M., Prosser, A., Tamoufe, U., et al., 2006. Patterns of bushmeat hunting and perceptions of disease risk among central African communities. Anim. Conserv. **9**, 357–363.

LeBreton, M., Yang, O., Tamoufe, U., et al., 2007. Exposure to wild primates among HIV-infected persons. Emerg. Infect. Dis. **13**, 1579–1582.

Leroy, E., Rouquet, P., Formenty, P., et al., 2004. Multiple Ebola virus transmission events and rapid decline of central African wildlife. Science **303**, 387–390.

Matsuda, H., Okada, K. Takahashi, K., et al., 2003. Severe hepatitis E virus infection after ingestion of uncooked liver from a wild boar. J. Infect. Dis. **188**, 944.

Monroe, M.C., Willcox, A.S., 2006. Could risk of disease change bushmeat-butchering behavior? Anim. Conserv. **9**, 368–369.

Mouinga-Ondémé, A., Caron, M., Nkoghé, D., et al., 2012. Cross-species transmission of simian foamy virus to humans in rural Gabon, Central Africa. J. Virol. **86**(2), 1255–1260.

Murphy, F., 2008. Emerging zoonoses: the challenge for public health and biodefense. Prev. Vet. Med. **86**, 216–223.

Ohl-Schacherer, J., Shepard, G.J., Kaplan, H., et al., 2007. The sustainability of subsistence hunting by Matsigenka native communities in Manu National Park, Peru. Conserv. Biol. **21**, 1174–1185.

Patz, J., Daszak, P., Tabor, G., et al., 2004. Unhealthy landscapes: policy recommendations on land use change and infectious disease emergence. Environ. Health Perspect. **112**, 1092–1098.

Peeters, M., 2004. Cross-species transmissions of simian retroviruses in Africa and risk for human health. Lancet **363**, 911–912.

Peeters, M., Courgnaud, V., Abela, B., et al., 2002. Risk to human health from a plethora of simian immunodeficiency viruses in primate bushmeat. Emerg. Infect. Dis. **8**, 451–457.

Peiris, J., Guan, Y., Yuen, K., 2004. Severe acute respiratory syndrome. Nat. Med. **10**, S88–S97.

Pike, B.L., Saylors, K.E., Fair, J.N., et al., 2010. The origin and prevention of pandemics. Clin. Infect. Dis. **50**, 1636–1640.

Poon, L., Guan, Y., Nicholls, J., et al., 2004. The aetiology, origins, and diagnosis of severe acute respiratory syndrome. Lancet Infect. Dis. **4**, 663–671.

Poulsen, J., Clark, C., Mavah, G., et al., 2009. Bushmeat supply and consumption in a tropical logging concession in northern Congo. Conserv. Biol. **23**, 1597–1608.

Reed, K.D., Melski, J.W., Graham, M.B., et al., 2004. The detection of monkeypox in humans in the Western Hemisphere. N. Engl. J. Med. **350**(4), 342–350.

Rimoin, A.W., Mulembakani, P.M., Johnston, S.C., et al., 2010. Major increase in human monkeypox incidence 30 years after smallpox vaccination campaigns cease in the Democratic Republic of Congo. Proc. Natl. Acad. Sci. U. S. A. **107**, 16262–16267.

Riolo, F., 2010. Web-Based GIS for Zoonoses Monitoring: European Food Safety Authority. HealthyGIS: ESRI HealthyGIS, p. 8.

Rosen, G.E., Smith, K.F., 2010. Summarizing the evidence on the international trade in illegal wildlife. Ecohealth **7**, 24–32.

Rowcliffe, J., Milner-Gulland, E., Cowlishaw, G., 2005. Do bushmeat consumers have other fish to fry? Trends Ecol. Evol. **20**, 274–276.

Sintasath, D.M., Wolfe, N.D., Lebreton, M., et al., 2009. Simian T-lymphotropic virus diversity among nonhuman primates, Cameroon. Emerg. Infect. Dis. **15**, 175–184.

Smith, K.M., Anthony, S.J., Switzer, W.M., et al., 2012. Zoonotic viruses associated with illegally imported wildlife products. PLoS One **7**(1), e29505.

Switzer, W.M., Garcia, A.D., Yang, C., et al., 2008. Coinfection with HIV-1 and simian foamy virus in West Central Africans. J. Infect. Dis. **197**, 1389–1393.

Takehisa, J., Kraus, M., Ayouba, A., et al., 2009. Origin and biology of simian immunodeficiency virus in wild-living western gorillas. J. Virol. **83**, 1635–1648.

Tei, S., Kitajima, N., Takahashi, K., et al., 2003. Zoonotic transmission of hepatitis E virus from deer to human beings. Lancet **362**, 371–373.

Ter Meulen, J., Lukashevich, I., Sidibe, K., et al., 1996. Hunting of peridomestic rodents and consumption of their meat as possible risk factors for rodent-to-human transmission of Lassa virus in the Republic of Guinea. Am. J. Trop. Med. Hyg. **55**, 661–666.

Toledo, L.F., Asmüssen, M.V., Rodríguez, J.P., 2012. Crime: track illegal trade in wildlife. Nature **483**(7387), 36.

Van Borm, S., Thomas, I., Hanquet, G., et al., 2005. Highly pathogenic H5N1 influenza virus in smuggled Thai eagles, Belgium. Emerg. Infect. Dis. **11**(5), 702–705.

Van Heuverswyn, F., Peeters, M., 2007. The origins of HIV and implications for the global epidemic. Curr. Infect. Dis. Rep. **9**, 338–346.

Vliet, N., Nasi, R., Taber, A., 2011. From the forest to the stomach: bushmeat consumption from rural to urban settings in Central Africa. In: Shackleton, S., Shackleton, C., Shanley, P., editors. Non-Timber Forest Products in the Global Context. Springer: Berlin Heidelberg; 129–145.

Waite, T., 2007. Revisiting evidence for sustainability of bushmeat hunting in West Africa. Environ. Manage. **40**, 476–480.

Weiss, R., McMichael, A., 2004. Social and environmental risk factors in the emergence of infectious diseases. Nat. Med. **10**, S70–S76.

Weiss, R., Wrangham, R., 1999. From Pan to pandemic. Nature **397**, 385–386.

Wolfe, N.D., Prosser, T.A., Carr, J.K., et al., 2004a. Exposure to nonhuman primates in rural Cameroon. Emerg. Infect. Dis. **10**, 2094–2099.

Wolfe, N.D., Switzer, W., Carr, J., et al., 2004b. Naturally acquired simian retrovirus infections in central African hunters. Lancet **363**, 932–937.

Wolfe, N.D., Heneine, W., Carr, J.K., et al., 2005a. Emergence of unique primate T-lymphotropic viruses among central African bushmeat hunters. Proc. Natl. Acad. Sci. U. S. A. **102**, 7994–7999.

Wolfe, N., Daszak, P., Kilpatrick, A., et al., 2005b. Bushmeat hunting, deforestation, and prediction of zoonoses emergence. Emerg. Infect. Dis. **11**, 1822–1827.

Wolfe, N.D., Dunavan, C., Diamond, J., 2007. Origins of major human infectious diseases. Nature **447**, 279–283.

Woodford, M.H., 2009. Veterinary aspects of ecological monitoring: the natural history of emerging infectious diseases of humans, domestic animals and wildlife. Trop. Anim. Health Prod. **41**(7), 1023–1033.

Worobey, M., Gemmel, M., Teuwen, D., et al., 2008. Direct evidence of extensive diversity of HIV-1 in Kinshasa by 1960. Nature **455**, 661–664.

Zheng, H., Wolfe, N.D., Sintasath, D.M., et al., 2010. Emergence of a novel and highly divergent HTLV-3 in a primate hunter in Cameroon. Virology **401**, 137–145.

Zinsstag, J., Schelling, E., Roth, F., et al., 2007. Human benefits of animal interventions for zoonosis control. Emerg. Infect. Dis. **13**(4), 527–531.

11

BIOLOGICAL SIGNIFICANCE OF BATS AS A NATURAL RESERVOIR OF EMERGING VIRUSES

Angela M. Bosco-Lauth and Richard A. Bowen

Colorado State University, Fort Collins, CO, USA

TABLE OF CONTENTS

11.1 INTRODUCTION

Over the past two decades, bats have been identified as the reservoir host for a number of high-impact zoonotic viruses that induce highly lethal disease in man and domestic animals, including Nipah and Hendra viruses, the severe acute respiratory syndrome coronavirus

Viral Infections and Global Change, First Edition. Edited by Sunit K. Singh.

(SARS CoV), and filoviruses such as Ebola and Marburg viruses. Additionally, although rabies virus has been known for many years as a pathogen transmitted by bats, a long list of other rabies-related lyssaviruses has recently been identified in bats. Speculation has thus intensified that bats are somehow special among vertebrates as a reservoir for dangerous pathogens. At present, there are no definitive answers to this question, and it seems imperative that we significantly expand our understanding of bat biology and ecology in order to address these issues.

11.2 BATS AS EXEMPLARS OF BIODIVERSITY

More than 1200 species of bats have been described, representing approximately 20% of vertebrates (Calisher et al., 2006; Teeling et al., 2005). One of the few descriptors all bats share is that they are flying mammals; aside from that, they manifest incredible diversity in lifestyle. Some bats subsist on insects, others on various fruits, and a number of species feed exclusively on the blood of other vertebrates. There are even nectar-eating and fish-catching bats. In terms of variations in body size, one only need to compare the diminutive bumblebee bat (*Craseonycteris thonglongyai*), with a body weight of approximately 2 g and wingspan of 12–13 cm, to several bats in the genus *Pteropus*, which have wingspans of 2 m. Further diversity in lifestyle is manifest by whether or not bats have the ability to echolocate, how they overwinter (hibernation vs. migration), and whether they tend to be solitary or roost in huge, high-density congregations. From the standpoint of formal taxonomy, bats belong to the order Chiroptera, which is composed of two suborders. The Megachiroptera group is composed of a single genus predominantly of fruit-eating bats, while bats of the Microchiroptera suborder are largely but by no means exclusively insectivorous. Although there has been speculation that extant bats represent two separate evolutionary lineages, recent evidence points to a monophyletic origin (Simmons et al., 2008). Except for the Arctic, Antarctic, and a few islands, bats enjoy a global distribution, likely established through their ability to fly.

Another characteristic of many temperate bats is their ability to enter a state of torpor and, in some species, to hibernate over winter. Torpor is defined as a metabolically depressed state with lower than normal body temperature and is often induced in response to scarcity of food. Periods of torpor are often interrupted by periods of normal body temperature. Such reductions in body temperature have significant suppressive effects on both innate and adaptive immune responses (Bouma et al., 2010). The influence of torpor and hibernation on the course of viral infection in bats is not well understood, but would be expected to modulate both viral replication and host immune responses. Rabies virus failed to replicate or replicated to only a limited extent in little brown bats (*Myotis lucifugus*) undergoing experimentally induced hibernation, but when such bats were warmed, virus was detected in several tissues (Sulkin et al., 1960). Similar effects on virus replication, resulting in very long periods of viremia, were demonstrated in both little brown bats and big brown bats (*Eptesicus fuscus*) infected with Japanese encephalitis (JE) virus (Sulkin and Allen, 1974; Sulkin et al., 1966). Recently, hibernation was identified as a critical factor in a mathematical model for the overwintering of rabies virus in big brown bats (George et al., 2011).

Most bats are highly social, often living in communal roosts where thousands of individuals come into contact daily. This high contact rate would certainly facilitate spread of viruses and also permit multiple sources of transmission to occur simultaneously, that is, oral, fecal, blood-borne, airborne, and shared vectors (Calisher et al., 2006). Once a colony has contracted a virus, or any transmissible pathogen, it is nearly impossible for them to get rid of it. In many instances, such as with Nipah virus and coronaviruses, bats are relatively unaffected by the disease itself, but in the case of white-nose syndrome, a fungal infection responsible for the deaths of millions of bats in North America, entire colonies are eliminated because they are never able to maintain the necessary level of immune versus susceptible individuals to establish a stable balance (Blehert et al., 2009).

11.3 BATS ARE RESERVOIR HOSTS FOR ZOONOTIC AND EMERGING PATHOGENS

From 1906 to 1908, several thousand cattle and horses died from rabies in one state of Brazil. It was noted that vampire bats were attempting to feed on these animals, and investigators hypothesized that the bats were somehow responsible for the disease. Subsequent investigations revealed the presence of Negri bodies in the brains of the bats, providing the first evidence for transmission of a viral disease from bats and setting the stage for future studies targeting bats as a source of emerging pathogens (Sulkin and Allen, 1974). Since that time, a large number of viruses, representing numerous genera, have been isolated from or detected in bats (Calisher et al., 2006). In most cases, the significance of such findings and the potential of the agent to emerge as a zoonotic pathogen are not known with any certainty. In the many cases where only antiviral antibody has been detected, it is likely that bats are only incidental hosts. Nonetheless, it has become clear that bats are the important reservoir host for a number of medically important viruses and, further, that some groups of viruses appear to have arisen in bats and diversified by host switching into other populations of vertebrates.

11.3.1 Lyssaviruses

Lyssaviruses belong to the family Rhabdoviridae and comprise rabies and rabies-related viruses. At this time, the *Lyssavirus* genus encapsulates 11 species, represented by rabies virus, Lagos bat virus, Mokola virus, Duvenhage virus, European bat lyssavirus types 1 and 2, Australian bat lyssavirus, Irkut virus, West Caucasian bat virus, Khujand virus, and Aravan virus. At least two additional species have been proposed for inclusion into this genus (Freuling et al., 2011; Kuzmin et al., 2010), and with increased surveillance efforts directed at bats, it is virtually certain that additional lyssaviruses will be identified. With the exception of Mokola virus, each of these viruses is maintained in a bat reservoir (Rupprecht et al., 2011), and most have been associated with spillover infections in humans and other animals.

The geographic distribution of lyssaviruses in bats is essentially global. Rabies virus itself has only been detected in New World bats, where it is endemic but also where it is the only lyssavirus known. Conversely, all of the other known lyssaviruses have come from the Old World where rabies virus itself has not been detected in bats. The greatest diversity of bat lyssaviruses has come from Africa (Warrel, 2010; Weyer et al., 2011), leading to a speculation that lyssaviruses originally evolved in African populations of bats.

Among the bat lyssaviruses, rabies virus has been studied most comprehensively. To a large extent, rabies virus has diversified in New World bat populations such that bats from different species harbor genetically distinct viruses (variants), which are readily distinguished from one another by nucleotide sequencing or application of panels of monoclonal antibodies. Application of such typing schemes has proven valuable in identifying the source of virus in cross species transmission events to humans or terrestrial carnivores. Moreover, the evolution of rabies virus variants seems to have been accompanied in some instances by the emergence of strains with increased infectivity or virulence. Specifically, virus variants associated with the silver-haired (*Lasionycteris noctivagans)* and tricolored (*Perimyotis subflavus*) bats have been shown responsible for roughly 70% of recent cases of human rabies in the United States, despite these two bats being uncommon relative to bats from other species and to be non-synanthropic (Messenger et al., 2003). Comparison of silver-haired bat rabies virus variants with virus strains from coyotes or dogs revealed that the bat variant displayed less neuronal tropism and more temperature sensitivity than the carnivore viruses, properties that may contribute to altered infectivity and hence virulence (Dietzschold et al., 2000; Morimoto et al., 1996). Considering these studies in light of the diversity of lyssaviruses known to be hosted by bats suggests that we have much to learn about the differences in biological properties of bat lyssaviruses and how such differences may influence such emergent events as host switching.

Lyssaviruses evolve through genomic point mutations promoted by lack of proofreading by its RNA polymerase (Davis et al., 2006; Hughes et al., 2005). However, it appears that host factors also contribute substantively to maintenance and evolution of virus variants. In temperate climates, seasonal factors such as migration or hibernation play a significant role in maintaining a stable enzootic cycle of bat rabies virus infection (George et al., 2011; Mondul et al., 2003). Moreover, evidence has been presented that the success of cross species transmission of rabies virus among bats is a consequence not only of virus evolution but of bat host restrictions (Streicker et al., 2010); although the nature of such restrictions is not known, host factors may greatly restrict the fraction of spillover events that result in true host-switching adaptations, although such phenomena have been documented (Leslie et al., 2006; Rupprecht et al., 2011).

Surveillance systems to detect emergence of new lyssaviruses are rudimentary, particularly in developing countries where we already know of multiple virus species. Coupled with the rich species diversity among bats and their known role as reservoirs for lyssaviruses, it is likely that many additional members of this virus genus will be detected, possibly when infections spillover into human populations.

11.3.2 Henipaviruses

Henipaviruses are RNA viruses comprising a genus within the family *Paramyxoviridae*; to date two species have been identified, Hendra and Nipah viruses, both of which utilize bats as the primary reservoir. In Brisbane, Australia, in 1994, an outbreak of severe influenza-like disease was observed in 21 Thoroughbred horses, and the horses either died or had to be euthanized. One week later, two caretakers of the diseased horses developed the same symptoms and one died of respiratory and renal failure (Murray et al., 1995; Selvey et al., 1995). Death was rapid after the onset of symptoms, and the clinical presentation did not point to anything specific, although airborne transmission was implicated. Over the course of the next decade, four more instances of equine infection occurred in Australia (Field et al., 2007), and multiple infections have been characterized in both horses and humans through 2011 in both Queensland and New South Wales. The agent responsible was a member of the *Paramyxoviridae* family and was originally deemed a morbillivirus, but later, genetic sequencing distinguished the virus as a novel pathogen, named Hendra after the location of the index case (Murray et al., 1995).

Due to its zoonotic nature, the Hendra virus outbreaks led to a search for a natural reservoir host. The first outbreaks could not be linked to one another through human or horse movement, so researchers turned to wildlife that had a large range and could easily move between locations as potential reservoirs. The obvious suspects were birds or bats due to their ability to fly, and because morbilliviruses had been previously isolated from bats (Henderson et al., 1995), researchers looked toward the local fauna and found that pteropid bats fit the image (Halpin et al., 2000). Serosurveillance showed that 40% of flying foxes had neutralizing antibodies against Hendra (Field et al., 2007), and repeated negative antibody response in other animals gave convincing evidence that pteropid bats were the natural reservoir (Halpin et al., 2011). Subsequent serosurveys linked a variety of species of flying foxes to Hendra virus outbreaks across much of Australia and into Papua New Guinea, and it appeared that flying foxes had harbored virus long before the outbreaks started (Chua et al., 2002b; Halpin et al., 2007; Mackenzie et al., 2001). To date, very little is known about Hendra and its enzootic cycle, even less about what sparks a spillover event with this virus.

In 1998, after a particularly bad drought, an outbreak of suspected JE in Malaysia killed 105 people and countless piglets, and none of the usual JE virus vaccines were effective. The causative agent was soon identified as a novel pathogen, much more closely related to Hendra virus than to any of the JE serogroup viruses (Chua et al., 2000). Like Hendra, this virus was transmitted through close contact with infected animals, except, instead of horses, pigs were the livestock hosts. Disease in pigs consisted of severe respiratory illness, while in humans influenza-like illness coupled with encephalitis was observed. Other species of animals, including dogs, tested antibody positive, but none were considered competent hosts (Looi and Chua, 2007). The similarities between these viruses led scientists to start testing bats for Nipah, and, as with Hendra virus, seropositive pteropid bats were found (Chua et al., 2002b; Halpin et al., 2011; Yob et al., 2001).

Originally, Hendra virus was grouped in the *Morbillivirus* genus of the family *Paramyxoviridae* (Murray et al., 1995), but further genetic analysis showed inconsistencies with known morbilliviruses, including a longer genome, indicating that this was a novel virus (Wang et al., 2000). Genome sequencing of Nipah virus showed closest homology to Hendra virus, and thus the two were deemed members of a new genus, *Henipavirus* (Chua et al., 2000; Wang et al., 2000). While these are not the first paramyxoviruses to be isolated from bats, they are the only ones described in *Pteropus* bats and are unique in their ability to cause severe, fatal disease in several host species, including humans (Chua et al., 2000; Field et al., 2007). Current understanding supports the theory that henipaviruses are novel pathogens, likely originating from pteropid bats, which have been transmitted across species barriers due to several anthropogenic emergence factors (Field et al., 2007). In particular, deforestation with loss of habitat, urbanization, and hunting have all contributed to the increasing contact between humans and flying foxes and, by doing so, augmented the likelihood of viral spillover events (Epstein et al., 2009).

The most likely chain of transmission in Malaysian Nipah outbreaks is that bats infected pigs and pigs infected humans; human-to-human transmission was not observed. In fact, it is assumed that perhaps only one or two introductions of bat Nipah virus into pigs were responsible for all the Malaysian human cases (AbuBakar et al., 2004; Pulliam et al., 2005). In Bangladesh, however, the story appears to be different. Since its recognition in 2001, no fewer than 23 separate introductions of Nipah virus into human hosts have occurred, and most if not all of these events were a direct result of bat-to-human infection by the bat *Pteropus giganteus* (Luby et al., 2009). What is perhaps more unique about the situation in Bangladesh is that more than half of the human cases were a direct result of person-to-person transmission. While the Malaysian outbreak was quelled by culling millions of pigs, in places where pig farming is less common, outbreaks in humans are more likely to come directly from bat reservoirs. The situation for Nipah is similar in India, where outbreaks in humans are linked to nearby seropositive bats (Epstein et al., 2008). One likely explanation for the increased frequency of bat-to-person transmission is the sharing of food sources, namely, date palm sap. In Bangladesh, the sweet sap is considered a delicacy and is frequently consumed raw by the natives. Fruit bats find it to their liking as well and have been observed lapping sap from the same trees people harvest. Bats shed the virus through their saliva, urine, and feces, so any tree where a bat has landed is likely to have been exposed to all three sources. Furthermore, the moist and sugary consistency of the sap allows the virus to persist for a few hours or longer, increasing the risk of spread (Fogarty et al., 2008; Rahman et al., 2012).

An intriguing finding for the study of bat-associated viruses comes from the sampling of nearly 5000 bats of diverse species from across the world and the use of RT-PCR to detect paramyxovirus sequences (Drexler et al., 2012). In addition to pointing toward an African origin for the henipaviruses, phylogenetic reconstructions based on detected sequences indicate that bats are the ancestral reservoir host for all paramyxoviruses extant today, including such human pathogens as mumps virus. Another way to view these data and analyses is that the diversity of paramyxoviruses we observe in mammals today is the result of host switching of viruses that originated in bats.

11.3.3 Filoviruses

Marburg and Ebola viruses are the two known species of the RNA virus family *Filoviridae*. Marburg virus was the first filovirus identified to cause human disease and is named after Marburg, Germany, the town in which laboratory workers contracted the disease from infected Ugandan primates (Gear et al., 1975). Since then, sporadic outbreaks have occurred every few years in Africa, with the largest and deadliest one in Angola in 2005, where 90% of the 252 cases were fatal (Hartman et al., 2010). Ebola virus was first identified in 1976 when two concurrent outbreaks took place with the same etiology, one in Sudan and one in the Democratic Republic of Congo (Heymann et al., 1980; Johnson et al., 1977). There are currently five distinct species of *Ebolavirus* recognized: Zaire, Sudan, Bundibugyo, Ivory Coast, and Reston, all named after the location of their respective discovery. The first three *Ebolavirus* species have all been linked to human epidemics where the case fatality rates vary from 42% to 100% (Hartman et al., 2010). Only one instance of human infection with the Ivory Coast ebolavirus has been reported, and that individual recovered (Le Guenno et al., 1995). The Reston ebolavirus is the only species that has not been associated with severe human disease, but antibodies against Reston virus were found in both humans and pigs, suggesting asymptomatic infection via aerosol exposure (Feldmann et al., 2004; Jahrling et al., 1990; Miranda et al., 1999).

Marburg and human-pathogenic Ebola viruses cause severe disease in both humans and nonhuman primates, but neither species is considered a virus reservoir (Leroy et al., 2011). Only recently have bats been implicated as reservoir hosts for filoviruses. Initially, studies conducted in Gabon, Uganda, and the Democratic Republic of Congo demonstrated antibodies specific to Ebola and Marburg viruses in several species of fruit and insectivorous bats (Leroy et al., 2005; Pourrut et al., 2007; Swanepoel et al., 2007; Towner et al., 2007, 2008). Importantly, Marburg virus RNA was detected in tissues, and virus-specific IgG was demonstrated in individual Egyptian fruit bats (*Rousettus aegyptiacus*) in Gabon (Towner et al., 2009). Antibodies against Reston ebolavirus have been also detected in *Rousettus amplexicaudatus* bats in the Philippines, but the presence of virus has not yet been demonstrated (Taniguchi et al., 2011).

A strong epidemiological link has been established between bats and human infection with Ebola virus (Leroy et al., 2009). A large outbreak of human disease occurred in 2007 in the Democratic Republic of Congo in association with stopover of migrating fruit bats. This migration involved especially high numbers of bats, and people from several villages shot large numbers of bats to feed their own villages and take to the surrounding markets to sell for food. The index case of the human outbreak was known to have purchased bats for consumption and apparently transmitted the virus to his infant daughter, who subsequently died. A woman who attended the infant's corpse later succumbed to hemorrhagic disease, and 11 contacts that had cared for that woman also developed disease and died. In total, 186 deaths out of 264 cases were reported during this outbreak. Of note, there are no established populations of chimpanzees or gorillas in this area, indicating that it is highly unlikely that the human infections were derived from primates and supporting the contention that human infection was initiated by contact with bats.

Marburg fever outbreaks have never been directly linked to bats, but the presence of live virus and antibodies in several bats suggests that they are indeed a reservoir. Ecologic niche modeling, while a relatively new science, provides a framework for geographic distribution of disease outbreaks. In the context of Marburg virus, this approach was used to identify the likely source of the 1975 Zimbabwe cases as near the Sinoia Caves region (Peterson et al., 2006). Several cave-dwelling bats species were identified as potential reservoirs of Marburg virus, including *Rhinolophus eloquens, R. aegyptiacus*, and *Miniopterus inflatus* (Swanepoel et al., 2007; Towner et al., 2009), and it is not unlikely that the Marburg virus infections could have come from infected bats roosting in caves. Seroprevalence of both Marburg and Ebola viruses co-circulating in bats in Gabon indicates a higher rate of infection among cave-dwelling bats compared to forest dwellers (Pourrut et al., 2009). The authors attribute this finding to the closer contact between bats in caves compared to those in the forest; for humans, this study implies that hunters who enter bat caves are probably at a higher risk for contracting one of these viruses.

Spatial modeling of Ebola virus outbreaks between 1994 and 2002 demonstrates that the outbreaks were all closely associated with drier conditions than normal at the end of the rainy season (Pinzon et al., 2004). By contrast, putative models of Marburg virus indicate that the virus is more likely to be found in areas with less annual temperature and precipitation variation and with a lower annual temperature mean (Peterson et al., 2006). By themselves, these findings do not provide conclusive evidence to implicate climatic or weather factors as directly responsible for filovirus outbreaks, but do offer theories on when and where an outbreak might be expected to occur. If a particular season is extremely dry, perhaps even undergoing drought, it may be expected that more Ebola cases will occur. Clearly, additional studies on bat-virus ecology are required to more fully understand the natural history of this disease. One interesting recent discovery is that *Myotis* species, along with some rodents and marsupials, have integrated copies of filovirus-like genes, specifically sequences related to the genes encoding VP35 and nucleoprotein (Taylor et al., 2010, 2011). The significance of these findings relative to the potential of bats to serve as filovirus reservoirs remains to be elucidated, but this observation does seem to suggest an ancient evolutionary relationship between bats and filoviruses.

11.3.4 Coronaviruses

Coronaviruses are large (~30 kb) negative-sense RNA viruses that infect a wide range of mammals and birds (Lai et al., 2007). Current classification by the International Committee on Taxonomy of Viruses (ictvonline.org) describes three genera of viruses within the subfamily *Coronavirinae*: alphacoronavirus, betacoronavirus, and gammacoronavirus, which have replaced the more classical antigenic groups 1, 2, and 3, respectively. Mammals are the host for alphacoronaviruses and betacoronaviruses, whereas gammacoronaviruses have avian hosts.

Diseases induced by coronaviruses have been known in livestock and laboratory rodents for many decades and include such syndromes as infectious bronchitis in chickens, transmissible gastroenteritis in pigs, and feline infectious peritonitis of cats, as well as mouse hepatitis in rodents. A majority of human coronaviruses are associated with mild respiratory or enteric disease.

In 2002, severe acute respiratory syndrome (SARS) emerged in Guangdong province in China, and over the following months, this disease spread explosively to 25 countries and resulted in an estimated 8000 infections with a case fatality rate of roughly 10% (Skowronski et al., 2005). Fortunately, SARS has not been recognized in human populations since 2004, and the prospects for its reemergence remain unknown. Within a short time of recognizing this newly emerged disease, its etiologic agent—now known as SARS CoV—was isolated and its nucleotide sequence determined (Drosten et al., 2003; Ksiazek et al., 2003; Marra et al., 2003; Peiris et al., 2003). Based on its genetic structure, SARS CoV was classified as a betacoronavirus and is only distantly related to other known human coronaviruses.

SARS was initially thought to have emerged by transmission of SARS CoV from one of several wild animals sold in wet markets, and viruses closely related to the human SARS CoV was isolated from palm civets and raccoon dogs in those markets (Guan et al., 2003). However, such viruses were not isolated from wild counterparts of these animals, and there was a large degree of sequence variability among the viruses isolated in markets (Kan et al., 2005), suggesting that they were in the process of adapting to a new host and solidifying the contention that these carnivores were not natural reservoir hosts of SARS CoV. In contrast, viruses closely related to SARS CoV were detected in species of horseshoe bats from the genus *Rhinolophus* (Lau et al., 2005; Li et al., 2005b), leading to the hypothesis that SARS CoV somehow emerged from SARS-like coronaviruses (SL CoV) harbored by bats. Although this remains a viable and appealing explanation for the origin of SARS CoV, questions remain and the mechanism for emergence is unclear. For example, the SARS CoV binds to human host cells via interactions between the viral spike protein and angiotensin-converting enzyme-2 (ACE2) protein (Li et al., 2005a), yet the spike proteins from bat-origin SL CoV contain a truncation that prevents them from binding to human ACE2 (Li et al., 2006). Conversely, the spike protein from SARS CoV does not bind to ACE2 protein from horseshoe bats (Ren et al., 2008). Collectively, these studies support the concept that bat SL CoV are ancestral to SARS CoV but that SARS CoV did not arise by direct transmission of a bat virus to humans. Among alternative hypotheses is that there may be bat SL CoV with spike protein that binds to ACE2 or that a bat SL CoV recombined in some intermediate host with another coronavirus capable of binding ACE2, leading to the generation of SARS CoV (Lau et al., 2010a). This latter hypothesis is appealing because coronaviruses are known to readily undergo recombination (Keck et al., 1988) and because different species of bats roosting together have been shown to carry different coronaviruses (Gloza-Rausch et al., 2008; Reusken et al., 2010; Tang et al., 2006) and a single bat may harbor multiple, genetically distinct coronaviruses (Lau et al., 2010b).

The identification of bat-origin SL CoV as a likely progenitor to SARS CoV potently stimulated a global search for other coronaviruses in bats, and sequences of viruses belonging to both alphacoronavirus and betacoronavirus genera have been detected in bats of multiple species from many regions of the world, including Asia and the Pacific Islands

(Chu et al., 2006; Poon et al., 2005; Tang et al., 2006; Watanabe et al., 2010; Woo et al., 2006), Africa (Pfefferle et al., 2009; Tong et al., 2009), Europe (Drexler et al., 2010; Gloza-Rausch et al., 2008; Reusken et al., 2010), North America (Dominguez et al., 2007; Misra et al., 2009; Osborne et al., 2011), and South America (Carrington et al., 2008). The SL CoV of bat origin have all come from either horseshoe bats or some bats from Africa (Lau et al., 2010b), and the zoonotic potential of bat coronaviruses more distantly related to SARS CoV remains speculative but clearly worthy of further study.

One frustration in the study of bat CoV infections is that none of the numerous viruses detected in bats by PCR have yet been isolated or propagated in cultured cells. The massive genetic dataset has, however, greatly increased our understanding of coronavirus evolution, and phylogenetic studies have pointed to bat CoV as ancestral to all of the alphacoronaviruses and betacoronaviruses (Vijaykrishna et al., 2007; Woo et al., 2009).

11.3.5 Arboviruses

Bats appear to become infected with a number of arboviruses, based on detection of antiviral antibody or, less commonly, isolation of the agents (Calisher et al., 2006). Prominent examples of such agents include JE, Rift Valley fever, and St. Louis encephalitis viruses. In the cases of both JE and St. Louis encephalitis viruses, it has been demonstrated that the viruses can persist in insectivorous bats through a period of hibernation (La Motte, 1958; Sulkin et al., 1963, 1966). However, bats have not yet been implicated as an epidemiologically important reservoir host for any vector-borne pathogen. West Nile virus (WNV) is an example in which bats were considered a possible reservoir host for an arbovirus. Many years ago, antibodies to WNV were demonstrated in apparently healthy *Rousettus* bats collected in Israel and Uganda (Constantine, 1970), and the virus was isolated from bats of the same genus in India (Paul et al., 1970). In 2000, shortly after its emergence in the United States, WNV infection was demonstrated in 2 of 150 bats (one *E. fuscus* and one *M. lucifugus*) submitted to the New York State Department of Health, but subsequent attempts over the next 2 years to demonstrate the virus in bats from endemic areas of New York failed. Further, experimental infection of both big brown and Mexican free-tailed bats revealed that WNV infection failed to induce clinical disease and viremia was low or undetectable, suggesting that they would not serve as amplifying hosts for mosquitoes (Davis et al., 2005). Although magnitude of viremia is often used to define reservoir host status for arboviruses, the possibility of mosquitoes becoming infected by feeding on infected hosts with undetectable viremia has been demonstrated, including for flying foxes infected with JE virus, raising the possibility that such viruses could be introduced into new territories by migrating bats (van den Hurk et al., 2009).

11.4 CONTACT RATE AS A DRIVER FOR EMERGENCE OF BAT-ASSOCIATED ZOONOSES

Contact rate among bats and between bats and humans or domestic animals is a major contributing factor in zoonotic virus transmission. As mentioned earlier, many species of bats are colonial, roosting in extremely close, almost overlapping proximity with large numbers of other bats of that species, a situation that likely has a significant impact on bat-to-bat transmission of pathogens. Interestingly, reports of large mortality events in bats are rare and usually ascribed to such events as pesticide poisoning. Also, the annual introduction of a new population of susceptible young is well known to influence the dynamics of pathogen maintenance for many species (Anderson and May, 1979). In the case of bats, colony formation and birth pulse were associated with major amplification of coronavirus and astrovirus in *Myotis myotis* bats over a 3-year period (Drexler et al., 2011). Cohabitation of roosts by multiple species of bats is also frequently observed, although there is typically some degree of spatial segregation among species in such situations. Nonetheless, multi-species roosting may be important in diversification of viruses within bats, as discussed earlier for lyssaviruses.

Throughout the world, bats are coming into an ever-closer contact with humans and their livestock. Indeed, in some cases, bat populations are flourishing as a result of enhanced roosting opportunities provided by human habitations. Increased contact rates between bats and humans or livestock can result from a number of factors, including encroachment of human operations into bat habitats (e.g., Nipah virus and piggeries in Malaysia) and anthropogenic changes such as deforestation, which can alter bat habitat and force alterations in their geographic distribution (Chua et al., 2002a; Wibbelt et al., 2010.) Human encroachment is one of the most likely explanations of increased contact with zoonotic reservoirs. The viruses causing AIDS, Ebola, Nipah, Hendra, SARS, and monkeypox are just a few of the pathogens that can be transmitted to humans by animals when direct contact occurs. Examples of encroachment include hunters venturing into the forest to catch prey, people moving into rural areas and building houses and communities, clearing land for agricultural purposes, and urbanization. The last two examples can have immediate effects on the local ecosystem and can also contribute to global climate change by increasing pollution and decreasing the amount of trees and foliage available to take up excess carbon. In farming, crop clearing is a necessary but destructive process that all but eradicates an ecosystem's "natural" environment. In this scenario, animals that formerly lived at some distance from humans are suddenly in much closer contact, allowing pathogen spillover events to occur. This phenomenon is likely responsible for Nipah virus movement from Sumatra, where the burning of vast swaths of forest to create palm plantations forced the flying foxes to migrate into Malaysia, thereby coming into contact with pigs and starting the transmission cycle that would lead to human disease (Breed et al., 2010; Chua et al., 2002a). In another example, reforestation of previously arable land in the Midwest region of the United States likely fostered Lyme disease emergence by bringing the reservoir hosts (white-tailed deer and white-footed mice) back into proximity with humans (Daszak et al., 2001). It is important to keep in mind that even seemingly beneficial environmental changes can have consequences on disease outcome.

Agricultural practices drastically alter the natural landscape and are a prime example of human encroachment, but sometimes, the human incursions are less physically destructive yet equally invasive. As an example, take the case of Ebola viruses. Likely, these viruses circulated for years in bats and occasionally spilled over into nonhuman primates, causing severe and often fatal disease. Humans then came into contact with infected primate populations while hunting them for food or trapping them and contracted the disease when handling the carcasses. This is one example that illustrates how changes in human behaviors in relation to their environment can lead to disease.

Many species of bats migrate, often on a transcontinental scale, and epizootic transmission of viral diseases can be a direct effect of bat migratory patterns. Plowright et al. (2011) examined the transmission dynamics of Hendra virus in Australian pteropid bats using a mathematical model parameterized with field and laboratory data. These models predicted that decreased migration of *Pteropus* bats in Australia, possibly due to urbanization and changes in food availability, could influence the intensity and duration of Hendra virus epidemics. Decreased migration gave rise to more intense but shorter outbreaks after local viral reintroduction. The mechanism proposed for this increase in epidemic intensity was that decreased bat migratory behavior could lead to a decline in transmission between colonies and, therefore, reduced inter-colony exposures and resulting immunity within colonies. This loss of immunity may lead to increased epidemic size and more rapid fade-out once infection was reintroduced into a susceptible colony. Thus, a reduction in migration was suggested to increase the amplitude of the seasonal outbreaks and increase the probability of spillover.

A final important pathway for increased contact between wildlife and humans is the procurement and trade in bushmeat (Wolfe et al., 2005). In many areas of the world, bats are not only hunted for food but also considered with favor in the diet. Handling, butchering, and preparing bats for cooking may be a significant mechanism for transmitting viruses to humans. A notable case in point is the strong association with hunting bats and the outbreak of *Ebolavirus* infection in Africa (Leroy et al., 2009). It follows that surveillance for pathogens in bats entering the human food chain may be a fruitful avenue for identifying new emerging viruses.

11.5 POTENTIAL IMPACT OF CLIMATE CHANGE ON VIRUSES TRANSMITTED BY BATS

Emerging infectious diseases are frequently linked to global changes, both natural and man-made. Most of the time, these pathogens are zoonotic, which is why any variation in their hosts' environment is likely to impact the virus as well. While human encroachment and other anthropogenic practices tend to have a significant and traceable impact on emerging zoonoses, factors like climate change and global warming may be playing a substantial role in emergence as well. In addition to affecting vector populations, weather and climatic variations can take a toll on susceptible hosts. Most mammals are relatively refractory to minor temperature variations and are not likely to drastically alter their habits or habitats, but creatures that migrate and/or hibernate are an exception. For one thing, insectivorous birds and bats are driven by their food source, so when the insects themselves change their distribution, their predators will follow. Additionally, severe weather changes can induce migration or hibernation, which can alter disease distribution. Migratory animals pose a unique challenge in dealing with disease epidemiology. For example, birds are the primary reservoirs for viruses including avian influenza, JE, West Nile, and Western equine encephalitis, and because of their motility, they permit these diseases to spread very rapidly. Likewise, bat migratory patterns are at least partially responsible for the spread of Nipah and Hendra viruses in Southeast Asia (Epstein et al., 2009), and because bats are capable of carrying other viruses, it is highly likely that there are other instances of viral translocation by bats that go undetected or unreported. It is difficult to say exactly what impact global climate change has on animal migration, but whether their patterns change due to seasonal weather or long-term temperature variations, the ultimate location of the host will determine the location of the pathogen and its proximity to humans.

Climate change may also influence transmission of bat-harbored viruses by inducing stress that alters the within-host dynamics of infection. A study on Australian flying foxes illustrated that hot temperature extremes (>42 °C) distressed bats and induced them to take measures to cool down, including wing fanning, shade seeking, panting, and saliva spreading. Even with these behavior modifications, 5–6% of the bats died (Welbergen et al., 2008). In terms of impact on the spread of disease, many bats excrete virus in their saliva, so excessive salivation could easily lead to increased transmission. Additionally, shade-seeking behavior could put bats in closer proximity to humans, thereby increasing the risk of exposure. At the very least, the induced stress of hyperthermia could alter the bat's immune system and modulate disease pathogenesis. Since 1994, temperatures in New South Wales Australia have exceeded 42 °C on 19 recorded occasions, causing the deaths of more than 30 000 flying foxes (Welbergen et al., 2008); there is no reason to believe that temperatures elsewhere in the world are not having similar effects. An additional concern is that increasing aridity associated with climate change will have significant effects on bat survival and reproduction (Adams and Hayes, 2008), possibly altering their distribution and, hence, transmission of their pathogens.

11.6 CONCLUSIONS

Over the past decade, interest in bats as reservoir hosts for zoonotic pathogens has increased dramatically as more and more infectious diseases are linked to these highly mobile and incredibly diverse mammals. Numerous investigators have asked the question of whether there is something special about bats that promotes their ability to serve this role and to tease out mechanisms that may explain such a propensity. Several attributes of chiropteran life history have been proposed to contribute to reservoir host abilities, including their highly mobile nature as flying mammals and the fact that they often live in dense population structures that likely facilitates transmission of pathogens and that bats are among the most numerous and diverse of vertebrates. Rodents are the only mammals that compete with bats in terms of species numbers and diversity, and when comparing their capacity to carry diseases, similarities are evident (Halpin et al., 2007; Mills and Child, 1998). It is perhaps useful to remember that many of the pathogens recently identified in bats have been detected only by PCR, which, if applied with equal intensity to rodents, might yield another large battery of pathogens.

An intriguing recent finding in virology is the verdict that ancestral members of several virus families (coronaviruses, paramyxoviruses, filoviruses, and lyssaviruses) appear to have arisen in bats and spread over time to other species by host switching. Additional confirmation of these deductions should solidify a prominent role of bats as viral reservoirs.

There can be little doubt that increased surveillance will identify additional viruses harbored by bats and hopefully prepare us for emergence of new disease threats to humans, livestock, and wildlife. This basic biology must be coupled with an enhanced understanding of bat biology and ecology. Finally, we cannot fail to account for the impact of human practices and behaviors as they relate to emergence of zoonotic diseases.

REFERENCES

AbuBakar S., Chang L.Y., Ali A.R., et al., 2004. Isolation and molecular identification of Nipah virus from pigs. *Emerg Infect Dis.* **10**(12), 2228–2230.

Adams R.A., Hayes M.A., 2008. Water availability and successful lactation by bats as related to climate change in arid regions of western North America. *J Anim Ecol.* **77**, 1115–1121.

Anderson R.M., May R.M., 1979. Population biology of infectious diseases: Part I. *Nature.* **280**(2), 361–367.

Blehert D.S., Hicks A.C., Behr M., et al., 2009. Bat white-nose syndrome: an emerging fungal pathogen? *Science.* **323**(9), 227.

Bouma H.J., Carey H.V., Kroese F.G.M., 2010. Hibernation: the immune system at rest? *J Leukocyte Biol.* **88**(4), 619–624.

Breed A.C., Field H.E., Smith C.S., et al., 2010. Bats without borders: long-distance movements and implications for disease risk management. *Ecohealth.* **7**(2), 204–212.

Calisher C.H., Childs J.E., Field H.E., et al., 2006. Bats: important reservoir hosts of emerging viruses. *Clin Microbiol Rev.* **19**(3), 531–545.

Carrington C.V., Foster J.E., Zhu H.C., et al., 2008. Detection and phylogenetic analysis of group 1 coronaviruses in South American bats. *Emerg Infect Dis.* **14**(12), 1890–1893.

Chu D.W., Poon L.M., Chan K., et al., 2006. Coronaviruses in bent-winged bats (*Miniopterus* spp.). *J Gen Virol.* **87**(9), 2461–2466.

Chua K.B., Bellini W.J., Rota P.A., et al., 2000. Nipah virus: a recently emergent deadly paramyxovirus. *Science.* **288**(5470), 1432–1435.

Chua K.B., Chua B.H., Wang C.W., 2002a. Anthropogenic deforestation, El Nino and the emergence of Nipah virus in Malaysia. *Malays J Pathol.* **24**(1), 15–21.

Chua K.B., Koh C.L., Hooi P.S., et al. 2002b. Isolation of Nipah virus from Malaysian Island flying foxes. *Microb Infect.* **4**(2), 145–151.

Constantine D.G., 1970. Bats in relation to the health, welfare, and economy of man. Wimsatt W.A., ed. Biology of Bats, Volume 2. New York: Academic Press, pp. 319–344.

Daszak P., Cunningham A.A., Hyatt A.D., 2001. Anthropogenic environmental change and the emergence of infectious diseases in wildlife. *Acta Trop.* **78**, 103–116.

Davis A., Bunning M., Gordy P., et al., 2005. Experimental and natural infection of North American bats with West Nile virus. *Am J Trop Med Hyg.* **73**(2), 467–469.

Davis P.L., Bourhy H., Holmes E.C., 2006. The evolutionary history and dynamics of bat rabies virus. *Infect Genet Evol.* **6**, 464–473.

Dietzschold B., Morimoto K., Hooper D.C., et al., 2000. Genotypic and phenotypic diversity of rabies virus variants involved in human rabies: implications for postexposure prophylaxis. *J Hum Virol.* **3**(1), 50–57.

Dominguez S.R., O'Shea T.J., Oko L.M., et al., 2007. Detection of group 1 coronaviruses in bats in North America. *Emerg Infect Dis.* **13**(9) 1295–1300.

Drexler J.F, Gloza-Rausch F., Glende J., et al., 2010. Genomic characterization of severe acute respiratory syndrome-related coronavirus in European bats and classification of coronaviruses based on partial RNA-dependent RNA polymerase gene sequences. *J Virol.* **84**(21), 11336–11349.

Drexler J.F., Corman V.M., Wegner T., et al., 2011. Amplification of emerging viruses in a bat colony. *Emerg Infect Dis.* **17**(3), 449–456.

Drexler J.F., Corman V.M., Muller M.A., et al., 2012. Bats host major mammalian paramyxoviruses. *Nat Commun.* **3**(796), 796–808.

Drosten C., Gunther S., Preiser W., et al., 2003. Identification of a novel coronavirus in patients with severe acute respiratory syndrome. *N Engl J Med.* **348**, 1967–1976.

Epstein J.H., Prakash V., Smith C.S., et al., 2008. Henipavirus infection in fruit bats (*Pteropus giganteus*), India. *Emerg Infect Dis.* **14**(8), 1309–1311.

Epstein J.H., Olival K.J., Pulliam J.R.C., et al., 2009. *Pteropus vampyrus*, a hunted migratory species with a multinational home-range and a need for regional management. *J Appl Ecol.* **46**, 991–1002.

Feldmann H., Walh-Jensen V., Jones S.M., et al., 2004. Ebola virus ecology: a continuing mystery. *Trends Microbiol.* **12**(10), 433–437.

Field H.E., Mackenzie J.S., Daszak P., 2007. Henipaviruses: emerging paramyxoviruses associated with fruit bats. *Curr Top Microbiol Immunol* **315**, 133–159.

Fogarty R., Halpin K., Hyatt A.D., et al., 2008. Henipavirus susceptibility to environmental variables. *Virus Res.* **132**(1–2), 140–144.

Freuling C.M., Beer M., Conraths F.J., et al., 2011. Novel lyssavirus in Natterer's bat, Germany. *Emerg Infect Dis.* **17**(8), 1519–1521.

Gear J.S., Cassel G.A., Gear A.J., et al., 1975. Outbreak of Marburg virus disease in Johannesburg. *Brit Med J.* **29**(4), 489–493.

George D.B., Webb C.T., Farnsworth M.L., et al., 2011. Host and viral ecology determine bat rabies seasonality and maintenance. *Proc Natl Acad Sci USA.* **108**(2), 10208–10213.

Gloza-Rausch F., Ipsen A., Seebens A., et al., 2008. Detection and prevalence patterns of group I coronaviruses in bats, Northern Germany. *Emerg Infect Dis.* **14**(4), 626–631.

Guan Y., Zheng B.J., He Y.Q., et al., 2003. Isolation and characterization of viruses related to the SARS coronavirus from animals in southern China. *Science.* **302**, 276–278.

Halpin K., Young P.L., Field H.E., et al., 2000. Isolation of Hendra virus from pteropid bats: a natural reservoir of Hendra virus. *J Gen Virol.* **81**(8), 1927–1932.

Halpin K., Hyatt A.D., Plowright R.K., et al., 2007. Emerging viruses: coming in on a wrinkled wing and a prayer. *Emerg Infect Dis.* **44**(5), 711–717.

Halpin K., Hyatt A.D., Fogarty R., et al., 2011. Pteropid bats are confirmed as the reservoir hosts of Henipaviruses: a comprehensive experimental study of virus transmission. *Am J Trop Med Hyg.* **85**(5), 946–951.

Hartman A.L., Towner J.S., Nichol S.T., 2010. Ebola and Marburg hemorrhagic fever. *Clin Lab Med.* **30**(1), 161–177.

Henderson G.W., Laird C., Dermott E., et al., 1995. Characterization of Mapuera virus: structure, proteins and nucleotide sequence of the gene encoding nucleocapsid protein. *J Gen Virol.* **76**(10), 2509–2518.

Heymann D.L., Weisfeld J.S., Webb P.A., et al., 1980. Ebola hemorrhagic fever: Tandala, Zaire, 1977–1978. *J Infect Dis.* **142**(3), 372–376.

Hughes G.J., Orciari L.A., Rupprecht C.E., 2005. Evolutionary timescale of rabies virus adaptation to North American bats inferred from the substitution rate of the nucleoprotein gene. *J Gen Virol.* **86**(5), 1467–1474.

Jahrling P.B., Geisbert T.W., Dalgard D.W., et al., 1990. Preliminary report: isolation of Ebola virus from monkeys imported to USA. *Lancet.* **335**(8688), 502–505.

Johnson K.M., Webb P.A., Lange J.V., et al., 1977. Isolation and partial characterization of a new virus causing acute haemorrhagic fever in Zaire. *Lancet.* **1**(8011), 569–571.

Kan B., Wang M., Jing H., et al., 2005. Molecular evolution analysis and geographic investigation of severe acute respiratory syndrome coronavirus-like virus in palm civets at an animal market and on farms. *J Virol.* **79**(18), 11892–11900.

Keck J.G., Matsushima G.K., Makino S., et al., 1988. In vivo RNA–RNA recombination of coronavirus in mouse brain. *J Virol.* **62**(3), 1810–1813.

Ksiazek T.G., Erdman D., Goldsmith C.S., et al., 2003. A novel coronavirus associated with severe acute respiratory syndrome. *N Engl J Med.* **348**, 1953–1966.

Kuzmin I.V., Mayer A.E., Niezgoda M., et al., 2010. Shimoni bat virus, a new representative of the Lyssavirus genus. *Virus Res.* **149**(2), 197–210.

La Motte L.C., 1958. Japanese B encephalitis in bats during simulated hibernation. *Am J Hyg.* **67**(1), 101–108.

Lai M.M.C., Perlman S., Anderson L.J., 2007. Coronaviridae. In: Fields Virology, Fifth Ed., DM Knipe and PM Howley (eds.). Philadelphia, PA: Lippincott.

Lau S.K., Woo P.C., Li K.S., et al., 2005. Severe acute respiratory syndrome coronavirus-like virus in Chinese horseshoe bats. *Proc Natl Acad Sci USA.* **102**(39), 14040–14045.

Lau S.K.P., Li K.S.M., Huang Y., et al., 2010a. Ecoepidemiology and complete genome comparison of different strains of severe acute respiratory syndrome-related Rhinolophus bat coronavirus in China reveal bats as a reservoir for acute, self-limiting infection that allows recombination events. *J Virol.* **84**(6), 2808–2818.

Lau S.K., Poon R.W., Wong B.H., et al., 2010b. Co-existence of different genotypes in the same bat and serological characterization of Rousettus bat coronavirus HKU9 belonging to a novel Betacoronavirus subgroup. *J Virol.* **84**(21), 11385–11394.

Le Guenno B., Formentry P., Wyers M., et al., 1995. Isolation and partial characterization of a new strain of Ebola virus. *Lancet.* **345**(8960), 1271–1274.

Leroy E.M., Kumulungui B., Pourrut X., et al., 2005. Fruit bats as reservoirs of Ebola virus. *Nature* **438**, 575–576.

Leroy E.M., Epelboin A., Mondonge V., et al., 2009. Human Ebola outbreak resulting from direct exposure to fruit bats in Luebo, Democratic Republic of Congo, 2007. *Vector-Borne Zoonot Dis.* **9**(6), 9723–9728.

Leroy E.M., Gonzalez J.P., Baize S., 2011. Ebola and Marburg haemorrhagic fever viruses: major scientific advances, but a relatively minor public health threat for Africa. *Clin Microbiol Infect.* **17**(7), 964–976.

Leslie M.J., Messenger S, Rohde R.E., et al., 2006. Bat-associated rabies virus in skunks. *Emerg Infect Dis.* **12**(8), 1274–1277.

Li F., Li W., Farzan M., et al., 2005a. Structure of SARS coronavirus spike receptor-binding domain complexed with receptor. *Science.* **309**(5742), 1864–1868.

Li W., Shi Z., Yu M., et al., 2005b. Bats are natural reservoirs of SARS-like coronaviruses. *Science.* **310**(5748), 676–679.

Li W., Wong S.K., Li F., et al., 2006. Animal origins of the severe acute respiratory syndrome coronavirus: insight from ACE2-S-protein interactions. *J Virol.* **80**(9), 4211–4219.

Looi L.M., Chua K.B., 2007. Lessons from the Nipah virus outbreak in Malaysia. *Malays J Pathol.* **29**(2), 63–67.

Luby S.P., Hossain M.J., Gurley E.S., et al., 2009. Recurrent zoonotic transmission of Nipah virus into humans, Bangladesh, 2001–2007. *Emerg Infect Dis.* **15**(8), 1229–1235.

Mackenzie J.S., Chua K.B., Daniels P.W., et al., 2001. Emerging viral diseases of Southeast Asia and the western Pacific. *Emerg Infect Dis.* **7**(3), 497–504.

Marra M.A., Jones S.J., Astell C.R., et al., 2003. The genome sequence of the SARS-associated coronavirus. *Science.* **300**, 1399–1404.

Messenger S.L., Smith J.S., Orciari L.A., et al., 2003. Emerging pattern of rabies deaths and increased viral infectivity. *Emerg Infect Dis.* **9**(2), 151–154.

Mills J.N., Childs J.E., 1998. Ecologic studies of rodent reservoirs: their relevance for human health. *Emerg Infect Dis.* **4**(4), 529–536.

Miranda M.E., Ksiazek T.G., Retuya T.J., et al., 1999. Epidemiology of Ebola (subtype Reston) virus in the Philippines, 1996. *J Infect Dis* **179**(Suppl 1), S115–S119.

Misra V., Dumonceaux T., Dubois J., et al., 2009. Detection of polyoma and corona viruses in bats of Canada. *J Gen Virol.* **90**, 2015–2022.

Mondul A.M., Krebs J.W., Childs J.E., 2003. Trends in national surveillance for rabies among bats in the United States (1993–2000). *J Am Vet Med Assoc.* **222**(5), 633–639.

Morimoto K., Patel M., Corisdeo S., et al., 1996. Characterization of a unique variant of bat rabies virus responsible for newly emerging human cases in North America. *Proc Natl Acad Sci USA.* **93**(11), 5653–5658.

Murray K., Selleck P., Hooper P., et al., 1995. A morbillivirus that caused fatal disease in horses and humans. *Science.* **268**(5207), 94–97.

Osborne C., Cryan P.M., O'Shea T.J., et al., 2011. Alphacoronaviruses in New World bats: prevalence, persistence, phylogeny, and potential for interaction with humans. *PLoS One.* **6**(5), e19156.

Paul S.D., Rajagopalan P.K., Screenivasan M.A., 1970. Isolation of the West Nile virus from the frugivorous bat, *Rousettus leschenaulti*. *Indian J Med Res.* **58**(9), 1–3.

Peiris J.S.M., Lai S.T., Poon L.L., et al., 2003. Coronavirus as a possible cause of severe acute respiratory syndrome. *Lancet.* **361**(9366), 1319–1325.

Peterson A.T., Lash R.R., Carroll D.S., et al., 2006. Geographic potential for outbreaks of Marburg hemorrhagic fever. *Am J Trop Med Hyg.* **75**(1), 9–15.

Pfefferle S., Oppong S., Drexler J.F., et al., 2009. Distant relatives of severe acute respiratory syndrome coronavirus and close relatives of human coronavirus 229E in bats, Ghana. *Emerg Infect Dis.* **15**(9), 1377–1384.

Pinzon J.E., Wilson J.M., Tucker C.J., et al., 2004. Trigger events: enviroclimatic coupling of Ebola hemorrhagic fever outbreaks. *Am J Trop Med Hyg.* **71**(5), 664–674.

Plowright R.K., Foley P., Field H.E., et al., 2011. Urban habituation, ecological connectivity and epidemic dampening: the emergence of Hendra virus from flying foxes (*Pteropus* spp.). *Proc R Soc B: Biol Sci.* **278**(1725), 3703–3712.

Poon L.L.M., Chu D.K.W., Chan K.H., et al., 2005. Identification of a novel coronavirus in bats. *J Virol.* **79**(4), 2001–2009.

Pourrut X., Delicat A., Rollin P.E., et al., 2007. Spatial and temporal patterns of *Zaire ebolavirus* antibody prevalence in the possible reservoir bat species. *J Infect Dis.* **196** (Suppl 2), S176–S183.

Pourrut X., Souris M., Towner J.S., et al., 2009. Large serological survey showing cocirculation of Ebola and Marburg viruses in Gabonese bat populations, and a high seroprevalence of both viruses in *Rousettus aegyptiacus*. *BMC Infect Dis.* **9**(159), 1–10.

Pulliam J.R., Field H.E., Olival K.J., 2005. Nipah virus strain variation. *Emerg Infect Dis.* **11**(12), 1978–1979.

Rahman M.A., Hossain M.J., Sultana S., et al., 2012. Date palm sap linked to Nipah virus outbreak in Bangladesh, 2008. *Vector-Borne Zoonot Dis.* **12**(1) 65–72.

Ren W., Qu X., Li W., et al., 2008. Difference in receptor usage between severe acute respiratory syndrome (SARS) coronavirus and SARS-like coronavirus of bat origin. *J Virol.* **82**(4), 1899–1907.

Reusken C.B., Lina P.H., Pielaat A., et al., 2010. Circulation of group 2 coronaviruses in a bat species common to urban areas in Western Europe. *Vector-Borne Zoonot Dis.* **10**(8), 785–791.

Rupprecht C.E., Turmelle A., Kuzmin I.V., 2011. A perspective on lyssavirus emergence and perpetuation. *Curr Opin Virol.* **1**(6), 662–670.

Selvey L.A., Wells R.M., McCormack J.G., et al., 1995. Infection of humans and horses by a newly described morbillivirus. *Med J Aust.* **162**(12), 642–645

Simmons N.B., Seymour K.L., Habersetzer J., et al., 2008. Primitive Early Eocene bat from Wyoming and the evolution of flight and echolocation. *Nature.* **451**, 818.

Skowronski D.M., Astell C., Brunham R.C., et al., 2005. Severe acute respiratory syndrome (SARS): a year in review. *Annu Rev Med.* **56**, 357–381.

Streicker D.G., Turmelle A.S., Vonhof M.J., et al., 2010. Host phylogeny constrains cross-species emergence and establishment of rabies virus in bats. *Science.* **329**(5992), 676–679.

Sulkin S.E., Allen R., 1974. Virus infections in bats. *Monogr. Virol.* **8**(0), 1–103.

Sulkin S.E., Allen R., Sims R., et al., 1960. Studies on the pathogenesis of rabies in insectivorous bats. II. Influence of environmental temperature. *J Exp Med.* **30**, 595–617.

Sulkin S.E., Allen R., Sims R., 1963. Studies of arthropod-borne virus infections in chiroptera. I. Susceptibility of insectivorous species to experimental infection with Japanese B and St. Louis encephalitis viruses. *Am J Trop Med Hyg.* **12**, 800–814.

Sulkin S.E., Allen R., Sims R., 1966. Studies of arthropod-borne virus infections in chiroptera. III. Influence of environmental temperature on experimental infection with Japanese B and St. Louis encephalitis viruses. *Am J Trop Med Hyg.* **15**(3), 406–417.

Swanepoel R., Smit S.B., Rollin P.E., et al., 2007. Studies of reservoir hosts for Marburg virus. *Emerg Infect Dis.* **13**(12), 1847–1851.

Tang X.C., Zhang J.X., Zhang S.Y., et al., 2006. Prevalence and genetic diversity of coronaviruses in bats from China. *J Virol.* **80**(5), 7481–7490.

Taniguchi S., Watanabe S., Masangkay J.S., et al., 2011. Reston Ebolavirus antibodies in bats, the Philippines. *Emerg Infect Dis.* **17**(8), 1559–1560.

Taylor D.J., Leach R.W., Bruenn J., 2010. Filoviruses are ancient and integrated into mammalian genomes. *BMC Evol Biol.* **10**, 193–203.

Taylor D.J., Dittman K., Ballinger M.J., et al., 2011. Evolutionary maintenance of filovirus-like genes in bat genomes. *BMC Evol Biol.* **11**, 336.

Teeling E.C., Springer M.S., Madsen O., et al., 2005. A molecular phylogeny for bats illuminates biogeography and the fossil record. *Science.* **307**, 580–584.

Tong S., Conrardy C., Ruone S., et al., 2009. Detection of novel SARS-like and other coronaviruses in bats from Kenya. *Emerg Infect Dis.* **15**(3), 482–485.

Towner J.S., Pourrut X., Albarino C.G., et al., 2007. Marburg virus infection detected in a common African bat. *PLoS One.* **8**, e764.

Towner J.S., Sealy T.K., Khristova M.L., et al., 2008. Newly discovered Ebola virus associated with hemorrhagic fever outbreak in Uganda. *PLoS Pathog.* **4**(11), 1–6.

Towner J.S., Amman B.R., Sealy T.K., et al., 2009. Isolation of genetically diverse Marburg viruses from Egyptian fruit bats. *PLoS Pathog.* **5**(7), 1–9.

van den Hurk A.F., Smith C.S., Field H.E., et al., 2009. Transmission of Japanese encephalitis virus from the black flying fox, *Pteropus alecto*, to *Culex annulirostris* mosquitoes, despite the absence of detectable viremia. *Am J Trop Med Hyg.* **81**(3), 457–462.

Vijaykrishna D., Smith G.J.D., Zhang J.X., et al., 2007. Evolutionary insights into the ecology of coronaviruses. *J. Virol.* **81**(8), 4012–4020.

Wang L.F., Yu M., Hansson E., et al., 2000. The exceptionally large genome of Hendra virus: support for creation of a new genus within the family Paramyxoviridae. *J Virol.* **74**(21), 9972–9979

Warrel M., 2010. Rabies and African bat lyssavirus encephalitis and its prevention. *Int J Antimicrob Agents.* **36S**, S47–S52.

Watanabe S., Masangkay, J.S., Nagata N., et al., 2010. Bat coronaviruses and experimental infection of bats, the Philippines. *Emerg Infect Dis.* **16**(8), 1217–1223.

Welbergen J.A., Klose S.M., Markus N., et al., 2008. Climate change and the effects of temperature extremes on Australian flying-foxes. *Proc R Soc B.* **275**(1633), 419–425.

Weyer J., Szmyd-Potapczuk A.V., Blumberg L.H., et al., 2011. Epidemiology of human rabies in South Africa, 1983–2007. *Virus Res.* **155**(1), 283–290.

Wibbelt G., Moore M.S., Schountz T., et al., 2010. Emerging diseases in Chiroptera: why bats? *Biol Lett.* **6**(4), 438–440.

Wolfe N.D., Daszak P., Kilpatrick A.M., et al., 2005. Bushmeat, hunting, deforestation and prediction of zoonoses emergence. *Emerg Infect Dis.* **11**(12), 1822–1827.

Woo P.C.Y., Lau S.K.P., Li K.S.M., et al., 2006. Molecular diversity of coronaviruses in bats. *Virology.* **351**(1), 180–187.

Woo P.C.Y., Lau S.K.P., Huang Y., et al., 2009. Coronavirus diversity, phylogeny and interspecies jumping. *Exp Biol Med.* **234**(10), 1117–1127.

Yob J.M., Field H., Rashdi A.M., et al., 2001. Nipah virus infection in bats (order Chiroptera) in peninsular Malaysia. *Emerg Infect Dis.* **7**(3), 439–441.

12

ROLE AND STRATEGIES OF SURVEILLANCE NETWORKS IN HANDLING EMERGING AND REEMERGING VIRAL INFECTIONS

Carlos Castillo-Salgado

Department of Epidemiology, Bloomberg School of Public Health, Johns Hopkins University, Baltimore, MD, USA

TABLE OF CONTENTS

Viral Infections and Global Change, First Edition. Edited by Sunit K. Singh.

12.1 INTRODUCTION

Infectious disease surveillance and health risk assessment are now major areas of concern in national and global public health. The renewed recognition of the importance of public health surveillance is due to the continuous and accelerated trend of newly emerging and reemerging infectious diseases and the constant cycles of infectious disease outbreaks in many countries and even the serious pandemics of viral infections such as AIDS, SARS, and influenza A (H1N1).

In the twenty-first century, the global relevance of the surveillance efforts to address the formidable challenge of new emerging and reemerging communicable diseases—most of them viral infections—is of outmost importance.

The aim of this chapter is to provide a description of the new vision of public health surveillance of viral infections and to review the main role, functions, and attributes of public health surveillance in a globalized world. The three sections of this chapter include (i) the accelerated global trend of viral infectious agents and diseases, (ii) the emergence of new global health networks of early warning and surveillance, and (iii) the role and strategies of the new global infectious disease surveillance networks.

12.2 GLOBAL TREND OF VIRAL INFECTIOUS AGENTS AND DISEASES

The majority of the global emerging infections are viral pathogens. During the last decades, the unprecedented speed in their global appearance and their virulence required the urgent need for a better understanding of their epidemiological distributions and trends.

In the early twentieth century, public health regulations contained disease control measures for 38 communicable diseases that were reportable in most countries. The first three diseases under global surveillance were cholera, plague, and yellow fever. Yellow fever is a viral disease still of great concern in many areas of the world. All of these diseases were of major importance in the economic globalization of the early twentieth century. The first International Health Regulations (IHR) originally designated in 1951 as "International Sanitary Regulations" included these three infectious diseases as part of the required national and global reporting for several decades.

Since the 1980s, the world experienced at least 30 previously unknown infectious disease agents, including human immunodeficiency virus (HIV); filoviruses such as Ebola hemorrhagic fever and Marburg hemorrhagic fever; adenoviruses such Lassa fever, hemorrhagic dengue, and Junin fever; and hepatitis C, Lyme disease, Nipah virus, West Nile virus, SARS, avian flu virus, and influenza A (H1N1) virus, for which no adequate treatments are available (Castillo-Salgado,

2010). Some of these viruses have been identified as potential bioterrorism agents. It is relevant to recognize that most of the new viral hemorrhagic fevers have been identified as some of the most deadly diseases and that they have limited interventions. Recognizing the role of globalization of trade and economy and the massive mobilization of goods and people across countries at unprecedented speed and the real potential for disease outbreaks and pandemics resulting from the emerging and reemerging viral and infectious agents and other public health risks (PHR) in 2005, the World Health Organization (WHO) issued the newly revised "IHR" (IHR-2005; WHO, 2005) as approved by all 194 WHO member states. These IHR entered into force in 2007 (WHO, 2008b). The IHR-2005 is the main regulatory source for global surveillance of priority infectious diseases, particularly from viral agents. It was acknowledged that it is possible to travel between most places in the world in less time than the incubation period for many infectious diseases (Choffnes, 2008). Jones et al. (2008) documented the emergence and trends of 335 new infectious diseases, called "emerging infectious disease (EID) events." They reported that most of the EIDs were zoonoses (60.3%), that the majority (71.8%) originated in wildlife, and that they were increasing significantly over time. They also mentioned that the peak incidence of these EIDs occurred in the 1980s concomitant with the HIV pandemic. Several of the viral infections of great public health importance include poliomyelitis and smallpox as well as childhood viral diseases such as measles, pertussis, varicella, rubella, mumps, diphtheria, and tetanus. For all of these conditions, vaccines are available. However, the reemergence of many of these conditions in many of developed countries reflects the incomplete vaccination schemes in many of those countries and the deterioration of their public health systems. Other important viral infections under surveillance in many countries are Haemophilus influenzae infection, rabies, herpes simplex, herpes zoster, rotavirus infection, parvovirus infection, human papillomavirus (HpV) infection, and hepatitis. In 2012, the U.S. Centers for Disease Control and Prevention (CDC) (CDC, 2012) launched an epidemiological alert recommending the screening for hepatitis C virus (HCV) for those born between 1945 and 1965 to address the current iceberg of hepatitis C prevalence and to identify and to treat the thousands of people with the infection without knowing they have it. The existence of HCV infection was confirmed in 1989 (Houghton, 2009). However, no vaccine against hepatitis C is currently available.

During the last decades the global trend of new emerging viral infections has presented an unprecedented challenge for global health, since it is occurring at the time of one of the most difficult economic global crises in recent history (Castillo-Salgado, 2010, p.12). International trade, commerce, and travel—as key dimensions of the economic globalization—have become key forces in transforming public health and disease surveillance.

12.3 RECOGNIZED IMPORTANCE OF PUBLIC HEALTH SURVEILLANCE

12.3.1 Public health surveillance as essential public health functions and core competencies

The public health community has recognized public health surveillance as one of the most important essential public health functions.

Based on the 10 essential public health functions framework (Public Health Foundation, 2010), public health surveillance has been included as one of the first and more critical functions of public health services.

To address this public health function, several core professional competencies have been identified. Professional competencies are defined as the "knowledge, skills and abilities demonstrated by the members of an organization that are critical to the effective and efficient function of that organization" (The Joint Task Group of Public Health Human

TABLE 12.1. Public Health Competencies. Public Health Professionals Working in Infectious Disease Surveillance Should Able to:

1. Identify key sources of data for infectious disease epidemiology
2. Identify the new emerging surveillance systems, including syndromic surveillance and real-time surveillance systems
3. Implement new or existing surveillance systems
4. Evaluate surveillance systems for infectious diseases
5. Explain how behavioral and social factors impact the transmission of infectious diseases
6. Describe the basic concepts in infectious disease epidemiology, including the reproductive number, herd immunity, virulence, and incubation, infectious, and latent periods
7. Describe how novel infectious agents emerge including changes from zoonotic infections to human diseases
8. Explain temporal and spatial patterns of infectious disease dynamics, including cyclical and seasonal patterns of transmission
9. Explain study designs used to evaluate control measures for infectious diseases, including vaccine efficacy and indirect effects

Source: Infectious Disease Epidemiology Tract. Department of Epidemiology, Bloomberg School of Public Health, Johns Hopkins University, Baltimore, MD. 2011. Working Internal Document.

Resources, 2005, p.24). The identified professional competencies are necessary for the practice of surveillance in the context of public health.

A list of important professional competencies for infectious disease surveillance is included in Table 12.1. Important institutions such as the Institute of Medicine, American Association of Schools of Public Health, U.S. CDC, Canadian CDC, European CDC, and the WHO have recommended professional competencies for public health and epidemiology professional practice (Council of State and Territorial Epidemiologists, 2010; ECDC, 2008; McNutt et al., 2008; Nelson et al., 2002).

Fundamental to infectious disease epidemiology is knowledge of the modes of transmission, risk factors for infection and disease, spatial and temporal patterns of transmission dynamics, and disease surveillance. An excellent source of relevant information about the epidemiology of most of the viral diseases is the Manual of Control of Communicable Diseases (Heymann, 2008). This important guidebook is in its 19th edition.

12.4 DEFINITION AND SCOPE OF PUBLIC HEALTH SURVEILLANCE

Public health surveillance has been defined as "Continuous analysis, interpretation and feedback of systematically collected data, generally using methods distinguished by their practicability, uniformity, and rapidity, rather than by accuracy or completeness" (Last, 1995, p.163). Key elements of this definition are the regular assessment, practicality, and uniformity of methods.

The International Epidemiological Association in the fifth edition of its "Dictionary of Epidemiology" emphasizes that "surveillance… is an essential feature of epidemiological and public health practice." "The final phase in the surveillance chain is the application of information to health promotion and to disease prevention and control" (Porta, 2008, p.239). Surveillance was defined by the IHR-2005 as the "systematic ongoing collection, collation and analysis of data for public health purposes and the timely dissemination of public health information for assessment and public health response as necessary" (WHO, 2008c, p.9). This definition expanded the traditional focus from monitoring three specific diseases—cholera, yellow fever, and plague—to expanding the efforts to prevent and control the international spread of diseases with high risk of a global pandemic and any major PHR including biological and chemical agents or nuclear radiation that affect global health under a multilateral public health response. Viral infections surveillance was identified as a main activity for public and global health surveillance. Public health surveillance is by nature an action-oriented process. Its main goal is the prompt initiation of steps to control or prevent a health problem, event, or risk of public health importance.

Public health surveillance is an integral research–action process providing public health authorities with important data for developing and implementing preventive and control programs and interventions. Also, it provides policy-makers useful information for planning public health actions. It is an integral component for monitoring and evaluating heath programs and assists for health prevention advocacy at local, national, and international level.

12.5 KEY FUNCTIONS AND USES OF DISEASE SURVEILLANCE

Most public health agencies have identified that the main schematic elements of surveillance are the type of data-gathering mechanisms, including the use of laboratory; central/local collection and analysis of the data; and their publication and dissemination and initiation of indicated action or response.

The surveillance community recognizes that the basic surveillance of communicable diseases involves five key functions: (i) the detection of cases of disease or epidemics in specific populations and reporting the information, (ii) analyzing and confirming reported case information to identify alerts and to early detect outbreaks, (iii) the monitoring of trends over time to assess the need for prevention measures, (iv) the evaluation of preventive measures, and (v) the generation of data and information for public health and health-care planning and resource allocation.

An illustrative review of the potential effectiveness of two types of surveillance (clinical recognition vs. syndromic surveillance) to recognize the main signs and characteristics of linking some of these infectious disease agents as bioterrorism-related epidemics is presented by Buehler et al. (2003). They concluded that the potential success of these surveillance systems depends on recognizing four characteristics: incubation period of the agent, duration of prodromal phase, presence of clinical signs, and the probability of making a diagnosis during the routine assessment. When effectively linked to specific infectious diseases or to appropriate syndromic events, surveillance provides key information for action against infectious disease risks and threats.

When public health surveillance is implemented, it is of great importance to incorporate the highest epidemiological quality control. Quality control in any surveillance system requires several basic conditions such as the use of a uniform case definition (CDC, 2012a) across the different entities of a national program; the collection of useful and sufficient information about the case under surveillance; the appropriate analysis by time, person, and space; and the timely dissemination of the information to assist in the development of proper intervention to reduce and control the preventable cases.

Case definition in infectious disease surveillance is a common methodological and operational problem. Case definition has been recognized as a set of objective criteria (symptoms, signs, and laboratory data) that allow a reliable and consistent report of the event, condition, or disease under surveillance. In most countries, it is recommended that a case definition should be a brief and straightforward account of the event to assist in the rapid and uniform reporting of the events under surveillance.

TABLE 12.2. Recognized Uses of Public Health Surveillance Early Warning Systems

Detecting outbreaks and epidemics
Identifying newly emerging agents and conditions
Recognizing unusual geospatial disease clusters
Definition of epidemic thresholds for priority health events and conditions
Recognizing changes in antibiotic-resistance patterns
Environmental health risk warnings
Risk factor surveillance
Recognizing health disparities among social, racial, ethnic, or gender groups

The need for standardized case definitions in global and national surveillance is promoted by all health agencies and the surveillance community. WHO and most national and international health agencies have included standard case definitions on their websites and documents for the known diseases and public health events. However, as mentioned earlier, for novel and emerging diseases and new public health events, there is a need for WHO global guidelines. There is a provision in the IHR-2005 to include surveillance and notification of public health events of international importance, even if the causative agent is not yet known.

Table 12.2 includes nine of the recognized uses of public health surveillance. In addition to providing early warning systems and institutional capacity for detecting outbreaks, epidemics, and pandemics, one key methodological consideration is to facilitate the identification of new EID agents and conditions.

The use of networks of national and global public health laboratories is of critical importance for the recognition of new viral agents and conditions.

12.6 NEW EXPANSION OF SURVEILLANCE BY THE IHR-2005

The IHR-2005 extended the types of events to be reported internationally. It introduced new key terms, such as "PHR" and "Public Emergency of International Concern" (PHEIC), as events under which the International Regulations apply. A global health emergency was defined as an "extraordinary event," which is determined by the following: (i) it constitutes a PHR to other states through the international spread of disease and (ii) it potentially requires a coordinated international response. In this regard, countries are required to notify the WHO of all events "that may constitute a public health emergency of international concern" according to the agreed framework of the Regulations. One of the key goals of the IHR is to monitor the emergence of any "public health emergency of international concern" (WHO, 2008c, p.32).

12.7 EMERGENCE OF NEW GLOBAL SURVEILLANCE NETWORKS

The entire global population is at risk of current and unexpected future diseases labeled emerging and reemerging viral infections because these biological pathogens do not respect geopolitical boundaries or international borders. Public health surveillance with enhanced early warning systems has been identified as a public health imperative both at national and global levels.

12.8 GLOBAL INFLUENZA SURVEILLANCE AND WHO'S PANDEMIC INFLUENZA PREPAREDNESS FRAMEWORK

An important global framework for influenza surveillance was adopted by the members of the WHO in 2011 (WHO, 2011) (http://whqlibdoc.who.int/publications/2011/9789241503082_eng.pdf). The aims of this important framework are to advance coordination and sharing of influenza viruses with pandemic potential and to achieve more effective and equitable access to vaccines and medical treatments during future pandemics.

The four major approaches used for influenza surveillance are described as (i) Viral Surveillance, (ii) Outpatient Illness Surveillance, (iii) Hospitalization Surveillance, and (iv) Mortality Surveillance.

WHO and most countries use regular weekly reports from national and collaborating laboratories on the number, percentage, and type of positive viral specimens for influenza as part of the viral surveillance. The percentage of visits for influenza-like illness reported by outpatient clinics and physician visits are used in Outpatient Illness Surveillance. The influenza laboratory-confirmed hospitalization rates of patients are used for the Hospitalization Surveillance. The influenza mortality and the influenza-associated deaths reported by vital statistics are used for the Mortality Surveillance.

For global influenza surveillance, the WHO developed the "Global Influenza Surveillance and Response System" (GISRS). The current name of this system was adopted following the 2011's Pandemic Influenza Preparedness Framework. The WHO GISRS system monitors the trends of influenza viruses and provides recommendations on laboratory diagnostics, vaccines, antiviral susceptibility, and epidemiological risk assessments. This global system serves as global alert mechanism for the emergence of influenza viruses with pandemic potential. The WHO GISRS system can be accessed at www.who.int/influenza/gisrs_laboratory/en/. Also, this global network is used for improving the pandemic influenza preparedness and response.

In June 2012, GISRS informed the global community that "Since their reemergence in 2003, highly pathogenic avian influenza A (H5N1) viruses continue to pose a serious threat to public health" (WHO/GIRRS, 2012, p.1). As part of its surveillance strengthening strategy for early detection of the emergencies of genetic changes of these viruses, an inventory of amino acid mutations in A (H5N1) viruses has been developed. This inventory can be accessed at http://www.cdc.gov/flu/avianflu/h5n1/inventory.htm.

This inventory will play an important role in assisting in the identification of the viral phenotypic characteristics of the genetic mutations. This inventory will be maintained and updated periodically by WHO and U.S. CDC.

In 2009, the WHO's Europe Region developed a regional web-based information platform for influenza surveillance in Europe named "EuroFlu" (http://www.euroflu.org/index.php). As described in the website, EuroFlu's objective is to assist countries to "reduce influenza morbidity and mortality in Europe by: (i) collecting and exchanging timely information; (ii) contributing to the annual determination of vaccine content; (iii) providing relevant information to health professionals and the general public; and (iv) contributing to the response to the influenza A (H1N1) pandemic. It was expected that clinicians, epidemiologists and virologists in the 53 countries of the WHO European Region will be part of a network reporting to EuroFlu. The laboratory network consists of several WHO-recognized national influenza centers, a WHO Collaborating Center for Reference and Research on influenza and two WHO H5 Reference Laboratories." A list of important laboratory links for this regional network is presented in the EuroFlu website.

Several countries have designed their own influenza surveillance systems, and they produce important alerts and trend assessments in their e-reports such as the Canada's FluWatch (www.phac-asp.gc.ca/fluwatch/index-eng.php) and the U.S. CDC FluView (www.cdc.gov/flu/weekly/fluactivitysurv.htm).

The U.S. CDC reported that in the United States the 2009 (H1N1) pandemic influenza A virus from April 2009 to March 13, 2010 infected between 43 and 88 million people, hospitalized between 192 000 and 398 000 people, and killed between 8720 and 18050 people (Reed et al., 2009). Estimates are available at http://www.cdc.gov/h1n1flu/estimates_2009_h1n1.htm.

The U.S. FluView report includes information from five different categories of Influenza Surveillance: Viral Surveillance, Surveillance of Novel Influenza A Viruses, Outpatient Illness Surveillance (ILINet), Mortality Surveillance, Hospitalization Surveillance (FluSurv_Net), and Geographic Spread of Influenza.

The viral infections surveillance in the United States is reported to be done through 80 U.S. WHO Collaborating Laboratories and 60 National Respiratory and Enteric Virus Surveillance System (NREVSS) laboratories located in all 10 regions of the country. The weekly total number of positive influenza tests, by virus type/subtype, and the percentage of positive specimens are presented in the FluView report.

Similar information is presented in the FluWatch summaries of Canada (Public Health Canada: FluWatch, 2012), available at www.phac-asp.gc.ca/fluwatch/index-eng.php. Depending on the epidemiological season, these reports are produced weekly or biweekly.

These reports summarize the influenza surveillance information and activities including the spread of flu and flu-like infections on an ongoing basis in their countries.

12.9 EARLY WARNING SURVEILLANCE SYSTEMS

Several public health agencies at national and international levels as well as governmental agencies linked to defense and homeland security and environment and food security have been funding, supporting, and implementing different *Early Warning Infectious Disease Surveillance* programs with the objective of addressing public health emergency preparedness activities in early detection, identification, and reporting of viral and infectious diseases associated with public health emergencies, potential bioterrorism agents, or other major threats to public health. Early Warning Surveillance Systems for viral agents and diseases are some of the most developed surveillance programs and systems. An inventory of global Early Warning Surveillance Systems was included in Castillo-Salgado (2010, pp. 6–7).

A list of selected Internet sites developed by WHO for monitoring and surveillance of viral global health events is included in Table 12.3.

A recent analysis by Wilson and Brownstein (2009) provides helpful examples of how Internet surveillance tools can assist in the early identification of disease outbreaks. The article concludes that web-based sources of information in addition to allowing timely detection of outbreaks will reduce cost and increase reporting transparency. These authors presented a useful list of major advantages and disadvantages of the "Internet-based surveillance."

Hitchcock et al. (2007) prepared a landmark review of 15 international surveillance and response programs (ISRPs). This review offers a useful classification of these key global systems. The classification is organized by the four basic components of surveillance and response programs: surveillance, reporting, verification, and response. These

TABLE 12.3. Selected WHO's Global Viral Infections Surveillance Sites

Selected WHO's global viral infections surveillance sites	Description and type of information under global surveillance
Strategic Health Operations (SHOC)	WHO's central platform of coordination of the response to infectious diseases outbreaks and public health crisis. Hub for alert and response operations http://www.who.int/csr/alertresponse/shoc/en/index.html
Global Alert and Response (GAR)	GOARN is the hub for information from the GPHIN and the official country sources. This is the main WHO's surveillance network with the collaboration of hundreds of health institutions http://www.who.int/csr/en/
FluNet	FluNet is the web-based data collection and reporting tool of the GISN. FluNet includes virological influenza information since 1995 from countries worldwide, provided by National Influenza Centres (NICs) and other national influenza reference laboratories collaborating actively with GISN. Data entry is restricted and data entry access password protected; data reports including tables, maps, and graphs are available to all public users
DengueNet	DengueNet, is part of the Global Health Atlas platform and functions as the WHO's central data management system for the global epidemiological and virological surveillance of dengue fever (DF) and dengue hemorrhagic fever (DHF) The network collects standardized data from all DengueNet partners worldwide and provides web access to key indicators such as incidence, case fatality rates (CFR), frequency and distribution of DF and DHF cases, number of fatalities, and circulating virus serotypes
Rabnet	"Rabnet version2" is a web-based data collection tool and interactive information system for the generation of graphs and maps with human and animal rabies data. The system produces yearly updates at country level and first administrative level (province or state)

Source: WHO website (http://www.who.int/en/).

authors concluded that only six ISRPs cover the four components of surveillance and response, five ISRPs include the surveillance and reporting components, and three ISRPs have the verification and response components. The six ISRPs are Global Polio Eradication Initiative, Regional Immunization Program of the Americas, Global Disease Detection (GDD) Program, Biological Threat Reduction Program (BTRP), and Epidemic and Pandemic Alert and Response (EPR). The five ISRPs are Global Public Health Intelligence Network (GPHIN), ProMED-mail, QFLU, European Influenza Surveillance Scheme (EISS), and the Global Influenza Surveillance Network (GISN). The three ISRPs are Outbreak Alert and Verification System (OAV), Global Outbreak Alert and Response Network (GOARN), and the Preparedness and Response Unit.

In 1997, a very original global surveillance initiative was proposed by the WHO in partnership with the Canadian Agency of Public Health to help identify significant disease outbreaks around the world, taking advantage of the existing globalized virtual communications. The GPHIN (Public Health Agency of Canada, 2004) is an Internet surveillance system, which gathers data and public health reports from diverse countries in seven languages, aimed at disseminating timely alerts to help control outbreaks, the spread of infectious disease, contamination of food and water, bioterrorism, natural disasters, and exposure to chemical agents and nuclear materials. This system also monitors questions

TABLE 12.4. Innovative Approaches for Viral and Infectious Disease Surveillance

Electronic and web-based information platforms for information reporting, sharing, and dissemination
Real-time and near real-time information
New updated statistical methods for tracking viral and infectious disease outbreaks
Use of GIS for mapping and geo-referencing public health events and risks of national and global importance
Using proxy and compiled web-based information from different sources
Incorporation of public–private partnerships in surveillance activities
Use of volunteer sentinel physicians
Improving guidelines and protocols for laboratory viral surveillance
Incorporating Health Situation Room or Command Centers for monitoring, analysis, and response in surveillance efforts

related to the safety of medications and medical products. The system assists WHO as an outbreak verification process. Between November 1999 and October 2000, WHO investigated 228 outbreak reports with 169 confirmed as outbreaks of global health significance. A very illustrative one is the 2009 declared pandemic level VI of influenza H1N1 (WHO, 2009b).

The GPHIN currently has hundreds of laboratories and disease notification systems, which provide immediate reports utilizing high technology such as systematic scanning of electronic resources—including websites, news, electronic public health services, and Internet discussion groups. GPHIN members can be governmental and nongovernmental agencies, which promote norm setting through wider use of this information.

12.10 INNOVATIVE APPROACHES FOR SURVEILLANCE

This section will highlight some of the most innovative approaches for enhancing public health surveillance in the first part of the twenty-first century. Table 12.4 lists nine useful innovative approaches for surveillance of viral and other infectious diseases.

12.11 ELECTRONIC AND WEB-BASED INFORMATION PLATFORMS FOR INFORMATION REPORTING, SHARING, AND DISSEMINATION

A highly information-oriented world and the occurrence of new emerging viral diseases and pathogens have produced a need for good laboratory and surveillance capacity but also efficient ways and channels of communication and sharing of surveillance information.

Virtual communication has transformed the way information is shared and used. In addition, the new vision of public health surveillance requires the incorporation of the new information technologies such as the Internet, web-based information systems, and the interoperability of national electronic information systems. The realities of a globalized world make possible the spatial diffusion of any infection or health condition. This diffusion could be completed in few hours or days by the intensified and continuous air transportation of thousands of people and goods from different continents in few hours.

During the last few years, the rapid expansion of health information technologies and the Internet have transformed every agency in the public health sector. Relevant ministries of health; health departments; public or private epidemiological research or operational centers; the United Nations Specialized Health Agencies such as WHO, UNICEF, and FAO; and private organizations expanded the use of electronic health and epidemiological records, web-based information platforms, mobile applications, and wireless devices for disease surveillance and dissemination of information and alerts in real or near real time.

WHO, USAID, CDC, Canadian CDC, European CDC, and other agencies provide in their websites useful practical descriptions of the different types of surveillance, including passive and active surveillance, facility-based routine surveillance, community-based surveillance, sentinel surveillance, and syndromic surveillance.

12.12 REAL-TIME AND NEAR REAL-TIME INFORMATION

The new health information technologies and the Internet are important drivers affecting the future directions of infectious disease surveillance communications, particularly facilitating the collection and transmission of information at speeds that allow for better emergency preparedness and response. Also, virtual information is immediately accessible to different public health stakeholders and the general public throughout the world.

With the Internet explosion and expanding virtual communications, surveillance information for the first time can be accessible in real time in any place in the world for those with virtual connectivity. In the new surveillance systems, infectious disease outbreaks are to be reported electronically 24 h a day by email to WHO and to the national epidemiology centers.

Web-based operated systems have had a dramatic impact on the new developments in infectious disease surveillance, including the eHealth networks for disease reporting and dissemination of information and alerts in real time.

One important change in the new IHR-2005 is the obligation of governments for improving their capacity of national and regional surveillance systems by incorporating real-time event management systems. In this regard, Internet and new health informatics methods are revolutionizing how real-time information about public health events and outbreaks is exchanged and communicated nationally and globally.

12.13 NEW UPDATED STATISTICAL METHODS FOR TRACKING VIRAL AND INFECTIOUS DISEASE OUTBREAKS

In the quest for better analytical tools for early detection of outbreaks, surveillance systems are implementing seasonal baselines and robust regression models to create better epidemic thresholds (epidemiological aberrations) to recognize significantly higher levels of cases than would be expected at different times and places.

A surveillance aberration is a statistical term used to describe deviations from a usual frequency distribution of health events. Several methods for detecting surveillance aberrations have been described recently (Fricker et al., 2008; Hutwagner et al., 2005; Lombardo et al., 2003). Aberration reporting systems have been implemented as statistical algorithms to give early warning of emerging viral or infectious diseases. These algorithms, entitled "C1," C2," and "C3" (www.cdc.gov/biosense), are implemented by the innovative syndromic surveillance systems. The C1 and C2 procedures use a moving sample average and sample standard deviation to standardize each reported case. The C1 uses the 7 days prior to the current reported case to calculate the sample average and the sample standard deviation (for more detail see www.bt.cdc.gov/surveillance/ears). The C3 compares the number of reported events in a period of 4 weeks with the average of these events from the previous 5 years. C2 and C3 are being proposed as useful temporal detection methods. U.S. CDC developed the Early Aberration Reporting System (EARS) as a free tool designed to facilitate the analysis of public health surveillance data using quality control charts. CDC suggests four potential uses of this tool: (i) temporary ("drop on") enhanced surveillance (e.g., G8 Summit, Presidential Inauguration, Olympic Games, Super-Bowls), (ii) routine surveillance for naturally occurring illness/diseases and bioterrorism events easily characterized by patient-reported chief complaints and physician diagnosis (e.g., influenza-like illness), (iii) post-disaster surveillance data analysis, and (iv) early detection of outbreaks identified by using nontraditional public health sources (e.g., school absenteeism rates, over-the-counter medication sales, 911 calls data, ambulance runs data).

In addition to providing early warning signals about "aberrations" in the frequency distributions of any health event of public health importance, surveillance is key for recognizing newly emerging conditions, such as Ebola, SARS, or influenza A (H1N1).

12.13.1 Use of Geographic Information Systems (GIS) for mapping and geo-referencing public health events and risks of national and global importance

A GIS is defined as "an integrated collection of computer software and data used to view and manage information connected with specific locations, analyze spatial relationships, and model spatial processes. GIS technology integrates common database operations, such as query and statistical analysis, with the unique visualization and geographic analysis benefits offered by maps" (ESRI, 2011, p.6).

Important consideration has been placed on recognizing unusual geospatial viral and infectious disease clusters (hot spots) at the local or national level (Castillo-Salgado, 2011).

A very dynamic new field in global health surveillance is represented by the incorporation of GIS (Castillo-Salgado, 2009). GIS provides new advanced analytical and technological tools for linking surveillance databases with spatial and map information. Environmental Systems Research Institute (ESRI), a recognized private company, has been developing advanced GIS systems and tools that are used in many of the current surveillance systems using real-time data over the world.

Another very active network of Internet-based surveillance was initiated by the International Society of Infectious Diseases's Program for Monitoring Emerging Diseases. ProMED-mail at present is considered to be one of the largest publicly available Internet-based reporting networks in the world (Madoff, 2004; Madoff, and Woodall, 2005).

Access to the enabling GIS technology has facilitated the deployment of informatics protocols aimed at automated classification and visualization of Internet media reports on disease outbreaks. HealthMap is a new global disease alert map website launched in September 2006 by the HealthMap Organization (HealthMap, 2009) with the sponsorship of Google Earth. Currently, this open-source platform is available in multiple languages, and it integrates outbreak data from news sources (such as Google News), curated personal accounts (such as ProMED), and validated official alerts (such as WHO). Through an automated text processing system, the data is aggregated by disease and displayed as thematic maps by location for user-friendly access to the original alert. HealthMap site indicates that it provides real-time information on EIDs that may have particular interest for public health officials and international travelers by integrating and filtering news from over 20 000 sources every hour.

In many reviews, Google has been recognized as a major driving force for providing technological advances in global health mapping accessible to Internet users all over the world. HealthMap as described in a previous section is a major open-source application for monitoring real-time global health events.

12.14 USING PROXY AND COMPILED WEB-BASED INFORMATION FROM DIFFERENT SOURCES

New terms linking surveillance and public health informatics are being proposed to analyze patterns of search and communication of surveillance information in the Internet. Eysenbach (2009) and Ginsberg et al. (2009) have described the fields of "*Infodemiology* and *Infoveillance*" as methods to be used for the analysis of queries from the Internet search engines to predict disease outbreaks. In addition, these informatics methods are proposed to be used to analyze search behaviors of people and navigation patterns in the Internet for health-related information, including how people communicate and share health and surveillance information.

Also, surveillance efforts have been reengineered to allow a systematic collection of data and information from different sources, including from other sectors not generated for health purposes.

As presented before, new web-based algorithms and systems are part of the uses of the U.S. CDC's EARS in the early detection of outbreaks identified by using nontraditional public health sources such as school absenteeism rates, over-the-counter medication sales, 911 calls data, and ambulance runs data.

Most health departments collect and analyze information from routine health information systems but also are being supplemented by pharmaceutics sales, web searches, press reports, and other health professional and consumer information inputs.

The WHO GOARN receives its information from different sources, including official and unofficial websites. The flow of outbreak information across national, regional, and subnational borders has improved the timely detection and the recognition of the geographical extent of potential outbreaks.

Using Google Search queries is being explored as a surveillance mechanism to identify trends of high proportion of health events or behaviors consistent with potential outbreaks. Search query surveillance has been mentioned as a valuable real-time, free, and public strategy that health institutions may adopt for complementing public health surveillance.

Electronic information systems reporting data with potential links with health risks or events of public health importance have been explored as novel strategies for surveillance. Reported daily absenteeism and data on visits to the school nurse for influenza-like illness has been reported as a reliable input for public health surveillance (Crawford et al., 2011; Paterson, 2011).

A recent online effort was developed with the goals of sharing and disseminating information and communications useful for flu surveillance in the United States using data provided by volunteer individuals. "Flu Near You" available at https://flunearyou.org/ is an initiative launched in November 2011 by HealthMap and the American Public Health Association to assist in tracking the spread of flu on a national level in the United States. Flu Near You is a web-based tool for people willing to report their flu symptoms weekly. This information is used in preparing real-time maps of influenza at local and national levels in the United States. Flu Near You is currently only available in the United States.

12.15 INCORPORATION OF PUBLIC–PRIVATE PARTNERSHIPS IN SURVEILLANCE ACTIVITIES

Networks of viral surveillance usually are developed after extensive consultations of public health and epidemiological government agencies, of experts in academic and research settings, and more frequently with more private health organizations. There is a call to encourage the growth of a public–private health technology infrastructure, including the development of surveillance systems.

Several of the networks for flu surveillance are developed by the private sector, and they may include information systems collecting viral infectious disease data using rapid flu testing from clinical sites, emergency departments, hospitals, private practices, laboratories, and health departments in a systematic manner and posted regularly on their websites. An example of these surveillance networks developed by the private sector is the National Flu Surveillance Network (NFSN) (http://www.fluwatch.com). Nevertheless, there is still a debate about the role of the private sector in surveillance activities (Hudson, 2001; Uyeki et al., 2002). However, all early warning systems generated by public–private partnerships are of critical importance in the recognition of potential outbreaks and pandemics. Another very useful and highly used surveillance information network is *Google Flu Trends* available at http://www.google.org/flutrends/.

12.16 USE OF VOLUNTEER SENTINEL PHYSICIANS

In addition to the reports generated from clinical and rapid diagnostic tools, health departments, U.S. CDC, Canadian CDC, European CDC, WHO, and several countries have developed surveillance networks for tracking types and strains of circulating influenza viruses using volunteer sentinel physicians. U.S. CDC in coordination with state and local health departments established both virus and disease surveillance for influenza in the United States using volunteer sentinel physicians for incidence information in 47 states and 122 participating cities for mortality data. Volunteer sentinel physicians report the percentage of their patient visit for influenza-like disease from October to May (CDC, 2006).

12.17 IMPROVING GUIDELINES AND PROTOCOLS FOR VIRAL SURVEILLANCE

It is recognized that all countries need to improve their disease surveillance systems to provide early detection of potential outbreaks and to establish guidelines and protocols to better respond to public health events of potential global importance. Also, they need to incorporate better monitoring mechanisms to enhance the safety and security of populations against terrorism and bioterrorism.

Conventional reporting in disease and viral surveillance uses a linear process initiated by a sick person contacting and being examined by his/her medical doctor followed by laboratory exams. If the results of laboratory exams are suggestive of a "reportable" condition or recognized as being unusual in trends or numbers, the doctor or laboratory notifies the local health authorities. The local health authorities inform regional and national health authorities of the surveillance findings, and these authorities notify the WHO and other international agencies particularly if the findings are of global concern. In general, this conventional reporting process takes a long time and frequently is subject to political clearances affecting the timely alerting and the effective initiation of public health interventions. A more effective process has been established with surveillance systems incorporating different reporting mechanisms and sources using list servers, websites, and Internet networks.

Validation protocols and quality control measures are needed for the recognition of epidemiological and health data and laboratory confirmation. These protocols are relevant for all sources of information: the traditional reporting systems, the syndromic surveillance, and the unofficial websites and other sources.

However, this important area of global surveillance is still developing. In many countries, there are still no clearly established guidelines for how they should conduct the different types of surveillance, including real-time surveillance, and which types of emerging disease syndromes should be confirmed and reported. Another important recognized challenge is the area of global surveillance enforcement. WHO as an intergovernmental agency has limited mechanisms of enforcement. However, the new IHR-2005 has incorporated recommendations to maximize global collaboration in times of severe outbreaks and pandemics such as the pandemic spread of H1N1 influenza.

WHO, with the assistance of International Collaborative Centers, and recognized academic and scientific leaders prepared specific guidelines for the surveillance of human infection with influenza A (H1N1) virus (WHO, 2009a). The use of networks and the Internet has facilitated in a few hours the rapid dissemination of these guidelines around the globe. This type of protocol includes information about the main objectives of the surveillance for this specific influenza virus including case definitions for confirmed and probable cases. Guidelines for the types of laboratory tests and reporting mechanisms are also now included for the designated national focal points of the IHR-2005. Countries without laboratory capacity for case confirmation are urged to contact WHO for coordinating the access to laboratories with this confirmation capacity.

Several guidelines have been proposed for the evaluation of surveillance systems (CDC, 2001; German et al., 2001; Romaguera et al., 2000; WHO, 1997). However, these evaluation guidelines were prepared before the implementation of the IHR-2005, and they need to incorporate the new information and reporting mechanisms and types of surveillance.

12.18 INCORPORATING HEALTH SITUATION ROOMS OR STRATEGIC COMMAND CENTERS FOR MONITORING, ANALYSIS, AND RESPONSE IN SURVEILLANCE EFFORTS

Health Situation Rooms are recognized by several ministries of health and WHO as their main strategic hubs for monitoring and surveillance activities and for coordination of their operations related to disasters and potential bioterrorist events. They have been extensively used in the epidemics of SARS, in the influenza pandemic of A (H1N1), and in the current monitoring of most relevant viral infections.

Several countries and international organizations have commissioned the creation of Health Situation Rooms, also known as Strategic Command Centers, as centralized coordination hubs for national and global monitoring and the assessment of epidemic and health emergencies. Many countries have been using these centers as the epidemiological intelligence units for coordination of the health sector with other public and private agencies required to integrate national responses to outbreaks, epidemics, and disasters. Leading examples of these centers are in United States at the DHHS in Washington (DHHS, 2006) and CDC in Atlanta (CDC_EOC, 2006) and in WHO in Geneva (WHO-SHOC, 2008a); at the Public Health Agency (PHA, 2009) in Ottawa, Canada; and at the Ministry of Health Brazil (CIEVS, 2009). Currently, a large number of ministries of health and local health departments around the world have developed these important coordinating hubs for surveillance, alert, and response operations.

12.19 CHALLENGES OF VIRAL AND PUBLIC HEALTH SURVEILLANCE

As presented by Castillo-Salgado (2010, p. 13), there is agreement among the different reviewed professional assessments that key constraints and challenges for global public health surveillance including viral surveillance are:

1. The development of core capacities for new viral and infectious disease surveillance and response systems for developing countries is affected by the lack or shortages of resources, limited trained national staff and officials, and weak networks of laboratories.
2. Many countries have multiple independent surveillance and health information systems with limited coordination and no interoperability.
3. Laboratory facilities in many developing countries are not familiar with quality assurance and control principles and regulations, and a large percentage of their equipment is obsolete or not functioning.
4. Changes in antibiotic-resistance patterns, particularly from hospital care, have been recognized as a major current use of infectious disease surveillance and nosocomial infection surveillance.
5. The global disease and viral monitoring through automated classification and visualization of events using electronic means is a limited option in many countries where the technological divide is extreme. Large numbers of countries or areas in the interior of the countries have no access to the Internet or to basic computerized systems.
6. Local health facilities in a large number of countries have limited operating telecommunications and transportation capabilities available.
7. Traditionally official surveillance systems are operated by staff not linked to the response teams, and the information collected is outdated and fragmented.
8. Many countries with severe human rights protection problems have difficulty maintaining the principles of fairness, objectivity, and transparency.
9. Compliance with global health regulations will require constant economic and technical cooperation with poorer countries.

The unique situation of viral and communicable disease surveillance requires closer collaboration between developed and developing countri es and the support of WHO and other international agencies such World Bank, UNICEF, FAO, and the private philanthropic organizations.

However, because of the comprehensive response needed to address disease and surveillance in society, risk factor surveillance and social determinants of health monitoring are emerging as a new orientation of surveillance in which not only the viral and health conditions are under surveillance but fundamentally the main risks and determinants of those infectious disease agents and conditions. As part of the recommendations from the WHO's Global Commission of Social Determinants of Health (WHO, 2008b), surveillance could play an important role in recognizing the enormous equity health gaps and disparities among social, racial, ethnic, and gender groups. Recognizing the health equity gaps will permit proposing more comprehensive societal responses linking the health sector with all other sectors, for example, housing, work and school, food, and transportation. Also, the civil society is of particular importance for the success of public health responses to the health threats identified by the viral and infectious disease surveillance.

REFERENCES

Buehler JW, Berkelman RL, Hartley DM, *et al.* Syndromic surveillance and bioterrorism-related epidemics. *Emerg Infect Dis.* 2003; **9**(10): 1197–1204. Available from: http://www.cdc.gov/ncidod/EID/vol9no10/03–0231.htm (accessed May 10, 2012).

Castillo-Salgado C. World Health Organization Forum Supports GIS. For Public Health in the Americas. HealthyGIS. ESRI, Spring 2009, p.2. Available from: http://www.esri.com/library/newsletters/healthygis/healthygis-spring2009.pdf (accessed May 10, 2012).

Castillo-Salgado C. Trends and directions of global public health surveillance. *Epidemiol Rev.* 2010; **32**(1): 93–109.

Castillo-Salgado, C. Urban health challenges in megacities: the case of Mexico City. In: Khan O, Pappas G (eds). Megacities and Global Health. Washington DC: American Public Health Association Press; 2011.

Centers for Disease Control (CDC). Updated guidelines for evaluating public health surveillance systems. *MMWR.* 2001; **50**(RR-13): 1–35.

CDC. U.S. Influenza Sentinel Provider Surveillance Network. Centers for Disease Control. 2006. Available from: http://www.doh.state.fl.us/disease_ctrl/epi/htopics/flu/FSPISN/RecruitmentCDCsys2006.pdf (accessed May 10, 2012).

CDC. Recommendations for the identification of chronic hepatitis C virus infection among persons born during 1945–1965. *MMWR Recomm Rep.* 2012a August; **61**(RR-4): 1–32.

CDC. Case Definitions: Nationally Notifiable Conditions Infectious and Non-Infectious Case. Atlanta, GA: Centers for Disease Control and Prevention; 2012b.

Choffnes ER. Improving infectious disease surveillance. In: Bulletin of the Atomic Scientists. Chicago, IL: University of Chicago; 2008.

Council of State and Territorial Epidemiologists. CDC/CSTE Development of Applied Epidemiology Competencies; 2010. Available from: www.cste.org/competencies.asp (accessed May 10, 2012).

Crawford GB, McKelvey S, Crooks J, *et al.* Influenza and school-based influenza-like illness surveillance: a pilot initiative in Maryland. *Public Health Rep.* 2011 July–August; **126**(4): 591–596.

ESRI. Geographic Information Systems and Pandemic Influenza Planning and Response. An ESRI White Paper. Redlands, CA: ESRI. February 2011, p.4.

Emergency Preparedness and Response. The Emergency Operating Center (EOC). Atlanta, GA: Centers for Disease Control and Prevention; 2006. Available from: http://www.bt.cdc.gov/cotper/eoc/ (accessed May 10, 2012).

European Centre for Disease Control (ECDC). Core Competencies for Public Health Epidemiologists Working in the Area of Communicable Disease Surveillance and Response, in the European Union. Stockholm: ECDC; January 2008. Available from: http://www.ecdc.europa.eu/en/Pages/home.aspx (accessed May 10, 2012).

Eysenbach G. Infodemiology and infoveillance: framework for an emerging set of public health informatics methods to analyze search, communication and publication behavior on the Internet. *J Med Int Res*. 2009; **11**(1): e11.

Fricker RD Jr, Hegler BL, Dunfee DA. Comparing syndromic surveillance detection methods: EARS' versus a CUSUM-based methodology. *Stat Med*. 2008 July; **27**(17): 3407–3429.

German RR, Lee LM, Horan JM, *et al.* Updated guidelines for evaluating public health surveillance systems: recommendations from the Guidelines Working Group. *MMWR Recomm Rep.* 2001; **50**(RR-13):1–35.

Ginsberg J, Mohebbi MH, Patel RS, *et al.* Detecting influenza epidemics using search engine query data. *Nature* 2009; **457**(7232): 1012–1014.

Heymann DL (ed.). Control of Communicable Diseases Manual, 19th Edition. Washington, DC: American Public Health Association Press; 2008.

Hitchcock P, Chamberlain A, Van Wagoner M, *et al.* Challenges to global surveillance and response to infectious disease outbreaks of international importance. *Biosecur Bioterror*. 2007; **5**(3): 206–227.

Houghton M. The long and winding road leading to the identification of hepatitis C virus. *J Hepatol.* 2009; **51**(5): 939–948.

Hudson RJ. Disease surveillance versus viral surveillance. *Clin Infect Dis.* 2001; **33**:265–266.

Hutwagner L, Browne T, Seeman GM, *et al.* Comparing aberration detection methods with simulated data. *Emerg Infect Dis*. 2005 February; **11**(2): 314–316.

Jones KE, Patel NG, Levy MA, *et al.* Global trends in emerging infectious diseases. *Nature* 2008 February; **451**: 990–993.

Last JM (ed.) A Dictionary of Epidemiology, 3rd Edition. New York: International Epidemiological Association; 1995, p. 163 (this definition was taken from: Eylensboch WJ, Noah ND (eds.) Surveillance in Health and Disease. Oxford: Oxford University Press; 1988).

Lombardo J, Burkom H, Elbert E, *et al.* A systems overview of the electronic surveillance system for the early notification of community-based epidemics (ESSENCE II). *J Urban Health* 2003 June; **80**(2 Suppl. 1): i32–i42.

Madoff LC. ProMED-mail: an early warning system for emerging diseases. *Clin Infect Dis.* 2004; **39**: 227–232.

Madoff LC, Woodall JP. The Internet and the global monitoring of emerging diseases: lessons learned from the first 10 years of ProMED-mail. *Arch Med Res*. 2005; **36**(6): 724–730.

McNutt L, Furner SE, Mose M, *et al.* Applied epidemiology competencies for governmental public health agencies: mapping content curriculum and the development of new curriculum. *Public Health Rep*. 2008; **123**(1): 13–18.

Ministerio de Salud. Brasil. Centro de Informações Estratégicas em Vigilância em Saúde – CIEVS (In Portuguese). Brasilia, Brazil: Ministry of Health; 2009. Available from: http://portal.saude.gov.br/portal/saude/Gestor/visualizar_texto.cfm?idtxt=22233 (accessed May 10, 2012).

Nelson JC, Essien JDK, Loudermilk R, *et al.* The Public Health Competency Handbook: Optimizing Individual & Organization Performance for Public's Health. Atlanta, GA: Center for Public Health Practice of the Rollins School of Public Health; 2002.

Paterson B, Caddis R, Durrheim D. Use of workplace absenteeism surveillance data for outbreak detection. *Emerg Infect Dis*. [serial on the Internet]. 2011 Oct. Avaialble from: http://dx.doi.org/10.3201/eid1710.110202HYPERLINK (accessed June 20, 2013).

Porta M (ed.) A Dictionary of Epidemiology, 5th Edition. International Epidemiological Association. New York: Oxford University Press; 2008, p. 239.

Public Health Agency of Canada. Global Public Health Intelligence Network (GPHIN). November 2004. Available from: http://www.phac-aspc.gc.ca/media/nr-rp/2004/2004_gphin-rmispbk-eng.php (accessed May 10, 2012).

Public Health Agency of Canada. Centre for Emergency and Response. Ottawa, Canada: Public Health Agency of Canada; 2009. Available from: http://www.phac-aspc.gc.ca/cepr-cmiu/index-eng.php (accessed May 10, 2012).

Public Health Canada: FluWatch. About FluWatch; 2012. Available from: http://www.phac-aspc.gc.ca/fluwatch/index-eng.php (accessed May 10, 2012).

Public Health Foundation: Council on Linkages Between Academia and Public Health Practice. Crosswalk of the Core Competencies for Public Health Professionals and the Essential Public Health Services; 2010. Available from: http://www.phf.org/resourcestools/Pages/Core_Public_Health_Competencies.aspx (accessed May 10, 2012).

Reed C, Angulo FJ, Swerdlow DL, *et al.* Estimates of the prevalence of pandemic (H1N1) 2009, United States, April–July 2009. *Emerg Infect Dis.* [serial on the Internet]. 2009 Dec. Available from: http://wwwnc.cdc.gov/eid/article/15/12/09-1413.htm (accessed May 10, 2012).

Romaguera RA, German RR, Klaucke DN. Evaluating public health surveillance. In: Teutsch SM, Churchill RE (eds.) Principles and Practice of Public Health Surveillance, 2nd Edition. New York: Oxford University Press; 2000.

The Joint Task Group on Public Health Human Resources (2005). Building the Public Health Workforce for the 21st Century. A Pan-Canadian Framework for Public Health Human Resources Planning. Appendix 1. Draft Core Public Health Competencies, p.24. Available from: http://www.phac-aspc.gc.ca/php-psp/pdf/building_the_public_health_workforce_fo_%20the-21stc_e.pdf (accessed May 10, 2012).

US DHHS Secretary's Command Center Fact Sheet. Washington, DC: Department of Health and Human Services; 2006. Available from: http://www.hhs.gov/news/factsheet/commandcenter.html (accessed May 10, 2012).

Uyeki TM, Fukuda K, Cox NJ. Influenza surveillance with rapid diagnostic tests. *Clin Infect Dis.* 2002; **34**(10): 142.

Wilson K, Brownstein JS. Early detection of disease outbreaks using the Internet. *CMAJ.* 2009; **180**(8): 829–831.

World Health Organization (WHO). Protocol for the Evaluation of Epidemiological Surveillance Systems. Geneva, Switzerland: World Health Organization; 1997; WHO/EMC/DIS/97.2.

WHO. International Health Regulations (IHR-2005). Geneva, Switzerland: World Health Organization; 2005. Available from: http://www.who.int/ihr/en/ (accessed May 12, 2012).

WHO. SHOC. Centre of Strategic Health Operation. Geneva, Switzerland: World Health Organization; 2008a. Available from: http://www.who.int/csr/alertresponse/shoc/en/index.html (accessed May 10, 2012).

WHO. Closing the gap in a generation: health equity through action on the social determinants of health. Commission on Social Determinants of Health – Final Report. Geneva, Switzerland: World Health Organization; 2008b. Available from: http://www.who.int/social_determinants/thecommission/finalreport/en/index.html (accessed May 10, 2012).

WHO. International Health Regulations (IHR-2005), 2nd Edition. Geneva, Switzerland: World Health Organization; 2008c. Available from: http://www.who.int/csr/ihr/IHRWHA58_3-en.pdf (accessed on May 22, 2012).

WHO. Interim WHO Guidance for the Surveillance of Human Infection with Swine Influenza A(H1N1) Virus. Geneva, Switzerland: World Health Organization; 27 April, 2009a. Available from: http://www.who.int/csr/disease/swineflu/WHO_case_definitions.pdf.

WHO. Weekly Epidemiological Record on Pandemic H1N1; 2009b. Available from: http://www.who.int/csr/disease/swineflu/wer/en/index.html (accessed May 10, 2012).

WHO. Pandemic Influenza Preparedness Framework for the Sharing of Influenza Viruses and Access to Vaccines and Other Benefits. Geneva, Switzerland: World Health Organization; 2011. Available from: http://www.who.int/influenza/resources/pip_framework/en/index.html (accessed May 10, 2012).

WHO. GISRS (2012) H5N1 Genetic Changes Inventory Influenza Knowledge Base for Surveillance and Preparedness, June 29, 2012, p.1. Available from: http://www.who.int/influenza/gisrs_laboratory/h5n1_genetic_changes_inventory/en/ (accessed May 10, 2012).

13

PREDICTIVE MODELING OF EMERGING INFECTIONS

Anna L. Buczak, Steven M. Babin, Brian H. Feighner, Phillip T. Koshute, and Sheri H. Lewis

Johns Hopkins University Applied Physics Laboratory, Laurel, MD, USA

TABLE OF CONTENTS

13.1 INTRODUCTION

Outbreaks of infectious disease, whether emerging or reemerging, result from a complicated interplay of the so-called epidemiological triad of host, pathogen, and environment (Wallace et al., 1998). This relationship is not a new concept, dating at least from the Aphorisms of Hippocrates in the fifth century BCE.:

Viral Infections and Global Change, First Edition. Edited by Sunit K. Singh.

> *"Whoever would study medicine aright must learn of the following subjects. First he must consider the effect of each season of the year and the differences between them. Secondly he must study the warm and the cold winds, both those that are common to every country and those peculiar to a particular locality. Lastly, the effect of water on the health must not be forgotten." (Lloyd, 1978)*

Traditional public health practice has relied on public health surveillance of disease in an effort to detect an outbreak early enough to mitigate its effects (Teutsch and Churchill, 1994; Wallace et al., 1998). For many emerging diseases, the ability to mitigate their effects is directly related to the rapidity of disease detection. The logical extension of this tenet is that mitigation might be maximized if the outbreak is accurately predicted before disease occurs.

In the past century several events have significantly increased interest in the prediction of emerging or reemerging diseases. First, many of the important infectious diseases facing mankind have been attributed to perturbations in the human–animal–environment interface (Dobson and Carper, 1996; Wolfe et al., 2007). Secondly, the exploding discipline of information technology now allows for the collection, ingestion, and analysis of unfathomable amounts of data. A direct result of this automated analysis of data of health importance is the vast expansion and improvement in computer modeling of epidemic disease (Grassly and Fraser, 2008). These two events, combined with advances in computational methods and the understanding of the natural history of diseases, have piqued the interest of scientists, health officers, and government officials in "early warning systems" for emerging or reemerging diseases (e.g., Committee on Climate, Ecosystems, Infectious Diseases, and Human Health, Board on Atmospheric Sciences and Climate, National Research Council. 2001; Wolfe et al., 2007). Coincident with advances in the understanding and modeling of infectious diseases, an important rearticulation of the epidemiological triad has emerged in the concept of "One Health." "One Health," originally termed "One Medicine" to describe the crucial interrelationship between human and animal health, has now expanded to describe the important interdependence between the ecologies of humans, animals, plants, and microbes (Atlas et al., 2010). Infectious disease modeling has a central, and growing, importance in the field of infectious disease epidemiology. Models are in use for a myriad of infectious diseases, ranging from influenza to foot-and-mouth disease, as well as for deliberate outbreaks intended as weapons of mass destruction (Grassly and Fraser, 2008).

13.2 TYPES OF MODELS

Modeling of infectious diseases can be done for at least two purposes: (i) to mathematically model the progress of an infectious disease or (ii) to predict when and where an outbreak of a given infectious disease is likely to occur.

For the first application, a popular type of model for simulating the progress of an infectious disease is the "agent-based" model. Agent-based models are a class of computational models that simulate the actions and interactions of individuals or groups in order to assess their effects on the system as a whole. Agent-based disease models combine elements of game theory, computational sociology, and biology to forecast the spread of disease by simulating the epidemiological triad of "host–pathogen–environment" (Epstein, 1999, 2009). Agent-based models were used extensively in the preparation and response to the 2009 H1N1 influenza pandemic (Committee on Modeling Community Containment for Pandemic Influenza, 2006). Classic epidemiological modeling of disease began in the

1920s with simple agent-based models (Epstein, 2009). These early models assumed all individuals and their behaviors to be identical, with perfect mixing of the population at risk. While useful, these models fail to represent the effect of individual behavior upon the spread of disease. Presently, agent-based models are the most popular for modeling the progress of disease, and they are mainly used to discover the likely outcomes of an epidemic or the effects of public health interventions on a given population. Agent-based models for a given disease in a given geographical region need to closely model the population of that region, and the dynamics of how a given disease is spread can make them very labor intensive when adapting them to a new geographical region.

For the second application, predictive models for an emerging disease attempt to predict when and where an outbreak of a given disease will occur. These are the models on which we will concentrate in this chapter. Predicting disease outbreaks is a very difficult task, and truly predictive models for emerging infectious diseases are still in their infancy.

13.3 REMOTE SENSING AND ITS USE IN DISEASE OUTBREAK PREDICTION

Detecting and predicting outbreaks of infectious disease have traditionally been performed using locally collected health data. In the case of dengue, for example, Focks et al. (1995) used the results of laboratory and field studies to develop transmission potential simulation models. Ambient temperature near the ground was used in their models because of its effects on biological parameters such as the extrinsic incubation period of the mosquito. Such field studies are extremely valuable but can be time-consuming, expensive, and technically difficult in remote regions of the world.

Recognition of the potential value of spaceborne platforms for studying environment-related disease risks dates back to the Apollo program when it was found that statistical multispectral pattern recognition techniques could distinguish different types of soils and crops (Anuta and Macdonald, 1971). Kanemasu et al. (1974) used data from the Earth Resources Technology Satellite (ERTS-1) to estimate crop yield and disease severity of wheat fields. Howard et al. (1979) described using satellite remote sensing to monitor agricultural disasters including insect infestations and disease. Linthicum et al. (1991) described how satellite data could be used to locate mosquito vector-breeding habitats. Linthicum also mentioned how satellite data could be used as a relatively inexpensive method of collecting long-term rainfall data that could be temporally correlated with changes in vector activity and thereby infer the risk for disease outbreaks. Linthicum et al. concluded that satellite remote sensing data would eventually be used for real-time prediction of disease outbreaks. The recognition that disease-related environmental conditions could be monitored by satellite indicated that remote sensing could be applied to vector-borne diseases in which the disease vector population varied with detectable environmental conditions.

Colwell (1996) emphasized how emerging infectious diseases are typically not new but are existing pathogens for which changes in environmental conditions have provided new opportunities to infect new host populations and accelerate the spread of disease. She described how pathogenic *Vibrio* spp. are adapted to certain environmental conditions in brackish water but also can survive by attachment to certain types of plankton. When environmental conditions are conducive to an increased plankton biomass, then there tends to be an increase in *Vibrio cholerae* in the environment. Colwell concluded that satellite remote sensing could be useful in predicting cholera outbreaks in coastal regions. Lobitz et al. (2000) described correlations of cholera outbreaks with coastal sea surface temperature

(SST) and sea surface height (related to human–plankton contact). They also mentioned the possibility of using satellite remote sensing of ocean color (i.e., sea surface chlorophyll) as an indicator of plankton biomass and a corresponding increase in the plankton reservoir of *V. cholerae*.

Goetz et al. (2000) reviewed advances in satellite remote sensing, especially in obtaining continuous fields of air temperature, land temperature, atmospheric water vapor, and soil moisture. They discussed application of multi-temporal maps of these data to the distribution and abundance of some common disease vectors and how this information could be applied to epidemiology. Curran et al. (2000) presented a framework for using remote sensing data in assessing risk of disease outbreaks: (i) remotely sensed data provides both temporal and spatial information on land cover and thus vector habitat, (ii) changes in this vector habitat are related to spatial and temporal changes in vector density, (iii) these remotely sensed data can thereby be used to estimate the spatiotemporal distribution of the vector-borne disease, and (iv) knowledge of this spatiotemporal relationship between the remote sensing data and vector-borne disease distribution may lead to an estimate of risk of disease outbreak now and in the future.

Kalluri et al. (2007) reviewed how advances in remote sensing techniques have shown promising results in assessing the risks for arthropod vector-borne infectious disease at different spatial scales. They discussed remote sensing uses for determining mosquito habitat related to land cover, leaf area index, soil moisture, and standing water. They mentioned the use of vegetation and temperature data for tick vectors. Black flies, which can transmit onchocerciasis, tend to live along forest fringes, so that remotely sensed land cover and forest cover data are useful.

While earlier disease prediction models (e.g., Focks et al., 1995) used ground-based temperature measurements, more recent ones often use satellite remote sensing as it has become a reliable way to estimate land surface and SST. Land surface temperature, after correction for atmospheric effects, can be derived from measurements of surface radiance at different wavelengths. For example, the Advanced Very High Resolution Radiometer (AVHRR) instrument uses infrared wavelengths to measure surface radiances in cloud-free conditions (Sobrino et al., 1991). The AVHRR instrument is present on a number of polar-orbiting satellites providing global coverage and a ground resolution as high as about 1 km (NOAA National Geophysical Data Center, 2012). When clouds are present, microwave radiometers measure microwave emissivity data, which are used to derive surface temperatures (Catherinot et al., 2011). The surface resolution of these radiometers varies but is typically around 10–25 km (Holmes et al., 2009).

Many disease vectors are typically adapted to a particular temperature range. Tun-Lin et al. (2000) studied development times, survival rates, and adult size of *Aedes aegypti* mosquitoes (vectors of dengue) at different temperatures. The elapsed time from first egg hatching to adult was shorter with increasing temperature. Focks et al. (1995) found that higher temperatures were associated with not only shorter gonotrophic cycles but also increases in infective bites. Because the mosquito vector requires water for completion of its life cycle, investigators have examined the use of rainfall data for disease outbreak prediction (Focks et al., 1995). The joint U.S.–Japan Tropical Rainfall Measuring Mission (TRMM) satellite data have been used to derive rainfall measurements (Kummerow et al., 1998) in remote and resource-limited regions, and these measurements have been used for outbreak predictions for diseases (Soebiyanto et al., 2010). The TRMM instruments include a precipitation radar, a microwave imager, visible and infrared scanners, a cloud and earth radiant energy sensor, and a lightning imaging sensor. Data from these sensors are used to derive a

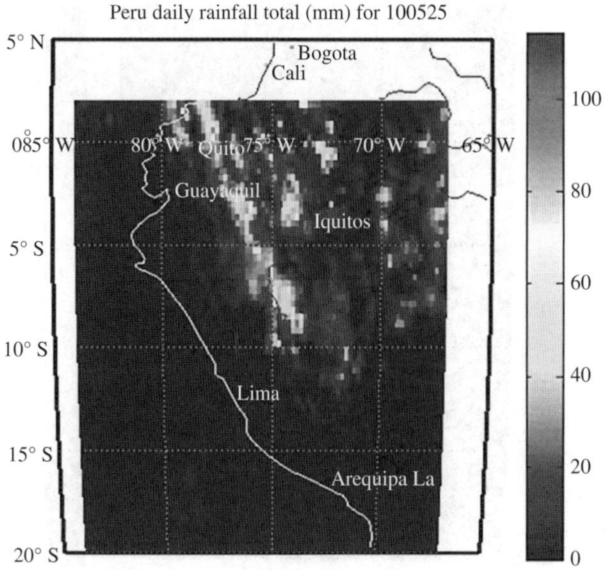

Figure 13.1. An example of the TRMM-derived rainfall for a single day in Peru. (Units are shown in mm). For color detail, please see color plate section.

variety of rainfall products such as hourly rainfall rates that averaged over 3-h intervals and may be obtained with 0.25 degree resolution. The TRMM rainfall data have been validated by ground-based observations at Kwajalein, Texas, Florida, Australia, Israel, Brazil, Guam, Taiwan, and Thailand. An example of TRMM rainfall data is shown in Figure 13.1.

Satellite measurements of leaf area indices have been used in studies of vector-borne disease. These indices are used to assess green leaf biomass, photosynthetic activity, and the effects of seasonal rainfall, which are then related to vector habitat characteristics and disease outbreaks (Anyamba et al., 2002). Photosynthesis utilizes solar radiation at particular wavelengths to produce energy for plant growth. Chlorophyll in green plants strongly absorbs at these wavelengths, but the leaves tend to reflect light at other wavelengths. Also, other types of surface cover such as snow and bare soil reflect at different wavelengths. One of the most commonly used leaf area indices is the Normalized Difference Vegetation Index (NDVI), which is defined as

$$\text{NDVI} = \frac{\rho_{\text{NIR}} - \rho_{\text{red}}}{\rho_{\text{NIR}} + \rho_{\text{red}}}$$

where ρ_{NIR} = surface reflectance for the near-infrared (NIR) wavelength band and ρ_{red} = surface reflectance for the red wavelength band. Each reflectance is a ratio, varying from 0 to 1, of the radiance at a particular wavelength channel received by the satellite and the extraterrestrial solar flux. Therefore, NDVI varies between −1 and +1. An area of dense vegetation has positive values, while clouds and snow have negative values. Oceans, rivers, and lakes have values close to 0, either slightly positive or negative. Bare soil has low positive values. However, soil often gets darker when wet, so NDVI may change due to soil moisture change rather than vegetation change. An example of NDVI data is shown in Figure 13.2.

NDVI is very sensitive to chlorophyll. Causes of lower than normal NDVI include very cold temperatures, clouds, and drought. NDVI works well when vegetation growth is limited by water so that vegetation density is an indicator of drought.

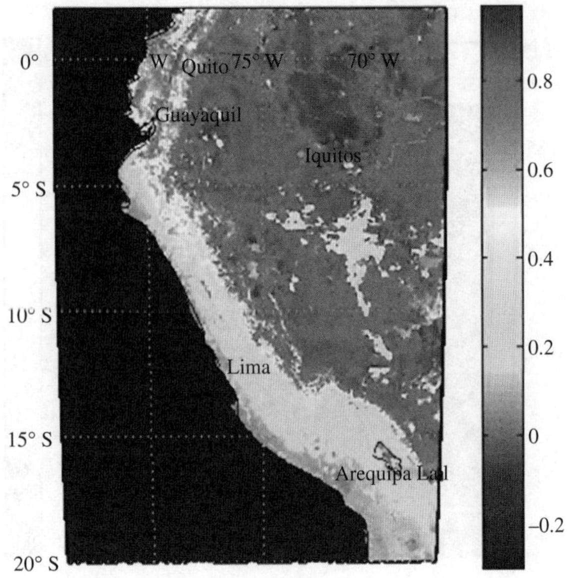

Figure 13.2. Example of NDVI values for a given 16-day interval in Peru. For color detail, please see color plate section.

Another useful vegetation index in the Enhanced Vegetation Index (EVI) defined as

$$\mathrm{EVI} = G \frac{\rho_{\mathrm{NIR}} - \rho_{\mathrm{red}}}{\rho_{\mathrm{NIR}} + C1\rho_{\mathrm{red}} - C2\rho_{\mathrm{blue}} + L}$$

where ρ_{NIR}, ρ_{red}, and ρ_{blue} are the surface reflectances for the NIR wavelength band, the red wavelength band, and the blue wavelength band, respectively. G is a gain factor used to limit the EVI values to range between −1 and +1, L is a canopy background calibration factor that normalizes differential red and NIR extinction through the canopy, and $C1$ and $C2$ are weighting factors for the aerosol resistance. For the Moderate Resolution Imaging Spectroradiometer (MODIS) satellite instrument, $L=1$, $C1=6$, $C2=7.5$, and $G=2.5$. EVI was adopted by NASA as a standard MODIS product in 2000. An example of EVI data is shown in Figure 13.3.

EVI is calculated similarly to NDVI but corrects for some distortions in the reflected light caused by particles in the air and ground cover below the vegetation. EVI does not become saturated as easily as NDVI when viewing rain forests and other areas with high chlorophyll levels. EVI tends to maintain a normal distribution of vegetation index values over high-biomass conditions, while the NDVI may saturate at high values. EVI has been found to reduce background and atmospheric noise. Like NDVI, EVI is also available from satellite sensors such as the AVHRR and MODIS instruments mentioned earlier. While NDVI is closely related to photosynthesis, EVI is closely related to leaf display (Fuller et al., 2009). Both NDVI and EVI are available at a spatial resolution of 0.05 degrees.

SST patterns influence rainfall patterns and are commonly derived from satellite measurements (e.g., AVHRR) at a spatial resolution of about 0.25 degrees. The presence of warmer than normal water, a positive sea surface temperature anomaly (SSTA), off the coast of a region increases the buoyancy of the air above it and enhances convection, leading to more rainfall downwind of this area. A negative SSTA would tend to decrease the potential for rainfall. El Niño refers to an abnormal warming

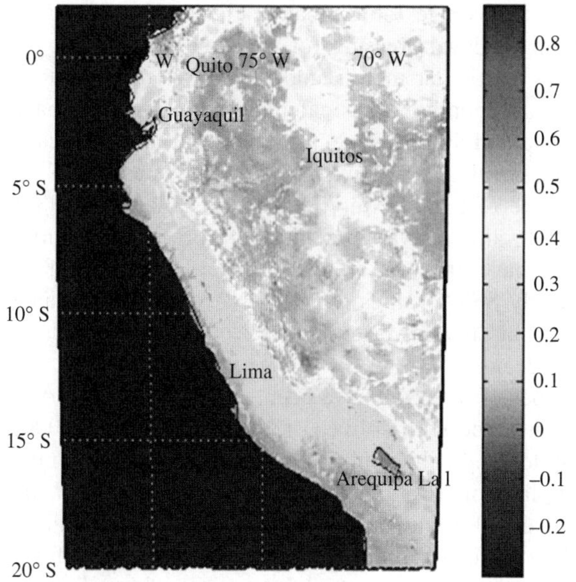

Figure 13.3. Example of EVI values for a given 16-day interval in Peru. For color detail, please see color plate section.

(positive SSTA) of the surface water in the eastern tropical Pacific. La Niña is the opposite (i.e., negative SSTA). El Niño/La Niña (also called the Southern Oscillation) is a climatic phenomenon that affects large regions of the Earth (Rasmussen and Carpenter, 1982). Southern Oscillation refers to the pattern of El Niño being followed eventually by a La Niña, and vice versa. There are also times in between when there is neither an El Niño nor a La Niña (no SST anomaly in the eastern tropical Pacific). The oscillation is not exactly periodic, with intervals as short as 2 years or as long as 7 years. A strong El Niño may last for 5 or more years. Normally, the trade winds blow from east to west along the equator. Southern Oscillation is the reversing monthly mean surface air pressure between the eastern and western tropical Pacific. If the surface pressure is high in the western part, then it tends to be low in the eastern part, and because wind blows from high to low pressure, this weakens the trade winds and allows warmer water to stay near the South American coast (resulting in a positive SSTA). If these pressures are reversed, then the trade winds blowing from east to west are enhanced, pushing water away from the coast and allowing it to be replaced by cooler water from below (upwelling) to fill the void (resulting in a negative SSTA). There is also a variable lag of several weeks between the establishment of an SSTA and the subsequent rainfall effects.

Because the aforementioned climate effects can indicate near-term future rainfall anomalies (increases and decreases), the Southern Oscillation Index (SOI) and SSTA values have been used as indicators of disease outbreaks (Fuller et al., 2009; Hales et al., 1999; Hu et al., 2010; Johansson et al., 2009). SOI is based on the pressure difference between Darwin (Australia) and Tahiti (French Polynesia) and gives an indication of the strength of the El Niño or La Niña. The SOI can be calculated as

$$\mathrm{SOI} = 10\frac{P_{\mathrm{diff}} - P_{\mathrm{diffavg}}}{\mathrm{SD_}P_{\mathrm{diff}}}$$

where P_{diff} is the difference between the average Tahiti mean sea level pressure for the month and the average Darwin mean sea level pressure for the same month, $P_{diffavg}$ is the long-term average of P_{diff} for that particular month, and SD_P_{diff} is the long-term standard deviation of P_{diff} for that same month. SOI is calculated on a monthly basis and quoted as a whole number. A sustained SOI less than a negative 8 usually indicates an El Niño, while a sustained SOI greater than a positive 8 usually indicates a La Niña. Thus, the SOI data provide a measure of the El Niño/Southern Oscillation (ENSO) climate effect. A single monthly SOI value is available and therefore is not location specific. Because the same SOI will have different effects in different regions of the world, local SSTAs are useful. Weekly SSTA values from the NASA Global Change Master Directory website (NASA Global Change Master Directory, 2012) are computed for different regions of the world's oceans.

As mentioned earlier, certain types of plankton have been shown to be a reservoir for *V. cholerae*, the bacteria that cause cholera (Colwell, 1996; Lobitz et al., 2000). Satellite remote sensing of ocean color (e.g., chlorophyll) has been used to determine increased concentrations of phytoplankton (Babin et al., 2004). For example, the MODIS instrument can measure ocean color with a resolution of 4 km. Phytoplankton are a food source for copepods, the type of plankton that serves as a reservoir for *V. cholerae*, and therefore may provide an indirect indicator of changes in the availability of copepods and thus *V. cholerae* (Lobitz et al., 2000). Magny et al. (2008) identified a 1-month lag between satellite-measured coastal ocean chlorophyll anomalies and cholera outbreaks in Kolkata, India.

These satellite-derived products, which offer great insight into local environmental conditions, are available for almost any region of the world. Not surprisingly, a number of epidemiological efforts have incorporated this remote sensing data into the investigation of the occurrence and distribution of diseases. A summary of some earlier studies follows, but note that these studies were primarily correlative, rather than predictive, in nature.

Examples include Bouma and Pascual (2001) who established a correlation between SST in the coastal Bay of Bengal and cholera spring deaths in Bangladesh; Mendelsohn and Dawson (2008) who ascertained a correlation between precipitation and cholera incidence in KwaZulu-Natal, South Africa; and Thomson et al. (2000, 2005) who determined rainfall and SST as drivers for interannual variability of malaria in Botswana. Thomson et al. (2005) also established that changes to population vulnerability to malaria are related to population dynamics, coinfection, and drug and insecticide resistance. Significant relationships between mean maximum daily temperatures (January–October) of the preceding season, total rainfall in the current summer months (November–March), and malaria case totals in KwaZulu-Natal, South Africa, have also been documented (Craig et al., 2004a, b). Wiwanitkit (2006) established a correlation between the rainfall and the prevalence of dengue in Thailand.

While useful, most of the studies mentioned earlier have not progressed much further than finding correlations. A truly useful prediction is one that can be made a priori, that is, all the data used are available before the date for which the final prediction is made. The accuracy of any prediction commonly decreases the further forward in time past the most recently inputted data. Having a prediction of no or increased risk of a disease outbreak at least 2-4 weeks in advance would be useful to the local public health professionals who may then have time to initiate an anticipatory response (e.g., intensive public outreach for draining standing water, proper usage of mosquito netting) that may mitigate the effects of the outbreak should it occur. Such predictive modeling will be described in more detail in Sections 13.4 and 13.5.

13.4 APPROACHES TO MODELING AND THEIR EVALUATION

Many of the so-called predictive techniques presently in use deal solely with modeling the statistical relationships between variables. Those methods include correlation analysis and various types of regression analysis. Correlation analysis often uses the Pearson correlation coefficient (Rodgers and Nicewander, 1988) as a measure of linear relationship between two variables. Values of the Pearson correlation coefficient are between −1 and 1. A correlation coefficient of 1 (−1) indicates that two variables are perfectly related in a positive linear sense (negative linear sense). A correlation coefficient of 0 indicates that there is no linear relationship between the two variables. Examples of correlations established between weather-related variables and outbreaks of disease were described in Section 13.2.

Regression analysis is a statistical tool for the investigation of relationships between variables. The goal of linear regression is to find the line that best predicts the dependent variable from the independent variables. Logistic regression (LR) (Agresti, 2007), employed extensively in medical and social sciences, can be used to predict a dependent binary variable on the basis of independent variables and to determine the percent of variance in the dependent variable explained by the independent ones. It can be also used to rank the relative importance of independent variables and to assess interaction effects. The impact of predictor variables is usually explained in terms of odds ratios.

R-squared (Steel and Torrie, 1960), also called the coefficient of determination, gives some information about the goodness of fit of a model; that is, it provides a measure of how well future outcomes are likely to be predicted by the model. It represents the proportion of variability in a data set that is accounted for by the statistical model. An *R*-squared of 1.0 indicates that the regression line perfectly fits the data. Often *R*-squared is computed only on exactly the same data on which the model was developed. As such it represents the goodness of the fit on the training data. Unfortunately even an *R*-squared of 1 computed on the training data does not guarantee that the prediction of the model on the testing data (i.e., data not used for model development) will have an acceptable accuracy. *R*-squared should be computed separately for the training data and the test data; however, this is rarely done in practice.

An example of the use of nonlinear regression for finding the best model for predicting dengue in Costa Rica was described by Fuller et al. (2009). The predictor variables used are the weekly ENSO SST indices and interpolated vegetation indices (EVI and NDVI). A simple additive model that includes lagged series as independent variables was used in the analysis. Models with different combinations of climate and vegetation index variables were analyzed using nonlinear regression, with *R*-squared as a measure of the goodness of the model. The best fit was from the model that included EVI, NDVI, and all four ENSO SST indices and resulted in an *R*-squared of 0.83, meaning that the model explained 83% of the variances in the weekly dengue fever (DF)/dengue hemorrhagic fever (DHF) cases in Costa Rica from 2003 to 2007. The problem with this model is that it used future data to predict past values of dengue (negative time lag) and therefore does not perform prediction. However, Fuller et al. also developed a predictive model (using past data to predict the future) for which the *R*-squared was 0.64. While the authors suggested that their approach might be used to predict outbreaks up to 40 weeks in advance, it appears that such validation studies were not completed. All the data were used to develop the model, and its performance in terms of *R*-squared was also assessed on all the data—a practice with which we take issue.

An example of the use of a regression model for dengue prediction is the work of Halide and Ridd (2008), who developed a multiple regression model for predicting monthly DHF cases in the city of Makassar, Indonesia. Model inputs included monthly temperature (maximum, minimum, mean), mean rainfall, mean relative humidity, and monthly El Niño 3.4 SSTA index (representing the monthly SSTA for the Pacific Ocean between 5 °N and 5 °S latitude and between 170 ° and 120 °W longitude). The data covered 1998 through 2005. A linear multiple regression model was developed for predicting DHF cases for a lead time between 1 and 12 months. This model was evaluated using a cross validation scheme in which one of the eight years was selected as the data for model testing and the remaining years were used for model training. The prediction was evaluated using R-squared, the root mean square error (RMSE), and the Pierce score (Manzato, 2007). The model performance degraded with increasing lead time, and according to the authors its predictions can be considered useful up to a lead time of 6 months. For these lead times the R-squared varied from 0.68 to 0.23. The model did not seem to do well in predicting high-incidence outbreaks and provided significant false alarms during 1999 and 2000. In contrast to other studies, the authors found no correlation with El Niño or rainfall.

Machine learning techniques, such as artificial neural networks (ANNs) (Bishop, 1995), are good candidates for developing prediction models. ANNs are computational models inspired by biological neural networks. ANNs consist of interconnected artificial neurons, and they process information using a connectionist approach to computation. ANNs are nonlinear statistical data modeling tools that are usually used to model complex relationships between inputs and outputs or to find patterns in data. The oldest and most widely used ANNs are feedforward neural networks with backpropagation learning rule (Rumelhart et al., 1986).

Husin et al. (2008) developed an ANN and a nonlinear regression model to predict dengue outbreaks in Malaysia and compared the results of those two methods. They examined weekly dengue cases in five districts for the years 2004–2005. Predictor variables included weekly dengue case data, location data, and weekly mean rainfall data. The prediction lead time was not specified in the paper. Their conclusion was primarily that the neural network model was better than the nonlinear regression model, based on RMSE of 0.0055 and 5.104 for the neural network model and the nonlinear regression model, respectively.

Another example of the use of neural network for disease prediction is the work of Cetiner et al. (2009) who developed a model using a feedforward neural network to predict the number of dengue cases in Singapore during 2001–2007. The inputs included weekly values of mean temperature, mean relative humidity, total rainfall, and the total number of confirmed dengue cases. The training set consisted of 2 years of data, and the rest of data was used for testing. The paper did not specify the temporal relationship between the input data and the predictions. Therefore, it is unclear if the output was lagged in comparison with the input, that is, if the network performed prediction or merely correlation (no time lag). The only validation was a time series correlation coefficient that was found to be as high as 0.76 for the test data, but it was noted that the model did not perform as well in 2005, with predicted cases being 2.2 times less than actual cases.

Another example of the use of a neural network (coupled with an entropy technique) for DHF prediction is the work of Rachata et al. (2008). The area of study was one province in northern Thailand, and the period of the study was 1999 through 2007. The method used an entropy technique to transform relevant weather data as input into a feedforward neural network. The predictor variables included weekly temperature (maximum, minimum, mean), weekly mean relative humidity, and weekly mean rainfall. The predictions were weekly and data at time T and earlier were used to predict outbreaks at time $T+1$. The network predicts either no outbreak risk or outbreak risk, with those quantities defined as 0–14 cases and greater than 14 cases, respectively. The accuracy achieved using the entropy technique was 85.92% and without it 78.16%.

Another method for developing prediction models is the use of Support Vector Machines (SVMs) (Vapnik, 1998). SVMs are based on statistical learning theory, and they construct a hyperplane in a high-dimensional space, which can be used for classification. A thorough description of an SVM-based method developed by two of the authors (ALB, SHL) for prediction of cholera is included in Section 13.5.2.

Fu et al. (2007) used a genetic algorithm (GA) and an SVM to predict increase or decrease in DF incidence from current week to next week. The data covered 2001–2005. The inputs to the SVM were chosen by a GA from the following possible inputs: daily temperature, relative humidity, cloudiness, rainfall, wind, and weekly DF cases. The daily meteorological variables were transformed into weekly maximum, minimum, and mean data to correspond to the weekly dengue case data. The GA was used to identify the important time lags to use for each variable (time lags under consideration were 0–12 weeks). The authors used each single year as a testing data set and the remaining 4 years as the training data set. The paper demonstrated how climatic factors, most importantly temperature and rainfall, influence trends in dengue incidence. The accuracy obtained was 77–83%.

When developing a disease outbreak prediction method, we believe it is important to follow the steps necessary for rigorous prediction:

1. All predictor variables need to be collected for the previous time period and used for prediction of outbreaks during a later time period. This ensures a realistic prediction: values of all the predictor variables can be obtained prior to performing the prediction.
2. The model must be validated on data that was not used in its development. When validating the model with data used in its development, it is relatively easy to obtain high accuracy (e.g., high R-squared), but that accuracy does not guarantee that the model will perform well on data it has not seen before. Rather it leads to "overfitting" of the model.
3. Prediction accuracy must be quantified (on the test set). This can be done using the commonly used metrics for measuring the validity of screening tests:
   - Positive Predictive Value (PPV): proportion of subjects with positive test results that are correctly identified.
   - Negative Predictive Value (NPV): proportion of subjects with negative test results that are correctly identified.
   - Sensitivity: proportion of Gold Standard positives that are correctly identified.
   - Specificity: proportion of Gold Standard negatives that are correctly identified.

It is important to note that sensitivity and specificity are inherent characteristics of the model or test, while predictive values are exquisitely sensitive to the prevalence of the outcome of interest. Also, a simple accuracy (the number of correct predictions divided by the number of all predictions) is not sufficient if the data set is skewed. Most of disease outbreak data are skewed: a very large number of normal cases exist and only a few outbreak cases. Often only 1% outbreak data is present in the whole data set. This means that if a method was to predict no outbreak all the time, its accuracy would be 99% (but sensitivity of that method and PPV would be 0%). This is why the measures of sensitivity, specificity, PPV, and NPV (collectively) are much better at describing the goodness of the method.

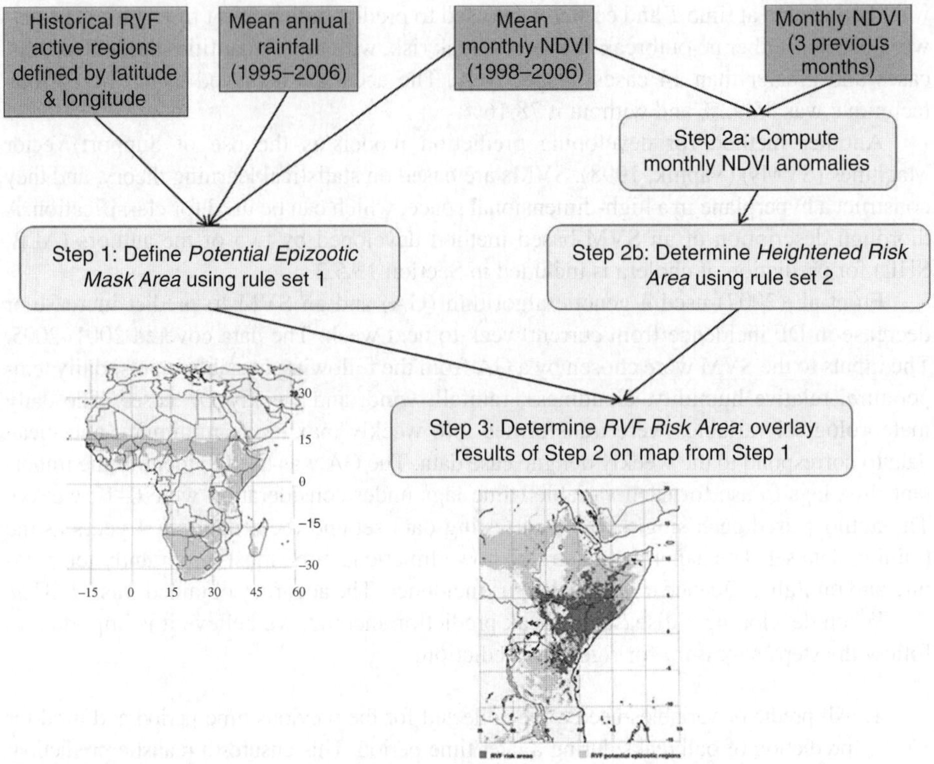

Figure 13.4. Major Steps of RVF Prediction (Method of Anyamba et al., 2009). For color detail, please see color plate section.

13.5 EXAMPLES OF PREDICTION MODELS

13.5.1 Rift Valley fever

Rift Valley fever (RVF) is a viral disease that predominately infects livestock but can spread to humans. The virus is primarily spread by *Aedes* and *Culex* mosquitoes. While infected humans often show no symptoms or mild fever, this illness can also cause a hemorrhagic fever and encephalitis. The mortality in humans is generally around 1% but is much higher in livestock, thereby causing significant economic hardship (Rich, 2010). Outbreaks of RVF have occurred frequently in sub-Saharan Africa, and in some outbreaks, hundreds to thousands of people have died.

One recent success story is Anyamba et al. (2009), who predicted an outbreak of RVF in humans and animals in the Horn of Africa during September 2006–May 2007. The RVF outbreak prediction method of Anyamba et al. has three major steps (Figure 13.4) described in the following text.

Step 1: Definition of the Potential Epizootic Area.

This area represents regions with a past history of RVF outbreaks and a past sensitivity to rainfall variations. Historical RVF active regions as defined by latitude and longitude, the mean annual rainfall, and the mean monthly NDVI constitute the inputs to Step 1. Using these inputs, an 8 km by 8 km grid cell is then included in the Potential Epizootic Area if either Rule 1 or Rule 2 is true:

Rule 1: (($25 \leq$ longitude ≤ 33) or (latitude ≤ 25 and longitude > 33) or (latitude ≤ 20 and longitude ≤ 25)) and ($0.15 \leq$ mean monthly NDVI ≤ 0.4) and ($100\,\text{mm} \leq$ mean annual rainfall $\leq 800\,\text{mm}$)

Rule 2: ($24 \leq$ latitude ≤ 36) and ($30 \leq$ longitude ≤ 35) and ($0.15 \leq$ mean monthly NDVI ≤ 0.4)

Rule 1 excludes the Sahara desert and Northern Africa, while Rule 2 includes the Nile Delta and western Arabian Peninsula that do not meet the precipitation criteria but where RVF has occurred in the past. NDVI thresholds identify areas of Africa with large interannual variability and where rainfall is directly related to vegetation growth.

Step 2: Determination of the Heightened Risk Area based on recent NDVI data.

The inputs to Step 2 are the historical mean monthly NDVI (1998–2006) and the monthly NDVI for the previous 3 months. Using these inputs, the monthly NDVI anomaly is computed as

$$\text{Monthly NDVI}\left(\text{year } y,\ \text{month } i\right) = \text{NDVI}\left(\text{year } y,\ \text{month } i\right) - \text{Monthly mean NDVI}\left(\text{year } y,\ \text{month } i\right)$$

where

$$\text{Monthly mean NDVI}\left(\text{year } y,\ \text{month } i\right) = \frac{1}{9}\sum_{k=1998}^{2006} \text{NDVI}\left(\text{year } k,\ \text{month } i\right)$$

Then, for each 1 km by 1 km grid cell to be included in the RVF Heightened Risk Area, the following conditions must both be met:

1. The monthly NDVI anomaly for each of the most recent 3 months must be greater than 0.025.
2. The mean of monthly NDVI anomaly for the most recent 3 months must be greater or equal to 0.1.

Step 3: Determination of the RVF Risk Area. The intersection of regions from Step 1 to Step 2 determines the final RVF Risk Area.

The results of Step 2 are overlaid on the map from Step 1. If a grid cell is both in the Potential Epizootic Risk Area and in the Heightened Risk Area, this cell is on the RVF Risk Area.

Using the aforementioned technique, Anyamba et al. (2009) detected anomalously elevated SSTs concurrently in the equatorial east Pacific and western Indian Oceans during July–October 2006, consistent with El Niño, followed by above normal rainfall from September to December and subsequent positive NDVI anomalies indicating elevated risk of RVF across the Horn of Africa. The early warning provided by this technique allowed entomological field investigations that confirmed virus activity and facilitated outbreak control activities.

13.5.2 Cholera

Cholera is an intestinal disease caused by the bacterium *V. cholerae*. As a result of drinking water or eating food containing these bacteria, this deadly disease presents with severe watery diarrhea and vomiting that rapidly result in dehydration and electrolyte imbalance. Although there are antibiotics that may shorten its duration and mitigate its severity, patients need rapid rehydration and restoration of their electrolyte balance if they are to survive. Vaccines are available but to date have only about a 50% efficacy. Although the microbial origin and transmission mechanism of cholera were first described in 1849–1855 by the pioneering physician John Snow, this disease continues to kill hundreds of thousands of people worldwide each year.

An example of predictive disease modeling is our (ALB, SHL) preliminary work on prediction of cholera outbreaks in Africa (Buczak et al., 2009; Chretien et al., 2009). In collaboration with the U.S. Armed Forces Health Surveillance Center and the Fogarty International Center, JHU/APL assessed demographic, economic, environmental, and climatic variables associated with *V. cholerae* infections across Africa during 1995–2005. The epidemiological data came from ProMED-mail as presented by Griffith et al. (2005). The main cited causes of outbreaks include water source contamination, poor sanitation, rainfall, flooding, and refugee settings. The predictor data sets that we identified as showing correlation with outbreaks included data from the World Health Organization (percent improved drinking water, percent improved sanitation, gross national income per capita, life expectancy at birth), National Aeronautics and Space Administration (rainfall), National Oceanic and Atmospheric Administration (SOI, SST), Gridded Population of the World (population density, percent urban constitution), and Dartmouth Flood Observatory (flood, flood severity, flood susceptibility). Another predictor that was used was the occurrence of a cholera outbreak in a neighboring province during the previous month.

We developed two approaches for predicting cholera outbreaks: hierarchical LR and SVM. Both methods performed predictions 1 month in advance, separately for each province of each country from Western, Eastern, Middle, and Southern Africa. The LR models were developed using 1997–2004 data and tested on 2005 data. They used the following predictors: population density, percent urban constitution, flood susceptibility, gross national income, occurrence of an outbreak in a neighboring province during previous month, region-specific random effects for rainfall 1 month prior, and country-specific random intercepts. There is no single predictor that can be used: heavy rainfall in several past months or the presence of an outbreak in a neighboring province during previous month does not constitute a certainty that there will be an outbreak next month in a given province. The SVM model was developed using 1996–2004 data and tested on 2005 data. There was only one model covering Western, Eastern, Middle, and Southern Africa. SVM used all the 14 predictors identified.

The Receiver Operating Characteristic (ROC) curves for the final LR and SVM models (Figure 13.5) show similar predictive accuracy, with a slight advantage by LR. The LR model has certain separate parameters for Eastern, Western, Middle, and Southern Africa. The SVM has at this point only one model covering all four regions of the continent. The preliminary results suggest that accurate forecasting of cholera outbreaks in Africa may be achievable with 1-month lead time using demographic, economic, environmental, and climatic predictors. Slightly better results are obtained by LR that uses region-specific rainfall effects for each African region. One method for improving the accuracy of the SVM is to develop separate SVM models for each of the regions (similarly to how the LR method was applied). Another possible technique for improving the overall accuracy is to fuse the results from both methods.

13.5.3 Dengue

Dengue is an acute febrile disease of humans caused by a single-stranded RNA flavivirus transmitted by *A. aegypti* mosquitoes. These mosquitoes thrive in tropical urban areas in

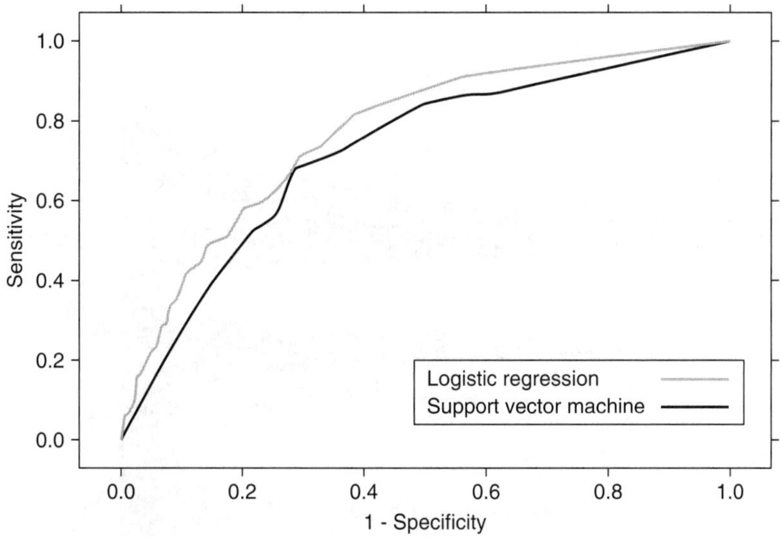

Figure 13.5. Receiver Operating Characteristic (ROC) curves for Logistic regression (LR) and Support Vector Machines (SVM). For color detail, please see color plate section.

uncovered containers holding rainwater, such as flower pots, buckets, and tires (Focks et al., 1995). Dengue is now the most common arboviral disease of humans in the world (Gibbons and Vaughn, 2002; TDR/WHO, 2009), with an estimated 50–100 million cases annually (Guzman and Kouri, 2002; Ranjit and Kissoon, 2010). Recent dengue outbreaks have occurred in the Philippines, Singapore, Thailand, Cambodia, Peru, Ecuador, and Brazil (Halsted, 2007). Dengue is endemic in Puerto Rico and recently made its resurgence in the Florida Keys in the United States (CDC, 2010).

Input variables for our model (Buczak et al., 2012) include previous dengue incidence, climatic data (rainfall, day and night temperature, NDVI, EVI, SSTA, SOI), and socioeconomic data (sanitation, water, and electricity). We obtained dengue case data from our collaborators from the Peru Ministry of Health, and we took into account cases marked as "probable" and "confirmed." The data set covered 2001–2009. The spatial resolution of the dengue data was a district and the temporal resolution, 1 week. We also obtained district population data from the 1993 and 2007 national censuses. This data was used to compute the dengue incidence rate per 1000 residents (Figure 13.6).

We developed a novel dengue prediction methodology (Buczak et al., 2012) comprised of the following steps:

1. Definition of spatiotemporal resolution and data preprocessing to fit that resolution.
2. Division of the data set into disjoint training, validation, and test subsets.
3. Rule extraction from training data using Fuzzy Association Rule Mining (FARM).
4. Automatic building of classifiers from the rules extracted in Step 3.
5. Choice of the best classifier based on its performance on the validation data set.
6. Computation of predictions on the test data using the classifier from Step 5. Computation of performance metrics.

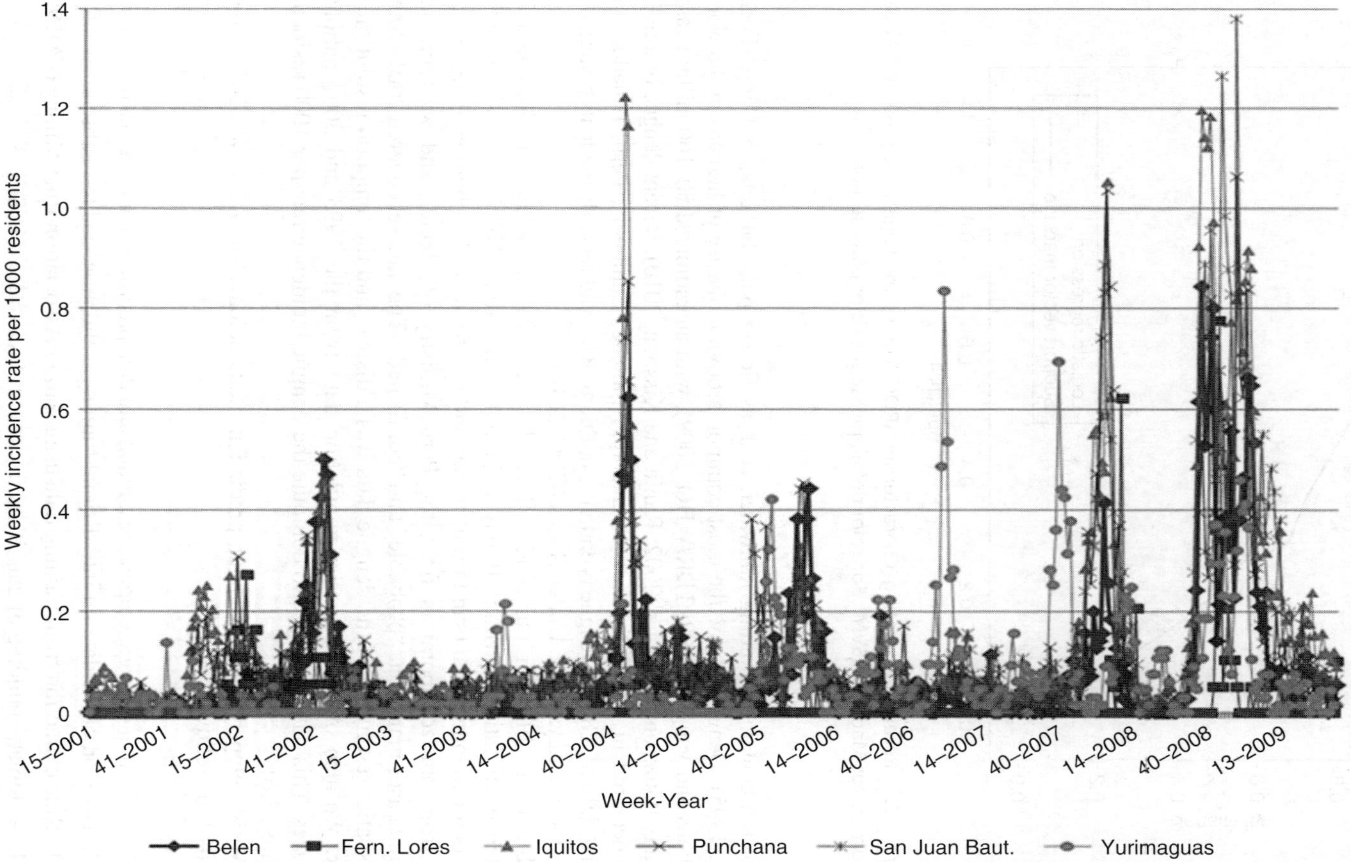

Figure 13.6. Dengue Weekly Incidence Rate. For color detail, please see color plate section.

Because different model input data come in disparate spatiotemporal scales, they were converted to one spatiotemporal scale to be used in the prediction method. The chosen temporal scale was 1 week and the chosen spatial distribution was one district. In Step 1 all the predictor and epidemiological data were converted into this spatiotemporal scale.

The second step was to divide the data set into disjoint training, validation, and test subsets. This is important as the model cannot be developed and tested on the same data.

In Step 3, FARM (Buczak and Gifford, 2010; Kuok et al., 1998) was executed on the training data to extract rules predicting future dengue incidence. FARM extracted rules of the form:

If (past incidence rate $T-1$ is *LOW*) and (past incidence rate $T-5$ is *LOW*) and (rainfall $T-3$ is *SMALL*) $\rightarrow$ Predicted incidence rate $T+4$ is *LOW*.

The preceding rule states that if the dengue incidence rate a week ago ($T-1$) was *LOW*, the dengue incidence rate 5 weeks ago ($T-5$) was *LOW*, and if the rainfall 3 weeks ago ($T-3$) was *SMALL*, then the predicted dengue incidence rate in 4 weeks ($T+4$) will be *LOW*.

FARM extracts a large number of rules (possibly hundreds or even thousands) from the training data set. When building a classifier, a subset of rules must be chosen; the subset chosen is the one that results in a smallest misclassification error for the training set. Step 4 involved the automatic building of classifiers (Liu et al., 1998) from the rules obtained in Step 3.

A separate, validation subset was used to choose the best performing classifier in Step 5. Finally, in Step 6, a third data subset (testing subset) was used to predict the dengue incidence and determine the accuracy of the method. The final step was the computation of predictions by the classifier. The outcome variable (predicted dengue incidence rate) was converted to a binary variable, either *HIGH* or *LOW* dengue incidence. This was performed on the test data, and the following performance metrics were used to assess the accuracy of the model prediction: PPV, NPV, sensitivity, and specificity.

We have developed two models, one performing a prediction for 3 weeks ahead ($T+3$) and one for 4 weeks ahead ($T+4$). The prediction generated by the classifier is a weekly prediction that specifies whether dengue incidence rate will be *LOW* or *HIGH*. In order for a dengue incidence prediction to fall exclusively into one class (*LOW* or *HIGH*), a threshold was set between *LOW* and *HIGH*. This was achieved by computing the mean (0.103) and standard deviation (0.175) of past weekly incidences. The threshold between *LOW* and *HIGH* was set at mean +2 standard deviations (rounded to 0.45).

The 3 weeks in advance incidence prediction achieved a PPV of 0.667, an NPV of 0.983, a sensitivity of 0.593, and a specificity of 0.987 on the test data set. This means that if the method predicted there would be a *HIGH* dengue incidence 3 weeks in the future, a *HIGH* dengue incidence occurred 66.7% of the time and 59.3% of the total *HIGH* dengue incidence rates 3 weeks in the future were captured. Similarly, if the method predicted a *LOW* dengue incidence 3 weeks in the future, a *LOW* dengue incidence occurred 98.3% of the time and 98.7% of the total *LOW* dengue incidence rates 3 weeks in the future were captured. The weekly, $T+4$ predictions were slightly less accurate with a PPV of 0.556, an NPV of 0.973, a sensitivity of 0.469, and a specificity of 0.981.

We also developed an LR model (Koshute et al., 2011) to check if this well-established method could yield satisfactory performance on the dengue data. The LR result gives the probability that a *HIGH* incidence rate of dengue will occur. If the estimate exceeds a predefined threshold (0.5 in our work), the model predicts a *HIGH* incidence rate; otherwise, a *HIGH* is not predicted.

For a prediction 3 weeks in advance ($T+3$), LR yielded a PPV of 0.5, an NPV of 0.962, a sensitivity of 0.25, and a specificity of 0.987. When predicting 4 weeks in advance ($T+4$), LR obtained a PPV of 0.583, an NPV of 0.961, a sensitivity of 0.219, and a specificity of 0.992. These results have very low sensitivity and are far inferior to the FARM results.

13.6 CONCLUSION

The ability to predict an infectious disease outbreak or its increased risk has often relied on studies of past incidence rates in the population and the prevalence of the pathogen. While some have developed mathematical models using only past incidence rates to forecast risks of future outbreaks (e.g., Choudhury et al., 2008), other studies have found value in adding socioeconomic factors of the population to the model (e.g., Bhandari et al., 2008). The interactions between climate and health have been recognized for millennia.

The RVF, cholera, and dengue prediction methods described in this chapter have found novel uses for satellite remote sensing data in predicting the risk of an outbreak. Remote sensing has the advantage of a continuous, or nearly continuous, monitoring of the environmental conditions that impact the vector population and thereby the risks of an outbreak.

For a disease outbreak prediction to become used by public health professionals, it has to be performed in a rigorous fashion. This means that (i) all predictor variables need to be collected for the previous time period and used for prediction of outbreaks during a later time period, (ii) the model must be validated on data that was not used in its development, and (iii) the prediction accuracy must be quantified so that the decision makers know what type of accuracy the method can provide.

If resources and effective interventions exist, the ability to anticipate an infectious disease outbreak is of great value to the public health community and the patients they serve. Such knowledge allows for early intervention by means of enhanced educational efforts (e.g., reemphasis on proper hand washing and cough covering), alerting of physicians and hospitals to be aware of diseases that may be spreading among their patients, vaccination programs, isolation and tracing contacts of symptomatic patients, etc. Early responses such as these may not only reduce economic costs compared with treating an infected populace, particular to countries whose resources are limited, but more importantly result in a reduction in the morbidity and mortality of their vulnerable populations.

REFERENCES

Agresti A. 2007. Building and applying logistic regression models. In *An Introduction to Categorical Data Analysis*, p. 138. Hoboken, NJ: Wiley.

Anuta P, Macdonald R. 1971. Crop surveys from multiband satellite photography using digital techniques. *Remote Sensing of Environment* **2**:53–67.

Anyamba A, Linthicum K, Mahoney R, et al. 2002. Mapping potential risk of Rift Valley fever outbreaks in African savannahs using vegetation index time series data. *Photogrammetric Engineering and Remote Sensing* **68**(2):137–145.

Anyamba A, Chretien JP, Small J, et al. 2009. Prediction of a Rift Valley fever outbreak. *Proceedings National Academy of Sciences* **106**(3):955–959.

Atlas R, Rubin C, Maloy S, et al. 2010. One Health – attaining optimal health for people, animals and the environment. *Microbe* **5**(9):383–389.

Babin S, Carton J, Dickey T, et al. 2004. Satellite evidence of hurricane-induced phytoplankton blooms in an oceanic desert. *Journal of Geophysical Research – Oceans* **109**:C03043.

Bhandari KP, Raju PLN, Sokhi BS. 2008. Application of GIS modeling for dengue fever prone area based on socio-cultural and environmental factors – a case study of Delhi city zone. *The International Archives of the Photogrammetry, Remote Sensing and Spatial Information Sciences*, Vol. XXXVII, Part B8, *Proceedings of the XXIst ISPRPS Congress Technical Commission VIII*, 3–11 July 2008, Beijing, China, pp. 165–170. *International Society for Photogrammetry and Remote Sensing.*

Bishop CM. 1995. *Neural Networks for Pattern Recognition*. Oxford: Clarendon Press.

Bouma MJ, Pascual M. 2001. Seasonal and interannual cycles of endemic cholera in Bengal 1891–1940 in relation to climate and geography. *Hydrobiologia* **460**:147–156.

Buczak AL, Gifford CM. 2010. Fuzzy association rule mining for community crime pattern discovery. In *Proceedings of the ACM SIGKDD Conference on Knowledge Discovery and Data Mining: Workshop on Intelligence and Security Informatics*, July, Washington, DC.

Buczak AL, Chretien JP, Philip TL, et al. 2009. Prediction of cholera epidemics in Africa. Presentation at the *8th Annual International Society for Disease Surveillance Conference*, 3–4 December, Miami, FL.

Buczak AL, Koshute PT, Babin SM, et al. 2012. A new epidemiological forecasting method for dengue outbreaks. *BMC Medical Informatics and Decision Making* **12**:124.

Catherinot J, Prigent C, Maurer R, et al. 2011. Evaluation of "all weather" microwave-derived land surface temperatures with in situ CEOP measurements. *Journal of Geophysical Research* **116**:D23105.

Cetiner B, Sari M, Aburas H. 2009. Recognition of dengue disease patterns using artificial neural networks. In *5th International Advanced Technologies Symposium (IATS'09)*, 13–15 May 2009, Karabuk, Turkey.

Choudhury Z, Banu S, Islam MA. 2008. Forecasting dengue incidence in Dhaka, Bangladesh: a time series analysis. *Dengue Bulletin* **32**:29–36.

Chretien JP, Buczak AL, Anyamba A, et al. 2009. Forecasting cholera epidemics in Africa. In *American Society of Tropical Medicine and Hygiene 58th Annual Meeting*, **717**:206, 18–22 November, Washington, DC.

Colwell R. 1996. Global climate and infectious disease: the cholera paradigm. *Science* **274**(5295):2025–2031.

Committee on Climate, Ecosystems, Infectious Diseases, and Human Health, Board on Atmospheric Sciences and Climate, National Research Council. 2001. Chap. 7: Toward the development of disease early warning systems. In: *Under the Weather: Climate, Ecosystems, and Infectious Disease*, 160 pp. Washington, DC: The National Academies Press. ISBN:0-309-51202-6.

Committee on Modeling Community Containment for Pandemic Influenza. 2006. *Modeling Community Containment for Pandemic Influenza: A Letter Report*, 47 pp. Washington, DC: Institute of Medicine, National Academies Press.

Craig MH, Kleinschmidt I, Le Sueur D, et al. 2004a. Exploring 30 years of malaria case data in KwaZulu-Natal, South Africa: Part II. The impact of non-climatic factors. *Tropical Medicine and International Health* **9**(12):1258–1266.

Craig MH, Kleinschmidt I, Nawn JB, et al. 2004b. Exploring 30 years of malaria case data in KwaZulu-Natal, South Africa: Part I. The impact of climatic factors. *Tropical Medicine and International Health* **9**(12):1247–1257.

Curran PJ, Atkinson PM, Foody GM, et al. 2000. Linking remote sensing, land cover and disease. *Advances in Parasitology* **47**: 37–78, IN1–IN4, 79–80.

Dobson AP, Carper ER. 1996. Infectious diseases and human history. *Bioscience* **46**(2):115–126.

Epstein JM. 1999. Agent-based computational models and generative social science. *Complexity* **4**(5):41–60.

Epstein JM. 2009. Modelling to contain pandemics. *Nature* **460**(6):687.

Focks DA, Daniels E, Haile DG, et al. 1995. A simulation model of the epidemiology of urban dengue fever: literature analysis, model development, preliminary validation, and samples of simulation results. *American Journal of Tropical Medicine and Hygiene* **53**(5):489–506.

Fu X, Liew C, Soh H, et al. 2007. Time-series infectious disease data analysis using SVM and genetic algorithm. In *IEEE Congress on Evolutionary Computation (CEC 2007)*, 25–28 September, Singapore, pp. 1276–1280.

Fuller D, Troyo A, Beier J. 2009. El Niño Southern Oscillation and vegetation dynamics as predictors of dengue fever cases in Costa Rica. *Environmental Research Letters* **4**:014011.

Gibbons RV, Vaughn DW. 2002. Dengue: an escalating problem. *British Medical Journal* **324**:1563.

Goetz S, Prince S, Small J. 2000. Advances in satellite remote sensing of environmental variables for epidemiological applications. *Advances in Parasitology* **47**:289–307.

Grassly NC, Fraser C. 2008. Mathematical models of infectious disease transmission. *Nature Reviews Microbiology* **6**:477–487.

Griffith DC, Kelly-Hope LA, Miller MA. 2005. Review of reported cholera outbreaks worldwide. *American Journal of Tropical Medicine and Hygiene* **75**:973–977.

Guzman MG, Kouri G. 2002. Dengue: an update. *Lancet Infectious Diseases* **2**:33–42.

Hales S, Weinstein P, Souares Y, et al. 1999. El Niño and the dynamics of vector-borne disease transmission. *Environmental Health Perspectives* **107**:99–102.

Halide H, Ridd P. 2008. A predictive model for dengue hemorrhagic fever epidemics. *International Journal of Environmental Health Research* **4**:253–265.

Halsted SB. 2007. Dengue. *Lancet Infectious Diseases* **370**:1644–1652.

Holmes T, De Jeu R, Owe M, et al. 2009. Land surface temperature from Ka band (37 GHz) passive microwave observations. *Journal of Geophysical Research* **224**:D04113.

Howard J, Barrett E, Heilkema J. 1979. The application of satellite remote sensing to monitoring of agricultural disasters. *Disasters* **4**:231–240.

Hu W, Clements A, Williams G, et al. 2010. Dengue fever and El Niño/Southern Oscillation in Queensland, Australia: a time series predictive model. *Occupational and Environmental Medicine* **67**:307–311.

Husin N, Salim N, Ahmad A. 2008. Modeling of dengue outbreak prediction in Malaysia: a comparison of neural network and nonlinear regression model. In *IEEE International Symposium on Information Technology (ITSim 2008),* 26–28 August 2008, Kuala Lumpur, Malaysia, Vol. **3**, pp. 1–4.

Johansson MA, Cummings DAT, Glass GE. 2009. Multiyear climate variability and dengue – El Niño Southern Oscillation, weather, and dengue incidence in Puerto Rico, Mexico, and Thailand: a longitudinal data analysis. *PLoS Medicine* **6**:e1000168.

Kalluri S, Gilruth P, Rogers D, et al. 2007. Surveillance of arthropod vector-borne infectious diseases using remote sensing techniques: a review. *PLoS Pathogens* **3**(10):1361–1371.

Kanemasu E, Niblett C, Manges H, et al. 1974. Wheat: its growth and disease severity as deduced from ERTS-1. *Remote Sensing of Environment* **3**:255–260.

Koshute PT, Buczak AL, Babin SM, et al. 2011. Dengue fever outbreak prediction. Presentation at the *10th Annual International Society for Disease Surveillance Conference*, Atlanta, GA. Abstract in *Emerging Health Threats* 4:11702.

Kummerow C, Barnes W, Kozu T, et al. 1998. The tropical rainfall measuring mission (TRMM) sensor package. *Journal of Atmospheric and Oceanic Technology* **15**:809–817.

Kuok CM, Fu A, Wong MH. 1998. Mining fuzzy association rules in databases. Association for Computing Machinery – Special Interest Group on Management of Data Record, Vol. **27**(1), pp. 41–46, New York.

Linthicum K, Bailey C, Tucker C, et al. 1991. Towards real-time prediction of Rift Valley fever epidemics in Africa. *Preventive Veterinary Medicine* **11**:325–334.

Liu B, Hsu W, Ma Y. 1998. Integrating classification and association rule mining. In *Knowledge Discovery and Data Mining (KDD)*, pp. 80–86.

Lloyd GER, ed. 1978. *Hippocratic Writings*. Harmondsworth: Penguin.

Lobitz B, Beck L, Huq A, et al. 2000. Climate and infectious disease: use of remote sensing for detection of *Vibrio cholerae* by indirect measurement. *Proceedings of the National Academy of Sciences of the United States of America* **97**(4):1438–1443.

Magny G, Murtugudde R, Sapiano M, et al. 2008. Environmental signatures associated with cholera epidemics. *Proceedings of the National Academy of Sciences of the United States of America* **105**(46):17676–17681.

Manzato A. 2007. A note on the maximum Pierce skill score. *Weather and Forecasting* **22**: 1148–1154.

Mendelsohn J, Dawson T. 2008. Climate and cholera in KwaZulu-Natal, South Africa: the role of environmental factors and implications for epidemic preparedness. *International Journal of Hygiene and Environmental Health* **211**:156–162.

NASA Global Change Mastery Directory website. 2012. http://gcmd.nasa.gov (accessed 5 March 2012).

NOAA National Geophysical Data Center. 2012. http://www.ngdc.noaa.gov/ecosys/cdroms/AVHRR97_d1/avhrr.htm (accessed 13 March 2012).

Rachata N, Charoenkwan P, Yooyativong T, et al. 2008. Automatic prediction system of dengue haemorrhagic fever outbreak risk by using entropy and artificial neural network. In: *IEEE International Symposium on Communications and Information Technologies (ISCIT 2008)*, 21–23 October 2008, Lao, pp. 210–214.

Ranjit S, Kissoon N. 2010. Dengue hemorrhagic fever and shock syndromes. *Pediatric Critical Care Medicine* **12**(1):90–100.

Rasmussen EM, Carpenter TH. 1982. Variations in tropical sea surface temperature and surface wind fields associated with the Southern Oscillation/El Niño. *Monthly Weather Review* **110**:354–384.

Rich KM. 2010. An assessment of the regional and national socio-economic impacts of the 2007 Rift Valley fever outbreak in Kenya. *American Journal of Tropical Medicine and Hygiene* **83**(Supplement 2):52–57.

Rodgers JL, Nicewander WA. 1988. Thirteen ways to look at the correlation coefficient. *The American Statistician* **42**(1):59–66.

Rumelhart DE, Hinton GE, Williams RJ. 1986. Learning representations by back-propagating errors. *Nature* **323**(6088):533–536.

Sobrino J, Coll C, Caselles V. 1991. Atmospheric correction for land surface temperature using NOAA-11 AVHRR channels 4 and 5. *Remote Sensing of Environment* **38**:19–34.

Soebiyanto R, Adimi F, Kiang R. 2010. Modeling and predicting seasonal influenza transmission in warm regions using climatological parameters. *PLoS One* **5**(3):e9450.

Steel RGD, Torrie JH. 1960. *Principles and Procedures of Statistics*, p. 187, 287. New York: McGraw-Hill.

Teutsch SM, Churchill RE, eds. 1994. *Principles and Practice of Public Health Surveillance*. New York: Oxford University Press.

Thomson MC, Palmer T, Morse AP, et al. 2000. Forecasting disease risk with seasonal climate predictions. *Lancet* **355**(9214):1559–1560.

Thomson MC, Mason SJ, Phindela T, et al. 2005. Use of rainfall and sea surface temperature monitoring for malaria early warning in Botswana. *American Journal of Tropical Medicine and Hygiene* **73**(1):214–221.

Tun-Lin W, Burkot T, Kay B. 2000. Effects of temperature and larval diet on development rates and survival of the dengue vector *Aedes aegypti* in north Queensland, Australia. *Medical and Veterinary Entomology* **14**:31–37.

US Centers for Disease Control and Prevention (CDC). 2010. Locally acquired dengue – Key West, Florida 2009–2010. *Morbidity and Mortality Weekly Report* **59**(19):577–581.

Vapnik VN. 1998. *Statistical Learning Theory*, 1st edition, 736pp. Series on *Adaptive and Learning Systems for Signal Processing, Communications, and Control*. New York: John Wiley & Son.

Wallace RB, Doebbeling BN, Last JM, eds. 1998. *Public Health and Preventive Medicine*, 14th edition. Stamford, CT: Appleton & Lange.

Wiwanitkit V. 2006. An observation on correlation between rainfall and the prevalence of clinical cases of dengue in Thailand. *Journal of Vector Borne Diseases* **43**:73–76.

Wolfe ND, Dunavan CP, Diamond J. 2007. Origins of major human infectious diseases. *Nature* **447**(17):279–283.

World Health Organization, Special Programme for Research and Training in Tropical Diseases (TDR/WHO). 2009. *Dengue Guidelines for Diagnosis, Treatment, Prevention and Control*, 160 pp. WHO Reference No. WHO/HTM/NTD/DEN/2009.1. Geneva: World Health Organization.

14

DEVELOPMENTS AND CHALLENGES IN DIAGNOSTIC VIROLOGY

Luisa Barzon

Department of Molecular Medicine, University of Padova, Padova, Italy
Microbiology and Virology Unit, Padova University Hospital, Padova, Italy

Laura Squarzon

Department of Molecular Medicine, University of Padova, Padova, Italy

Monia Pacenti

Microbiology and Virology Unit, Padova University Hospital, Padova, Italy

Giorgio Palù

Department of Molecular Medicine, University of Padova, Padova, Italy
Microbiology and Virology Unit, Padova University Hospital, Padova, Italy

TABLE OF CONTENTS

Viral Infections and Global Change, First Edition. Edited by Sunit K. Singh.

14.1 INTRODUCTION

In the last 50 years, sanitation improvement, childhood immunization, and increasing available antibiotics had contributed to reduce the morbidity and mortality of infectious diseases, at least in the developed countries and in most of the regions of the world (Reingold, 2000). By the mid to late 1970s, it was believed that infectious diseases soon would have been a problem of the past; however, novel emerging and reemerging pathogens have been reported continuously, representing one of the major problems in public health (Lederberg, 1993; Murphy, 2008).

Infectious agents that are introduced in an area different other than the region where they are endemic and that undergo a series of genetic mutations are classified as emerging, while reemerging agents are those that are able to reactivate in quiescent *réservoirs* and in areas where they seemed to have disappeared.

Climate change is a current global concern that affects and will continue to affect the incidence and prevalence of indigenous and imported infections. While climate restricts the range of infectious diseases, weather affects the timing and intensity of outbreaks. The largest health impact from climate change worldwide seems to occur from vector-borne infectious diseases. The ranges of several key diseases or their vectors are already changing in altitude due to warming. In addition, intense and costly weather events create conditions conducive to outbreaks of infectious diseases, such as heavy rains that leave insect breeding sites, drive rodents from burrows, and contaminate clean water systems. Mosquito-borne diseases, including malaria, dengue, and viral encephalitis, are among those diseases most sensitive to climate. Climate changes directly affect disease transmission by shifting vector geographic range and increasing reproductive and biting rates and by shortening the pathogen incubation period. Combined with climate change, globalization of the economy, high-speed international travel, transports, urbanization, and deforestation give a strong contribution to the increasing spread of emerging and reemerging infections (Jones et al., 2008; Kilpatrick, 2011; Matthews and Woolhouse, 2005; Pang and Guindon, 2004; Weaver and Reisen, 2010; Weissenböck et al., 2010; Wolfe et al., 2007). In the last decades, many countries have faced a succession of several viral outbreaks, including those caused by avian influenza viruses, Nipah virus, Rift Valley fever virus, Marburg and Ebola viruses, Japanese encephalitis virus, hantaviruses, Crimean–Congo hemorrhagic fever virus (CCHFV), dengue virus (DENV), chikungunya virus (CHIKV), Usutu virus, and West Nile virus (WNV) (Barzon et al., 2009a, 2012; Ergonul et al., 2004; Ergonul and Whitehouse, 2007; Murphy, 2008; Papa et al., 2002, 2004; Rezza et al., 2007; Sirbu et al., 2011; Vaheri et al., 2012; Weissenböck et al., 2010).

WNV appeared for the first time in the Americas in August 1999 and within 7 years had caused approximately 10 000 cases of neuroinvasive disease in North America, with a 10% fatality rate among clinically apparent encephalitic cases (Petersen et al., 2012). WNV outbreaks have been recorded in birds, horses, and humans in Europe and Russia since the 1960s, where both lineage 1 and lineage 2 are circulating and causing diseases. Large outbreaks in humans have been reported in Hungary (2005), Italy (since 2008), Romania (2010), and Greece (since 2010), where the virus has established endemic cycles (Barzon et al., 2009a, b, 2011b, 2012; Danis et al., 2011; Krisztalovics et al., 2008; Sirbu et al., 2011).

For many years, CHIKV has been associated with outbreaks of chikungunya fever in India and Africa. In 2004, an outbreak was reported in Kenya, followed by other cases in 2005 in Comoros, Seychelles, and Mauritius islands. The majority of cases were observed in La Réunion (~266 000 cases). In 2006, another outbreak occurred in India (Bonn, 2006; Chastel et al., 2005). From April 2005 to June 2006, 766 imported cases of chikungunya were identified in metropolitan France (Krastinova et al., 2006), and several imported cases were recorded also in other European countries and in the United States. In Italy, a large outbreak with 205 autochthonous cases occurred in the Emilia-Romagna Region in 2007 (Rezza et al., 2007). The outbreak was triggered by a CHIKV-infected traveler going to Italy from India. Molecular investigation during WNV and CHIKV late outbreaks showed the peculiar propensity of these arboviruses to adapt to new vectors by mutation, reassortment, or recombination that confer them a selective advantage. In particular, CHIKV acquired the A226V mutation in the E1 sequence, which improves its transmission competence for *Aedes albopictus* (Tsetsarkin and Weaver, 2011).

Among arboviruses, DENV is the most important in terms of morbidity and mortality. Similar vector transmission by *Aedes* mosquitoes and similar geographic distribution can make it sometimes difficult to distinguish between chikungunya and dengue fever. During the past century, the four serotypes (DENV 1 to DENV 4) have spread to about a hundred countries in the tropical and subtropical world including Asia, Africa, the Americas, and the Pacific. Each year, an estimated 50 million people contract dengue fever with at least 500 000 cases of dengue hemorrhagic fever or dengue shock syndrome leading to 25 000 deaths (Simmons et al., 2012). In 2010, autochthonous cases of DENV infection were reported in France and Croatia, thus suggesting a local transmission of DENV (Gjenero-Margan et al., 2011; La Ruche et al., 2010). These events were not entirely unexpected, taking into account the increase in imported cases from the French West Indies and other endemic–epidemic areas (Gautret et al., 2010), as well as the presence of competent vectors for transmitting this flavivirus.

CCHFV is geographically the second most widespread of all medically important arboviruses after dengue and circulates in nature in a tick–vertebrate–tick cycle. Crimean hemorrhagic fever was firstly described as a clinical entity in 1944–1945, when about 200 Soviet military personnel were infected while assisting peasants in devastated Crimea after Nazi invasion (Hoogstraal, 1979). By the year 2000, new outbreaks have been reported in Pakistan (Athar et al., 2003), Iran (Mardani et al., 2003), Senegal (Nabeth et al., 2004b), Albania (Papa et al., 2002), Kosovo (Drosten et al., 2002), Bulgaria (Papa et al., 2004), Turkey (>2500 cases) (Ergonul et al., 2004; Maltezou and Papa, 2010), Greece (Maltezou et al., 2009), Kenya (Dunster et al., 2002), Mauritania (Nabeth et al., 2004a), and recently India (Mishra et al., 2011; Patel et al., 2011). The serologic evidence for CCHFV was documented in Egypt, Portugal, Hungary, France, and Benin, although no human case has been reported yet (Ergonul and Whitehouse, 2007; Hoogstraal, 1979; Yilmaz et al., 2008).

In the past century, two major outbreaks of disease led to the discovery of hantaviruses that are carried by rodents. The first outbreak occurred during the Korean War (1950–1953), wherein more than 3000 United Nations troops fell ill with hemorrhagic fever with renal syndrome (HFRS) (Lee et al., 1978). The second outbreak occurred in the Four Corners region of the United States in 1993 and was referred as hantavirus pulmonary syndrome (HPS) or hantavirus cardiopulmonary syndrome (HCPS). These viruses can cause serious diseases in humans and have reached mortality rates of 12% (HFRS) and 60% (HPS) in some outbreaks. Approximately 150 000 HFRS cases are estimated to occur worldwide annually, more than 90% being reported from Asia where the most severe cases with fatality rates reaching 15% are recorded (Kariwa et al., 2007). The most common European hantavirus is Puumala virus (PUUV) that causes a mild form of HFRS often called nephropathia epidemica. Severe HFRS is also seen in Europe and is caused by Dobrava–Belgrade virus (DOBV) (Heyman et al., 2009; Klempa, 2009; Klempa et al., 2008). More than 10 000 individuals in Europe are affected by HFRS annually (Vaheri et al., 2012). The epidemiological pattern has a particular temporal cyclicity and can change geographically. In 2005, Finland had about 2500 serologically diagnosed HFRS cases and in 2008, a record year, 3259 PUUV–HFRS cases. Belgium had peak years in 2007 (298 cases) and 2008 (336 cases), Sweden in 2007 (2195 cases), and Germany in 2007 (1688 cases), in 2010 (>2000 cases), and in 2011–2012 (852 cases) (Boone et al., 2012; Heyman et al., 2011, 2009; Makary et al., 2010; Vaheri et al., 2011). It is reasonable to suggest that climate change will impact on the distribution of these and other disease vectors and viruses, even in the future (Gould and Higgs, 2009). Extreme climatic events, such as flooding, will have an impact on the local disease incidence, and moreover, flooding will provide water corridors for the distribution of insect vectors over large distances. Rapid adaptation to cooler climates or the adaptation of viruses to new vectors will represent also two critical issues that must be considered, as well as the need for continued adaptation of our diagnostic tools (Barnard et al., 2011).

14.2 PREPAREDNESS

To identify, prevent, and moderate the spread of emerging infectious diseases, epidemiological vigilance, appropriate diagnostic capacity, risk assessment, scientific developments, and regulatory measures are required at both the national and international level. Research is also required if effective new disease prevention, control, and policy tools are to be developed and translated into concrete risk management measures and policies. Accurate classification of pathogens with epidemic potential by using new rapid genotyping tools can be useful to follow the evolution of genetic and geographic shifts and to implement effective prevention strategies and vaccines, as well as to optimize communicable disease control and reduce associated costs. Thereafter, a closer interplay between epidemiological surveillance and disease management strategies must be established (Jones et al. 2008; Matthews and Woolhouse, 2005; Wolfe et al., 2007). An efficient integration of surveillance results with public health decision-making, especially in unexpected outbreaks, is required. As an example, in November 2011, public health authorities in the European Union (EU) member states were alerted about the outbreak of Schmallenberg virus (SBV) in ruminants. Since then, animal and human health authorities, both at national and EU level, have been closely collaborating on this topic to ensure the rapid detection of changes in the epidemiology in animals and humans, particularly among people who had been in close contact with infected animals. Epidemiological and microbiological studies conducted by the Robert Koch Institute in Germany and the National Institute of Public Health and the Environment in the Netherlands confirmed that the zoonotic potential of SBV was absent or very low (Ducomble et al., 2012). This outbreak revealed how surveillance could be a powerful and excellent tool at each level, among vectors, *réservoirs*, and humans. Surveillance has two main objectives, aimed at the increase of the detection of imported diseases from areas to which novel emerging or reemerging viruses are endemic and at the prompt identification of new potential autochthonous cases. At European level, the ECDC is responsible for the surveillance of infectious diseases in the EU and maintains the databases for epidemiological surveillance. As other institutions involved worldwide in surveillance programs, ECDC (available at website: http://www.ecdc.europa.eu/en/Pages/home.aspx) promotes search, collection, comparison, evaluation, and dissemination of relevant scientific and technical data, as well as the coordination and the integration of the dedicated surveillance networks. In order to maintain an updated database for surveillance activities, integrated data collection systems and all notifiable communicable diseases are constantly updated. Applied scientific studies and projects for the feasibility, development, and preparation of surveillance activities are also promoted, as well as a close cooperation between the organizations operating in the field of data collection. More specifically, in order to strengthen public health work force, surveillance activities first aim to define strategy, tools, and guidelines to enhance the preparedness of different countries for the prevention and control of communicable diseases, then to ensure an appropriate and timely response in case of public health threats, and, after all, to support countries in planning preparedness activities, making them operational, testing them by exercises, and refining the existing plans, as well as providing support to strengthen and optimize the response capacity.

14.3 CHALLENGES IN DIAGNOSIS OF EMERGING VIRAL INFECTIONS

A challenging aspect in preparedness activities for new infectious disease emergency is the establishment of diagnostic assays and tools able to detect them. Development of such tests needs to address a number of fundamental issues, which include sensitivity and specificity for the target and validation in comparison with existing technologies. Cost and speed to deliver desirable benefits from new adoption must encourage competition between old and new methodologies. Moreover, potential and real applications of a test need to be considered. For example, some tests may be applied to surveillance. In this case the test needs to be amenable to cost-effective delivery of high volumes of samples. For some technologies, the cost of individual tests is prohibitive for application to large numbers of samples or in resource-poor areas. In these situations, molecular detection techniques have been very competitive, because they are faster and less time consuming as compared with traditional detection methods such as virus isolation and plaque-reduction neutralization testing (Johnson et al., 2012). Virus isolation is very useful when a new isolate must be investigated in terms of viral replication rate, pathogenicity, and virulence. Anyway, high biosafety-level laboratory (i.e., BL3 and BL4), suitable cell lines, positive and negative controls, and personnel with highly specific technical expertise are strongly required to manage these cell culture-based assays. These are some of the reasons why rapid identification of newly emerging viruses through the use of modern genomics tools is strongly evolving and represents one of the major challenges for the near future (Haagmans et al., 2009).

Accurate and timely viral diagnosis is fundamental to optimize patient management, to appropriately use antiviral drugs, to reduce unnecessary tests and superfluous antibiotics, and to implement hospital precautions to limit nosocomial spread. Early and accurate identification and genotyping of viral pathogens has a great impact in public health initiatives, as was demonstrated by identification of the coronavirus associated with severe acute respiratory syndrome (SARS) and the 2009 pandemic H1N1 influenza A virus. Prompt antigenic and genomic characterization of the novel influenza A strain permitted the rapid development of accurate diagnostic tests and an effective vaccine (Shinde et al., 2009; Trifonov, 2009).

A hypothesis-driven diagnostic approach is generally adopted when facing a suspected viral disease, since most of currently available diagnostic methods for viral infections are based on the specific testing of a single type or genus of viruses (e.g., molecular methods based on nucleic acid amplification or probe hybridization, antigen detection, detection of antibody response, and virus isolation in cell culture). Diagnostic suspicion is based on signs, symptoms, and epidemiological data, and to be successful, such an approach requires skill in presumptive diagnosis.

In the context of a changing disease environment, the conventional hypothesis-based diagnosis has limitations in the identification of novel, unpredictable, or mutated viruses.

Novel viruses are continuously discovered with surveillance programs and with the application of new molecular techniques, and the number of newly discovered viruses will continue to exponentially increase with the contribution coming from metagenomic studies. Among the new viruses discovered in the last 15 years are avian influenza H5N1, Hendra and Nipah viruses, SARS-associated coronavirus, human bocavirus, novel human parvoviruses, human polyomaviruses, human metapneumoviruses, flaviviruses, etc. Besides viruses to be discovered, new introductions of viruses in areas where they had never been detected before represent a challenge for diagnostic virology. New introductions may occur as a consequence of changes in vector distribution or to adaptation of viruses to new vectors, such as in the case of CHIKV and WNV, which spread in higher latitudes as a consequence of adaptation of mosquitoes to cooler climates and adaptation of viruses to new vectors (Romi et al., 2006; Tsetsarkin and Weaver, 2011). Movements of infected humans are also a cause of appearance of new viruses or virus variants, like emergence and spread of pandemic H1N1 influenza A virus (Neumann et al., 2009), the spread of influenza virus with oseltamivir resistance due to H275Y mutation in the neuraminidase gene (Meijer, 2009), or the occurrence of a large outbreak with over 200 cases of chikungunya fever in Italy originated from a CHIKV-infected traveler from India (Rezza et al., 2007).

Mutation and recombination in viral genomes pose problems for diagnostic virology, mainly for molecular testing, even in the diagnosis of autochthonous viruses. Viruses, especially RNA viruses, are characterized by a high mutation rate and some exist as quasispecies. These sequence variations may lead to false-negative results in both commercial and in-house target-focused molecular assays (Whiley et al., 2008). Design of primer and probes in conserved regions of the viral genome, the use of at least two different assays targeting different viral genes, or multiplexed assays might improve the sensitivity of diagnostic tests.

The implications of all these aspects of emerging viral infection in a changing environment require that diagnostic tests be rapidly adaptable to, and capable of, detecting and identifying both known and unknown, and even variant or novel, viruses in clinical and field samples (Barnard et al., 2011).

14.4 APPROACHES TO THE DIAGNOSIS OF EMERGING VIRAL INFECTIONS

There are different approaches to the diagnosis of emerging viral diseases, based on the level of confidence regarding the suspected etiological agent, that is, if the specific identity of the etiological agent is suspected, or if it is suspected only at genus or family level or group of viruses that cause a disease, or if the identity of the etiological agent is completely unknown.

When a specific virus is suspected, targeted molecular testing, antigen detection, serology assays, or culture isolation may be applied. If the suspect is a broad number of agents, virus isolation in different cell culture systems may be attempted, and multiplex molecular tests or PCR amplification with degenerate primers followed by sequencing or microarray detection may be used. Metagenomic approaches may be helpful when focused assays have failed and the etiological agent is completely unknown.

The principal viral diagnostic methods that are used in the laboratory are culture, antigen detection, nucleic acid detection, and serology. Culture and antigen detection require viable virus or relatively intact viral fragments; specimens must be collected during active viral replication. Nucleic acid-based methods have the highest sensitivity and throughput time and outpace timelines for standard culture. Nucleic acid-based tests are invaluable for identifying viruses that propagate slowly or not at all in cell culture. In addition, viral nucleic acid fragments may persist into the recovery phase of illness, long after viable virions have been eliminated by the immune response. Since mutations responsible for antiviral drug resistance have been defined in many cases, mutation detection is the method of choice for appropriately targeting antiviral therapy.

Traditional serologic diagnosis requires paired sera from the acute and convalescent stages of infection to demonstrate a significant rise in specific viral antibody titer. Detection of virus-specific IgM antibody can diagnose acute illness from a single serum obtained during the acute or early convalescent period for some viral infections. Virus-specific IgG antibody tests are typically helpful to document immune status for selected viruses.

14.4.1 Specimen collection

Proper specimen collection and transportation are key factors for the success of viral diagnostic testing. Specimens for culture, antigen assays, or molecular testing should be obtained early in acute illness, when virus shedding is highest.

Specimens (e.g., swab specimens, nasopharyngeal aspirates, and tissue biopsies) should be collected in viral transport medium (VTM) for viral isolation and antigen detection. VTMs contain a buffered salt solution, protein and saccharide nutrients, pH indicator, and antibiotics to inhibit bacterial and fungal contaminants and are formulated to stabilize virions and infected cells for approximately 48 h. Body fluids, lavages, blood, and bone marrow samples do not need VTM. All specimens except blood samples should be transported and stored until setup at 4 °C. Blood samples for virus isolation should be collected aseptically and held at room temperature and immediately processed for culture or placed on dry ice or other suitable deep-freezing agent if virus isolation procedures cannot be set up immediately. Viral cultures are best set up on the day specimens arrive in the laboratory, preferably within 36 h of collection. For molecular assays, specimens are stored at 4 °C for short periods or frozen at −80 °C for periods longer than 24 h, especially for RNA viruses. Serum specimens to be tested only for antibodies can be shipped at ambient temperature for brief periods, provided they are collected aseptically and kept free of contaminating microorganisms.

Transport of infectious samples should be done according to current national and international regulations for the shipment of dangerous goods (Kabrane and Kallings, 2012; WHO, 2010). Laboratory procedures should be carried on according to international guidelines on laboratory biosafety and security practices (WHO, 2004).

14.4.2 Viral culture

Virus isolation and propagation in tissue culture was developed more than 60 years ago and remains a versatile, cost-effective, and comprehensive diagnostic approach. However, culture is labor intensive, involves technical proficiency, requires at least 1-day turnaround for positive results, and frequently is less sensitive than nucleic acid-based detection methods. The laboratory should select a panel of cell lines to cover the usual viral pathogens in its patient population and should vary the number and type to meet seasonal demands. A single cell line may support growth of several viruses, each producing distinctive cytopathic effect (CPE) or other measurable identifying features (hemagglutination, hemadsorption, interference). CPE may develop in 1–3 days (HSV) or may not be recognized for 5–20 days (RSV, VZV, CMV); therefore, cultures may need extended incubation for 2 weeks or longer. With the shell vial technique, the cell monolayer rests on a round coverslip in a shell vial. The specimen is inoculated into the vial, which then is centrifuged to enhance viral attachment to the cell surface. After 1–3 days of incubation, the monolayer is tested for specific viral antigens, usually with a fluorescent antibody stain, or the supernatant is subcultured for viral isolation.

Some arboviruses (DENV, CHIKV, Rift Valley virus) produce viremia for approximately a week after symptom onset allowing virus isolation on early blood samples. Other arboviruses, like WNV and tick-borne encephalitis virus, produce a brief, low-level viremia and are generally no longer detectable when neurological symptoms occur. Since these viruses can be recovered from blood only before the development of immune response, they are generally isolated from blood donors identified by screening tests or from immunodeficient subjects. Viral isolates can be recovered also from other biological samples collected from patients with acute infection, according to the tropism and tissue distribution of viruses.

There is no single virus isolation system adequate for all arboviruses; therefore, it is recommended that specimens are inoculated into different cell culture types, including C6/36, AP61, or TR-284 mosquito cells and Vero or LLC-MK2 monkey kidney cell lines. To avoid inhibition effect, serum specimens should be diluted to 1:10 or 1:20. Mosquito cells are generally more sensitive than mammalian cells and mice for isolation of many arboviruses, but they do not produce a clear CPE, so subculture in mammalian cells or virus detection by immunofluorescence assays (IFA) or molecular techniques is required. In some cases, viruses can be isolated from biological specimens only in 2–4-day-old suckling mice.

Risk group classification of medically important arboviruses ranges from BSL2 to BSL4 and most are assigned to BSL3. Since many arboviruses and hemorrhagic fever viruses have caused infections in laboratory workers, it is recommended that virus cultivation is performed under BSL3/4 conditions, consistent with the classification of the infectious agent (WHO, 2004).

14.4.3 Viral antigen detection

Viral antigens may be detected by both direct fluorescent antibody (DFA) and enzyme immunoassay (EIA) methods. The methods have comparable sensitivity and specificity. DFA has the advantage of microscopic visualization of the specimen's cell content so sample adequacy can be easily verified, but DFA procedures are labor intensive and may be unfeasible during large viral outbreaks with high specimen demand. EIA methods require less subjective interpretation than DFA, but assessment of sample quality is lost.

Many EIA procedures are engineered for individual specimen testing with positive and negative controls included into a single-use device. Sensitivity is highly dependent on adequate specimen collection, as illustrated by the variable sensitivity seen with these assays in the recent outbreak of 2009 influenza A (H1N1) (Ginocchio et al., 2009a).

Dengue nonstructural protein 1 (NS1) antigen detection by ELISA is a sensitive test for diagnosis of acute dengue infection (Blacksell et al., 2011; Duong et al., 2011). Sensitive immunochromatographic rapid test for dengue NS1 antigen detection is also available and can be used as point-of-care (POC) test (Dussart et al., 2008; Fry et al., 2011; Najioullah et al., 2011). For other arboviruses, direct antigen detection methods are generally applied to specimens collected from mosquitoes or in postmortem examination of infected hosts but not for routine clinical diagnosis. Antigen detection assay for some arboviruses, like Rift Valley virus, is under development and has shown good analytical performance (Fukushi et al., 2012; Jansen et al., 2009).

14.4.4 Molecular detection

Molecular methods, characterized by high sensitivity and specificity and possibility of automation for high-throughput testing, are replacing cell culture procedures for virus isolation. Molecular tests are available to identify many common viruses and can also be used to measure viral load. Highly sensitive molecular methods that can consistently detect as few as 10 virions ml^{-1} of specimen have been developed for screening blood and organ donations. Sensitivity is also a key factor in the detection of viruses, like WNV, which replicate at very low viral load in the human host.

Nucleic acid hybridization and amplification-based tests detect virus by targeting specific regions of the viral RNA or DNA genome. *In situ* hybridization (ISH) is a direct probe technique that can be used in anatomic pathology to detect a variety of viral pathogens and may be used in postmortem examination of organs for the presence of the virus. Polymerase chain reaction (PCR) can also be used for the diagnosis of viral infection in formalin-fixed, paraffin-embedded tissue biopsies.

Nucleic acid amplification technologies (NAAT) have raised the sensitivity of viral diagnosis beyond that of antigen assays or culture. Commonly employed target amplification technologies include PCR, real-time PCR, strand displacement amplification (SDA), nucleic acid sequence-based amplification (NASBA), and transcription-mediated amplification (TMA). Signal amplification technologies include hybrid capture, branched DNA (bDNA), invader chemistry, and gold nanoparticle probe technology. The clinical utility of PCR was greatly expanded with the introduction of real-time PCR, in which amplification and detection take place simultaneously in a closed tube or plate; crossing threshold (cycle number in which amplicon is detectable) is proportionate to viral copy, permitting accurate quantitation. Automation of nucleic acid-based testing, including extraction methods and incorporation of internal controls, has increased the precision of quantitative methods, reduced the risk of amplicon contamination, reduced the rate of false-positive or false-negative errors, and accelerated throughput.

The number of available Food and Drug Administration (FDA)- and/or European Union-cleared commercial kits is still limited and includes assays for HIV, HCV, HBV, HPV, CMV, enterovirus, respiratory viruses, and WNV. Most nucleic acid-based viral assays, and especially those for emerging viruses, are laboratory developed and utilize in-house designed primers and probes. Most molecular virology methods currently require specialized facilities (separation of areas for reagents, specimen processing and amplification steps, and appropriate air handling) and highly skilled technologists.

Many viruses contain RNA genomes, and sample collection and processing requires proper handling of specimens and storage of extracted nucleic acids to ensure accurate and sensitive diagnosis. Analytical and clinical validation and verification of in-house methods are the responsibilities of the individual laboratories, and an appropriate level of oversight and monitoring is required. Increasing availability of automation has greatly influenced the precision and ease of use of nucleic acid-based testing, particularly for quantitative analyses and screening tests (e.g., HIV, HCV, HBV viral loads, HPV detection systems).

Real-time PCR is the method of choice in clinical virology laboratories. It is routinely used in target-focused tests for many viruses, including HIV, HBV, HCV, WNV, DENV, influenza A virus, influenza B virus, respiratory viruses, and human herpesviruses. Multiplexing may be used to enable co-amplification of internal and external controls along with the target viral nucleic acid or to include different targets in one reaction tube. However, multiplexing may decrease PCR sensitivity. Real-time PCR assays, either single-plex or multiplex, are dependent on exact-match primers and probes directed to the target genes, so they are susceptible to sequence variation and "dropout of signal."

Multiplexing can be also achieved by using microspheres, microbeads, or microarrays to capture and identify different PCR products. Commercially available assays that rely on targeted multiplex PCR (e.g., ResPlex II v2.0 (Qiagen), MultiCode®-PLx (EraGen Biosciences), xTAG® (Luminex), FilmArray (Idaho Technologies Inc.)) are available for the identification of panels of respiratory viruses and virus subtypes, or gastrointestinal pathogens, and are under development for the diagnosis of different disease conditions. Fully automated multiplex systems (e.g., FilmArray technology) may be used for sensitive and rapid POC testing. Since even these methods are based on target-specific primers and probes, they are susceptible to dropout or failure to detect variant viruses (Ginocchio et al., 2009b).

Pan-genus diagnostic approaches are less prone to dropout failure than target gene-based approaches and are useful when the identity of the etiological agent is suspected at the level of genus or group of genera. A range of broad-spectrum RT-PCR assays have been developed for Flaviviruses (Johnson et al., 2010; Maher-Sturgess et al., 2008; Moureau et al., 2007; Scaramozzino et al., 2001), Bunyaviridae (Lambert and Lanciotti, 2009), Phleboviruses (Sanchez-Seco et al., 2003), Lyssaviruses (Hayman et al., 2011), and a variety of arboviruses (Kuno, 1998). After PCR amplification with broad-spectrum primers, virus species are identified by sequencing of the PCR product, by hybridization with oligonucleotide probes on a microarray or conjugated to Luminex microbeads (Balada-Llasat et al., 2011; Chen et al., 2011a; Huguenin et al., 2012; Leveque et al., 2011; Ohrmalm et al., 2012; Yu et al., 2011), or by mass spectrometry analysis (Chen et al., 2011b; Ecker et al., 2008; Gijavanekar et al., 2012; Grant et al., 2010; Grant-Klein et al., 2010). Sequencing allows to identify novel viruses within a genus. The use of broad-spectrum PCR enabled the identification of Usutu virus in patients with encephalitis in Italy (Pecorari et al., 2009), the diagnosis of cases of Alkhurma hemorrhagic fever in travelers returning in Italy from Egypt (Carletti et al., 2010), and the discovery of new Phleboviruses (Anagnostou et al., 2011; Papa et al., 2011) and could be useful to detect a variety of unexpected viral pathogens.

When the identity of the etiological agent is completely unknown and there is no hypothesis even at the genus level, metagenomic approaches may be helpful to discover the virus present in clinical samples, as discussed in the section on virus discovery.

14.4.5 Viral serology

Serologic diagnosis of viral infection is often determinant for emerging viral infections. In fact, several viruses responsible for emerging infections cause transient and low-level viremia and are generally no longer detectable when symptoms occur, while specific IgM antibodies are already present. Evidence of current/recent infection requires demonstration of virus-specific IgM during the acute stage of illness or demonstration of a significant rise in virus-specific IgG titer between acute and convalescent sera. Specific IgM is usually found in blood within the first week of primary infection and typically becomes undetectable within 1–3 months. IgM antibodies are tested in serum and may be tested also in CSF in the case of neuroinvasive disease. Virus-specific IgG is normally produced 1–2 weeks after primary infection, peaks at 4–8 weeks, and then declines but usually remains detectable indefinitely at low titer. The secondary immune response following viral reinfection or reactivation produces a different serologic profile: IgM may reappear transiently and in low titer, and IgG rapidly increases in titer. These are only general response patterns; intensity, specificity, timing, and class of antibody are influenced by the infecting virus, site of infection, and immune status of the host.

Serology is a very useful diagnostic approach for viruses that are no longer present when clinical symptoms develop, such as in the case of WNV and other arboviruses, which are often cleared from CSF at the onset of clinical encephalitis, or for viruses that require complex isolation procedures or animal inoculation (arboviruses, some Coxsackie A viruses), are a biohazardous risk (HIV, arboviruses, hemorrhagic fever viruses), or have no readily available nucleic acid amplification protocols.

The problem in the diagnosis of arbovirus and other emerging virus infections is the cross-reactivity among antigenically related viruses, mainly of the Alphavirus, Flavivirus, and Bunyavirus genera. Therefore, the presence of antibodies detected by IFA or EIA needs confirmation by detection of arboviral-specific neutralizing antibodies using the plaque-reduction neutralization tests or other specific neutralization assays (WHO, 2009). In some cases, immunoblotting may be also used as confirmation test.

Serology assay may not be commercially available for emerging viruses. In these cases, EIA and IFA may be developed in-house by experienced laboratories that can handle and inactivate infectious virus.

14.4.6 Metagenomics and virus discovery

The advent of modern laboratory techniques for the detection of novel viruses has been stimulated because there are many clinical syndromes in which viruses are suspected to play a role, but for which traditional microbiology techniques routinely fail in uncovering the etiologic agent. Broad-range molecular methods for microbial discovery were introduced two decades ago. Now, new molecular techniques such as high-throughput sequencing, mRNA expression profiling, and array-based single nucleotide polymorphism analysis are able to provide ways to rapidly identify emerging pathogens and to analyze the diversity of their genomes, as well as the host responses against them. The use of genomic approaches for both research and routine clinical applications, especially for phylogenetic assessments and for characterization of previously unrecognized pathogens, is now recommended for the robustness, reliability, and portability of molecular sequence-based data. Approaches for targeting differentially abundant or phylogenetically informative molecules have been joined by less efficient but more powerful methods for broad sequence surveys of clinical and environmental samples with the use of high-density DNA microarrays and shotgun sequencing (Barzon et al., 2011a). The advantages of DNA microarrays include the simultaneous detection of diverse sequences with widely varying relative abundance and recovery of captured sequences of interest directly from the microarray. A pan-viral DNA microarray with oligonucleotides designed from all known viral genera was used to characterize the novel causative agent of SARS (DeRisi et al., 2003; Palacios et al., 2007; Wang et al., 2003) and to detect viruses in nasopharyngeal aspirates from children with a variety of acute respiratory syndromes (Chiu et al., 2008). On the other hand, the disadvantages of DNA microarrays include their insensitivity to rare microbial sequences in the presence of highly abundant host sequences (i.e., those obtained from host tissues) and their reliance on previous knowledge of microbial sequence diversity (Palacios et al., 2008). These limits have been overcome by subtractive techniques and shotgun sequencing approaches.

Subtractive techniques, such as representational difference analysis or random sequencing of plasmid libraries of nuclease resistant fragments of viral genomes, have led in the past to the discovery of several viruses, including human herpesvirus type 8 (Chang et al., 1994), human GB virus (Simons et al., 1995), Torque teno virus (Nishizawa et al., 1997), bocavirus (Allander et al., 2005), human parvovirus 4 (Jones et al., 2005), WU polyomavirus (Gaynor et al., 2007), and KI polyomavirus (Allander et al., 2007). These traditional techniques, however, are poorly sensitive and time consuming and thus are unsuitable for large-scale analysis. High-throughput next-generation sequencing (NGS) techniques represent a powerful tool that can be applied to metagenomics-based strategies for the detection of unknown disease-associated viruses and for the discovery of novel human viruses (Barzon et al., 2011a; MacConaill et al., 2008; Tang and Chiu, 2010). Compared with microarray-based assays, NGS methods offer the advantage of higher sensitivity and the potential to detect the full spectrum of viruses, including unknown and unexpected viruses.

Next-generation high-throughput sequencing technologies (Schadt et al., 2010) include 454 FLX pyrosequencing, Ion Torrent PGM, Illumina, and SOLiD platforms. This field is in rapid expansion and novel and improved platforms, defined as third- and fourth-generation sequencing platforms, are continuously being developed and released, like Heliscope by Helicos, a real-time sequencing platform by Pacific Biosciences, and nanopore sequencing technology (Schneider and Dekker, 2012). These NGS methods have different underlying biochemistries and differ in sequencing protocol (sequencing by synthesis for 454 pyrosequencing, Illumina GA, Ion Torrent PGM, and Heliscope and sequencing by ligation for SOLiD), throughput, and sequence length. Thus, the SOLiD system may be more suitable for applications that require a very high throughput of sequences, but not long reads, such as whole-genome re-sequencing or RNA-sequencing projects, while both 454 and Illumina provide data suitable for de novo assembly, and the relative long length of 454 FLX (and its smaller version GS Junior) reads allows deep sequencing of amplicons, with applications in microbial and viral metagenomics and analysis of viral quasispecies (Barzon et al., 2011a; Relman, 2011).

Up to now, exciting results have been achieved in the detection of viruses responsible for disease that could not be detected by other techniques, including microarray analysis (Palacios et al., 2008), for the investigation of seasonal influenza (Yang et al., 2011; Yongfeng et al., 2011), norovirus outbreaks (Nakamura et al., 2009), acute respiratory tract infections (Cheval et al., 2011; Yang et al., 2011; Yozwiak et al., 2012), encephalitis (Quan et al., 2010a, b), and hemorrhagic fever (Briese et al., 2009; Xu et al., 2011; Yozwiak et al., 2012). In several cases, new or unexpected viruses were discovered, such as H1N1 pandemic influenza (Greninger et al., 2010; Kuroda et al., 2010), enterovirus 109 (Yozwiak et al., 2010), and viral genomes that could be assembled de novo using metagenomic data.

Vector-borne viruses and zoonotic viruses represent an important and challenging field for virus discovery. The feasibility of detecting arthropod-borne viruses was explored in *Aedes aegypti* mosquitoes experimentally infected with DENV (Bishop-Lilly et al., 2010) in small RNA libraries obtained from invertebrate hosts (Wu et al., 2010), while a new coronavirus was discovered in gastrointestinal tissues obtained from bats (Quan et al., 2010a).

14.4.7 POC testing

POC testing is defined as analytical testing performed outside the central laboratory using devices that can be easily transported to the vicinity of the patient. The value of near-patient testing for routine infectious disease diagnosis is well recognized given that real-time test results can direct timely therapeutic interventions and improve patients' clinical outcomes (Clerc and Greub, 2010; Peeling and Mabey, 2010). With the increasing threat of epidemic-to-pandemic transitions of new or reemerging infectious disease outbreaks owing to globalization, decentralizing diagnostic testing closer to frontline clinical settings can facilitate earlier implementations of public health responses to contain and mitigate such events. Moreover, in developing countries where high infectious disease burden is compounded by diagnostic challenges due to poor clinical laboratory infrastructure and cost constraints, the potential utility for POC testing is even greater (Park et al., 2011). Today, the field of POC testing technology is quickly developing and producing instruments that are increasingly reliable, while their size is being gradually reduced (Olasagasti and Ruiz de Gordoa, 2012). Proteins are a common target for many POC analysis methods. They can be detected and quantified in a number of ways, but lateral flow immunochromatography (LFI) is the most common detection method due to its specificity. It has been used for detection of antigens related to infectious diseases, inflammation and sepsis proteins, cardiac markers, tumor markers, glycosylated hemoglobin, and some hormones. Regarding emerging diseases, immunochromatographic tests have been developed for the detection of DENV NS1 antigen and IgM, IgG, and IgA antibodies by a number of commercial companies and have found wide application because of their ease of use (no specialized equipment or training is required) and rapidity of results (~10–15 min) (Blacksell, 2012). Despite their high value for helping in rapid diagnosis in the early phase of infection, these devices are not suitable when late clinical presentations occur. Results coming from these technologies must be combined with conventional diagnostic tools, such as PCR detection, serologic assays, and viral isolation, when available. Anyhow, other microfluidic devices, such as on-chip PCR devices, and microarrays have been employed for similar detection. The advantage of these systems comes from the possibility to detect multiple analytes simultaneously (Park et al., 2011).

14.5 CONCLUSIONS

During the last years, emerging and reemerging infections were characterized by a dramatic increase of human cases. This situation is attributable to several factors, including climate change, international travels, commercial transports, urbanization, deforestation, and adaptation of vectors to different conditions. Public health policies should increasingly focus on surveillance programs, vector control measures, and education of health-care workers to improve their skills in clinical and laboratory management of emergencies. Emerging infections are challenging not only for clinicians but also for researchers, which are continuously engaged to develop new diagnostic tools and to investigate pattern of increased pathogenicity and virulence. Up to know, many advances have been done, in particular in molecular techniques field. Starting from traditional methods that rely on virus culture, serology, and PCR followed by genome sequencing, new generation molecular diagnostic systems have been developed that enable not only to detect viruses from a specific family but also to understand if the causative agent of an outbreak is entirely novel or is an unknown sequence variant. Very recent high-throughput sequencing techniques have also given a boost to the discovery of novel viruses. These techniques are still expensive, poorly sensitive, and time consuming and thus are unsuitable for large-scale analysis. But while much has been done, much remains to be done, especially to make commercial tests available and less expensive for developing countries, which face more frequently than developed countries with epidemics and unexpected outbreaks. Finally, international networks that provide information and help are crucial for outbreak management. Surveillance activities, continuous monitoring, and timely actions should be the priorities of public health services, supported also by trained task forces that are able to prevent, diagnose, and manage new emergencies.

ACKNOWLEDGEMENT

Funding from the EU (FP7 project WINGS, grant no. 261426) is acknowledged.

REFERENCES

Allander, T., Tammi, M.T., Eriksson, M., et al., 2005. Cloning of a human parvovirus by molecular screening of respiratory tract samples. *Proc. Natl. Acad. Sci. U.S.A.* **102**, 12891–12896.

Allander, T., Andreasson, K., Gupta, S., et al., 2007. Identification of a third human polyomavirus. *J. Virol.* **81**, 4130–4136.

Anagnostou, V., Pardalos, G., Athanasiou-Metaxa, M., et al., 2011. Novel phlebovirus in febrile child, Greece. *Emerg. Infect. Dis.* **17**, 940–941.

Athar, M.N., Baqai, H.Z., Ahmad, M., et al., 2003. Short report: Crimean–Congo hemorrhagic fever outbreak in Rawalpindi, Pakistan, February 2002. *Am. J. Trop. Med. Hyg.* **69**, 284–287.

Balada-Llasat, J.-M., LaRue, H., Kelly, C., et al., 2011. Evaluation of commercial ResPlex II v2.0, MultiCode®-PLx, and xTAG® respiratory viral panels for the diagnosis of respiratory viral infections in adults. *J. Clin. Virol.* **50**, 42–45.

Barnard, R.T., Hall, R.A., Gould, E.A., 2011. Expecting the unexpected: nucleic acid-based diagnosis and discovery of emerging viruses. *Expert. Rev. Mol. Diagn.* **11**, 409–423.

Barzon, L., Franchin, E., Squarzon, L., et al., 2009a. Genome sequence analysis of the first human West Nile virus isolated in Italy in 2009. *Eurosurveillance* **14**(44), pii: 19384.

Barzon, L., Squarzon, L., Cattai, M., et al., 2009b. West Nile virus infection in Veneto region, Italy, 2008–2009. *Eurosurveillance* **14**(31), pii: 19289.

Barzon, L., Lavezzo, E., Militello, V., et al., 2011a. Applications of next-generation sequencing technologies to diagnostic virology. *Int. J. Mol. Sci.* **12**, 7861–7884.

Barzon, L., Pacenti, M., Cusinato, R., et al., 2011b. Human cases of West Nile Virus infection in north-eastern Italy, 15 June to 15 November 2010. *Eurosurveillance* **16**(33), pii: 19949.

Barzon, L., Pacenti, M., Franchin, E., et al., 2012. New endemic West Nile virus lineage 1a in northern Italy, July 2012. *Eurosurveillance* **17**(31), pii: 20231.

Bishop-Lilly, K.A., Turell, M.J., Willner, K.M., et al., 2010. Arbovirus detection in insect vectors by rapid, high-throughput pyrosequencing. *PLoS Negl. Trop. Dis.* **4**(11), e878.

Blacksell, S.D., 2012. Commercial dengue rapid diagnostic tests for point-of-care application: recent evaluations and future needs? *J. Biomed. Biotechnol.* **151967**.

Blacksell, S.D., Jarman, R.G., Bailey, M.S., et al., 2011. Evaluation of six commercial point-of-care tests for diagnosis of acute dengue infections: the need for combining NS1 antigen and IgM/IgG antibody detection to achieve acceptable levels of accuracy. *Clin. Vaccine Immunol.* **18**(12), 2095–2101.

Bonn, D., 2006. How did the chikungunya reach the Indian Ocean? *Lancet Infect. Dis.* **6**, 543.

Boone, I., Wagner-Wiening, C., Reil, D., et al., 2012. Rise in the number of notified human hantavirus infections since October 2011 in Baden-Wurttemberg, Germany. *Eurosurveillance* **17**(21), pii: 20180.

Briese, T., Paweska, J.T., McMullan, L.K., et al., 2009. Genetic detection and characterization of Lujo Virus, a new hemorrhagic fever-associated arenavirus from Southern Africa. *PLoS Pathog.* **5**(5), e1000455.

Carletti, F., Castilletti, C., Di Caro, A., et al., 2010. Alkhurma hemorrhagic fever in travelers returning from Egypt, 2010. *Emerg. Infect. Dis.* **16**, 1979–1982.

Chang, Y., Cesarman, E., Pessin, M.S., et al., 1994. Identification of herpesvirus-like DNA sequences in AIDS-associated Kaposi's sarcoma. *Science* **266**, 1865–1869.

Chastel, C., 2005. Chikungunya virus: its recent spread into the southern Indian Ocean and Reunion Island (2005–2006). *Bull. Acad. Natl. Med.* **9**, 1827–1835.

Chen, E.C., Miller, S.A., DeRisi, J.L., et al., 2011a. Using a pan-viral microarray assay (Virochip) to screen clinical samples for viral pathogens. *J. Vis. Exp.* **50**, pii: 2536.

Chen, K.F., Rothman, R.E., Ramachandran, P., et al., 2011b. Rapid identification viruses from nasal pharyngeal aspirates in acute viral respiratory infections by RT-PCR and electrospray ionization mass spectrometry. *J. Virol. Methods* **173**, 60–66.

Cheval, J., Sauvage, V., Frangeul, L., et al., 2011. Evaluation of high throughput sequencing for identifying known and unknown viruses in biological samples. *J. Clin. Microbiol.* **49**, 3268–3275

Chiu, C.Y., Urisman, A., Greenhow, T.L., et al., 2008. Utility of DNA microarrays for detection of viruses in acute respiratory tract infections in children. *J. Pediatr.* **153**, 76–83.

Clerc, O., Greub, G., 2010. Routine use of point-of-care tests: usefulness and application in clinical microbiology. *Clin. Microbiol. Infect.* **16**, 1054–1061.

Danis, K., Papa, A., Theocharopoulos, G., et al., 2011. Outbreak of West Nile virus infection in Greece, 2010. *Emerg. Infect. Dis.* **17**, 1868–1872.

DeRisi, J.L., Wang, D., Urisman, A., et al., 2003. Viral discovery and sequence recovery using DNA microarrays. *PLoS Biol.* **1**, 257–260.

Drosten, C., Minnak, D., Emmerich, P., et al., 2002. Crimean–Congo hemorrhagic fever in Kosovo. *J. Clin. Microbiol.* **40**, 1122–1123.

Ducomble, T., Wilking, H., Stark, K., et al., 2012. Lack of evidence for Schmallenberg virus infection in highly exposed persons, Germany, 2012. *Emerg. Infect. Dis.* **18**, 1333–1335.

Dunster, L., Dunster, M., Ofula, V., et al., 2002. First documentation of human Crimean–Congo hemorrhagic fever, Kenya. *Emerg. Infect. Dis.* **8**, 1005–1006.

Duong, V., Ly, S., Lorn Try, P., et al., 2011. Clinical and virological factors influencing the performance of a NS1 antigen-capture assay and potential use as a marker of dengue disease severity. *PLoS Negl. Trop. Dis.* **5**, e1244.

Dussart, P., Petit, L., Labeau, B., et al., 2008. Evaluation of two new commercial tests for the diagnosis of acute dengue virus infection using NS1 antigen detection in human serum. *PLoS Negl. Trop. Dis.* **2**, e280.

Ecker, D.J., Sampath, R., Massire, C., et al., 2008. Ibis T5000: a universal biosensor approach for microbiology. *Nat. Rev. Microbiol.* **6**, 553–558.

Ergonul, O., Whitehouse, C.A., 2007. Introduction. In: Crimean Congo Hemorrhagic Fever: A Global Perspective. Ergonul, O., Whitehouse, C.A. (eds.). Dordrecht: Springer, pp. 3–11.

Ergonul, O., Celikbas, A., Dokuzoguz, B., et al., 2004. Characteristics of patients with Crimean–Congo hemorrhagic fever in a recent outbreak in Turkey and impact of oral ribavirin therapy. *Clin. Infect. Dis.* **39**, 284–287.

Fry, S.R., Meyer, M., Semple, M.G., et al., 2011. The diagnostic sensitivity of dengue rapid test assays is significantly enhanced by using a combined antigen and antibody testing approach. *PLoS Negl. Trop. Dis.* **5**, e1199.

Fukushi, S., Nakauchi, M., Mizutani, T., et al., 2012. Antigen-capture ELISA for the detection of Rift Valley fever virus nucleoprotein using new monoclonal antibodies. *J. Virol. Methods* **180**, 68–74.

Gautret, P., Botelho-Nevers, E., Charrel, R.N., et al., 2010. Dengue virus infections in travelers returning from Benin to France, July–August 2010. *Eurosurveillance* **15**(36), pii: 19657.

Gaynor, A.M., Nissen, M.D., Whiley, D.M., et al., 2007. Identification of a novel polyomavirus from patients with acute respiratory tract infections. *PLoS Pathog.* **3**(5), e64.

Gijavanekar, C., Drabek, R., Soni, M., et al., 2012. Detection and typing of viruses using broadly sensitive cocktail-PCR and mass spectrometric cataloging: demonstration with dengue virus. *J. Mol. Diagn.* **14**(4), 402–407.

Ginocchio, C.C., Lotlikar, M., Falk, L., et al., 2009a. Clinical performance of the 3M Rapid Detection Flu A+B Test compared to R-Mix culture, DFA and BinaxNOW Influenza A&B Test. *J. Clin. Virol.* **45**, 146–149.

Ginocchio, C.C., St George, K., 2009b. Likelihood that an unsubtypeable influenza A result in the Luminex xTAG respiratory virus panel is indicative of novel H1N1 (swinelike) influenza. *J. Clin. Microbiol.* **47**, 2347–2348.

Gjenero-Margan, I., Aleraj, B., Krajcar, D., et al., 2011. Autochthonous dengue fever in Croatia, August–September 2010. *Eurosurveillance* **16**(9), pii: 19805.

Gould, E.A., Higgs, S., 2009. Impact of climate change and other factors on emerging arbovirus diseases. *Trans. R. Soc. Trop. Med. Hyg.* **103**, 109–121.

Grant, R.J., Baldwin, C.D., Nalca, A., et al., 2010. Application of the Ibis-T5000 pan-Orthopoxvirus assay to quantitatively detect monkeypox viral loads in clinical specimens from macaques experimentally infected with aerosolized monkeypox virus. *Am. J. Trop. Med. Hyg.* **82**, 318–323.

Grant-Klein, R.J., Baldwin, C.D., Turell, M.J., et al., 2010. Rapid identification of vector-borne flaviviruses by mass spectrometry. *Mol. Cell. Probes* **24**, 219–228.

Greninger, A.L., Chen, E.C., Sittler, T., et al., 2010. A metagenomic analysis of pandemic influenza A (2009 H1N1) infection in patients from North America. *PLoS One* **5**(10), e13381.

Haagmans, B.L., Andeweg, A.C., Osterhaus, A.D., 2009. The application of genomics to emerging zoonotic viral diseases. *PLoS Pathog.* **5**(10), e1000557.

Hayman, D.T., Banyard, A.C., Wakeley, P.R., et al., 2011. A universal real-time assay for the detection of Lyssaviruses. *J. Virol. Methods* **177**, 87–93.

Heyman, P., Vaheri, A., Lundqvist, Å., et al., 2009. Hantavirus infections in Europe: from virus carriers to a major public-health problem. *Expert. Rev. Anti. Infect. Ther.* **7**, 205–217.

Heyman, P., Ceianu, C., Christova, I., et al., 2011. A five year perspective on the situation of haemorrhagic fever with renal syndrome and status of the hantavirus reservoirs in Europe, 2005–2010. *Eurosurveillance* **16**(36), pii: 19961.

Hoogstraal, H., 1979. The epidemiology of tick-borne Crimean–Congo hemorrhagic fever in Asia, Europe, and Africa. *J. Med. Entomol.* **15**, 307–417.

Huguenin, A., Moutte, L., Renois, F., et al., 2012. Broad respiratory virus detection in infants hospitalized for bronchiolitis by use of a multiplex RT-PCR DNA microarray system. *J. Med. Virol.* **84**, 979–985.

Jansen van Vuren, P., Paweska, J.T., 2009. Laboratory safe detection of nucleocapsid protein of Rift Valley fever virus in human and animal specimens by a sandwich ELISA. *J. Virol. Methods* **157**, 15–24.

Johnson, N., Wakeley, P.R., Mansfield, K.L., et al., 2010. Assessment of a novel real-time pan-flavivirus RT-polymerase chain reaction. *Vector Borne Zoonotic Dis.* **10**, 665–671.

Johnson, N., Voller, K., Phipps, L.P., et al., 2012. Rapid molecular detection methods for arboviruses of livestock of importance to northern Europe. *J. Biomed. Biotechnol.* **2012**, 719402.

Jones, M.S., Kapoor, A., Lukashov, V.V., et al., 2005. New DNA viruses identified in patients with acute viral infection syndrome. *J. Virol.* **79**(13), 8230–8236.

Jones, K.E., Patel, N.G., Levy, M.A., et al., 2008. Global trends in emerging infectious diseases. *Nature* **451**(7181), 990–993.

Kabrane, Y., Kallings, I., 2012. Transportation of biological samples international regulations. In: European Manual of Clinical Microbiology. 1st ed. Cornaglia, G., Courcol, R., Herrmann, J.-L., Kahlmeter, G., Peigue-Lafeuille, H., Vila, J., ESCMID, SFM (eds.), pp. 87–99.

Kariwa, H., Yoshimatsu, K., Arikawa, J., 2007. Hantavirus infection in East Asia. *Comp. Immunol. Microbiol. Infect. Dis.* **30**, 341–356.

Kilpatrick, A.M., 2011. Globalization, land use, and the invasion of West Nile virus. *Science* **334**(6054), 323–327.

Klempa, B., 2009. Hantaviruses and climate change. *Clin. Microbiol. Infect.* **15**, 518–523.

Klempa, B., Tkachenko, E.A., Dzagurova, T.K., et al., 2008. Haemorrhagic fever with renal syndrome caused by 2 lineages of Dobrava hantavirus, Russia. *Emerg. Infect. Dis.* **14**, 617–625.

Krastinova, E., Quatresous, I., Tarantola, A., 2006. Imported cases of chikungunya in metropolitan France: update to June 2006. *Eurosurveillance* **11**(8), E060824.1

Krisztalovics, K., Ferenczi, E., Molnar, Z., et al., 2008. West Nile virus infections in Hungary, August–September 2008. *Eurosurveillance* **13**(45), pii: 19030.

Kuno, G., 1998. Universal diagnostic RT-PCR protocol for arboviruses. *J. Virol. Methods* **72**(1), 27–41.

Kuroda, M., Katano, H., Nakajima, N., et al., 2010. Characterization of quasispecies of pandemic 2009 influenza A virus (A/H1N1/2009) by de novo sequencing using a next-generation DNA sequencer. *PLoS One* **5**(4), e10256.

La Ruche, G., Souarès, Y., Armengaud, A., et al., 2010. First two autochthonous dengue virus infections in metropolitan France, September 2010. *Eurosurveillance* **15**(39), 19676.

Lambert, A.J., Lanciotti, R.S., 2009. Consensus amplification and novel multiplex sequencing method for S segment species identification of 47 viruses of the Orthobunyavirus, Phlebovirus, and Nairovirus genera of the family Bunyaviridae. *J. Clin. Microbiol.* **47**, 2398–2404.

Lederberg, J., 1993. Emerging infections: microbial threats to health. *Trends Microbiol.* **1**, 43–44.

Lee, H.W., Lee, P.W., Johnson, K.M., 1978. Isolation of the etiologic agent of Korean hemorrhagic fever. *J. Infect. Dis.* **137**, 298–308.

Leveque, N., Van Haecke, A., Renois, F., et al., 2011. Rapid virological diagnosis of central nervous system infections by use of a multiplex reverse transcription-PCR DNA microarray. *J. Clin. Microbiol.* **49**, 3874–3879.

MacConaill, L., Meyerson, M., 2008. Adding pathogens by genomic subtraction. *Nat. Genet.* **40**, 380–382.

Maher-Sturgess, S.L., Forrester, N.L., Wayper, P.J., et al., 2008. Universal primers that amplify RNA from all three flavivirus subgroups. *Virol. J.* **5**, 16.

Makary, P., Kanerva, M., Ollgren, J., et al., 2010. Disease burden of Puumala virus infections, 1995–2008. *Epidemiol. Infect.* **138**, 1484–1492.

Maltezou, H.C., Papa, A., 2010. Crimean-Congo hemorrhagic fever: risk for emergence of new endemic foci in Europe? *Travel Med. Infect. Dis.* **8**, 139–143.

Maltezou, H.C., Papa, A., Tsiodras, S., et al., 2009. Crimean–Congo hemorrhagic fever in Greece: a public health perspective. *Int. J. Infect. Dis.* **3**, 713–716.

Mardani, M., Jahromi, M.K., Naieni, K.H., et al., 2003. The efficacy of oral ribavirin in the treatment of Crimean–Congo hemorrhagic fever in Iran. *Clin. Infect. Dis.* **36**, 1613–1618.

Matthews, L., Woolhouse, M., 2005. New approaches to quantifying the spread of infection. *Nat. Rev. Microbiol.* **3**, 529–536.

Meijer, A., Lackenby, A., Hungnes, O., et al., 2009. Oseltamivir-resistant influenza virus A (H1N1), Europe, 2007–2008 Season. *Emerg. Infect. Dis.* **15**(4), 552–560.

Mishra, A.C., Mehta, M., Mourya, D.T., et al., 2011. Crimean–Congo haemorrhagic fever in India. *Lancet* **378**, 372.

Moureau, G., Temmam, S., Gonzalez, J.P., et al., 2007. A real-time RT-PCR method for the universal detection and identification of flaviviruses. *Vector Borne Zoonotic Dis.* **7**, 467–477.

Murphy, F.A., 2008. Emerging zoonoses: the challenge for public health and biodefense. *Prev. Vet. Med.* **86**, 216–223.

Nabeth, P., Cheikh, D.O., Lo, B., et al., 2004a. Crimean–Congo hemorrhagic fever, Mauritania. *Emerg. Infect. Dis.* **10**, 2143–2149.

Nabeth, P., Thior, M., Faye, O., et al., 2004b. Human Crimean–Congo hemorrhagic fever, Senegal. *Emerg. Infect. Dis.* **10**, 1881–1882.

Najioullah, F., Combet, E., Paturel, L., et al., 2011. Prospective evaluation of nonstructural 1 enzyme-linked immunosorbent assay and rapid immunochromatographic tests to detect dengue virus in patients with acute febrile illness. *Diagn. Microbiol. Infect. Dis.* **69**, 172–178.

Nakamura, S., Yang, C.S., Sakon, N., et al., 2009. Direct metagenomic detection of viral pathogens in nasal and fecal specimens using an unbiased high-throughput sequencing approach. *PLoS One*, **4**, e4219.

Neumann, G., Noda, T., Kawaoka, Y., 2009. Emergence and pandemic potential of swine-origin H1N1 influenza virus. *Nature* **459**(7249), 931–939.

Nishizawa, T., Okamoto, H., Konishi, K., et al., 1997. A novel DNA virus (TTV) associated with elevated transaminase levels in posttransfusion hepatitis of unknown etiology. *Biochem. Biophys. Res. Commun.* **241**(1), 92–97.

Ohrmalm, C., Eriksson, R., Jobs, M., et al., 2012. Variation-tolerant capture and multiplex detection of nucleic acids: application to detection of microbes. *J. Clin. Microbiol.* **50**(10), 3208–3215..

Olasagasti, F., Ruiz de Gordoa, J.C., 2012. Miniaturized technology for protein and nucleic acid point-of-care testing. *Transl. Res.* **160**(5), 332–345.

Palacios, G., Quan, P.L., Jabado, O.J., et al., 2007. Panmicrobial oligonucleotide array for diagnosis of infectious diseases. *Emerg. Infect. Dis.* **13**(1), 73–81.

Palacios, G., Druce, J., Du, L., et al., 2008. A new arenavirus in a cluster of fatal transplant-associated diseases. *N. Engl. J. Med.* **358**(10), 991–998.

Pang, T., Guindon, G.E., 2004. Globalization and risk to health. EMBO Reports. **5**, S11–S16.

Papa, A., Velo, E., Bino, S., 2011. A novel phlebovirus in Albanian sandflies. *Clin. Microbiol. Infect.* **17**(4), 585–587.

Papa, A., Bino, S., Llagami, A., et al., 2002. Crimean–Congo hemorrhagic fever in Albania, 2001. *Eur. J. Clin. Microbiol. Infect. Dis.* **21**, 603–606.

Papa, A., Christova, I., Papadimitriou, E., et al., 2004. Crimean–Congo hemorrhagic fever in Bulgaria. *Emerg. Infect. Dis.* **10**, 1465–1467.
Papa, A., Velo, E., Bino, S., 2011. A novel phlebovirus in Albanian sandflies. *Clin. Microbiol. Infect.* **17**(4), 585–587.
Park, S., Zhang, Y., Lin, S., et al., 2011. Advances in microfluidic PCR for point-of-care infectious disease diagnostics. *Biotechnol. Adv.* **29**(6), 830–839.

Patel, A.K., Patel, K.K., Mehta, M., et al., 2011. First Crimean–Congo hemorrhagic fever outbreak in India. *J. Assoc. Physicians India* **59**, 585–589.

Pecorari, M., Longo, G., Gennari, W., et al., 2009. First human case of Usutu virus neuroinvasive infection, Italy, August–September 2009. *Eurosurveillance* **14**(50), pii: 19446.

Peeling, R.W., Mabey, D., 2010. Point-of-care tests for diagnosing infections in the developing world. *Clin. Microbiol. Infect.* **16**(8), 1062–1069.

Petersen, L.R., Carson, P.J., Biggerstaff, B.J., et al., 2012. Estimated cumulative incidence of West Nile virus infection in US adults, 1999–2010. *Epidemiol. Infect.* **28**, 1–5.

Quan, P.L., Firth, C., Street, C., et al., 2010a. Identification of a severe acute respiratory syndrome coronavirus-like virus in a leaf-nosed bat in Nigeria. *mBio* **1**(4), pii: e00208-10.

Quan, P.L., Wagner, T.A., Briese, T., et al., 2010b. Astrovirus encephalitis in boy with X-linked agammaglobulinemia. *Emerg. Infect. Dis.* **16**(6), 918–925.

Reingold, A., 2000. Infectious disease epidemiology in the 21st century: will it be eradicated or will it reemerge? *Epidemiol. Rev.* **22**, 57–63.

Relman, D.A., 2011. Microbial genomics and infectious diseases. *N. Engl. J. Med.* **365**, 347–357.

Rezza, G., Nicoletti, L., Angelini, R., et al., 2007. Infection with chikungunya virus in Italy: an outbreak in a temperate region. *Lancet* **370**(9602), 1840–1846.

Romi, R., Severini, F., Toma, L., 2006. Cold acclimation and overwintering of female *Aedes albopictus* in Roma. *J. Am. Mosq. Control. Assoc.* **22**, S149–S151.

Sánchez-Seco, M.P., Echevarría, J.M., Hernandez, L., et al., 2003. Detection and identification of Toscana and other phleboviruses by RT-nested-PCR assays with degenerated primers. *J. Med. Virol.* **71**, 140–149.

Scaramozzino, N., Crance, J.M., Jouan, A., et al., 2001. Comparison of flavivirus universal primer pairs and development of a rapid, highly sensitive heminested reverse transcription-PCR assay for detection of flaviviruses targeted to a conserved region of the NS5 gene sequences. *J. Clin. Microbiol.* **39**, 1922–1927.

Schadt, E.E., Turner, S., Kasarskis, A., 2010. A window into third-generation sequencing. *Hum. Mol. Genet.* **19**, R227–R240.

Schneider, G.F., Dekker, C., 2012. DNA sequencing with nanopores. *Nat. Biotechnol.* **30**, 326–328.

Shinde, V., Bridges, C.B., Uyeki, T.M., et al., 2009. Triple-reassortant swine influenza A (H1) in humans in the United States, 2005–2009. *N. Engl. J. Med.* **360**, 2616–2625.

Simmons, C.P., Farrar, J.J., Nguyen, V.V., et al., 2012. Dengue. *N. Engl. J. Med.* **366**, 1423–1432.

Simons, J.N., Pilot-Matias, T.J., Leary, T.P., et al., 1995. Identification of two flavivirus-like genomes in the GB hepatitis agent. *Proc. Natl. Acad. Sci. U.S.A.* **92**(8), 3401–3405.

Sirbu, A., Ceianu, C.S., Panculescu-Gatej, R.I., et al., 2011. Outbreak of West Nile virus infection in humans, Romania, July to October 2010. *Eurosurveillance* **16**(2), pii: 19762.

Tang, P., Chiu, C., 2010. Metagenomics for the discovery of novel human viruses. *Future Microbiol.* **5**(2), 177–189.

Trifonov, V., Khiabanian, H., Rabadan, R., 2009. Geographic dependence, surveillance, and origins of the 2009 influenza A (H1N1) virus. *N. Engl. J. Med.* **361**(2), 115–119.

Tsetsarkin, K.A., Weaver, S.C., 2011. Sequential adaptive mutations enhance efficient vector switching by Chikungunya virus and its epidemic emergence. *PLoS Pathog.* **7**(12), e1002412.

Vaheri, A., Mills, J.N., Spiropoulou, C.F., et al., 2011. Hantaviruses. In: Oxford Textbook of Zoonoses – Biology, Clinical Practice and Public Health Control. 2nd edn. Palmer, S.R., Soulsby, L., Torgerson, P.R., Brown, D.W.G. (eds). Oxford: Oxford University Press, pp. 307–322.

Vaheri, A., Henttonen, H., Voutilainen, L., et al., 2012. Hantavirus infections in Europe and their impact on public health. *Rev. Med. Virol.*

Wang, D., Urisman, A., Liu, Y.T., et al., 2003. Viral discovery and sequence recovery using DNA microarrays. *PLoS Biol.* **1**(2), E2.

Weaver, S.C., Reisen, W.K., 2010. Present and future arboviral threats. *Antiviral Res.* **85**(2), 328–345.

Weissenböck, H., Hubálek, Z., Bakonyi, T., et al., 2010. Zoonotic mosquito-borne flaviviruses: worldwide presence of agents with proven pathogenicity candidates of future emerging diseases. *Vet. Microbiol.* **140**, 271–280.

Whiley, D.M., Lambert, S.B., Bialasiewicz, S., et al., 2008. False-negative results in nucleic acid amplification tests – do we need to routinely use two genetic targets in all assays to overcome problems caused by sequence variation? *Crit. Rev. Microbiol.* **34**, 71–76.

Wolfe, N.D., Dunavan, C.P., Diamond, J., 2007. Origins of major human infectious diseases. *Nature* **447**(7142), 279–283.

World Health Organization (WHO), 2004. Laboratory Biosafety Manual. 3rd edn. Geneva. Available: http://www.who.int/csr/resources/publications/biosafety/en/Biosafety7.pdf (accessed July 6, 2013).

WHO, 2009. Dengue Guidelines for Diagnosis, Treatment, Prevention and Control. World Health Organization. Available: http://whqlibdoc.who.int/publications/2009/9789241547871_eng.pdf (accessed July 6, 2013)

WHO, 2010. Guidance on Regulations for the Transport of Infectious Substances, 2011–2012. World Health Organization. Available: http://whqlibdoc.who.int/hq/2010/WHO_HSE_IHR_2010.8_eng.pdf (accessed July 6, 2013)

Wu, Q., Luo, Y., Lu, R., et al., 2010. Virus discovery by deep sequencing and assembly of virus-derived small silencing RNAs. *Proc. Natl. Acad. Sci. U.S.A.* **107**(4), 1606–1611.

Xu, B., Liu, L., Huang, X., et al., 2011. Metagenomic analysis of fever, thrombocytopenia and leukemia syndrome (FTLS) in Henan province, China: discovery of a new Bunyavirus. *PLoS Pathog.* **7**(11), e1002369.

Yang, J., Yang, F., Ren, L., et al., 2011. Unbiased parallel detection of viral pathogens in clinical samples using a metagenomic approach. *J. Clin. Microbiol.* **49**, 3463–3469.

Yilmaz, G.R., Buzgan, T., Irmak, H., et al., 2008. The epidemiology of Crimean–Congo hemorrhagic fever in Turkey, 2002–2007. *Int. J. Infect. Dis.* **13**, 380–386.

Yongfeng, H., Fan, Y., Jie, D., et al., 2011 Direct pathogen detection from swab samples using a new high-throughput sequencing technology. *Clin. Microbiol. Infect.* **17**(2), 241–244.

Yozwiak, N.L., Skewes-Cox, P., Gordon, A., et al., 2010. Human enterovirus 109: a novel interspecies recombinant enterovirus isolated from a case of acute pediatric respiratory illness in Nicaragua. *J. Virol.* **84**, 9047–9058.

Yozwiak, N.L., Skewes-Cox, P., Stenglein, M.D., et al., 2012. Virus identification in unknown tropical febrile illness cases using deep sequencing. *PLoS Negl. Trop. Dis.* **6**(2), e1485.

Yu, D., Wu, S., Wang, B., et al., 2011. Rapid detection of common viruses using multi-analyte suspension arrays. *J. Virol. Methods* **177**, 64–70.

15

ADVANCES IN DETECTING AND RESPONDING TO THREATS FROM BIOTERRORISM AND EMERGING VIRAL INFECTIONS

Stephen A. Morse and Angela Weber

Division of Foodborne, Waterborne, and Environmental Diseases, National Center for Emerging and Zoonotic Infectious Diseases, Centers for Disease Control and Prevention, Atlanta, GA, USA

TABLE OF CONTENTS

Viral Infections and Global Change, First Edition. Edited by Sunit K. Singh.

15.1 INTRODUCTION

A comprehensive literature review identified 1415 microbial species that were pathogenic for humans (Taylor et al., 2001); many more are capable of causing infections in animals and plants. Approximately 61% of the 1415 species are transmitted by animals, for which the human represents a dead-end host (Weiss and McMichael, 2004). Occasionally a zoonotic infection adapts to human-to-human transmission and diversifies away from its animal origin. Epidemic diseases are generally caused by infectious agents that are directly transmissible between humans. Human immunodeficiency virus (HIV) is a recent example of a human infection initiated by a switch of host species. Fortunately, only a few are capable of affecting human, animal, or plant health on a large scale. Nevertheless, according to the World Health Organization's *2004 World Health Report*, infectious diseases accounted for about 26% of the 57 million deaths worldwide in 2002 (World Health Organization, 2004). Collectively, infectious diseases are the second leading cause of death globally, following cardiovascular disease, but among persons under the age of 50, infections are overwhelmingly the leading cause of death. Infectious diseases account for nearly 30% of all disability-adjusted life years (DALYs), which reflect the number of healthy years lost to illness (World Health Organization, 2004). A significant proportion of these deaths are due to infections caused by viruses (World Health Organization, 2010). Addressing the global problem of infectious diseases is a complex endeavor, requiring new approaches to public health surveillance and rapid laboratory diagnostic capabilities. This chapter will focus on detecting and responding to threats from infectious diseases of viral etiology as described in the next section.

15.2 EMERGING, REEMERGING, AND INTENTIONALLY EMERGING DISEASES

Among the infectious diseases throughout the world, there is a baseline group of infectious diseases of viral etiology that constitutes an ongoing threat. Human demographics and behavior, technology and industry, economic development and land use, international travel and commerce, microbial evolution and change, and the breakdown of public health measures are important factors that change patterns of infectious disease (Lederberg et al., 1992). Diseases that have newly appeared in a population or have existed previously but are now rapidly increasing in incidence or geographic range are referred to as *emerging* or *reemerging* diseases, respectively (Morse, 1995). Examples of emerging viral diseases, that is, diseases that have never been recognized before, are HIV/AIDS, severe acute respiratory syndrome (SARS) (Peiris et al., 2004), and Nipah virus encephalitis (Table 15.1). Reemerging diseases are those that have been around for decades or centuries, but have come back in a different form or in a different location. Examples of reemerging viral diseases are West Nile virus in the western hemisphere and monkeypox in the United States (Table 15.1). Reemergence can also be deliberate resulting from an act of terrorism (see following text). However, whether the disease is newly emerging, reemerging, or deliberately emerging, it is treated in much the same way from a public health and scientific perspective (Morens et al., 2004). Although some of the apparent increase in infectious disease may be attributed to better diagnostic methods and surveillance, there seems little doubt that more incidents are occurring and have the potential to spread more widely than 50 years ago, as outbreaks and spread of infections like Nipah virus and SARS would not have passed unnoticed (Weiss and McMichael, 2004).

TABLE 15.1. Examples of Newly Emerging and Reemerging Viral Diseases

Agent	Mode of transmission	Cause(s) of emergence
Newly emerging		
Enterovirus 71	Direct contact with infectious nasal discharge, saliva, stools, or blister fluid	A cause of hand-foot-and-mouth disease in children; more recently, a cause of severe central nervous system disease
H5N1 influenza virus	Aerosol from infected birds	
Hantavirus pulmonary syndrome	Inhalation of aerosolized rodent urine and feces	Human invasion of virus ecological niche
Hendra virus	Probably respiratory route	Newly recognized
Hepatitis C	Percutaneous exposure to contaminated blood or plasma, sexual transmission	Recognition through molecular virology applications
Hepatitis E	Contaminated water	Newly recognized
HIV-1/HIV-2	Sexual contact, exposure to blood or tissues of an infected person, vertical transmission	Urbanization, changes in lifestyles and mores, increased IV drug use, international travel, medical technology
Human herpesvirus 6	Unknown; possibly respiratory spread	Newly recognized
Human papillomavirus	Direct contact (sexual contact/contact with contaminated surfaces)	Newly recognized
Human parvovirus B19	Contact with respiratory secretions of infected person, vertical transmission	Newly recognized
Lassa fever virus	Contact with urine or feces of infected rodents	Urbanization/conditions favoring infestation by rodents
Nipah virus	Direct transmission from infected pigs	Newly recognized
SARS coronavirus	Respiratory route	Newly recognized
Whitewater arroyo virus	Contact with urine or feces of infected rodents	Newly recognized
Reemerging		
Chikungunya	Bite of infected mosquito	Unknown
Crimean–Congo hemorrhagic fever	Bite of infected adult tick	Ecological changes favoring increased human exposure to ticks on sheep and small wild animals
Dengue	Bite of infected mosquito	Poor mosquito control, increased urbanization in tropics, increased air travel
Filoviruses (Ebola and Marburg)	Direct contact with infected blood, organs, secretions, and semen	Unknown; possibly contact with infected bats
Human monkeypox	Direct exposure to infected animals	Importation of wild rodents from West Africa
Japanese encephalitis	Bite of infective mosquito	Changing agricultural practices
Measles	Airborne, direct contact with respiratory secretions of infected persons	Deterioration of public health infrastructure supporting immunization
Norwalk and Norwalk-like agents	Fecal–oral	Increased recognition
Rabies	Bite of a rabid animal	Introduction of infected reservoir hosts to new area
Rift Valley fever	Bite of an infective mosquito	Importation of infected mosquitoes, development (dams, irrigation)
Rotavirus	Primarily fecal–oral	Increased recognition
Venezuelan equine encephalitis	Bite of an infective mosquito	Movement of mosquitoes and amplification hosts (horses)
West Nile virus	Bite of an infective mosquito	Unknown
Yellow fever	Bite of an infective mosquito	Lack of effective mosquito control and widespread vaccination, urbanization in tropics, increased air travel

Sources: Adapted from references Geisbert and Jahrling (2004), Lederberg et al. (1992), Mackenzie et al. (2004), Morens et al. (2004), and Murray et al. (1998).

15.3 BIOTERRORISM

Biological organisms and toxins were useful as weapons of war long before the germ theory of disease was understood. But as the twentieth century came to a close, the perceived difficulties in production, weaponization, and deployment of these biological weapons, as well as a belief that moral restraints would preclude the use of these weapons, gave many a false sense of security. Recently, a number of events have served to focus attention on the threat of terrorism and the potential for the use of biological, chemical, or nuclear weapons against the military, civilian populations, or agriculture for the purpose of causing illness, death, or economic loss (Falkenrath et al., 1998). This potential became a reality in October 2001 when spores of *Bacillus anthracis* were sent through the U.S. mail to media companies in New York City and Boca Raton, Florida, resulting in five deaths and considerable panic throughout the United States and other countries (Jernigan et al., 2001).

Viruses that could be used as weapons against humans, animals, or plants generally possess traits including ease of production and dissemination, transmissibility, environmental stability, and high morbidity and mortality rates. The use of biological agents is often characterized by the manner in which they are used. For the purposes of this chapter, *biological warfare* is defined as a special type of warfare conducted by a government against a target; *bioterrorism* is defined as the threat or use of a biological agent (or toxin) against humans, animals, or plants by individuals or groups motivated by political, religious, ecological, or other ideological objectives. Terrorists can be distinguished from other types of criminals by their motivation and objective; however, criminals may also be driven by psychological pathologies and may use biological agents. When criminals use biological agents for murder, extortion, or revenge, it is called a *biocrime*.

The use of viral agents for biological warfare has a long history, which predates their recognition and isolation by culture (Meyer and Morse, 2008; Morse, 2006). Their early use is consistent with what, at the time, was known about infectious diseases, particularly smallpox caused by the virus Variola major (Hopkins, 1983). Several examples will suffice. In the sixteenth century, the Spanish explorer, Francisco Pizarro, presented the indigenous peoples of South America with Variola-contaminated clothing, resulting in widespread epidemics of smallpox. During the French and Indian War (1745–1767), Sir Jeffrey Amherst, commander of the British forces in North America, suggested the deliberate use of smallpox to "reduce" Native American tribes hostile to the British. Captain Ecuyer (one of Amherst's subordinates), fearing an attack on Ft. Pitt from Native Americans, acquired two Variola-contaminated blankets and a handkerchief from a smallpox hospital and, in a gesture of good will, distributed them to the Native Americans. As a result, several outbreaks of smallpox occurred in various tribes in the Ohio River valley. In 1775, during the Revolutionary War, the British attempted to spread smallpox among the Continental forces by inoculating (variolation) civilians fleeing Boston. In the Southern colonies, there is evidence that the British were going to distribute slaves who had escaped during hostilities, and were sick with smallpox, back to the rebel plantations in order to spread the disease.

The use of viruses other than Variola major is a more recent phenomenon and reflects an increased knowledge of how to grow and stabilize viruses for delivery purposes. Allegations have been made by the government of Cuba that the CIA was responsible for the massive outbreak of dengue fever in 1980 that ravaged the country (Christopher et al., 1999). However, subsequent investigations have failed to find substantive proof of CIA involvement in these outbreaks. The Aum Shinrikyo, a religious cult responsible for the 1995 release of sarin gas in the Tokyo subway system, was also involved in developing biological weapons and sent a team of 40 people to Zaire to acquire Ebola virus (Carus, 2002; Kaplan and Marshall, 1996; Olsen, 1999). Fortunately, they were unsuccessful in this endeavor. In 1997, unknown farmers in New Zealand deliberately and illegally introduced rabbit hemorrhagic disease virus (a calicivirus) onto the south island as an animal control tool to kill feral rabbits (Carus, 2002).

Over the past two decades, HIV has been involved in a number of biocrimes (Carus, 2002). This most likely reflects the availability of HIV-contaminated blood as a source of this virus. For example, in 1990, Graham Farlow, an asymptomatic HIV-positive inmate at a prison in New South Wales, Australia, injected a guard with HIV-contaminated blood (Jones, 1991). The guard became infected with HIV; Farlow subsequently died of AIDS. In 1992, Brian T. Stewart, a phlebotomist at a St. Louis, MO, hospital, injected his 11-month-old son with HIV-contaminated blood during a fight over payment of child support (Metzker et al., 2002). In 1993, Iwan E. injected his former girlfriend with 2.5 ml of HIV-contaminated blood after she broke up with him (Carus, 2002; Veenstra et al., 1995). In 1994, Dr. Richard J. Schmidt, a married Louisiana gastroenterologist, injected a former lover with HIV-contaminated blood (Carus, 2002). Molecular typing of the HIV strains demonstrated that her strain was very similar to the strain of HIV found in the blood of one of Dr. Schmidt's patients (Harmon, 2005). In perhaps the most famous case, Dr. David Acer, a Florida dentist infected with HIV, transmitted the disease to six of his patients between 1987 and 1990 (Carus, 2002) as confirmed by molecular analyses (Crandall, 1995; Ou et al., 1992). The intentional infection of these patients is a possibility although there is no direct evidence. In spite of these incidents, HIV has not been included on lists of threat agents for public health bioterrorism preparedness (Rotz et al., 2002) in spite of the contention that HIV has great weapon potential if the goal is to destabilize a society (Casadevall and Pirofsky, 2004).

Viruses have also been involved in suspected incidents or hoaxes. In 1999, an article appeared suggesting that the CIA was investigating whether Iraq was responsible for causing the outbreak of West Nile fever in the New York City area. The story relied heavily on a previous story written by an Iraqi defector, claiming that Saddam Hussein planned to use West Nile virus strain SV 1417 to mount an attack. An investigation concluded that there was no evidence of bioterrorism involved in the spread of West Nile virus (Lanciotti et al., 1999). A fictional "virus" was also involved in large bioterrorism hoaxes in 2000 (Carus, 2002). According to e-mail messages widely circulated on the Internet, an organization known as the Klingerman Foundation was mailing blue envelopes containing sponges contaminated with a fictional pathogen called the "Klingerman virus." According to the e-mail alert, 23 people had been infected with the virus, including 7 who died.

15.4 VIRUSES AS BIOWEAPONS

Advances in viral culture and virus stabilization made during the second half of the twentieth century have facilitated the large-scale production of viral agents for aerosol dissemination. A report for the United Nations on chemical and biological weapons and the effects of their possible use gave estimates on the numbers of casualties produced by a hypothetical biological attack (Table 15.2). Three viruses (Rift Valley fever virus, tick-borne encephalitis virus, and Venezuelan equine encephalomyelitis (VEE) virus) were modeled in a scenario in which 50 kg of the agent was released by aircraft along a 2 km line upwind of a population center of 500 000. Not surprisingly, the viral agents produced fewer casualties and impacted a smaller area than the bacterial agents used in this hypothetical

TABLE 15.2. Estimates of Casualties Produced by Hypothetical Biological Attack

Agent	Downwind Reach (km)	Dead	Incapacitated
Rift Valley fever	1	400	35 000
Tick-borne encephalitis	1	9500	35 000
VEE	1	200	19 800
Francisella tularensis	>20	30 000	125 000
B. anthracis	>20	95 000	125 000

Source: World Health Organization (WHO). 1970. *Health Aspects of Chemical and Biological Weapons*. Geneva: WHO.

Note: These estimates are based on the following scenario: release of 50 kg of agent by aircraft along a 2 km line upwind of a population center of 500 000.

model. Of note, smallpox was not evaluated in this study because it had not yet been eradicated and the level of vaccine-induced immunity in the population was high.

Viral agents were part of the biological weapons arsenal of both the Soviet Union and the United States (Table 15.3). VEE virus was stockpiled by both countries as an incapacitating agent; Variola major and Marburg viruses were stockpiled as lethal agents by the Soviet Union. The Soviet Union reportedly conducted a live field test of Variola major virus on Vozrozhdeniye Island in the Aral Sea in the 1970s in which 400 g of the virus was released into the atmosphere by explosion (Esrink, 2002). Unfortunately, a laboratory technician who was collecting plankton samples from an oceanographic research vessel 15 km from the island became infected. It was reported that after returning home to Aralsk, she transmitted the infection to several people including children. All those infected died. A number of other viruses that infect humans (e.g., Ebola virus, Lassa fever virus, enterovirus 70) or livestock (e.g., foot-and-mouth disease virus (FMDV), rinderpest, Newcastle disease virus) have also been studied for their offensive capabilities or for the development of medical and veterinary countermeasures.

Today, with the increased level of concern, a number of viruses have been cited as possible weapons for use against humans or animals (Table 15.3). Many of these viruses are associated with naturally occurring emerging or reemerging diseases (Table 15.1). These viruses share one or more criteria that would make them an ideal biological warfare agent including availability, ease of production, stability after production, a susceptible population, absence of specific treatment, ability to incapacitate or kill the host, appropriate particle size in aerosol so that the virus can be carried long distances by prevailing winds and inhaled deeply into the lungs of unsuspecting victims, ability to be disseminated via food or water, and the availability of a vaccine to protect certain groups. Other factors such as the economic and psychological impact of a biological attack on animal agriculture with a viral agent must also be considered.

Variola major is considered to be the major viral threat agent for humans because of its aerosol infectivity, relatively high mortality, and stability (Harper, 1961; World Health Organization, 1970). Thus, considerable effort has been expended toward preparing the public health and medical communities for the possibility that this agent will be employed by a terrorist (Henderson et al., 1999). Variola major is considered to be an ideal terrorist weapon because it is highly transmissible by the aerosol route from infected to susceptible persons, the civilian populations of most countries contain a high proportion of susceptible persons, and the

TABLE 15.3. Classification of Viral Agents That Are Considered to Be of Concern for Bioterrorism and Biowarfare and Those That Have Been Weaponized or Studied for Offensive or Defensive Purposes as Part of Former or Current National Biological Weapons Programs

Nucleic acid	Family	Genus	Species
Negative-sense single-stranded RNA	Arenaviridae	Arenaviruses	Lassa fever[a,b]
			Junin[a,b]
			Machupo[a,b]
			Sabia
			Guanarito
	Bunyaviridae	Phlebovirus	Rift Valley fever[b]
		Nairovirus	Crimean–Congo HF
		Hantavirus	Hantaan and related viruses[b]
			Sin Nombre
	Orthomyxoviridae	Influenzaviruses	Influenza A[b]
	Filoviridae	Filovirus	Ebola[a]
			Marburg[c]
	Paramyxoviridae	Henipavirus	Nipah virus
		Morbillivirus	Rinderpest[a,b,c,d,e,f]
		Avulavirus	Newcastle disease virus[b]
Positive-sense single-stranded RNA	Flaviviridae	Flavivirus	Yellow fever[a,b,d]
			Dengue[b]
			Tick-borne encephalitis virus[a]
			Kyasanur forest disease virus
			Japanese encephalitis virus[a]
			Omsk hemorrhagic fever virus
	Togaviridae	Alphavirus	VEE virus[c,g]
			EEE virus[b]
			WEE virus[b]
			Chikungunya virus[b]
	Picornaviridae	Enterovirus	Enterovirus 70[h]
		Hepatovirus	Hepatitis A virus
		Aphthovirus	FMDV[f,i]
Double-stranded DNA	Poxviridae	Orthopoxvirus	Variola major[b,c,j]
			Camelpox[h]
	Asfarviridae	Asfivirus	African swine fever virus[a]

Source: Adapted from reference Meyer and Morse (2008).
[a] Studied by the Soviet Union BW program.
[b] Studied by the U.S. BW program.
[c] Weaponized by the Soviet Union BW program.
[d] Studied by the Canada BW program.
[e] Studied by the France BW program.
[f] Studied by the Germany BW program.
[g] Weaponized by the U.S. BW program.
[h] Studied by the Iraq BW program.
[i] Studied by the Iran BW program.
[j] Studied by the North Korea BW program.

disease is associated with a high morbidity and about 30% mortality; initially, the diagnosis of a disease that has not been seen for almost 30 years would be difficult, and other than the vaccine, which may be effective in the first few days post infection, there is no proven treatment available (Daum et al., 2007).

VEE, eastern equine encephalomyelitis (EEE), and western equine encephalomyelitis (WEE) (Table 15.3) are also of concern because they can be produced in large amounts in inexpensive and unsophisticated systems, they are relatively stable and highly infectious for humans as aerosols, and strains are available that produce incapacitating (e.g., VEE) or lethal infections (EEE case fatality rates range from 50% to 75%) (Smith et al., 1997). Furthermore, the existence of multiple serotypes of VEE and EEE viruses, as well as the inherent difficulties of inducing efficient mucosal immunity, makes defensive vaccine development difficult.

The filoviruses, arenaviruses, and other viruses that cause hemorrhagic fever have also been considered as agents that might be used by terrorists because of their high virulence and capacity for causing fear and anxiety (Bray, 2003; Charrel and de Lamballerie, 2003; Cleri et al., 2006; Sidwell and Smee, 2003). The filoviruses, Ebola and Marburg, can also be highly infectious by the airborne route. Humans are generally susceptible to infection with these viruses with fatality rates greater than 80%, and infection can be transmitted between humans through direct contact with virus-containing body fluids. There are five species of arenaviruses (Lassa fever, Junin, Machupo, Guanarito, and Sabia) that can cause viral hemorrhagic fevers with a case fatality rate of about 20%. Large quantities of these viruses can be produced by propagation in cell culture. Infection occurs via the respiratory pathway, suggesting that dissemination via aerosol might be used by a terrorist. Human-to-human transmission has also been reported with aerosol transmission, the most likely route for at least some of the secondary cases. The filoviruses and arenaviruses discussed earlier are biosafety level 4(BSL-4) agents, and diagnostic capacities for infections caused by these viruses are limited.

From an economic perspective, FMDV (Table 15.3) is an important biothreat agent for attacks on animal agriculture in countries that are free of foot-and-mouth disease (FMD). For example, after being free of FMD for more than 20 years, the United Kingdom suffered its worst FMD epidemic ever in 2001. In just 11 months, at least 6 million animals were destroyed and losses were calculated to exceed £12 billion (Martin et al., 2001; Thompson et al., 2002).

15.5 IMPACT OF BIOTECHNOLOGY

Because the nucleic acid of many viruses, including some that are currently not threats, can be manipulated in the laboratory, the potential for genetic engineering remains a serious threat. The use of a genetically engineered virus has profound implications for public health surveillance systems as well as laboratory diagnosis. Biotechnology, which has had a tremendous impact on the development of medicines and vaccines and in the technologies needed to counter the threat of naturally occurring disease, can also be used to modify viruses with unintended consequences or for the development of novel biological agents. Several examples involving viruses are presented in the succeeding text.

15.5.1 Mousepox virus

An Australian research group (Jackson et al., 2001) was investigating virally vectored immunocontraceptive vaccines based on ectromelia virus, the causative agent of the disease termed mousepox. They created a recombinant virus, which expressed the mouse cytokine

IL-4 in order to enhance the antibody-mediated response to other recombinant antigens carried on the virus vector. Unexpectedly, the ectromelia virus vector expressing IL-4 altered the host's immune response to this virus resulting in lethal infections in the normally genetically resistant C57BL/6 mice. Additionally, this virus also caused lethal infections in mice previously immunized against infection with ectromelia virus. The creation of this "supermousepox" virus led to speculation that similar genetic engineering could be performed on Variola major leading to a biological weapon that would be effective against an immunized population.

15.5.2 Influenza A

The influenza pandemic of 1918–1919, which followed World War I, was uniquely severe, causing an estimated 20–40 million deaths globally (Crosby, 1989). This pandemic happened before the advent of viral culture and very little was known about this virus until the discovery of the polymerase chain reaction (PCR). Recently, the complete coding sequences of all eight viral RNA segments have been determined by using reverse transcription-PCR (RT-PCR) to amplify the viral RNA sequences from formalin-fixed and frozen tissue samples from individuals who died during this pandemic in an effort to shed light on both the reasons for its extraordinary virulence and evolutionary origin (Reid et al., 1999, 2000; Taubenberger et al., 1997). More recently, researchers reconstructed the 1918 Spanish influenza pandemic virus using reverse genetics and observed that this reconstructed virus exhibited exceptional virulence in several model systems and that the 1918 hemagglutinin and polymerase genes (H1N1) were essential for optimal virulence (Tumpey et al., 2005).

H5N1 avian influenza virus is a seasonal occurrence in at least 63 countries (Editorial, 2012). It is rarely transmitted (i.e., requires a large inoculum) from animal hosts to humans (Webby et al., 2004). But when it does, it causes severe infections—with overall mortality rates around 60% (World Health Organization, 2013). Recently, Ron Fouchier of the Erasmus Medical Center in Rotterdam, the Netherlands, and Yoshihiro Kawaoka of the University of Wisconsin, Madison, and the University of Tokyo's Institute of Medical Science have identified the genetic changes needed for A/H5N1 virus to be efficiently transmitted between ferrets, a surrogate for human-to-human transmission. The publication of these studies in their entirety has been very controversial due to concerns that the results could lead to the development of a highly transmissible, highly lethal strain of A/H5N1 by terrorists, or that the strain may be accidently released from the laboratory into the general population (Le Duc and Franz, 2012).

15.5.3 Synthetic genomes

A full-length poliovirus complementary DNA (cDNA) (c. 7500 B.P.) has been synthesized in the laboratory by assembling oligonucleotides of plus and minus strand polarity (Cello et al., 2002). The synthetic poliovirus cDNA was transcribed by RNA polymerase into viral RNA, which was translated and subsequently replicated in a cytoplasmic extract of uninfected HeLa S3 cells, resulting in the de novo synthesis of infectious poliovirus. The publication of this research raised concerns that more complicated viruses (e.g., Variola major or Ebola) could be synthesized from scratch based on publicly available sequences, or that viruses could be created that do not exist in the wild.

15.6 DETERRENCE, RECOGNITION, AND RESPONSE

An effective defense requires a comprehensive approach that includes prevention of access to viral stocks, improved means of detecting deliberately induced disease outbreaks, rapid medical recognition of specific syndromes (e.g., hemorrhagic fever syndrome), rapid laboratory identification of viruses in patient specimens, prevention of person–person transmission, reliable decontamination procedures, development of effective vaccines, and development of effective antiviral therapy.

15.6.1 Deterrence

In the United States, the possession, use, and transfer of microorganisms and toxins deemed to have the potential to pose a severe threat to public health and safety, animal health or animal products, and plant health and plant products are regulated by the CDC and Animal and Health Inspection Service (APHIS) Select Agent Programs (Morse and Weirich, 2011). A number of viruses capable of infecting humans and/or animals are regulated by these programs. A complete list can be found at http://www.selectagents.gov.

15.6.2 Laboratory Response Network

Rapid and accurate detection is crucial for detecting, identifying, and lessening the impact of an infectious threat. A major effort to enhance diagnostic capacity is the Laboratory Response Network (LRN), established by the CDC in collaboration with the Association of Public Health Laboratories, the Federal Bureau of Investigation, and the U.S. Army Medical Research Institute for Infectious Diseases (USAMRIID) to respond quickly to acts of chemical and biological terrorism, emerging infectious diseases, and other public health threats and emergencies (Morse et al., 2003). The diagnostic network became operational in 1999 and now links about 170 federal, state, and local reference laboratories. The LRN is a unique asset in the nation's growing preparedness for chemical and biological terrorism as well as emerging infectious diseases. In the years since its creation, the LRN has played an instrumental role in improving the public health infrastructure by helping to enhance laboratory capacity. Laboratories are better equipped, their staff levels are increasing, and they are employing advanced technologies for the identification of threat agents or emerging infectious diseases.

LRN laboratories are designated as either national, reference, or sentinel (Rotz and Hughes, 2004). This designation depends on the types of tests a laboratory can perform and how it handles infectious agents to protect workers and the public. National laboratories (CDC and USAMRIID) have BSL-4 facilities to handle highly infectious viruses, such as Ebola and Variola major, for which other laboratories have insufficient containment facilities and unvaccinated staff. These national laboratories also maintain extensive culture collections of critical agents, against which an isolate from an intentional or a naturally occurring outbreak can be compared using molecular methods to determine its likely origin.

Reference laboratories can perform tests to detect and confirm the presence of a threat agent. These laboratories ensure a timely local response in the event of a terrorist incident or for the recognition of an emerging infectious disease. Rather than having to rely on confirmation from laboratories at the CDC, reference laboratories are capable of producing conclusive results using standard protocols and reagents for the identification and confirmation of many threat agents. This ability allows local authorities to respond quickly to emergencies. Reference laboratories are primarily local and state public health laboratories with BSL-2 facilities where BSL-3 practices are employed and public health laboratories with full BSL-3 facilities as well as laboratories with certified animal facilities necessary for performing some of the tests.

Sentinel laboratories constitute the thousands of hospital-based and commercial clinical laboratories that routinely process human specimens for the presence of microbial agents. Formal recognition of these facilities as an integral part of the LRN is the responsibility of each state (or in some cases local) LRN reference laboratory in partnership with the CDC. In an unannounced or covert terrorism attack, patient specimens are collected during routine patient care. Thus, sentinel laboratories may be the first facilities to spot suspicious specimen results (Bush et al., 2001). A sentinel laboratory's responsibility is to refer a suspicious culture to the correct reference laboratory with the capabilities to test for the suspect agent. To facilitate this process, the CDC, in collaboration with the American Society for Microbiology, has developed protocols and algorithms for clinical laboratories. These algorithms can be found on the Internet at either organization's website (www.bt.cdc.gov or www.asm.org).

The LRN has strengthened the U.S. public health response to new and emerging infectious diseases. In 2002, the LRN played a vital role in the development and deployment of a new rapid test for a previously unknown disease, SARS. CDC laboratories identified the unique DNA sequence of the coronavirus that causes SARS (Rota et al., 2003); the CDC/LRN subsequently developed nucleic acid amplification assays and reagents and provided member laboratories access to these materials. More recently, human infections with influenza A/H5N1 have heightened concern that this strain might become pandemic. To facilitate the diagnosis of patients with influenza/H5N1, CDC developed a new test for the detection of this virus, which was subsequently approved by the Food and Drug Administration (FDA) and distributed to LRN laboratories nationwide. The LRN also played an important role in the identification and public health response to an outbreak of monkeypox in the midwestern United States (Reed et al., 2004) (Figure 15.1). What was unique about this outbreak was the appearance of individuals with smallpox-like lesions, which on first blush may have indicated the deliberate release of smallpox virus (Figure 15.2a and Figure 15.2b). However, a thorough integrated epidemiologic and laboratory investigation identified the disease as monkeypox, a previously unrecognized disease in the United States, which in itself could suggest bioterrorism (Morse and Khan, 2005). Affected individuals were infected by prairie dogs purchased as pets, which had acquired their infection while co-housed with infected giant Gambian rats that had recently been imported from Ghana, and not from deliberate dissemination (Centers for Disease Control and Prevention, 2003a, b, c, d, e).

The National Animal Health Laboratory Network (NAHLN) was established by the APHIS, United States Department of Agriculture (USDA), to protect the nation from bioterrorist attacks that would impact animal health and agriculture (http://www.aphis.usda.gov/animal_health/nahln/). The NAHLN consists of 57 laboratories with the ability to diagnose infections caused by avian influenza virus, classical swine fever virus, FMDV, Newcastle disease virus, pseudorabies virus, swine influenza virus, and vesicular stomatitis virus. Many of these viruses are exotic and are not found in the United States. Thus, their presence through either an intentional or accidental introduction could have severe economic consequences.

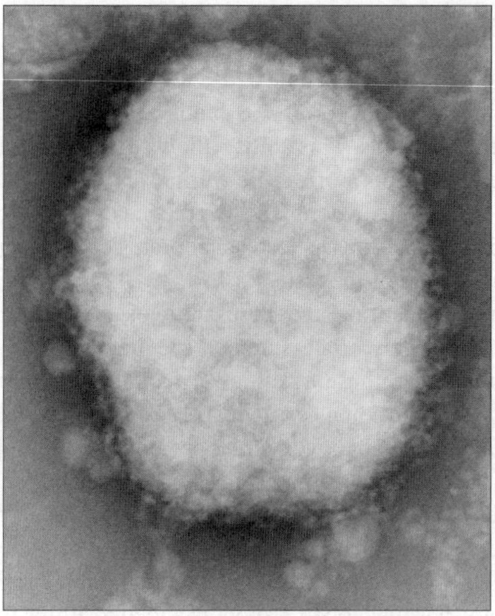

Figure 15.1. Negative stain electron micrograph reveals an "M" (mulberry type) monkeypox virion in human vesicular fluid. Courtesy CDC Public Image Library. For color detail, please see color plate section.

(a) (b)

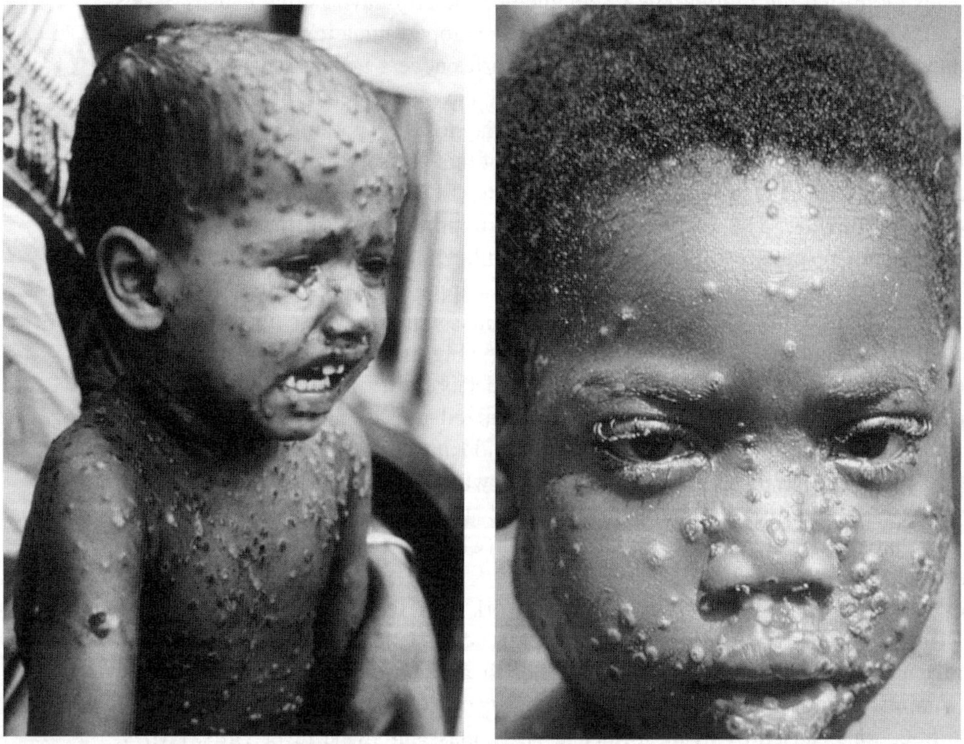

Figure 15.2. Clinical appearance of (**a**) smallpox and (**b**) monkeypox is similar. (**a**) The face of a boy infected with smallpox with facial lesions in various stages of resolution. At this point in time, the patient is still highly contagious. (**b**) The face of a young boy from the DRC exhibiting the characteristic maculopapular cutaneous rash of monkeypox. Courtesy CDC Public Image Library and World Health Organization. For color detail, please see color plate section.

15.6.3 Advances in diagnostics

Innovations in existing technologies (Gubala et al., 2012) as well as new technologies have been used to develop diagnostics for the detection of infections caused by previously unknown, newly emerging, or intentionally released viral agents.

15.6.4 Point-of-care diagnostics

Point-of-care (POC) diagnostics are *in vitro* diagnostic tests that do not involve the use of laboratory staff and facilities to provide the result (Bissonnette and Bergeron, 2010; Gubala et al., 2012; Peeling and Mabey, 2010). The most common format of a POC device is a lateral flow immunoassay that, depending on the test, can detect either an antigen or a specific antibody. Many POC tests have been developed for the identification of viral infections (Clerc and Greub, 2010). Specimens for these tests include blood, saliva, urine, stool, or other bodily fluids. POC tests can be used "near patient" in a hospital, clinic, or doctor's office or administered at home or in the field. POC diagnostic tests provide results rapidly and are generally less expensive than tests performed in the laboratory. For example, POC tests used to screen for infection with HIV have a sensitivity and specificity of 99–100%, which is comparable to that of standard tests (Branson, 2007). In 2009, a newly emerged influenza A/H1N1 virus of swine lineage was detected in humans in the United States and in just over a month infected over 10 000 people in more than 40 countries. In an effort to slow the spread of the virus, countries endeavored to achieve early detection of infected patients and implement quarantine and contact-tracing measures. The fastest diagnostic tool for the detection of influenza viruses was commercially available POC tests that could generate a result in 15 min or less. Hurt et al. (2009) evaluated three commercial rapid POC tests for influenza virus (BinaxNow Influenza A&B (Inverness Medical, Waltham, MA), Quidel QuickVue Influenza A+B Test (Quidel, San Diego, CA), and BD Directigen EZ Flu A+B (BD EZ; Sparks, MD)) for their ability to detect the recently emerged swine lineage A/H1N1 virus. All three of the POC tests detected the swine lineage A/H1N1 at high virus concentrations (10^3–10^5 $TCID_{50}ml^{-1}$ (Tissue Culture Infectious $Dose_{50}$)), with the BD Directigen test having a marginally greater sensitivity than the other two tests. Quantitative comparison using viral infectivity and RNA load data at the detection limit of the tests suggested that both the Quidel and Binax tests were less sensitive for the detection of the swine lineage A/H1N1 viruses than for human seasonal strains. In contrast, the BD Directigen test exhibited similar sensitivity for the swine lineage A/H1N1 and human seasonal viruses. Early diagnosis of infection when viral titers are low can assist in rapid treatment; however, these POC tests are significantly less sensitive than PCR assays, and as such, negative results should be confirmed by a laboratory test.

Innovations to improve the sensitivity of POC tests to detect antigens or antibodies include the use of alternate materials to improve fluid flow rates and decrease nonspecific binding, the use of quantum dots for highly visible labels, and the use of fluorescent labels and FRET for increasing sensitivity (Gubala et al., 2012).

15.6.5 PCR

Real-time PCR has become an effective methodology for viral detection, typing, and subtyping (Le Duc and Franz, 2012); platforms for performing these assays can be laboratory based or portable. For example, Daum et al. (2007) developed a real-time reverse transcriptase PCR assay for type and subtype detection of influenza A and B viruses. The portable, ruggedized instrument with field deployable assays will provide rapid, POC screening needs arising from a pandemic influenza outbreak.

MassTag-PCR, a multiplex PCR technology, is a relatively recent advance that has been used for pathogen discovery, surveillance, outbreak investigation, and epidemiologic investigations. It uses primer pairs that are tagged with molecules of known masses called MassCodes. The chemical bond between the MassCode and the DNA is photolabile and can be cleaved by exposure to UV light. Unlike conventional multiplex PCR, more than 15 primer pairs can be used at the same time. After PCR, products are purified, subjected to UV light, and the released MassCodes detected with a mass spectrometer. MassTag-PCR has been shown to enhance the detection of viral pathogens in clinical specimens where conventional diagnostic tests are negative as well as to identify previously unrecognized viruses (Dominguez et al., 2008).

Electrospray ionization mass spectrometry (a time-of-flight mass spectrometer) has been used in conjunction with PCR or reverse transcriptase PCR using broad-range or universal primers to identify DNA and RNA viruses, respectively. Base composition of the amplicon(s) is determined using a time-of-flight mass spectrometer and the virus identified from the amplicon database. This method has been used to identify vector-borne flaviviruses in ticks and mosquitoes (Grant-Klein et al., 2010). This technology provides a new analytical tool for viral diagnosis and epidemiologic surveillance.

Trade in live animals and animal products has led to the emergence of several zoonotic pathogens, of which RNA viruses are the most common (e.g., SARS). PCR using degenerate primers coupled with high-throughput sequencing was shown to be effective in discovering a new filovirus in an insectivorous bat in Europe (Negredo et al., 2011) as well as in identifying zoonotic viruses associated with illegally imported wildlife products (Smith et al., 2012).

Sequencing machines are becoming faster, smaller, and cheaper—spreading beyond big centers into clinics and small labs (Kupferschmidt, 2011). Unbiased high-throughput sequencing can be used to identify viruses in complex clinical specimens through partial sequences (Nakamura et al., 2009). Future advances in technology have encouraged scientists to envision a global system to share and mine genomic data from microorganisms, which can give better faster answers (Kupferschmidt, 2011).

15.7 PUBLIC HEALTH SURVEILLANCE

15.7.1 Passive surveillance

Historically, public health surveillance has been passive and voluntary. For example, most routine notifiable disease surveillance in the United States is passive. A notifiable disease is one for which regular, frequent, timely information on individual cases is considered to prevent and control the disease (Centers for Disease Control and Prevention, 2011). The list of nationally notifiable diseases is reviewed and modified by the Council of State and Territorial Epidemiologists (CSTE) and CDC once each year. There are currently 36 infectious conditions listed that are associated with viruses (Centers for Disease Control and Prevention, 2012). The healthcare system at the local level is at the core of public health surveillance. Primary care physicians, infection control professionals, laboratories, and emergency departments are in every community in the United States working on a daily basis, so while passive surveillance has known disadvantages, few of the alternatives have the geographic reach of this system. While passive surveillance systems are simple and more affordable to operate, they can dramatically underreport the occurrence of a disease or condition.

15.7.2 Active surveillance

When detecting and responding to an urgent, epidemiologic investigation, routine monitoring efforts can be intensified to meet the needs of a more timely response. Active surveillance systems can be used to validate passive data and assure more complete reporting of disease conditions. While active surveillance may allow for accurate and complete reporting, it is a complex, time-intensive process. For emerging, reemerging, or deliberately emerging diseases, a combination of both active and passive surveillance systems will be utilized to collect health-related information.

15.7.3 Syndromic surveillance

Strategies used for the early detection of a biological incident have been designed around the increasing availability of electronic medical data (Bravata et al., 2004). Syndromic surveillance tools are used to collect, aggregate, and evaluate data from health-related sources in order to identify aberrations (i.e., significant changes) from the expected numbers of pre-identified syndromes. Syndromic surveillance systems can evaluate chief complaints and primary diagnoses from medical facilities and first-responders, prescription, or over-the-counter medication sales; veterinary data; or environmental data. Aberrations in the data can signal, with sufficient probability, the occurrence of an outbreak that warrants further public health investigation.

While syndromic surveillance systems cannot definitively identify the occurrence of a biological incident, they can still provide timely access to health data for immediate analysis and feedback. Syndromic surveillance compliments traditional passive surveillance systems by providing incident-specific information that may assist in defining the scope and progression of the outbreak. There are numerous syndromic surveillance systems available, many of which have not been thoroughly evaluated (Bravata et al., 2002). One concern about syndromic surveillance is its potentially low specificity that would result in false alarms. For example, differentiating between a bioterrorism-related incident and a naturally occurring epidemic may be difficult because early symptoms may not be distinguishable from those of common diseases (Buehler et al., 2003).

15.7.4 Detection of viral infections

Generally, state and local health agencies track relevant environmental, animal, and human health data for specific exposures and diseases, which may be indicative of a biological incident in their jurisdiction. Identification of a natural or intentional biological incident may occur through notification of (i) a probable/confirmed case of a notifiable illness; (ii) a probably/confirmed case of a highly infectious, life-threatening illness; (iii) a death from unknown causes; (iv) an aberration in an illness indicator using a syndromic surveillance system; and (v) a highly credible threat involving a bioterrorism agent. Additional clues are presented in Table 15.4.

Viruses are constantly changing. Many lessons were learned from the 2009 novel influenza A/H1N1 virus, which emerged and caused the first pandemic of the twenty-first century (Blanton et al., 2011). Influenza A and B viruses both undergo gradual but

TABLE 15.4. Epidemiologic Clues That May Signal a Bioterrorist Attack

- Single case of disease caused by an uncommon agent (e.g., smallpox, viral hemorrhagic fever, inhalation or cutaneous anthrax, glanders, FMD) without adequate epidemiologic explanation
- Unusual, atypical, genetically engineered, or antiquated strain of an agent (or antibiotic resistance pattern)
- Higher morbidity and mortality in association with a common disease or syndrome or failure of such patients (or animals or plants) to respond to usual therapy (or treatments)
- Unusual disease presentation (e.g., inhalation anthrax, pneumonic plague)
- Disease with an unusual geographic or seasonal distribution (e.g., plague in a non-endemic area, influenza occurring in the northern hemisphere in the summer)
- Stable endemic disease with an unexplained increase in incidence (e.g., tularemia, plague)
- Atypical disease transition through aerosols, food, or water, in a mode suggesting sabotage (i.e., no other possible physical explanation)
- No illness in persons who are not exposed to common ventilation system (have separate closed ventilation systems) when illness is seen in patients in close proximity who have a common ventilation system
- Several unusual or unexplained diseases coexisting in the same patient (or animal, or plant)
- Unusual illness that affects a large, disparate population (e.g., respiratory disease in a large heterogeneous population may suggest exposure to an inhaled biological agent)
- Illness that is unusual (or atypical) for a given population or age group (e.g., outbreak of measles-like rash in adults)
- Unusual pattern of death or illness among animals (which may be unexplained or attributed to an agent of bioterrorism) that precedes or accompanies illness or death in humans
- Unusual pattern of illness or death in humans that precedes or accompanies illness or death in animals (which may be unexplained or attributed to an agent of bioterrorism)
- Ill persons who seek treatment at about the same time (point source with compressed epidemic curve)
- Similar genetic type among agents isolated from temporally or spatially distinct sources
- Simultaneous clusters of similar illness in noncontiguous areas—domestic or foreign
- Large numbers of cases of unexplained diseases or deaths

Source: Modified from reference Treadwell et al. (2003).

continuous change in proteins (i.e., antigenic drift). Both virologic and disease surveillance are necessary to identify new virus variants, to monitor their health impact in populations, and to provide data necessary for selection of vaccine components each year. Influenza surveillance in the United States consists of five categories of information: (i) viral surveillance by the U.S. WHO collaborating laboratories, National Respiratory and Enteric Virus Surveillance System (NREVSS), and novel influenza A reporting; (ii) outpatient illness surveillance by the U.S. Outpatient Influenza-like Illness Surveillance Network (ILINet); (iii) mortality surveillance by 122 Cities Mortality Reporting System and influenza-associated pediatric mortality reporting; (iv) surveillance of hospitalization data by the Influenza Hospitalization Network (FluSurv-NET) and Aggregate Hospitalization and Death Reporting Activity (AHDRA); and (v) summary of the geographic spread of influenza by the State and Territorial Epidemiologists' reports of influenza activity level. Thus, there are multiple agent-specific surveillance systems for influenza used throughout the United States, which is also true for other viral diseases.

Is it possible that there could be a pandemic that is not detected through currently available surveillance systems? Torque teno virus (TTV) and flavivirus GBV-C are highly prevalent human viral infections that have not yet been associated with disease (Bernardin et al., 2010). If TTV or GBV were to enter humans and spread, through, for example, transfusions, it may not be detected as most surveillance systems rely on the detection of symptoms. It is important to keep in mind that even if a new virus appears harmless, we may need to monitor its potential exposures pathways as, moving through populations quickly, viruses can change through mutation, recombine with other viruses, and, through the mixing of genetic material, create a novel virus.

The majority of pandemics caused by a virus almost always begin with the zoonotic transmission of a virus from an animal to a human (Wolfe et al., 2007). Surveillance for novel diseases like monkeypox in some of the most rural regions in the world highlights the importance of global surveillance systems. Study authors, through the use of surveillance data, have indicated that human monkeypox incidence has dramatically increased in rural Democratic Republic of the Congo (DRC) (Rimoin et al., 2010). Mass smallpox vaccination campaigns ended in this region in 1977 after the global eradication of the disease, so there is now limited immunity in the population to monkeypox. Comparison of active surveillance data in the same health zone from 1981 to 1986 (0.72 per 10 000) and 2006–2007 (14.42 per 10 000) suggests a 20-fold increase in human monkeypox incidence. Between 1986 and 2006, there was no consistent surveillance program in place. Authors note that given the limited resources, it was not possible for their surveillance team to reach all reported cases because of their remote locations and inconsistent access to healthcare providers. The increase in cases offers additional opportunities for a unique virus to emerge, which is why surveillance of this disease is so important. Of additional concern, in 2003, the first report of human monkeypox outside of the African continent occurred in the midwestern United States and was associated with direct contact with ill prairie dogs that were being kept or sold as pets (Reed et al., 2004). There were a total of 93 human cases of monkeypox in six midwestern states and New Jersey associated with this single event. This highlights the gaps in surveillance capabilities in developing countries and also demonstrates this as a global health concern.

15.8 CONCLUSION

For the terrorist, the use of a viral agent would pose a challenge due to problems associated with acquisition, cultivation, and dissemination. The target for an attack with a viral agent can range from humans to animals and plants. Therefore, agricultural targets are also a major concern. Nature has provided many challenges to combating viral diseases. Viral agents are much more prone to genetic variation and mutation and can be manipulated or created in the laboratory to take on desired characteristics. Differentiating between natural and intentional viral disease outbreaks relies on surveillance systems and epidemiologic investigations and can be challenging. Unlike bacterial diseases, many of which are treatable, there are fewer medical countermeasures to employ when dealing with viral infections. Laboratory diagnostic methods and reagents must continuously be refined to account for genetic changes and variants. Fortunately, advances in technology will facilitate the detection of known and novel viruses.

REFERENCES

Bernardin, F., E. Operskalski, M. Busch, et al. 2010. Transfusion transmission of highly prevalent human viruses. *Transfusion* **50**(11):2474–2483.

Bissonnette, L., M.G. Bergeron. 2010. Diagnosing infections – current and anticipated technologies for point-of-care diagnostics and home-based testing. *Clinical Microbiology and Infectious Diseases* **16**:1044–1053.

Blanton, L., L. Brammer, L. Finelli, et al. 2011. Influenza. In: *Manual for the Surveillance of Vaccine-Preventable Diseases*, 5th edition. Atlanta: Centers for Disease Control and Prevention. www.cdc.gov/vaccines/pubs/surv-manual/chpt06-influenza.html (accessed on March 6, 2012).

Branson, B.M. 2007. State of the art for diagnosis of HIV infection. *Clinical Infectious Diseases* **45**:S221–S225.

Bravata, D.M., K. McDonald, D.K. Owens, et al. 2002. Bioterrorism preparedness and response: use of information technologies and decision support systems. Evidence Reports/Technology Assessment No. 59. Rockville: Agency for Healthcare Research and Quality. www.ahrq.gov/clinic/bioitinv.htm (accessed on March 6, 2012).

Bravata, D.M., K.M. McDonald, W.M. Smith, et al. 2004. Systematic review: surveillance systems for early detection of bioterrorism-related diseases. *Annals of Internal Medicine* **140**:910–922.

Bray, M. 2003. Defense against filoviruses used as biological weapons. *Antiviral Research* **57**:53–60.

Buehler, J.W., R.L. Berkelman, D.M. Hartley, et al. 2003. Syndromic surveillance and bioterrorism-related epidemics. *Emerging Infectious Diseases* **9**(10):1197–1204.

Bush, L.M., B.H. Abrams, A. Beall, et al. 2001. Index case of fatal inhalational anthrax due to bioterrorism in the United States. *New England Journal of Medicine* **345**:1607–1610.

Carus, W.S. 2002. *Bioterrorism and Biocrimes. The Illicit Use of Biological Agents Since 1900.* Amsterdam: Fredonia Books.

Casadevall, A., L. Pirofsky. 2004. The weapon potential of a microbe. *Trends in Microbiology* **12**:259–263.

Cello, J., A.V. Paul, E. Wimmer. 2002. Chemical synthesis of poliovirus cDNA: generation of infectious virus in the absence of natural template. *Science* **297**:1016–1018.

Centers for Disease Control and Prevention. 2003a. Multistate outbreak of monkeypox – Illinois, Indiana, and Wisconsin, 2003. *Morbidity and Mortality Weekly Report* **52**:537–540.

Centers for Disease Control and Prevention. 2003b. Update: multistate outbreak of monkeypox—Illinois, Indiana, Kansas, Missouri, Ohio, and Wisconsin, 2003. *Morbidity and Mortality Weekly Report* **52**:561–564.

Centers for Disease Control and Prevention. 2003c. Update: multistate outbreak of monkeypox—Illinois, Indiana, Kansas, Missouri, Ohio, and Wisconsin, 2003. *Morbidity and Mortality Weekly Report* **52**:589–590.

Centers for Disease Control and Prevention. 2003d. Update: multistate outbreak of monkeypox—Illinois, Indiana, Kansas, Missouri, Ohio, and Wisconsin, 2003. *Morbidity and Mortality Weekly Report* **52**:616–618.

Centers for Disease Control and Prevention. 2003e. Update: multistate outbreak of monkeypox—Illinois, Indiana, Kansas, Missouri, Ohio, and Wisconsin, 2003. *Morbidity and Mortality Weekly Report* **52**:642–646.

Centers for Disease Control and Prevention. 2011. *National Notifiable Diseases Surveillance System.* http://wwwn.cdc.gov/nndss

Centers for Disease Control and Prevention. 2012. 2012 *Case Definitions: National Notifiable Diseases and Conditions Infectious and Non-infectious Case*. Atlanta, GA: CDC.

Charrel, R.N., X. de Lamballerie. 2003. Arena viruses other than Lassa virus. *Antiviral Research* **57**:89–100.

Christopher, G.W., T.J. Cieslak, J.A. Pavlin, et al. 1999. Biological warfare: a historical perspective. In: J. Lederberg (ed.), *Biological Weapons: Limiting the Threat*, pp. 17–35. Cambridge: MIT Press.

Clerc, O., G. Greub. 2010. Routine use of point-of-care tests: usefulness and application in clinical microbiology. *Clinical Microbiology and Infectious Diseases* **16**:1054–1061.

Cleri, D.J., A.J. Ricketti, R.B. Porwancher, et al. 2006. Viral hemorrhagic fevers: current status of endemic disease and strategies for control. *Infectious Disease Clinics of North America* **20**:359–393.

Crandall, K. 1995. Interspecific phylogenetics: support for dental transmission of human immunodeficiency virus. *Journal of Virology* **69**:2351–2356.

Crosby, A. 1989. *America's Forgotten Pandemic: The Influenza of 1918*. New York: Cambridge University Press.

Daum, L.T., L.C. Canas, B.P. Arulanandam, et al. 2007. Real-time RT-PCR assays for type and subtype detection of influenza A and B viruses. *Influenza and Other Respiratory Viruses* **1**:167–175.

Dominguez, S.R., T. Briese, G. Palacios, et al. 2008. Multiplex Mass Tag-PCR for respiratory pathogens in pediatric nasopharyngeal washes negative by conventional diagnostic testing shows a high prevalence of viruses belonging to a newly recognized picornavirus clade. *Journal of Clinical Virology* **43**:219–222.

Editorial. 2012. Why genies don't go back into bottles. *Nature Biotechnology* **30**:117.

Esrink, M. 2002. Did bioweapons test cause a deadly smallpox outbreak? *Science* **296**:2116–2117.

Falkenrath, R.A., R.D. Newman, B.A. Thayer. 1998. *America's Achilles' Heel. Nuclear, Biological, and Chemical Terrorism and Covert Attack*. Cambridge: MIT Press.

Geisbert, T.W, P.B. Jahrling. 2004. Exotic emerging viral diseases: progress and challenges. *Nature Medicine* **10**:S110–S121.

Grant-Klein, R.J., C.D. Baldwin, M.J. Turell, et al. 2010. Rapid identification of vector-borne flaviviruses by mass spectrometry. *Molecular and Cellular Probes* **24**:219–228.

Gubala, V, L.F. Harris, A.J. Ricco, et al. 2012. Point of care diagnostics: status and future. *Analytical Chemistry* **84**:487–515.

Harmon, R. 2005. Admissibility standards for scientific evidence. In: R.G. Breeze, B. Budowle, S.E. Schutzer (eds.), *Microbial Forensics*, pp. 381–392. San Diego: Elsevier Academic Press.

Harper, G.J. 1961. Airborne microorganisms. *Journal of Hygiene* **59**:479–486.

Henderson, D.A., T.V. Inglesby, J.G. Gartlett, et al. 1999. Smallpox as a biological weapon. *Journal of the American Medical Association* **281**:2127–2137.

Hopkins, D.R. 1983. *Princes and Peasants. Smallpox in History*. Chicago: University of Chicago Press.

Hurt, A.C., C. Bass, Y.-M. Deng, et al. 2009. Performance of influenza rapid point-of-care tests in the detection of swine lineage A (H1N1) influenza viruses. *Influenza and Other Respiratory Viruses* **3**:171–176.

Jackson, R.J., A.J. Ramsay, C.D. Christensen, et al. 2001. Expression of mouse interleukin-4 by a recombinant ectromelia virus suppresses cytolytic lymphocyte responses and overcomes genetic resistance to mousepox. *Journal of Virology* **75**:1205–1210.

Jernigan, J.A., D.S. Stephens, D.A. Ashford, et al. 2001. Bioterrorism-related inhalational anthrax: the first 10 cases reported in the United States. *Emerging Infectious Diseases* **7**:933–944.

Jones, P.D. 1991. HIV transmission by stabbing despite zidovudine prophylaxis. *Lancet* **338**:884.

Kaplan, D.E., A. Marshall. 1996. *The Cult at the End of the World*. New York: Crown Publishers.

Kupferschmidt, K. 2011. Outbreak detectives embrace the genome era. *Science* **333**:1818–1819.

Lanciotti, R.S., J.T. Roehrig, V. Duebel, et al. 1999. Origin of the West Nile virus responsible for an outbreak of encephalitis in the northeastern United States. *Science* **286**:2333–2337.

Le Duc, J.W., D.R. Franz. 2012. Genetically engineered transmissible influenza A/H5N1: a call for laboratory safety and security. *Biosecurity and Bioterrorism: Biodefense Strategy, Practice and Science* **10**:1–2.

Lederberg, J., R.E. Shope, S.C. Oaks, Jr (eds). 1992. *Emerging Infections, Microbial Threats to Health in the United States*. Washington, DC: National Academy Press.

Mackenzie, J.S., D.J. Gubler, L.R. Petersen. 2004. Emerging flaviviruses: the spread and resurgence of Japanese encephalitis, West Nile and dengue viruses. *Nature Medicine* **10**:S98–S109.

Martin, M.J., F. Gonzales-Candelas, F. Sobrino, et al. 2001. It has been a difficult year for the British Ministry of Agriculture. *Nature Immunology* **2**:565.

Metzker, M.L., D.P. Mindell, X. Liu, et al. 2002. Molecular evidence of HIV-1 transmission in a criminal case. *Proceedings of the National Academy of Sciences of the United States of America* **99**:14292–14297.

Meyer, R., S.A. Morse. 2008. Viruses and bioterrorism. In: B.W.J. Mahy, M.H.V. Van Regenmortel (eds.), *Encyclopedia of Virology*, 3rd edition, Vol. **5**, pp. 406–411. San Diego: Academic Press.

Morens, D.M., G.K. Folkers, A.S. Fauci. 2004. The challenge of emerging and re-emerging infectious diseases. *Nature* **430**:242–249.

Morse, S.S. 1995. Factors in the emergence of infectious diseases. *Emerging Infectious Diseases* **1**:7–15.

Morse, S.A. 2006. Historical perspectives of microbial bioterrorism. In: H. Friedman, B. Anderson, M. Bendinetti (eds.), *Microorganisms and Bioterrorism*, pp. 15–29. New York: Springer.

Morse, S.A., A.S. Khan. 2005. Epidemiologic investigation for public health, biodefense, and forensic microbiology. In: R.G. Breeze, B. Budowle, S.E. Schutzer (eds.), *Microbial Forensics*, pp. 157–171. San Diego: Elsevier Academic Press.

Morse, S.A., E. Weirich. 2011. Select agent regulations. In: B. Budowle, S.E. Schutzer, R.G. Breeze, et al. (eds.), *Microbial Forensics*, 2nd edition, pp. 199–220. San Diego: Elsevier Academic Press.

Morse, S.A., R.B. Kellogg, S. Perry, et al. 2003. Detecting biothreat agents: the Laboratory Response Network. *ASM News* **69**:433–437.

Murray, K., B. Eaton, P. Hooper, et al. 1998. Flying foxes, horses, and humans: a zoonosis caused by a new member of the *Paramyxoviridae*. In: W.M. Scheld, D. Armstrong, J.M. Hughes (eds.), *Emerging Infections*, Vol. **1**, pp. 43–58. Washington, DC: ASM Press.

Nakamura, S., C.-S. Yang, N. Sakon, et al. 2009. Direct metagenomic detection of viral pathogens in nasal and fecal specimens using an unbiased high-throughput sequencing approach. *PLoS One* **4**(1):e4219.

Negredo, A., G. Palacios, S. Vasquez-Moron, et al. 2011. Discovery of an ebolavirus-like filovirus in Europe. *PLoS Pathogens* **7**(10):e1002304.

Olsen, K.B. 1999. Aum Shinrikyo: once and future threat? *Emerging Infectious Diseases* **5**:513–516.

Ou, C.Y., C.A. Ciesielski, G. Myers, et al. 1992. Molecular epidemiology of HIV transmission in a dental practice. *Science* **256**:1165–1171.

Peeling, R.W., D. Mabey. 2010. Point-of-care tests for diagnosing infections in the developing world. *Clinical Microbiology and Infectious Diseases* **16**:1062–1069.

Peiris, J.S.M., Y. Guan, K.Y. Yuen. 2004. Severe acute respiratory syndrome. *Nature Medicine* **10**:S88–S97.

Reed, K.D., J.W. Melski, M.B. Graham, et al. 2004. The detection of monkeypox in humans in the Western Hemisphere. *New England Journal of Medicine* **350**:341–350.

Reid, A.H., T.G. Fanning, J.V. Hultin, et al. 1999. Origin and evolution of the "Spanish" influenza virus hemagglutinin gene. *Proceedings of the National Academy of Sciences of the United States of America* **96**:1651–1656.

Reid, A.H., T.G. Fanning, T.A. Janczewski, et al. 2000. Characterization of the 1918 "Spanish" influenza virus neuraminidase gene. *Proceedings of the National Academy of Sciences of the United States of America* **97**:6785–6790.

Rimoin, A.W., P.M. Mulembakani, S.C. Johnston, et al. 2010. Major incidence in human monkeypox incidence 30 years after smallpox vaccination campaigns cease in the Democratic Republic of Congo. *Proceedings of the National Academy of Sciences of the United States of America* **107**(37):16262–16267.

Rota, P.A., M.S. Oberste, S.S. Monroe, et al. 2003. Characterization of a novel coronavirus associated with severe acute respiratory syndrome. *Science* **300**:1399–1404.

Rotz, L.D., J.M. Hughes. 2004. Advances in detecting and responding to threats from bioterrorism and emerging infectious diseases. *Nature Medicine* **10**:S130–S136.

Rotz, L.D., A.S. Khan, S.R. Lillibridge, et al. 2002. Public health assessment of potential biological terrorism agents. *Emerging Infectious Diseases* **8**:225–229.

Sidwell, R.W., D.F. Smee. 2003. Viruses of the Bunya- and Togaviridae families: potential as bioterrorism agents and means of control. *Antiviral Research* **57**:101–111.

Smith, J.F., K. Davis, M.K. Hart, et al. 1997. Viral encephalitides. In: F.R. Sidell, E.T. Takafuji, D.R. Franz (eds.), *Medical Aspects of Chemical and Biological Warfare, Textbook of Military Medicine*, pp. 561–589. Washington, DC: Office of the Surgeon General, TMM Publications.

Smith, K.M., S.J. Anthony, W.M. Switzer, et al. 2012. Zoonotic viruses associated with illegally imported wildlife products. *PLoS One* **7**(1):e29505.

Taubenberger, J.K., A.H. Reid, A.E. Krafft, et al. 1997. Initial characterization of the 1918 "Spanish" influenza virus. *Science* **275**:1793–1796.

Taylor, L.H., S.M. Latham, M.E.J. Woolhouse. 2001. Risk factors for human disease emergence. *Philosophical Transactions of the Royal Society London B* **356**:983–989.

Thompson, D., P. Muriel, D. Russel, et al. 2002. Economic costs of the foot and mouth disease outbreak in the United Kingdom in 2001. *Review of Science and Technology* **21**:675–687.

Treadwell, T.A., D. Koo, K. Kuker, et al. 2003. Epidemiologic clues to bioterrorism. *Public Health Reports* **118**:92–98.

Tumpey, T.M., C.F. Basler, P.V. Aguilar, et al. 2005. Characterization of the reconstructed 1918 Spanish influenza pandemic virus. *Science* **310**:77–80.

Veenstra, J., R. Schurrman, M. Cornelissen, et al. 1995. Transmission of zidovudine-resistant human immunodeficiency virus type 1 variants following deliberate injection of blood from a patient with AIDS: characteristics and natural history of the virus. *Clinical Infectious Diseases* **21**:556–560.

Webby, R., E. Hoffmann, R. Webster. 2004. Molecular constraints to interspecies transmission of viral pathogens. *Nature Medicine* **10**:S77–S81.

Weiss, R.A., A.J. McMichael. 2004. Social and environmental risk factors in the emergence of infectious diseases. *Nature Medicine* **10**:S70–S76.

Wolfe, N.D., C.P. Dunavan, J. Diamond. 2007. Origins of major infectious diseases. *Nature* **447**:279–283.

World Health Organization. 1970. *Health Aspects of Chemical and Biological Weapons*. Geneva: WHO Press.

World Health Organization. 2004. The world health report 2004 – changing history. http://www.who.int/whr/2004/en/report04_en.pdf (accessed on February 17, 2012).

World Health Organization. 2010. *World Health Statistics*. Geneva: WHO Press.

World Health Organization. 2013. Influenza at the human-animal interface. Human infections with H5N1. http://www.who.int/influenza/human_animal_interface/Influenza_Summary_IRA_HA_int (accessed on August 28, 2013).

MOLECULAR AND EVOLUTIONARY MECHANISMS OF VIRAL EMERGENCE

Juan Carlos Saiz

Departamento de Biotecnología, INIA, Ctra. Coruña, Madrid, Spain

Francisco Sobrino

Centro de Biología Molecular "Severo Ochoa" (CSIC-UAM), Consejo Superior de Investigaciones Científicas (CSIC), Campus de Cantoblanco, Madrid, Spain

Noemí Sevilla and Verónica Martín

Centro de Investigación en Sanidad Animal, Instituto Nacional de Investigación Agraria y Alimentaria, Valdeolmos, Madrid, Spain

Celia Perales and Esteban Domingo

Centro de Biología Molecular "Severo Ochoa" (CSIC-UAM), Consejo Superior de Investigaciones Científicas (CSIC), Campus de Cantoblanco, Madrid, Spain
Centro de Investigación Biomédica en Red de Enfermedades Hepáticas y Digestivas (CIBERehd), Barcelona, Spain

TABLE OF CONTENTS

16.1	Introduction: Biosphere and virosphere diversities	298
16.2	Virus variation as a factor in viral emergence: A role of complexity	299
16.3	High error rates originate quasispecies swarms	300
16.4	Evolutionary mechanisms that may participate in viral disease emergence	302
16.5	Ample genetic and host range variations of FMDV: A human epidemic to be?	304

Viral Infections and Global Change, First Edition. Edited by Sunit K. Singh.
© 2014 John Wiley & Sons, Inc. Published 2014 by John Wiley & Sons, Inc.

16.1 INTRODUCTION: BIOSPHERE AND VIROSPHERE DIVERSITIES

Genome analyses have confirmed an immense genetic diversity in natural populations of viruses, in particular the RNA viruses. Viral diversity is but a manifestation of the remarkable complexity of the biological world whose robustness in the face of climatic change remains an open question. Understanding the origins and maintenance of biological diversity stands as a key scientific question that captivates theoreticians and experimentalists alike. Viruses are extremely attractive biological systems because, despite being totally dependent on cells for replication, they are endowed with a genetic program that can be expressed in short times (minutes to hours) and can potentially produce a large amount of infectious progeny. Compared to differentiated organisms, viruses can undergo evolutionary events in an accelerated manner.

The International Committee on Taxonomy of Viruses (ICTV) tries to cope with diversity and newly discovered viruses with periodic updates of virus classification (King et al., 2011). A clear division among viruses—which has found no exceptions so far—is between those that use DNA and those that use RNA as genetic material. This is an important difference that affects virus biology in ways that are highly relevant to viral disease emergence, the focus of this chapter. All present-day cellular genomes known have DNA genomes, although according to some theories an RNA-based cellular world existed in early phases of the evolution of cellular life on Earth (reviewed in Forterre, 2010). Be descendants of an early RNA world or not, about 70% of the known viruses in our biosphere, referred to as the "virosphere," are RNA viruses. Together with subviral replicating entities, such as the plant viroids, and some defective, virus-dependent elements, such as the animal hepatitis delta agent, they are unique in using RNA as the repository of inheritable information. A number of theoretical and experimental considerations suggest that evolution toward more complex genomes (those encoding more information) required a transition from RNA to DNA, in part due to the higher physical resistance of DNA to environmental aggressions, particularly hydrolytic reactions. This line of thinking is consistent with the fact that all present-day cells that have been characterized contain a DNA genome, albeit of widely different complexity: 4.7 million (or mega) base pairs (Mbp) for the bacterium *Escherichia coli*, 33 Mbp for the parasites of the genus *Leishmania*, 170 Mbp for the fly *Drosophila*, and 3200 Mbp for man. The same is true of viruses. The more complex viruses have a DNA genome: up to 0.4 Mbp for the important pathogens poxviruses and herpesviruses and 1.2 Mbp for the amoeba mimiviruses. In sharp contrast, the RNA viruses characterized to date have genomes of 3000–33 000 nucleotides, the largest being the coronaviruses.

The evolution of genome size paralleled an evolution of the machinery involved in replication of the corresponding genomes. Cells evolved at least two distinct and complementary mechanisms for correction of mistakes (incorporation of incorrect bases) during genome replication: a proofreading-repair 3′–5′ exonuclease activity and a number of post-replicative DNA repair pathways (Friedberg et al., 2006). The 3′–5′ exonuclease activity can excise an incorrect nucleotide incorporated at the 3′-end of the growing chain of DNA. This activity is present in a distinct protein domain of replicative cellular DNA-dependent DNA polymerases (Bernad et al., 1989), but it is absent in most viral RNA-dependent RNA polymerases and RNA-dependent DNA polymerases (or reverse transcriptases). The known post-replicative repair pathways that are active on DNA are inactive (or very inefficient) on RNA. In consequence, the RNA viruses show several orders of magnitude higher mutation rates (meaning the rate of incorporation of incorrect nucleotides during RNA synthesis) than their host cells (Batschelet et al., 1976; Drake and Holland, 1999; Sanjuan et al., 2010). Thus, an increase of genome complexity necessitated recruiting resources to preserve the integrity of the genetic material. It required a trade-off between copying fidelity and biochemical expenditures. It has been estimated that about 130 human genes are directly or indirectly involved in DNA repair, emphasizing the importance of maintaining the coding accuracy of the cellular genetic material. Consistently, the only RNA viruses known to encode a proofreading-repair activity are the coronaviruses (Denison et al., 2011; Eckerle et al., 2007, 2010), the largest among the RNA viruses. Smaller Nidovirales (the family to which the coronaviruses belong) do not have an equivalent repair activity. Thus, a number of theoretical and experimental studies converge to indicate that only the small RNA viruses can successfully replicate and persist in nature with an error-prone replication.

Several lines of evidence suggest that high mutation rates are not an accidental (unselected) feature of RNA genomes. Rather, error-prone replication confers a selective advantage to RNA viruses because it favors their adaptability. This was elegantly shown using a mutant of poliovirus with an amino acid substitution in its polymerase that increased about fivefold its template copying fidelity, relative to the wild-type virus. This mutant was less pathogenic for some strains of mice than the corresponding wild-type virus. The difference was due to the lower error rate, and ensuing lower virus population diversity, that impeded adaptation to the complex environment of the animal, including replication in the brain (Pfeiffer and Kirkegaard, 2005; Vignuzzi and Andino, 2010; Vignuzzi et al., 2006). These observations *in vivo* must be added to many others during clinical practice, or obtained in experiments designed in cell culture, that indicate the presence of many classes of escape mutants in viral populations (Domingo et al., 2012). The presence of escape mutants to allow a virus population to replicate despite the presence of selective constraints is a consequence of high mutation rates, again reinforcing the concept that error-prone replication, at least within the range of error rates observed in nature, confers a selective advantage to viruses. High adaptability associated with the standard error rates operating during RNA virus replication contributes to the frequent emergence of new RNA viruses as agents of human disease (Domingo, 2010), as discussed in the next sections.

16.2 VIRUS VARIATION AS A FACTOR IN VIRAL EMERGENCE: A ROLE OF COMPLEXITY

The emergence of new viral diseases is an important threat for public health. This is now amply recognized and has been dramatically demonstrated by the AIDS pandemic that rates as one of the major health problems worldwide. From all indications, unless new paradigms in vaccinology or chemotherapy are discovered, human immunodeficiency virus type 1 (HIV-1) infections are bound to continue for decades. However, in the words of Howard Temin, HIV-1 is "not unique but merely different," when compared with other variable viruses (Temin, 1989). Viral emergences in the human population occur at a rate of about one per year, involving different types of viral pathogens, frequently of a zoonotic origin (Calisher et al., 2006; Domingo, 2010; Krauss et al., 2003; Morse, 1994a; Nakamura et al., 2009; Negredo et al., 2011; Peters, 2007; Woo et al., 2006) (see other chapters of this volume).

In a report of the U.S. Academy of Medicine, a total of 13 major influences of a very different nature were identified as being involved in infectious disease emergence (Smolinski et al., 2003). One of the factors in emergences—which applies not only to viruses but also to cellular prokaryotic and eukaryotic pathogens—is microbial adaptability. Other influences relate to environmental changes that affect the traffic of viruses, of their vectors, and of potential new host species. Yet another group of influences consist of sociopolitical circumstances (i.e., malnutrition, poor hygienic conditions, social unrest, war) that can enhance human susceptibility to disease. Very frequently no single factor can explain by itself the emergence or reemergence of a viral pathogen. When a final outcome is the result of many individual influences, but none of the influences individually nor the sum of them can explain the outcome, we enter the domain of the sciences of complexity. A viral emergence can be regarded as a change in the state of a virus which is highly unpredictable. The words of Herbert A. Simon on adaptive systems are relevant: "Feedback mechanisms,…, by continually responding to discrepancies between a system's actual and desired states, adapt it to long-range fluctuations in the environment without forecasting" (Simon, 1996, p. 149). Or, closer to the situation we are dealing with, viral emergence can be viewed as complexity in the sense of the definition proposed by Solé and Goodwin: "the study of those systems in which there is no simple and predictable relationship between levels, between the properties of parts and of wholes" (Solé and Goodwin, 2000). The connection of viral emergences with complexity was suggested in a previous article by one of us (Domingo, 2010). Here we elaborate it further and expand in the light of recent developments, in particular evidence of intrapopulation complexity derived from the application of ultra-deep sequencing methodologies. A justification of how the mutual influences between virus genetics and socio-environmental factors lead to different levels of complexity requires that we examine, in turn, these different features.

16.3 HIGH ERROR RATES ORIGINATE QUASISPECIES SWARMS

Mutation and recombination stand as the more widespread mechanisms of virus genome variation. In the case of RNA viruses, these mechanisms can be further subdivided in standard mutagenesis, hypermutational events triggered by cellular editing activities, replicative and nonreplicative recombination, genome segment reassortment (in the case of viruses with segmented genomes such as the influenza viruses), some example of internal deletions that give rise to viruses that infect and kill cells by complementation, and lateral gene transfers (Figure 16.1) (see Domingo et al., 2008; Ehrenfeld et al., 2010 for review). Due to absence of error-correcting activities, the mutation rates during RNA virus replication lie in the range of 10^{-3}–10^{-5} misincorporations per nucleotide copied (Batschelet et al., 1976; Drake and Holland, 1999; Sanjuan et al., 2010). These error rates imply that, on average of 0.1–10, mutations will occur upon replication of a 10 000-residue RNA genome. Expressed in other terms, even in a single replicative unit within an infected cell (i.e., the replication complex at a membrane site), it is not possible to maintain unaltered the nucleotide sequence of a replicating RNA genome. The mutant distributions generated during RNA genome replication are termed viral quasispecies, and their appearance is schematically depicted in Figure 16.1b. Quasispecies was first developed theoretically as a mathematical

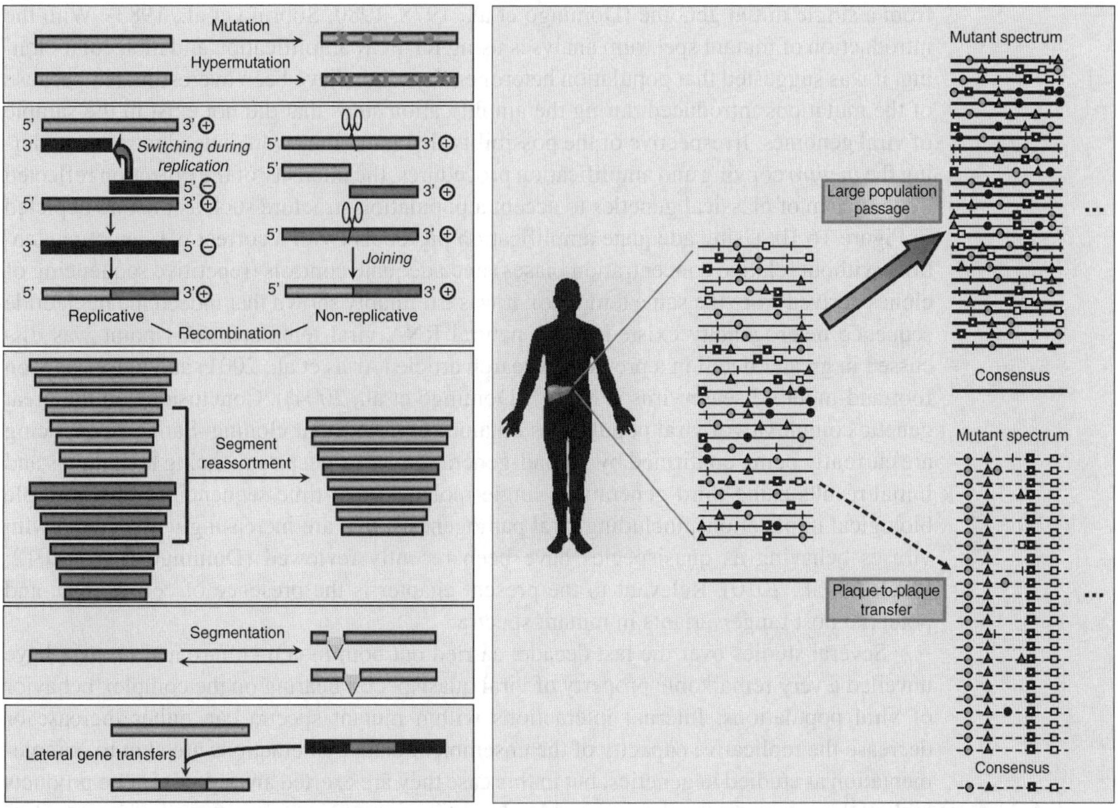

Figure 16.1. Different types of genetic variation of viruses. (**a**) Mutation, hypermutation (associated with cellular editing factors), molecular recombination (replicative and nonreplicative), reassortment in segmented genomes, genome segmentation (as observed in FMDV (Ojosnegros et al., 2011)), and lateral gene transfers (such as in some virulent forms of influenza virus (Khatchikian et al., 1989)) are examples of genetic variations that can potentially contribute to viral disease emergence. (**b**) High mutation rates are a general feature of RNA viruses that have as a consequence the generation of complex mutant spectra (swarms or clouds) termed viral quasispecies. Evolution is exquisitely dependent on population size, with large population passages leading to fitness gain in a given environment and plaque-to-plaque transfers leading to mutation accumulation in the consensus sequence and fitness decrease (see text for biological implications and literature references). For color detail, please see color plate section.

formulation to describe self-organization and adaptability of primitive replicons that might have populated the Earth at an early phase of life development (Eigen, 1992; Eigen and Schuster, 1979).

When viruses were shown to follow the mutant generation and competition dynamics implicit in quasispecies theory, it became clear that they were endowed with a unique mechanism for adaptation and long-term survival. This unique mechanism can be described as a realm of ever-changing, heterogeneous viral subpopulations, which give up a precise identity for a better chance of survival (Domingo et al., 1978, 2001, 2012). Indeed, mutant spectra (Figure 16.1b) constitute reservoirs of genetic and phenotypic variants, ready to be selected (i.e., to increase in frequency) in response to an environmental demand. Early analyses of mutant spectra based on sampling of biological clones that did not require any *in vitro* genome copying or amplification (processes that could be a source of artifactual mutations) indicated the extensive heterogeneity of viral populations, even those grown from a single initial genome (Domingo et al., 1978, 1980; Sobrino et al., 1983). With the introduction of mutant spectrum analyses using RT-PCR amplification and molecular cloning, it was suggested that population heterogeneity might have been overestimated because of the mutations introduced during the amplification steps that did not exist in the sample of viral genomes. Irrespective of the possibility that mutations were indeed introduced during the *in vitro* copying and amplification procedures, the intensity of the objection reflected a skepticism of classical genetics to accept a population structure such as the one depicted in Figure 16.1b. Using adequate amplification protocols (with a correct pH, intact nucleotides without relative concentration biases) and adequate controls (repetitive sequencing of clones derived from the same template), it was irrefutably shown that indeed the nucleotide sequence heterogeneity existed in the natural RNA viral templates (this point was discussed in greater detail in a previous research article (Arias et al., 2001) and in a review on foot-and-mouth disease virus (FMDV) (Domingo et al., 2004)). Conclusions on the great genetic complexity of viral populations obtained by molecular cloning–Sanger sequencing are currently being confirmed by second-generation, ultra-deep sequencing techniques and initial results using third-generation, single-molecule real-time sequencing. The multiple biological implications, including viral pathogenesis, that are increasingly associated with viruses behaving as quasispecies have been recently reviewed (Domingo et al., 2012; Perales et al., 2010). Relevant to the present chapter is the presence of cell tropism and potential host range variants in mutant spectra.

Several studies over the last decade, carried out both in cell culture and *in vivo*, have unveiled a very remarkable property of viral quasispecies bearing on the complex behavior of viral populations. Internal interactions within mutant spectra can either increase or decrease the replicative capacity of the ensemble. Positive interactions are akin to complementation as studied in genetics, but in this case they are exerted among viral gene products within the same replicative ensemble. Negative interactions become prominent when defective, but RNA replication-competent genomes increase in frequency to the point that replication of the entire population is compromised (Domingo et al., 2012; Grande-Pérez et al., 2005). Such internal interactions that can shift from being positive to being negative with relative ease render an ensemble with a biological behavior, which is unpredictable from the traits of the individual components, themselves rather indeterminate. It is not surprising that when such a volatile genetic structure meets with environmental unknowns, the realm of complexity is fully manifested (Domingo, 2010; Solé and Goodwin, 2000).

In the emergence of viral pathogens, three steps have been distinguished: introduction, establishment, and dissemination or propagation (Domingo, 2010; Morse, 1994b; Peters, 2007; Pulliam, 2008). In addition to the molecular events that lead to quasispecies swarms, several evolutionary mechanisms can affect each of the three steps involved in an emergence.

16.4 EVOLUTIONARY MECHANISMS THAT MAY PARTICIPATE IN VIRAL DISEASE EMERGENCE

A successful establishment and dissemination of an infection may be influenced by a number of molecular and evolutionary mechanisms that affect the virus, once environmental factors have prompted the initial encounter between a virus and a potential new host. These mechanisms (Figure 16.1 and Table 16.1) are of a general nature, and here we summarize those evolutionary features that we consider particularly relevant for a viral

TABLE 16.1. Mechanisms That May Be Involved in a Viral Emergence

Molecular
Genetic modifications of the virus[a]
Mutation
Hypermutation
Recombination
Genome segment reassortment
Lateral gene transfers

Evolutionary
Positive selection (i.e., triggered by receptor availability)
Coevolution of antigenicity and cell receptor specificity
Rate of evolution and capacity to explore sequence space
Bottleneck events

[a] See also Figure 16.1.

emergence. Positive selection is the process by which a viral genotype becomes dominant in a population, as a result of evaluation of the phenotype that it expresses. Particularly relevant for disease emergence are the cases in which a minority genome in a viral population expresses a surface amino acid sequence that has the capacity to recognize a receptor in a cell type belonging to an organism that is not the standard host for the virus. A critical issue is that one or few amino acid substitutions may be often sufficient to permit a modification of host cell tropism (several examples have been previously reviewed (Baranowski et al., 2003; Domingo et al., 2008, 2012)).

Positive selection events that tend to render a viral population increasingly competent to replicate in a new organism—through the replacement of some viral subpopulations by others—are critical for the process of establishment of the infection. Dissemination may occur as a result of a general increase of viral load, or it may require additional, transmission-favoring mutations. In the latter case, statistical reasons will increase the probability of occurrence of the transmission-associated mutations when the replicative load is high.

Receptor-recognition sites often overlap with antigenic sites on the virion surface. This implies that an antigenic change in a virus may have as a consequence a potential change in receptor specificity (Baranowski et al., 2003; Domingo et al., 2012 and references therein). Mutual effects in antigenic structure and receptor recognition fall within the general concept of coevolution, which implies mutual interactions among biological entities that exert evolutionary influences on one another (Woolhouse et al., 2002). In this case, coevolution is due to residues on the virion surface that participate in two important traits subjected to different selective pressures. There is now some concern that the extensive use of some antiviral vaccines may act as a new selective agent to favor a gradual drift of sequences that may enhance the chances of cell tropism and host range modifications (Domingo et al., 2012).

The emphasis on genetic variation as a factor of viral emergence is often put at the intra-host level. However, long-term evolution of virus, the one that results from multiple, successive host-to-host transmissions (with multiple quasispecies swarms in each of the infected hosts), is increasingly recognized as relevant in viral pathogenesis and, consequently, in disease emergence. One example is provided by an evolutionary lineage of norovirus that displayed a higher polymerase error rate, generated ample genetic diversity, and displayed pandemic potential (Bull et al., 2011). Likewise, some rapidly evolving enterovirus lineages may be prone to epidemic potential (Zhang et al., 2011).

Finally, bottleneck events (Figure 16.1b) introduce an important element of stochasticity (chance) in virus evolution and, therefore, in the probability that a specific variant from a viral quasispecies reaches a potential new host. Model studies in cell culture, in particular using FMDV, have unveiled the molecular basis of fitness decrease that is associated with repeated bottleneck events (as key references and reviews, see Domingo et al., 2012; Escarmís et al., 1996, 2006, 2009; Lázaro et al., 2003). These studies revealed a key implication for viral emergence: that minority genomes hidden in mutant spectra, and that may be endowed with atypical phenotypic traits, may become dominant — and, thus, available to the external environment — as a result of bottleneck events. Rather than being exceptional, current evidence suggests that viruses undergo frequent bottleneck events in the course of their life cycles inside the infected host individual, not only during host-to-host transmission (several examples reviewed in Domingo et al., 2012). This is true of arboviruses during their spread in insect vectors, as discussed in a later section of this chapter and in other chapters of this book.

Therefore, two groups of mechanisms, molecular and evolutionary (Table 16.1), are intimately interwoven to produce a highly unpredictable outcome of which variant of which virus is going to emerge in which new host, how, and when. The mechanisms involved in short-term and long-term evolution of a few viral systems on which we have some direct expertise are described to underline some of the general points we have made.

16.5 AMPLE GENETIC AND HOST RANGE VARIATIONS OF FMDV: A HUMAN EPIDEMIC TO BE?

FMDV has been one of the model systems for understanding variation of RNA viruses, in particular that antigenic variation is a consequence of a general genetic variability that affects the entire genome. The studies over the past decades have provided experimental and conceptual support to the notion that populations of RNA viruses consist of multiple variants collectively termed quasispecies, as described in the previous sections. A complex equilibrium between a high mutation rate inherent to RNA replication and the competitive fitness in response to the selective pressures takes place continuously, as part of the lifestyle of FMDV and RNA viruses in general.

Genetic and antigenic heterogeneity of FMDV populations, as well as high rates of evolution, has been observed in populations derived from cloned viruses upon a limited number of acute or persistent infections in cell culture (Sobrino et al., 1983). The antigenic heterogeneity of FMDV populations was revealed by the high frequency of isolation (around 2×10^{-5}) of monoclonal antibody-resistant (MAR) mutants, estimated after minimal amplification of cloned viruses (Martínez et al., 1991). Propagation of these heterogeneous populations can result in a rapid emergence of antigenic variants even in the absence of immune pressure (Bolwell et al., 1989; Borrego et al., 1993; Domingo et al., 1993). Such an emergence may constitute a problem not only for the implications in antigenic drift in the field but also for the evaluation of vaccine efficacy. This is because serial passages of vaccine and challenge strains are usually required for vaccine production and testing (Gonzalez et al., 1992). Similar frequencies of isolation, about 10^{-5}, have been reported for mutants with increased resistance to acidic pH (Martin-Acebes et al., 2011), while frequencies as high as 8×10^{-2} have been found for mutants with increased sensitivity to acidic pH (Martin-Acebes et al., 2010). Acid-sensitive mutants are present at similar frequencies in lesions from infected animals (A. Vázquez-Calvo et al., manuscript in preparation). These observations support the view that FMDVs, as other RNA viruses, vary at many genomic sites and that our awareness of variation is limited by the availability of phenotypic tests to detect it. In this case, pH sensitivity or reactivity with antibodies revealed the occurrence of variations at sites that happened to be amenable to analysis.

Infections by FMDV can be either acute and cytolytic or persistent with little cell damage both in cell culture and *in vivo*. The analysis of viruses recovered from serial passages of BHK-21 cells persistently infected with FMDV demonstrated a gradual accumulation of nucleotide substitutions (de la Torre et al., 1985) and the rapid generation of heterogeneity and the occurrence of remarkable phenotypic changes (i.e., virion stability, virulence for the host cells) (de la Torre et al., 1988; Díez et al., 1990). A coevolution of both the virus (which showed increased virulence for the parental BHK-21 cells) and cells (which became progressively resistant to the initial virus) was documented in great detail (de la Torre et al., 1988). Virus–cell coevolution has been evidenced with several viruses (Ahmed et al., 1981; Chen and Baric, 1996; Ron and Tal, 1985) and more recently with the newly established cell culture system for hepatitis C virus (HCV) (Zhong et al., 2006).

Model studies with FMDV in cell culture have documented that viral quasispecies may possess a memory of their past evolutionary history in the form of minority components of the mutant spectrum (Ruiz-Jarabo et al., 2000). This observation may be relevant to the response of RNA viruses to fluctuating selective constraints during prolonged infections *in vivo*. Memory was extended to HIV-1 and its implications in virus adaptability have been reviewed (Briones and Domingo, 2008).

The potential for variation of FMDV has also been observed in host animals. Sequence heterogeneity among individual cloned viruses recovered from a single animal has been documented (Domingo et al., 1980; King et al., 1981; Rowlands et al., 1983; Wright et al., 2011). Genetic and phenotypic heterogeneity has been found in viral populations recovered upon infection of swine with plaque-purified viruses (Carrillo et al., 1990, 1998).

During centuries of evolution of FMDV in the field, repeated opportunities for variation have led to the viral diversification that is observed nowadays, reflected in seven serotypes that coexist in the world. Despite having been a subject of controversy, it is now well accepted that both random (chance) events, a continuously acting negative selection (the process to keep unfit genomes at low frequencies) and episodes of positive Darwinian selection, all have contributed to shape the FMDV populations we see today. Genetic heterogeneity and diversification may result in amino acid substitutions in virtually any FMDV protein, but particularly in exposed surface loops of capsid proteins which critically affect the antigenic properties of the virus but do not jeopardize virion stability (Lea et al., 1994). On the basis of cross-protection studies with convalescent and vaccinated animals, seven serotypes of the virus have been described: O, A, C, Asia1 and SAT1, SAT2, SAT3 (Bachrach, 1968; Pereira, 1981). Viruses of a different serotype do not manifest cross-protection. In addition, many subtypes within a serotype—which evoke partially cross-protective immune responses—were identified by classical immunological techniques. The use of monoclonal antibodies documented that FMDV populations are composed of a continuum of antigenic variants with amino acid substitutions affecting virtually each of the several antigenic sites that have been defined on viral particles (Mateu, 1995).

The most frequent mutational events observed during FMDV evolution are point mutations. Insertions and deletions appear to be fixed at lower frequencies. *In vitro* recombination in RNA viruses was first described with FMDV (King et al., 1982), and it occurs at high frequency among highly homologous strains in cell culture (reviewed in King, 1988). Recent reports indicate that recombination is also found in natural FMDV isolates (Abdul-Hamid et al., 2011; Yang et al., 2011), and it appears to be a widespread feature of all picornaviruses (reviewed in Agol, 2010; Simmonds, 2010).

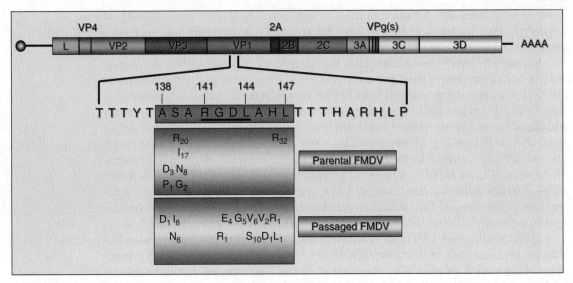

Figure 16.2. Specific example of alteration of a receptor recognition site as a result of viral evolution. The picornavirus FMDV genome (top) encodes a number of structural and nonstructural viral proteins (indicated in the boxes along the genome). Capsid protein VP1 includes a major antigenic determinant, and one of the epitopes (residues 138–147) is boxed. Within the epitope, the RGDL sequence is critical for integrin recognition, the major cellular receptor for FMDV. Monoclonal antibody-escape mutants of the parental FMDV (with a limited number of passages in cell culture) mapped around the RGDL but not within the RGDL (box labeled as parental FMDV). In contrast, the escape mutants from passaged FMDV (subjected to 100 serial passages in cell culture) affected both the RGDL and the region around the RGDL (bottom box). The difference in escape mutant repertoire was highly significant statistically. Scheme modified from Perales et al. (2005), with permission. For color detail, please see color plate section.

Flexibility in receptor usage by FMDV has also important implications for the evolution of virus antigenicity. FMDV uses integrins as its major cellular receptors, and integrin recognition occurs via an RGD triplet located at an exposed, mobile loop in protein VP1 (Acharya et al., 1989). Since an RGDL sequence is also a key part of several epitopes recognized by neutralizing antibodies, dispensability of the RGDL integrin-binding motif for cell entry greatly expanded the repertoire of antigenic FMDV variants. The consequence was the isolation of viable mutants with profoundly altered antigenicity that included mutations at the RGDL sequence (Figure 16.2).

Accessibility of integrin-binding sites to neutralizing antibodies suggests the possibility of a coevolution of FMDV antigenicity and receptor usage (Baranowski et al., 2001, 2003), as discussed in the previous section (Table 16.1). The genomic changes that can endow FMDV with the capacity to use alternative mechanism of cell recognition are minimal (Baranowski et al., 2000), and viruses with unusual receptor-binding specificities are likely to be present in the mutant spectra of FMDV replicating in animal hosts. Evidence came from experiments with cattle immunized with synthetic peptides representing B-cell and T-cell epitopes, including the RGD-containing G-H loop sequence (Taboga et al., 1997). Peptides conferred only partial protection, and animals challenged with virulent virus developed lesions. Direct sequencing using viral genomes extracted from lesions revealed unusual amino acid replacements affecting the RGD motif (R141G) or positions +1 or +4 relative to RGD (L144P and L147P) known to be critical for binding to some RGD-dependent integrins. The antigenic alterations produced by these replacements suggest that the emergence of these particular FMDV mutants *in vivo* is the result of selection of antigenic variants that escaped neutralization by anti-FMDV antibodies in peptide-vaccinated cattle (Taboga et al., 1997).

Little is known of changes in receptor specificity and antigenicity accompanying host range alterations in nature, but there is no reason to rule out that emerging and reemerging FMDVs could be endowed with unusual biological properties. A study of the genetic changes selected during adaptation of FMDV to guinea pig documented the progressive dominance of an unusual amino acid replacement (L147P) affecting the antigenic structure, although other substitutions in nonstructural proteins were the main responsible of the expanded host range. Specifically, a single amino acid substitution in protein 3A, which is involved in RNA replication, can confer the ability to produce clinical symptoms in the guinea pig (Núñez et al., 2001).

The severity of the symptoms (virulence) caused by FMDV may vary depending on the species affected as well as the dose and genomic characteristics of the infecting viral isolate (Bachrach, 1968; Burrows et al., 1981; Pereira, 1981). The concept of host range is mostly based on the capacity of FMDV to induce clinical symptoms, and absence of those symptoms does not exclude the potential of FMDV to replicate in a given species. Adult outbred mice do not develop vesicular lesions, albeit they show fever, and infective virus can be isolated from circulating blood. Little is known of virulence and host range determinants of FMDV *in vivo*. As discussed previously, interaction of the RGD triplet of VP1 with integrins is considered a requirement for viral entry. Nevertheless, integrins are expressed in a variety of tissues and species, including those that are not permissive for FMDV.

Nonstructural proteins can be also involved in FMDV virulence and host tropism. Deletions in nonstructural protein 3A were associated with attenuation for cattle of FMDV serotypes O and C, achieved upon virus passage in chicken embryos (Giraudo et al., 1990). An overlapping 10-amino acid deletion, together with different point mutations in 3A, has been shown to contribute to the low virulence for cattle of a variant of FMDV serotype O, isolated during the Taiwan 1997 epizootic (Beard and Mason, 2000). Interestingly, this virus was highly virulent for swine, a fact that illustrates the species specificity often associated with FMDV attenuation. It reflects also the complex set of interactions that are established between the host and the virus that determine the progress of the infection toward the appearance of lesions and clinical symptoms (Núñez et al., 2001).

FMD has been restricted to *Artiodactyla* (even-toed ungulates) for centuries. Nevertheless, due to the high potential of FMDV for exploring new biological niches and to the increasing trade globalization and modification of traditional ecosystems and animal production practices, the selection of virus mutants with expanded tropism and host range cannot be ruled out. As commented earlier, integrin receptors that can mediate virus infection in cultured cells are expressed in different "nonsusceptible" species, and limited genetic changes can increase the virulence of FMDV infection. Indeed, a debate exists on whether the few descriptions of FMD in humans can justify considering FMDV as a potential zoonotic agent (David and Brown, 2001; Schinazi et al., 1997; Schrijver et al., 1998). Despite difficulties derived from the socioeconomic situation in several countries where FMD is enzootic, surveillance of the disease and means to restrict human contacts with diseased animals would appear as a valid anticipatory action to prevent the emergence of a new human disease.

16.6 THE ARBOVIRUS HOST ALTERNATIONS: HIGH EXPOSURE TO ENVIRONMENTAL MODIFICATIONS

Arthropod-borne viruses (arboviruses), which include a taxonomic diverse group (King et al., 2011), are single-stranded RNA viruses, many of which are zoonotic. Most arbovirus infections are asymptomatic, but they can also cause diseases of the central nervous system that may result in death. Arboviruses replicate to high titers in their natural hosts, but they do it to a much lesser extent in occasional hosts that are considered dead-end hosts, as they hardly transmit the virus or serve as reservoir (Weaver and Barrett, 2004) (Figure 16.3).

Arboviruses seem to evolve more slowly compared with other single-stranded RNA viruses, as their consensus sequences often remain highly conserved in nature. This probably reflects their unique lifestyle because they are maintained in cycles that require replication on disparate hosts (hematophagous arthropod vectors and vertebrates). As a consequence, it has been proposed that constraints should be expected in both its evolution (stasis) and host-specific adaptation (fitness trade-off), which is generally attributed to the differential selective pressure exerted by vertebrate and invertebrate hosts (Ciota and Kramer, 2010).

Data regarding the trade-off hypothesis (i.e., that sustained replication in a single host results in rapid adaptation to the specialized host, with fitness loss in the bypassed host) are inconsistent, and results depend on the viral model (i.e., flaviviruses vs. alphaviruses) or experimental systems (cell culture vs. animals) used. However, many of the experimental data come from cell culture studies that may not represent the course of events in natural infections. Understanding the mechanism by which arboviral hosts shape the viral evolution would greatly contribute to develop rational strategies of control (see other chapters of this volume); however, no generalization on the acting mechanisms seems to be possible with our current knowledge (Ciota and Kramer, 2010; Weaver, 2006; Weaver and Barrett, 2004).

Many experimental evidences do not support that the slow accumulation of changes is a result of host cycling alone; rather, the changes are more likely due to selective pressures acting on individual hosts. However, in most of the studies, only consensus sequences have been evaluated. For instance, dengue virus (DENV) serially passaged in mosquito or mammalian cells showed no, or limited, consensus sequence changes, and while fitness gain was reported in mammalian-specialized virus, mosquito-derived population showed lower titers in bypassed invertebrate cells (Chen et al., 2003). A later study in which DENV was sequentially passaged in mosquito and human cells showed adaptation to the cell type in which the virus was passaged, and fitness loss in the bypassed cells, but fitness gain was recorded after alternate passages. These results do not support that host-specific trade-offs are a consequence of cycling events, but, rather, that they are more probably due to purifying selection acting in vertebrates that, in some cases, leads to convergent evolution (Vasilakis et al., 2009). West Nile virus (WNV) and St. Louis encephalitis virus (SLEV) sequentially passaged in mosquito cells showed host specialization with little effect in bypassed host and modest consensus sequences changes (Ciota et al., 2007).

In vivo studies in which WNV was serially passaged in mosquitoes showed fitness gain in mosquitoes, without fitness loss in chicks and limited accumulation of changes in the consensus sequences (Ciota et al., 2008). A further study (Ciota et al., 2009) showed that SLEV specialization in chicks displayed increased fitness in chicks, but not in mosquitoes. Mosquito-specialized viruses did not display fitness variation and no accumulation of changes in the consensus sequences in either host. Thus, it seems that flavivirus evolution is controlled by alternating cycles of genetic expansion in permissive vectors (mosquitoes) and restricted in vertebrate hosts where purifying selection is dominant and intra-host genetic variation is low (Deardorff et al., 2011). However, fitness gains *in vivo* are difficult to analyze, because of several intervening factors such as immune responses, multiple environmental conditions, and activation of specific host genes.

Even though limited consensus sequence variation after WNV and SLEV passaged in mosquito cells has been observed (Ciota et al., 2007), genetic changes were shown in the

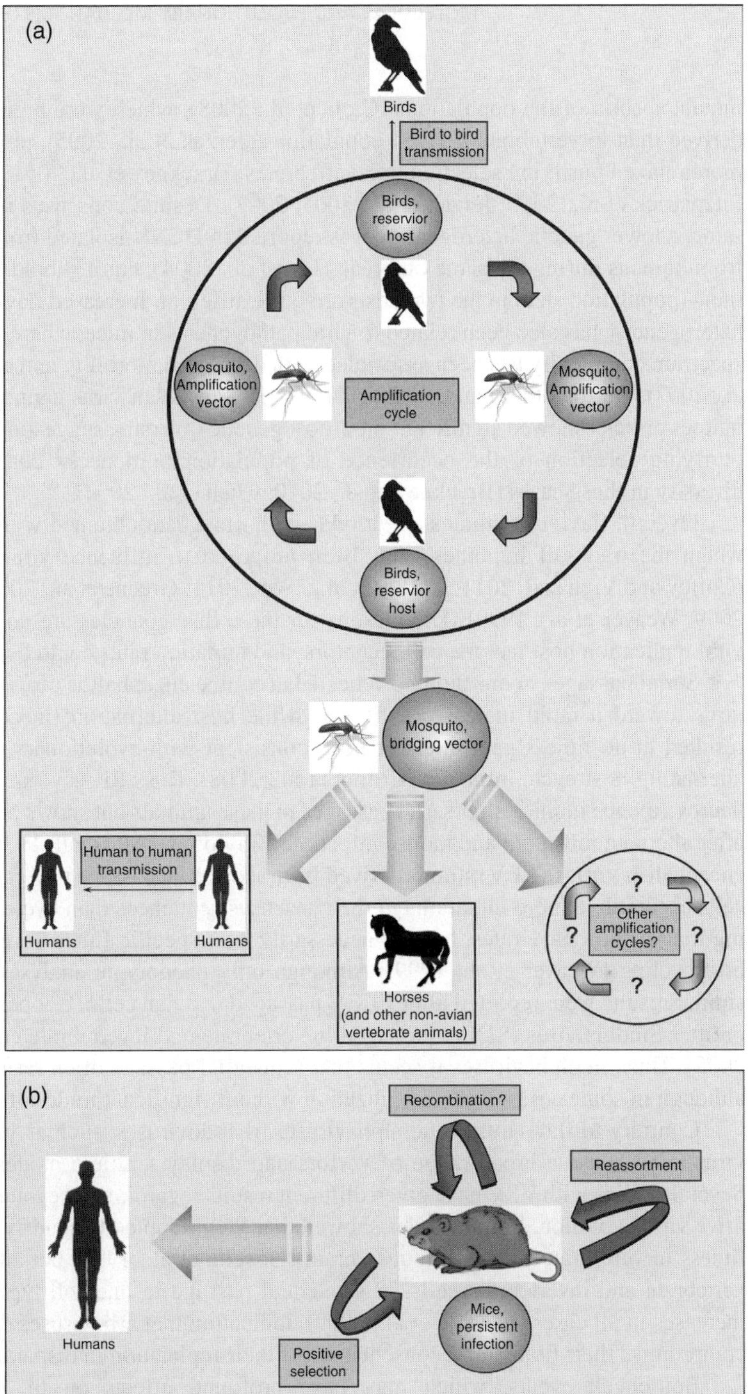

Figure 16.3. Simplified view of the influence of multiple hosts in the evolution of some viruses. (**a**) The amplification and transmission cycle of WNV involves reservoir hosts, amplification, and bridging vectors, as depicted schematically. Human-to-human transmission appears to be restricted to blood transfusion, organ transplantation, intrauterine virus spread, or breast milk feeding, although other means cannot be excluded. Likewise, there might be amplification cycles other than birds and mosquitoes in maintenance of this pathogen in nature. (**b**) Rodents are an extensive reservoir of many RNA viruses that have become (or have the potential to become) a zoonotic threat for humans. Arenaviruses constitute an example of rodent viruses that have emerged as human pathogens. Compare this scheme with the molecular mechanisms and the evolutionary forces described in Figure 16.1 and Table 16.1 (see text for additional implications and references). For color detail, please see color plate section.

mutant spectra of the populations (Ciota et al., 2008), which were higher in invertebrate-derived than in vertebrate-derived populations (Jerzak et al., 2005), probably reflecting a more relaxed purifying selection in invertebrates (Brackney et al., 2011; Ciota et al., 2009; Fitzpatrick et al., 2010; Jerzak et al., 2005, 2007). Despite consensus sequence being the same, a lower genetic heterogeneity was reported in DENV isolated from mosquitoes than from humans during the same outbreak (Lin et al., 2004), but it should be noted that viral titers (population size) in the two hosts are quite different. Increased flaviviral quasispecies heterogeneity has also been related to viral pathogenesis in mice, where increase in mutant spectrum complexity has been associated with decreased morbility and mortality (Jerzak et al., 2007). On the other hand, neither DENV nor Powassan virus naturally infecting mosquitoes or ticks showed significant intra-host genetic diversity, suggesting that either strong purifying selection or the occurrence of population bottlenecks constrains population diversity in these cases (Brackney et al., 2010; Chen et al., 2003).

Overall, flavivirus studies are at odds with most data obtained with alphaviruses, for which the trade-off hypothesis has been proposed to influence viral–host interactions (Coffey and Vignuzzi, 2011; Coffey et al., 2008, 2011; Greene et al., 2005; Vasilakis et al., 2009; Weaver et al., 1999). The reasons for these discrepancies are not known, although viral replication host factors, cell receptors, and mutation rates could be among them.

Serial passages in mosquitoes Venezuelan equine encephalitis virus (VEEV) drove the virus toward a rapid increase of fitness, while host alternation (mosquitoes and mice) resulted in no fitness gain in either host, consistent with evolutionary constraints due to alternating host cycle in nature (Coffey et al., 2008). Ross River virus exhibits increased neurovirulence in mice after serial passages in these animals but shows phenotypic stability after alternate mosquito and mouse infections (Taylor and Marshall, 1975). Eastern equine encephalitis virus (EEEV) strains derived from sequential passage in vertebrate or mosquito cells accumulate more mutations in the consensus sequences than cycled strains, suggesting that evolutionary rates, but not necessarily host-specific fitness, were constrained by host cycling (Weaver et al., 1999). Although only phenotypic analyses were conducted, similar results were reported after EEEV passaged in avian cells (Cooper and Scott, 2001), or after Sindbis virus (SINV) passaged in vertebrate and invertebrate cells (Greene et al., 2005). Thus, in alphaviruses it seems that sequential passage often results in fitness cost, although in some cases, host specialization without significant trade-offs can be observed.

Contrary to flaviviruses and alphaviruses, rhabdoviruses, such as vesicular stomatitis virus (VSV), use a broad range of vectors and display multiple modes of transmission. Several studies with VSV have given different results regarding the evolutionary forces that drive viral evolution. Initial studies showed that VSV adapted to sandfly cells had reduced fitness in vertebrate cells and mouse brain (Novella et al., 1995) but alternate passage in vertebrate and invertebrate cells, or sequential passage in one cell type, drove to fitness increases in all cases (Novella et al., 1999), indicating that arboviruses did not necessarily compromise their fitness as a consequence of their replication in disparate hosts.

Bottleneck events, which may have profound effects on the mutant spectrum (population size), are probably a common occurrence in the arbovirus replication cycle (Elena et al., 2001) (see also Section 16.4). For instance, mosquitoes usually transmit low amounts of virus, and thus, population size during transmission is low (Weaver, 2006). In any case, contradictory results have been published regarding bottlenecks effects in arbovirus evolution (Brackney et al., 2011; Ciota and Kramer, 2010), and further studies are needed to clarify this important point.

Recombination is generally considered low in arboviruses and difficult to predict. Superinfections of hosts by closely related alphaviruses are usually excluded once the primary infection has began (Karpf et al., 1997), but within a few hours, multiple sequential infections could occur and mosquitoes could feed in several hosts infected by different viruses, which may favor recombination (Weaver, 2006). Intra-serotype recombination in DENV has been documented (Craig et al., 2003; Holmes and Twiddy, 2003) and related to the feeding habits of its main vector, *Aedes aegypti*, that takes multiple blood meals from different human and animal hosts (Harrington et al., 2001). In any case, recombination does not seem to be essential for arbovirus evolution, even though an ancient recombination event has been suggested to be responsible for the emergence of WEEV (Hahn et al., 1988; Weaver et al., 1997). Recently, flavivirus sequences have been found in the DNA genome of mosquitoes, which are, thus, conserved during replication (Crochu et al., 2004). Recombination of mosquito cell transcripts and RNA from an infecting virus could result in restoration of ancestral viral RNA sequences and in the generation of new genomic types (Weaver, 2006).

Besides the limited accumulation of changes observed during arboviral evolution and the role that the population dynamics may play in it, as described above for FMDV, single mutations may have a pivotal role in viral behavior. A single mutation on the E2 gene of VEEV has been associated to increased vector competence (Brault et al., 2001, 2004) and equine virulence (Anishchenko et al., 2006). A dominant strain of WNV that originated from the original NY-99 strain, with just two synonymous and one nonsynonymous changes, was transmitted earlier and more efficiently by *Culex* mosquitoes (Moudy et al., 2007). Likewise, recent outbreaks of chikungunya virus in different geographical regions have been associated with viruses with a single common substitution (A226V) in the E1 glycoprotein (Powers et al., 2000; Schuffenecker et al., 2006), which replicate and disseminate more efficiently (de Lamballerie et al., 2008; Tsetsarkin et al., 2007; Vazeille et al., 2007), without compromising its fitness in *A. aegypti* (Ng and Hapuarachchi, 2010). Thus, despite overall slow evolutionary rates, arboviruses are able to produce variants that can exploit new environments.

Adaptation to new vector hosts and ecosystems has also important consequences on arboviral evolution and spread. For instances, infection of humans by DENV probably was the consequence of viral evolution from sylvatic ancestors in combination with change in host range (from nonhuman primates to human) and with increase in human urban density (Weaver and Barrett, 2004).

Similarly, the abovementioned substitution (A226V) in the E1 glycoprotein of chikungunya virus, which appeared in isolates from *Aedes albopictus* mosquitoes years after a huge outbreak in La Reunion (de Lamballerie et al., 2008), contributes to replicate and disseminate the virus more efficiently (Tsetsarkin et al., 2007; Vazeille et al., 2007). This E1 substitution has been related to the adaptation of chikungunya virus to a new mosquito host and to the consequent expansion of the virus to new geographical regions (Parola et al., 2006). Thus, chikungunya virus has adapted to *A. albopictus* without compromising its fitness in *A. aegypti* and humans (Ng and Hapuarachchi, 2010). Changing vectors has important consequences, as *A. aegypti* is mainly urban and highly anthropophilic, while *A. albopictus* covers a wider ecosystem (urban and rural) with expanded host range, thus facilitating intraspecies transmission. The fact that the A226V substitution has been detected in different regions and host species as a result of three separate events suggests that the same biological solution (convergent evolution) has been independently applied in response to similar requirements, transmission by *A. albopictus*. These mutations might have been selected from mutant spectra of different viral population (Gould and Higgs, 2009). Thus, mosquito diversity seems to be associated with arbovirus geographical range (Hemmerter et al., 2007).

Although WNV remains as a relatively homogeneous population, new WNV strains, closely related to the NY-99 isolate, were detected in western United States 2 years after the introduction of the virus in the country. They have now replaced the original virus and have become dominant (Beasley et al., 2002; Davis et al., 2005; Lanciotti et al., 2002), probably due to genetic drift of virus during dissemination (Kilpatrick et al., 2006). It remains to be determined whether WNV is evolving toward reduced virulence as it adapts to new hosts and ecosystems, but Mexican strains have been reported to be less virulent for North American avian species (Brault et al., 2011). By reverse genetic, two point mutations located in structural genes have been confirmed to be responsible for attenuation in cell culture and in susceptible avian species (Langevin et al., 2011).

Even though the WNV isolated in New York for the first time in 1999 was genetically very close to the strains circulating in the Mediterranean Basin at that time, its introduction in the United States had devastating consequences. This might have been a consequence of the introduction of virulent strains into a naïve host population in the presence of competent amplifying vectors (Martín-Acebes and Saiz, 2012; Weaver and Barrett, 2004). However, the initial high-efficient amplification and virulence in the United States seems to have declined with time. This could be due to a strong selection for resistance to WNV resulting from high bird mortality rates, which may lead to decreased viral spillover. Likewise, preexisting immunity to WNV due to previous infection with other related flaviviruses may also explain a decline in WNV cases in recent years, as well as the different behavior of the disease observed in Central and South America, where few cases have been reported in human or horses and very low avian mortality has been observed (Martín-Acebes and Saiz, 2012; Weaver and Barrett, 2004).

Climate (temperature, humidity, rainfall rate, etc.) and landscape conditions (altitude, latitude, water sources, forestation, etc.) directly affect mosquito populations, density, and stability, thus playing a pivotal role in arboviral dissemination and evolution (Gould and Higgs, 2009; Weaver and Barrett, 2004; see other chapters of this volume). Climate change affects the geographical and seasonal distribution of arthropod vectors, a fact that has many consequences. Expansion of mosquito species to more extreme latitudes and elevations allows them to reach new host populations to feed on and to share new ecological niches with other vector species. Heavy rainfall and warm temperatures benefit the increase of mosquito populations and have been positively correlated with arbovirus transmission (Martin-Acebes and Saiz, 2011; Weaver and Barrett, 2004). Likewise, overabundance of birds that share urban and rural areas with ubiquitous fastidious mosquitoes that feed on birds and mammals plays an important role in WNV transmission. In fact, in two of the major urban WNV outbreaks reported, which took place in Bucharest (Romania) and Volgograd and Volzhsky (Russia), the regions were heavily infested by potential mosquito vectors. Epidemics of chikungunya virus have been reported to be also dependent on mosquito population density, which increases following rainfall periods (Lumsden, 1955). However, epidemics have been also documented during drought periods, probably due to the infrequent replacement of water stores where mosquitoes can survive (Chretien et al., 2007). Thus, arboviral epidemics are probably more related to the concomitant profusion of hosts and vector populations in the same area, generally wastelands and mashes where migratory birds are abundant, than to other factors. Selection for alternative vectors that occur as a result of deforestation has also been proposed (Brault et al., 2004).

An important factor in arboviral evolution is socioeconomic development since it has led to increased and faster travelling, urbanization, deforestation, irrigation, and migratory activities leading to overpopulation. For instance, increase in the geographical range of some natural JEV hosts (cattle) during the nineteenth century has been implicated in viral spread. Likewise, and because JEV outbreaks were mainly restricted to Japan before the Second World War, the conflict seems to have also favored virus spread (Weaver and Barrett, 2004). Hence, new human and animal behaviors and climate changes are facilitating contact between vectors and hosts that, together with viral genetic mechanisms, may lead to the emergence or reemergence of new arbovirus with expanded tropism and host range and, thus, represents an important threat for human and animal health.

16.7 ARENAVIRUSES AS AN EMERGING THREAT

The *Arenaviridae* family encompasses a large group of viruses that merit attention for being clinically important human pathogens and excellent experimental model systems to study viral pathology (Buchmeier et al., 2007; Gonzalez et al., 2007; Oldstone, 2002). Thus, this family includes the highly human pathogenic Lassa fever virus (LFV) that causes thousands of deaths by severe hemorrhagic fever (McCormick and Fisher-Hoch, 2002) and lymphocytic choriomeningitis virus (LCMV), the prototypic arenavirus, and a neglected human pathogen. The study of LCMV infection in rodents has led to major advances in virology and immunology (Oldstone, 2002; Zinkernagel, 2002). They are enveloped viruses with a bisegmented negative strand RNA genome, comprising a large (L) and small (S) RNA segment, each encoding two proteins in an ambisense coding strategy (Buchmeier et al., 2007). Based on serological and phylogenetic data, arenaviruses have been classified into two groups: the Old World (OW) arenaviruses, found in Africa and Europe, and the New World (NW) arenaviruses, endemic in America (Clegg, 2002; Charrel et al., 2008). LFV and LCMV are considered representative members of the OW arenaviruses. Regarding the NW arenaviruses, Junin virus (JUNV), Machupo virus (MACV), Guanarito virus (GTOV), and Sabia virus (SABV) have emerged in different areas of South America as causative agents of severe hemorrhagic fevers (Peters, 2002). Each arenavirus species has as natural reservoir one or a limited number of closely related rodent species from the family Muridae, except Tacaribe virus (TCRV) that has been isolated from bats (Salazar-Bravo et al., 2002). Arenaviruses cause a persistent and asymptomatic infection that is vertically transmitted within the natural reservoir population while transmission to humans through mucosal exposure to aerosols can cause severe disease (Geisbert and Jahrling, 2004).

Currently, there are 22 recognized species into the arenavirus family, but still novel arenavirus members are identified nowadays as a result of the evolution of arenaviruses in different areas (Briese et al., 2009; Coulibaly-N'Golo et al., 2011), indicating that new viral isolates emergence is a real threat for humans. The different mechanisms implicated in arenavirus variation and evolution are not very well understood, but some data have provided evidence of some mechanism operating in the field, which will be discussed in this section.

The segmentation of the arenavirus genome confers the potential for the reassortment of their segments when different viruses coinfect the same cell (Figure 16.1a and 16.3b). Although the possibility to coinfect cultured cells has been demonstrated *in vitro* (Lukashevich, 1992; Lukashevich et al., 2005; Riviere and Oldstone, 1986), coinfections in nature might be unlikely. A coinfection event implies that two different arenaviruses coinfect the same rodent, and this would be against the superinfection exclusion concept in arenaviruses (cells infected with a given arenavirus become highly resistant to infection by other related arenavirus) (Ellenberg et al., 2004, 2007). The basis of the exclusion remains poorly understood. In addition, the new generated reassortant needs to have a higher fitness that both parental viruses in order to be maintained and to spread. Thus, it does not appear as an easy occurrence in nature, and in fact, there is no genetic data supporting the generation of new arenavirus species by reassortment (Cajimat and Fulhorst, 2004; Cajimat et al., 2011; Charrel et al., 2003). Nevertheless, Weaver et al. (2000) have described a rodent infected by two different genotypes of Pirital virus, indicating that the superinfection exclusion is not absolute and reassortment event may occur between members of arenavirus quasispecies replicating *in vivo*.

The recombination events in arenaviruses have been a controversial issue. Initially, the hypothesis of genetic recombination was proposed as the cause of discrepancy between the nucleoprotein (NP) and the glycoprotein precursor (GPC)-based phylogenetic trees (Charrel et al., 2001). The NW arenaviruses are phylogenetically divided in three clades: A, B, and C. There are two North American arenaviruses, Catarina virus and Skinner Tank virus, that show phylogenetic incongruence using NP or GPC sequences (Cajimat et al., 2007, 2008), giving place to a new lineage, referred to as the A/Rec lineage (Charrel et al., 2003). In addition, there are a number of amino acid domains shared between arenaviruses from clade A and B. Taken all together, the data point to the consistency of a character among closely related viruses, which suggest an heritable trait, supporting the idea that a recombination event may be the origin of the A/Rec lineage (Archer and Rico-Hesse, 2002; Charrel et al., 2008). The generation of recombinant viruses is subjected to the same constraints as those described for the generation of reassortant arenaviruses: different viruses have to coinfect the same cell, and this coinfection might be subjected to the superinfection exclusion principle. From a geographical point of view, some arenaviruses share a geographical localization such as Junin and Oliveros in Argentina (Mills et al., 1996), Guanarito and Pirital in Venezuela (Fulhorst et al., 1999; Weaver et al., 2000), Machupo, Latino, and Chapare in Bolivia (Delgado et al., 2008), and Lassa and Mopeia in Mozambique (Wulff et al., 1977), rendering a possible, in principle, coinfection event. In addition, some arenaviruses share the host like Guanarito and Pirital viruses that were isolated both from cotton rat *Sigmodon alstoni* and from cane mouse *Zygodontomys brevicauda* (Weaver et al., 2000), although a recombination event between Guanarito and Pirital has not been documented (Cajimat and Fulhorst, 2004; Fulhorst et al., 2008). More research *in vitro* (cell culture systems) with marked variants is needed to determine the occurrence of true recombination of arenaviruses.

Among different evolutionary mechanisms that may be involved in viral emergence (Table 16.1), positive selection of new viral variants in response to different selection pressures has been broadly studied in arenaviruses. Attachment of a virus to its receptor molecules on cell membrane is the first step of virus infection, and it represents an important determinant for the cellular tropism, host range, and pathogenesis. Within the arenavirus family, α-dystroglycan (α-DG) is the primary receptor for most isolates of LCMV and LFV and the clade C of New World arenaviruses (Cao et al., 1998; Sevilla et al., 2000; Spiropoulou et al., 2002). Differences in binding affinity to α-DG have been documented among arenaviruses, and such differences have been linked to differences in virus–host interaction. In the arenavirus family, LCMV has been used as a model to study various aspects of arenaviruses pathology and the emergence of viral variants. The infection of immunocompetent mice with LCMV results in viral clearance mediated by a robust class-I-restricted cytotoxic $CD8^+$ T cell (CTL) response (Byrne and Oldstone, 1984; Fung-Leung et al., 1991). If neonates are infected, mice remain persistently infected for life. In the course of a persistent infection, distinct variants can be isolated from brain and lymphoid tissue (Ahmed et al., 1984). These viral variants show different biological properties: those isolated from brain are similar to the parental virus (Arm), inducing a potent CTL response (Evans et al., 1994), whereas isolates from lymphoid tissue cause chronic infection in adult mice associated with the absence of CTL response (Ahmed and Oldstone, 1988; Borrow and Oldstone, 1992). The prototypic virus of the immunosuppressive variants is named clone 13. These specific viral variants arising in the entire animal can be explained because the different organs and cell types represent a rich environment for the selection of viral variants (Sevilla et al., 2000). Thus, those viruses with a growth advantage in certain cells will be selected. Sequence analyses of a large number of LCMV variants consistently showed an amino acid exchange F260L or F260I in the GP1 of the immunosuppressive variants, which correlated with a 200- to 500-fold enhanced binding affinity for α-DG (Kunz et al., 2001, 2003). The anatomical distribution of immunosuppressive variants showed a tropism for cells of the marginal zone and white pulp of spleen, whereas nonimmunosuppressive variants were found in cells of the red pulp (Borrow et al., 1995; Sevilla et al., 2000; Smelt et al., 2001). In addition, immunosuppressive variants infect dendritic cells that are the major cell population expressing α-DG in spleen (Sevilla et al., 2004). Thus, viral variants are being selected *in vivo* based on the affinity for the cellular receptor, α-DG.

Other evidence of the association of arenaviruses variants arising during the course of a persistent infection can be found in the neonatal infection of C3H/St mice with LCMV that induce growth hormone (GH) deficiency syndrome (GHDS) (Oldstone et al., 1982). This disease is characterized by marked growth retardation and development of severe hypoglycemia. The development of GHDS is associated with high viral load in the GH-producing cells of the anterior pituitary (Oldstone et al., 1985). Studies using reassortant viruses between strains of LCMV, which do (Arm) or do not (WE) cause GHDS, mapped the ability to cause disease to the S segment (Riviere et al., 1985). The molecular basis of the differences between viral variants that do cause GHDS and those that do not cause GHDS has been mapped to the amino acid change S153F in the viral GP1 (Teng et al., 1996). These variants replicated in the GH-producing cells and in liver cells, suggesting that viral replication in the liver may favor a preferential accumulation of variants that cause GHDS (Buesa-Gomez et al., 1996). Taking together all these findings, this is an illustration of arenaviruses being subjected to positive selection for replicating in a cell type, arising new virus that may cause new diseases.

In summary, new arenaviruses are being nowadays detected, suggesting that many others will be uncovered in the future. These new detections are clear proof of the emergent nature of this important group of viruses. The natural inhabit of arenaviruses, rodents, will fluctuate in response to environmental changes. Thus, in the regions where pathogenic arenaviruses now circulate, significant effects are likely to occur as a result of the global climate change, affecting more likely the geographical location and the incidence of arenaviruses. The combination of a dynamic RNA virus and a dynamic host species promotes encounters between humans and new arenaviruses, with high risk for humans.

16.8 CONCLUSION

The emergence of viral disease has some common underlying mechanisms that have been summarized in general terms in the early sections of this chapter and illustrated with three very dissimilar viral systems: the picornavirus FMDV, several arboviruses, and the arenaviruses. These examples should complement others described in different chapters of the book. The incessant generation of diversity in our virosphere, in particular among the RNA viruses, acts as the fuel for these parasitic genetic elements to find new ecological niches where to thrive. Viral disease emergence is just a manifestation of the highly opportunistic nature of genetic elements whose major dictum is replicate and survive. And all might be due to the fact that life evolved thanks to genetic transfers promoted by these and related genetic parasites.

ACKNOWLEDGEMENT

Work at CBMSO was supported by grants BFU2011-23604 (to E.D.) and BIO2011-24351 (to F.S. and Fundación R. Areces). CIBERehd is funded by Instituto de Salud Carlos III. Work at INIA was supported by grants RTA2011-00036 and the Network for Animal Disease Infectology and Research—European Union NADIR-EU-228394. Research at CISA was supported by grants RYC-2010-06516, AGL 2011–25025, RTA2011-00036, and the Network for Animal Disease Infectology and Research—European Union NADIR-EU-228349, AGL2009-07353, RYC-2010-06516, and AGL2011-25025.

REFERENCES

Abdul-Hamid, N. F., Firat-Sarac, M., Radford, A. D., et al. (2011). Comparative sequence analysis of representative foot-and-mouth disease virus genomes from Southeast Asia. *Virus Genes* **43**(1), 41–45.

Acharya, R., Fry, E., Stuart, D., et al. (1989). The three-dimensional structure of foot-and-mouth disease virus at 2.9Å resolution. *Nature* **337**(6209), 709–716.

Agol, V. I. (2010). Picornaviruses as a model for studying the nature of RNA recombination. In: "The Picornaviruses" (E. Ehrenfeld, E. Domingo, and R. P. Roos, Eds.). ASM Press, Washington, DC, pp. 239–252.

Ahmed, R., and Oldstone, M. B. (1988). Organ-specific selection of viral variants during chronic infection. *J Exp Med* **167**(5), 1719–1724.

Ahmed, R., Canning, W. M., Kauffman, R. S., et al. (1981). Role of the host cell in persistent viral infection: coevolution of L cells and reovirus during persistent infection. *Cell* **25**(2), 325–332.

Ahmed, R., Salmi, A., Butler, L. D., et al. (1984). Selection of genetic variants of lymphocytic choriomeningitis virus in spleens of persistently infected mice. Role in suppression of cytotoxic T lymphocyte response and viral persistence. *J Exp Med* **160**(2), 521–540.

Anishchenko, M., Bowen, R. A., Paessler, S., et al. (2006). Venezuelan encephalitis emergence mediated by a phylogenetically predicted viral mutation. *Proc Natl Acad Sci USA* **103**(13), 4994–4999.

Archer, A. M., and Rico-Hesse, R. (2002). High genetic divergence and recombination in Arenaviruses from the Americas. *Virology* **304**(2), 274–281.

Arias, A., Lázaro, E., Escarmís, C., et al. (2001). Molecular intermediates of fitness gain of an RNA virus: characterization of a mutant spectrum by biological and molecular cloning. *J Gen Virol* **82**(Pt 5), 1049–1060.

Bachrach, H. L. (1968). Foot-and-mouth disease virus. *Annu Rev Microbiol* **22**, 201–244.

Baranowski, E., Ruíz-Jarabo, C. M., Sevilla, N., et al. (2000). Cell recognition by foot-and-mouth disease virus that lacks the RGD integrin-binding motif: flexibility in aphthovirus receptor usage. *J Virol* **74**(4), 1641–1647.

Baranowski, E., Ruiz-Jarabo, C. M., and Domingo, E. (2001). Evolution of cell recognition by viruses. *Science* **292**(5519), 1102–1105.

Baranowski, E., Ruiz-Jarabo, C. M., Pariente, N., et al. (2003). Evolution of cell recognition by viruses: a source of biological novelty with medical implications. *Adv Virus Res* **62**, 19–111.

Batschelet, E., Domingo, E., and Weissmann, C. (1976). The proportion of revertant and mutant phage in a growing population, as a function of mutation and growth rate. *Gene* **1**(1), 27–32.

Beard, C. W., and Mason, P. W. (2000). Genetic determinants of altered virulence of Taiwanese foot-and-mouth disease virus. *J Virol* **74**(2), 987–991.

Beasley, D. W., Li, L., Suderman, M. T., et al. (2002). Mouse neuroinvasive phenotype of West Nile virus strains varies depending upon virus genotype. *Virology* **296**(1), 17–23.

Bernad, A., Blanco, L., Lazaro, J. M., et al. (1989). A conserved 3′→5′ exonuclease active site in prokaryotic and eukaryotic DNA polymerases. *Cell* **59**(1), 219–228.

Bolwell, C., Brown, A. L., Barnett, P. V., et al. (1989). Host cell selection of antigenic variants of foot-and-mouth disease virus. *J Gen Virol* **70**(Pt 1), 45–57.

Borrego, B., Novella, I. S., Giralt, E., et al. (1993). Distinct repertoire of antigenic variants of foot-and-mouth disease virus in the presence or absence of immune selection. *J Virol* **67**(10), 6071–6079.

Borrow, P., and Oldstone, M. B. (1992). Characterization of lymphocytic choriomeningitis virus-binding protein(s): a candidate cellular receptor for the virus. *J Virol* **66**(12), 7270–7281.

Borrow, P., Evans, C. F., and Oldstone, M. B. (1995). Virus-induced immunosuppression: immune system-mediated destruction of virus-infected dendritic cells results in generalized immune suppression. *J Virol* **69**(2), 1059–1070.

Brackney, D. E., Brown, I. K., Nofchissey, R. A., et al. (2010). Homogeneity of Powassan virus populations in naturally infected Ixodes scapularis. *Virology* **402**(2), 366–371.

Brackney, D. E., Pesko, K. N., Brown, I. K., et al. (2011). West Nile virus genetic diversity is maintained during transmission by Culex pipiens quinquefasciatus mosquitoes. *PLoS One* **6**(9), e24466.

Brault, A. C., Powers, A. M., Medina, G., et al. (2001). Potential sources of the 1995 Venezuelan equine encephalitis subtype IC epidemic. *J Virol* **75**(13), 5823–5832.

Brault, A. C., Powers, A. M., Ortiz, D., et al. (2004). Venezuelan equine encephalitis emergence: enhanced vector infection from a single amino acid substitution in the envelope glycoprotein. *Proc Natl Acad Sci USA* **101**(31), 11344–11349.

Brault, A. C., Langevin, S. A., Ramey, W. N., et al. (2011). Reduced avian virulence and viremia of West Nile virus isolates from Mexico and Texas. *Am J Trop Med Hyg* **85**(4), 758–767.

Briese, T., Paweska, J. T., McMullan, L. K., et al. (2009). Genetic detection and characterization of Lujo virus, a new hemorrhagic fever-associated arenavirus from southern Africa. *PLoS Pathog* **5**(5), e1000455.

Briones, C., and Domingo, E. (2008). Minority report: hidden memory genomes in HIV-1 quasispecies and possible clinical implications. *AIDS Rev* **10**(2), 93–109.

Buchmeier, M. J., de la Torre, J. C., and Peters, C. J. (2007). Arenaviridae. In: "Fields Virology," 5th edition. (D. M. Knipe, and P. M. Howley, Eds.). Lippincott, Williams and Wilkins, Philadelphia, pp. 1791–1828.

Buesa-Gomez, J., Teng, M. N., Oldstone, C. E., et al. (1996). Variants able to cause growth hormone deficiency syndrome are present within the disease-nil WE strain of lymphocytic choriomeningitis virus. *J Virol* **70**(12), 8988–8992.

Bull, R. A., Luciani, F., McElroy, K., et al. (2011). Sequential bottlenecks drive viral evolution in early acute hepatitis C virus infection. *PLoS Pathog* **7**(9), e1002243.

Burrows, R., Mann, J. A., Garland, A. J., et al. (1981). The pathogenesis of natural and simulated natural foot-and-mouth disease infection in cattle. *J Comp Pathol* **91**(4), 599–609.

Byrne, J. A., and Oldstone, M. B. (1984). Biology of cloned cytotoxic T lymphocytes specific for lymphocytic choriomeningitis virus: clearance of virus in vivo. *J Virol* **51**(3), 682–686.

Cajimat, M. N., and Fulhorst, C. F. (2004). Phylogeny of the Venezuelan arenaviruses. *Virus Res* **102**(2), 199–206.

Cajimat, M. N., Milazzo, M. L., Bradley, R. D., et al. (2007). Catarina virus, an arenaviral species principally associated with *Neotoma micropus* (southern plains woodrat) in Texas. *Am J Trop Med Hyg* **77**(4), 732–736.

Cajimat, M. N., Milazzo, M. L., Borchert, J. N., et al. (2008). Diversity among Tacaribe serocomplex viruses (family Arenaviridae) naturally associated with the Mexican woodrat (*Neotoma mexicana*). *Virus Res* **133**(2), 211–217.

Cajimat, M. N., Milazzo, M. L., Haynie, M. L., et al. (2011). Diversity and phylogenetic relationships among the North American Tacaribe serocomplex viruses (family Arenaviridae). *Virology* **421**(2), 87–95.

Calisher, C. H., Childs, J. E., Field, H. E., et al. (2006). Bats: important reservoir hosts of emerging viruses. *Clin Microbiol Rev* **19**(3), 531–545.

Cao, W., Henry, M. D., Borrow, P., et al. (1998). Identification of alpha-dystroglycan as a receptor for lymphocytic choriomeningitis virus and Lassa fever virus. *Science* **282**(5396), 2079–2081.

Carrillo, C., Plana, J., Mascarella, R., et al. (1990). Genetic and phenotypic variability during replication of foot-and-mouth disease virus in swine. *Virology* **179**(2), 890–892.

Carrillo, C., Borca, M., Moore, D. M., et al. (1998). In vivo analysis of the stability and fitness of variants recovered from foot-and-mouth disease virus quasispecies. *J Gen Virol* **79**(Pt 7), 1699–1706.

Ciota, A. T., and Kramer, L. D. (2010). Insights into arbovirus evolution and adaptation from experimental studies. *Viruses* **2**(12), 2594–2617.

Ciota, A. T., Lovelace, A. O., Ngo, K. A., et al. (2007). Cell-specific adaptation of two flaviviruses following serial passage in mosquito cell culture. *Virology* **357**(2), 165–174.

Ciota, A. T., Lovelace, A. O., Jia, Y., et al. (2008). Characterization of mosquito-adapted West Nile virus. *J Gen Virol* **89**(Pt 7), 1633–1642.

Ciota, A. T., Jia, Y., Payne, A. F., et al. (2009). Experimental passage of St. Louis encephalitis virus in vivo in mosquitoes and chickens reveals evolutionarily significant virus characteristics. *PLoS One* **4**(11), e7876.

Clegg, J. C. (2002). Molecular phylogeny of the arenaviruses. *Curr Top Microbiol Immunol* **262**, 1–24.

Coffey, L. L., and Vignuzzi, M. (2011). Host alternation of chikungunya virus increases fitness while restricting population diversity and adaptability to novel selective pressures. *J Virol* **85**(2), 1025–1035.

Coffey, L. L., Vasilakis, N., Brault, A. C., et al. (2008). Arbovirus evolution in vivo is constrained by host alternation. *Proc Natl Acad Sci USA* **105**(19), 6970–6975.

Coffey, L. L., Beeharry, Y., Borderia, A. V., et al. (2011). Arbovirus high fidelity variant loses fitness in mosquitoes and mice. *Proc Natl Acad Sci USA* **108**(38), 16038–16043.

Cooper, L. A., and Scott, T. W. (2001). Differential evolution of eastern equine encephalitis virus populations in response to host cell type. *Genetics* **157**(4), 1403–1412.

Coulibaly-N'Golo, D., Allali, B., Kouassi, S. K., et al. (2011). Novel arenavirus sequences in *Hylomyscus* sp. and Mus (Nannomys) setulosus from Cote d'Ivoire: implications for evolution of arenaviruses in Africa. *PLoS One* **6**(6), e20893.

Craig, S., Thu, H. M., Lowry, K., et al. (2003). Diverse dengue type 2 virus populations contain recombinant and both parental viruses in a single mosquito host. *J Virol* **77**(7), 4463–4467.

Crochu, S., Cook, S., Attoui, H., et al. (2004). Sequences of flavivirus-related RNA viruses persist in DNA form integrated in the genome of Aedes spp. mosquitoes. *J Gen Virol* **85**(Pt 7), 1971–1980.

Charrel, R. N., de Lamballerie, X., and Fulhorst, C. F. (2001). The Whitewater Arroyo virus: natural evidence for genetic recombination among Tacaribe serocomplex viruses (family Arenaviridae). *Virology* **283**(2), 161–166.

Charrel, R. N., de Lamballerie, X., and Emonet, S. (2008). Phylogeny of the genus Arenavirus. *Curr Opin Microbiol* **11**(4), 362–368.

Charrel, R. N., Lemasson, J. J., Garbutt, M., et al. (2003). New insights into the evolutionary relationships between arenaviruses provided by comparative analysis of small and large segment sequences. *Virology* **317**(2), 191–196.

Chen, W., and Baric, R. S. (1996). Molecular anatomy of mouse hepatitis virus persistence: coevolution of increased host cell resistance and virus virulence. *J Virol* **70**(6), 3947–3960.

Chen, W. J., Wu, H. R., and Chiou, S. S. (2003). E/NS1 modifications of dengue 2 virus after serial passages in mammalian and/or mosquito cells. *Intervirology* **46**(5), 289–295.

Chretien, J. P., Anyamba, A., Bedno, S. A., et al. (2007). Drought-associated chikungunya emergence along coastal East Africa. *Am J Trop Med Hyg* **76**(3), 405–407.

David, W., and Brown, G. (2001). Foot and mouth disease in human beings. *Lancet* **357**(9267), 1463.

Davis, C. T., Ebel, G. D., Lanciotti, R. S., et al. (2005). Phylogenetic analysis of North American West Nile virus isolates, 2001–2004: evidence for the emergence of a dominant genotype. *Virology* **342**(2), 252–265.

Deardorff, E. R., Fitzpatrick, K. A., Jerzak, G. V. S., et al. (2011). West Nile virus experimental evolution *in vivo* and the trade-off hypothesis. *PLoS Pathog* **7**(11), e1002335.

de Lamballerie, X., Leroy, E., Charrel, R. N., et al. (2008). Chikungunya virus adapts to tiger mosquito via evolutionary convergence: a sign of things to come? *Virol J* **5**, 33.

de la Torre, J. C., Davila, M., Sobrino, F., et al. (1985). Establishment of cell lines persistently infected with foot-and-mouth disease virus. *Virology* **145**(1), 24–35.

de la Torre, J. C., Martínez-Salas, E., Diez, J., et al. (1988). Coevolution of cells and viruses in a persistent infection of foot-and-mouth disease virus in cell culture. *J Virol* **62**(6), 2050–2058.

Delgado, S., Erickson, B. R., Agudo, R., et al. (2008). Chapare virus, a newly discovered arenavirus isolated from a fatal hemorrhagic fever case in Bolivia. *PLoS Pathog* **4**(4), e1000047.

Denison, M. R., Graham, R. L., Donaldson, E. F., et al. (2011). Coronaviruses: an RNA proofreading machine regulates replication fidelity and diversity. *RNA Biol* **8**(2), 270–279.

Díez, J., Dávila, M., Escarmís, C., et al. (1990). Unique amino acid substitutions in the capsid proteins of foot-and-mouth disease virus from a persistent infection in cell culture. *J Virol* **64**(11), 5519–5528.

Domingo, E. (2010). Mechanisms of viral emergence. *Vet Res* **41**(6), 38.

Domingo, E., Davila, M., and Ortin, J. (1980). Nucleotide sequence heterogeneity of the RNA from a natural population of foot-and-mouth-disease virus. *Gene* **11**(3–4), 333–346.

Domingo, E., Sabo, D., Taniguchi, T., et al. (1978). Nucleotide sequence heterogeneity of an RNA phage population. *Cell* **13**(4), 735–744.

Domingo, E., Díez, J., Martínez, M. A., et al. (1993). New observations on antigenic diversification of RNA viruses. Antigenic variation is not dependent on immune selection. *J Gen Virol* **74**, 2039–2045.

Domingo, E., Biebricher, C., Eigen, M., et al. (2001). "Quasispecies and RNA Virus Evolution: Principles and Consequences". Landes Bioscience, Austin.

Domingo, E., Ruiz-Jarabo, C. M., Arias, A., et al. (2004). Quasispecies dynamics and evolution of foot-and-mouth disease virus. In: "Foot-and-Mouth Disease" (F. Sobrino, and E. Domingo, Eds.). Horizon Bioscience, Wymondham, UK.

Domingo, E., Parrish, C., and Holland, J. J. E. (2008). Origin and Evolution of Viruses, 2nd edition. Elsevier, Oxford.

Domingo, E., Sheldon, J., and Perales, C. (2012). Viral quasispecies evolution. *Microbiol Mol Biol Rev* **76**, 159–216.

Drake, J. W., and Holland, J. J. (1999). Mutation rates among RNA viruses. *Proc Natl Acad Sci USA* **96**, 13910–13913.

Eckerle, L. D., Lu, X., Sperry, S. M., et al. (2007). High fidelity of murine hepatitis virus replication is decreased in nsp14 exoribonuclease mutants. *J Virol* **81**(22), 12135–12144.

Eckerle, L. D., Becker, M. M., Halpin, R. A., et al. (2010). Infidelity of SARS-CoV Nsp14-exonuclease mutant virus replication is revealed by complete genome sequencing. *PLoS Pathog* **6**(5), e1000896.

Ehrenfeld, E., Domingo, E., and Ross, R. P. (2010). "The Picornaviruses." ASM Press, Washington, D.C.

Eigen, M. (1992). "Steps Towards Life." Oxford University Press, Oxford.

Eigen, M., and Schuster, P. (1979). "The Hypercycle. A Principle of Natural Self-Organization." Springer, Berlin.

Elena, S. F., Sanjuan, R., Borderia, A. V., et al. (2001). Transmission bottlenecks and the evolution of fitness in rapidly evolving RNA viruses. *Infect Genet Evol* **1**(1), 41–48.

Ellenberg, P., Edreira, M., and Scolaro, L. (2004). Resistance to superinfection of Vero cells persistently infected with Junin virus. *Arch Virol* **149**(3), 507–522.

Ellenberg, P., Linero, F. N., and Scolaro, L. A. (2007). Superinfection exclusion in BHK-21 cells persistently infected with Junin virus. *J Gen Virol* **88**(Pt 10), 2730–2739.

Escarmís, C., Dávila, M., Charpentier, N., et al. (1996). Genetic lesions associated with Muller's ratchet in an RNA virus. *J Mol Biol* **264**, 255–267.

Escarmís, C., Lázaro, E., and Manrubia, S. C. (2006). Population bottlenecks in quasispecies dynamics. *Curr Top Microbiol Immunol* **299**, 141–170.

Escarmis, C., Perales, C., and Domingo, E. (2009). Biological effect of Muller's Ratchet: distant capsid site can affect picornavirus protein processing. *J Virol* **83**(13), 6748–6756.

Evans, C. F., Borrow, P., de la Torre, J. C., et al. (1994). Virus-induced immunosuppression: kinetic analysis of the selection of a mutation associated with viral persistence. *J Virol* **68**(11), 7367–7373.

Fitzpatrick, K. A., Deardorff, E. R., Pesko, K., et al. (2010). Population variation of West Nile virus confers a host-specific fitness benefit in mosquitoes. *Virology* **404**(1), 89–95.

Forterre, P. (2010). The universal tree of life and the last universal cellular ancestor: revolution and counterrevolutions. In: "Evolutionary Genomics and Systems Biology" (G. Caetano-Anollés, Ed.). Wiley-Blackwell, New Jersey, pp. 43–62.

Friedberg, E. C., Walker, G. C., Siede, W., et al. (2006). "DNA Repair and Mutagenesis." American Society for Microbiology, Washington, DC.

Fulhorst, C. F., Bowen, M. D., Salas, R. A., et al. (1999). Natural rodent host associations of Guanarito and pirital viruses (Family Arenaviridae) in central Venezuela. *Am J Trop Med Hyg* **61**(2), 325–330.

Fulhorst, C. F., Cajimat, M. N., Milazzo, M. L., et al. (2008). Genetic diversity between and within the arenavirus species indigenous to western Venezuela. *Virology* **378**(2), 205–213.

Fung-Leung, W. P., Kundig, T. M., Zinkernagel, R. M., et al. (1991). Immune response against lymphocytic choriomeningitis virus infection in mice without CD8 expression. *J Exp Med* **174**(6), 1425–1429.

Geisbert, T. W., and Jahrling, P. B. (2004). Exotic emerging viral diseases: progress and challenges. *Nat Med* **10**(12 Suppl.), S110–S121.

Giraudo, A. T., Beck, E., Strebel, K., et al. (1990). Identification of a nucleotide deletion in parts of polypeptide 3A in two independent attenuated aphthovirus strains. *Virology* **177**(2), 780–783.

Gonzalez, M., Mateu, M. G., Martinez, M. A., et al. (1992). Comparison of capsid protein VP1 of the viruses used for the production and challenge of foot-and-mouth disease vaccines in Spain. *Vaccine* **10**(11), 731–734.

Gonzalez, J. P., Emonet, S., de Lamballerie, X., et al. (2007). Arenaviruses. *Curr Top Microbiol Immunol* **315**, 253–288.

Gould, E. A., and Higgs, S. (2009). Impact of climate change and other factors on emerging arbovirus diseases. *Trans R Soc Trop Med Hyg* **103**(2), 109–121.

Grande-Pérez, A., Lazaro, E., Lowenstein, P., et al. (2005). Suppression of viral infectivity through lethal defection. *Proc Natl Acad Sci USA* **102**(12), 4448–4452.

Greene, I. P., Wang, E., Deardorff, E. R., et al. (2005). Effect of alternating passage on adaptation of sindbis virus to vertebrate and invertebrate cells. *J Virol* **79**(22), 14253–14260.

Hahn, C. S., Lustig, S., Strauss, E. G., et al. (1988). Western equine encephalitis virus is a recombinant virus. *Proc Natl Acad Sci USA* **85**(16), 5997–6001.

Harrington, L. C., Edman, J. D., and Scott, T. W. (2001). Why do female *Aedes aegypti* (Diptera: Culicidae) feed preferentially and frequently on human blood? *J Med Entomol* **38**(3), 411–422.

Hemmerter, S., Slapeta, J., van den Hurk, A. F., et al. (2007). A curious coincidence: mosquito biodiversity and the limits of the Japanese encephalitis virus in Australasia. *BMC Evol Biol* **7**, 100.

Holmes, E. C., and Twiddy, S. S. (2003). The origin, emergence and evolutionary genetics of dengue virus. *Infect Genet Evol* **3**(1), 19–28.

Jerzak, G., Bernard, K. A., Kramer, L. D., et al. (2005). Genetic variation in West Nile virus from naturally infected mosquitoes and birds suggests quasispecies structure and strong purifying selection. *J Gen Virol* **86**(Pt 8), 2175–2183.

Jerzak, G. V., Bernard, K., Kramer, L. D., et al. (2007). The West Nile virus mutant spectrum is host-dependant and a determinant of mortality in mice. *Virology* **360**(2), 469–476.

Karpf, A. R., Lenches, E., Strauss, E. G., et al. (1997). Superinfection exclusion of alphaviruses in three mosquito cell lines persistently infected with Sindbis virus. *J Virol* **71**(9), 7119–7123.

Khatchikian, D., Orlich, M., and Rott, R. (1989). Increased viral pathogenicity after insertion of a 28S ribosomal RNA sequence into the haemagglutinin gene of an influenza virus. *Nature* **340**(6229), 156–157.

Kilpatrick, A. M., Kramer, L. D., Jones, M. J., et al. (2006). West Nile virus epidemics in North America are driven by shifts in mosquito feeding behavior. *PLoS Biol* **4**(4), e82.

King, A. M. Q. (1988). Genetic recombination in positive strand RNA viruses. In: "RNA Genetics" (E. Domingo, J. J. Holland, and P. Ahlquist, Eds.), Vol. II. CRC Press Inc., Boca Raton, FL.

King, A. M., Underwood, B. O., McCahon, D., et al. (1981). Biochemical identification of viruses causing the 1981 outbreaks of foot and mouth disease in the UK. *Nature* **293**(5832), 479–480.

King, A. M., McCahon, D., Slade, W. R., et al. (1982). Recombination in RNA. *Cell* **29**(3), 921–928.

King, A. M. Q., Lefkowitz, E. J., Adams, M. J., et al. (2011). "Virus Taxonomy. Ninth Report of the International Committee on Taxonomy of Viruses." Academic Press, Elsevier, San Diego, CA.

Krauss, H., Weber, A., Appel, M., et al. (2003). "Zoonoses. Infectious Diseases Transmissible from Animals to Humans." ASM Press, Washington, DC.

Kunz, S., Sevilla, N., McGavern, D. B., et al. (2001). Molecular analysis of the interaction of LCMV with its cellular receptor [alpha]-dystroglycan. *J Cell Biol* **155**(2), 301–310.

Kunz, S., Edelmann, K. H., de la Torre, J. C., et al. (2003). Mechanisms for lymphocytic choriomeningitis virus glycoprotein cleavage, transport, and incorporation into virions. *Virology* **314**(1), 168–178.

Lanciotti, R. S., Ebel, G. D., Deubel, V., et al. (2002). Complete genome sequences and phylogenetic analysis of West Nile virus strains isolated from the United States, Europe, and the Middle East. *Virology* **298**(1), 96–105.

Langevin, S. A., Bowen, R. A., Ramey, W. N., et al. (2011). Envelope and pre-membrane protein structural amino acid mutations mediate diminished avian growth and virulence of a Mexican West Nile virus isolate. *J Gen Virol* **92**(Pt 12), 2810–2820.

Lázaro, E., Escarmis, C., Perez-Mercader, J., et al. (2003). Resistance of virus to extinction on bottleneck passages: study of a decaying and fluctuating pattern of fitness loss. *Proc Natl Acad Sci USA* **100**(19), 10830–10835.

Lea, S., Hernández, J., Blakemore, W., et al. (1994). The structure and antigenicity of a type C foot-and-mouth disease virus. *Structure* **2**(2), 123–139.

Lin, B., Vora, G. J., Thach, D., et al. (2004). Use of oligonucleotide microarrays for rapid detection and serotyping of acute respiratory disease-associated adenoviruses. *J Clin Microbiol* **42**(7), 3232–3239.

Lukashevich, I. S. (1992). Generation of reassortants between African arenaviruses. *Virology* **188**(2), 600–605.

Lukashevich, I. S., Patterson, J., Carrion, R., et al. (2005). A live attenuated vaccine for Lassa fever made by reassortment of Lassa and Mopeia viruses. *J Virol* **79**(22), 13934–13942.

Lumsden, W. H. (1955). An epidemic of virus disease in Southern Province, Tanganyika Territory, in 1952–53. II. General description and epidemiology. *Trans R Soc Trop Med Hyg* **49**(1), 33–57.

Martin-Acebes, M. A., and Saiz, J. C. (2011). A West Nile virus mutant with increased resistance to acid-induced inactivation. *J Gen Virol* **92**(Pt 4), 831–840.

Martín-Acebes, M. A., and Saiz, J. C. (2012). West Nile virus: a re-emerging pathogen revisited. *World J Virol* **1**, 51–70.

Martin-Acebes, M. A., Rincon, V., Armas-Portela, R., et al. (2010). A single amino acid substitution in the capsid of foot-and-mouth disease virus can increase acid lability and confer resistance to acid-dependent uncoating inhibition. *J Virol* **84**(6), 2902–2912.

Martin-Acebes, M. A., Vazquez-Calvo, A., Rincon, V., et al. (2011). A single amino acid substitution in the capsid of foot-and-mouth disease virus can increase acid resistance. *J Virol* **85**(6), 2733–2740.

Martínez, M. A., Carrillo, C., Gonzalez-Candelas, F., et al. (1991). Fitness alteration of foot-and-mouth disease virus mutants: measurement of adaptability of viral quasispecies. *J Virol* **65**(7), 3954–3957.

Mateu, M. G. (1995). Antibody recognition of picornaviruses and escape from neutralization: a structural view. *Virus Res* **38**(1), 1–24.

McCormick, J. B., and Fisher-Hoch, S. P. (2002). Lassa fever. *Curr Top Microbiol Immunol* **262**, 75–109.

Mills, J. N., Barrera Oro, J. G., Bressler, D. S., et al. (1996). Characterization of Oliveros virus, a new member of the Tacaribe complex (Arenaviridae: Arenavirus). *Am J Trop Med Hyg* **54**(4), 399–404.

Morse, S. S., Ed. (1994a). "The Evolutionary Biology of Viruses." Raven Press, New York.

Morse, S. S. (1994b). The viruses of the future? Emerging viruses and evolution. In: "The Evolutionary Biology of Viruses" (S. S. Morse, Ed.). Raven Press, New York, pp. 325–335.

Moudy, R. M., Meola, M. A., Morin, L. L., et al. (2007). A newly emergent genotype of West Nile virus is transmitted earlier and more efficiently by Culex mosquitoes. *Am J Trop Med Hyg* **77**(2), 365–370.

Nakamura, S., Yang, C. S., Sakon, N., et al. (2009). Direct metagenomic detection of viral pathogens in nasal and fecal specimens using an unbiased high-throughput sequencing approach. *PLoS One* **4**(1), e4219.

Negredo, A., Palacios, G., Vazquez-Moron, S., et al. (2011). Discovery of an ebolavirus-like filovirus in europe. *PLoS Pathog* **7**(10), e1002304.

Ng, L. C., and Hapuarachchi, H. C. (2010). Tracing the path of Chikungunya virus—evolution and adaptation. *Infect Genet Evol* **10**(7), 876–885.

Novella, I. S., Clarke, D. K., Quer, J., et al. (1995). Extreme fitness differences in mammalian and insect hosts after continuous replication of vesicular stomatitis virus in sandfly cells. *J Virol* **69**(11), 6805–6809.

Novella, I. S., Quer, J., Domingo, E., et al. (1999). Exponential fitness gains of RNA virus populations are limited by bottleneck effects. *J Virol* **73**(2), 1668–1671.

Núñez, J. I., Baranowski, E., Molina, N., et al. (2001). A single amino acid substitution in nonstructural protein 3A can mediate adaptation of foot-and-mouth disease virus to the guinea pig. *J Virol* **75**(8), 3977–3983.

Ojosnegros, S., Garcia-Arriaza, J., Escarmis, C., et al. (2011). Viral genome segmentation can result from a trade-off between genetic content and particle stability. *PLoS Genet* **7**(3), e1001344.

Oldstone, M. B. (2002). Biology and pathogenesis of lymphocytic choriomeningitis virus infection. In: "Arenaviruses" (M. B. Oldstone, Ed.), Vol. **263**, pp. 83–118. Springer, Berlin.

Oldstone, M. B., Sinha, Y. N., Blount, P., et al. (1982). Virus-induced alterations in homeostasis: alteration in differentiated functions of infected cells in vivo. *Science* **218**(4577), 1125–1127.

Oldstone, M. B., Ahmed, R., Buchmeier, M. J., et al. (1985). Perturbation of differentiated functions during viral infection in vivo. I. Relationship of lymphocytic choriomeningitis virus and host strains to growth hormone deficiency. *Virology* **142**(1), 158–174.

Parola, P., de Lamballerie, X., Jourdan, J., et al. (2006). Novel chikungunya virus variant in travelers returning from Indian Ocean islands. *Emerg Infect Dis* **12**(10), 1493–1499.

Perales, C., Martin, V., Ruiz-Jarabo, C. M., et al. (2005). Monitoring sequence space as a test for the target of selection in viruses. *J Mol Biol* **345**(3), 451–459.

Perales, C., Lorenzo-Redondo, R., López-Galíndez, C., et al. (2010). Mutant spectra in virus behavior. *Future Virol* **5**(6), 679–698.

Pereira, H. G. (1981). Foot-and-mouth disease virus. In: "Virus Diseases of Food Animals" (R. P. G. Gibbs, Ed.), Vol. **2**, pp. 333–363. Academic Press, New York.

Peters, C. J. (2002). Human infection with arenaviruses in the Americas. *Curr Top Microbiol Immunol* **262**, 65–74.

Peters, C. J. (2007). Emerging viral diseases. In: "Fields Virology," 5th edition (D. M. Knipe, and P. M. Howley, Eds.). Lippincott Williams & Wilkins, Philadelphia, PA, pp. 605–625.

Pfeiffer, J. K., and Kirkegaard, K. (2005). Increased fidelity reduces poliovirus fitness under selective pressure in mice. *PLoS Pathog* **1**, 102–110.

Powers, A. M., Brault, A. C., Tesh, R. B., et al. (2000). Re-emergence of Chikungunya and O'nyong-nyong viruses: evidence for distinct geographical lineages and distant evolutionary relationships. *J Gen Virol* **81**(Pt 2), 471–479.

Pulliam, J. R. (2008). Viral host jumps: moving toward a predictive framework. *Ecohealth* **5**(1), 80–91.

Riviere, Y., and Oldstone, M. B. (1986). Genetic reassortants of lymphocytic choriomeningitis virus: unexpected disease and mechanism of pathogenesis. *J Virol* **59**(2), 363–368.

Riviere, Y., Ahmed, R., Southern, P., et al. (1985). Perturbation of differentiated functions during viral infection in vivo. II. Viral reassortants map growth hormone defect to the S RNA of the lymphocytic choriomeningitis virus genome. *Virology* **142**(1), 175–182.

Ron, D., and Tal, J. (1985). Coevolution of cells and virus as a mechanism for the persistence of lymphotropic minute virus of mice in L-cells. *J Virol* **55**(2), 424–430.

Rowlands, D. J., Clarke, B. E., Carroll, A. R., et al. (1983). Chemical basis of antigenic variation in foot-and-mouth disease virus. *Nature* **306**(5944), 694–697.

Ruiz-Jarabo, C. M., Arias, A., Baranowski, E., et al. (2000). Memory in viral quasispecies. *J Virol* **74**, 3543–3547.

Salazar-Bravo, J., Ruedas, L. A., and Yates, T. L. (2002). Mammalian reservoirs of arenaviruses. *Curr Top Microbiol Immunol* **262**, 25–63.

Sanjuan, R., Nebot, M. R., Chirico, N., et al. (2010). Viral mutation rates. *J Virol* **84**(19), 9733–9748.

Schinazi, R. F., Larder, B. A., and Mellors, J. W. (1997). Mutations in retroviral genes associated with drug resistance. *Int Antivir News* **5**, 129–142.

Schrijver, R. S., van Oirschot, J. T., Dekker, A., et al. (1998). Foot and mouth disease is not a zoonosis. *Tijdschr Diergeneeskd* **123**(24), 750.

Schuffenecker, I., Iteman, I., Michault, A., et al. (2006). Genome microevolution of chikungunya viruses causing the Indian Ocean outbreak. *PLoS Med* **3**(7), e263.

Sevilla, N., Kunz, S., Holz, A., et al. (2000). Immunosuppression and resultant viral persistence by specific viral targeting of dendritic cells. *J Exp Med* **192**(9), 1249–1260.

Sevilla, N., McGavern, D. B., Teng, C., et al. (2004). Viral targeting of hematopoietic progenitors and inhibition of DC maturation as a dual strategy for immune subversion. *J Clin Invest* **113**(5), 737–745.

Simmonds, P. (2010). Recombination in the evolution of picornaviruses. In: "The Picornaviruses" (E. Ehrenfeld, E. Domingo, and R. P. Roos, Eds.). ASM Press, Washington, DC, pp. 229–238.

Simon, H. A. (1996). "The Sciences of the Artificial," 3rd edition. The MIT Press, Cambridge, MA.

Smelt, S. C., Borrow, P., Kunz, S., et al. (2001). Differences in affinity of binding of lymphocytic choriomeningitis virus strains to the cellular receptor alpha-dystroglycan correlate with viral tropism and disease kinetics. *J Virol* **75**(1), 448–457.

Smolinski, M. S., Hamburg, M. A., and Lederberg, J., Eds. (2003). Microbial Threats to Health. Emergence, Detection and Response. The National Academies Press, Washington, DC.

Sobrino, F., Dávila, M., Ortín, J., et al. (1983). Multiple genetic variants arise in the course of replication of foot-and-mouth disease virus in cell culture. *Virology* **128**, 310–318.

Solé, R., and Goodwin, B. (2000). "Signs of Life. How Complexity Pervades Biology." Basic Books, New York.

Spiropoulou, C. F., Kunz, S., Rollin, P. E., et al. (2002). New World arenavirus clade C, but not clade A and B viruses, utilizes alpha-dystroglycan as its major receptor. *J Virol* **76**(10), 5140–5146.

Taboga, O., Tami, C., Carrillo, E., et al. (1997). A large-scale evaluation of peptide vaccines against foot-and-mouth disease: lack of solid protection in cattle and isolation of escape mutants. *J Virol* **71**(4), 2606–2614.

Taylor, W. P., and Marshall, I. D. (1975). Adaptation studies with Ross River virus: laboratory mice and cell cultures. *J Gen Virol* **28**(1), 59–72.

Temin, H. M. (1989). Is HIV unique or merely different? *J AIDS* **2**(1), 1–9.

Teng, M. N., Borrow, P., Oldstone, M. B., et al. (1996). A single amino acid change in the glycoprotein of lymphocytic choriomeningitis virus is associated with the ability to cause growth hormone deficiency syndrome. *J Virol* **70**(12), 8438–8443.

Tsetsarkin, K. A., Vanlandingham, D. L., McGee, C. E., et al. (2007). A single mutation in chikungunya virus affects vector specificity and epidemic potential. *PLoS Pathog* **3**(12), e201.

Vasilakis, N., Deardorff, E. R., Kenney, J. L., et al. (2009). Mosquitoes put the brake on arbovirus evolution: experimental evolution reveals slower mutation accumulation in mosquito than vertebrate cells. *PLoS Pathog* **5**(6), e1000467.

Vazeille, M., Moutailler, S., Coudrier, D., et al. (2007). Two Chikungunya isolates from the outbreak of La Reunion (Indian Ocean) exhibit different patterns of infection in the mosquito, *Aedes albopictus*. *PLoS One* **2**(11), e1168.

Vignuzzi, M., and Andino, R. (2010). Biological implications of picornavirus fidelity mutants. In: "The Picornaviruses" (E. Ehrenfeld, E. Domingo, and R. F. Roos, Eds.). ASM Press, Washington, DC, pp. 213–228.

Vignuzzi, M., Stone, J. K., Arnold, J. J., et al. (2006). Quasispecies diversity determines pathogenesis through cooperative interactions in a viral population. *Nature* **439**, 344–348.

Weaver, S. C. (2006). Evolutionary influences in arboviral disease. *Curr Top Microbiol Immunol* **299**, 285–314.

Weaver, S. C., and Barrett, A. D. (2004). Transmission cycles, host range, evolution and emergence of arboviral disease. *Nat Rev Microbiol* **2**(10), 789–801.

Weaver, S. C., Kang, W., Shirako, Y., et al. (1997). Recombinational history and molecular evolution of western equine encephalomyelitis complex alphaviruses. *J Virol* **71**(1), 613–623.

Weaver, S. C., Brault, A. C., Kang, W., et al. (1999). Genetic and fitness changes accompanying adaptation of an arbovirus to vertebrate and invertebrate cells. *J Virol* **73**(5), 4316–4326.

Weaver, S. C., Salas, R. A., de Manzione, N., et al. (2000). Guanarito virus (Arenaviridae) isolates from endemic and outlying localities in Venezuela: sequence comparisons among and within strains isolated from Venezuelan hemorrhagic fever patients and rodents. *Virology* **266**(1), 189–195.

Woo, P. C., Lau, S. K., and Yuen, K. Y. (2006). Infectious diseases emerging from Chinese wet-markets: zoonotic origins of severe respiratory viral infections. *Curr Opin Infect Dis* **19**(5), 401–407.

Woolhouse, M. E., Webster, J. P., Domingo, E., et al. (2002). Biological and biomedical implications of the co-evolution of pathogens and their hosts. *Nat Genet* **32**(4), 569–577.

Wright, C. F., Morelli, M. J., Thebaud, G., et al. (2011). Beyond the consensus: dissecting within-host viral population diversity of foot-and-mouth disease virus by using next-generation genome sequencing. *J Virol* **85**(5), 2266–2275.

Wulff, H., McIntosh, B. M., Hamner, D. B., et al. (1977). Isolation of an arenavirus closely related to Lassa virus from Mastomys natalensis in south-east Africa. *Bull World Health Org* **55**(4), 441–444.

Yang, X., Zhou, Y. S., Wang, H. N., et al. (2011). Isolation, identification and complete genome sequence analysis of a strain of foot-and-mouth disease virus serotype Asia1 from pigs in southwest of China. *Virol J* **8**, 175.

Zhang, Y., Wang, J., Guo, W., et al. (2011). Emergence and transmission pathways of rapidly evolving evolutionary branch C4a strains of human enterovirus 71 in the Central Plain of China. *PLoS One* **6**(11), e27895.

Zhong, J., Gastaminza, P., Chung, J., et al. (2006). Persistent hepatitis C virus infection in vitro: coevolution of virus and host. *J Virol* **80**(22), 11082–11093.

Zinkernagel, R. M. (2002). Lymphocytic choriomeningitis virus and immunology. *Curr Top Microbiol Immunol* **263**, 1–5.

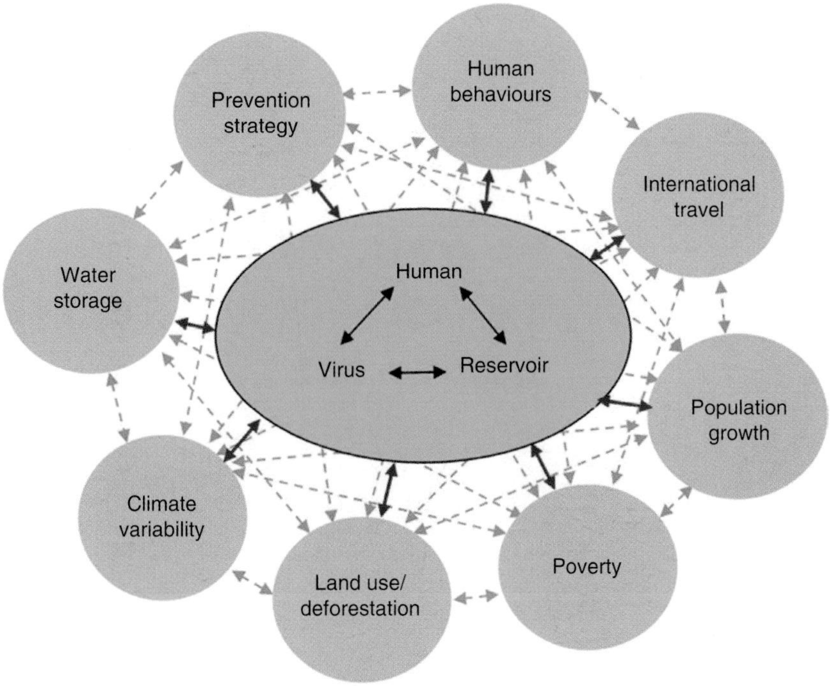

Figure 1.1. The VVD episystem showing interactions with influencing factors.

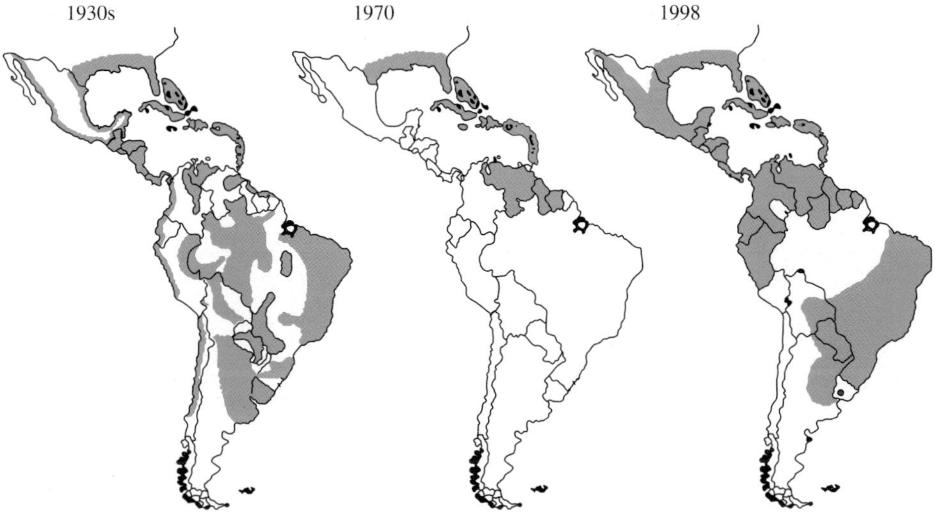

Figure 1.2. Reemergence of *Ae. aegypti* and dengue in the Americas.

Viral Infections and Global Change, First Edition. Edited by Sunit K. Singh.

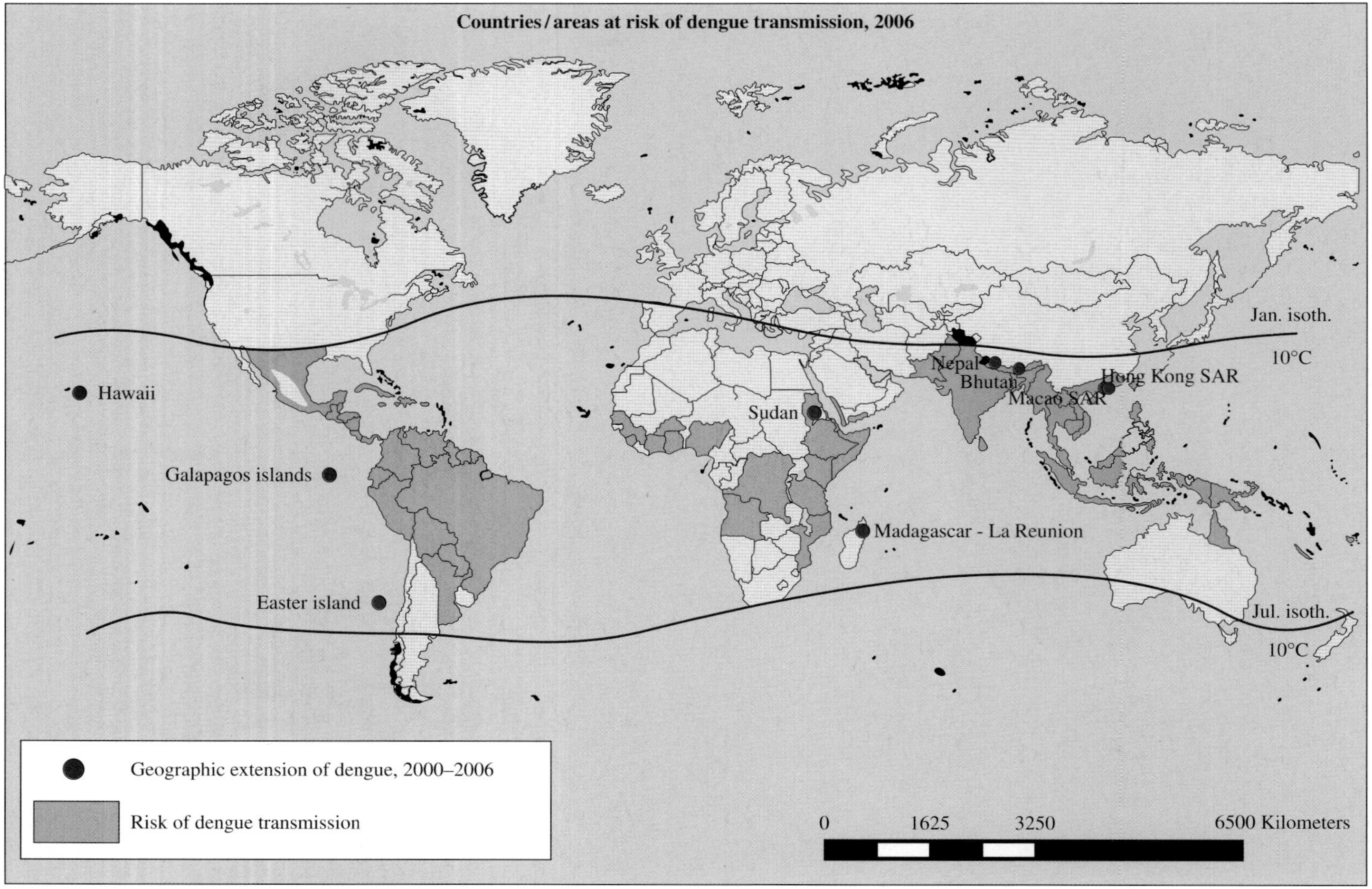

Figure 1.3. Areas at risk of dengue transmission. Source: Image from WHO

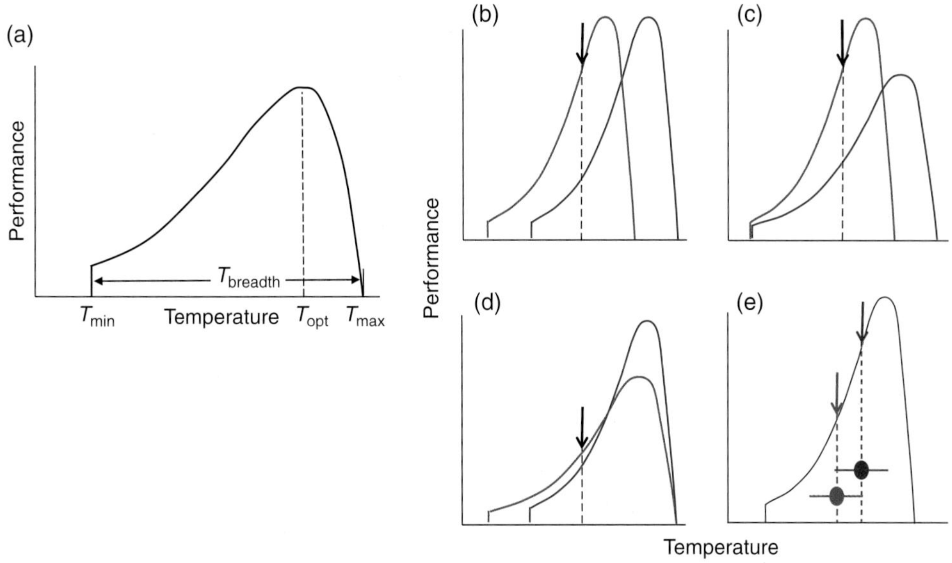

Figure 3.1. Hypothetical performance curve. A hypothetical performance curve illustrating the relationship between a performance-based measure of fitness and environmental temperature (**a**) and scenarios for performance curve or thermal preference evolution in response to climate warming (**b–e**) (redrawn from Gilchrist and Folk, 2008). The optimal temperature (T_{opt}), maximum limit of performance (T_{max}), minimum limit of performance (T_{min}), and degree of temperature specialization as indicated by the performance breadth ($T_{breadth}$) are depicted. For (**b**) through (**d**), a constant area under the performance curve constrains performance curve evolution. Current conditions are shown in blue and future conditions under global warming are shown in red. The black arrows indicate thermal preference prior to selection. Selection by climate warming generally decreases performance at the preference point. (**b**) The performance curve shifts horizontally with climate warming. (**c**) Evolution of the maximum thermal performance limit and the minimum thermal limit is constrained. (**d**) Evolution of the minimum thermal limit and the maximum thermal limit is constrained. (**e**) The performance curve does not change with global warming, but the temperature preference changes (from the blue arrow to the red arrow). The mean and variation in the temperature regimes of two hypothetical climate states are shown under the curve. A warming climate increases the risk of thermal damage as the population resides in an environment that is closer to the collapse in performance at high environmental temperature.

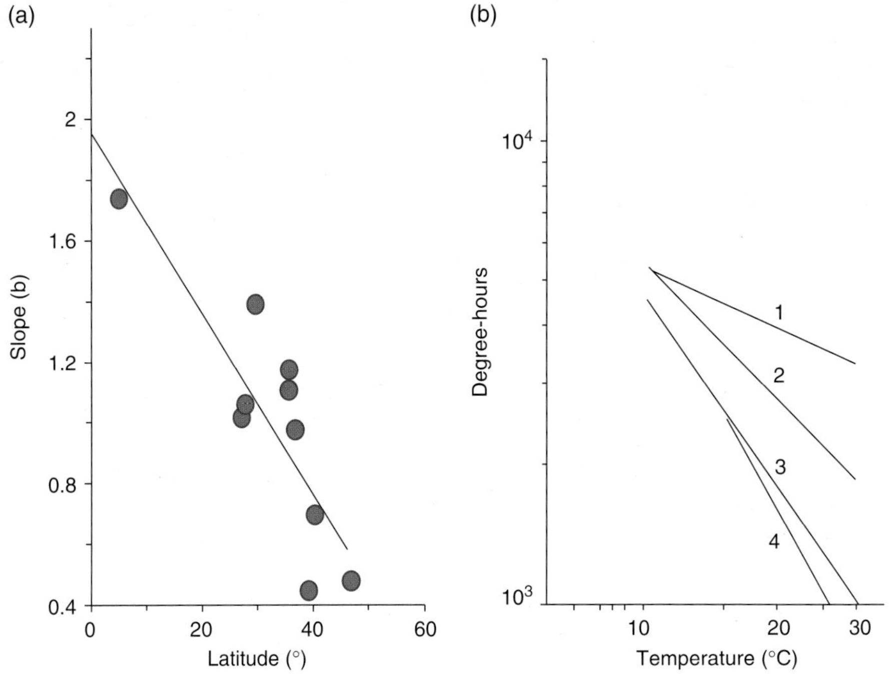

Figure 3.4. The relationship between the slope for developmental rate and temperature. The relationship between the slope for developmental rate with temperature for mosquitoes living at different latitudes (**a**) and the thermal requirement for development of four mosquito species (**b**) (redrawn from Pritchard and Mutch, 1985). The slope of the log transformation of the relationship $D=aT^b$ is the dependent variable in panel (**a**). D is development time. T is temperature (°C, uncorrected for developmental zero). The numbers in panel (**b**) correspond to *Ae. sticticus* from 50 N (1), *Aedes vexans* from 40 N (2), *Anopheles quadrimaculatus* from 32.5 N (3), and *Toxorhynchites brevipalpis* from 7S (4).

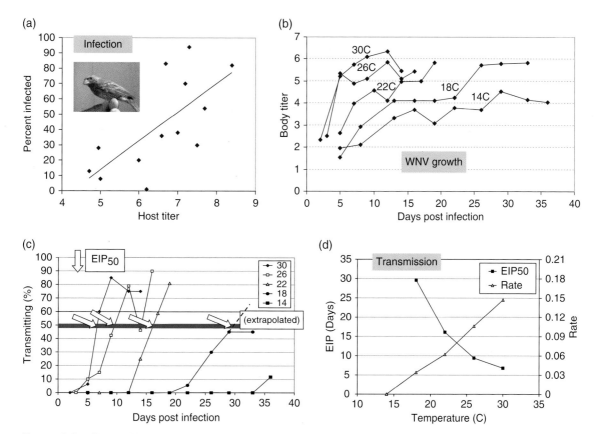

Figure 3.6. Effects of host titer and temperature on the vector competence of *Cx. tarsalis* for WNV. (a) Percent infected as a function of host viremia in \log_{10} plague forming units [PFU] of WNV per ml, (b) virus growth in PFU per mosquito as a function of days, (c) percent transmission as a function of time after infection, and time to 50% transmission and (d) rate of the EIP plotted as a function of temperature (redrawn from Reisen et al., 2006).

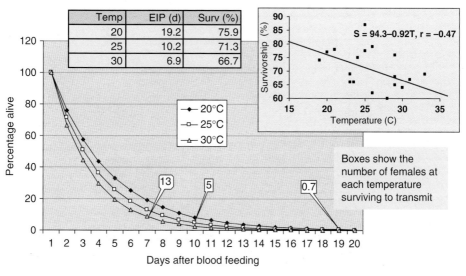

Figure 3.7. Effects of temperature on a number of infected females alive to transmit at the end of the EIP. Percentage of females surviving each day plotted as a function of temperature and the percentage alive at the end of the EIP at three temperatures.

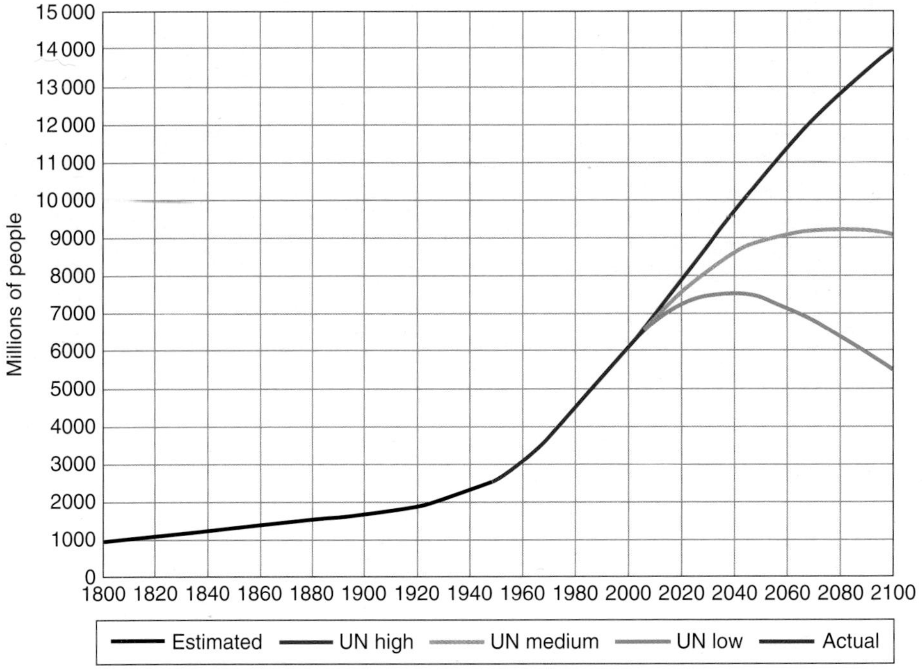

Figure 4.1. Global human population size. From: United Nations, Department of Economic and Social Affairs, Population Division (2011). World Population Prospects: The 2010 Revision, CD-ROM Edition.

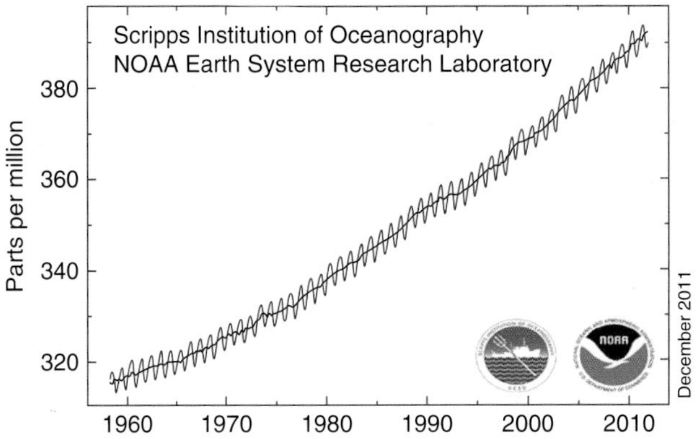

Figure 4.2. Change in atmospheric CO_2 concentrations measured at Mauna Loa, Hawaii (from P. Tans (www.esrl.noaa.gov/gmd/ccgg/trends/) and R. Keeling (scrippsco2.ucsd.edu/); accessed 1 Jan 2012).

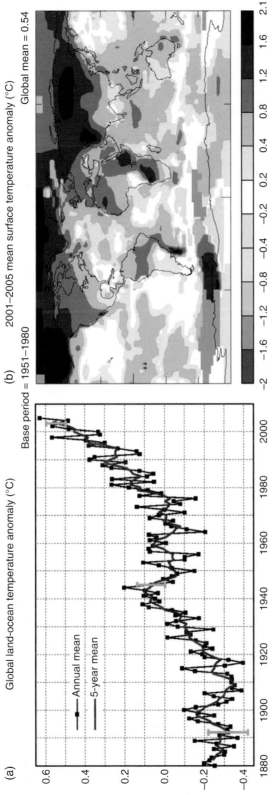

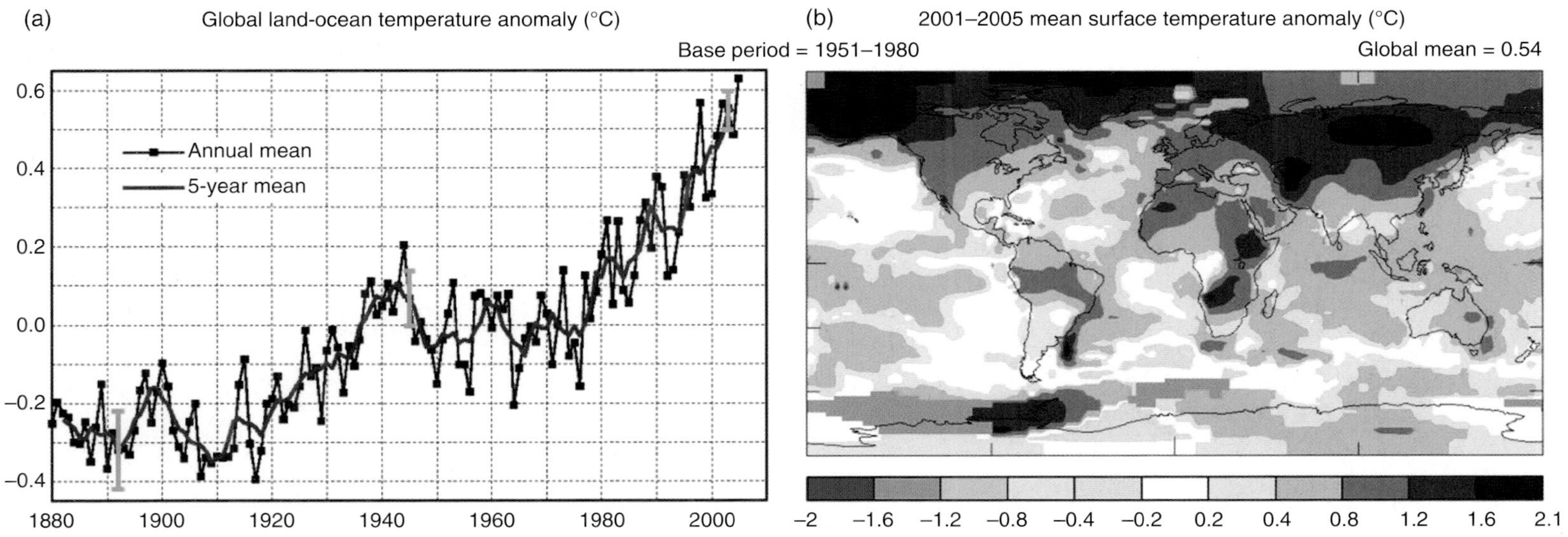

Figure 4.3. Surface temperature anomalies relative to 1951–1980 from surface air measurements at meteorological stations and ship and satellite SST measurements: (**a**) global annual mean anomalies and (**b**) temperature anomaly for the first half decade of the twenty-first century (Hansen et al., 2006).

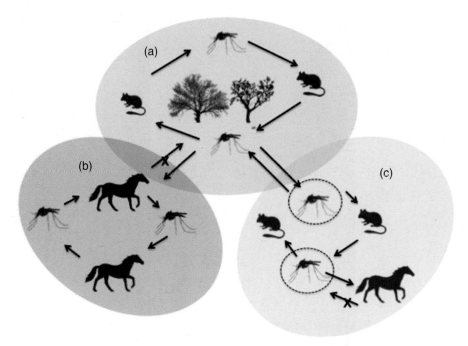

Figure 4.4. (a) Depicts an enzootic VEEV transmission cycle in which small rodents serve as reservoir hosts and *Culex* (*Mel.*) spp mosquitoes as vectors. (b) Depicts classical epizootic emergence in which viruses acquire the capacity for elevated replication in equids and viruses are transmitted by epizootic vectors capable of being infected with these high eqid titers. (c) Depicts the novel epizootic emergence evevt in 1993/96 in which epizootic emergence was mediated by adaptation to an epizootic mosquito vector (*Ae. taeniorhynchus*) as depicted by the dashed circle.

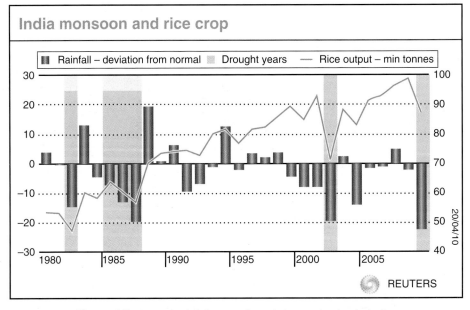

Figure 4.7. Annual rainfall anomaly and rice production in India.

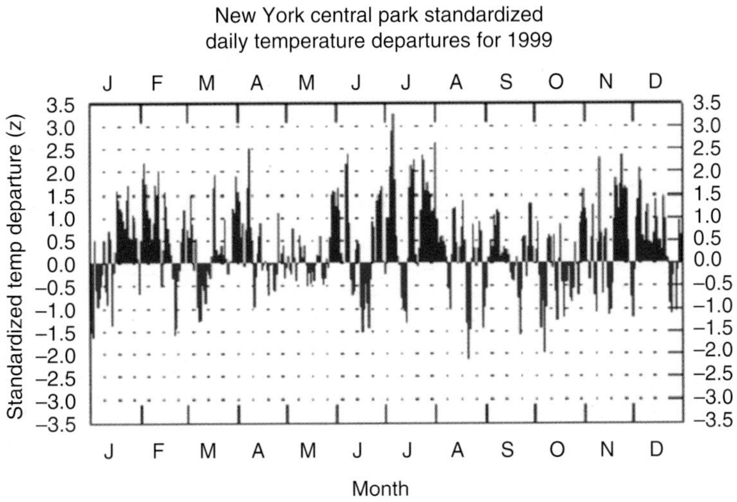

Figure 4.8. Departures from 50-year average temperature in Central Park, New York City, during 1999 (data from http://www.climatestations.com/new-york-city/).

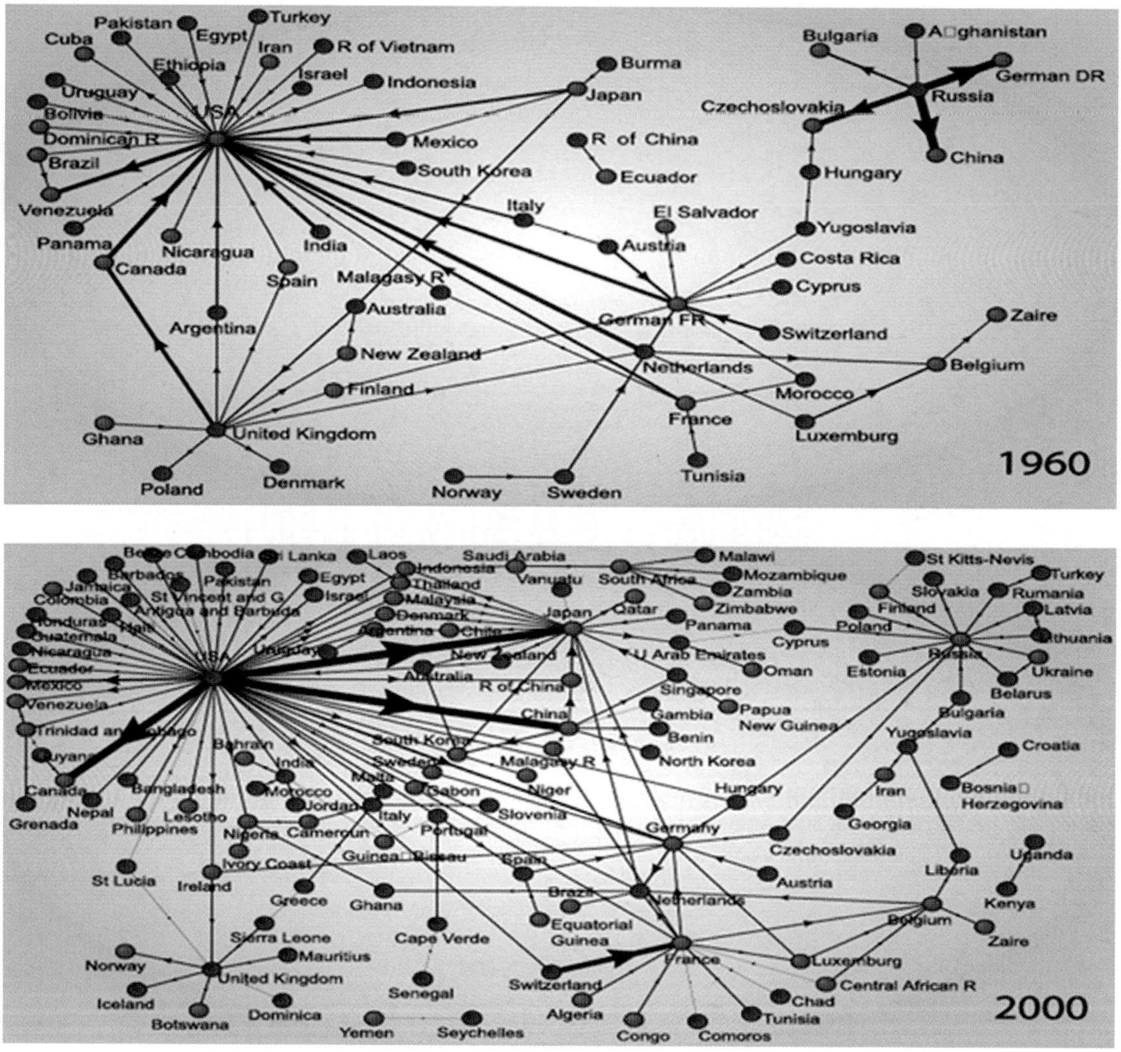

Figure 7.3. Evolution of modern trade patterns (Source: Patterns of dominant flows in the world trade web. Serrano MA, Bogunia M, Vespignani A J. Econ. Interac. Coor. 2007;2:111. Available from URL: http://arxiv.org/pdf/0704.1225v1.pdf. Accessed February 3, 2012).

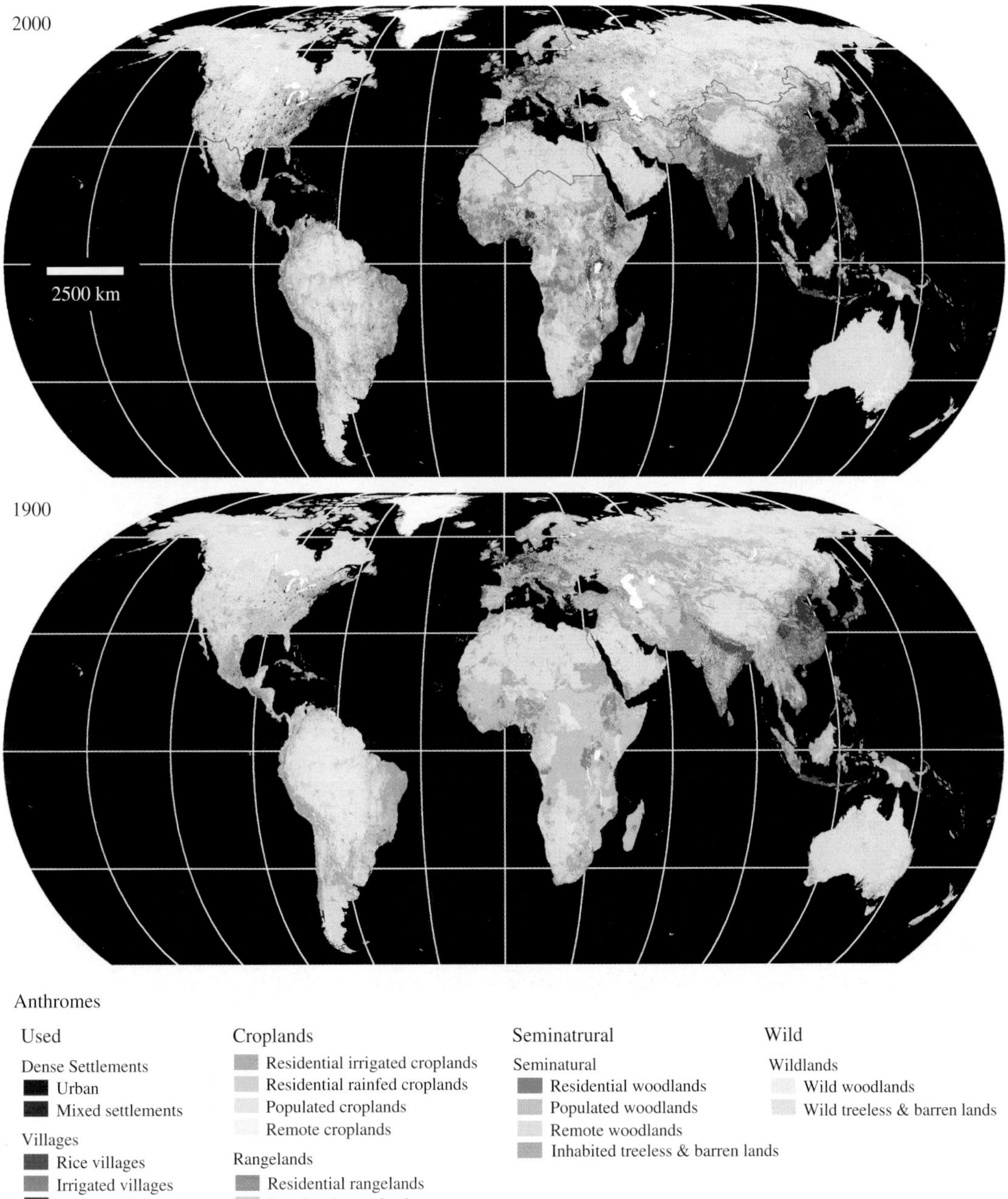

Figure 8.1. Anthropogenic biomes, 1900–2000, reproduced from Ellis et al. (2010).

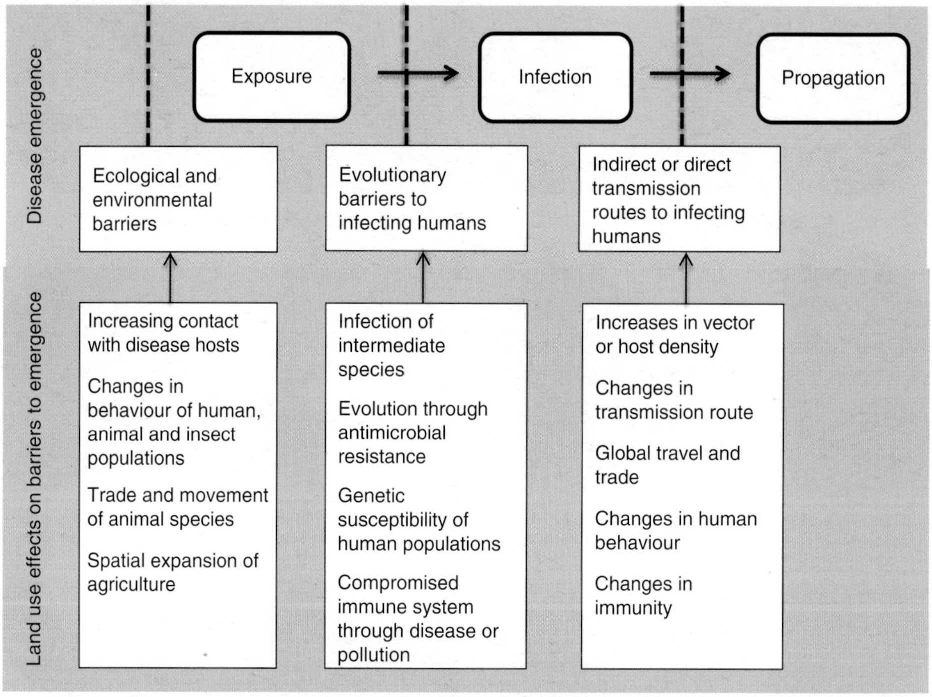

Figure 8.2. Effects of land-use changes on the process of disease emergence.

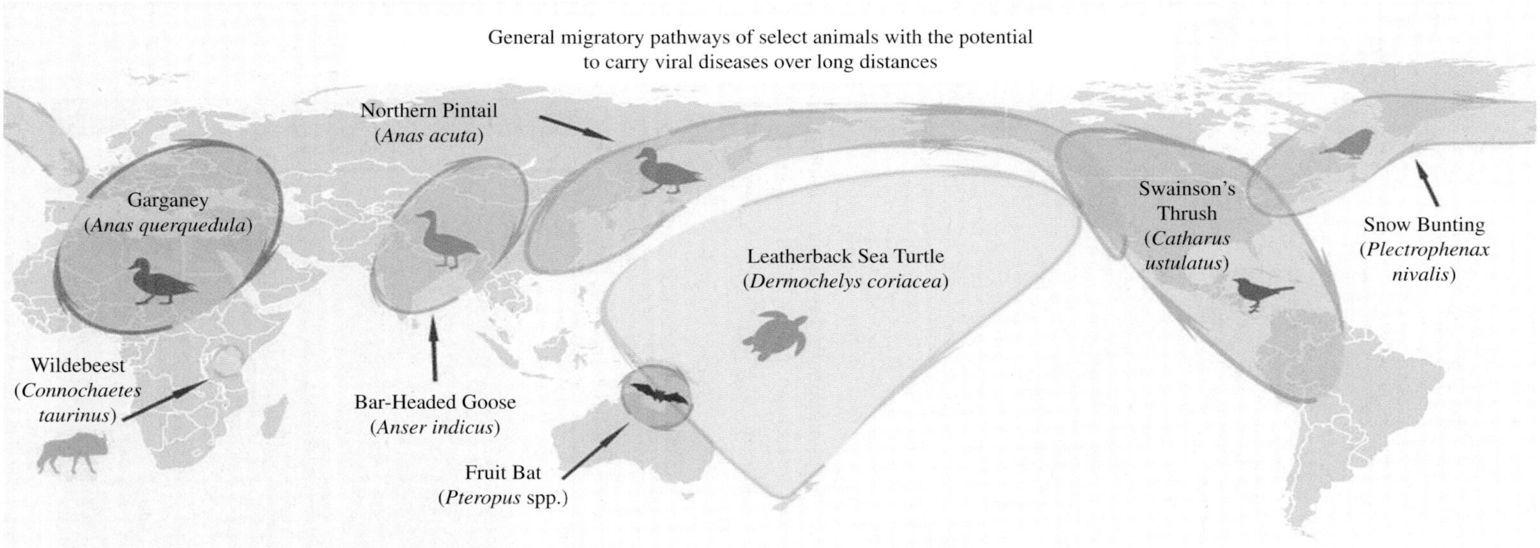

Figure 9.1. Examples of general migratory pathways for select animals that may contribute to the dispersal of viral diseases over long distances. Migratory routes for each species mentioned earlier were obtained and modified from the following sources: garganey (Gaidet et al., 2008), wildebeest (Serneels and Lambin, 2001), bar-headed goose (Takekawa et al., 2009), northern pintail (Miller et al., 2005; Yamaguchi et al., 2010), fruit bat (Breed et al., 2010), leatherback sea turtle (Benson et al., 2011), Swainson's thrush (Mack and Yong, 2000), and snow bunting (Lyngs, 2003). Pathways shown are examples of reported pathways from the literature but do not necessarily represent all known pathways for each species.

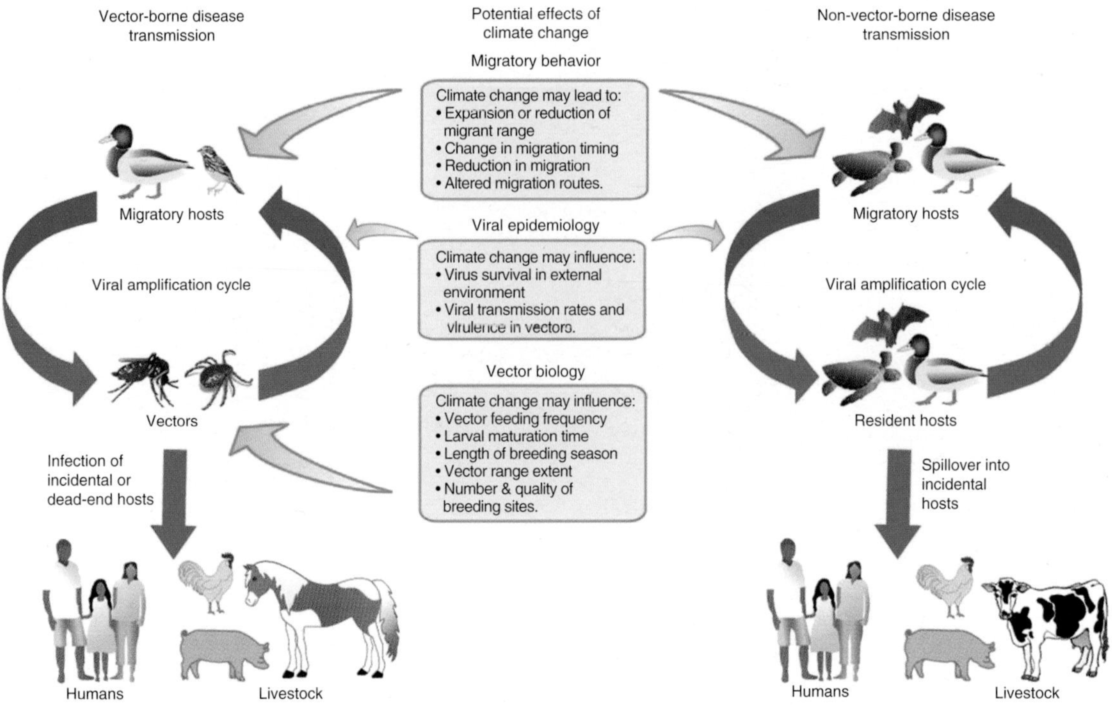

Figure 9.2. Potential effects of climate change on disease transmission pathways of vector-borne (left) and non-vector-borne (right) viruses associated with migratory animals. Symbols used in this Figure were provided courtesy of the Integration and Application Network, University of Maryland Center for Environmental Science (ian.umces.edu/symbols/).

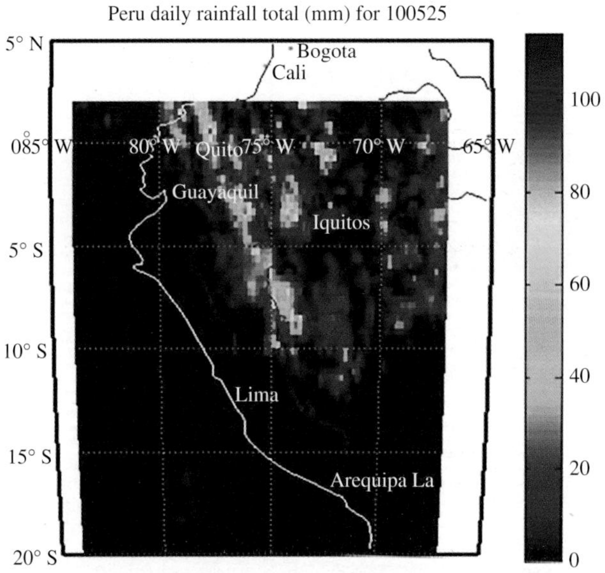

Figure 13.1. An example of the TRMM-derived rainfall for a single day in Peru. (Units are shown in mm.)

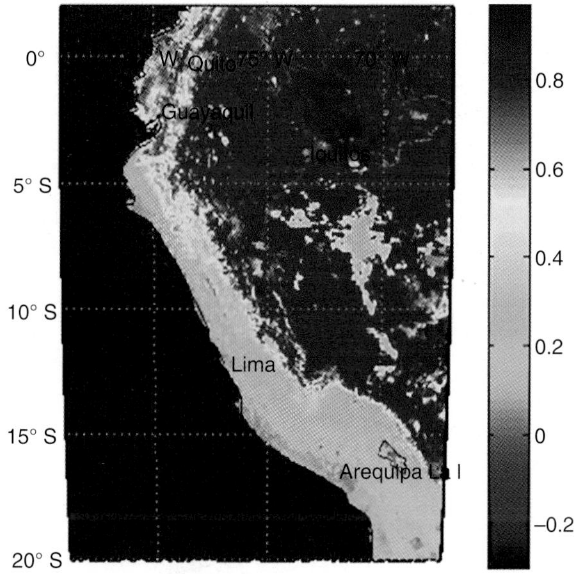

Figure 13.2. Example of NDVI values for a given 16-day interval in Peru.

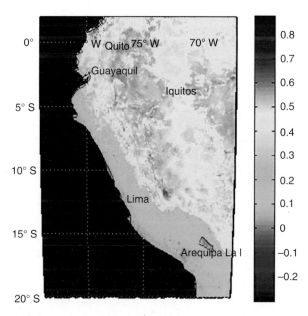

Figure 13.3. Example of EVI values for a given 16-day interval in Peru.

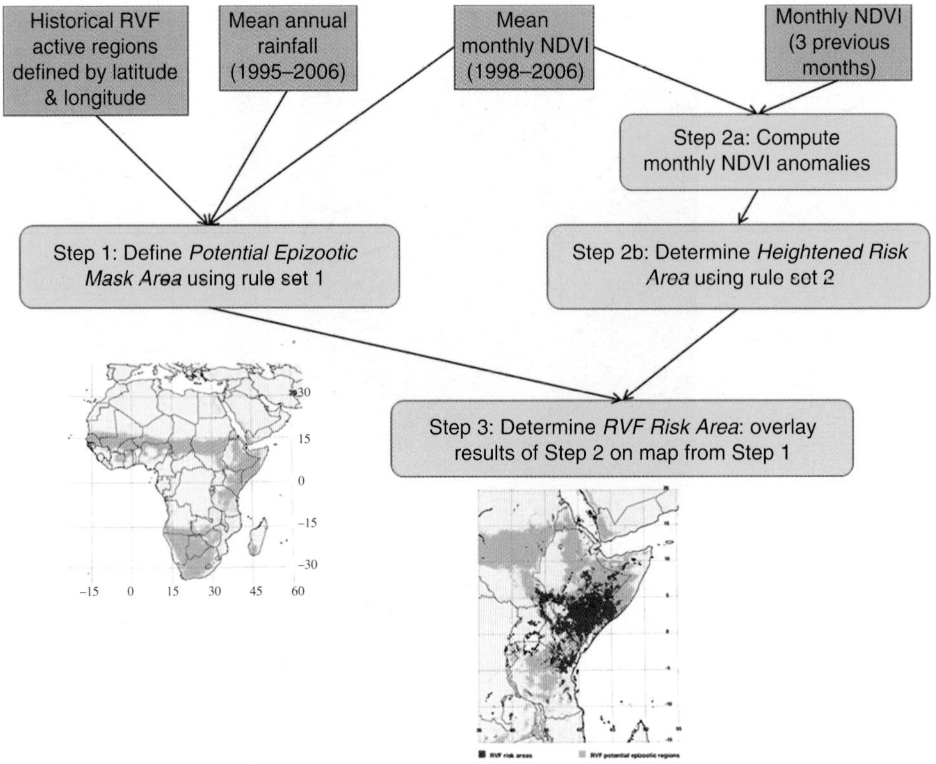

Figure 13.4. Major Steps of RVF Prediction (Method of Anyamba et al., 2009).

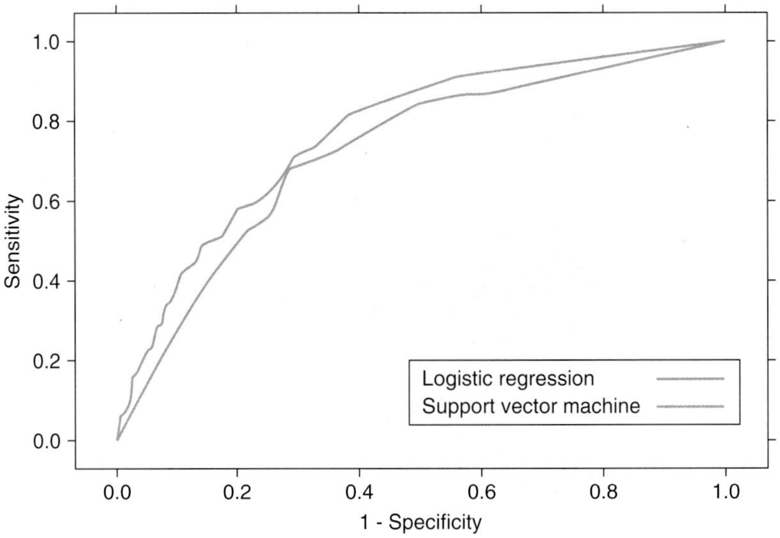

Figure 13.5. Receiver Operating Characteristic (ROC) curves for Logistic regression (LR) and Support Vector Machines (SVM).

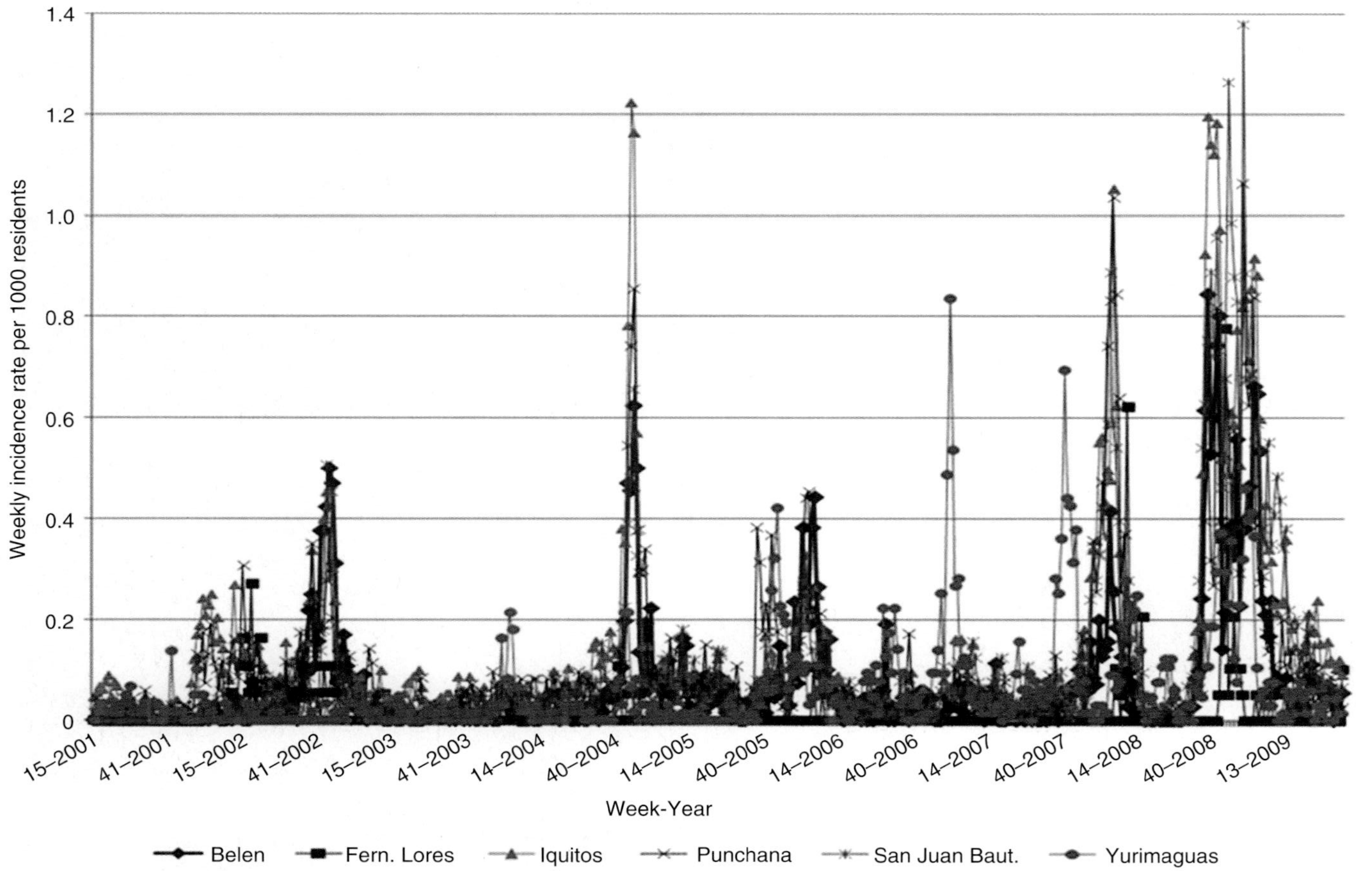

Figure 13.6. Dengue Weekly Incidence Rate.

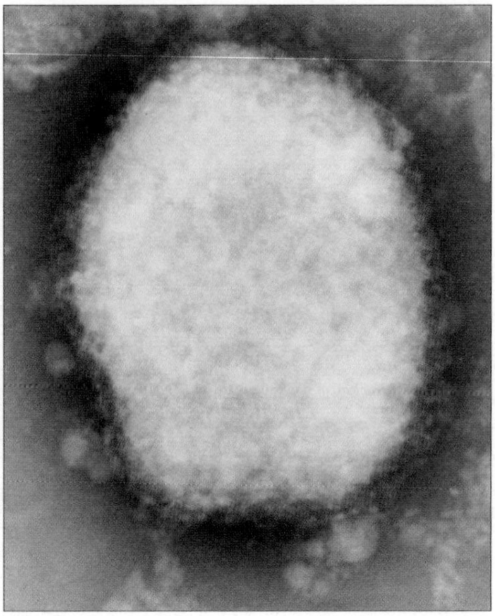

Figure 15.1. Negative stain electron micrograph reveals an "M" (mulberry type) monkeypox virion in human vesicular fluid. Courtesy CDC Public Image Library.

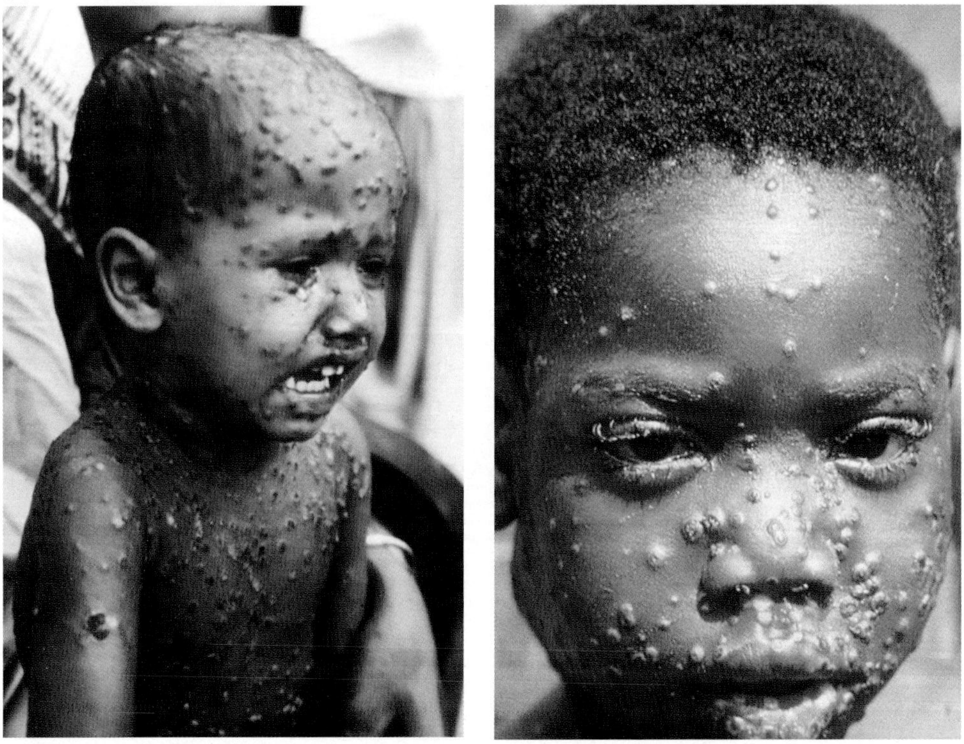

Figure 15.2. Clinical appearance of (**a**) smallpox and (**b**) monkeypox is similar. (**a**) The face of a boy infected with smallpox with facial lesions in various stages of resolution. At this point in time, the patient is still highly contagious. (**b**) The face of a young boy from the DRC exhibiting the characteristic maculopapular cutaneous rash of monkeypox. Courtesy CDC Public Image Library and World Health Organization.

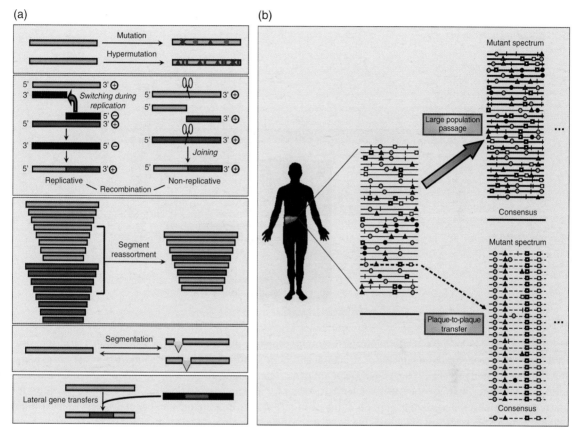

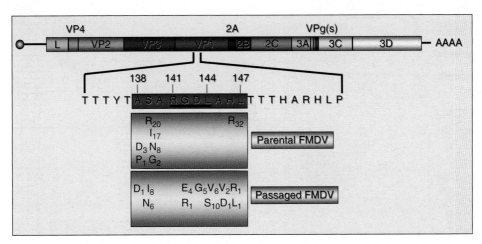

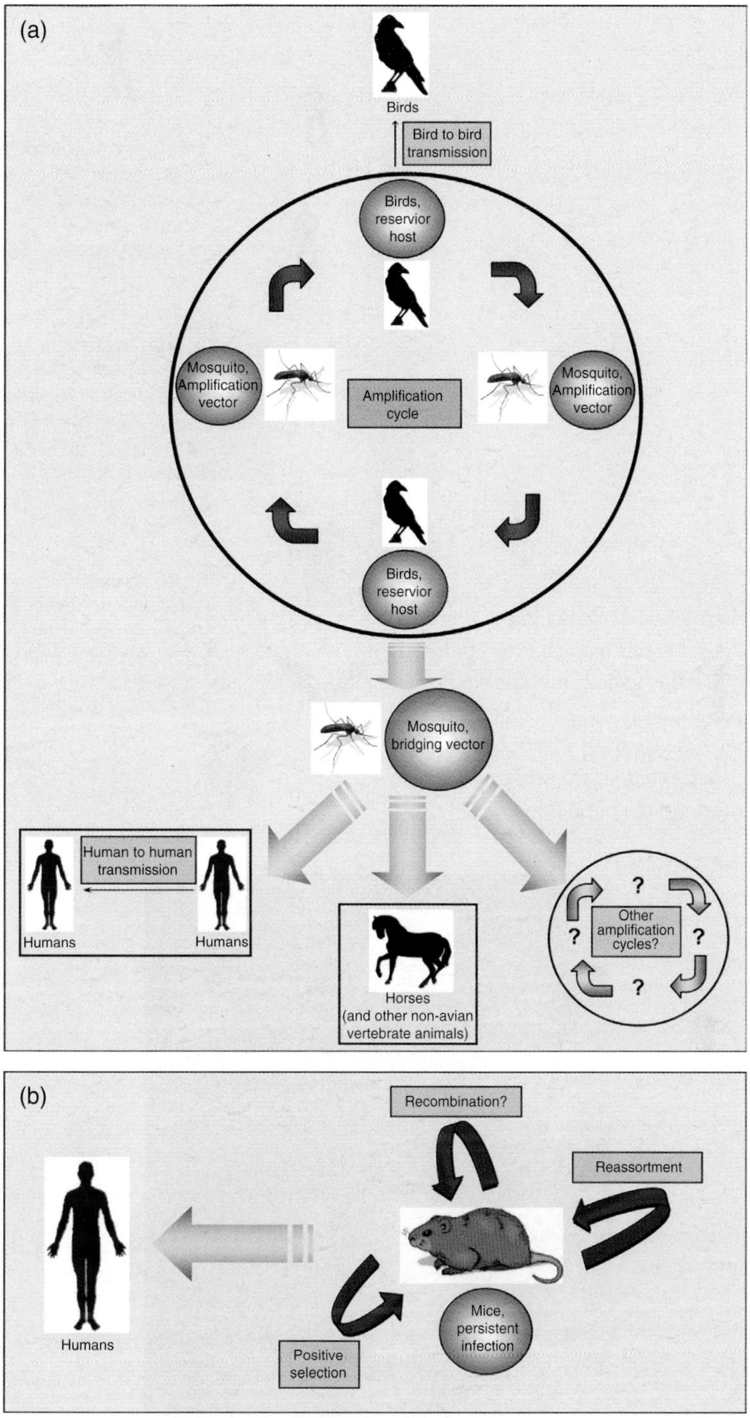

Figure 16.3. Simplified view of the influence of multiple hosts in the evolution of some viruses. (**a**) The amplification and transmission cycle of WNV involves reservoir hosts, amplification, and bridging vectors, as depicted schematically. Human-to-human transmission appears to be restricted to blood transfusion, organ transplantation, intrauterine virus spread, or breast milk feeding, although other means cannot be excluded. Likewise, there might be amplification cycles other than birds and mosquitoes in maintenance of this pathogen in nature. (**b**) Rodents are an extensive reservoir of many RNA viruses that have become (or have the potential to become) a zoonotic threat for humans. Arenaviruses constitute an example of rodent viruses that have emerged as human pathogens. Compare this scheme with the molecular mechanisms and the evolutionary forces described in Figure 16.1 and Table 16.1 (see text for additional implications and references).

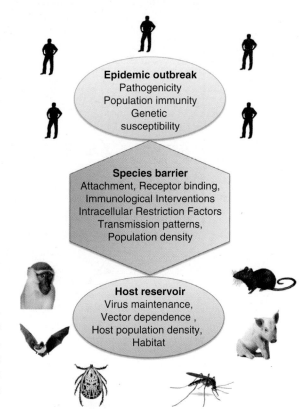

Figure 18.1. Steps involved in host switching of viruses. The zoonotic viruses have to cross many barriers before infecting human beings. Attachment, receptor binding, immunological interventions, intracellular host restriction factors, transmission patterns, and population density are the major species barriers to be crossed by a virus to successfully infect human population.

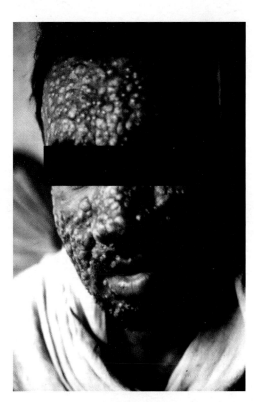

Figure 20.1. Cutaneous lesion in smallpox patient (Courtesy of Dr. Takeshi Kurata, National Institute of Infectious Diseases, Japan).

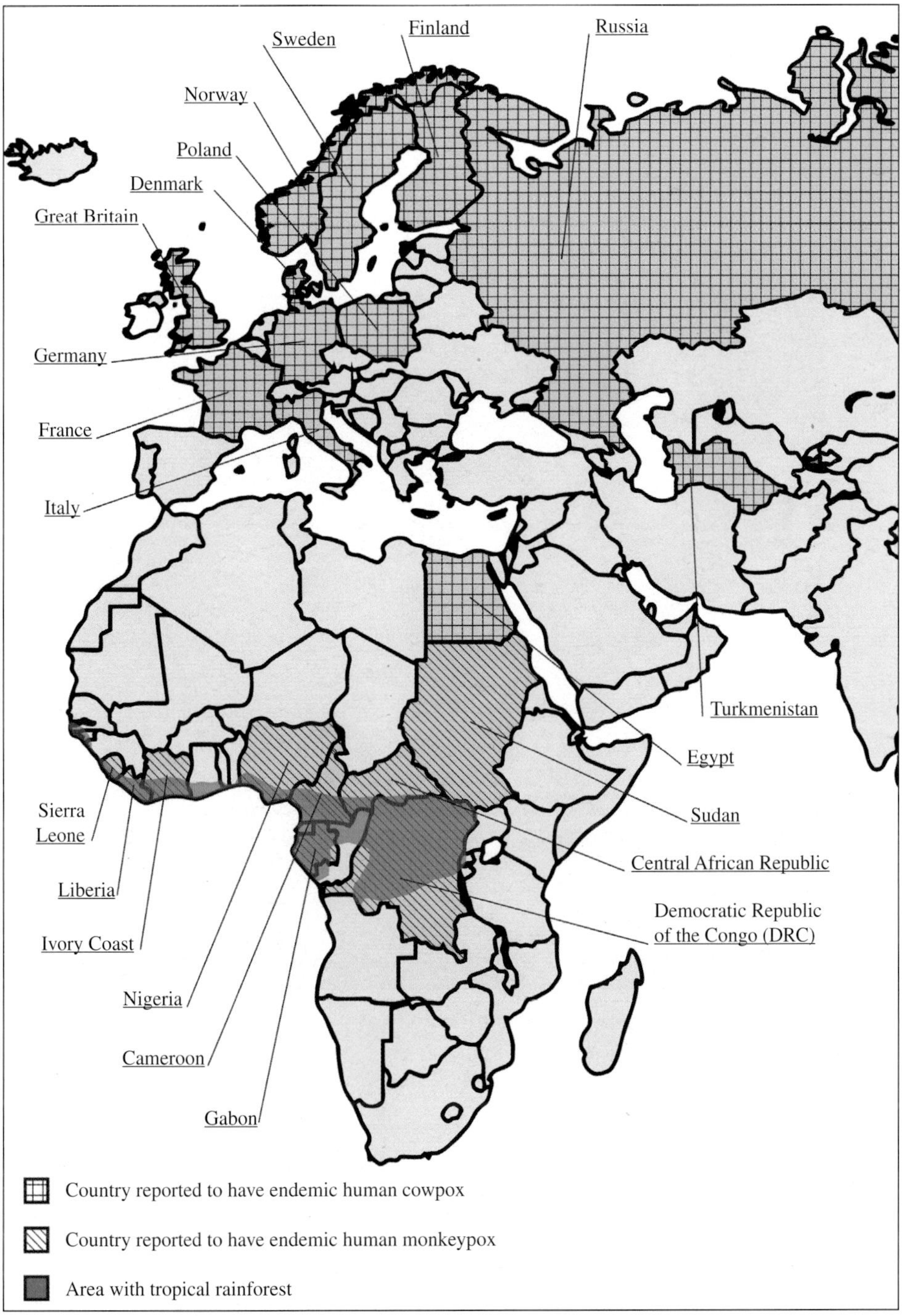

Figure 20.2. Geographical distribution of zoonotic poxvirus infections.

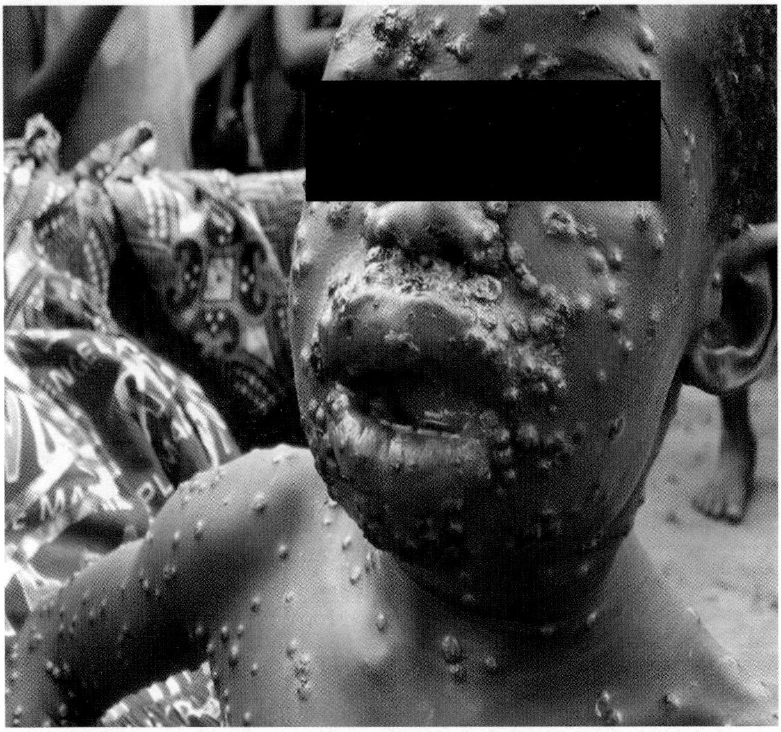

Figure 20.3. Cutaneous lesion in monkeypox patient (Courtesy of Dr. Muyembe-Tamfum JJ, National Institute of Biomedical Research, Democratic Republic of the Congo).

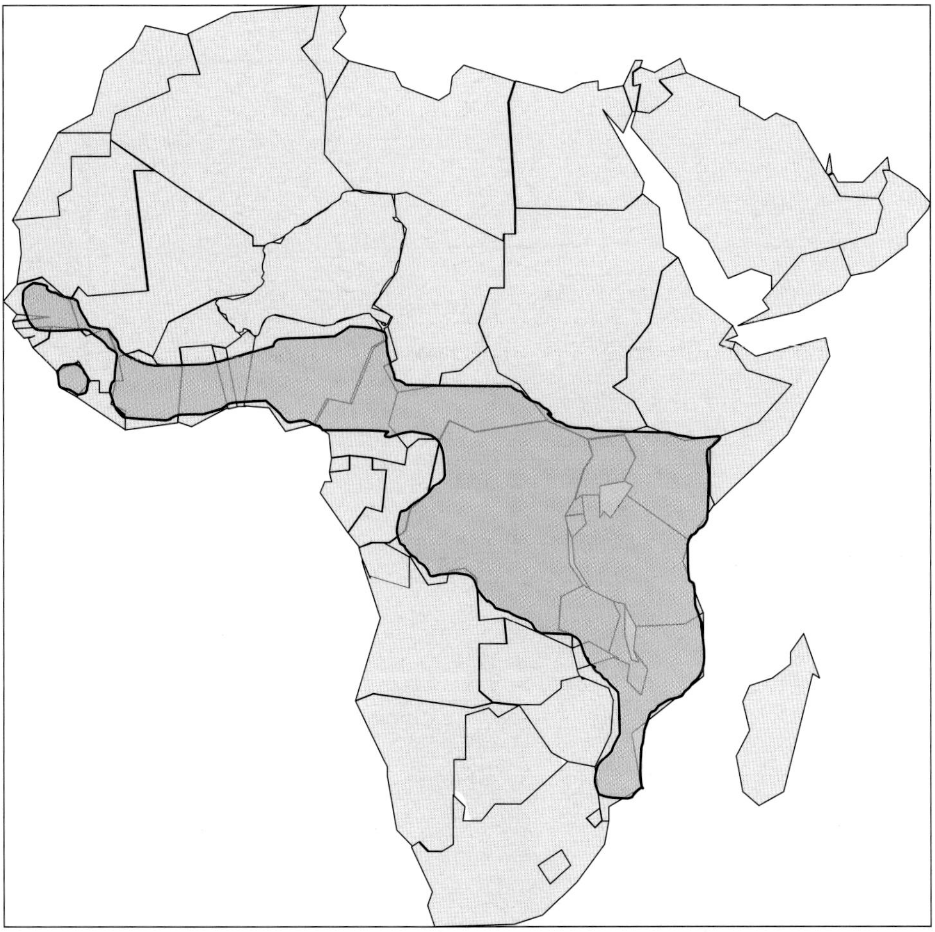

Figure 23.1. ONNV distribution map. Approximate known geographic distribution of ONNV. Positive areas determined by human febrile cases, human and/or animal serosurveys, and mosquito collections.

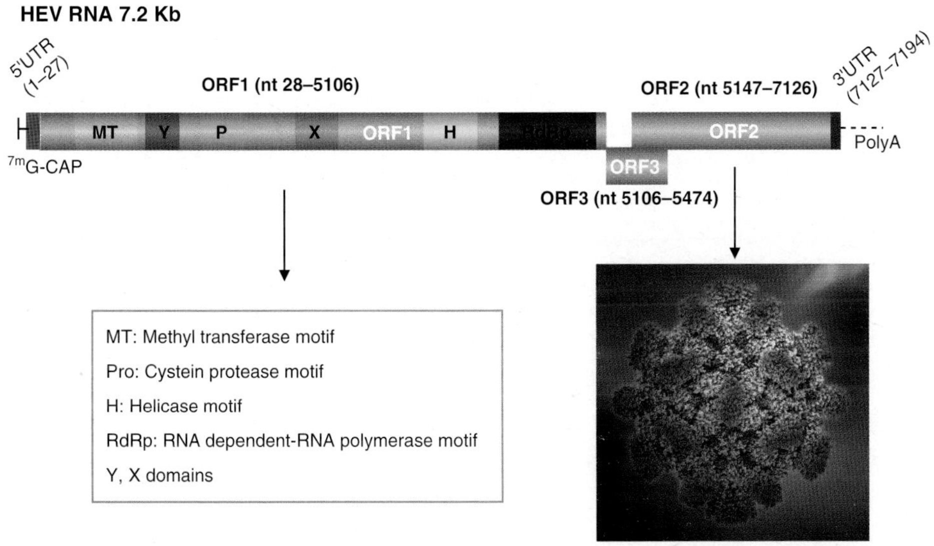

Figure 24.1. Hepatitis E virus genome organization and schematic structure.

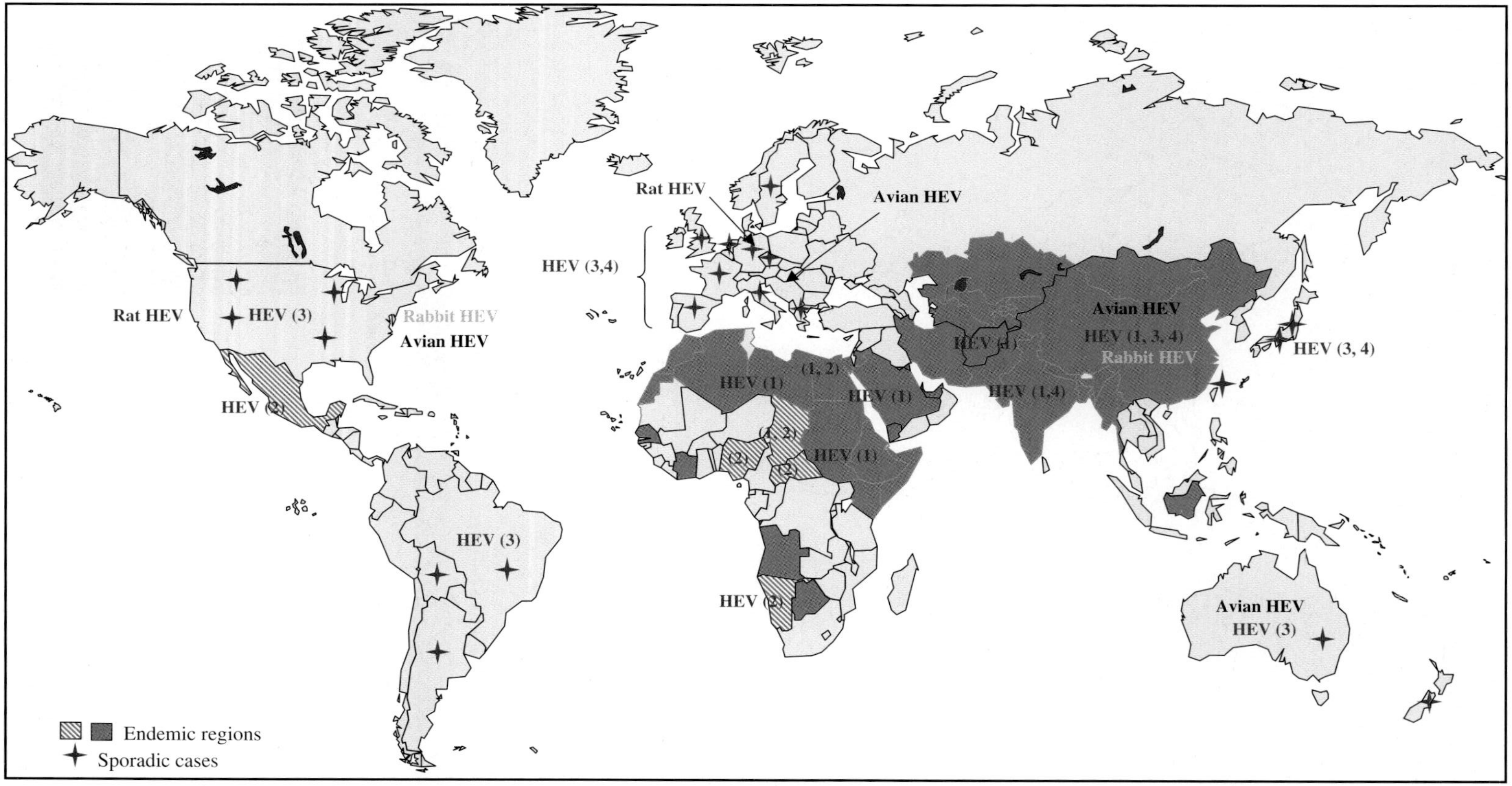

Figure 24.3. Geographic distribution of HEV and HEV-like viruses.

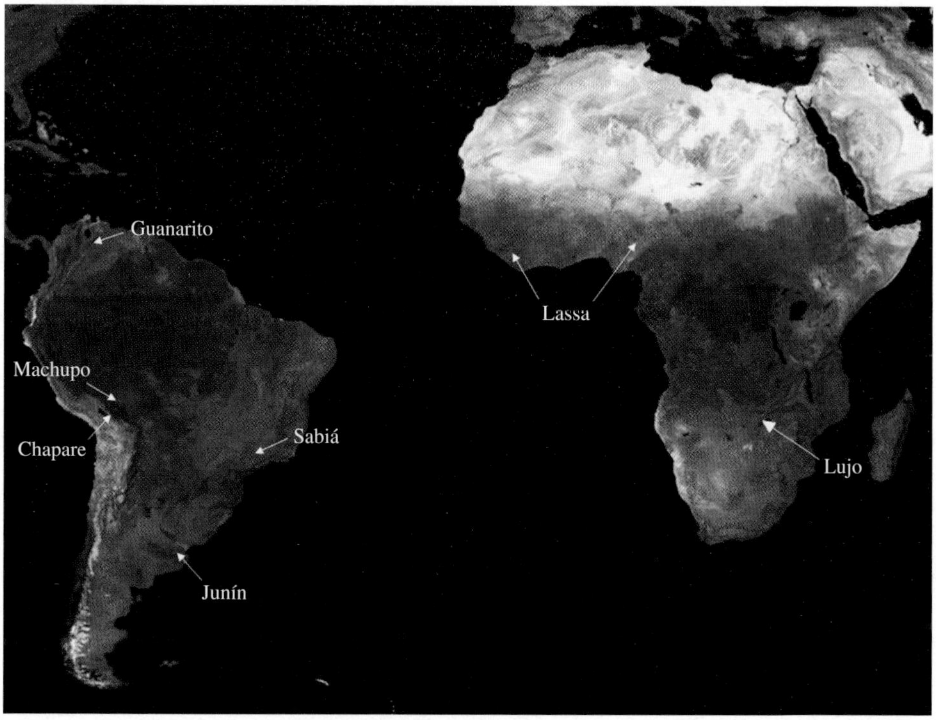

Figure 25.1. Geographic locations of arenaviruses associated with human hemorrhagic fevers. Map provided by NASA Visible Earth. Available at http://visibleearth.nasa.gov/.

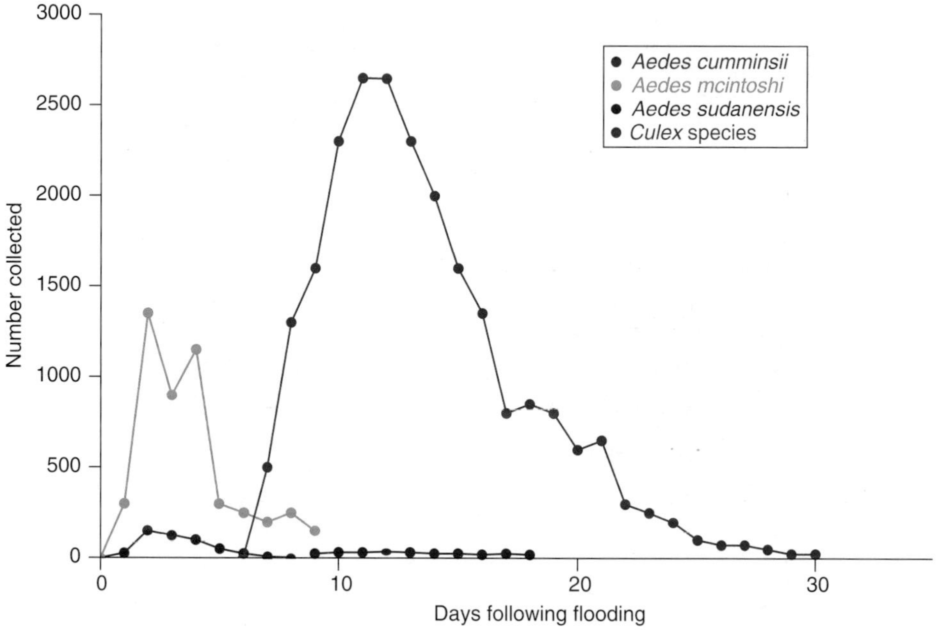

Figure 26.1. Mosquito population succession in a *dambo* habitat (reconstructed after Linthicum et al., 1983)

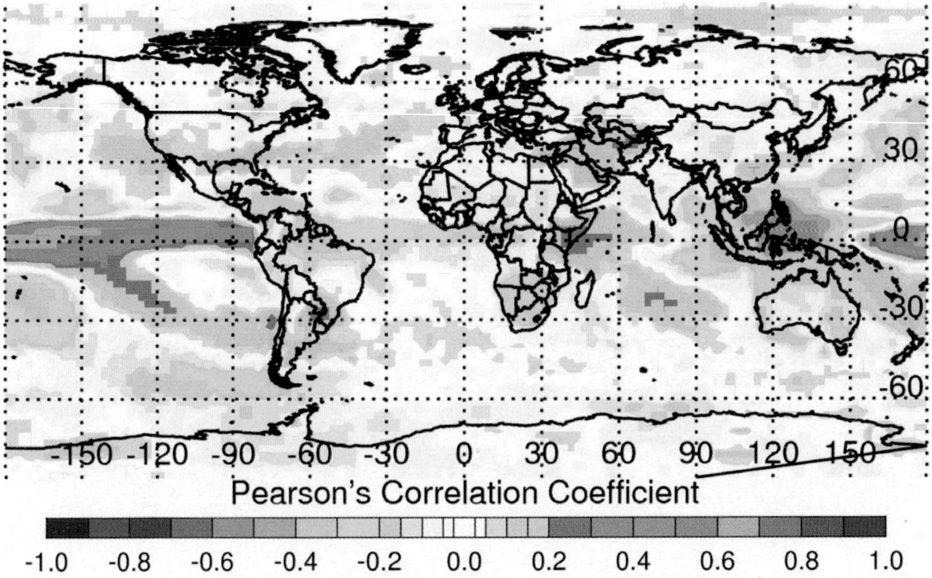

Figure 26.2. Correlation of SST and rainfall anomalies illustrates ENSO teleconnection patterns. There is a tendency for above (below)-normal rainfall during *El Niño* (*La Niña*) events over East Africa (Southern Africa, Southeast Asia). Similar differential anomaly patterns were observed for other regions, especially within the global tropics. These extremes (above or below) in rainfall influence regional ecology and consequently dynamics of mosquito disease vector populations and patterns of mosquito-borne disease outbreaks (Anyamba et al., 2012).

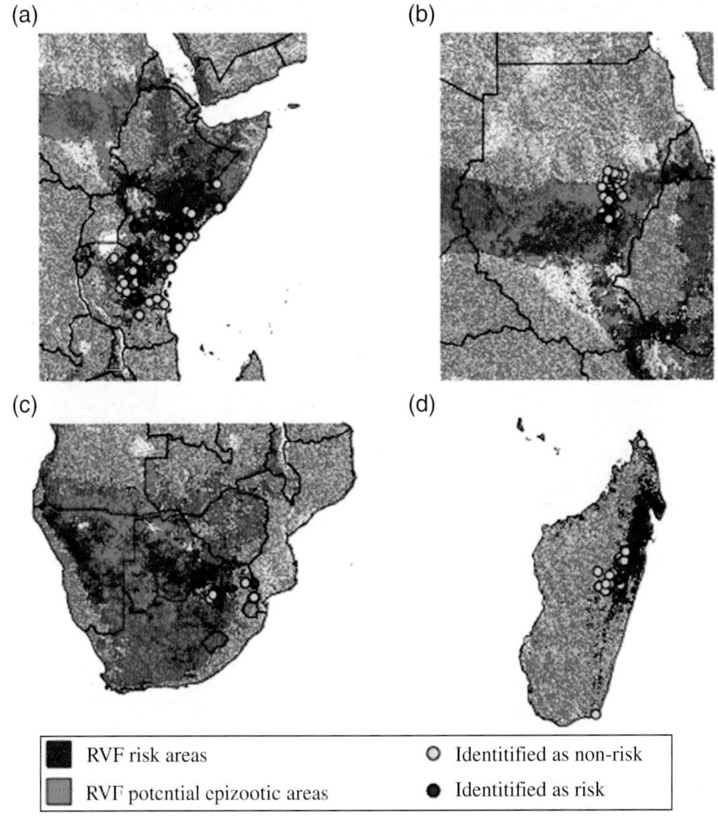

Figure 26.3. Summary RVF risk map, of Eastern Africa (September, 2006–May, 2007). Areas shown in green represent RVF potential epizootic areas, areas shown in red represent pixels that were mapped by the prediction system to be at risk for RVF activity during the respective time periods, blue dots indicate human cases identified to be in the RVF risk areas, and yellow dots represents human cases in areas not mapped to be at risk.

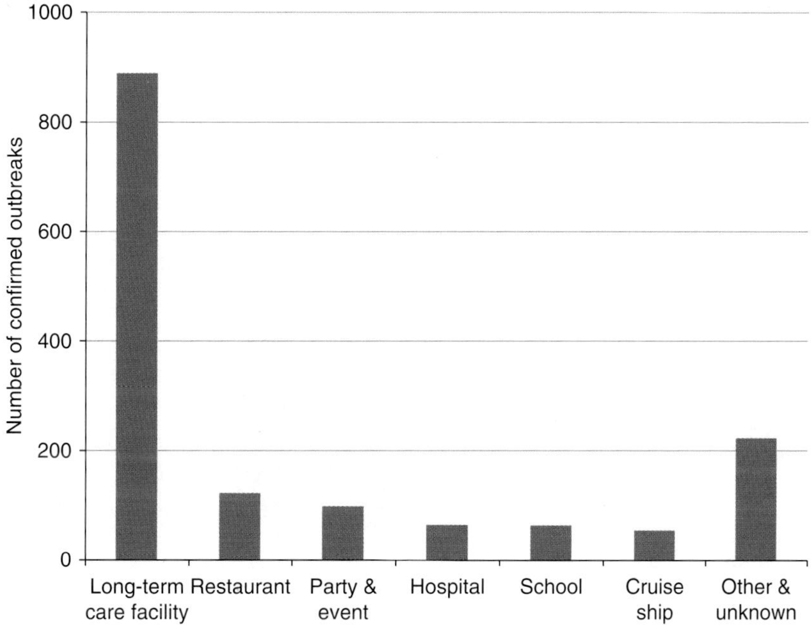

Figure 28.1. Setting of 1518 confirmed norovirus outbreaks in the United States from 2010–2011. Data are reproduced from http://www.cdc.gov/features/dsnorovirus/figure2.html.

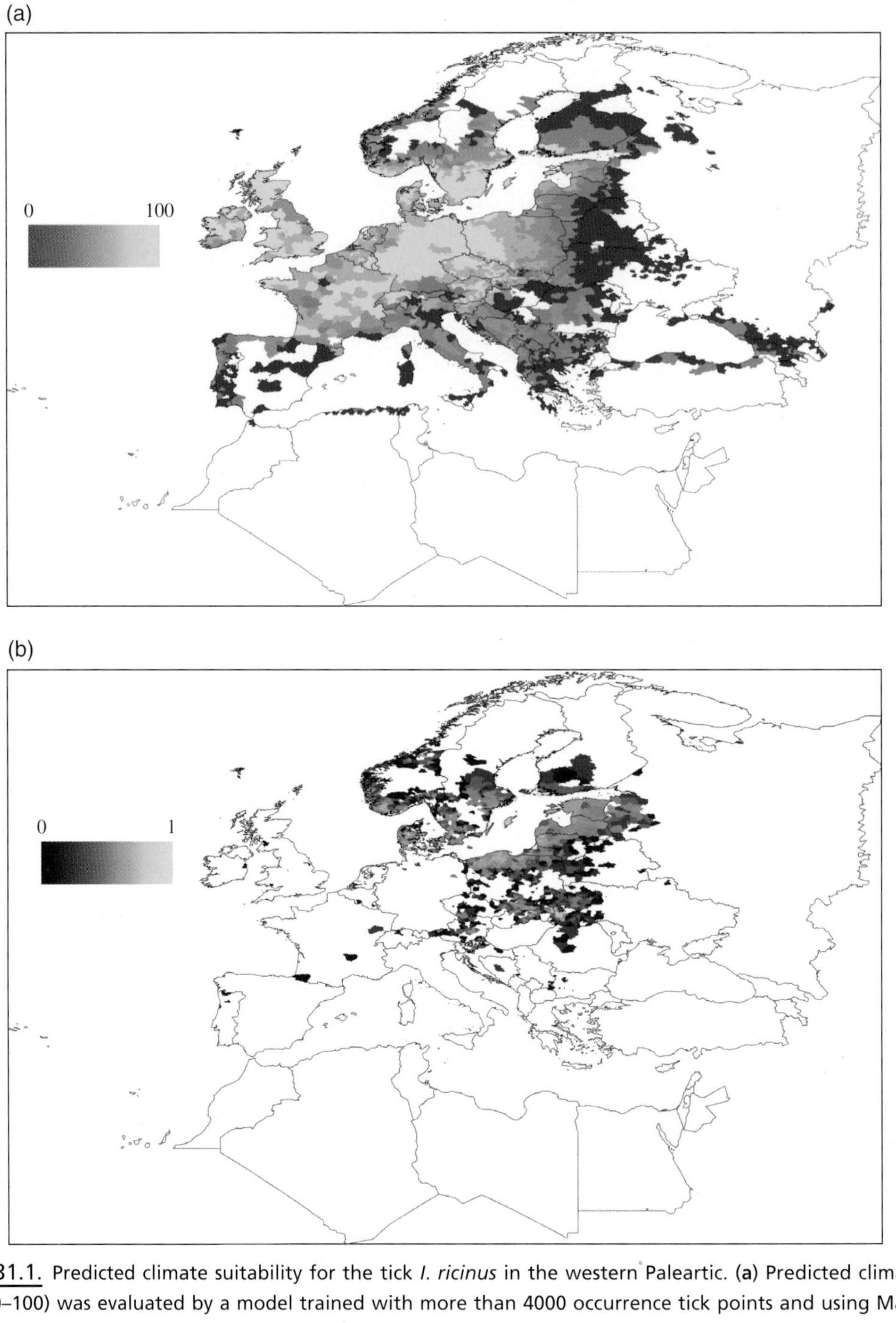

Figure 31.1. Predicted climate suitability for the tick *I. ricinus* in the western Paleartic. (**a**) Predicted climate suitability (0–100) was evaluated by a model trained with more than 4000 occurrence tick points and using Maxent as modeling software. The map is based on previous developments by Estrada-Peña et al. (2006). The ramp of colors shows the probability to find permanent populations of the tick, as driven only by climate conditions, including a set of remotely sensed monthly average temperature and monthly average vegetation stress (NDVI, a proxy for tick water stress) between the years 2000 and 2010. (**b**) Changes in climate suitability for *I. ricinus* in the period 2000–2010 (from 0, the minimum, to 1, the maximum) based in the same model. It is based on the modeling of climate suitability separately for each year, then evaluating the trend of such an index along the period 2000–2010. Both maps (**a** and **b**) are not a depiction of tick abundance but of the appropriateness of the climate for the development of the tick (**a**) and how such a factor evolved in time (**b**). Figures reproduced from Estrada-Peña et al. (2012).

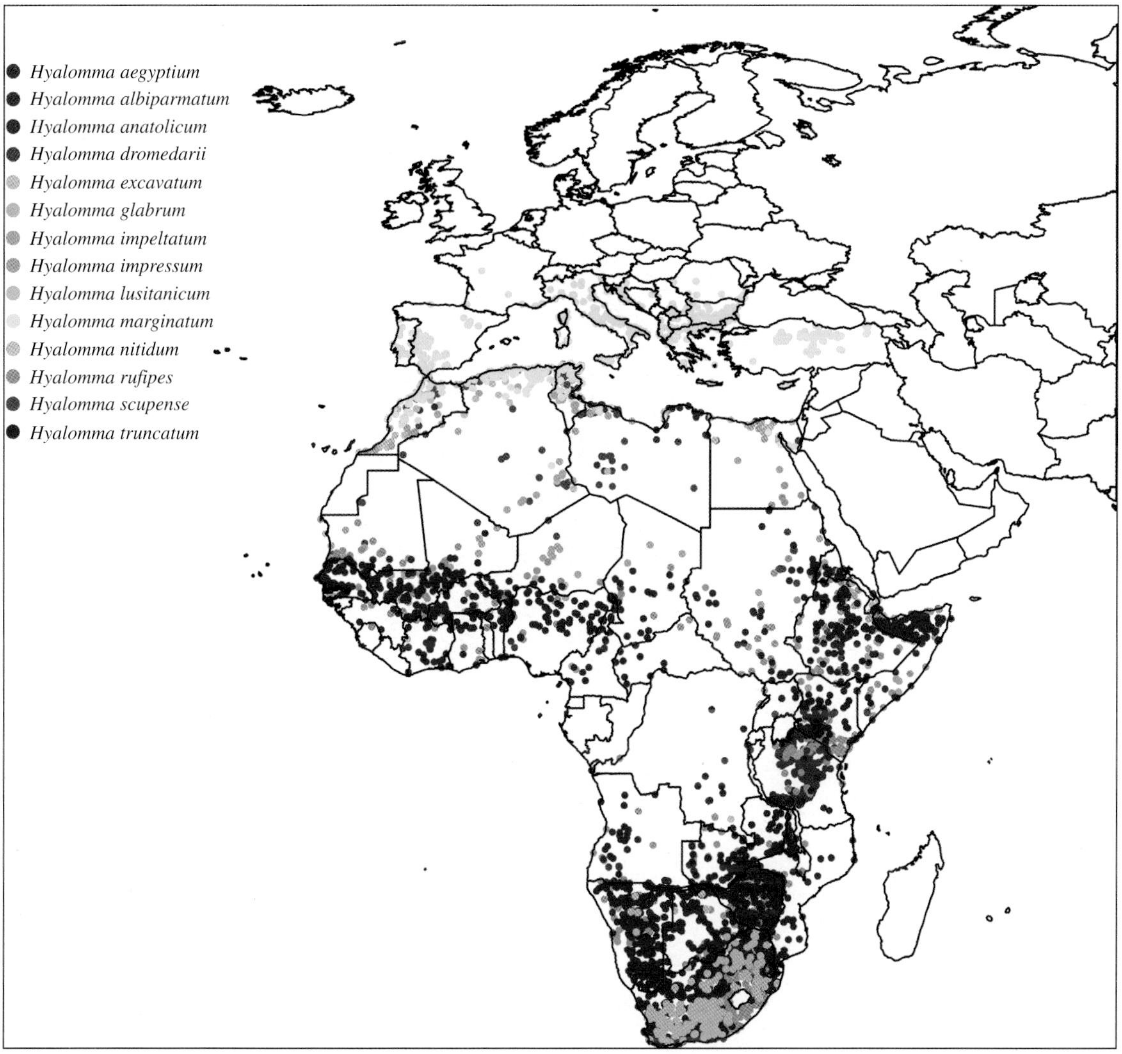

Figure 31.2. The distribution of the most prominent species of the genus *Hyalomma* in Africa. Only those records with accurate georeferences where included, that is, with a pair of coordinates or with an unambiguous name of a locality. The compilation is not intended to be exhaustive, and it provides only general information on the distribution of those species.

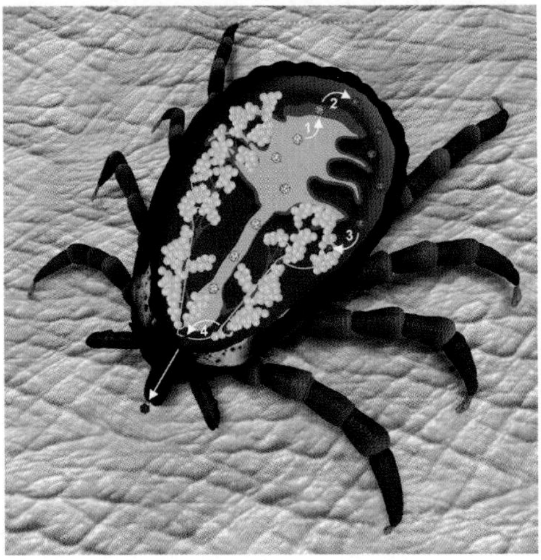

Figure 32.1. Barriers to virus infection of a tick vector. Virus enters the tick with the blood meal during feeding and is imbibed into the midgut (light virions). Entry into midgut cells represents the first barrier to infection of the tick (arrow 1) and likely defines vector competence. Specific interactions between virus and midgut cell receptors have not been identified. Virus must then replicate in midgut cells and overcome the second barrier to infection, exiting the midgut (arrow 2), for spread throughout the tick. Virus must then enter the salivary glands (arrow 3) and replicate. Finally, virus must exit the salivary glands (arrow 4) via secretion into the saliva for transmission to a host (darkest particles). Green, salivary glands; blue, midgut; virions range in color from light pink to dark red; arrows represent barriers to infection.

17

DRIVERS OF EMERGENCE AND SOURCES OF FUTURE EMERGING AND REEMERGING VIRAL INFECTIONS

Leslie A. Reperant and Albert D.M.E. Osterhaus

Department of Viroscience, Erasmus Medical Center, Rotterdam, The Netherlands

TABLE OF CONTENTS

Viral Infections and Global Change, First Edition. Edited by Sunit K. Singh.

17.1 INTRODUCTION

Infectious diseases have plagued the human species since its first evolutionary steps in and out of Africa. Some were caused by so-called heirloom pathogens, which coevolved or were vertically transmitted by ancestral hominins (Van Blerkom, 2003). A classic example is that of the herpesviruses, which infected hominins long before the emergence of *Homo sapiens* (McGeoch et al., 2006). Others originated from more or less ancient horizontal cross species transmission of animal pathogens that subsequently adapted to the human species, eventually becoming human pathogens with relatively strict species specificity. Among these are the viruses that cause smallpox, measles, and mumps, the major scourges of humans since prehistoric times (Wolfe et al., 2007). Yet others arose and continue to arise following repeated cross species transmission of animal pathogens, with no or limited ability to further transmit between humans. Rabies virus is likely one of the most ancient animal viruses with limited onward transmission in humans, which continues to bear high mortality burdens, especially in developing countries. Among the latter pathogens, however, some may eventually adapt to sustained transmission between humans and evolve towards new human pathogens, not unlike those ancient animal pathogens now long established in the human species. A devastating example is that of the human immunodeficiency virus type 1 (HIV-1), which initiated the acquired immunodeficiency syndrome (AIDS) pandemic in the past decades: HIV-1 colonized the human species about 100 years ago in Africa upon multiple cross species transmissions of simian immunodeficiency viruses (SIV), most likely during butchering of chimpanzee meat (Van Heuverswyn et al., 2006). A century later, it is distributed globally, annually killing more than two million individuals. A more recent example is that of the severe acute respiratory syndrome (SARS) coronavirus, which emerged in humans in 2002, following a cross species transmission event that took place at a live-animal market in China (Peiris et al., 2003). Within a few months, the virus had spread to 26 countries, causing more than 8000 diagnosed cases of SARS, of which approximately 10% were fatal. In contrast to the human immunodeficiency virus (HIV) pandemic, however, the SARS unfolding pandemic was nipped in the bud through a global, concerted and exhaustive public health effort, effectively bringing this newly emerging human virus to extinction.

The advances of modern medicine since the nineteenth century and a fortiori during the twentieth to twenty-first century, as well as the implementation of intervention strategies in recent times, such as movement restrictions, case isolation, and disinfection, together with the development of antimicrobial and antiviral drugs and vaccines, resulted in the effective control of a number of infectious diseases in both human and animal populations. Many infectious diseases have largely receded in industrialized countries and to a lesser extent in developing countries. In 1980, the human smallpox virus was formally declared eradicated, after millennia of unyielding spread in the human population (World Health Organization, 1980). This was the first successful global eradication of an infectious agent, followed in 2011 by the formal declaration of the eradication of rinderpest virus, a morbillivirus of cattle (Office International des Epizooties (OIE), 2011). The latter provides much welcome hope for the future eradication of measles—the human counterpart of rinderpest—still responsible for several hundred thousands of deaths among children every year, mainly in developing countries (de Swart et al., 2012).

Despite medical, technological, and scientific progress, however, dramatic ecological and anthropogenic changes associated with the unrelenting development of the human

TABLE 17.1. Drivers of Disease Emergence Listed by Smolinski et al. (2003)

Drivers of disease emergence
Microbial adaptation and change
Human susceptibility to infection
Climate and weather
Changing ecosystems
Human demographics and behaviors
Economic development and land use
International travel and commerce
Technology and industry
Breakdown of public health measures
Poverty and social inequality
War and famine
Lack of political will
Intent to harm

species also led to the emergence and reemergence of various infectious pathogens in human and animal populations. Emergence is defined by the occurrence of a novel pathogen in a particular host species, while reemergence is defined as the occurrence of a pathogen that is already known to infect a particular host species and displays an expanding geographical range or increasing incidence. Currently, approximately 75% of human emerging infectious diseases are zoonotic of origin, that is, are horizontally transmitted from animal reservoirs, in particular wildlife, to humans (Taylor et al., 2001; Woolhouse and Gowtage-Sequeria, 2005). As such, the global pool of animal pathogens represents a virtually inexhaustible source of emerging pathogens for the human species. To date, in average three new pathogens capable of infecting humans are discovered every year (Woolhouse and Gaunt, 2007).

To better face the threat of emerging and reemerging infectious pathogens, factors associated with their emergence in humans have been studied. These drivers of emergence represent key targets to better control and anticipate the rise of novel and reemerging pathogens in the human population. In 2003, the Institute of Medicine (IOM) officially listed the main drivers of disease emergence (Table 17.1) (Smolinski et al., 2003). In this chapter, we will briefly review the prehistoric and historic unfolding of these drivers in association with the development of human societies. We will also present a simplifying framework reducing the categories of drivers of emergence, based on the theory of island biogeography (MacArthur and Wilson, 1967). This framework may help to identify the animal species that will become the source of a newly emerging infectious disease (Reperant, 2010).

17.2 PREHISTORIC AND HISTORIC UNFOLDING OF THE DRIVERS OF DISEASE EMERGENCE

Although the drivers of disease emergence listed by the IOM (Table 17.1) apply principally to modern times, a vast majority of them have been present since prehistoric or ancient historic times. They catalyzed disease emergence in humans then, not unlike today, yet on a typically more restricted or slower scale.

17.2.1 Prehistory and before: Microbial adaptation and change

The human species (*Homo sapiens sapiens*), emerging some 200 thousand years ago, inherited a plethora of pathogens that infected ancestral hominins (Cockburn, 1971; Van Blerkom, 2003). These included DNA and RNA viruses, such as members of the *Herpesviridae*, *Papillomaviridae*, *Polyomaviridae*, and *Retroviridae* families, of which the phylogeny largely mirrors that of their respective host species. Geographical or behavioral isolation of hosts led to isolation of pathogen populations and bottlenecks, followed by further diversification, co-speciation, and, in some instances, coevolution of pathogens and their hosts. The relatively slow rate of evolution of vertebrate host species allowed for the co-speciation of pathogens with slow rates of evolution, such as DNA viruses. As such, *microbial adaptation* was the primary driver of the evolution of these so-called human heirloom pathogens.

Of all drivers of disease emergence, *microbial adaptation and change* has undoubtedly influenced the epidemiology and evolution of infectious pathogens since the dawn of parasitism. *Microbial adaptation* is an intrinsic property of infectious pathogens, allowing them to persist in the face of changes, such as the effects of host immune responses, changing environmental conditions, host speciation, and host switch. Adaptation upon host switch was recently well illustrated with the SARS coronavirus upon its cross species transmission from animals to humans, with the selection in humans of a variant that could efficiently bind to the human angiotensin-converting enzyme 2 (ACE2) receptor (Li et al., 2005a; Song et al., 2005).

17.2.2 Prehistoric human migrations and international travel

The human species migrated out of Africa 100 000 years ago (Oppenheimer, 2012). It colonized Australasia, Europe, and eventually the Americas, as small groups of nomadic hunter-gatherers followed coastlines or tracked the migration of large game species. The colonization of new geographical areas resulted in contacts with species of these new environments and the acquisition of new pathogens, representing the most ancient instances of disease emergence associated with large-scale migration and travel (*international travel*). A number of pathogens, including papillomaviruses, polyomaviruses, and human T-lymphotropic viruses (HTLV), can be used as valuable tools for the study of these ancient human migration patterns, as their phylogeography reflects that of prehistoric humans (de The, 2007; Van Blerkom, 2003). While HTLV-II emerged following simian-to-hominin transmission in Africa some 400 thousand years ago, long before the rise of modern humans, a new lineage of HTLV (HTLV-I) was acquired in Asia at the time of the migrations of modern humans out of Africa (Vandamme et al., 2000). Subsequently, repeated cross species transmission of HTLV between humans and other primates occurred in both directions, as humans migrated back to Africa. It resulted in the current diversity of primate T-lymphotropic viruses over these two continents. This example not only illustrates the emergence of a human pathogen during prehistoric migrations but also demonstrates the early role of humans in spreading pathogens to new worlds and eventually new species.

As such, the unequaled mobility of the human species is among the most ancient drivers of disease emergence. As mobility further developed and accelerated to reach the unprecedented—and ever increasing—levels of current times, it continues to drive disease emergence, from isolated cases of bat lyssavirus or filovirus infection of travelers to endemic countries (Timen et al., 2009; van Thiel et al., 2007) to the cross continental spread of the SARS epidemic (Peiris et al., 2003) and the global spread of pandemic and seasonal influenza (Brownstein et al., 2006; Fraser et al., 2009; Taubenberger and Morens, 2006).

17.2.3 Domestication, demographic, and behavioral changes

One of the major bursts of infectious disease emergence in humans was associated with the transition from hunter-gatherer economies to farming, about 10 000 years ago (Diamond, 2002; Wolfe et al., 2007). The emergence of major scourges of the human species during this transition time, and persisting in the millennia to come, was driven by several factors newly characterizing the prehistoric human species. Particularly these were changes in *human demographics*, *behavior*, and *susceptibility to infection*.

Gradual behavioral changes led hunter-gatherers to settle, cultivate plants, and domesticate animals in several regions of the world, eventually replacing most hunter-gatherer economies worldwide. Contact with domestic animals kept in crowded conditions favored cross species transmission of pathogens between animals and humans. Smallpox, measles, and mumps viruses are believed to have emerged from animal viruses that likely flourished in dense populations of domesticated species in those times (Wolfe et al., 2007). Environmentally transmitted pathogens, such as rotaviruses and caliciviruses, also must have frequently crossed the species barrier between livestock species, such as cattle, swine, and humans, in both directions (Van Blerkom, 2003). New behavioral characteristics associated with a sedentary and crowded lifestyle, at the time of animal and plant domestication and culture, created unprecedented conditions for disease emergence that would continue to develop and expand up to this day.

Since the earliest attempts of domestication, the animal bestiary associated with the human species has continually widened, culminating in the current massive trade of conventional and exotic pets, wildlife, and wildlife products worldwide. Similarly to the few livestock species at the origin of some of human's most ancient pathogens, these have become the sources of numerous emerging and reemerging pathogens in the recent past. Wildlife species used for bushmeat frequently act as sources for pathogens crossing to the human species, such as HIV, SARS, and filoviruses (Wolfe et al., 2005). Such cross species transmission events typically occur initially at a local scale, although *international trade* and interspecies contacts may well introduce such pathogens far from their homeland. Exotic pet species in particular may act as Trojan horses at a much wider scale, as illustrated by the first outbreak of monkeypox virus outside Africa, triggered by the international trade of exotic rodents that infected local exotic pets and their new owners in North America in 2003 (Peiris et al., 2003).

Poorer health quality conditions and malnutrition characterized developing farming communities, compared to the health and nutritional status of hunter-gatherers, and resulted in increased *human susceptibility to infection* (Larsen, 2006). The transition from foraging to farming economies, about 10000 years ago, is thought to have resulted in less varied food, reduced meat consumption, and reduced access to key micronutrients, such as iron, zinc, vitamin A, and vitamin B_{12}, in early farming communities. Periods of malnutrition were evidenced by reduced growth rate and more frequent signs of growth disruption and anemia, probably associated with parasitic infections, in early farmers than in hunter-gatherers. This likely favored the colonization of humans by new emerging pathogens in this crucial era of transition.

Although the nutritional status of the human population has drastically improved, especially since the nineteenth-century industrial revolution, malnutrition continues to take a painful toll on the human population in less developed countries, going hand in hand with parasitic and other infectious diseases. Malnutrition severely affecting *human susceptibility to infection* is a pernicious driver of disease emergence, afflicting the human population since prehistory and against which a vast part of humanity has remained at struggle to date. More recently, *human susceptibility to infection* has further been affected by the AIDS pandemic, particularly affecting certain developing countries, due to HIV-1 infection-associated immunosuppression (Morens et al., 2004).

While the initial transition from hunter-gatherer economies to farming may have led to poorer health and nutritional status, settlements and food production also prompted the explosive *demographic growth* of the human species, still relentlessly continuing to this day. These new demographic conditions permitted the further evolution of acute and more virulent infections and the eventual maintenance of acute and virulent pathogens, such as smallpox, measles, and mumps viruses (Dobson and Carper, 1996; King et al., 2009; Wolfe et al., 2007). A critical community size is required for acute and immunizing pathogens, such as measles virus, to circulate sustainably in the human population (Grenfell et al., 2001). Measles virus has been estimated to require a human population size of 200 000–500 000 individuals to be maintained. Although measles virus' critical community size may have been smaller during ancient history, it is believed that the networks of cities that developed at that time, for example, in Mesopotamia, provided the ground for recurring epidemics (Dobson and Carper, 1996), not unlike the recurrent morbillivirus epidemics seen in fragmented seal populations in recent decades (Jensen et al., 2002; Swinton et al., 1998). Since then, the complexity of *human demographics and behavior* has not ceased to increase, evolve, and develop, leading to more and more complex dynamics of social contact and movement patterns, paving the way for the efficient spread of emerging pathogens.

17.2.4 Human settlements and changing ecosystems

Although in a primitive way, farming and the prehistoric development of settlements initiated human society's *economic development and changes in land use* and eventually resulted in *changing ecosystems*. Early farmers cleared forests and plowed land, using slash-and-burn techniques (Nikiforuk, 1991). This contributed to the emergence and evolution of commensal species, such as mice, rats, and anthropophilic insects, which thrive in anthropogenic environments (McCormick, 2003). These animals were not only predating on harvests and animal feed but also transmitted infectious pathogens to domestic animals and humans, further contributing to the pool of emerging infections that arose during prehistoric and ancient historic times. Major diseases, plaguing humans until medieval history and further, include the plague, caused by *Yersinia pestis*, and typhus fever, caused by *Rickettsia* bacteria, both transmitted to humans by insect vectors ingesting infected blood from rodent reservoir hosts (Perry and Fetherston, 1997).

Agricultural development and destruction of natural habitat continue to favor adaptable species to thrive as commensals in many parts of the world. In South America, agricultural development led to population explosion and expansion of various rodent species, eventually resulting in the emergence of several arenaviruses and hantaviruses in the past decades (Charrel and Lamballerie, 2003; Zeier et al., 2005). In Malaysia and Australia, the destruction of natural habitats led to colonization of plantations and managed parks by fruit bats, leading to the emergence of infections with henipaviruses (Nipah and Hendra viruses) in domestic animals and humans (Field et al., 2001). As such, development and environmental impact have accompanied the human species since its very first settlements, leaving heavier environmental footprints as human societies relentlessly grew and expanded into most of the world's ecosystems.

17.2.5 Ancient and medieval times: Commerce, warfare, poverty, and climate

Ancient historic and medieval times were characterized among others by the development of *geographically dispersed commerce* and *organized wars* associated with large-scale movements of humans, animals, and goods. The first major pandemics afflicting the human species must have emerged at these times, further fueled by crowding conditions, poor sanitation, *poverty*, *social inequality*, and *famines*. The plague ravaged Europe upon two devastating pandemics lasting several centuries each, during the sixth to eighth and fourteenth to seventeenth centuries (Perry and Fetherston, 1997). It was introduced in Europe from Central Asia via the Silk Road, a major route for *international commerce* at these times. Among the first instances of biological warfare, the plague was used with the *intent to harm* in 1346 by the Mongol army, which catapulted infected corpses over the city walls of Caffa in Crimea (Wheelis, 2002). This violent reemergence of the disease at the doorstep of Europe preceded the entry of the pathogen into Sicily, sparking the infamous Black Death epidemics that ravaged Europe through 1351. Typhus fever likewise swept through Europe and Asia during the fifteenth century. It was introduced into the Americas upon their discovery, together with smallpox and measles (Acemoglu et al., 2003). Because the indigenous Amerindians had never been exposed to these pathogens, they proved *highly susceptible to these infections*, and their communities were ravaged by virgin-soil epidemics. It has been documented that smallpox was further used as a biological weapon in the New World, not only by the Spanish Conquistadores against the Inca and Aztec empires during the fifteenth century but also by the English during the French and Indian war of the eighteenth century (Christopher et al., 1997). Not unlike these times, biological warfare and the bioterrorist's *intent to harm* remain a threat today, with origins as ancient as most drivers of disease emergence.

The colonization of new worlds also resulted in emerging infections among European settlers in new territories. In particular, Europeans suffered high morbidity and mortality in Africa, due to malaria and yellow fever, flourishing under tropical and subtropical *climate* (Acemoglu et al., 2003). It has been suggested that these diseases have largely hampered institutional and economic development at this continent and likely contributed to the slave trade. These pathogens were introduced into the Americas decimating indigenous Amerindians and European settlers alike. *Climate and weather*, as such, have long been important determinants of the epidemiology of numerous infectious pathogens. The Little Ice Age of the sixteenth to eighteenth centuries, characterized by *weather instability* and cold temperatures, may have further contributed to severe famines and consequently large epidemics in Europe (Appleby, 1980).

Changes affecting *global climate* today nevertheless represent an unprecedented footprint of the human population on the global environment. Although the consequences of *global climate change* on infectious diseases are still unfolding, they likely will play an increasing role in disease emergence in the future (Gould and Higgs, 2009; Patz et al., 1996). In particular, the effect of *global warming* on the geographical distribution and life cycle of arthropod vectors is anticipated to drive the emergence of arbovirus diseases in currently temperate climate regions; such effect may already be seen for a number of zoonotic pathogens, including tick-borne encephalitis virus in Europe (Randolph, 2001).

17.2.6 Recent past and modern times: Technology and industry

The industrial revolution of the nineteenth century likely contributed to initiate the changes affecting global climate today. It also initiated the unabated progress in *technology and industry* that would allow for the unprecedented growth of food production animal populations. This resulted in major changes in farming practices worldwide, including intensive farming and *worldwide trade*. These conditions undoubtedly favored the emergence of zoonotic infectious diseases in poultry and livestock and their cross species transmission to humans. These include the emergence of human metapneumovirus (hMPV) about 100 years ago, following the cross species transmission of avian metapneumovirus from poultry (de Graaf et al., 2008). Likewise, the emergence of influenza A virus in humans resulted in the past 100 years from several instances of cross species transmission of avian or swine influenza viruses and rapid adaptation of these viruses to humans (Taubenberger and Morens, 2006). The latest influenza pandemic occurred in 2009, after cross species transmission of an influenza virus of porcine origin, with genetic elements from different lineages circulating in America and Eurasia (Smith et al., 2009). It further illustrated the globalization of livestock populations, via *global trade practices* that fuel the exchange and emergence of a wide diversity of pathogens in animal and eventually human populations.

In conclusion, most of the recognized drivers of disease emergence have unfolded in parallel to changes associated with the human species development since prehistoric and ancient historic times. However, the dramatic and accelerating changes affecting the modern human population worldwide today, as well as its relationship with and impact on the global environment and climate, have undoubtedly taken domestication, agriculture, urbanization, industrialization, and colonization to new and unprecedented levels, contributing to the current rise in emerging and reemerging infectious diseases.

17.3 PROXIMAL DRIVERS OF DISEASE EMERGENCE AND SOURCES OF FUTURE EMERGING AND REEMERGING VIRAL INFECTIONS

Although the IOM list of drivers provides valuable insights into the evolutionary, ecological, and anthropogenic nature of the factors favoring the emergence of infectious diseases, their multifactorial and complex interactive nature limits our ability to effectively predict or act against disease emergence. Since the majority of emerging infectious diseases have a zoonotic origin, the possibility to predict the animal species from which future pathogens may be expected to emerge would be crucial. However, such endeavor was deemed impossible, because of the range, scope, and interactive nature of the drivers involved (WHO/FAO/OIE, 2004).

Most drivers of disease emergence have led and continue to lead to changes in population dynamics of human and animal species sources of emerging infectious diseases. These critical population dynamic changes can be readily identified as direct factors for disease emergence, by use of the theory of island biogeography, developed by McArthur and Wilson more than 50 years ago (MacArthur and Wilson, 1967). These proximal factors of emergence provide new means for the identification of the most probable animal sources of future emerging infectious pathogens while simplifying the categories of drivers of disease emergence (Reperant, 2010).

Biogeography is the study of the distribution and abundance of living organisms—or biodiversity—in space and time (MacArthur and Wilson, 1967). Because islands provide ideal conditions for such a study, the theory of island biogeography was developed as a general approach for the prediction of biodiversity patterns. McArthur and Wilson proposed that a biotic equilibrium of the number of species present on an island is reached when the rate of immigration and the rate of extinction (species/unit time) equal each other (Figure 17.1a). The immigration rate increases as the island is located nearer to the source of immigrating species and decreases as the number of species present on the island increases. Conversely, the extinction rate increases as the size of the island decreases and as the number of species present on the island increases. It has long been recognized that this theory can be applied to parasites and pathogens, when viewing their hosts as islands (Kuris et al., 1980; Poulin, 2004). Parasite and pathogen species richness in a host species is thought to result from a balance between parasite and pathogen colonization and extinction rates (Figure 17.1b). Interactions between species correspond to the distance from source to island and influence parasite and pathogen colonization rate. The host body size and lifespan, as well as the population size, density, and intraspecific interactions of a species, correspond to the size of the island and impact on parasite and pathogen extinction rate. Several studies have attempted to define general relationships between parasite species richness and species characteristics in direct line of the theory of island biogeography. For example, it has been shown for several taxa that parasite species richness increases with host species body size (Poulin, 2004). While most of these studies have focused on linking parasite species richness with constant characteristics of species across taxa (e.g., species median body size, social behavior, geographical range), evaluating the effect of short- to medium-term changes in species characteristics may provide useful clues on proximal drivers of disease emergence.

Based on the theory of island biogeography, one may predict that changes in interspecies interactions ("distance of island to source") and changes in species population size, density, or intraspecific interactions ("island size") likely will result in changes in parasite and pathogen colonization and extinction rates, favoring parasite and pathogen emergence or extinction in that species (Reperant, 2010). An increase in interspecies interactions, for example, following novel geographical overlap with other species or behavioral changes, likely will result in an increase in parasite and pathogen colonization rate. Likewise, an increase in a species population size, density, or geographical range or the intensification of intraspecific interactions likely will result in a decrease in parasite and pathogen extinction rate. In fact, most changes affecting the human population since prehistory (*demographic and behavioral changes* as well as *international travel*) have led to increased interspecies interactions (with domesticated, hunted, and traded animals) and increased population size, density, and intraspecific interactions (via demographic growth and complex mixing dynamics associated with ever-evolving human mobility and social behavior). These changes, which have not ceased to accelerate and expand since the emergence of the human species, undoubtedly favored parasite and pathogen colonization rate and hindered parasite and pathogen extinction rate in the human species. For example, the establishment of acute pathogens, such as measles, mumps, and smallpox viruses, during prehistory represents a clear example of a decrease in pathogen extinction rate in humans associated with human behavioral changes and the increase in population size and density. In sum, all drivers of emergence associated with changes in human demographics and behavior that result in closer proximity to sources of pathogens and in bigger "island size" in terms of population size, density, and intraspecific interactions can be grouped in a set of proximal drivers of disease emergence that characterize the human (receiver) host species.

From the standpoint of the source of species populating an island, the theory of island biogeography predicts that size and distance from source to island impact the rate and probability of propagules leaving the source and reaching the island (MacArthur and

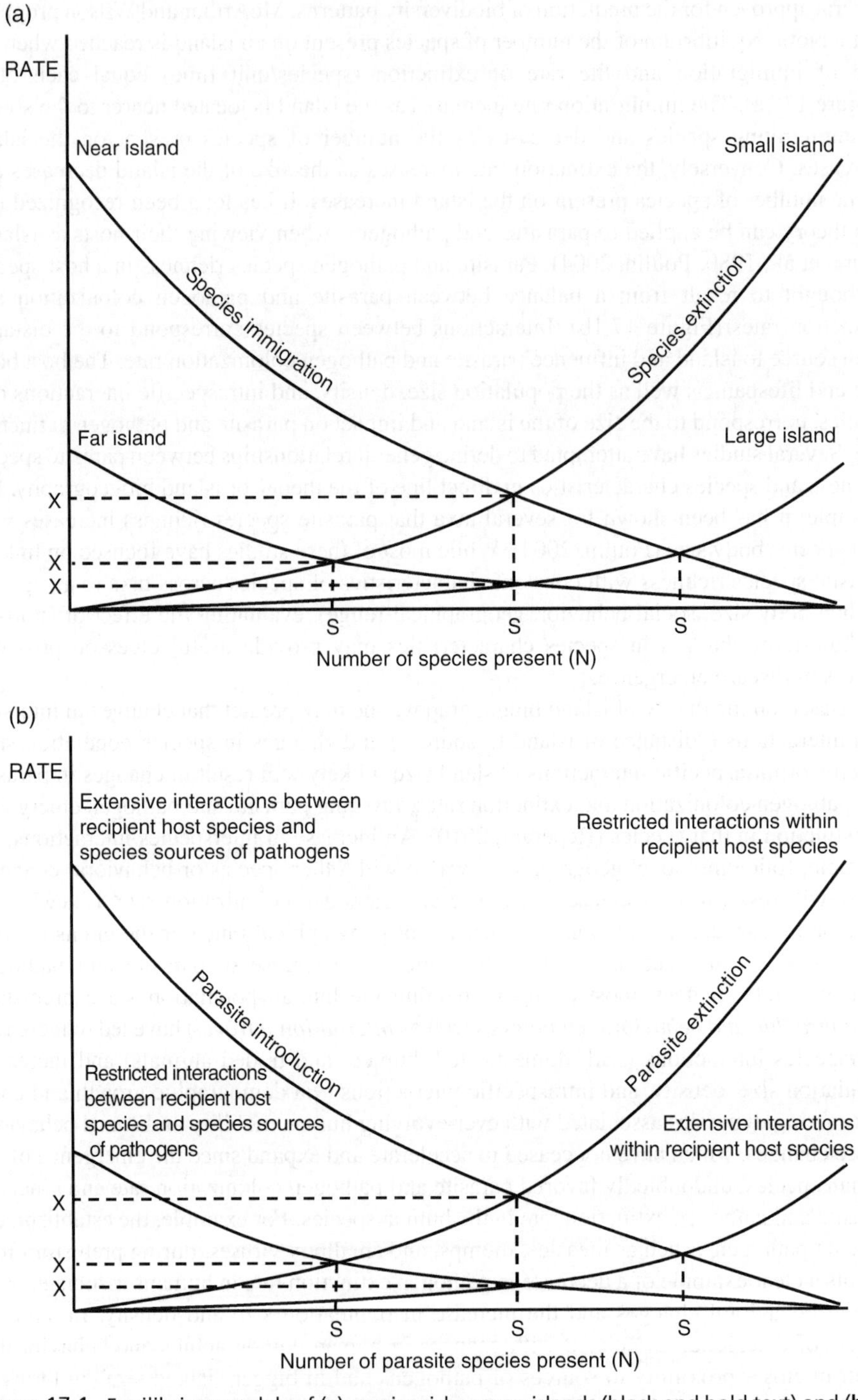

Figure 17.1. Equilibrium models of (**a**) species richness on islands (black and bold text) and (**b**) of parasite species richness within host species (grey and non-bold text). S, number of species at equilibrium; X, species turnover rate at equilibrium. Adapted from MacArthur and Wilson (1967), and modified from Reperant (2010).

Wilson, 1967). A large area at the source favors propagules' emigration rate, while close distance to island favors their colonization rate. Consequently, one can predict from the theory of island biogeography that species with large population size, high densities, or intensive intraspecific interactions ("source size") may represent main sources of pathogens for other species that are in contact through geographical or behavioral overlap (Reperant, 2010). An increase in population size, density, or geographical range of source species thus may lead to an increase in parasite emigration rate and disease emergence in contact species. Most ecological drivers listed by the IOM are known to have led to a significant increase in the population size, density, or geographical range of animal host or reservoir species of emerging infectious pathogens. Most of the examples of zoonotic pathogens emerging in the human population listed previously can be attributed to such changes. For example, domestication, *technology and industry*, and *international trade* led to increased animal host population size, density, and/or geographical range, resulting in the colonization of humans by smallpox, measles, and mumps viruses during prehistory and by hMPV, influenza virus, HIV, filoviruses, SARS coronavirus, or monkeypox virus in more recent times. *International travel and commerce* and *climate change and weather* also typically drive demographic changes or geographical expansion of animal reservoirs or insect vectors of current emerging pathogens.

Likewise, animal species displaying changes in behavior that result in an increase in interspecies interactions with humans ("distance of source to island") may also be considered major sources of emerging infectious diseases, due to their closer proximity to humans. While domestication contributes to increased population size, density, and geographical range, it also brings animal species in close contact with humans, further promoting the emergence of infectious diseases in humans since prehistory. *Changing ecosystems, economic development*, and *land use* have been driving animal species with a commensal proclivity closer to the human species since the time of domestication and up to this day. This promoted, for example, the emergence of arenaviruses, hantaviruses, and henipaviruses in humans in the more recent past. *Wars and famines*, *poverty*, and *social inequality* not only favor commensal species, due to poor sanitation and limited options for the control of their populations, but also powerfully fuel the demand for bushmeat hunting, bringing an increasing number of hunted species closer to humans. This favored the emergence of HIV, filoviruses, and SARS coronavirus in humans. In sum, all drivers of emergence associated with changes in animal populations that mirror those in humans, as listed earlier, namely, changes in animal species demographics and behavior that result in closer proximity to humans and in bigger "island size" in terms of population size, density, and intraspecific interactions, can be grouped in a set of proximal drivers of disease emergence that characterize the animal (source) host species.

In conclusion, while the first IOM driver of disease emergence (*microbial adaptation and change*) likely shapes the evolution of emerging infectious pathogens in most if not all cases since times immemorial, the ecological and anthropogenic drivers may be dissected into proximal causes of disease emergence, characterizing the receiver and source host species populations. These can be divided into three categories: (i) an increase in human population size, density, or intraspecific interactions (associated with *human demographic and behavioral changes* and *international travel*); (ii) an increase in interspecies interactions between humans and other animal species (e.g., *domestication, changing ecosystems, economic development and land use, war and famines, poverty and social inequality*); and (iii) an increase in animal (or vector) population size, density, or intraspecific interactions (*domestication, technology and industry, international trade, climate and weather*). An important consequence of the application of the theory of island biogeography to emerging pathogens is the simplification and reduction of the range of drivers involved, facilitating predictions, in particular with regard to the most probable sources of future emerging infections in humans. Because changes in animal species demographics and behavior in association with anthropogenic factors and climate change may be anticipated to a large extent, the animal species sources of future emerging infectious diseases may largely be identified. In particular, flourishing species in close proximity to humans and notably domestic and exotic pet species, hunted and traded species, commensal species, and species favorably affected by climate change likely represent the most likely sources of future emerging pathogens (Reperant, 2010). Targeted surveillance of these species and characterization of the parasites and pathogens they harbor are therefore warranted to improve human preparedness against emerging scourges from the animal world.

17.4 FURTHER INSIGHTS FROM THE THEORY OF ISLAND BIOGEOGRAPHY

The theory of island biogeography formed the basis of a powerful framework for the general study of biodiversity patterns. While it provides useful clues on more proximal drivers of emergence associated with human and animal population dynamics, it further supports two important predictions that may be applied to pathogens. These predictions may further provide new means to anticipate the rise of future emerging diseases.

First, it was shown that the presence of stepping-stone islands connecting the source of immigrating species to the final island under study significantly increases the rate of species immigration (MacArthur and Wilson, 1967). This may be paralleled by the role of intermediary hosts, such as domesticated species, in the emergence of a number of pathogens, such as henipaviruses (transmitted from bats to swine and horses before further transmission to humans) and influenza A viruses (transmitted from wild birds to poultry and swine before further transmission to humans). The role of stepping-stones may also be played by more exotic or wild species. Prairie dogs were exotic pets that served as stepping-stones for monkeypox virus during the human outbreak of 2003 in the United States, following the importation of infected African rodent hosts. In 2003, the proximal source of the SARS coronavirus was traced to civet cats kept at live-animal markets (Song et al., 2005). However, bats are now generally believed to be the main reservoir hosts of SARS and SARS-like coronaviruses (Li et al., 2005b), whereas carnivores at the live-animal markets were likely stepping-stones in the chain of transmission of this new pathogen to humans. Likewise, primates are typically considered the proximal sources of filoviruses in Africa, while bats are considered the most likely reservoir hosts (Monath, 1999). These examples illustrate the broad nature of stepping-stone species and their possible involvement in disease emergence in humans.

Second, the theory of island biogeography states that at equilibrium, the turnover of species present on an island constantly occurs but the number of species remains unchanged (MacArthur and Wilson, 1967). For example, the number of species recolonizing defaunated mangrove islands, which were inhabited by arboreal arthropods, was shown to reach and oscillate around the initial number of species recorded before treatment. However, the species composition was always different from the initial composition (Simberloff, 1974). The eradication of infectious diseases from human populations may result in a relative increase in the rate of pathogen colonization in humans, due to similar dynamics. The waning immunity against eradicated pathogens may open niches for colonization by novel pathogens, in the same way defaunated mangrove islands offered open niches for colonization by novel arthropod species. The threat of exotic zoonotic poxviruses to humans, like monkeypox and cowpox viruses, may be perceived as increasing due to waning immunity against smallpox virus following its eradication and the subsequent abrogation of vaccination (Di Giulio and Eckburg, 2004; Vorou et al., 2008). Similarly, animal morbilliviruses may colonize human populations should vaccination be discontinued after the eventual eradication of measles virus (de Swart et al., 2012; Stittelaar and Osterhaus, 2001).

In conclusion, the dynamic nature of disease emergence is paralleled at least in part by changes in the dynamics of receiver and source host populations. These changes include demographic and behavioral changes that characterize the human species and animal species in contact with humans since prehistoric times. Nonetheless, these changes are accelerating and expanding in breadth and scope. They affect receiver and source host ecology, demographics, and behavior in direct association with the unrelenting development of the human species and its global impact on its environment, from climate change to the eradication of infectious pathogens. Understanding the proximal effect of these changes on the dynamics of infectious pathogens is essential to better predict and anticipate disease emergence.

REFERENCES

Acemoglu, D., Robinson, J., and Johnson, S. (2003). Disease and development in historical perspective. *J Eur Econ Assoc* **1**(2/3), 397–405.

Appleby, A. B. (1980). Epidemics and famine in the little ice age 1550–1700. *J Interdiscip Hist* **10**, 643–663.

Brownstein, J. S., Wolfe, C. J., and Mandl, K. D. (2006). Empirical evidence for the effect of airline travel on inter-regional influenza spread in the United States. *PLoS Med* **3**(10), e401.

Charrel, R. N., and Lamballerie, X. de. (2003). Arenaviruses other than Lassa virus. *Antiviral Res* **57**(1–2), 89–100.

Christopher, G. W., Cieslak, T. J., Pavlin, J. A., et al. (1997). Biological warfare. A historical perspective. *J Am Med Assoc* **278**(5), 412–417.

Cockburn, T. A. (1971). Infectious diseases in ancient populations. *Curr Anthropol* **12**, 45–62.

de Graaf, M., Osterhaus, A. D., Fouchier, R. A., et al. (2008). Evolutionary dynamics of human and avian metapneumoviruses. *J Gen Virol* **89**(Pt 12), 2933–2942.

de Swart, R. L., Duprex, W. P., and Osterhaus, A. D. M. E. (2012). Rinderpest eradication: lessons for measles eradication? *Curr Opin Virol* **2**(3), 330–334.

de The, G. (2007). Microbial genomes to write our history. *J Infect Dis* **196**(4), 499–501.

Di Giulio, D. B., and Eckburg, P. B. (2004). Human monkeypox: an emerging zoonosis. *Lancet Infect Dis* **4**(1), 15–25.

Diamond, J. (2002). Evolution, consequences and future of plant and animal domestication. *Nature* **418**(6898), 700–707.

Dobson, A. P., and Carper, E. R. (1996). Infectious diseases and human population history. *Bioscience* **46**(2), 115–126.

Field, H., Young, P., Yob, J. M., et al. (2001). The natural history of Hendra and Nipah viruses. *Microbes Infect* **3**(4), 307–314.

Fraser, C., Donnelly, C. A., Cauchemez, S., et al. (2009). Pandemic potential of a strain of influenza A (H1N1): early findings. *Science* **324**(5934), 1557–1561.

Gould, E. A., and Higgs, S. (2009). Impact of climate change and other factors on emerging arbovirus diseases. *Trans R Soc Trop Med Hyg* **103**(2), 109–121.

Grenfell, B. T., Bjornstad, O. N., and Kappey, J. (2001). Travelling waves and spatial hierarchies in measles epidemics. *Nature* **414**(6865), 716–723.

Jensen, T., van de Bildt, M., Dietz, H. H., et al. (2002). Another phocine distemper outbreak in Europe. *Science* **297**(5579), 209.

King, A. A., Shrestha, S., Harvill, E. T., et al. (2009). Evolution of acute infections and the invasion-persistence trade-off. *Am Nat* **173**(4), 446–455.

Kuris, A. M., Blaustein, A. R., and Alio, J. J. (1980). Hosts as islands. *Am Nat* **116**(4), 570–586.

Larsen, C. S. (2006). The agricultural revolution as environmental catastrophe: implications for health and lifestyle in the Holocene. *Quat Int* **150**, 12–20.

Li, F., Li, W., Farzan, M., et al. (2005a). Structure of SARS coronavirus spike receptor-binding domain complexed with receptor. *Science* **309**(5742), 1864–1868.

Li, W., Shi, Z., Yu, M., et al. (2005b). Bats are natural reservoirs of SARS-like coronaviruses. *Science* **310**(5748), 676–679.

MacArthur, R. H., and Wilson, E. O. (1967). "The Theory of Island Biogeography." Princeton Landmarks in Biology. Princeton University Press, Princeton, NJ.

McCormick, M. (2003). Rats, communications, and plague: toward an ecological history. *J Interdiscip Hist* **34**(1), 1–25.

McGeoch, D. J., Rixon, F. J., and Davison, A. J. (2006). Topics in herpesvirus genomics and evolution. *Virus Res* **117**(1), 90–104.

Monath, T. P. (1999). Ecology of Marburg and Ebola viruses: speculations and directions for future research. *J Infect Dis* **179**(s1), S127–S138.

Morens, D. M., Folkers, G. K., and Fauci, A. S. (2004). The challenge of emerging and re-emerging infectious diseases. *Nature* **430**(6996), 242–249.

Nikiforuk, A. (1991). "The Fourth Horseman." Penguin Books, Canada.

Office International des Epizooties (OIE) (2011). "Declaration of Global Eradication of Rinderpest and Implementation of Follow-Up Measures to Maintain World Freedom from Rinderpest." Resolution no. 18. Office International des Epizooties, Paris.

Oppenheimer, S. (2012). Out-of-Africa, the peopling of continents and islands: tracing uniparental gene trees across the map. *Philos Trans R Soc Lond B—Biol Sci* **367**(1590), 770–784.

Patz, J. A., Epstein, P. R., Burke, T. A., et al. (1996). Global climate change and emerging infectious diseases. *J Am Med Assoc* **275**(3), 217–223.

Peiris, J. S., Yuen, K. Y., Osterhaus, A. D., et al. (2003). The severe acute respiratory syndrome. *N Engl J Med* **349**(25), 2431–2441.

Perry, R. D., and Fetherston, J. D. (1997). *Yersinia pestis*—etiologic agent of plague. *Clin Microbiol Rev* **10**(1), 35–66.

Poulin, R. (2004). Macroecological patterns of species richness in parasite assemblages. *Basic Appl Ecol* **5**, 423–434.

Randolph, S. E. (2001). The shifting landscape of tick-borne zoonoses: tick-borne encephalitis and Lyme borreliosis in Europe. *Philos Trans R Soc B* **356**(1411), 1045–1056.

Reperant, L. A. (2010). Applying the theory of island biogeography to emerging pathogens: toward predicting the sources of future emerging zoonotic and vector-borne diseases. *Vector Borne Zoonotic Dis* **10**(2), 105–110.

Simberloff, D. S. (1974). Equilibrium theory of island biogeography and ecology. *Annu Rev Ecol Syst* **5**, 161–182.

Smith, G. J., Vijaykrishna, D., Bahl, J., et al. (2009). Origins and evolutionary genomics of the 2009 swine-origin H1N1 influenza A epidemic. *Nature* **459**(7250), 1122–1125.

Smolinski, M. S., Hamburg, M. A. and Lederberg, J. (2003) Microbial Threats to Health: Emergence, Detection, and Response. The National Academies Press, Washington, DC.

Song, H. D., Tu, C. C., Zhang, G. W., et al. (2005). Cross-host evolution of severe acute respiratory syndrome coronavirus in palm civet and human. *Proc Natl Acad Sci USA* **102**(7), 2430–2435.

Stittelaar, K. J., and Osterhaus, A. D. M. E. (2001). MVA: a cuckoo in the vaccine nest? *Vaccine* **19**(27), V–Vi.

Swinton, J., Harwood, J., Grenfell, B. T., et al. (1998). Persistence thresholds for phocine distemper virus infection in harbour seal *Phoca vitulina* metapopulations. *J Anim Ecol* **67**(1), 54–68.

Taubenberger, J. K., and Morens, D. M. (2006). 1918 influenza: the mother of all pandemics. *Emerg Infect Dis* **12**(1), 15–22.

Taylor, L. H., Latham, S. M., and Woolhouse, M. E. J. (2001). Risk factors for human disease emergence. *Philos Trans R Soc Lond Ser B—Biol Sci* **356**(1411), 983–989.

Timen, A., Koopmans, M. P., Vossen, A. C., et al. (2009). Response to imported case of Marburg hemorrhagic fever, the Netherlands. *Emerg Infect Dis* **15**(8), 1171–1175.

Van Blerkom, L. M. (2003). Role of viruses in human evolution. *Am J Phys Anthropol* **37**(Suppl.), 14–46.

Van Heuverswyn, F., Li, Y., Neel, C., et al. (2006). Human immunodeficiency viruses: SIV infection in wild gorillas. *Nature* **444**(7116), 164.

Van Thiel, P-P. A. M., de Bie, R. M. A., Eftimov, F., et al. (2009). Fatal human rabies due to Duvenhage virus from a bat in Kenya: failure of treatment with coma-induction,ketamine, and antiviral drugs. *PLoS Negl Trop Dis* **3**(7), e428.

Vandamme, A. M., Bertazzoni, U., and Salemi, M. (2000). Evolutionary strategies of human T-cell lymphotropic virus type II. *Gene* **261**(1), 171–180.

Vorou, R. M., Papavassiliou, V. G., and Pierroutsakos, I. N. (2008). Cowpox virus infection: an emerging health threat. *Curr Opin Infect Dis* **21**(2), 153–156.

Wheelis, M. (2002). Biological warfare at the 1346 siege of Caffa. *Emerg Infect Dis* **8**(9), 971–975.

WHO/FAO/OIE (2004). "Report of the WHO/FAO/OIE Joint Consultation on Emerging Zoonotic Diseases." World Health Organization, Geneva.

Wolfe, N. D., Daszak, P., Kilpatrick, A. M., et al. (2005). Bushmeat hunting, deforestation, and prediction of zoonoses emergence. *Emerg Infect Dis* **11**(12), 1822–1827.

Wolfe, N. D., Dunavan, C. P., and Diamond, J. (2007). Origins of major human infectious diseases. *Nature* **447**(7142), 279–283.

Woolhouse, M., and Gaunt, E. (2007). Ecological origins of novel human pathogens. *Crit Rev Microbiol* **33**(4), 231–242.

Woolhouse, M. E. J., and Gowtage-Sequeria, S. (2005). Host range and emerging and reemerging pathogens. *Emerg Infect Dis* **11**(12), 1842–1847.

World Health Organization (WHO) (1980). The global eradication of smallpox: final report of the global commission for the certification of smallpox eradication. World Health Organization, Geneva.

Zeier, M., Handermann, M., Bahr, U., et al. (2005). New ecological aspects of hantavirus infection: a change of a paradigm and a challenge of prevention—a review. *Virus Genes* **30**(2), 157–180.

Song, H.-D., Tu, C.-C., Zhang, G.-W., et al. (2005) Cross-host evolution of severe acute respiratory syndrome coronavirus in palm civet and human. Proc Natl Acad Sci USA 102(7): 2430–2435.

[illegible] (2001) [illegible]

Swinton, J., Harwood, J., Grenfell, B. T., et al. (1998) Persistence thresholds for phocine distemper virus infection in harbour seal Phoca vitulina metapopulations. J Anim Ecol 67: [illegible]

Taubenberger, J. K., and Morens, D. M. (2006) 1918 Influenza: the mother of all pandemics. Emerg Infect Dis 12(1): 15–22.

Taylor, L. H., Latham, S. M., and Woolhouse, M. E. J. (2001) Risk factors for human disease emergence. Philos Trans R Soc Lond B Biol Sci 356(1411): 983–989.

[illegible] (2009) [illegible]

Van Blerkom, L. M. (2003) Role of viruses in human evolution. Am J Phys Anthropol Suppl [illegible]

Van Heuverswyn, F., Li, Y., Neel, C., et al. (2006) Human immunodeficiency viruses: SIV infection in wild gorillas. Nature 444(7116): 164.

[illegible] (2006) [illegible]

Vandamme, A.-M., Bertazzoni, U., and Salemi, M. (2000) Evolutionary strategies of human T-cell lymphotropic virus type II. Gene 261(1): 171–180.

[illegible] (2007) [illegible]

[illegible] (2002) [illegible]

WHO/UNAIDS (2006) [illegible] World Health Organization, Geneva.

Wolfe, N. D., [illegible] (2005) [illegible]

[illegible] (2007) [illegible] Nature 447(7142): 279–283.

[illegible] (2002) [illegible]

World Health Organization (WHO) (1980) The global eradication of smallpox. Final report of the Global Commission for the Certification of Smallpox Eradication. World Health Organization, Geneva.

[illegible] (2008) [illegible]

18

SPILLOVER TRANSMISSION AND EMERGENCE OF VIRAL OUTBREAKS IN HUMANS

Sunit K. Singh

Laboratory of Neurovirology and Inflammation Biology, Centre for Cellular and Molecular Biology (CCMB), Hyderabad, India

TABLE OF CONTENTS

18.1 INTRODUCTION

Humans have been prone to infection by pathogens throughout their evolutionary history. Human society has undergone a series of major transitions that has affected pattern of infectious disease acquisition and dissemination. These transitions illustrate the interrelationship between environmental and behavioral influences on the

Viral Infections and Global Change, First Edition. Edited by Sunit K. Singh.

emergence of viral outbreaks. As humans encroach further into previously uncultivated environments, new contacts between wild fauna and humans and their livestock increase the risk of cross species infection.

Host switch of animal viruses to human hosts is increasing, and there are not many tools to handle such emergence of viral outbreaks. Many recent outbreaks (avian and swine influenza, West Nile virus, Japanese encephalitis virus, Chikungunya virus, Ebola virus, Nipah virus (NiV), and hantavirus) and past introductions (human immunodeficiency virus (HIV)) are believed to be zoonotic in their origin. These are reported as spillover event from animals to humans. Spillover transmission of viruses from animals is a major concern for human health. These viruses could gain efficiency in human transmission in due course of time.

18.2 MAJOR ANTHROPOGENIC FACTORS RESPONSIBLE FOR SPILLOVER

The limited opportunity for animal–human contact is an important barrier to host transmitted diseases. This barrier is formed by the geographical, ecological, and behavioral separation of the animal and human hosts. Anthropogenic changes are the major player for transmission of viruses from animal to human hosts. Viral transmission from one host to another can take place in distantly or closely related species, but transmission might be easier to a distant host because closely related species may develop cross immunity toward related pathogens. The transmission of a virus to a new host cell can be regulated at many different levels in a cell. Mutations in receptor binding can play a big role in the host switching of virus. Besides receptor barriers, viruses face host intracellular defenses as well. These include interferon release, macrophages, and other cell-mediated and humoral immune responses. Viral transmission is also affected by physical barriers to the virus along with host factors.

Cross species transmission is more common in rapidly evolving viruses. RNA viruses have a more error-prone replication rate and therefore have a higher variability. High variability helps the virus transmission and establishment by improving the ability to adapt to the new host mechanisms. The mode of transmission of a virus can determine the ability of the virus to spread.

Contact between host and virus, host immune defenses, viral factors are major determinants responsible for efficient transmission of virus to hosts. The determinants of viral spread in population are the major factors for the spread in human population.

Wolfe et al. (2007) classified pathogenic infection in five stages, which can lead to successful host switching in viruses and the emergence of viral outbreaks: (i) pathogens exclusively infecting animals, (ii) pathogens transmitting from animals to humans to cause primary infection but do not exhibit secondary infection (human to human), (iii) pathogens spilling over to human populations from animal reservoirs and can cause limited cycles of human-to-human transmission (iv) pathogens originating and persisting in animal reservoirs but can cause self-sustaining chains of transmission in human populations, and (v) pathogens exclusively infecting humans.

It is important to understand the mechanisms of viral entry and spread in new hosts, including factors such as demographics, host and cellular properties, and the controls of virus transmission (Figure 18.1).

The geographical distribution of a virus is limited by geographical distribution of its reservoir host, whereas the geographical distribution of a viral disease is a result of the

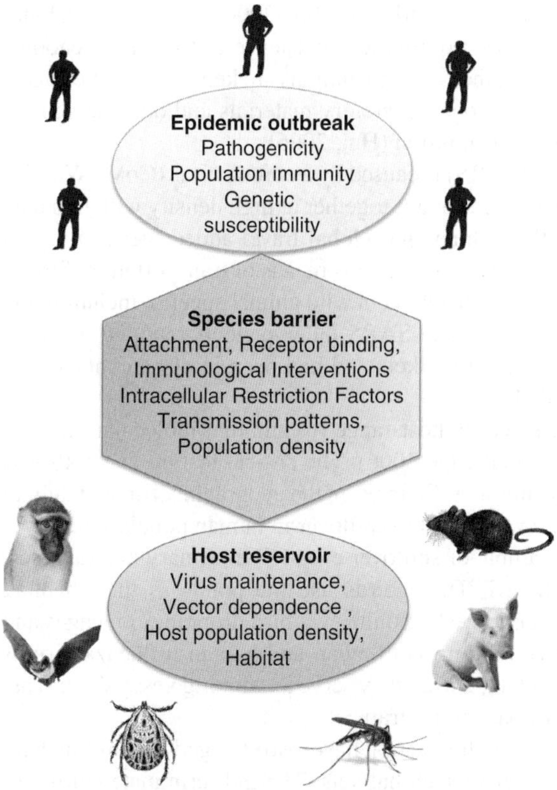

Figure 18.1. Steps involved in host switching of viruses. The zoonotic viruses have to cross many barriers before infecting human beings. Attachment, receptor binding, immunological interventions, intracellular host restriction factors, transmission patterns, and population density are the major species barriers to be crossed by a virus to successfully infect human population. For color detail, please see color plate section.

combined effect of the reservoir's host distribution, the location of spillover host, and the environment in which they are brought together (Figure 18.1).

Various anthropogenic factors are responsible for the contact between host and viruses such as urbanization, changes in agricultural practices, deforestation, human travel and trade, vector travel, hunting, and pasture practices (Figure 18.1). In Africa, close contact between nonhuman primates and humans occurs frequently due to the traditional habits of hunting animals for food. In Cameroon, more than 60% of the population is directly exposed to fresh nonhuman primate blood and bodily fluids from hunting, butchering, or petting (Wolfe et al., 2004, 2005).

An important example of a viral infection resulting from host switching is HIV. HIV is believed to emerge from primates in Africa between late 1940 and the early 1950. The earliest known case was in 1959, in a man from the Democratic Republic of Congo (Zhu et al., 1998). The closest relative of HIV-1 and HIV-2 found infecting other primates are termed as simian immunodeficiency viruses (SIVs). HIV-1 is more closely related to SIV_{CPZ}, whereas HIV-2 is related to sooty mangabeys SIV_{SM}. Chimpanzees as well as sooty mangabeys and many other species are often hunted and butchered for food and sold as bushmeat in market. Direct exposure of humans to animal blood or mucosal secretions as a result of hunting and field dressing provides an easy way for cross species transmission of zoonotic viruses to humans. However, iatrogenic (human-induced) modes of cross species transmission have also been proposed (Sharp et al., 2001).

Pasturage practices are the major cause of proximity of animals to humans in developing countries, which may result into recombination events in viruses. Wild animals are commonly sold in live animal markets (wet markets) (Hui, 2006) in southern China. Recombination and reassortment of viruses require the coexistence of two viral genomes in the same cell. Before being slaughtered, the captured animals are kept in the same cages, and this leads to contamination of cages through excretory materials and other secretions and which can result into cross species transmission (Hui, 2006).

Severe acute respiratory syndrome (SARS) is caused by a corona virus (CoV). Wildlife trade, in which many species of animals are brought together at high density with humans, which led to the emergence of SARS in China, and global travel added fuel in fire and expanded the geographical area of a virus through various means of transportation. Strong evidence exists that SARS-CoV was present in China in wild animal species, including the Himalayan palm civet (Guan et al., 2003). Many SARS index patients reported in China worked as restaurant staff in the live animal markets and had been in close contact with wild animals (Hui, 2006; Xu et al., 2004).

It appears that SARS-CoV broadened its host range by minimal molecular changes and then started infecting human beings (Riedel, 2006). The critical feature in SARS was its capability of human-to-human transmission. Changes in the geographical area of animal reservoirs provide new opportunities for the virus to spillover in human population.

The influenza virus is another example of spillover event. The primary natural reservoir of influenza A viruses is waterfowl. These birds use the ponds as their habitat. Occasionally, influenza A viruses infect domestic poultry via bird droppings on cages and feed (Hui, 2006). Since pigs are susceptible to both avian and human influenza viruses (Hui, 2006; Ito et al., 1998; Ninomiya et al., 2002), they serve as "mixing vessels" for reassortment, responsible for generation of pandemic strains.

Changes in agricultural practices and deforestation have also changed the wildlife habitat and brought human population closer to such habitats. The indiscriminate cutting of forests to develop new agricultural lands exposes workers to arthropod and rodent vectors. Argentine hemorrhagic fever (AHF), caused by Junin virus, spilled over to human population through changes in agricultural practices (Hui, 2006; Maiztegui, 1975). The rodent *Calomys musculinus*, a natural reservoir of Junin virus, increased in numbers after clearing the pampas for maize cultivation. This rodent species shed virus through their urine and solid excreta. Agricultural workers working in the fields got infected by inhalation of virus-contaminated aerosols produced from rodent excreta (Hui, 2006). A similar trend appeared with the emergence of Bolivian hemorrhagic fever and Lassa fever, caused by Machupo virus and Lassa fever virus, respectively, via their natural respective host rodents *Calomys callosus* and *Mastomys natalensis* (Richmond and Baglole, 2003; Vainrub and Salas, 1994).

The genus *Henipavirus* represents two viruses, Hendra virus (HeV) and NiV. These viruses are nonsegmented negative-strand RNA viruses. Both of these viruses have caused zoonotic diseases affecting a broad range of hosts. HeV has been associated with deaths of horses and humans in Australia. This HeV outbreak emerged as a spillover event from horses (equines) having respiratory infection to the people who came in contact with the infected horses (O'Sullivan et al., 1997; Selvey et al., 1995). NiV was isolated in an outbreak in Malaysia in 1999 and was responsible for many deaths. People working in very close contact with pigs or engaged in large-scale pig farming became infected through NiV-infected pigs. Since then, the distribution of NiV extends beyond Southeast Asia.

The dengue virus (DENV) infection has also been reported as a consequence of cross species transmission. Humans got repeatedly exposed to sylvatic DENV, which resulted in four sustained transmission chains in humans represented by DENV-1 to DENV-4 (Vasilakis et al., 2011). It is difficult to say where and when the cross species transmission of DENV took place that lead to the DENV infection in humans. It is expected that the DENV arose in the forests of Central and West Africa. Two monkey species are known to serve as amplification or reservoir hosts for sylvatic DENV in Africa: African green monkey (*Chlorocebus sabaeus*) and Guinea baboon (*Papio papio*) limited to West Africa (Vasilakis et al., 2011). The third primate host of sylvatic DENV is the patas monkey (*Erythrocebus patas*).

The Rift Valley fever (RVF) epidemic was reported in Mauritania in 1987, following the dam construction on the Senegal River, that resulted into 200 human deaths (Nabeth et al., 2001). A similar kind of RVF outbreak has been reported in Egypt after the opening of Aswan Dam in 1993. Three mosquito species (*Aedes cumminsil, Aedes circumlocutious, and Aedes mcintoshi*) are known vectors for RVF virus (Hui, 2006). Commercial farming resulted into the deforestation and brought the humans and their livestock in very close contact with the wild animals. In many instances, the cleared land has been used for irrigated agriculture, which resulted into an increase in vector-borne viral infections such as Japanese encephalitis virus (JEV); as mosquitoes and water birds are brought in close proximity to domestic pigs in nearby villages (Keiser et al., 2005; Morse, 1995).

18.3 MAJOR VIRAL FACTORS PLAYING A ROLE IN SPILLOVER

18.3.1 Reassortment

Reassortment is restricted to RNA viruses having segmented genomes and involves packaging of segments with different ancestry into a single virion. Reassortment requires that a cell should be infected with more than one virus (McDonald and Patton, 2011; Nibert et al., 1996; Simon-Loriere and Holmes, 2011; Urquidi and Bishop, 1992). It is believed that segmented viruses arose following the coinfection of a single cell by two or more viruses, which then evolved to function together through complementation (Simon-Loriere and Holmes, 2011). Extensive variation has been reported in the reassortment among viruses. Some segmented viruses (hantaviruses and Lassa viruses) exhibit relatively low levels of reassortment (Simon-Loriere and Holmes, 2011). However, others (influenza viruses and rotaviruses) show higher degree of reassortment (Iturriza-Gomara et al., 2001; Watanabe et al., 2001). Viral evolution may play an important role in successful host transfer through recombination and reassortment in viruses having segmented genomes. The emergence of the H2N2 and H3N2 influenza A viruses in 1957 and 1958 is the best example of reassortment. Recombination and reassortment play a role in host switching but can also be helpful in the adaption of the virus during intermediate host transfer.

18.3.2 Recombination

RNA viruses are having enormous genetic variability due to the combination of high mutation rates and large population sizes. Recombination leads to major impact on the evolution, emergence, and epidemiology of RNA viruses. It has been reported to be associated with the expansion of viral host range (Brown, 1997; Gibbs and Weiller, 1999; Simon-Loriere and Holmes, 2011), increases in virulence (Khatchikian et al., 1989), evasion of host immunity (Malim and Emerman, 2001), and resistance to antivirals (Nora et al., 2007). Recombination can occur in all RNA viruses, whether their genomes are composed of single segments or multiple segments. Recombination events in viruses can be responsible for efficient transfer through infection in new hosts and lead to viral outbreaks. The widely accepted model of RNA recombination is copy-choice recombination. In this process, the RNA polymerase (RNA-dependent RNA polymerase (RdRP)) in most RNA viruses and reverse transcriptase (RT) in retroviruses switch from one RNA molecule (the donor template) to another (acceptor template) during synthesis while remaining bound to nascent nucleic acid chain, thereby generating an RNA molecule with mixed ancestry (Breyer and Matthews, 2001; Von Hippel et al., 1994). The extent of local sequence identity between RNA templates, the kinetics of transcription, and the secondary structures in the RNA influence template switching (Baird et al., 2006; Simon-Loriere and Holmes, 2011). Recombination can be of two types: homologous and nonhomologous or illegitimate. Homologous recombination occurs most often between regions of high sequence similarity. Nonhomologous recombination takes place between different and hence genetically dissimilar genomic regions or between unrelated molecules (Simon-Loriere and Holmes, 2011). Recombination could assist in the process of cross species transmission of viruses because it enables the viruses to explore a greater proportion of the sequence space compared to mutation, thereby increasing the likelihood of finding genetic configuration that facilitates host adaptation. Viruses leading to persistent rather than acute infections, such as HIV, may have higher recombination rates due to increased chance of acquiring mixed infections in a single host.

Certain genome structures also facilitate recombination. Retrovirus virions are having two RNA molecules, which make them "pseudodiploid." As a consequence, the viruses with different ancestries that simultaneously infect a host cell may be packaged together, producing "heterozygous" virions and thereby enabling the production of genetically distinct progeny through copy-choice recombination (Simon-Loriere and Holmes, 2011). However, not all retroviruses recombine with the frequency of HIV. A high rate of recombination has been reported in CoV (Lai and Cavanagh, 1997), eastern equine encephalitis virus (EEEV), and western equine encephalitis virus (WEEV) (Hui, 2006).

18.3.3 Mutation

The common adaptation strategy for virus evolution is point mutation. RNA viruses mutate at the maximum error rate compatible with maintaining the integrity of genetic information (error threshold) because this would allow them to quickly find the beneficial mutations required for the adaptation (Domingo, 2000; Domingo and Holland, 1997; Holland et al., 1982; Novella et al., 1995). The high mutation rate of RNA viruses is due to poor fidelity of RdRP (i.e., RT), in which there is a lack of the 3'–5' proofreading exonuclease. Thus, the point mutation rates among RNA viruses is approximately 10^{-4}–10^{-5}, whereas DNA viruses is having 10^{-8}–10^{-11} (Hui, 2006). Usually viral proteins differing by as little as 1 or 2 amino acid residues confer striking differences in the phenotype. Such changes could lead to the development of antiviral drug resistance and host switching phenomenon.

Several changes are required for a virus to be able to switch hosts. When a virus undergoes any change in order to successfully adapt in the new host, it loses its fitness level or ability to infect the reservoir. In order for a virus to regain its fitness level in both the host and the reservoir, it must undergo through many changes. This can create complications for many viruses that require a different fitness mechanism in each host it infects when there is a vector and two or three other alternative hosts involved. Influenza A virus completed a successful host transfer and fitness change by going from an enteric virus in the avian population to a respiratory virus in the new human host. Host switching is rare phenomenon and virus adaptability to a new host takes time; it is not impossible for a virus like the H5N1 influenza A to gain mutations for efficient human-to-human transmission and at the same time retain its deadly pathogenicity. The amount of exposure the alternative host has to a virus is a key component in deciding whether or not it will have a successful host transfer. However, new host may not get infected due to the failure of the virus to adapt in the new host species.

18.4 INTERMEDIATE HOSTS AND SPECIES BARRIERS IN VIRAL TRANSMISSION

Intermediate hosts or amplifier hosts also have ways of helping a virus by facilitating the transmission in a new host. Intermediate hosts may play a critical role in disease emergence by bringing animal viruses into close contact with recipient hosts. The outbreak of NiV emerged from the NiV primary host, fruit bat (*Pteropus* sp.), to the pigs inhabiting near or around the fruit farms. The fruit farms were a major attraction to the bats.

Similarly SARS-CoV infection appears to have originated in bats and then infected humans along with civet cats (*Paguma larvata*). The exact pathway of transfer is uncertain; it is possible that the infection of the domesticated animals resulted in increased human exposures (Vijaykrishna et al., 2007; Wang and Eaton, 2007). Viruses oftentimes have to infect an intermediate host before infecting a new host. This suggests that some important changes take place in intermediate hosts, which are an important part of host transfer.

Host cells have many barriers (Figure 18.1) that a virus must cross before having a successful host transfer. These barriers include cell receptor and binding, immune responses to viruses, and skin barriers. Prominent changes in the viral sequences have been reported among the same viruses extracted from two different hosts affected over a long period of time. Complete adaptation of a virus in a new host takes a lot of time, and it involves a number of mutations. This suggests that the virus may have had to undergo a series of adaptations before it creates an emergence of disease outbreak (Figure 18.1). Patterns in human interaction with the environment and the emergence of new diseases help in predicting a pathway of emergence.

18.5 CONCLUSION

In order to minimize these spillover events, we would have to avoid the man-made disturbances in the ecological systems that would minimize the contact between reservoirs and hosts, which would lead to the decrease in the rate of transmission of zoonotic viral infections. Along with minimizing the human interferences in the natural ecological habitats, we should have the vaccination of animals and development of antiviral drugs to deal with such viral outbreaks. We must have a generic strategy in order to combat the emergence of new diseases because it is impossible to predict which virus will switch to human hosts and how it would affect humans. The animal viruses should be studied more closely to understand the molecular mechanisms involved in host switching, and effective strategies should be designed to combat the emergence of viral outbreaks due to host switching.

REFERENCES

Baird, H. A., Galetto, R., Gao, Y., et al. (2006). Sequence determinants of breakpoint location during HIV-1 intersubtype recombination. *Nucleic Acids Res* **34**(18), 5203–5216.

Breyer, W. A., and Matthews, B. W. (2001). A structural basis for processivity. *Protein Sci* **10**(9), 1699–1711.

Brown, D. W. (1997). Threat to humans from virus infections of non-human primates. *Rev Med Virol* **7**(4), 239–246.

Domingo, E. (2000). Viruses at the edge of adaptation. *Virology* **270**(2), 251–253.

Domingo, E., and Holland, J. J. (1997). RNA virus mutations and fitness for survival. *Annu Rev Microbiol* **51**, 151–178.

Gibbs, M. J., and Weiller, G. F. (1999). Evidence that a plant virus switched hosts to infect a vertebrate and then recombined with a vertebrate-infecting virus. *Proc Natl Acad Sci USA* **96**(14), 8022–8027.

Guan, Y., Zheng, B. J., He, Y. Q., et al. (2003). Isolation and characterization of viruses related to the SARS coronavirus from animals in southern China. *Science* **302**(5643), 276–278.

Holland, J., Spindler, K., Horodyski, F., et al. (1982). Rapid evolution of RNA genomes. *Science* **215**(4540), 1577–1585.

Hui, E. K. (2006). Reasons for the increase in emerging and re-emerging viral infectious diseases. *Microbes Infect* **8**(3), 905–916.

Ito, T., Couceiro, J. N., Kelm, S., et al. (1998). Molecular basis for the generation in pigs of influenza A viruses with pandemic potential. *J Virol* **72**(9), 7367–7373.

Iturriza-Gomara, M., Isherwood, B., Desselberger, U., et al. (2001). Reassortment in vivo: driving force for diversity of human rotavirus strains isolated in the United Kingdom between 1995 and 1999. *J Virol* **75**(8), 3696–3705.

Keiser, J., Maltese, M. F., Erlanger, T. E., et al. (2005). Effect of irrigated rice agriculture on Japanese encephalitis, including challenges and opportunities for integrated vector management. *Acta Trop* **95**(1), 40–57.

Khatchikian, D., Orlich, M., and Rott, R. (1989). Increased viral pathogenicity after insertion of a 28S ribosomal RNA sequence into the haemagglutinin gene of an influenza virus. *Nature* **340**(6229), 156–157.

Lai, M. M., and Cavanagh, D. (1997). The molecular biology of coronaviruses. *Adv Virus Res* **48**, 1–100.

Maiztegui, J. I. (1975). Clinical and epidemiological patterns of Argentine haemorrhagic fever. *Bull World Health Organ* **52**(4–6), 567–575.

Malim, M. H., and Emerman, M. (2001). HIV-1 sequence variation: drift, shift, and attenuation. *Cell* **104**(4), 469–472.

McDonald, S. M., and Patton, J. T. (2011). Assortment and packaging of the segmented rotavirus genome. *Trends Microbiol* **19**(3), 136–144.

Morse, S. S. (1995). Factors in the emergence of infectious diseases. *Emerg Infect Dis* **1**(1), 7–15.

Nabeth, P., Kane, Y., Abdalahi, M. O., et al. (2001). Rift Valley fever outbreak, Mauritania, 1998: seroepidemiologic, virologic, entomologic, and zoologic investigations. *Emerg Infect Dis* **7**(6), 1052–1054.

Nibert, M. L., Margraf, R. L., and Coombs, K. M. (1996). Nonrandom segregation of parental alleles in reovirus reassortants. *J Virol* **70**(10), 7295–7300.

Ninomiya, A., Takada, A., Okazaki, K., et al. (2002). Seroepidemiological evidence of avian H4, H5, and H9 influenza A virus transmission to pigs in southeastern China. *Vet Microbiol* **88**(2), 107–114.

Nora, T., Charpentier, C., Tenaillon, O., et al. (2007). Contribution of recombination to the evolution of human immunodeficiency viruses expressing resistance to antiretroviral treatment. *J Virol* **81**(14), 7620–7628.

Novella, I. S., Duarte, E. A., Elena, S. F., et al. (1995). Exponential increases of RNA virus fitness during large population transmissions. *Proc Natl Acad Sci USA* **92**(13), 5841–5844.

O'Sullivan, J. D., Allworth, A. M., Paterson, D. L., et al. (1997). Fatal encephalitis due to novel paramyxovirus transmitted from horses. *Lancet* **349**(9045), 93–95.

Richmond, J. K., and Baglole, D. J. (2003). Lassa fever: epidemiology, clinical features, and social consequences. *BMJ* **327**(7426), 1271–1275.

Riedel, S. (2006). Crossing the species barrier: the threat of an avian influenza pandemic. *Proc (Bayl Univ Med Cent)* **19**(1), 16–20.

Selvey, L. A., Wells, R. M., McCormack, J. G., et al. (1995). Infection of humans and horses by a newly described morbillivirus. *Med J Aust* **162**(12), 642–645.

Sharp, P. M., Bailes, E., Chaudhuri, R. R., et al. (2001). The origins of acquired immune deficiency syndrome viruses: where and when? *Philos Trans R Soc Lond B: Biol Sci* **356**(1410), 867–876.

Simon-Loriere, E., and Holmes, E. C. (2011). Why do RNA viruses recombine? *Nat Rev Microbiol* **9**(8), 617–626.

Urquidi, V., and Bishop, D. H. (1992). Non-random reassortment between the tripartite RNA genomes of La Crosse and snowshoe hare viruses. *J Gen Virol* **73**(Pt 9), 2255–2265.

Vainrub, B., and Salas, R. (1994). Latin American hemorrhagic fever. *Infect Dis Clin North Am* **8**(1), 47–59.

Vasilakis, N., Cardosa, J., Hanley, K. A., et al. (2011). Fever from the forest: prospects for the continued emergence of sylvatic dengue virus and its impact on public health. *Nat Rev Microbiol* **9**(7), 532–541.

Vijaykrishna, D., Smith, G. J., Zhang, J. X., et al. (2007). Evolutionary insights into the ecology of coronaviruses. *J Virol* **81**(8), 4012–4020.

Von Hippel, P. H., Fairfield, F. R., and Dolejsi, M. K. (1994). On the processivity of polymerases. *Ann N Y Acad Sci* **726**, 118–131.

Wang, L. F., and Eaton, B. T. (2007). Bats, civets and the emergence of SARS. *Curr Top Microbiol Immunol* **315**, 325–344.

Watanabe, M., Nakagomi, T., Koshimura, Y., et al. (2001). Direct evidence for genome segment reassortment between concurrently-circulating human rotavirus strains. *Arch Virol* **146**(3), 557–570.

Wolfe, N. D., Dunavan, C. P., and Diamond, J. (2007). Origins of major human infectious diseases. *Nature* **447**(7142), 279–283.

Wolfe, N. D., Switzer, W. M., Carr, J. K., et al. (2004). Naturally acquired simian retrovirus infections in central African hunters. *Lancet* **363**(9413), 932–937.

Wolfe, N. D., Heneine, W., Carr, J. K., et al. (2005). Emergence of unique primate T-lymphotropic viruses among central African bushmeat hunters. *Proc Natl Acad Sci USA* **102**(22), 7994–7999.

Xu, R. H., He, J. F., Evans, M. R., et al. (2004). Epidemiologic clues to SARS origin in China. *Emerg Infect Dis* **10**(6), 1030–1037.

Zhu, T., Korber, B. T., Nahmias, A. J., et al. (1998). An African HIV-1 sequence from 1959 and implications for the origin of the epidemic. *Nature* **391**(6667), 594–597.

II

SPECIFIC INFECTIONS

19

NEW, EMERGING, AND REEMERGING RESPIRATORY VIRUSES

Fleur M. Moesker, Pieter L.A. Fraaij, and Albert D.M.E. Osterhaus

Department of Viroscience, Erasmus Medical Center, Rotterdam, The Netherlands

TABLE OF CONTENTS

Viral Infections and Global Change, First Edition. Edited by Sunit K. Singh.

19.1 INTRODUCTION

19.1.1 History

Infectious diseases have been a major cause of mortality in human history. However, in the last century, this picture has changed remarkably with the introduction of hygienic and sanitary as well as veterinary control measures and new medical intervention strategies, like vaccination and the use of antibiotics and antivirals. Vaccination even resulted in the complete eradication of smallpox, a virus disease that has claimed the lives of millions of humans over the centuries. Recently vaccination has also contributed to the final eradication of rinderpest that has likewise devastated cattle populations (Morens et al., 2011; Roeder, 2011). It is reasonable to expect that the eradication of other human infectious diseases like polio and measles will follow in the decades to come (Aylward and Tangermann, 2011; Centers for Disease Control and Prevention, 2012; Moss and Griffin, 2012). In the 1970s of the last century, these developments prompted influential policymakers and scientists to speculate that infectious diseases would become a minor health issue in the future. However, this assumption proved completely wrong in the decades thereafter when the world was confronted with an unexpected number of new or reemerging infectious diseases. Most of these were viral diseases derived from the animal world, like pandemic influenza, severe acute respiratory syndrome (SARS), Ebola, Nipah, and dengue (Kuiken et al., 2011; Osterhaus, 2001, 2008; Reperant et al., 2012a, b).

Probably the most striking example is the still ongoing HIV/AIDS pandemic that started by multiple introductions of a simian lentivirus into the human population in the early decades of the twentieth century (Keele et al., 2006). Currently, almost 60 million people have been infected with HIV and approximately one-third of them have died from AIDS (WHO, 2012b). To date we continue to be confronted with new, emerging, and reemerging infectious diseases causing mild to severe illnesses in humans and animals. Emerging and reemerging infections are defined as "infections that have newly appeared in a susceptible population or have existed previously but are rapidly increasing in incidence and/or geographic range" (Morens et al., 2004). They also include infections of which the incidence in humans threatens to increase in the near future (Lederberg et al., 1992). These two characteristics set them aside from newly discovered pathogens, which may have been circulating for a long time, but may have gone unnoticed. In this chapter we will focus on newly discovered, emerging, and reemerging respiratory viruses causing acute respiratory tract infections (ARTIs). In the population at large, each person experiences on average two to three ARTIs every year. Furthermore, it is the most frequent reason for emergency department visits and hospital admissions of infants (Brodzinski and Ruddy, 2009). The disease burden is estimated at more than 94 million disability-adjusted life years lost globally (WHO, 2011). Since ARTIs may aggravate and lead to pneumonia, their clinical impact is considerable, as pneumonia is one of the leading causes of death among children under the age of five. The World Health Organization (WHO) estimates that 1.4 million children die annually from pneumonia (WHO, 2012c).

19.1.2 Newly discovered human respiratory viruses that recently crossed the species barrier

In the past two decades, several viruses have been described as the cause of human respiratory tract infections for the first time. Among the clinically most significant ones are avian, swine, and pandemic influenza viruses; Hendra and Nipah virus; human metapneumovirus (HMPV); three human coronaviruses (HCoV), including the virus that caused SARS; and human bocavirus (HBoV) (Allander et al., 2005; Fouchier et al., 2005; Fraaij and Osterhaus, 2010; van den Hoogen et al., 2001; Ksiazek et al., 2011).

All these viruses have crossed animal–human species barriers, from either wild or domestic animals, in the recent or more distant past. Some of these have just manifested themselves as zoonotic viruses, while others have subsequently adapted to become real human viruses, spreading efficiently from human to human (Lederberg et al., 1992; Morens et al., 2004). The WHO definition of a zoonosis is any disease or infection that is naturally transmissible from vertebrate animals to humans. Although not called zoonosis, all living organisms can be affected by such an event of interspecies transmission (Osterhaus, 2001; WHO, 2013). A complex interplay of predisposing factors governs the occurrence of emerging infectious diseases. These include behavioral changes such as air travel, close contact with animals, and progressive deforestation—exposing humans to pathogens never encountered on a large scale before. In addition environmental changes will affect ecological factors contributing to the spread of viruses in new susceptible populations (Harvell et al., 1999; Kuiken et al., 2003a,b; Osterhaus, 2001).

A changing local climate as the result of global warming may destroy but also create new habitats for animals. For instance, rodents seeking new ground after flooding are more likely to come in close contact with humans. This encounter may give rise to rodent-borne infections, like in 1993, the rodent-borne Sin Nombre virus (SNV) that was detected in the United States of America (Duchin et al., 1994). It causes hantavirus pulmonary syndrome, which until today continues to be a health problem in the USA (Carver et al., 2010; Klempa, 2009; Macneil et al., 2011). Virus spread appeared to follow El Niño events, creating a good environment for the reproduction of the deer mouse, which harbors SNV, and increased human–mouse interaction (Carver et al., 2010; Klempa, 2009). An additional example of this phenomenon is the change in habitats of mosquitoes that may function as vectors for several viral pathogens. This was illustrated in Italy in 2007 where the combination of lack of adequate garbage disposal and warm temperatures led to an

TABLE 19.1 Virus characteristics

Virus	Influenza viruses	HMPV	HCoV	HBoV	KI and WU polyomaviruses	NiV and HeV
Order	Unassigned	Mononegavirales	Nidovirales	Unassigned	Unassigned	Mononegavirales
Family	Orthomyxoviridae	Paramyxoviridae	Coronaviridae	Parvoviridae	Polyomaviridae	Paramyxoviridae
Subfamily		Pneumovirinae	Coronavirinae	Parvovirinae		Pneumovirinae
Genus	*Influenzavirus A*, *Influenzavirus B*, *Influenzavirus C*, *Thogotovirus* *Isavirus*	*Metapneumovirus*	*Alpha-* *Beta-* *Delta-* *Gammacoronavirus*	*Bocavirus*	*Polyomavirus*	*Henipavirus*
Species	Influenza A virus Influenza B virus Influenza C virus Dhori virus Thogoto virus	HMPV	Group 1 (HCoV- 229E, HCoV-NL63) Group 2 (SARS-CoV, HCoV-OC43) Group 3	Human bocavirus Bovine parvovirus Canine minute virus	Human polyomavirus	Hendra virus Nipah virus
Subtypes	Influenza A virus: H1N1, H3N2, H5N1, H7N3, H7N7, H9N2	A1, A2, B1, B2	—	HBoV type 1–4	KIPyV, WUPyV	—
Genome	– ssRNA eight segments	– ssRNA non-segmented	+ ssRNA non-segmented	– ssDNA	circular dsDNA	– ssRNA non-segmented
Baltimore group[a]	V	V	IV	II	I	V

Source: *Field's Virology* Fifth edition 2007, www.ictvonline.org, viralzone.expasy.org.

–, negative sense; +, positive sense; ss, single stranded; ds, double stranded; RNA, ribonucleic acid; DNA, deoxyribonucleic acid.

Note: We depicted the viruses according to their taxonomy. On the left side we sorted order, family, subfamily, genus, species, and subtypes.

[a]The Baltimore group is a classification system according to viral genome replication.

unprecedented spread of the newly introduced mosquito *Aedes albopictus* harboring *Chikungunya* virus (Carrieri et al., 2011).

In the following sections, we will further highlight emerging and reemerging human respiratory viruses in relation to climate change that (may) have a great impact on public health. Their characteristics have been summarized in Table 19.1.

19.2 INFLUENZA VIRUSES

Influenza viruses belong to the *Orthomyxoviridae* family, which encompasses five genera: *Influenzavirus A*, *Influenzavirus B*, *Influenzavirus C*, *Thogotovirus* (a tick-borne virus of mammals), and *Isavirus* (infectious salmon anemia virus). Influenza A viruses are further divided into subtypes, characterized by their surface glycoproteins, hemagglutinin ((HA), $n=17$ (Tong et al., 2012)), and neuraminidase ((NA), $n=9$) (see Figure 19.1). Influenza A viruses are able to infect a wide array of birds and mammals, including humans. Influenza B virus infections are generally considered to be restricted to the human host, although recently, infections in seals were also demonstrated (Osterhaus et al., 2000). Influenza C viruses seem to have a somewhat broader host range, with the virus being isolated from humans, pigs, and dogs (Wright et al., 2007). *Thogotovirus* and *Isavirus* are of limited or no consequences for the human population (Fraaij and Osterhaus, 2010).

In moderate climate zones, influenza A and B virus infections are the cause of the annually recurring seasonal influenza outbreaks. Such seasonality has not been reported for influenza C viruses, which generally cause mild upper respiratory tract infections in young children (Moriuchi et al., 1991). In addition, influenza A viruses can also cause worldwide outbreaks, or pandemics of influenza as recently happened with the 2009 "swine flu" or "Mexican flu." Pandemics can occur after the introduction of novel influenza A viruses, antigenically distinct from their seasonal counter parts, into the global population. Obviously the population at large will largely lack immune-mediated protection, which allows widespread infection. An antigenically new influenza A subtype virus can emerge in the human population through an event called "antigenic shift." This can be the result of

(a) (b)

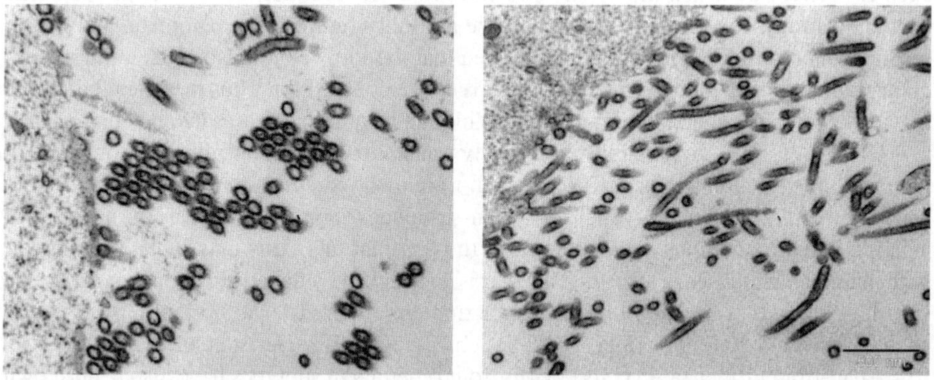

Figure 19.1. **H5N1**. Electron microscopy of viruses budding from human 293T cells. (**a**) Shows viral aggregates released in clusters of spherical particles of virus 1, while (**b**) shows single virus 2 particles being released from the infected cell as both spherical and filamentous virions (Sorrell et al., 2011). Picture courtesy: Ton de Jong.

either of two mechanisms that can also happen simultaneously. The first is the reshuffling or reassortment among the eight RNA genome segments of different influenza A viruses that may simultaneously infect a cell of an avian or mammalian host. This may result in the introduction of a reassortant virus with completely new surface glycoproteins. The second mechanism by which a new influenza A virus subtype can be introduced into the human population is by gradual adaptation of an avian influenza virus upon zoonotic transmission, through sequential mutation (Fraaij and Osterhaus, 2010). The most dramatic influenza pandemic known today started in 1918 and was called "the Spanish flu," causing over 50 million deaths in a relatively short period of time (Johnson and Mueller, 2002). The 1918 virus originated in birds (Taubenberger et al., 2005). Whether it had directly crossed the species barrier to humans, or an intermediate host like the pig was involved before it started to circulate among humans, remains unclear (Gibbs and Gibbs, 2006). Still as such the "Spanish flu" is an ominous well-documented example in recent history on how a zoonosis can eventually develop into a major human health catastrophe.

19.2.1 Transmission of avian influenza to humans

Until 1997, infections of humans with avian influenza viruses were sporadic reported and thought to be of minimal public health risk. However, this all changed in 1997 when highly pathogenic avian influenza (HPAI) A (H5N1) virus was first encountered in Hong Kong (Claas et al., 1998a, b; de Jong et al., 1997; Osterhaus et al., 2002). It was first described upon fatal disease in a 3-year-old boy, and subsequently 17 other patients were identified, of these 5 also died. This high mortality rate was at that time atypical, but, as we know now, not uncharacteristic for human infections with HPAI A (H5N1) (Fraaij and Osterhaus, 2010). The human infections coincided with an epizootic caused by the same virus infection in domestic poultry in the same geographical area (Claas et al., 1998a; de Jong et al., 1997). The Hong Kong authorities responded by having 1.6 million birds culled at the live bird marketplaces in order to eradicate the epizootic from the birds and prevent further zoonotic infections. Although this did stop the initial outbreak, it did not prevent the virus from returning, albeit in different genetic constellations. From 1997 on multiple infections with HPAI A (H5N1) in animals and humans, derived from wild birds, have been reported in an increasing number of countries. Up to now more than 600 human hospitalized cases have been registered, of which more than 350 were fatal in 15 countries (WHO, 2012a). One should however be cautious to interpret these mortality figures. A recent meta-analysis on sero-evidence for H5N1 infections concluded that the virus could also cause mild or subclinical infections in humans that are not currently accounted for. Thus the true fatality rate may be lower than 50%, although it remains high in hospitalized cases (Wang et al., 2012).

HPAI A (H5N1) virus is most commonly transmitted to humans through contact with infected birds (Aditama et al., 2012). Human-to-human infections happen only very rarely and apparently inefficiently and in a non-sustained way. The increasing number of avian-to-human infections does give rise to the fear that this virus may mutate or reassort with mammalian influenza viruses. This fear is augmented by data from animal experiments showing that only a limited number of mutation are required for airborne transmission to occur (Chu et al., 2012; Fouchier, 2012; Fouchier et al., 2012; Herfst et al., 2012a, b).

Apart from influenza A (H5N1) virus, four other avian subtypes have also been shown to be pathogenic to humans: H5N1, H7N3, H7N7, and H9N2 (Fraaij and Osterhaus, 2010). In this respect, of note is an outbreak of HPAI A virus (H7N7) in domestic poultry in the Netherlands. Eventually the same virus was also detected in 86 humans who were involved in the mass culling of about 30 million chickens. Three additional cases were found in family members of infected poultry workers indicating human-to-human transmission. Most patients developed self-limiting conjunctivitis; however, several patients developed more severe respiratory disease, including one veterinarian who died because of acute respiratory distress syndrome (ARDS) (Fouchier et al., 2004; Koopmans et al., 2004).

19.2.2 Clinical manifestations

Influenza virus infection causes influenza or the "flu." According to the WHO, seasonal influenza symptoms are characterized by a combination of the following clinical manifestations: sudden onset of high fever, (dry) cough, sore throat, nasal congestion, muscle and joint pain, fatigue, severe malaise, and headaches (WHO, 2009d). Most of the clinical cases are mild and self-limiting. However, complications leading to more severe disease do occur. Especially feared, mostly associated with H5N1 and pandemic (p)H1N1, but also to a lesser extent with seasonal influenza, is influenza-related ARDS. This distinct clinical manifestation is classically associated with multiorgan failure and disseminated intravascular coagulation (Fraaij and Osterhaus, 2010). Another potential fatal complication is secondary bacterial pneumonia. Influenza virus infection can also cause disease outside the respiratory tract including rhabdomyolysis, pericarditis, and meningoencephalitis (Fraaij and Heikkinen, 2011).

19.2.3 Diagnosis

Diagnosis on the basis of clinical signs and symptoms can be made by an experienced physician—when influenza is in the community—with about 60–80% sensitivity and specificity, depending on experience and adherence to the predefined criteria (Monto et al., 2000).Therefore, clinical diagnosis should be followed by laboratory confirmation, if indeed a definitive diagnosis is required. Current laboratory diagnosis of influenza virus infections may be based on any of the following methods: reverse transcriptase polymerase chain reaction (RT-PCR) (conventional PCR, real-time RT-PCR, and multiplex PCR), direct and indirect specimen immunofluorescence, rapid diagnostic tests (including antigen-detecting (EIA) and neuraminidase detection assay), viral culture, and serological testing (HA inhibition, enzyme-linked immunosorbent assay (ELISA), complement fixation, and neutralization), each with their own characteristic advantages and disadvantages (Fraaij and Osterhaus, 2012).

19.2.4 Treatment, prognosis, and prevention

Specific antiviral medication for the treatment of influenza is available. Medication includes neuraminidase inhibitors (oseltamivir, zanamivir, and in certain countries also peramivir) and for some influenza A viruses M2-channel blockers (amantadine and rimantadine). Due to rapid development of antiviral resistance, M2-channel blockers have currently become virtually obsolete in regular clinical practice.

The most effective and cost-effective way to combat influenza is vaccination if indeed a vaccine is available. Seasonal influenza vaccines are therefore updated twice a year to accommodate for the never-ending antigenic drift. The efficiency of seasonal vaccination to prevent seasonal influenza has been estimated to be between almost nonexisting to over 80%, depending on age and risk groups of the population vaccinated but also on the match of the vaccine strains with the actually circulating seasonal influenza viruses (Fiore et al., 2009; Fraaij et al., 2011; Michiels et al., 2011; Osterholm

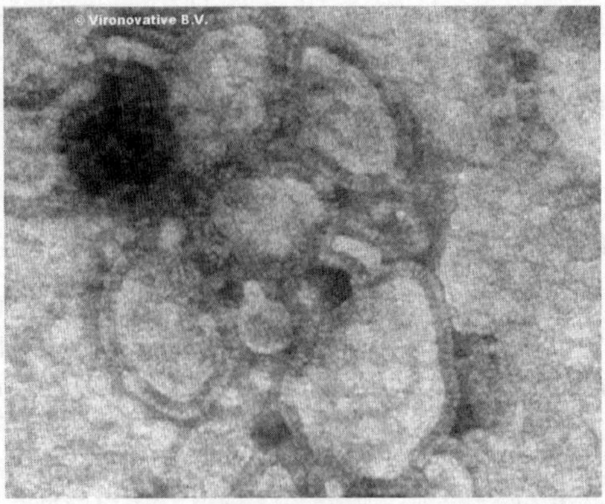

Figure 19.2. **HMPV**. Electron Micrograph of HMPV showing a typical Paramyxovirus morphology (www.vironovative.com).

et al., 2012). For the prevention of human infections with avian H5N1, so-called pre-pandemic candidate vaccines have been developed, which have not been used in the field yet (WHO, 2012d). For pandemics with a previously unknown influenza virus, enormous collaborative endeavors are required from scientists, clinicians, health authorities, and the pharmaceutical industry to manufacture a "new" vaccine in sufficient quantities within a record period of time. The difficulties of this process are probably best illustrated by the course of events during the 2009 H1N1 pandemic. Early in the pandemic, huge efforts were made to develop a pandemic vaccine, initially to vaccinate the highest-risk subjects. Pharmaceutical industry had to produce millions of vaccine doses in a relatively short period of time, which proved to take more than 6 months from the onset of the first efforts to develop the vaccine. Therefore, at the southern hemisphere, the first vaccine produced could only be used after the first wave of the pandemic outbreak was over and at the northern hemisphere well into the autumn wave of the pandemic (Butler, 2010).

19.3 HUMAN METAPNEUMOVIRUS

In 2001 HMPV was first described as the cause of respiratory tract infections of unknown etiology in children (van den Hoogen et al., 2001). They discovered HMPV, a negative single-stranded RNA virus genetically similar to avian metapneumovirus (AMPV) subgroup C, which is a causative agent of respiratory tract illnesses in poultry. Despite its close relationship to AMPV, HMPV does not readily infect birds. Currently, humans are the only known natural hosts. Still several animal models have been identified including hamsters, guinea pigs, ferrets, and nonhuman primates. HMPV has been categorized as a member of the *Metapneumovirus* genus in the subfamily of Pneumovirinae and family Paramyxoviridae (Schildgen et al., 2011). Van den Hoogen identified two lineages of the virus based on sequence homology (HMPV A and B). Both can be further divided into subtypes called A1, A2, B1, and B2 (Schildgen et al., 2011). Co-circulation of these subtypes during outbreaks is common, although predominance of the subtypes may change (Agapov et al., 2006) (see Figure 19.2). People have been shown to be infected several times during their lifetime with this virus (Schildgen et al., 2011).

19.3.1 Epidemiology

After its discovery HMPV has been shown to be a ubiquitous virus. On the northern hemisphere in the period between December and April, "winter" epidemics occur. Serological studies show that most individuals experience their first infection before the age of 1 year (Schildgen et al., 2011). Hereafter reinfections occur repeatedly and frequently (Schildgen et al., 2011). Children (especially <2 years) are at risk for complicated HMPV infection. It causes approximately 5–10% of the lower respiratory tract infections (LRTIs) in infants and is the second leading cause of bronchiolitis after RSV. In addition, patients with underlying medical conditions like neoplasmata and chronic lung disease, immunocompromised patients, and the elderly are at risk for severe disease. In the latter, especially in institutionalized elderly, HMPV infections have been shown to be associated with significant mortality (Schildgen et al., 2011; Te Wierik et al., 2012).

19.3.2 Clinical manifestations

HMPV is a respiratory pathogen causing a wide spectrum of disease ranging from mild ARTI to profound LRTIs including pneumonia. The clinical spectrum of disease is very similar to that of RSV (Brodzinski and Ruddy, 2009). Common symptoms are rhinorrhea, cough, and fever. Other respiratory symptoms can (infrequently) occur and include conjunctivitis, vomiting, diarrhea, and rash. The lower respiratory illnesses caused by HMPV mostly include bronchiolitis, pneumonia, and asthma exacerbation (Schildgen et al., 2011).

19.3.3 Diagnosis

RT-PCR-based techniques are generally the methods of choice for the detection of HMPV, but other assays, such as isothermal real-time nucleic acid sequence-based amplification (NASBA), have also been used. Virus isolation by culture proved relatively difficult for HMPV although shell vial cultures could offer an option (Crowe, 2011; Landry et al., 2005). ELISAs have been developed for antibody detection in plasma and serum.

19.3.4 Treatment, prognosis, and prevention

Overall, HMPV is mostly a self-limiting disease of the respiratory tract, although severe infections do occur (Jartti et al., 2012b). Currently there are no antivirals available for the treatment of HMPV infection. In case of severe disease, treatment consists of supportive care. Bronchodilators and corticosteroids have been used empirically, but there are no controlled trials to support or refute their efficacy. The only currently licensed antiviral drug for a related virus, RSV, is ribavirin, but no well-controlled clinical trials have been performed on the treatment of HMPV infection, and its efficacy has been debated (Schildgen et al., 2011). There is no vaccine available yet (Feuillet et al., 2012).

19.4 HUMAN CORONAVIRUSES: SARS AND NON-SARS

HCoV are viruses with a large positive RNA genome that belong to the genus *Coronavirus*. Together with the viruses grouped in the genus *Torovirus*, they constitute the family of *Coronaviridae*, which in turn are part of the order of *Nidovirales* (Abdul-Rasool and Fielding, 2010). Five HCoV have been identified to date. The first two were discovered in the 1960s, HCoV-OC43 and HCoV-229E. Both were found to cause common cold

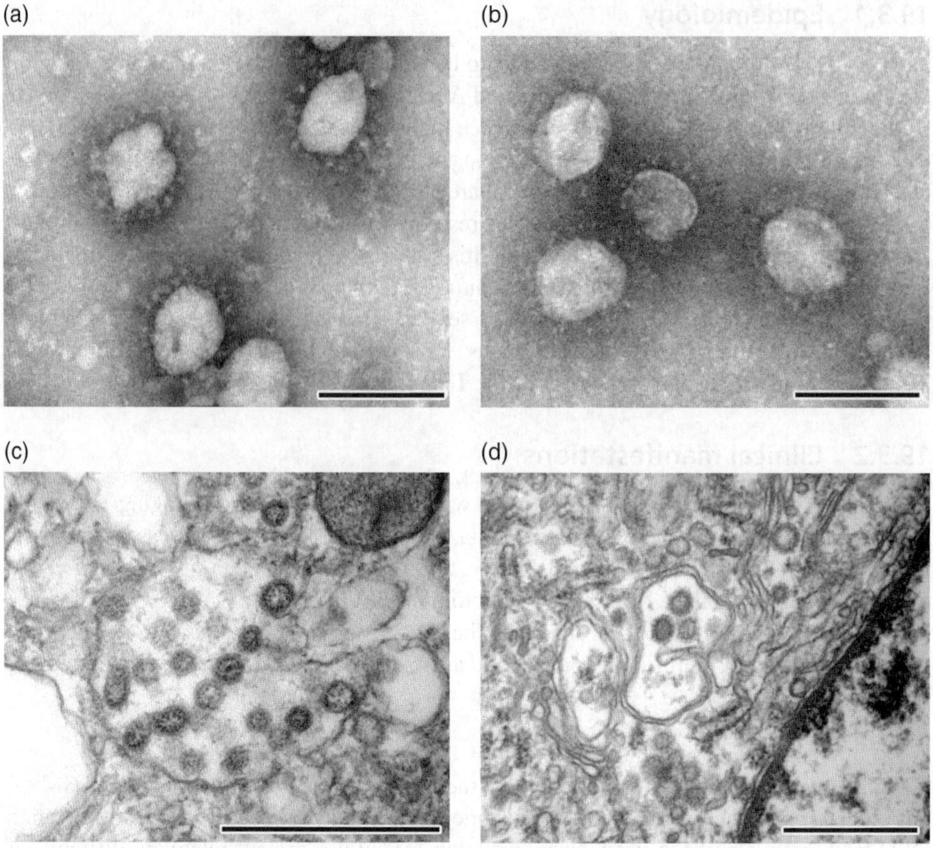

Figure 19.3. **SARS-CoV**. Electron microscopy of SARS-CoV in inoculum, clinical samples, and tissue samples of experimentally infected cynomolgus macaques. (**a**) Negative-contrast electron microscopy of virus stock used to inoculate cynomolgus macaques shows the typical club-shaped surface projections of coronavirus particles; negatively stained with phosphotungstic acid, bar = 100 nm. (**b**) Morphologically identical particles isolated from nasal swabs of infected macaques; negatively stained with phosphotungstic acid, bar = 100 nm. (**c**) Transmission electron microscopy of infected Vero 118 cell shows viral nucleocapsids with variably electron-dense and electron-lucent cores in smooth-walled vesicles in the cytoplasm; stained with uranyl acetate and lead citrate, bar = 500 nm. (**d**) Morphologically similar particles occur in pulmonary lesions of infected macaques, within vesicles of the Golgi apparatus of pneumocytes; stained with uranyl acetate and lead citrate; bar = 500 nm (Kuiken et al., 2003a,b).

(Bradburne et al., 1967). Recently, three new coronaviruses were described. The first was discovered in 2003 and caused the SARS outbreak (Drosten et al., 2003; Rota et al., 2003) (Figure 19.3). Soon thereafter the two other coronaviruses were discovered: human coronavirus NL63 (HCoV-NL63) in 2004 and human coronavirus HKU1 (HCoV-HKU1) in 2005 (Fouchier et al., 2004; van der Hoek et al., 2004; Woo et al., 2005).

19.4.1 The SARS-CoV outbreak

The 2002–2003 SARS outbreak was again an impressive example of how a zoonosis can eventually develop into a pandemic threat to the human population. During this outbreak, approximately 8000 cases were reported from 29 countries, with an overall staggering mortality of approximately 10% (Pang et al., 2003; Svoboda et al., 2004; WHO, 2003). Due to a quick and well-coordinated international response, the epidemic could be stopped in its wake (Peiris et al., 2004). However, the impact of the outbreak was massive because of reduced air travel and the impact on our globalized economy (Stockman et al., 2006). Since 2004 no new human cases of SARS-CoV have been reported. The masked palm civet (*Paguma larvata*) was initially believed to be the reservoir for SARS-CoV, and elimination was necessary from markets where they were sold for human consumption. However, later the presence of the virus was also demonstrated in other civet species and raccoon dogs, but there reservoir is now thought to be fruit bats. Therefore, ongoing vigilance for reappearance needs to be upheld (Graham and Baric, 2010; Kuiken et al., 2005).

19.4.2 Clinical manifestations of SARS-CoV

The WHO and the Center of Disease Control (CDC) made the following clinical case definition during the outbreak of SARS-CoV: patients with a history of (documented) fever and one or more symptoms of lower respiratory tract illness (cough, difficulty breathing, shortness of breath) and radiographic evidence or autopsy findings of lung infiltrates consistent of pneumonia or ARDS and no alternative diagnosis fully explaining the illness. Most patients with SARS have no upper respiratory symptoms during the first 3–7 days (the prodrome phase), the respiratory phase starts with a nonproductive cough, and dyspnea and respiratory failure occur as the disease progresses. The case description is still applicable if there is suspicion of a new SARS-CoV case; obviously it must always be combined with laboratory confirmation (WHO, 2009a).

19.4.3 Diagnosis of SARS-CoV

The diagnosis is based on the clinical picture and the following laboratory tests: RT-PCR assay for detecting viral RNA and eventually virus isolation from a clinical specimen. To confirm an infection, an additional sample should be tested. ELISA and/or immunofluorescence assay (IFA) can be performed for serum antibody detection (Kenneth McIntosh, 2012).

19.4.4 Epidemiology of non-SARS coronaviruses

HCoV OC43 and 229E circulate throughout the year and have a worldwide distribution. In temperate climates, there is an incidence peak in winter, while in other geographical areas such seasonality is not observed (Bastien et al., 2005; Brodzinski and Ruddy, 2009). The prevalence of the two newly discovered HCoV NL63 and HKU1 varies greatly. In both out- and inpatient populations, NL63 virus may be found in as many as 1 to almost 10% in patients with ARTIs. The virus is mostly detected in children, elderly, and immunocompromised patients (Pyrc et al., 2007).

19.4.5 Clinical manifestations of non-SARS coronaviruses

HCoV NL63, HKU1, 229E, and OC43 mostly cause common cold and are usually associated with a relatively mild clinical picture. Still hospitalization as a result of infection with one of the two viruses may occur. In this respect, HCoV-NL63 is often associated with pseudocroup in young children (Abdul-Rasool and Fielding, 2010). Moreover, it has also been found to cause wheezing and bronchiolitis and is associated with pneumonia. A relation with Kawasaki disease in children has been suggested, but could not be confirmed in two other study groups (Esper et al., 2005). HKU1 was first described among elderly patients with major underlying disease, in particular of the respiratory and cardiovascular systems.

19.4.6 Diagnosis of non-SARS coronaviruses

RT-PCR techniques are used to detect these coronaviruses (Gaunt et al., 2010). Complement-fixing and ELISA antibody assays for coronaviruses 229E and OC43 are available (Gerna et al., 2006). Recently also a novel double-antibody sandwich ELISA based on specific monoclonal antibodies allows detection and differentiation of HCoV-NL63 and HCoV-229E infections (Greenberg, 2011).

19.4.7 Treatment, prognosis, and prevention

Many treatment options for SARS-CoV have been suggested; ribavirin, corticosteroids, lopinavir and ritonavir, type 1 interferon (IFN), intravenous immunoglobulin, and SARS convalescent plasma have been implemented in battling SARS. However, none of these approaches were found to be uniformly effective (Sastre et al., 2011). Especially of note is the use of pegylated IFN-α in macaque studies (Haagmans et al., 2004). Despite all efforts, the prognosis of SARS-CoV remained to be poor during the epidemic with an overall mortality of 10%, which was however largely restricted to older-age categories (Chan et al., 2004; Liang et al., 2004; Liu et al., 2006).

Treatment options for the other coronaviruses are also limited, but in most cases unnecessary since infections are mostly self-limiting. There is no vaccine available to prevent for coronavirus infection.

19.5 HUMAN BOCAVIRUS

HBoV was discovered in 2005 (Allander et al., 2005). It has a single-stranded DNA genome and was classified in the *Bocavirus* genus (family Parvoviridae, subfamily Parvovirinae). To date, four strains have been identified: HBoV 1–4 (Jartti et al., 2012a). The closest relatives of this new parvovirus are *bo*vine parvovirus and minute virus of *ca*nines, hence its name: human bocavirus. It is a parvovirus that is only distantly related to parvovirus B19, the agent of fifth disease and other human disorders (Allander, 2008).

19.5.1 Epidemiology

HBoV has been detected in all continents of the world. It is not clear yet whether there is true seasonal variability for HBoV, but most studies indicate a higher incidence during winter (Chow and Esper, 2009). Seroprevalence of specific antibodies increases with age from >64% at 2–4 years of age to 100% in children 7 years of age (Jartti et al., 2012a).

19.5.2 Clinical manifestations

The clinical impact of this newly discovered virus remains topic for debate, although most papers report a relation between HBoV and ARTI (Allander, 2008). As for most respiratory viruses, the clinical picture includes fever, cough, and rhinorrhea. In addition, episodes of wheezing, acute obstructive bronchitis, bronchiolitis, and pneumonia have been found to be associated with HBoV infection. Furthermore, HBoV2 has been implicated in some cases of acute gastroenteritis (Jartti et al., 2012a). It has been suggested that most clinical manifestations of HBoV infections are associated with coinfection with other respiratory viruses (Guo et al., 2011).

19.5.3 Diagnosis

RT-PCR is widely used for the detection of HBoV in study setups. Serology for HBoV detection in serum is also used to diagnose acute infection (Chow and Esper, 2009; Jartti et al., 2012a).

19.5.4 Treatment, prognosis, and prevention

Most HBoV-induced disease is non-severe and self-limiting. This renders treatment unnecessary. However, in more severe cases, often in patients with underlying illness, supportive care should be given (Jartti et al., 2012a). The overall prognosis of HBoV infection is good.

19.6 KI AND WU POLYOMAVIRUSES

In the last decade, several new *polyomaviruses* (PyV) have been identified. Two of them were detected in samples taken from children with ARTIs and named KIPyV (Karolinska Institute) and WUPyV (Washington University); another was detected in a rare skin tumor and was named MCPyV (Merkel cell carcinoma) (Babakir-Mina et al., 2011; Jartti et al., 2012b). Whether or not the polyomaviruses KI and WU are indeed true pathogens remains a matter of debate. Still, in some patients with acute respiratory symptoms, they were the only pathogen detected (Han et al., 2007; Mourez et al., 2009).

19.6.1 Epidemiology

Serological studies show that both viruses occur in many geographical areas (Kean et al., 2009). Respiratory tract samples obtained during childhood show that primary exposure occurs early in life (Babakir-Mina et al., 2011). In the adult population, the seroprevalence of KIPyV and WUPyV are 55% and 69%, respectively (Kean et al., 2009).

19.6.2 Clinical manifestations

KIPyV and WUPyV have been detected in children with complaints of the respiratory and gastrointestinal tract. The majority of patients have the following symptoms: rhinitis, cough, bronchiolitis, or pneumonia (Jartti et al., 2012b).

19.6.3 Diagnosis

Most research on the occurrence of KI and WU was performed using PCR-based techniques in different samples including nasopharyngeal aspirates, sera, stool, and solid tissues (Babakir-Mina et al., 2011).

19.6.4 Prognosis

The prognosis is good without intervention.

19.7 NIPAH AND HENDRA VIRUSES

Two new viruses, Nipah virus (NiV) and Hendra virus (HeV), both *paramyxoviruses* derived from animals were discovered in 1994 and 1998. Both belong to the recently defined genus *Henipavirus* (Chua et al., 1999; Young et al., 1996). Fruit bats were found to be the main reservoir for both viruses. HeV was found to cause human infections upon exposure to affected horses in Australia, whereas most human cases caused by NiV took place upon contacts with affected pigs or bats in Southeast Asia (Ksiazek et al., 2011).

19.7.1 Outbreaks

The spread of NiV and HeV is reliant on the presence of the Pteropid fruit bats. Until now 13 outbreaks have been reported for NiV infections, all of which happened in South Asia (WHO, 2009c). Isolated human cases of HeV infection, derived from occupational contacts with diseased horses, have been reported in Australia from the mid-1990s onward and apparently continue to cause severe and fatal cases (Mendez et al., 2012; WHO, 2009b). The recently (2011) reported increase in the incidence of severe HeV infections in horses and isolation of HeV from a dog in Australia are a sharp reminder that the problems with HeV are far from over (Mendez et al., 2012; Young et al., 2011), as an unwanted fear for infection with HeV has caused Australian veterinarians to cease equine practice (Mendez et al., 2012).

19.7.2 Clinical manifestations and prognosis

NiV infection can cause both respiratory and neurological signs and symptoms. Disease onset is characterized by fever, headache, myalgia, and dizziness. Within a few days, drowsiness, confusion, and other neurological signs like hyporeflexia or areflexia, segmental myoclonus, gaze palsy, and limb weakness may develop. NiV is known to cause encephalitis and virus can be detected in the cerebrospinal fluid (Ksiazek et al., 2011). During the outbreak in Malaysia, the fatality rate was 30–40% and during several outbreaks in Bangladesh over 70% (Goh et al., 2000; Hossain et al., 2008). HeV has a high fatality rate of almost 60% (4 out of 7 patients identified have died). Disease symptoms initially are influenza-like with fever, myalgia, headache, lethargy, sore throat, nausea, and vomiting. Within days the central nervous system gets involved with meningitis and (late-onset) encephalitis (Tulsiani et al., 2011).

19.7.3 Diagnosis

PCR is performed on urine, respiratory, and cerebrospinal fluid samples. In addition serology can be performed. Virus isolation should only be performed in biosafety level 4 laboratories (the highest biosafety classification). The "gold standard" for HeV detection is the serum neutralization test, but during an outbreak the other methods are more commonly used. Currently, new methods are being developed like antigen-capture ELISA and Luminex-based tests (Tulsiani et al., 2011).

19.7.4 Treatment and prevention

Supportive care is given to patients infected with NiV and HeV. The use of antivirals such as ribavirin showed inconclusive outcomes. New and promising developments include the use of humanized monoclonal antibodies (Bossart et al., 2009). A number of vaccines have been developed for animal use only. Although their use may be beneficial in livestock, also with regard to animal welfare, surveillance and culling remain to be most cost-effective (Ksiazek et al., 2011; Tulsiani et al., 2011).

19.8 CONCLUSION

As discussed earlier, new, emerging, and re-emerging infectious diseases will continue to pose a challenge for health authorities worldwide. Frequently they are the direct result of ecological changes that may at least in part be climate driven. Research efforts to mitigate their effects on the human population concentrate on four main areas: improved surveillance, improved diagnostic methods, the development of vaccines, and the development of antiviral agents. Detection of a new, emerging, or reemerging virus starts with its first detection in animals or humans; therefore, vigilance of clinicians and veterinarians to recognize newly emerging disease entities is of great importance. We should realize that viruses not only cross species barriers but also geographical borders. Therefore, it is in the best interest of public and animal health programs to invest in international and preferably global surveillance programs. Diagnostic methods combining old and new advanced techniques such as random RT-PCR assays, nucleotide sequencing, and phylogenetic analysis have led to fast detection of new or modified known viruses. As discussed earlier, recent outbreaks have shown that there is an urgent need for (antiviral) treatment, as for most viral infections safe and effective specific treatment usually is lacking. Furthermore, public awareness programs concerning viral threats should be available and easy to access worldwide. In our current society, information can be spread fast via the Internet and social media. This may give us a head start facing new viral threats by early detection of a starting epidemic, early alerts, and easy access to this information by efficient communication. This may then result in extensive collaboration between expert laboratories, as shown previously with the successful identification and elimination of SARS-CoV (Peiris et al., 2004).

19.9 LIST OF ABBREVIATIONS

Viruses:

AMPV	avian metapneumovirus
CDV	canine distemper virus
HBoV	human bocavirus
HeV	Hendra virus
HIV	human immunodeficiency virus
HMPV	human metapneumovirus
HPAI	highly pathogenic avian influenza virus
HCoV	human coronavirus
KIPyV	Karolinska Institute polyomavirus
NiV	Nipah virus
PyV	polyomavirus

RSV	respiratory syncytial virus
SNV	Sin Nombre virus
WUPyV	Washington University polyomavirus
Others:	
AIDS	acquired immunodeficiency syndrome
ARDS	acute respiratory distress syndrome
ARTI	acute respiratory tract infection
CDC	centers for disease control and prevention
DNA	deoxyribonucleic acid
EIA	enzyme immunoassay
ELISA	enzyme-linked immunosorbent assay
HA	hemagglutinin
IFA	immunofluorescence assay
IFN	interferon
LRTI	lower respiratory tract infection
NA	neuraminidase
NASBA	nucleic acid sequence-based amplification
p	pandemic
RNA	ribonucleic acid
RT-PCR	reverse transcriptase polymerase chain reaction
SARS	severe acute respiratory syndrome
URTI	upper respiratory tract infection
WHO	World Health Organization

REFERENCES

Abdul-Rasool, S. & Fielding, B.C., 2010. Understanding human coronavirus HCoV-NL63. *The open virology journal*, **4**, pp. 76–84.

Aditama, T.Y., Samaan, G., Kusriastuti, R. et al., 2012. Avian influenza H5N1 transmission in households, Indonesia. *PLoS one*, **7**(1), p. e29971.

Agapov, E., Sumino, K.C., Gaudreault-Keener, M. et al., 2006. Genetic variability of human metapneumovirus infection: evidence of a shift in viral genotype without a change in illness. *The journal of infectious diseases*, **193**(3), pp. 396–403.

Allander, T., 2008. Human bocavirus. *Journal of clinical virology*, **41**(1), pp. 29–33.

Allander, T., Tammi, M.T., Eriksson, M. et al., 2005. Cloning of a human parvovirus by molecular screening of respiratory tract samples. *Proceedings of the national academy of sciences of the United States of America*, **102**(36), pp. 12891–6.

Aylward, B. & Tangermann, R., 2011. The global polio eradication initiative: lessons learned and prospects for success. *Vaccine*, **29**(Suppl 4), pp. D80–5.

Babakir-Mina, M., Ciccozzi, M., Perno, C.F. et al., 2011. The novel KI, WU, MC polyomaviruses: possible human pathogens? *The new microbiologica*, **34**(1), pp. 1–8.

Bastien, N., Robinson, J.L., Tse, A. et al., 2005. Human coronavirus NL-63 infections in children: a 1-year study. *Journal of clinical microbiology*, **43**(9), pp. 4567–73.

Bossart, K.N., Zhu, Z., Middleton, D. et al., 2009. A neutralizing human monoclonal antibody protects against lethal disease in a new ferret model of acute nipah virus infection. *PLoS pathogens*, **5**(10), p. e1000642.

Bradburne, A.F., Bynoe, M.L. & Tyrrell, D.A., 1967. Effects of a "new" human respiratory virus in volunteers. *British medical journal*, **3**(5568), pp. 767–9.

Brodzinski, H. & Ruddy, R.M., 2009. Review of new and newly discovered respiratory tract viruses in children. *Pediatric emergency care*, **25**(5), pp. 352–60; quiz 361–3.

Butler, D., 2010. Portrait of a year-old pandemic. *Nature*, **464**(7292), pp. 1112–3.

Carrieri, M., Angelini, P., Venturelli, C. et al., 2011. *Aedes albopictus* (Diptera: Culicidae) population size survey in the 2007 chikungunya outbreak area in Italy. I. Characterization of breeding sites and evaluation of sampling methodologies. *Journal of medical entomology*, **48**(6), pp. 1214–25.

Carver, S., Kilpatrick, A.M., Kuenzi, A. et al., 2010. Environmental monitoring to enhance comprehension and control of infectious diseases. *Journal of environmental monitoring*, **12**(11), pp. 2048–55.

Centers for Disease Control and Prevention, 2012. Progress toward interruption of wild poliovirus transmission – worldwide, January 2011–March 2012. *Morbidity and mortality weekly report*, **61**, pp. 353–7.

Chan, T.Y., Miu, K.Y., Tsui, C.K. et al., 2004. A comparative study of clinical features and outcomes in young and older adults with severe acute respiratory syndrome. *Journal of the American geriatrics society*, **52**(8), pp. 1321–5.

Chow, B.D.W. & Esper, F.P., 2009. The human bocaviruses: a review and discussion of their role in infection. *Clinics in laboratory medicine*, **29**(4), pp. 695–713.

Chu, C., Fan, S., Li, C. et al., 2012. Functional analysis of conserved motifs in influenza virus PB1 protein. L. L. M. Poon, ed. *PLoS one*, **7**(5), p. e36113.

Chua, K.B., Goh, K.J., Wong, K.T. et al., 1999. Fatal encephalitis due to Nipah virus among pig-farmers in Malaysia. *Lancet*, **354**(9186), pp. 1257–9.

Claas, E.C., Osterhaus, A.D., van Beek, R. et al., 1998a. Human influenza A H5N1 virus related to a highly pathogenic avian influenza virus. *Lancet*, **351**(9101), pp. 472–7.

Claas, E.C., de Jong, J.C., van Beek, R. et al., 1998b. Human influenza virus A/HongKong/156/97 (H5N1) infection. *Vaccine*, **16**(9–10), pp. 977–8.

Crowe, J.E. Jr., 2011. Human metapneumovirus infections. *Uptodate.com.*

Drosten, C., Günther, S., Preiser, W. et al., 2003. Identification of a novel coronavirus in patients with severe acute respiratory syndrome. *The new England journal of medicine*, **348**(20), pp. 1967–76.

Duchin, J.S., Koster, F.T., Peters, C.J. et al., 1994. Hantavirus pulmonary syndrome: a clinical description of 17 patients with a newly recognized disease. The Hantavirus Study Group. *The new England journal of medicine*, **330**(14), pp. 949–55.

Esper, F., Shapiro, E.D., Weibel, C. et al., 2005. Association between a novel human coronavirus and Kawasaki disease. *The journal of infectious diseases*, **191**(4), pp. 499–502.

Feuillet, F., Lina, B., Rosa-Calatrava, M. et al., 2012. Ten years of human metapneumovirus research. *Journal of clinical virology*, **53**(2), pp. 97–105.

Fiore, A.E., Bridges, C.B. & Cox, N.J., 2009. Seasonal influenza vaccines. *Current topics in microbiology and immunology*, **333**, pp. 43–82.

Fouchier, R., 2012. H5N1. Ron Fouchier: in the eye of the storm. *Science*, **335**(6067), pp. 388–9.

Fouchier, R.A.M., Hartwig, N.G., Bestebroer, F.W. et al., 2004. A previously undescribed coronavirus associated with respiratory disease in humans. *Proceedings of the national academy of sciences of the United States of America*, **101**(16), pp. 6212–6.

Fouchier, R.A., Rimmelzwaan, G.F., Kuiken, T. et al., 2005. Newer respiratory virus infections: human metapneumovirus, avian influenza virus, and human coronaviruses. *Current opinion in infectious diseases*, **18**(2), pp. 141–6.

Fouchier, R.A.M., Herfst, S. & Osterhaus, A.D.M.E., 2012. Public health and biosecurity. Restricted data on influenza H5N1 virus transmission. *Science*, **335**(6069), pp. 662–3.

Fraaij, P.L.A. & Osterhaus, A.D.M.E., 2010. The epidemiology of influenza viruses in humans. *Respiratory medicine*, **5**(13), pp. 7–14.

Fraaij, P.L.A. & Heikkinen, T., 2011. Seasonal influenza: the burden of disease in children. *Vaccine*, **29**(43), pp. 7524–8.

Fraaij, P.L. & Osterhaus, A.D., 2012. Influenza virus. In: G. Cornaglia, R. Courcol, J.-L. Herrmann, G. Kahlmeter, editors. *European Manual of Clinical Microbiology*. Paris: Société Française de Microbiologie, pp. 363–8.

Fraaij, P.L.A., Bodewes, R., Osterhaus, A.D. et al., 2011. The ins and outs of universal childhood influenza vaccination. *Future microbiology*, **6**(10), pp. 1171–84.

Gaunt, E.R., Hardie, A., Claas, E.C. et al., 2010. Epidemiology and clinical presentations of the four human coronaviruses 229E, HKU1, NL63, and OC43 detected over 3 years using a novel multiplex real-time PCR method. *Journal of clinical microbiology*, **48**(8), pp. 2940–7.

Gerna, G., Campanini, G., Rovida, F. et al., 2006. Genetic variability of human coronavirus OC43-, 229E-, and NL63-like strains and their association with lower respiratory tract infections of hospitalized infants and immunocompromised patients. *Journal of medical virology*, **78**(7), pp. 938–49.

Gibbs, M.J. & Gibbs, A.J., 2006. Molecular virology: was the 1918 pandemic caused by a bird flu? *Nature*, **440**(7088), pp. E8; discussion E9–10.

Goh, K.J., Tan, C.T., Chew, N.K. et al., 2000. Clinical features of Nipah virus encephalitis among pig farmers in Malaysia. *The new England journal of medicine*, **342**(17), pp. 1229–35.

Graham, R.L. & Baric, R.S., 2010. Recombination, reservoirs, and the modular spike: mechanisms of coronavirus cross-species transmission. *Journal of virology*, **84**(7), pp. 3134–46.

Greenberg, S.B., 2011. Update on rhinovirus and coronavirus infections. *Seminars in respiratory and critical care medicine*, **32**(4), pp. 433–46.

Guo, L., Gonzalez, R., Xie, Z. et al., 2011. Bocavirus in children with respiratory tract infections. *Emerging infectious diseases*, **17**(9), pp. 1775–7.

Haagmans, B.L., Kuiken, T., Martina, B.E. et al., 2004. Pegylated interferon-alpha protects type 1 pneumocytes against SARS coronavirus infection in macaques. *Nature medicine*, **10**(3), pp. 290–3.

Han, T.H., Chung, J.Y., Koo, J.W. et al., 2007. WU polyomavirus in children with acute lower respiratory tract infections, South Korea. *Emerging infectious diseases*, **13**(11), pp. 1766–8.

Harvell, C.D., Kim, K., Burkholder J.M. et al., 1999. Emerging marine diseases—climate links and anthropogenic factors. *Science*, **285**(5433), pp. 1505–10.

Herfst, S., Schrauwen, E.J.A., Linster, M. et al., 2012a. Airborne transmission of influenza A/H5N1 virus between ferrets. *Science*, **336**(6088), pp. 1534–41.

Herfst, S., Osterhaus, A.D.M.E. & Fouchier, R.A.M., 2012b. The future of research and publication on altered H5N1 viruses. *The journal of infectious diseases*, **205**(11), pp. 1628–31.

Hossain, M.J., Gurley, E.S., Montgomery, J.M. et al., 2008. Clinical presentation of nipah virus infection in Bangladesh. *Clinical infectious diseases: an official publication of the infectious diseases society of America*, **46**(7), pp. 977–84.

Lederberg, J., Shope, R.E., Oaks, C.S., Jr., editors; Committee on Emerging Microbial Threats to Health, Institute of Medicine, 1992. *Emerging Infections: Microbial Threats to Health in the* United States, Washington, DC: National Academy Press.

Jartti, T., Hedman, K., Jartti, L. et al., 2012a. Human bocavirus-the first 5 years. *Reviews in medical virology*, **22**(1), pp. 46–64.

Jartti, T., Jartti, L., Ruuskanen, O. et al., 2012b. New respiratory viral infections. *Current opinion in pulmonary medicine*, **18**(3), 271–8.

Johnson, N.P.A.S. & Mueller, J., 2002. Updating the accounts: global mortality of the 1918–1920 "Spanish" influenza pandemic. *Bulletin of the history of medicine*, **76**(1), pp. 105–15.

de Jong, J.C., Claas, E.C., Osterhaus, A.D. et al., 1997. A pandemic warning? *Nature*, **389**(6651), p. 554.

Kean, J.M., Rao, S., Wang, M. et al., 2009. Seroepidemiology of human polyomaviruses. *PLoS pathogens*, **5**(3), p. e1000363.

Keele, B.F., Van Heuverswyn, F., Li, Y. et al., 2006. Chimpanzee reservoirs of pandemic and nonpandemic HIV-1. *Science*, **313**(5786), pp. 523–6.

Kenneth McIntosh, M., 2012. Severe acute respiratory syndrome (SARS). *Uptodate*. Available at: http://www.uptodate.com/contents/severe-acute-respiratory-syndrome-sars?source=search_result&search=SARS-CoV&selectedTitle=1~6#H19 [Accessed June 20, 2012].

Klempa, B., 2009. Hantaviruses and climate change. *Clinical microbiology and infection: the official publication of the European society of clinical microbiology and infectious diseases*, **15**(6), pp. 518–23.

Koopmans, M., Wilbrink, B., Conyn, M. et al., 2004. Transmission of H7N7 avian influenza A virus to human beings during a large outbreak in commercial poultry farms in the Netherlands. *Lancet*, **363**(9409), pp. 587–93.

Ksiazek, T.G., Rota, P.A. & Rollin, P.E., 2011. A review of Nipah and Hendra viruses with an historical aside. *Virus research*, **162**(1–2), pp. 173–83.

Kuiken, T., Fouchier, R., Rimmelzwaan, G. et al., 2003a. Emerging viral infections in a rapidly changing world. *Current opinion in biotechnology*, **14**(6), pp. 641–6.

Kuiken, T., Fouchier, R.A.M., Schutten, M., et al., 2003b. Newly discovered coronavirus as the primary cause of severe acute respiratory syndrome. *Lancet*, **362**(9380), 263–70.

Kuiken, T., Leighton, F.A., Fouchier, R.A. et al., 2005. Public health. Pathogen surveillance in animals. *Science*, **309**(5741), pp. 1680–1.

Kuiken, T., Fouchier, R., Rimmelzwaan, G. et al., 2011. Pigs, poultry, and pandemic influenza: how zoonotic pathogens threaten human health. *Advances in experimental medicine and biology*, **719**, pp. 59–66.

Landry, M.L., Ferguson, D., Cohen, S. et al., 2005. Detection of human metapneumovirus in clinical samples by immunofluorescence staining of shell vial centrifugation cultures prepared from three different cell lines. *Journal of clinical microbiology*, **43**(4), pp. 1950–2.

Liang, W., Zhu, J., Guo, J. et al., 2004. Severe acute respiratory syndrome, Beijing, 2003. *Emerging infectious diseases*, **10**(1), pp. 25–31.

Liu, M., Liang, W.N., Chen, Q. et al., 2006. Risk factors for SARS-related deaths in 2003, Beijing. *Biomedical and environmental sciences: BES*, **19**(5), pp. 336–9.

Macneil, A., Nichol, S.T. & Spiropoulou, C.F., 2011. Hantavirus pulmonary syndrome. *Virus research*, **162**(1–2), pp. 138–47.

Mendez, D.H., Judd, J. & Speare, R., 2012. Unexpected result of Hendra virus outbreaks for veterinarians, Queensland, Australia. *Emerging infectious diseases*, **18**(1), pp. 83–5.

Michiels, B., Govaerts, F., Remmen, R. et al., 2011. A systematic review of the evidence on the effectiveness and risks of inactivated influenza vaccines in different target groups. *Vaccine*, **29**(49), pp. 9159–70.

Monto, A.S., Gravenstein, S., Elliott, M. et al., 2000. Clinical signs and symptoms predicting influenza infection. *Archives of internal medicine*, **160**(21), pp. 3243–7.

Morens, D.M., Folkers, G.K. & Fauci, A.S., 2004. The challenge of emerging and re-emerging infectious diseases. *Nature*, **430**(6996), pp. 242–9.

Morens, D.M., Holmes, E.C., Davis, A.S. et al., 2011. Global rinderpest eradication: lessons learned and why humans should celebrate too. *The journal of infectious diseases*, **204**(4), pp. 502–5.

Moriuchi, H., Katsushima, N., Nishimura, H. et al., 1991. Community-acquired influenza C virus infection in children. *The journal of pediatrics*, **118**(2), pp. 235–8.

Moss, W.J. & Griffin, D.E., 2012. Measles. *Lancet*, **379**(9811), pp. 153–64.

Mourez, T., Bergeron, A., Ribaud, P. et al., 2009. Polyomaviruses KI and WU in immunocompromised patients with respiratory disease. *Emerging infectious diseases*, **15**(1), pp. 107–9.

Osterhaus, A., 2001. Catastrophes after crossing species barriers. *Philosophical transactions of the royal society of London. Series B: biological sciences*, **356**(1410), pp. 791–3.

Osterhaus, A.D., 2008. New respiratory viruses of humans. *The pediatric infectious disease journal*, **27**(10 Suppl), pp. S71–4.

Osterhaus, A.D., Rimmelzwaan, G.F., Martina, B.E. et al., 2000. Influenza B virus in seals. *Science*, **288**(5468), pp. 1051–3.

Osterhaus, A.D.M.E., de Jong, J.C., Rimmelzwaan, G.F. et al., 2002. H5N1 influenza in Hong Kong: virus characterizations. *Vaccine*, **20**(Suppl 2), pp. S82–3.

Osterholm, M.T., Kelley, N.S., Sommer, A. et al., 2012. Efficacy and effectiveness of influenza vaccines: a systematic review and meta-analysis. *The Lancet infectious diseases*, **12**(1), pp. 36–44.

Pang, X., Zhu, Z., Xu, F. et al., 2003. Evaluation of control measures implemented in the severe acute respiratory syndrome outbreak in Beijing, 2003. *The journal of the American medical association*, **290**(24), pp. 3215–21.

Peiris, J.S.M., Guan, Y. & Yuen, K.Y., 2004. Severe acute respiratory syndrome. *Nature medicine*, **10**(12 Suppl), pp. S88–97.

Pyrc, K., Berkhout, B. & van der Hoek, L., 2007. The novel human coronaviruses NL63 and HKU1. *Journal of virology*, **81**(7), pp. 3051–7.

Reperant, L.A., Kuiken, T. & Osterhaus, A.D.M.E., 2012a. Adaptive pathways of zoonotic influenza viruses: from exposure to establishment in humans. *Vaccine*, (Chu), **30**(30), pp. 4419–34.

Reperant, L.A., Kuiken, T. & Osterhaus, A.D.M.E., 2012b. Influenza viruses: from birds to humans. *Human vaccines & immunotherapeutics*, **8**(1), pp. 7–16.

Roeder, P.L., 2011. Rinderpest: the end of cattle plague. *Preventive veterinary medicine*, **102**(2), pp. 98–106.

Rota, P.A., Oberste, M.S., Monroe, S.S. et al., 2003. Characterization of a novel coronavirus associated with severe acute respiratory syndrome. *Science*, **300**(5624), pp. 1394–9.

Sastre, P., Dijkman, R., Camuñas, A. et al., 2011. Differentiation between human coronaviruses NL63 and 229E using a novel double-antibody sandwich enzyme-linked immunosorbent assay based on specific monoclonal antibodies. *Clinical and vaccine immunology*, **18**(1), pp. 113–8.

Schildgen, V., van den Hoogen, B., Fouchier, R. et al., 2011. Human Metapneumovirus: lessons learned over the first decade. *Clinical microbiology reviews*, **24**(4), pp. 734–54.

Sorrell, E., Schrauwen, E., Linster, M., et al., 2011. Predicting "airborne" influenza viruses: (trans-) mission impossible? *Current opinion in virology*, **1**(6), 635–42.

Stockman, L.J., Bellamy, R. & Garner, P., 2006. SARS: systematic review of treatment effects. *PLoS medicine*, **3**(9), p. e343.

Svoboda, T., Henry, B., Shulman, L. et al., 2004. Public health measures to control the spread of the severe acute respiratory syndrome during the outbreak in Toronto. *The new England journal of medicine*, **350**(23), pp. 2352–61.

Taubenberger, J.K., Reid, A.H., Lourens, R.M. et al., 2005. Characterization of the 1918 influenza virus polymerase genes. *Nature*, **437**(7060), pp. 889–93.

Te Wierik, M.J., Nguyen, D.T., Beersma, M.F. et al., 2012. An outbreak of severe respiratory tract infection caused by human metapneumovirus in a residential care facility for elderly in Utrecht, the Netherlands, January to March 2010. *Euro surveillance: bulletin europêensur les maladies transmissibles = European communicable disease bulletin*, **17**(13), pp. 1–7.

Tong, S., Li, Y., Rivailler, P. et al., 2012. A distinct lineage of influenza A virus from bats. *Proceedings of the national academy of sciences of the United States of America*, **109**(11), pp. 4269–74.

Tulsiani, S.M., Graham, G.C., Moore, P.R. et al., 2011. Emerging tropical diseases in Australia. Part 5. Hendra virus. *Annals of tropical medicine and parasitology*, **105**(1), pp. 1–11.

van den Hoogen, B.G., de Jong, J.C., Groen, J. et al., 2001. A newly discovered human pneumovirus isolated from young children with respiratory tract disease. *Nature medicine*, **7**(6), pp. 719–24.

Van der Hoek, L., Pyrc, K., Jebbink, M. F. et al. (2004). Identification of a new human coronavirus. *Nature medicine*, **10**(4), pp. 368–73.

Wang, T.T., Parides, M.K. & Palese, P., 2012. Seroevidence for H5N1 influenza infections in humans: meta-analysis. *Science*, **335**(6075), p. 1463.

WHO, 2003. WHO | Summary of probable SARS cases with onset of illness from 1 November 2002 to 31 July 2003. Available at: http://www.who.int/csr/sars/country/table2004_04_21/en/index.html [Accessed March 29, 2012].

WHO, 2009a. Case definitions for the 4 diseases requiring notification to WHO in all circumstances under the IHR (2005). *Relevéépidémiologiquehebdomadaire/Section d'hygiène du Secrétariat de la Société des Nations = Weekly epidemiological record/Health Section of the Secretariat of the League of Nations*, **84**(7), pp. 52–6.

WHO, 2009b. Hendra, Fact sheet N°329. Available at: http://www.who.int/mediacentre/factsheets/fs329/en/ [Accessed June 5, 2013].

WHO, 2009c. Nipah virus, Fact sheet N°262. Available at: http://www.who.int/mediacentre/factsheets/fs329/en/ [Accessed June 5, 2013].

WHO, 2009d. WHO Influenza (Seasonal), Fact sheet N°211. Available at: http://www.who.int/mediacentre/factsheets/fs211/en/index.html [Accessed June 5, 2013].

WHO, 2011.World Health Organization, Pneumonia, Fact sheet N°331. Available at: http://www.who.int/mediacentre/factsheets/fs331/en/index.html [Accessed June 5, 2013].

WHO, 2012a. Cumulative number of confirmed human cases for avian influenza A(H5N1) reported to WHO, 2003–2012. *WHO*. Available at: http://www.who.int/influenza/human_animal_interface/EN_GIP_20120312uCumulativeNumberH5N1cases.pdf [Accessed June 5, 2013].

WHO, 2012b. WHO Global Health Observatory HIV. Available at: http://www.who.int/gho/hiv/en/ [Accessed July 8, 2013]

WHO, 2012c. WHO I Acute respiratory infections. Available at: http://www.who.int/vaccine_research/diseases/ari/en/ [Accessed March 23, 2012].

WHO, 2012d. WHO I FAQs: H5N1 influenza. Available at: http://www.who.int/influenza/human_animal_interface/avian_influenza/h5n1_research/faqs/en/ [Accessed June 20, 2012].

WHO, 2013. WHO I Zoonoses. Geneva: World Health Organization. Available at: http://www.who.int/topics/zoonoses/en/ [Accessed June 13, 2013].

Woo, P.C.Y., Lau, S.K.P., Chu, C., et al. (2005). Characterization and complete genome sequence of a novel coronavirus, coronavirus HKU1, from patients with pneumonia. *Journal of virology*, **79**(2), pp. 884–95.

Wright P.F., Neumann, G. & Kawaoka, Y., 2007. Orthomyxoviruses. In: D.M. Knipe, P.M. Howley, editors. *Field's Virology*, 5th ed., Philadelphia: Lippincott Williams & Wilkins.

Young, P.L., Halpin, K., Selleck, P.W. et al., 1996. Serologic evidence for the presence in Pteropus bats of a paramyxovirus related to equine morbillivirus. *Emerging infectious diseases*, **2**(3), pp. 239–40.

Young, J.R., Selvey, C.E. & Symons, R., 2011.Hendra virus. *The medical journal of Australia*, **195**(5), pp. 250–1.

20

EMERGENCE OF ZOONOTIC ORTHOPOX VIRUS INFECTIONS

Tomoki Yoshikawa and Masayuki Saijo

Department of Virology I, National Institute of Infectious Diseases, Tokyo, Japan

Shigeru Morikawa

Department of veterinary science, National Institute of Infectious Diseases, Tokyo, Japan

TABLE OF CONTENTS

20.1 SMALLPOX, A REPRESENTATIVE ORTHOPOXVIRUS INFECTION: THE ERADICATED NON-ZOONOTIC ORTHOPOXVIRUS

Although this chapter is focused on zoonotic poxviruses, this introduction focuses on smallpox virus, which is not zoonotic, because it is one of the most well-known human viruses. Smallpox is caused by the variola virus (a member of the *Orthopoxvirus* genus). It had been endemic to populous areas, largely China and India in ancient times, appearing in Europe in the sixth

Viral Infections and Global Change, First Edition. Edited by Sunit K. Singh.

century and in America in 1520 (Barquet and Domingo, 1997). From the historical datasets, which show the peak of smallpox transmission to be January, a relationship between smallpox transmission and seasonal humidity was suggested (Nishiura and Kashiwagi, 2009). The transmission is apparently increased by dry weather. In the early 1950s, an estimated 50 million cases of smallpox occurred in the world each year. Fortunately, smallpox has been eradicated through a global smallpox eradication campaign comprising a broad array of measures, such as mass vaccination, containment (patient isolation), and surveillance, coordinated by the World Health Organization (WHO). The last patient with endemic smallpox was recorded in Somalia in 1977, and the eradication of smallpox was certified in December 1979 and subsequently endorsed by the World Health Assembly in 1980. Three main factors made the eradication feasible (Fenner et al., 1988). First, the host of the variola virus is only humans. Second, a very effective smallpox vaccine, including the vaccinia virus, was used in the eradication campaign, and ring vaccination, a strategic method of vaccination that targets all susceptible individuals in a prescribed area around an epidemic, was effective. Third, all of the susceptible subjects infected with variola virus showed symptoms associated with smallpox, making it possible to identify the areas in which the variola virus was present.

Vaccinia virus, also known as the Edward Jenner virus, is a member of the genus *Orthopoxvirus*, as well as the variola virus. There are no known natural hosts. Routine use of the vaccinia virus as a smallpox vaccine was terminated in most countries between 1980 and 1981, by 1985 at the latest, and no countries have continued to give the vaccine on a large scale. However, it is still used for some military forces and laboratory workers potentially exposed to orthopoxviruses, because the vaccine provides cross protection against several other orthopoxviruses (e.g., cowpox virus and monkeypox virus). In the early 1980s, a fatal vaccinia infection occurred in an HIV-infected adult following vaccination in the U.S. military, an event that accelerated the rate of termination of routine vaccination programs for military personnel (Redfield et al., 1987). Approximately 30 years have passed since the termination of the routine smallpox vaccination in the mid-1980s. However, while no endemic infections have occurred, there is a serious concern that the variola virus could be used as a biological weapon for terrorism (Breman and Henderson, 1998).

20.1.1 Clinical features

Two clinical forms of smallpox are known, variola major and variola minor.

Variola major is the severe form of smallpox caused by variola virus major, causing a more extensive rash and having a high mortality rate. There are four subtypes of smallpox variola major: ordinary (the most common type, accounting for 90 % or more of the cases), modified (mild and occurring in previously vaccinated persons), flat, and hemorrhagic (both rare and very severe). The overall lethality rate of variola major is approximately 30 %, but the patients who develop either flat or hemorrhagic smallpox generally die.

Variola minor is the other form of smallpox and is a much less severe disease than variola major. The lethality rate is approximately 1 % or less. In general, the incubation period of smallpox after exposure is 12–14 days. The symptoms in the early phase of smallpox include fever, malaise, head and body aches, and sometimes vomiting. At this stage, people are usually too sick to carry on their ordinary activities. This is called the prodromal phase and may last for 2–4 days. Next, a rash emerges in the oral cavity and on the skin surface. By 3 days after the onset, the rash becomes raised bumps, as shown in Figure 20.1, and then scabs. About 6 days after the initial symptoms develop, the scabs begin to fall off and become pitted scars. Most scabs will have fallen off three weeks after the rash appears. The patient is infectious to others until all of the scabs have fallen off.

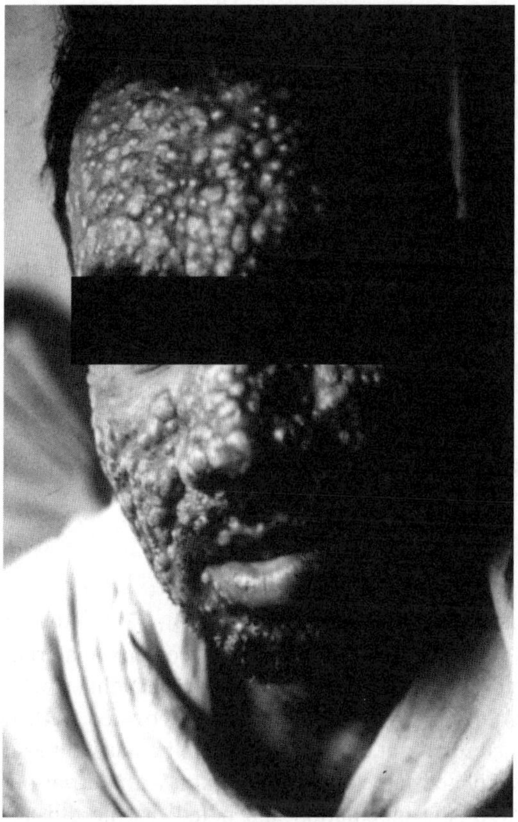

Figure 20.1. Cutaneous lesion in smallpox patient (Courtesy of Dr. Takeshi Kurata, National Institute of Infectious Diseases, Japan). For color detail, please see color plate section.

20.2 ZOONOTIC ORTHOPOXVIRUSES

Orthopoxviruses are some of the largest viruses with regard to size. The virus particle, which is brick shaped and measures around 200–300 nm in diameter, contains a linear, double-stranded genome of 140–300 kb in length that encodes several hundred polypeptides. The released virion is enveloped (external enveloped virion (EEV)), and inside the cell, the virion (intracellular mature virion (IMV)), which is also infectious, often has a double membrane. The lateral bodies contain various enzymes essential for virus replication (Damon, 2006). Orthopoxviruses are not difficult to isolate and grow in a variety of cell culture media *in vitro* and produce pocks on the chick chorioallantoic membrane (CAM) (Moss, 2006). The genus *Orthopoxvirus* is one of the viruses of the subfamily *Chordopoxvirinae*, which consists of 9 genera, 4 of which are known to be pathogenic to humans: *Molluscipoxvirus*, *Orthopoxvirus*, *Parapoxvirus*, and *Yatapoxvirus* (Moss, 2006). These viruses are capable of infecting vertebrate hosts. All of them, with the exception of the variola virus and *Molluscipoxvirus*, are transmitted to humans from other animals, in a process called zoonosis. Most cases of zoonotic orthopoxvirus infection are occupational and sporadic, thus resulting in skin lesions often associated with systemic illness. These lesions are painful blisters and crusted eschars and usually are self-limiting. These infections are rarely fatal. Exceptionally, the morbidity and childhood mortality of monkeypox virus infections (known as human monkeypox), whose symptoms resemble those of smallpox, are high.

Molecular characterization using polymerase chain reaction (PCR) amplification and other methods provides accurate phylogenetic identification and suggests that a cowpox-like virus is the probable ancestor of the variola virus and other zoonotic poxviruses (Lewis-Jones, 2004). The reservoirs for the cowpox virus, and possibly monkeypox virus, are believed to be rodents (Moss, 2006). Antibodies against vaccinia virus have been documented in several rodents (bank voles, wood mice, short-tailed field voles) and carnivores (cats, red foxes, and lynx) in ELISA and/or immunofluorescence studies (Baxby, 1977; Bennett et al., 1990; Chantrey et al., 1999; Czerny et al., 1996; Henning et al., 1995; Pelkonen et al., 2003; Tack and Reynolds, 2011; Tryland et al., 1998a). In recent years, the incidence of zoonotic poxvirus infections has increased outside their usual geographical range. Three of the major factors contributing to this increase in the incidence of zoonotic poxviruses include the end of routine smallpox vaccination, the increase of rodent export and import, and advancing urbanization, which leads to more frequent encounters between humans and nondomesticated animals.

This chapter focuses on two representative zoonotic poxvirus infections, namely, cowpox and monkeypox infections.

20.2.1 Cowpox

20.2.1.1 *Cowpox as the First Basis of Vaccination.* On May 14, 1796, Edward Jenner inoculated James Phipps, an 8-year-old male subject, with cowpox material, which was obtained from the vesicular lesions present on the hand of Sarah Nelmes, who was a local dairymaid. James became mildly ill a few days later, but recovered well. It was confirmed that cowpox could be transmitted not only from cows to humans but also from humans to humans. Next, Jenner variolated James, inoculating him with the variola virus, on July 1 to confirm the efficacy of the inoculation and noted that he did not develop smallpox. Although earlier accounts of cowpox, and even vaccination, exist, Jenner published all of his research on smallpox in a book entitled *An Inquiry into the Causes and Effects of the Variolae Vaccinae, A Disease Discovered in Some of the Western Counties of England, Particularly Gloucestershire, and Known by the Name of Cow Pox* in 1798 (Jenner, 1798; Riedel, 2005). The following year, he published the results of further experiments, which contained the first thorough descriptions of bovine and human cowpox. For two centuries, these reports formed the basis of our understanding of the natural history of cowpox.

20.2.1.2 *Characteristics of Cowpox.* Cowpox virus infection in humans (human cowpox) is much milder than smallpox infections. Cowpox is sometimes confused with pseudocowpox, which is an alternative name for milker's nodules, which are caused by the pseudocowpox virus, a member of the parapoxvirus family. The cowpox virus is distributed throughout Europe, Russia, the western states of the former Soviet Union, and adjacent areas of northern and central Asia (Damon, 2006). On the other hand, it is not found in Ireland, the United States, Australasia, or the Middle or Far East. Human cowpox usually causes localized skin lesions; however, immunocompromised patients are at risk of developing severe generalized skin infections (Baxby et al., 1994; Eis-Hubinger et al., 1990).

The virus exhibits low infectivity for humans and is transmitted only by direct contact with the skin lesions of infected animals. Despite widespread disease reported in domestic and wild animals, human infections are very rare (Bennett et al., 1989; Chantrey et al., 1999). Human-to-human transmission has not been reported. However, patients should be careful, since, as was mentioned earlier, their lesions contain infectious viruses, which were used by Jenner for the smallpox vaccination.

20.2.1.3 ***Natural Reservoirs.*** Cowpox virus is so named because it can infect the teats of cows. It must be kept in mind that, despite its name, the cowpox virus does not have cattle as its reservoir host. In fact, rodents are the reservoirs. In Europe, bank voles, wood mice, and short-tailed field voles constitute the main reservoirs (Baxby, 1977; Chantrey et al., 1999), whereas the cowpox virus was only sporadically detected in rats (*Rattus norvegicus*) (Marennikova and Shelukhina, 1976; Wolfs et al., 2002). Therefore, the direct transmission of cowpox virus from rodents to humans is naturally a major transmission route (Chantrey et al., 1999; Wolfs et al., 2002). Domestic cats, which are likely the most common host for cowpox virus because they hunt rodents, can also transmit cowpox virus to humans (Baxby and Bennett, 1997; Baxby et al., 1994). Notably, cat-to-human transmission usually occurs during the autumn season, which correlates with the peak rodent population size and activity. The cats probably become infected while hunting rodents, through a bite, scratch, or possibly via ingestion (Bennett et al., 1990; Pfeffer et al., 2002). Rats and other nondomestic animals, such as monkeys and elephants, that are kept as pets have also been associated with human infection (Becker et al., 2009; Campe et al., 2009; Elsendoorn et al., 2011; Girling et al., 2011; Hemmer et al., 2010; Kurth et al., 2008, 2009; Ninove et al., 2009; Wolfs et al., 2002). Transmission to humans likely occurs via direct contact with an affected animal, resulting in the implantation of the virus into non-intact skin or mucous membranes (Baxby et al., 1994). Children and those involved in animal care appear to be at risk (Baxby and Bennett, 1997).

20.2.1.4 ***Clinical Features.*** Human cowpox usually causes painful pustular lesions located on the hands or face, and the patients usually complain of flu-like symptoms. The lesions are large, ulcerative with inflammation, and edematous. The crust formed by the lesions is thick, hard, and black. Mucosal lesions have also been reported (Dina et al., 2011). Local lymphadenopathy and systemic symptoms such as pyrexia, lethargy, sore throat, and general malaise are common and often severe enough for people to miss school or work. Infections are typically self-limiting, and most people recover within 6–8 weeks (Baxby and Bennett, 1997; Baxby et al., 1994). Generalized infections have only been reported in those with predisposing factors, such as eczema, atopic dermatitis, and Darier's disease, and are fatal in a limited number of cases (Baxby and Bennett, 1997; Baxby et al., 1994; Haase et al., 2011; Pelkonen et al., 2003).

20.2.1.5 ***Epidemiology.*** Human cowpox outbreaks have occurred sporadically in Europe, in the western part of Russia, and in northern and central Asia, as shown in Figure 20.2 (Amer et al., 2001; Baxby and Bennett, 1997; Baxby et al., 1994; Becker et al., 2009; Campe et al., 2009; Cardeti et al., 2011; Hemmer et al., 2010; Honlinger et al., 2005; Kurth et al., 2008; Mancaux et al., 2011; Marennikova et al., 1978, 1984; Ninove et al., 2009; Pelkonen et al., 2003; Postma et al., 1991; Tryland et al., 1998b; Vestergaard et al., 2008; Vogel et al., 2011; Vorou et al., 2008; Wolfs et al., 2002). Although a global outbreak has never been reported, cowpox virus infections seem to be increasing (Vorou et al., 2008). One of the major reasons is the increase in the populations that have not been vaccinated with smallpox vaccine. The other factor is the increase in the number of people keeping wild and exotic animals as pets. Continued active surveillance is required to assess the risk of infections and to prevent humans from developing cowpox virus infections.

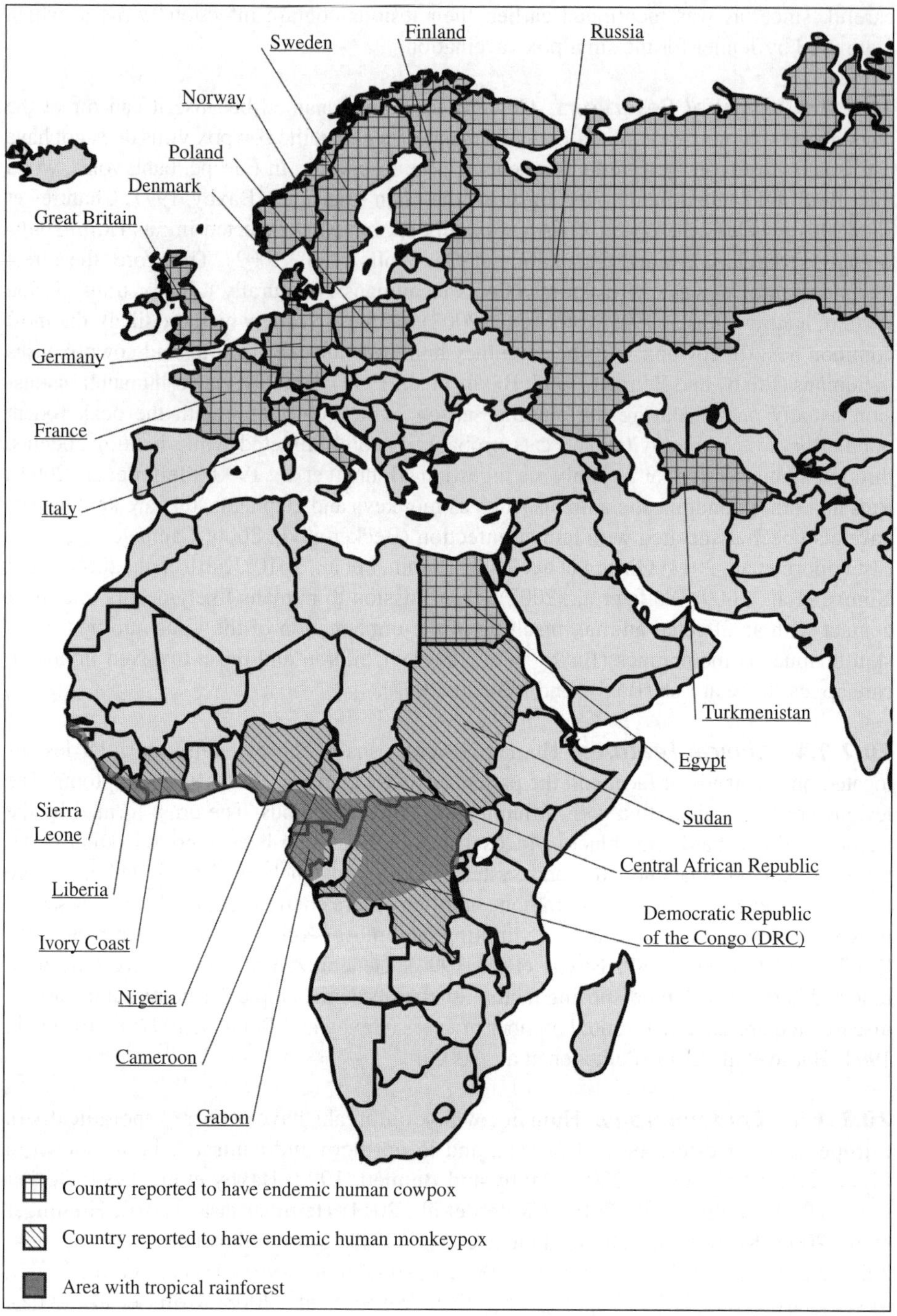

Figure 20.2. Geographical distribution of zoonotic poxvirus infections. For color detail, please see color plate section.

20.2.2 Monkeypox

20.2.2.1 ***The Increase in Zoonosis after the Cessation of Smallpox Vaccination.*** Human monkeypox (monkeypox virus infections in humans) leads to clinical signs similar to those seen in smallpox patients (Di Giulio and Eckburg, 2004; Heymann et al., 1998; Ladnyj et al., 1972). It was first isolated from sick animals in a colony of captive cynomolgus monkeys in Copenhagen Zoo in 1958 (von Magnus et al., 1959). Despite the high case fatality rate in monkeys, it had not been considered a threat to humans as long as the smallpox vaccination continued. This was because vaccination with the smallpox vaccine using the vaccinia virus led to cross protection against various orthopoxviruses, including monkeypox virus. In addition, the virus showed no evidence of sustained transmissibility in humans. Although person-to-person spread was reported (Arita et al., 1972, 1985; Jezek et al., 1983, 1988a, b; Khodakevich et al., 1988), the longest chain of documented human-to-human transmission was only five generations (four serial transmissions) (Jezek et al., 1986), and a stochastic model for the spread of human monkeypox indicated that the monkeypox virus was highly unlikely to be maintained permanently in humans (Jezek et al., 1987a). Accordingly, the Global Commission for the Certification of Smallpox Eradication concluded in its final report in 1979 that continued smallpox vaccination to prevent human monkeypox was not justified. Moreover, in addition to the known adverse events associated with smallpox vaccination even in immunocompetent subjects, the emergence of AIDS in the 1980s raised further concerns about the use of the vaccine (Heymann et al., 1998). The Global Commission did, however, recommend that measures be taken to assess the public health significance of this emerging zoonosis more accurately (WHO, 1980). Notably, in comparison with the incidence of human monkeypox (0.72 per 10 000 population) in the Democratic Republic of the Congo (DRC, formerly called Zaire) in the 1980s, the rate from 2006 to 2007 (14.42 per 10 000 population) suggests an approximately 20-fold increase in infections (Rimoin et al., 2010). This data suggests that the increase in the incidence may have been related to the cessation of smallpox vaccination programs in the human monkeypox-endemic regions.

20.2.2.2 ***Natural Reservoirs.*** Monkeypox virus has a wide range of hosts, including ground squirrels and rats, which are both reservoirs (Gispen, 1975). However, the main animal reservoir for monkeypox virus has not yet been conclusively determined (Mutombo et al., 1983; The current status of human monkeypox, 1984). Several epidemiological studies conducted in the DRC have implicated squirrels (especially *Funisciurus anerythrus*) inhabiting agricultural areas as the primary candidates for the infection of humans with monkeypox virus in the endemic areas (Khodakevich et al., 1986, 1988). In these epidemiological surveys, *Funisciurus* spp. squirrels had a higher rate of monkeypox virus seropositivity (24 %) than any of the other animals tested, including *Heliosciurus* spp. squirrels (15 %) and nonhuman primates (8 %) (Khodakevich et al., 1987a, b, 1988). A subsequent seroprevalence study conducted as part of the investigation of an outbreak in February 1997 in the DRC showed even higher positivity rates in these squirrels (40 % in *Funisciurus* spp. and 50 % in *Heliosciurus* spp.) (Hutin et al., 2001). Furthermore, 16 % of Gambian giant rats in the study that were tested showed a positive reaction in the serological analysis, suggesting that they had been exposed to the monkeypox virus.

After a monkeypox outbreak occurred in 2003 due to its accidental importation into the United States, which had been a virus-free area (CDC 2003a, b; Reed et al., 2004; Sejvar et al., 2004), there was concern that the virus would be established throughout the country. Although it is unclear whether monkeypox virus has established an enzootic

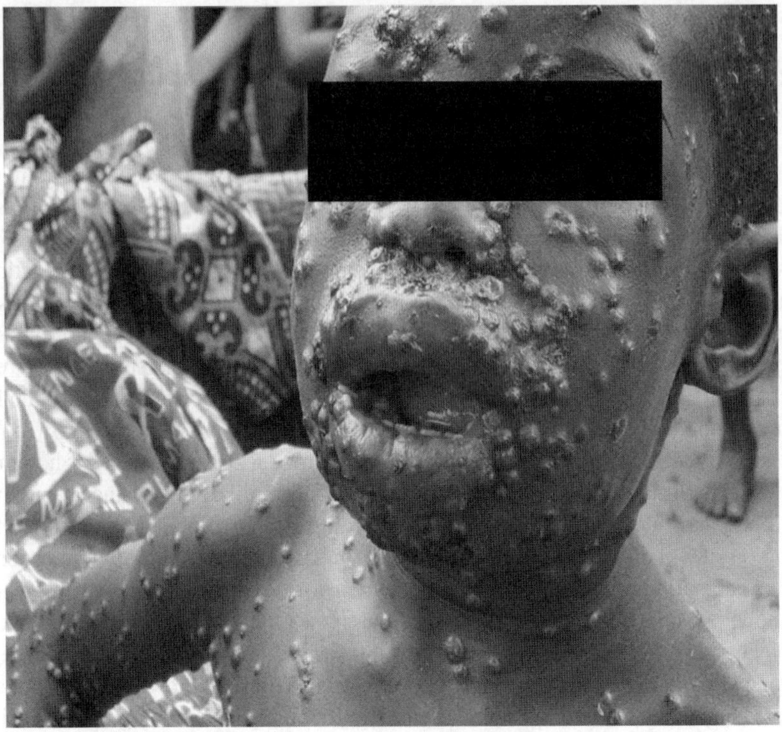

Figure 20.3. Cutaneous lesion in monkeypox patient (Courtesy of Dr. Muyembe-Tamfum JJ, National Institute of Biomedical Research, Democratic Republic of the Congo). For color detail, please see color plate section.

reservoir in the United States, the infection of a rabbit (family *Leporidae*) after exposure to a diseased prairie dog at a veterinary clinic confirmed the transmissibility of the virus between mammal species common in North America. This rabbit was implicated as the source of primary infection in one U.S. case (CDC 2003a; Hutson et al., 2007).

20.2.2.3 ***Clinical Features.*** The incubation period of human monkeypox varies from 6 to 16 days. Infected humans present with a generalized rash and lymphadenopathy, which is preceded by a 2–4-day prodrome of fever, headache, backache, myalgia, lethargy, and general malaise (Jezek et al., 1983, 1987b; Reed et al., 2004). The presence of lymphadenopathy is a characteristic feature of human monkeypox infection. The skin eruption appears on the face (in 95 % of cases), on the palms of the hands and soles of the feet (75 %), and on the body nearly simultaneously. The evolution of the rash from maculopapules (lesions with flat bases) to vesicles (small fluid-filled blisters) and pustules, followed by crusts, occurs in approximately 10 days (Figure 20.3). Three weeks might be necessary before the complete elimination of the crusts occurs. The symptoms of human monkeypox usually last for 14–21 days.

The number of lesions varies from a few to several thousand, affecting oral mucous membranes (in 70 % of cases), genitalia (30 %), and conjunctivae (eyelid, 20 %), as well as the cornea (eyeball). Unlike the typical presentation of human cowpox, a disseminated rash is commonly observed even in immunocompetent individuals. Affected individuals also suffer from more severe systemic disease and have a compressed incubation period (Reed et al., 2004; Reynolds et al., 2007). The case fatality rate has varied widely between

TABLE 20.1. Two Clades of Monkeypox Virus

Genetic clade		Congo Basin	West African
Outbreak in		Congo Basin (Central Africa)	West Africa, United States
Virulence to	Human (Likos et al., 2005) Cynomolgus monkey (Chen et al., 2005; Saijo et al., 2009) Prairie dog (Hutson et al., 2010b) Mouse (Hutson et al., 2010a) Ground squirrel (Sbrana et al., 2007)	Virulent	Avirulent

epidemics. One likely reason for this is that there are two distinct clades of monkeypox viruses, typed West African and Congo Basin monkeypox viruses, which display different levels of virulence in humans, monkeys, and rodents (Chen et al., 2005; Hutson et al., 2010a, b; Likos et al., 2005; Saijo et al., 2009; Sbrana et al., 2007) (Table 20.1).

20.2.2.4 *Epidemiology*

20.2.2.4.1 The Latter Part of the Twentieth Century. As the number of cases in Africa accumulated in the 1970s, human monkeypox was concluded to resemble smallpox in terms of symptoms, severity, and mortality (Gispen, 1975). However, in contrast to smallpox, the transmissibility of the monkeypox virus among human beings was lower than that of the variola virus. Therefore, the epidemiology of human monkeypox differs significantly from that of smallpox. Human monkeypox is endemic to the central and western parts of Africa, while smallpox had been endemic worldwide. Until early 2003, sporadic human monkeypox cases were reported only in the rainforest areas of Central and West Africa, as shown in Figure 20.2 (Arita et al., 1985; Breman et al., 1980; Damon et al., 2006; Monkeypox, 1991; The current status of human monkeypox, 1984).

The first documented patient with human monkeypox occurred in 1970 in a child in the DRC (Breman et al., 1980; Ladnyj et al., 1972; Marennikova et al., 1972). Of the total of 59 patients with human monkeypox reported in Cameroon, Cote d'Ivoire, Liberia, Nigeria, Sierra Leone, and the DRC from 1970 to 1980, 47 patients lived in the tropical rainforest region of the DRC (Breman et al., 1980). Twenty-three of these 47 patients had severe disease, and 8 of these patients died, resulting in a case fatality rate of 17 %. Human monkeypox outbreaks occurred in clusters in some instances, and 4 of the 47 reported patients appeared to have acquired the infection through person-to-person transmission, bringing the secondary transmission rate to 9 %, whereas no suggestion of tertiary spread was found. The four patients infected with monkeypox virus through person-to-person transmission occurred among 123 non-vaccinated contacts (a 3.3 % secondary attack rate) (Arita et al., 1985; Breman et al., 1980).

The Global Commission recommended that measures should be taken to assess the public health significance of this emerging zoonosis more accurately (WHO, 1980). Based on these recommendations, an active surveillance program for human monkeypox was established in the DRC by national authorities, with the support of the WHO from 1981 to 1986 (Jezek and Fenner, 1988). Of the 404 patients reported in Central and West Africa between 1981 and 1986, 386 were identified in the DRC. The majority of the patients were from the tropical rainforest areas, in which intensive surveillance was conducted, such as the regions of Bandundu, Equateur, and Kasai Oriental. For the period from 1981 to 1986, the annual incidence was estimated to be 0.63 per 10 000 population in one health zone where about one-third of the cases had occurred. The incidence was much lower than for other infectious diseases passively reported in the same zone and during the same period (e.g., malaria, 32.1 per 10 000; helminthiasis, 27.6 per 10 000). The 8-fold increase in the number of patients during the period from 1981 to 1986 (compared to 1970–1980) was more likely a result of the improved surveillance system and case identification than an actual increase in the incidence (Heymann et al., 1998). Interestingly, about one-third of the patients were diagnosed in June, July, or August, coinciding with the period of most intensive human outdoor activity, such as agricultural work and hunting in the regions.

The potential for human-to-human transmission of monkeypox virus is considerably lower than that of variola virus, with the secondary attack rates for monkeypox virus in 1970–1986 being 4–12% among unvaccinated household contacts, in contrast with the rate of 37–88% for smallpox (Jezek and Fenner 1988; Jezek et al., 1988b; Mack et al., 1972; Mukherjee et al., 1974; Rao et al., 1968). The attack rate of monkeypox virus was thought to be much lower than those of the other viral diseases in humans maintained naturally through human-to-human spread (e.g., measles, mumps, pertussis, and varicella), including smallpox. After the intensified surveillance project ended in 1986, the WHO resources in the country were directed towards AIDS, which by then had emerged as a serious public health threat. The number of reported human monkeypox patients from Western and Central Africa declined to six in 1987, one in 1990, five in 1991, and one in 1992. All of them were children, and nine occurred in two clusters, which independently occurred in 1987 and 1991, involving siblings in neighboring villages in Gabon. Six patients were diagnosed to have human monkeypox by the WHO Collaborating Centers in Atlanta and Moscow. After 1992, no additional patients with human monkeypox had been reported until 1996, when human monkeypox outbreaks occurred in the DRC.

The first known patient in those outbreaks, presumably the index case, was a 35-year-old male, who became ill on February 15th (Human monkeypox—Kasai Oriental, 1997a, b; Mukinda et al., 1997). In addition to the 92 patients identified during the initial investigation in February 1997, a further 419 cases were identified by a follow-up investigation in October of that year. The epidemiological analyses of these 419 patients indicated that there were some differences from the previously reported outbreaks. The 1.5 % case fatality rate was much lower than the rate in the previous outbreaks. The secondary attack rate was finally concluded from the follow-up investigation to be an estimated 8 %, which was similar to that estimated during the monkeypox surveillance in the DRC during the early 1980s (4–12 %) (Human monkeypox—Kasai Oriental, 1997b; Jezek and Fenner 1988), although the rate seemed to be substantially higher during the initial investigation (Cohen, 1997; Human monkeypox—Kasai Oriental, 1997a; Mukinda et al., 1997).

20.2.2.4.2 In the Twenty-First Century. In the twenty-first century, endemic human monkeypox infections have still been reported in the DRC (Rimoin et al., 2007). In 2003, the prospect of the global spread of monkeypox virus infections in humans was seen following cases of human monkeypox infection that occurred in the previously virus-free Western Hemisphere due to the accidental importation of infected animals (CDC 2003a, b; Reed et al., 2004; Sejvar et al., 2004). The outbreak of human monkeypox in North America occurred in the Midwest, with 71 patients eventually becoming infected (CDC, 2003b). Unlike the smallpox-like disease described in the DRC, the severity of human monkeypox in the United States appeared to be milder. None of the cases were attributed to secondary transmission, and none resulted in death. These patients, who developed fever and a rash, were linked to skin exposure to pets, notably prairie dogs (Kile et al., 2005). Trace back investigations identified an international shipment of about 800 small mammals from

Ghana to Texas that contained 762 African rodents as the probable source for the introduction of monkeypox virus into the United States (CDC, 2003c). These imported animals were six genera of African rodents: rope squirrels (*Funisciurus* spp.), tree squirrels (*Heliosciurus* spp.), Gambian giant rats (*Cricetomys* spp.), brush-tailed porcupines (*Atherurus* spp.), dormice (*Graphiurus* spp.), and striped mice (*Hybomys* spp.). Laboratory testing of some of the potential animals by virus isolation and PCR amplification at the CDC in Atlanta, GA, revealed that at least one Gambian giant rat, two rope squirrels, and three dormice were infected with monkeypox virus. Gambian giant rats from this shipment were transported from Texas via an Iowa animal vendor to a pet distributor in Illinois, where they were co-housed with prairie dogs (*Cynomus* spp.). These prairie dogs were infected with the monkeypox virus from the rats and then were transported from the distributor to a vendor in Wisconsin, where they were sold to the index patient and others.

In addition to routine sales by animal vendors, these infected animals also were sold or traded at "swap meets" (i.e., gatherings of animal traders, exhibitors, and buyers) in Illinois, Indiana, and Ohio (CDC 2003a, b, c, d). According to genomic sequencing analyses, the virus isolated in the United States was most closely linked to the West African clade (Chen et al., 2005; Esposito and Knight, 1985; Likos et al., 2005; Reed et al., 2004). West African clade isolates are suggested to be less virulent in humans and less apt to spread person-to-person than Congo Basin variants of monkeypox virus. It was reported that Congo Basin strains Zr79 and Zr-599 were more virulent than the West African strain US03, which was the strain that caused the outbreaks in the United States and Liberia, by animal studies using a ground squirrel model or cynomolgus monkeys, respectively (Saijo et al., 2009; Sbrana et al., 2007). Currently, the United States prohibits the importation of African rodents and their sale, distribution, transport, and release into the environment.

In November 2005, a human monkeypox outbreak occurred in Bentiu, a wetland area of the Unity State Sudan, although human monkeypox had usually been reported in the dry savannah area in Africa (Damon et al., 2006). By a retrospective analysis of the outbreak, 10 laboratory-confirmed and 9 probable patients with human monkeypox were reported to have occurred from September to December across the rainy season (May–October) to dry season (November–April). Up to five generations of human-to-human transmission were reported. None of the affected patients died. Although the source of the outbreak was not fully determined, the virological data indicated that the monkeypox virus isolated during this outbreak was a novel virus belonging to the Congo Basin clade (Formenty et al., 2010). These reports clearly show that active surveillance and epidemiological analyses of monkeypox virus infections are indispensable to better assess the public health burden and develop effective strategies for reducing the risk of outbreaks.

ACKNOWLEDGEMENT

This chapter was supported in part by grants from the Japan Health Sciences Foundation (KHC1204).

REFERENCES

Amer, M., El-Gharib, I., Rashed, A., et al., *Human cowpox infection in Sharkia Governorate, Egypt.* Int J Dermatol, 2001. **40**(1): p. 14–7.

Arita, I., Gispen, R., Kalter, S.S., et al., *Outbreaks of monkeypox and serological surveys in nonhuman primates.* Bull World Health Organ, 1972. **46**(5): p. 625–31.

Arita, I., Jezek, Z., Khodakevich, L., et al., *Human monkeypox: a newly emerged orthopoxvirus zoonosis in the tropical rain forests of Africa*. Am J Trop Med Hyg, 1985. **34**(4): p. 781–9.

Barquet, N. and P. Domingo, *Smallpox: the triumph over the most terrible of the ministers of death*. Ann Intern Med, 1997. **127**(8 Pt 1): p. 635–42.

Baxby, D., *Is cowpox misnamed? A review of 10 human cases*. Br Med J, 1977. **1**(6073): p. 1379–81.

Baxby, D. and Bennett, M., *Cowpox: a re-evaluation of the risks of human cowpox based on new epidemiological information*. Arch Virol Suppl, 1997. **13**: p. 1–12.

Baxby, D., Bennett, M., and Getty, B., *Human cowpox 1969–93: a review based on 54 cases*. Br J Dermatol, 1994. **131**(5): p. 598–607.

Becker, C., Kurth, A., Hessler, F., et al., *Cowpox virus infection in pet rat owners: not always immediately recognized*. Dtsch Arztebl Int, 2009. **106**(19): p. 329–34.

Bennett, M., Gaskell, R.M., Gaskell, C.J., et al., *Studies on poxvirus infection in cats*. Arch Virol, 1989. **104**(1–2): p. 19–33.

Bennett, M., Gaskell, C.J., Baxbyt, D., et al., *Feline cowpox virus infection*. J Small Anim Pract, 1990. **31**(4): p. 167–73.

Breman, J.G. and Henderson, D.A., *Poxvirus dilemmas—monkeypox, smallpox, and biologic terrorism*. N Engl J Med, 1998. **339**(8): p. 556–9.

Breman, J.G., Kalisa-Ruti, Steniowski, M.V., et al., *Human monkeypox, 1970–79*. Bull World Health Organ, 1980. **58**(2): p. 165–82.

Campe, H., Zimmermann, P., Glos, K., et al., *Cowpox virus transmission from pet rats to humans, Germany*. Emerg Infect Dis, 2009. **15**(5): p. 777–80.

Cardeti, G., Brozzi, A., Eleni, C., et al., *Cowpox virus in llama, Italy*. Emerg Infect Dis, 2011. **17**(8): p. 1513–15.

Centers for Disease Control and Prevention, *Update: multistate outbreak of monkeypox—Illinois, Indiana, and Wisconsin*, 2003. Morb Mortal Wkly Rep, 2003a. **52**(23): p. 537–40.

CDC, *Update: multistate outbreak of monkeypox—Illinois, Indiana, Kansas, Missouri, Ohio, and Wisconsin, 2003. Morb* Mortal Wkly Rep, 2003b. **52**(27): p. 642–6.

CDC, *Update: multistate outbreak of monkeypox—Illinois, Indiana, Kansas, Missouri, Ohio, and Wisconsin, 2003. Morb* Mortal Wkly Rep, 2003c. **52**(24): p. 561–4.

CDC, *Update: multistate outbreak of monkeypox—Illinois, Indiana, Kansas, Missouri, Ohio, and Wisconsin*, 2003. Morb Mortal Wkly Rep, 2003d. **52**(26): p. 616–8.

Chantrey, J., Meyer, H., Baxby, D., et al., *Cowpox: reservoir hosts and geographic range*. Epidemiol Infect, 1999. **122**(3): p. 455–60.

Chen, N., Li, G., Liszewski, M.K., et al., *Virulence differences between monkeypox virus isolates from West Africa and the Congo basin*. Virology, 2005. **340**(1): p. 46–63.

Cohen, J., *Is an old virus up to new tricks?* Science, 1997. **277**(5324): p. 312–3.

Czerny, C.P., Wagner, K., Gessler, K., et al., *A monoclonal blocking-ELISA for detection of orthopoxvirus antibodies in feline sera*. Vet Microbiol, 1996. **52**(3–4): p. 185–200.

Damon, I.K., Poxviruses. In: D.M. Knipe, P.M. Howley, D.E. Griffin, R. A. Lamb, M.A. Martin, R. Barnard, S.E. Straus et al., Editors. *Fields virology*, 2006, Philadelphia: Lippincott Williams & Wilkins. p. 2947–75.

Damon, I.K., Roth, C.E. and Chowdhary, V., *Discovery of monkeypox in Sudan*. N Engl J Med, 2006. **355**(9): p. 962–3.

Di Giulio, D.B. and Eckburg, P.B., *Human monkeypox: an emerging zoonosis*. Lancet Infect Dis, 2004. **4**(1): p. 15–25.

Dina, J., Lefeuvre, P.F., Bellot, A., et al., *Genital ulcerations due to a cowpox virus: a misleading diagnosis of herpes*. J Clin Virol, 2011. **50**(4): p. 345–7.

Eis-Hubinger, A.M., Gerritzen, A., Schneweis, K.E., et al., *Fatal cowpox-like virus infection transmitted by cat*. Lancet, 1990. **336**(8719): p. 880.

Elsendoorn, A., Agius, G., Le Moal, G., et al., *Severe ear chondritis due to cowpox virus transmitted by a pet rat.* J Infect, 2011. **63**(5): p. 391–3.

Esposito, J.J. and Knight, J.C., *Orthopoxvirus DNA: a comparison of restriction profiles and maps.* Virology, 1985. **143**(1): p. 230–51.

Fenner, F., Henderson, D. A., Arita, I., et al., *Smallpox and its eradication*, 1988, History of International Public Health, No. 61988, Geneva: World Health Organization.

Formenty, P., Muntasir, M.O., Damon, I., et al., *Human monkeypox outbreak caused by novel virus belonging to Congo Basin clade, Sudan, 2005.* Emerg Infect Dis, 2010. **16**(10): p. 1539–45.

Girling, S.J., Pizzi, R., Cox, A., et al., *Fatal cowpox virus infection in two squirrel monkeys (Saimiri sciureus).* Vet Rec, 2011. **169**(6): p. 156.

Gispen, R., *Relevance of some poxvirus infections in monkeys to smallpox eradication.* Trans R Soc Trop Med Hyg, 1975. **69**(3): p. 299–302.

Haase, O., Moser, A., Rose, C., et al., *Generalized cowpox infection in a patient with Darier disease.* Br J Dermatol, 2011. **164**(5): p. 1116–8.

Hemmer, C.J., Littmann, M., Löbermann, M., et al., *Human cowpox virus infection acquired from a circus elephant in Germany.* Int J Infect Dis, 2010. **14**(Suppl 3): p. e338–40.

Henning, K., Czerny, C.-P., Meyer, H., et al., *A seroepidemiological survey for orthopox virus in the red fox (Vulpes vulpes).* Vet Microbiol, 1995. **43**(2–3): p. 251–9.

Heymann, D.L., Szczeniowski, M., and Esteves, K., *Re-emergence of monkeypox in Africa: a review of the past six years.* Br Med Bull, 1998. **54**(3): p. 693–702.

Honlinger, B., Huemer, H.P., Romani, N., et al., *Generalized cowpox infection probably transmitted from a rat.* Br J Dermatol, 2005. **153**(2): p. 451–3.

Human monkeypox—Kasai Oriental, Democratic Republic of Congo, February 1996–October 1997. Morb Mortal Wkly Rep, 1997a. **46**(49): p. 1168–71.

Human monkeypox—Kasai Oriental, Zaire, 1996–1997. Morb Mortal Wkly Rep, 1997b. **46**(14): p. 304–7.

Hutin, Y.J., Williams, R.J., Malfait, P., et al., *Outbreak of human monkeypox, Democratic Republic of Congo, 1996 to 1997.* Emerg Infect Dis, 2001. **7**(3): p. 434–8.

Hutson, C.L., Lee, K.N., Abel, J., et al., *Monkeypox zoonotic associations: insights from laboratory evaluation of animals associated with the multi-state US outbreak.* Am J Trop Med Hyg, 2007. **76**(4): p. 757–68.

Hutson, C.L., Abel, J.A., Carroll, D.S., et al., *Comparison of West African and Congo Basin monkeypox viruses in BALB/c and C57BL/6 mice.* PLoS One, 2010a. **5**(1): p. e8912.

Hutson, C.L., Carroll, D.S., Self, J., et al., *Dosage comparison of Congo Basin and West African strains of monkeypox virus using a prairie dog animal model of systemic orthopoxvirus disease.* Virology, 2010b. **402**(1): p. 72–82.

Jenner, E., *An inquiry into the causes and effects of variolae vaccinae, a disease discovered in some western counties of England.* 1798, London: Sampson Low.

Jezek, Z. and F. Fenner, *Human monkeypox.* Monographs in Virology. Vol. **17**, 1988. Basel: Karger. p. 1–142.

Jezek, Z., Gromyko, A.I. and Szczeniowski, M.V., *Human monkeypox.* J Hyg Epidemiol Microbiol Immunol, 1983. **27**(1): p. 13–28.

Jezek, Z., Arita, I., Mutombo, M., et al., *Four generations of probable person-to-person transmission of human monkeypox.* Am J Epidemiol, 1986. **123**(6): p. 1004–12.

Jezek, Z., Grab, B. and Dixon, H., *Stochastic model for interhuman spread of monkeypox.* Am J Epidemiol, 1987a. **126**(6): p. 1082–92.

Jezek, Z., Szczeniowski, M., Paluku, K.M., et al., *Human monkeypox: clinical features of 282 patients.* J Infect Dis, 1987b. **156**(2): p. 293–8.

Jezek, Z., Grab, B., Szczeniowski, M., et al., *Clinico-epidemiological features of monkeypox patients with an animal or human source of infection*. Bull World Health Organ, 1988a. **66**(4): p. 459–64.

Jezek, Z., Grab, B., Szczeniowski, M.V., et al., *Human monkeypox: secondary attack rates*. Bull World Health Organ, 1988b. **66**(4): p. 465–70.

Khodakevich, L., Jezek, Z. and Kinzanzka, K., *Isolation of monkeypox virus from wild squirrel infected in nature*. Lancet, 1986. **1**(8472): p. 98–9.

Khodakevich, L., Szczeniowski, M., Nambu-ma-Disu, et al., *Monkeypox virus in relation to the ecological features surrounding human settlements in Bumba zone, Zaire*. Trop Geogr Med, 1987a. **39**(1): p. 56–63.

Khodakevich, L., Szczeniowski, M., Nambu-ma-Disu, et al., *The role of squirrels in sustaining monkeypox virus transmission*. Trop Geogr Med, 1987b. **39**(2): p. 115–22.

Khodakevich, L., Jezek, Z., and Messinger, D., *Monkeypox virus: ecology and public health significance*. Bull World Health Organ, 1988. **66**(6): p. 747–52.

Kile, J.C., Fleischauer, A.T., Beard, B., et al., *Transmission of monkeypox among persons exposed to infected prairie dogs in Indiana in 2003*. Arch Pediatr Adolesc Med, 2005. **159**(11): p. 1022–5.

Kurth, A., Wibbelt, G., Gerber, H.P., et al., *Rat-to-elephant-to-human transmission of cowpox virus*. Emerg Infect Dis, 2008. **14**(4): p. 670–1.

Kurth, A., Nitsche, A., Straube, M., et al., *Cowpox virus outbreak in banded mongooses* (Mungos mungo) *and jaguarundis* (Herpailurus yagouaroundi) *with a time-delayed infection to humans*. PLoS One, 2009. **4**(9): p. e6883.

Ladnyj, I.D., Ziegler, P., and Kima, E., *A human infection caused by monkeypox virus in Basankusu Territory, Democratic Republic of the Congo*. Bull World Health Organ, 1972. **46**(5): p. 593–7.

Lewis-Jones, S., *Zoonotic poxvirus infections in humans*. Curr Opin Infect Dis, 2004. **17**(2): p. 81–9.

Likos, A.M., Sammons, S.A., Olson, V.A., et al., *A tale of two clades: monkeypox viruses*. J Gen Virol, 2005. **86**(Pt 10): p. 2661–72.

Mack, T.M., Thomas, D.B. and Muzaffar Khan, M., *Epidemiology of smallpox in West Pakistan. II. Determinants of intravillage spread other than acquired immunity*. Am J Epidemiol, 1972. **95**(2): p. 169–77.

Mancaux, J., Vervel, C., Bachour, N., et al., *Necrotic skin lesions caused by pet rats in two teenagers*. Arch Pediatr, 2011. **18**(2): p. 160–4.

Marennikova, S.S. and Shelukhina, E.M., *White rats as source of pox infection in carnivora of the family Felidae*. Acta Virol, 1976. **20**(5): p. 442.

Marennikova, S.S., Seluhina, E.M., Mal'ceva, N.N., et al., *Isolation and properties of the causal agent of a new variola-like disease (monkeypox) in man*. Bull World Health Organ, 1972. **46**(5): p. 599–611.

Marennikova, S.S., Ladnyj, I.D., Ogorodinikova, Z.I., et al., *Identification and study of a poxvirus isolated from wild rodents in Turkmenia*. Arch Virol, 1978. **56**(1–2): p. 7–14.

Marennikova, S.S., Shelukhina, E.M. and Efremova, E.V., *New outlook on the biology of cowpox virus*. Acta Virol, 1984. **28**(5): p. 437–44.

Monkeypox, 1991. Gabon. Wkly Epidemiol Rec, 1992. **67**(14): p. 101–2.

Moss, B., Poxviridae: the viruses and their replication. In: D.M. Knipe, P.M. Howley, D.E. Griffin, R. A. Lamb, M.A. Martin, R. Barnard, S.E. Straus, Editors. *Fields virology*, 2006, Philadelphia: Lippincott Williams & Wilkins. p. 2905–45.

Mukherjee, M.K., Sarkar, J.K. and Mitra, A.C., *Pattern of intrafamilial transmission of smallpox in Calcutta, India*. Bull World Health Organ, 1974. **51**(3): p. 219–25.

Mukinda, V.B., Mwema, G., Kilundu, M., et al., *Re-emergence of human monkeypox in Zaire in 1996. Monkeypox Epidemiologic Working Group*. Lancet, 1997. **349**(9063): p. 1449–50.

Mutombo, M., Arita, I., and Jezek, Z., *Human monkeypox transmitted by a chimpanzee in a tropical rain-forest area of Zaire*. Lancet, 1983. **1**(8327): p. 735–7.

Ninove, L., Domart, Y., Vervel, C., et al., *Cowpox virus transmission from pet rats to humans, France*. Emerg Infect Dis, 2009. **15**(5): p. 781–4.

Nishiura, H. and Kashiwagi, T., *Smallpox and season: reanalysis of historical data*. Interdiscip Perspect Infect Dis, 2009. **2009**: p. 591935.

Pelkonen, P.M., Tarvainen, K., Hynninen, A., et al., *Cowpox with severe generalized eruption, Finland*. Emerg Infect Dis, 2003. **9**(11): p. 1458–61.

Pfeffer, M., Pfleghaar, S., von Bomhard, D., et al., *Retrospective investigation of feline cowpox in Germany*. Vet Rec, 2002. **150**(2): p. 50–1.

Postma, B.H., Diepersloot, R.J.A., Niessen, G.J.C.M., et al., *Cowpox-virus-like infection associated with rat bite*. Lancet, 1991. **337**(8743): p. 733–4.

Rao, A.R., Jacob, E.S., Kamalakshi, S., et al., *Epidemiological studies in smallpox. A study of intra-familial transmission in a series of 254 infected families*. Indian J Med Res, 1968. **56**(12): p. 1826–54.

Redfield, R.R., Wright, D.C., James, W.D., et al., *Disseminated vaccinia in a military recruit with human immunodeficiency virus (HIV) disease*. N Engl J Med, 1987. **316**(11): p. 673–6.

Reed, K.D., Melski, J.W., Graham, M.B. et al., *The detection of monkeypox in humans in the Western Hemisphere*. N Engl J Med, 2004. **350**(4): p. 342–50.

Reynolds, M.G., Davidson, W.B., Curns, A.T., et al., *Spectrum of infection and risk factors for human monkeypox, United States, 2003*. Emerg Infect Dis, 2007. **13**(9): p. 1332–9.

Riedel, S., *Edward Jenner and the history of smallpox and vaccination*. Proc (Bayl Univ Med Cent), 2005. **18**(1): p. 21–5.

Rimoin, A.W., Kisalu, N., Kebela-Ilunga et al., *Endemic human monkeypox, Democratic Republic of Congo, 2001–2004*. Emerg Infect Dis, 2007. **13**(6): p. 934–7.

Rimoin, A.W., Mulembakani, P.M., Johnston, S.C., et al., *Major increase in human monkeypox incidence 30 years after smallpox vaccination campaigns cease in the Democratic Republic of Congo*. Proc Natl Acad Sci U S A, 2010. **107**(37): p. 16262–7.

Saijo, M., Ami, Y., Suzaki, Y., et al., *Virulence and pathophysiology of the Congo Basin and West African strains of monkeypox virus in non-human primates*. J Gen Virol, 2009. **90**(Pt 9): p. 2266–71.

Sbrana, E., Xiao, S.Y., Newman, P.C., et al., *Comparative pathology of North American and central African strains of monkeypox virus in a ground squirrel model of the disease*. Am J Trop Med Hyg, 2007. **76**(1): p. 155–64.

Sejvar, J.J., Chowdary, Y., Schomogyi, M., et al., *Human monkeypox infection: a family cluster in the midwestern United States*. J Infect Dis, 2004. **190**(10): p. 1833–40.

Tack, D.M. and Reynolds, M.G., *Zoonotic poxviruses associated with companion animals*. Animals, 2011. **1**(4): p. 377–95.

The current status of human monkeypox: memorandum from a WHO meeting. Bull World Health Organ, 1984. **62**(5): p. 703–13.

Tryland, M., Sandvik, T., Hansen, H., et al., *Characteristics of four cowpox virus isolates from Norway and Sweden*. APMIS, 1998a. **106**(6): p. 623–35.

Tryland, M., Sandvik, T., Mehl, R., et al., *Serosurvey for orthopoxviruses in rodents and shrews from Norway*. J Wildl Dis, 1998b. **34**(2): p. 240–50.

Vestergaard, L., Vestergaard, L., Vinner, L., et al., *Identification of cowpox infection in a 13-year-old Danish boy*. Acta Derm Venereol, 2008. **88**(2): p. 188–90.

Vogel, S., Sárdy, M., Glos, K., et al., *The Munich outbreak of cutaneous cowpox infection: transmission by infected pet rats*. Acta Derm Venereol, 2011. **92**(2): p. 126–31

von Magnus, P., Andersen, E.K., Petersen, K.B., et al., *A pox-like disease in cynomolgus monkeys*. Acta Pathol Microbiol Scand, 1959. **46**(2): p. 156–76.

Vorou, R.M., Papavassiliou, V.G. and Pierroutsakos, I.N., *Cowpox virus infection: an emerging health threat*. Curr Opin Infect Dis, 2008. **21**(2): p. 153–6.

WHO, *The global eradication of smallpox: final report of the Global Commission for the Certification of Smallpox Eradication*, 1980, History of International Public Health, No. 41980, Geneva: World Health Organization.

Wolfs, T.F., Wagenaar, J.A., Niesters, H.G., et al., *Rat-to-human transmission of cowpox infection*. Emerg Infect Dis, 2002. **8**(12): p. 1495–6.

BIOLOGICAL ASPECTS OF THE INTERSPECIES TRANSMISSION OF SELECTED CORONAVIRUSES

Anastasia N. Vlasova and Linda J. Saif

Food Animal Health Research Program, The Ohio State University, Wooster, OH, USA

TABLE OF CONTENTS

21.1	Introduction	393
21.2	Coronavirus classification and pathogenesis	397
21.3	Natural reservoirs and emergence of new coronaviruses	399
21.4	Alpha-, beta- and gamma coronaviruses: cross-species transmission	404
	21.4.1 Alpha-coronaviruses cross-species transmission	404
	21.4.2 Beta-coronaviruses cross-species transmission	405
	21.4.3 Gamma-coronaviruses cross-species transmission	407
21.5	Anthropogenic factors and climate influence on coronavirus diversity and outbreaks	407
21.6	Conclusion	410
	References	410

21.1 INTRODUCTION

Coronavirinae subfamily members are enveloped viruses with a helical capsid, and a positive-stranded nonsegmented RNA (27–32 kb) genome (Spaan et al., 1988; Tyrrell et al., 1975). The 5′ and 3′ends of coronavirus (CoV) genomes contain short untranslated regions (UTRs). For the coding regions, the genome organization of all CoVs is similar,

Viral Infections and Global Change, First Edition. Edited by Sunit K. Singh.
© 2014 John Wiley & Sons, Inc. Published 2014 by John Wiley & Sons, Inc.

with the characteristic gene order 5′-replicase ORF1ab, spike (S), envelope (E), membrane (M) and nucleocapsid (N)-3′, although variable numbers of additional ORFs are present in each subgroup of coronaviruses (Table 21.1). A transcription regulatory sequence (TRS) motif is present at the 3′ end of the leader sequence preceding most ORFs. Like other members of the *Nidovirales* order, CoVs produce a set of 3′ nested transcripts with a common short leader sequence at the 5′ terminus (Cavanagh, 1997; Gorbalenya et al., 2006; Spaan et al., 1988).

Coronavirus (CoV) genetic diversity is maintained through accumulation of point mutations in genes (genetic drift) due to low fidelity of the RNA-dependent RNA polymerase and homologous RNA recombination (genetic shift) (Domingo, 1998; Domingo et al., 1998a, b, 2006). Recombination is facilitated by a unique template switching "copy-choice" mechanism during RNA replication with the transcription-regulating sequence (TRS) motifs believed to direct it (Lai, 1992; Lai et al., 1985). Additionally, because CoVs possess the largest RNA genomes, their capacity for accommodating gene rearrangements and modifications (sometimes significant: such as in the porcine respiratory coronavirus (PRCV) spike gene deletion) is highest among all RNA viruses. This genetic plasticity allows CoVs to generate remarkable diversity in emergence of new strains and species and to adapt to new hosts and ecological niches without employing common biological vectors such as ticks, mosquitoes etc. Utilization of mechanical vehicles is not well documented, but is less likely to play a major role in CoV spread due to CoV instability in the environment (Sizun et al., 2000). An exception may be enhanced CoV stability when frozen permitting its increased transmission in winter.

Feline infectious peritonitis (FIP), first described in 1912 was presumably the earliest report of a CoV associated disease, whereas infectious bronchitis virus (IBV) was the first CoV isolated from chickens in 1937 (Beaudette and Hudson, 1937). This was followed by identification and characterization of murine hepatitis virus (MHV) and other mammalian CoVs in 1940s (Cheever and Daniels et al., 1949; Doyle and Hutchings, 1946).

Another two decades elapsed before CoV was recognized as the etiological agent of common colds in humans in 1965 (Hamre and Procknow, 1966; Tyrrell and Bynoe, 1966). Later it was estimated that CoV infections contribute to as much as 35% of the total viral respiratory disease load during epidemics (Fielding, 2011). Overall, the proportion of adult colds caused by CoVs was estimated at 5% (McIntosh et al., 1970). Prior to the severe acute respiratory syndrome CoV (SARS-CoV) emergence and global pandemic in 2002–2003, it was commonly accepted that in humans CoVs cause mainly mild upper respiratory tract infections (Fielding, 2011), with the exception of human enteric CoV (HECV-4408) isolated from a child with acute diarrhea (Zhang et al., 1994). In contrast, in animals, CoVs cause a wide spectrum of clinical conditions including respiratory, enteric, hepatic and neurological diseases, with clinical outcomes ranging from mild symptomatology to lethal. The SARS epidemic has substantially advanced CoV research efforts, especially studies of CoV biodiversity and genomics. Since the discovery of SARS-CoV, numerous novel animal CoVs have been identified and characterized revealing a remarkable diversity of animal CoVs. The SARS-CoV was postulated to be of animal origin, with horseshoe bats as a potential natural reservoir (Lau et al., 2005; Li et al., 2005a). Besides SARS-CoV, bats are known to be reservoirs of important zoonotic viruses (including Ebola, Marburg, Nipah, Hendra, rabies and influenza) and viruses that can infect man or other animals (Calisher et al., 2006; Tong et al., 2012). Being abundant, diverse and geographically widespread, various species of bats, which are flying mammals equivalent to mosquitoes as insect vectors, were also recently shown to be natural hosts to a variety of CoVs (Calisher et al., 2006; Dominguez et al., 2007; Donaldson et al., 2010; Tang et al., 2006). Additionally,

TABLE 21.1. Coronavirus Genome Organization Comparison of Different Genera

Coronavirus	No. of nsp[a] in ORF1ab	No. of papain-like proteases in ORF1ab	No. of small ORFs between ORF1ab and N	No. of small ORFs downstream to N[b]	Conserved S[b] cleavage site presence	HE[b] gene presence
Alphacoronavirus						
Alphacoronavirus 1						
Transmissible gastroenteritis virus (TGEV)	16	2	2	1	N	N
Porcine respiratory coronavirus (PRCV)	16	2	1	1	N	N
Feline coronavirus (FCoV, FECV, FIPV)	16	2	4	2	N	N
Canine coronavirus (CCoV)	16	2	4	2	N	N
Human coronavirus NL63 (HCoV-NL63)	16	2	1	–	N	N
Human coronavirus 229E (HCoV-229E)	16	2	2	–	N	N
Porcine Epidemic diarrhea virus (PEDV)	16	2	1	–	N	N
Mink coronavirus (MCoV)	16	2	1	3	N	N
Ferret coronavirus (FeCoV)	16	2	1	2	N	N
Rhinolophus bat coronavirus HKU2 (BtCoV-HKU2)	16	2	1	1	N	N
Miniopterus bat coronavirus 1A/1B (BtCoV-1A/1B)	16	2	1	–	N	N
Betacoronavirus						
Subgroup A						
Human coronavirus HKU1 (HCoV-HKU1)	16	2	1	–	Y	Y
Human coronavirus OC43 (HCoV-OC43)	16	2	1	–	Y	Y
Canine respiratory CoV (CRCoV)	16	2	3	–	Y	Y
Mouse hepatitis virus (MHV)	16	2	2	–	Y	Y/N
Sialodacryoadenitis virus (SDAV)	16	2	2	1(?)	Y	Y
Bovine coronavirus (BCoV)	16	2	3	–	Y	Y
Porcine hemagglutinating encephalomyelitis virus (PHEV)	16	2	2	–	Y	Y
Equine coronavirus (ECoV)	16	2	2	–	Y	Y

(*Continued*)

TABLE 21.1. (*Continued*)

Coronavirus	No. of nsp[a] in ORF1ab	No. of papain-like proteases in ORF1ab	No. of small ORFs between ORF1ab and N	No. of small ORFs downstream to N[b]	Conserved S[b] cleavage site presence	HE[b] gene presence
Human enteric coronavirus 4408 (HECV-4408)	16	2	3	–	Y	Y
Subgroup B						
Severe acute respiratory syndrome coronavirus (SARS-CoV)	16	1	7	–	N	N
SARS-related Rhinolphus bat coronavirus HKU3 (Bat SARS-CoV)	16	1	5	–	N	N
Subgroup C						
Tylonycteris bat coronavirus HKU4 (BtCoV-HKU4)	16	1	4	–	N	N
Pipistrellus bat coronavirus HKU5 (BtCoV-HKU5)	16	1	4	–	N	N
Subgroup D						
Rousettus bat coronavirus HKU9 (BtCoV-HKU9)	16	1	1	2	N	N
Gammacoronavirus						
Infectious bronchitis virus (IBV)	15	1	4	–	Y	N
Turkey coronavirus (TCoV)	15	1	5	–	Y	N
Beluga whale coronavirus	15	1	8	–	N	N
Deltacoronavirus						
Bulbul coronavirus HKU11	15	1	1	3	N	N
Thrush coronavirus HKU11	15	1	1	3	N	N
Munia coronavirus HKU11	15	1	1	3	N	N

[a] *nsp, non-structural protein*
[b] *N, nucleoprotein; S, spike protein; HE, haemagglutinin esterase*

a novel highly divergent CoV was reported recently in a captive deceased beluga whale (Mihindukulasuriya et al., 2008) emphasizing once more the great adaptability and widespread prevalence of *Coronavirinae* subfamily members.

In this chapter, we first summarize recent studies by us and others on coronavirus biodiversity and genomics. Secondly we discuss the biologic factors that contribute to novel CoV emergence and interspecies jumping with emphasis on animal CoVs.

21.2 CORONAVIRUS CLASSIFICATION AND PATHOGENESIS

Coronaviruses (CoVs) belong to the order *Nidovirales*, family *Coronaviridae*. Classification into three (1 to 3) antigenic groups was initially established based on antigenic cross-reactivity, and was further confirmed by phylogenetic analysis (Lai and Holmes, 2001). Historically, within each group, subgroups (provisional subgroups) were sequentially established: in group 1—1a and 1b were defined; in group 2—2a, 2b, 2c and 2d; and in group 3—3a, 3b and 3c; however, some of these taxonomic units were contentious. The abundance and complexity of new data on novel CoV phylogeny post-SARS epidemic, and especially the availability of complete genomic sequences, revealed a need to revise the existing CoV taxonomy. Therefore, within the *Coronaviridae* family, two subfamilies were recently defined: *Coronavirinae* and *Torovirinae* (International Committee on Taxonomy of Viruses (ICTV; 2009). The *Coronavirinae* subfamily is further subdivided into three genera: *alphacoronavirus* (formerly CoVs group 1), *betacoronavirus* (formerly group 2) and *gammacoronavirus* (formerly group 3), with *alphacoronavirus 1 species* corresponding to former subgroup 1a (transmissible gastroenteritis virus (TGEV), PRCV, canine coronavirus (CCoV) and feline coronavirus (FCoV)) and other alphacoronavirus species (formerly subgroup 1b) each represented by different CoVs (human coronavirus NL63 (HCoV-NL63), human coronavirus 229E (HCoV-229E), porcine epidemic diarrhea virus (PEDV) etc.) from various hosts (Gonzalez et al., 2003) (Figure 21.1). No new species (or other relevant taxonomic units) were established to replace the former subgroups 2a–2d. Therefore, the majority of prototype mammalian and human CoVs are currently distributed between the *alpha*- and *betacoronavirus* (subgroup 2a) genera; while the *betacoronavirus* subgroup 2b is represented by the SARS- and SARS-like CoV species and *betacoronavirus* subgroups 2c and 2d include bat CoV (BtCoV) species. Gammacoronaviruses are detected primarily in domestic birds or related avian species. Gough and colleagues have recently identified a parrot CoV that is genetically distinct from alpha-, beta-, and gammacoronaviruses (Gough et al., 2006). Additional novel CoVs that are genetically similar to the parrot coronavirus were subsequently detected in terrestrial birds (Woo et al., 2009). Therefore, CoVs of this novel lineage recently have been proposed to form a new genus, provisionally named *deltacoronavirus* (de Groot et al., 2011) that included some species from the former provisional subgroup 3c. Additionally, findings from other studies suggested that there is much diversity in CoVs circulating in wild birds (Hughes et al., 2009; Muradrasoli et al., 2010). Recently conducted phylogenetic analyses of diverse avian CoVs demonstrated that there are various gammacoronaviruses and deltacoronaviruses circulating in birds. Gammacoronaviruses were found predominantly in *Anseriformes* birds, whereas deltacoronaviruses could be detected in *Ciconiiformes*, *Pelecaniformes*, and *Anseriformes* birds (Chu et al., 2011). Chu et al. (2011) also suggested that there is frequent interspecies transmission of gammacoronaviruses between duck species; whereas deltacoronaviruses may have more stringent host specificities (Chu et al., 2011). Furthermore, a novel virus from a beluga whale that died of generalized pulmonary disease and terminal acute liver failure was suggested to be a highly divergent *Coronavirinae* member most closely related to *gammacoronavirus* genus members (former subgroup 3b) (Mihindukulasuriya et al., 2008). If so, this finding could be the first evidence of mammalian non-alpha-, non-betacoronaviruses. It also suggests the possibility of intriguing CoV circulation between birds and aquatic mammals.

Genomic organization within each CoV genus has some unique features and provides data that support most of the newly established taxonomic units (Table 21.1). However, the

(a)

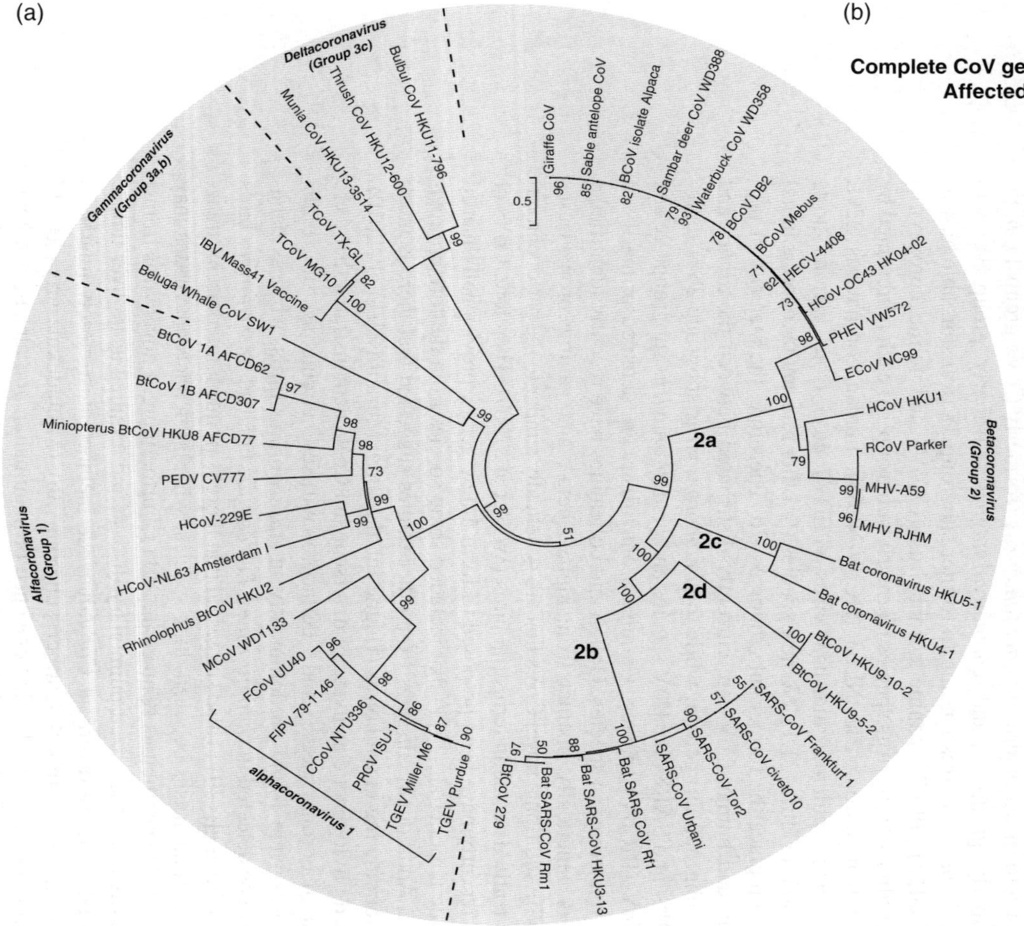

(b)

	2002	2012
Complete CoV genomic sequences	17	503
Affected species	7	>55

Figure 21.1. (**a**) Neighbour-joining phylogenetic tree based on the complete genome sequences of representative alpha-, beta-, gamma- and deltacoronaviruses. The genera are named as described in the 9th report of the International Committee on Taxonomy of Viruses with the former group names in brackets. Additionally, *alphacoronavirus 1* species and subgroups 2a-2d are indicated. Bootstrap support values are indicated. Scale bar indicates number of nucleotide substitutions per site. ClustalW algorithm was used to generate the tree in MEGA 5.05 software. (**b**) The summary indicates the approximate numbers of the complete genomic sequences released and species affected by those CoVs prior (2002) to after (2012) the SARS outbreak.

rapidly accumulating novel and vast CoV genomic data may result in further revisions of the *Coronaviridae* nomenclature.

Coronavirus infection can be asymptomatic or can cause respiratory, intestinal, hepatic and neurological disease with a wide range of clinical manifestations. The molecular mechanisms of the pathogenesis of several CoVs have been actively studied since the 1970s (Weiss and Navas-Martin, 2005). Some CoVs of animal origin, such as TGEV, bovine CoV (BCoV), and avian IBV, are of veterinary importance. Other animal CoVs warrant careful monitoring because they infect domestic animals (pets) living in close proximity to humans (such as CCoV and feline infectious peritonitis virus (FIPV)) or may serve as natural reservoirs for emerging CoVs persisting in wild animals (bats, wild ruminants or wild carnivores). Feline CoV which causes inapparent-to-mild enteritis in cats and sometimes leads to fatal effusive abdominal FIP is a pathogenesis paradigm for the scientific community (Pedersen et al., 2008). Additionally, preexisting immunity was shown to accelerate and exacerbate the severity of this disease (Pedersen et al., 1984b). A range of MHV strains are recognized with different tissue tropisms and levels of virulence (Navas-Martin et al., 2005). The MHV and PRCV are studied as models for human disease. The MHV infection of the mouse is regarded as one of the best animal models for the study of demyelinating diseases such as multiple sclerosis (Weiss and Navas-Martin, 2005). Because the pulmonary pathology of PRCV in pigs resembles that of SARS patients, the effect of dexamethasone treatment and a pre-existing respiratory viral infection on PRCV pathogenesis and immune responses was studied in detail as a model for SARS-CoV (Jung et al., 2007, 2009; Zhang et al., 2008). The SARS-CoV pathogenesis is unique among HCoVs which normally cause only mild upper respiratory tract disease (Ding et al., 2004; Drosten et al., 2003a, b; Ksiazek et al., 2003; Sims et al., 2008). The SARS-CoV which causes acute respiratory distress, especially in the elderly, provides a demonstration of how CoVs acquired via recent interspecies transmission events can cause excessive and sometimes dysregulated innate immune responses in new hosts (Perlman and Dandekar, 2005; Smits et al., 2010; Yoshikawa et al., 2009). Undoubtedly, the variety of clinical diseases in animals and the mainly respiratory pathology in humans caused by CoVs requires further detailed studies of CoV pathogenesis. The accumulated data indicate that CoV tissue tropism and virulence are not defined solely by the cellular receptor exploited or the genetic relatedness among strains, but rather result from the interplay of individual viral gene products and the host immune responses (Table 21.2) (de Groot et al., 2011). There are numerous reports indicating that S, N and some accessory non-structural proteins (nsps) may govern or affect CoV tissue tropism and virulence (Cowley and Weiss, 2010; Cowley et al., 2010; Frieman et al., 2012; Herrewegh et al., 1995; Rottier et al., 2005; Sanchez et al., 1999; Yeager et al., 1992). It is also believed that pathogenesis is determined by viral quasispecies rather than by the action of a single genotype (Vignuzzi et al., 2006). Another significant feature of CoV pathogenesis, possibly related to the existence of quasispecies, is the often dual pneumoenteric tropism, even in case of one dominating syndrome (Leung et al., 2003; Park et al. 2007; Shi et al., 2005; Zhang et al., 2007).

21.3 NATURAL RESERVOIRS AND EMERGENCE OF NEW CORONAVIRUSES

An estimated 75% of emerging diseases arise from zoonotic sources (Taylor et al., 2001). There are two essential conditions for emergence of novel viruses and sustainability in a new host: within-natural host population growth and between-host transmission efficiency

TABLE 21.2. Major Coronavirus Associated Diseases and Pathotypes In Different Hosts.

Pathotype	Coronavirus	Cellular receptor (Chu et al., 2006; Madu et al. 2007; Oh et al., 2003; Weiss and Navas-Martin, 2005)	Genus	Disease/syndrome	Host/age	Severity
Respiratory	HCoV-NL63	ACE2	*Alphacoronavirus*	Common cold, croup	Human/any	Mild
	HCoV-229E	Human APN	*Alphacoronavirus*	Common cold, upper respiratory tract disease	Human/any	Mild
	PRCV	Porcine APN	Alphacoronavirus	Respiratory disease	Pig/any	Mild
	HCoV-HKU1	Neu5,9Ac2-containing moiety	Betacoronavirus	Common cold, upper respiratory tract disease	Human/any	Mild
	HCoV-OC43	Neu5,9Ac2-containing moiety	*Betacoronavirus*	Common cold, upper respiratory tract disease	Human/any	Mild
	CRCoV	ND	*Betacoronavirus*	Respiratory disease	Dog/any	Mild
	SDAV	ND	*Betacoronavirus*	Respiratory disease	Rats/any	Mild
	BRCV[a]	Neu5,9Ac2-containing moiety	*Betacoronavirus*	Respiratory disease, anorexia, shipping fever	Cow/any	Mild
	SARS-CoV	ACE2	*Betacoronavirus*	Acute Respiratory distress, pneumonia	Human/adults and elderly	Mild-severe (may be lethal in immuno-compromised and elderly)
	IBV	α2,3-linked sialic acid + heparan sulfate	*Gammacoronavirus*	Respiratory disease, reproductive disorder	Chicken/any	Mild-severe
Enteric	TGEV	Porcine APN	*Alphacoronavirus*	Enteritis	Pig/any, more severe in young (<2–3 weeks)	Moderate-severe (may be lethal in young piglets)
	PEDV	Porcine APN	*Alphacoronavirus*	Enteritis	Pig/any	Moderate-severe (may be lethal in young piglets)
	FCoV	Feline APN	*Alphacoronavirus*	Enteritis	Cat/any	Asymptomatic-Mild
	CCoV	Canine APN	*Alphacoronavirus*	Enteritis	Dog/puppies	Mild-severe-lethal

Category	Virus	Receptor	Genus	Disease	Host/Age	Severity
	BECV[b]	Neu5,9Ac2-containing moiety	Betacoronavirus	Winter dysentery	Cow/any	Moderate-severe
	ECoV	ND	Betacoronavirus	Enteritis	Horse/any	Moderate
	PHEV	Neu5,9Ac2-containing moiety	Betacoronavirus	Enteritis, fever Vomiting and wasting disease[c]	Pig/<1 month	Severe-deadly
	HECV-4408	ND	Betacoronavirus	Acute diarrhea	Human/children	Moderate (Zhang et al., 1994)
	SARS-CoV	ACE2	Betacoronavirus	Diarrhea[d]	Human/adults and elderly	Moderate
	TCoV	ND	Gammacoronavirus	Enteritis	Turkey/poults	Moderate-severe
Hepatic	MHV	Murine CEACAM1	Betacoronavirus	Hepatatis	Mouse/any	Severe-lethal
Neurological	MHV	Murine CEACAM1	Betacoronavirus	Encephalitis, CNS demyelination	Mouse/any	Severe-lethal
	PHEV	Neu5,9Ac2-containing moiety	Betacoronavirus	Hemagglutinating encephalomyelitis	Pig/<1 month	Severe-lethal
Reproductive	IBV	α2,3-linked sialic acid + heparan sulfate	Gammacoronavirus	Reproductive disorder (egg production drop)	Chicken/adults	Mild
	TCoV	ND	Gammacoronavirus	Reproductive disorder (egg production drop), poor growth	Turkey/adults	Mild
Systemic	FIPV	Feline APN	Alphacoronavirus	Infectious peritonitis with systemic granulomatous-necrotizing lesions	Cat/any	Severe-lethal
Other	SDAV	ND	Betacoronavirus	Conjunctivitis	Rats/any	Asymptomatic-mild
	BtCoV	ND	Alphacoronavirus Betacoronavirus	Asymptomatic	Bat(various species)/any	Asymptomatic-mild

[a] BRCV, Bovine respiratory CoV
[b] BECV, Bovine enteric CoV
[c] Suggested to be induced by vagus nerve damage, therefore may be not a true enteropathogenic effect
[d] Although reported, diarrhea was not a major symptom in SARS patients, but with the exceptions noted in Hong Kong (Amoy Gardens) and among medical care workers at Chang Gung Memorial Hospital, Kaohsiung Medical Center (Chiu et al., 2004; Peiris et al., 2003a).

(Dennehy et al., 2006). Viral fitness is critical in source (reservoir, original) and sink (new) hosts and may be poor initially in a new host with subsequent genetic adaptation to establish a persistent population and/or epidemiological spread as in the case of SARS-CoV (Dennehy et al., 2006; Li et al., 2005b; Qu et al. 2005).

The natural reservoirs of CoVs and their precise emergence pathways remain largely unknown. For some CoVs, persistently or chronically infected (asymptomatic) animals, birds and humans may serve as a source of infection and in favorable conditions may lead to disease outbreaks (An et al., 2011; Che et al., 2006; Dominguez et al., 2007; King et al., 2011; Lee et al., 2003; Tang et al., 2006; Vogel et al., 2010; Walsh et al., 1999; Watanabe et al., 2010; Wilder-Smith et al., 2005) http://www.savsnet.co.uk/canine-enteric-coronavirus/. Increasing evidence indicates that CoV quasispecies lead to the selection of new viral forms and to the sporadic emergence of new viral species with virulent phenotypes (Domingo, 1998; Domingo et al., 1998a, b, 2006; Holland et al., 1982). For some CoVs, there is evidence that they have emerged as a result of recombination between existing CoVs. For instance FIPV, is closely related to TGEV and CCoV and may have initially emerged as a result of recombination between CCoV and FCoV type I (Herrewegh et al., 1998). The hypothesis that FIPV is a relatively common natural mutant of FCoV—the within-host spontaneous mutation theory—was suggested by Vennema et al (Vennema et al., 1998). The PRCV is another example of a new CoV emergence due to a natural mutation in TGEV within the swine host (Laude et al. 1993; Pensaert et al., 1986; Wesley et al., 1990). Other CoVs (HECV-4408, HCoV-OC43, SARS-CoV, etc.), may have evolved as a result of inter-species transmission events and adaptation to new hosts due to close cohabitation, translocation or handling/consumption of infected animal by-products (Guan et al., 2003; Lau et al., 2005; Li et al., 2005a; Vijgen et al., 2005; Zhang et al., 1994).

Since the emergence of SARS, knowledge of CoV molecular epidemiology and genomics has increased greatly. However, identification of the CoV common ancestors or source remains a challenging task requiring a comprehensive understanding of CoV ecology and the factors affecting it. It is noteworthy that according to the current *Coronavirinae* subfamily classification, all known mammalian CoVs belong to *alpha-* and *betacoronavirus* genera, while *gamma-* and newly emerging *deltacoronavirus* genera comprise exclusively (or predominantly) avian CoVs (Woo et al., 2010, 2012). Furthermore, while CoVs from bat species, omnivores and carnivores are uniformly distributed between *alpha-* and *betacoronavirus* genera, CoVs from herbivores appear to be strictly confined at present to the *betacoronavirus* genus. These observations indicate that the type of feeding and digestive system may affect the ecology and evolution of CoVs along with host species mobility and prevalence. Bats constitute 20% of the mammalian population on Earth being the most divergent and widely distributed nonhuman mammalian species (Dominguez et al., 2007). Numerous recent studies indicate that bats harbor a much wider diversity of CoVs than any other mammalian/avian species (Donaldson et al., 2010; Vijaykrishna et al., 2007). Molecular clock analysis indicates that BtCoVs are evolutionarily older than CoVs from any other animals, with analysis of population dynamics indicating that CoVs in bats have constant population growth and that viruses from all other hosts show epidemic-like increases in population (Vijaykrishna et al., 2007). This indicates that diverse CoVs are endemic in different bat species, and that they may account for repetitive CoV introductions into and occasionally permanent establishment in other species (Vijaykrishna et al., 2007). It is also interesting that BtCoVs are found in both *alpha-* and *betacoronavirus* genera, but not yet in *gammacoronavirus* genus, suggesting that BtCoVs may be ancestral to all mammalian CoVs but not to CoVs from avian species (Figure 21.2). A recent report documented mammalian CoVs in the *deltacoronavirus* genus and suggested an avian

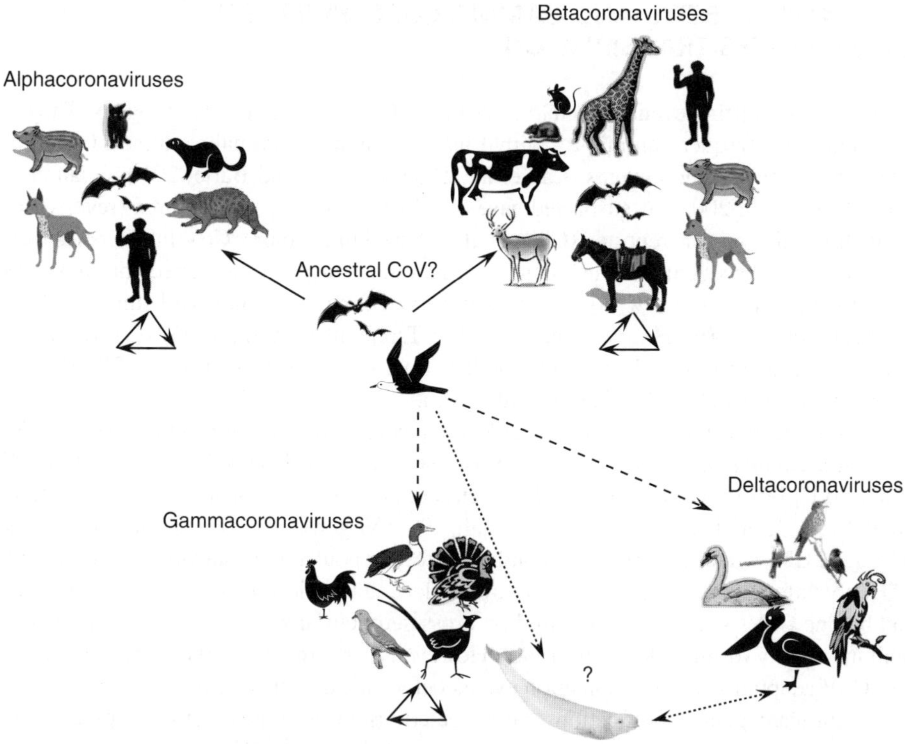

Figure 21.2. Tentative schematic of CoV ecology in diverse hosts. Solid and dashed lines represent confirmed and hypothetical interspecies transmission pathways, respectively; and arrow triangles indicate confirmed interspecies transmission within each genera; "?" means suggested but not confirmed.

ancesteral CoV as the source of genes for *gamma-* and *deltacoronavirus* (Woo et al., 2012) (Figure 21.2). Alternatively, avian CoVs could have split off earlier from the ancestral CoV and diverged significantly over time evolving separately from mammalian CoVs.

Recent studies of BtCoVs in China demonstrated that genetically divergent CoVs are present in and are species-specific in different bats (Tang et al., 2006). Moreover, the same group demonstrated that there is an additional isolated sublineage of CoVs (possibly non-*alpha-*, *beta-* or *gammacoronavirus*) consisting of exclusively BtCoVs, further emphasizing their extraordinary genetic diversity that affects the entire CoV taxonomy (Tang et al., 2006). Overall, BtCoV abundance, absence of epidemic patterns with only asymptomatic cases identified to date and the remarkable viral-host fitness indicate that bats can play a central role in CoV ecology and persistence (Dominguez et al., 2007; Vijaykrishna et al., 2007). Therefore, bat species and other animals sharing ecological niches with bats should be carefully monitored for CoV persistence, diversity and interspecies adaptation.

Highly diverse avian CoVs, also suggested to form new and separate (from mammalian CoVs) genera, do not seem to account for new CoV emergence and the current alpha- and betacoronavirus diversity in mammalian species including humans (Chu et al., 2011; Gough et al., 2006; Hughes et al., 2009; Muradrasoli et al., 2010; Woo et al., 2012). However, they may affect the entire CoV evolution and ecology through unknown pathways and, therefore, should also be under surveillance (Woo et al., 2012).

21.4 ALPHA-, BETA- AND GAMMA CORONAVIRUSES: CROSS-SPECIES TRANSMISSION

Frequent host-shifting events are characteristic for CoV evolution and phylogeny. They are represented by frequent animal-to-animal and occasionally animal-to-human (zoonosis) interspecies transmission events (Lau et al., 2005; Graham and Baric, 2010; Guan et al., 2003; Rota et al., 2003). A suspected case of SARS-CoV human-to-animal (reverse zoonosis) transmission was reported (Chen et al., 2005). Furthermore, CoV interspecies transmission and subsequent establishment in a new host may be facilitated through direct spread (original host → new host) as was observed for bovine and wild ruminant CoVs (Alekseev et al. 2008; Hasoksuz et al., 2007; Tsunemitsu et al., 1995) or involving an intermediate host (original host → intermediate host → new host) in the case of SARS-CoV (Peiris et al., 2003a, b, 2004; Zhong et al. 2006).

There are multiple examples of CoV interspecies transmission within each *alpha-*, *beta-* and *gammacoronavirus* genera (Alekseev et al., 2008; Hasoksuz et al., 2007; Jin et al., 2007; Lorusso et al., 2008, 2009; Peiris et al., 2003b, 2004; Perlman and Netland, 2009; Pfefferle et al., 2009; Tsunemitsu et al., 1995; Vijgen et al., 2006; Zhong et al., 2006) (Figure 21.2). Under experimental conditions turkey poults were shown to be susceptible to BCoV infection (Ismail et al., 2001), and close antigenic relationship between turkey and bovine CoVs was suggested based on cross-seroneutralization and hemagglutination-inhibition assay results (Dea et al., 1990). However, there are no reports of natural interspecies CoV adaptation or recombination events between CoVs from different genera.

Significant gene modifications (often deletions in accessory nsps or spike protein genes) or rearrangements (deletion or insertion of the whole ORFs encoding accessory nsps) are often involved in or follow host-shift or tissue tropism change events (Hasoksuz et al., 2007; Laude et al., 1993; Lorusso et al., 2008; Vaughn et al., 1995; Vennema et al., 1998; Vijgen et al., 2005); while scattered genome-wide point mutations normally reflect within host evolution of CoVs (Zhang et al., 2006, 2007) (Table 21.3).

21.4.1 Alpha-coronaviruses cross-species transmission

The pleiotropic molecular mechanisms that govern cross-species transmission of alphacoronaviruses are not completely understood but homologous recombination following co-infection with different CoVs appears to be a common route of generation of new sustainable CoV populations in heterologous hosts (Decaro et al., 2008a, 2010; Lorusso et al., 2008). Recombinants between TGEV, CCoV and FCoV, that represent host range variants of the same CoV species, are known to occur (Benetka et al., 2006; Decaro et al., 2009, 2010; Herrewegh et al., 1998; Pedersen et al., 1981; Wesley, 1999). Canine CoVs illustrate the genetic evolution and complexity of alphacoronaviruses. There are two CCoV genotypes identified to date, CCoV-I and CCoV-II (Decaro et al., 2008a). Based on the high genetic relatedness between the CCoV-II and TGEV genomes and on the presence of ORF3 remnants in both genomes, it has been hypothesized earlier that TGEV originated from CCoV-II through cross-species transmission (Lorusso et al., 2008). More recently, novel CCoV-II strains have been isolated, which likely originated from a double recombination event with TGEV, occurring in the 5′ end of the spike protein gene (Decaro et al., 2009; Erles and Brownlie, 2009).

While FCoV replication is primarily restricted to the mature intestinal epithelial cells (Pedersen et al., 1981, 1984a), virulent FIPV strains exhibit a prominent tropism for macrophages (Pedersen, 1976; Petersen and Boyle, 1980) with a rapid dissemination of the

TABLE 21.3. Coronaviruses that Emerged as a Result of Interspecies Transmission or Tissue Tropism Changes and Suggested Associated Genomic Modifications

Resulting CoV/host	Suspected original CoV/host	Genomic modification	References
TGEV/pig	CCoV-II/dog	ORF3 insertion	Decaro et al. (2007)
CCoV-II/dog	TGEV/pig	Recombination in the 5′ end of the spike gene	Decaro et al. (2009)
CCoV-II/dog	CCoV-I/dog and unknown CoV	Recombinant spike gene	Lorusso et al. (unpublished)
FIPV/cat	FCoV/cat and CCoV/dog	Substitutions in M and ORF7b genes and FCoV-CCV recombinations in spike and pol genes	Brown et al., (2009), Herrewegh et al. (1998)
PRCV/pig	TGEV/pig	621–681-nt deletion in the 5′ end of the spike gene; deletions in ORF3	Wesley et al. (1991)
HCoV-OC43/human	BCoV/cow	290-nt deletion (corresponding to the absence of BCoV nsp 4.9 kDa and nsp 4.8 kDa)	Vijgen et al. (2005)
HECV-4408/human	BCoV/cow	?	
GiCoV/giraffe	BCoV/cow	Deletion in the S1 subunit (amino acid 543–547) of the spike protein	Hasoksuz et al. (2007)
SARS-CoV/human	Bat and civet SARS-CoV/ horseshoe bat and civet cats	29-nt deletion in ORF8 and substitutions in spike gene and ORF3	Lau et al. (2005)

? means not known.

virus throughout the body. Feline coronavirus (FCoV or FIPV) naturally infects domestic and wild *Felidae*. People are not susceptible, but dogs and swine can be experimentally infected with FIPV (Pedersen, 2009). A mild to moderately severe TGE-like syndrome occurs in baby pigs post FIPV infection (Woods et al., 1981). It is noteworthy that FIP appeared within a decade of the initial descriptions of TGE in pigs in North America in 1946 (Doyle and Hutchings, 1946; Haelterman, 1962; Haelterman and Hutchings, 1956; Pedersen, 2009). At least one strain of CCoV can induce mild enteritis in cats and enhance a subsequent infection with FIPV, indicating a special closeness to FCoVs (McArdle et al., 1992). Therefore, empirical data indicate that CCoV may be a parent of FCoV in this scenario (Pedersen et al., 2008, 2009).

21.4.2 Beta-coronaviruses cross-species transmission

Among betacoronaviruses, interspecies transmission events appear to be very common even between genetically distant hosts. There are at least two human CoVs of suspected bovine origin—HCoV-OC43 and HECV-4408 (Vijgen et al., 2005; Zhang et al., 1994). Bovine CoVs and HCoV-OC43 show remarkable genetic and antigenic similarities and utilize the same cellular receptor (Table 21.2). Interestingly, while HCoV-OC43 is fairly common in humans, causing mild respiratory disease (Huang et al., 2009), human enteric CoV 4408 which also shares high genetic and antigenic similarity with BCoV (Han et al., 2006), appears to be an isolated case scenario (Zhang et al., 1994). This may indicate that fecal-oral transmission and therefore enteric pathology is not very common among HCoVs or that HCoV-OC43 was introduced into the human population earlier in the past (Vijgen et al., 2005) and now possesses better host fitness properties than HECV-4408. There are also multiple examples of BCoV (or its host range variants) interspecies transmission between related species of ruminants including white tail deer, waterbuck, sable antelope, giraffe, and water buffalo (Decaro et al., 2008c; Hasoksuz et al., 2007; Tsunemitsu et al., 1995).

Persistent MHV infection *in vitro* resulted in the emergence of host range variants capable of efficient replication in normally nonpermissive cell lines derived from non-mouse hosts including Chinese hamster ovary (CHO), human hepatocellular liver carcinoma (HepG2), and in human breast adenocarcinoma (MCF7) (Baric et al., 1999). In another study, host range MHV mutants were isolated from mixed cultures containing progressively increasing concentrations of nonpermissive Syrian baby hamster kidney (BHK) cells and decreasing concentrations of permissive murine astrocytoma (DBT) cells (Hensley et al., 1998).

Emergence of SARS-CoV is the most recent, most significant and best documented event of betacoronavirus interspecies transmission (Drosten et al., 2003b; Ksiazek et al., 2003; Peiris et al., 2004). Although the discovery of SARS-CoV in palm civets and raccoon dogs from live animal markets in China had initially identified them as an immediate source of the SARS-CoV infection in humans, subsequent surveillance in wild areas suggested that they may have only served as an amplification host for SARS-CoV (Lau et al., 2005). Genomic analyses, efficiency of civet ACE2 receptor binding to SARS-CoV S proteins from different stages of the SARS epidemic (Li et al., 2005b) and lack of widespread SARS-CoV infection among farmed civets (Tu et al., 2004) revealed that they were likely not a natural reservoir of human SARS-CoV (Table 21.3). The prevalence of SARS-like CoVs (~40%) and SARS-CoV specific antibodies (>60%) among Chinese horseshoe bats in Hong Kong, together with the high genetic similarities between the bat and human SARS-CoVs provided evidence for bats as a potential natural host of SARS-CoV infection in humans (Lau et al., 2005). Later epidemiological investigations suggested categorizing SARS outbreaks into three groups: interspecies, early-mid epidemic in humans and late epidemic, with each one revealing successive steps of SARS-CoV adaptive evolution in humans. Twelve amino acid sites in the S protein (with a larger proportion (24%) of them located in the receptor-binding domain) were identified as being under positive selective pressure and potentially responsible for SARS-CoV adaptation to new hosts (Zhang et al., 2006). Therefore, SARS-CoV emergence followed a common pattern for zoonoses: natural reservoir (bats) → intermediate host (civets) → new host (humans) → adaptation (viral-host fitness optimization) within the new host (humans).

It is interesting that cats, ferrets, several inbred mouse species (BALB/c, C57BL/6 (B6), 129S) and palm civets were all susceptible to natural or experimental infection with SARS-CoV Urbani strain (Glass et al., 2004; Hogan et al., 2004; Martina et al., 2003; Roberts et al., 2005; Subbarao et al., 2004; Wu et al., 2005). Additionally, a mink lung cell line (Mv1Lu) was permissive to SARS-CoV, expressing a functional ACE2 receptor for viral entry (Gillim-Ross et al., 2004; Heller et al., 2006; Mossel et al., 2005). This represents a wide range of susceptible species which may be a result of the relatively recent interspecies transmission event, ongoing adaptation and incomplete host fitness. The HCoV-OC43 according to molecular clock dating was transmitted from the bovine species to humans around 1890 (Vijgen et al., 2005). These betacoronaviruses share high genetic similarities and possibly a common ancestor with canine respiratory CoV (CRCoV) and PHEV, all of which appear to be species-specific CoVs due to a prolonged evolution and host fitness optimization (Vijgen et al., 2006).

21.4.3 Gamma-coronaviruses cross-species transmission

There is no indisputable evidence for interspecies transmission of known avian CoVs to phylogenetically distant hosts. Frequent interspecies transmissions of gammacoronaviruses between duck species were recently reported by Chu et al. (2011). However, in this case it is hard to establish the boundary between true interspecies transmission and host range CoV variants circulating among related host species. This scenario is similar to the close genetic relatedness among betacoronaviruses circulating among various ruminant species (Alekseev et al. 2008; Hasoksuz et al., 2007; Tsunemitsu et al., 1995). Isolation of avian IBV from domestic peafowl (*Pavo cristatus*) and teal (*Anas*) in China (Liu et al., 2005) reveals a potential for interspecies transmission of gammacoronaviruses or may be indicative of a wider natural host range. The latter is supported by the fact that TCoV strains have been successfully propagated in embryonated chicken and turkey eggs by inoculation of the amniotic cavity (Nagaraja and Pomeroy, 1997).

A suspected interspecies transmission event in the case of a captive beluga whale CoV, provisionally classified as a gammacoronavirus (Mihindukulasuriya et al., 2008), is supported by genetic data and sharing of ecological niches between aquatic mammalian and aquatic avian species. Whether the captive beluga whale initially acquired a CoV from a cohabitating avian species or if they share a common CoV ancestor is unclear (Mihindukulasuriya et al., 2008). However, the widespread presence of gammacoronaviruses in aquatic avian species (Chu et al., 2011) and the clinical severity (generalized pulmonary disease and acute liver failure) of the CoV infection in the beluga whale (Mihindukulasuriya et al., 2008) may be indicative of bird-to-whale CoV transmission. Sharing of a common ancestral CoV between avian species and the deceased beluga whale remains a possibility. Disease severity and death can be attributed to a number of factors including within host spontaneous CoV mutation resulting in higher virulence or immune incompetence.

The provisional *deltacoronavirus* genus includes parrot, thrush, bulbul and other terrestrial bird CoVs (Chu et al., 2011; Gough et al., 2006; Woo et al., 2009). Deltacoronaviruses infect and co-circulate with gammacoronaviruses in a variety of *Galloanserae*, *Neoaves* and *Passeriformes* (Chu et al., 2011). Although Chu and colleagues (Chu et al., 2011) suggested that deltacoronaviruses may have more stringent host specificities, reports by Woo et al. (Woo et al., 2009, 2012) indicate possible interspecies transmission of deltacoronaviruses from birds to an Asian leopard and pig.

21.5 ANTHROPOGENIC FACTORS AND CLIMATE INFLUENCE ON CORONAVIRUS DIVERSITY AND OUTBREAKS

With the world's population exceeding seven billion and the ability to travel long distances within a short timeframe, humans are one of the most significant modifying influences on viral ecology, including that of CoVs. Besides the contributions of population density and mobility of humans, to CoV emergence and spread as exemplified by the SARS-CoV epidemic, there are numerous sociodemographic influences that can affect CoV ecology and evolution. These include: global disease control efforts, improvements in public health infrastructure (public health training, emergency response, and prevention and control programs), improvement in veterinary care, alterations of natural animal habitats and land-use influences (deforestation, agricultural development, water projects, and urbanization) and shifting ecological niches (farming and exotic animal farming, animal transportation, selection and breeding, preservation of endangered species in captivity, introduction of feral animals back into wild habitats, etc.) (Patz et al., 2003).

There was a dramatic shift in the status, keeping, and breeding of cats and dogs as pets in the second half of the twentieth century. The numbers of pet cats/dogs greatly increased, purebreeding and cattery/kennel rearing became increasingly popular, and more cats and dogs were placed in shelters. These large multiple cat/dog indoor environments are known to favor feline and canine enteric CoV infections and FIP (Pedersen, 2009). The increasing demand for farm animals (such as pigs, cows, horses) that are often kept in close proximity to cats and dogs or exposed to wild rodents, bats or carnivores creates favorable conditions for genetic exchange between different species of CoVs. This can result in continuous emergence of new (recombinant) or mutant CoVs, thereby increasing their diversity as previously described for TGEV and CCoV (Decaro et al., 2009; Lorusso et al., 2008). Captive exotic wild ruminants (giraffe, waterbuck and Sable antelope) from Africa were transported and kept in captivity in the US in wildlife parks where they may have acquired BCoV from domestic ruminants resulting in similar diarrheal disease outbreaks (Alekseev et al. 2008; Hasoksuz et al., 2007; Tsunemitsu et al., 1995). It is unknown, whether two human CoVs—HCoV-OC43 and HECV-4408—derived from BCoV (Vijgen et al., 2005; Zhang et al., 1994) were acquired through handling/consumption of dairy or beef products or through a direct contact with infected cattle.

Identification and characterization of novel CoVs in farmed mink and ferrets in the US (Gorham et al., 1990; Vlasova et al., 2011; Wise et al., 2006, 2010) reveals a need for extensive wildlife surveillance to rule out anthropogenic influences in emergence of these pathogens. The SARS-CoV emergence and pandemic provide a vivid illustration of how modern anthropogenic activities can facilitate CoV introduction and spread in humans. On the other hand, the timely applied strict control and intervention measures curtailed the infection (Drosten et al., 2003a; Ksiazek et al., 2003; Peiris et al., 2003b). Crowded housing of the exotic civet cats intended for human consumption in animal markets in China and within restaurants fostered the spread of SARS-like CoV in this intermediate host and its initial spill-over into the human population. International travel facilitated the spread of this "atypical pneumonia" worldwide (Drosten et al., 2003b; Ksiazek et al., 2003; Lau et al., 2005; Peiris et al., 2003b). After the danger was recognized, the prompt search for the etiological agent, its discovery and detailed scientific characterization, together with a WHO issued travel advisory and extensive medical treatment of affected patients, may have restrained the wider expansion of this new virus before its adaptation to the human host was complete (Guan et al., 2004; Lau et al., 2005; Li et al., 2005a; Peiris et al., 2003a). Interestingly, although civet cats from Chinese animal markets were commonly positive for SARS-like CoV; farmed civet cats across the country were predominantly free of the infection (Tu et al., 2004). Thus, even the initial spread of the SARS-like CoV from bat species to civet cats (Lau et al., 2005) may have been due to human activity. The captive beluga whale that died of gammacoronavirus associated disease (Mihindukulasuriya et al., 2008) provides another possible example of human influence on the ecology of CoVs, and the possibility of a similar scenario occurring in nature. Collectively, these examples emphasize the importance of anthropogenic influences on CoV biodiversity and evolution.

Climate remains one of several important factors influencing the incidence of infectious diseases. There are multiple reviews on association of climate changes and vector-borne/water-borne diseases (Berberian and Rosanova, 2012; Dobson, 2009; Greer et al., 2008; Hales et al., 2002; Harley et al., 2011; Lafferty, 2009; Morillas-Marquez et al., 2010; Ostfeld, 2009; Patz et al., 2003; Rosenthal, 2009; Shuman, 2010; Wilson et al., 2011); however, the information for other infectious disease is insufficient. In this era of global development and human domination in most ecosystems, climate change effects can substantially affect the ecology of CoVs and modify the extent to which humans can control and respond to the outcomes of multiple disease modifying influences. The methods of CoV transmission and their natural host reservoirs are two critical features that can be affected by climate changes. The MHV and TGEV remained infectious in sewage and water for up to 3 weeks (Casanova et al., 2009) indicating that floods can contribute to increased spread of CoVs through contaminated water sources. Additionally, BCoV was shown to survive on lettuce surfaces, retaining infectivity for up to at least 14 days (Mullis et al., 2012) demonstrating that contaminated vegetables (due to contaminated water sources) may serve as potential vehicles for CoV transmission to humans. Numerous data indicate that humidity is important for CoV survival (Casanova et al., 2009, 2010a, b; Sizun et al., 2000). A faulty sewage system, initially contaminated by the excreta of the SARS index case, with an aerosol route of transmission was suggested to be responsible for a cluster of SARS cases in 2003 in Hong Kong (Peiris et al., 2003a). Interestingly, recent experimental data revealed that MHV and TGEV survivability on environmental surfaces was greater at high (80%) or low (20%) than at moderate (~50%) relative humidity (Casanova et al., 2010b). Another study, however, demonstrated that TGEV remains viable in the airborne state at low relative humidity longer than at high (Kim et al., 2007). This indicates that both floods and extreme droughts can directly affect CoV viability and transmission. Additionally, they may affect survival or migration of wildlife species (avian and mammalian) which serve as natural reservoirs for these CoVs, imposing an indirect effect on CoV circulation and preservation. Furthermore, experimental data confirm that CoV are inactivated faster at higher temperatures: TGEV and MHV remained infectious for up to 28 days at +4 °C, while at +20 °C the viruses persisted for only 5–20 days, with the fastest inactivation at +40 °C (Casanova et al., 2010b). Therefore, global warming and subsequent droughts could negatively affect CoV survivability. On the other hand, such significant climate changes may also increase the number of and density among environmental refugees (animals as well as humans) migrating to new areas thereby creating more favorable conditions for CoV circulation, recombination and spread.

The existing data regarding seasonality of known CoVs are incomplete and sometimes contradictory. Although in temperate climates, most respiratory CoV infections in humans occur more often in the winter and spring (cold season) (Bastien et al., 2005a, b; Dowell and Ho, 2004; Vabret et al. 2005), there are data that HCoV-NL63 and HKU1 occurrence in sub-tropical (Hong Kong, China) or in humid continental (Beijing, China) climates was higher in spring, summer and fall (Cui et al., 2011; Leung et al., 2009). However, the peak infection for another human respiratory CoV, HCoV-OC43 in Hong Kong was detected in December–January (Leung et al., 2009). Interestingly, there are also variable observations on the seasonality of animal CoVs. Winter dysentery in cows occurs in winter in temperate climates, but also occurs at other times in non-temperate climates (Decaro et al., 2008b; Fukutomi et al., 1999; Park et al. 2006; Saif et al., 1991) suggesting that diverse mechanisms may promote BCoV spread in different seasons and environments. There are also data indicating that TGEV and PRCV in pigs are more prevalent in winter (Pensaert et al., 1986, 1993; Saif and Sestak, 2006) suggesting that better preservation of these viruses in frozen state in feces or at cold temperatures assists with CoV transfer. Additionally, starlings (Pilchard, 1965) and flies (Gough and Jorgenson, 1983) were indicated as potential mechanical vehicles for TGEV transmission which emphasizes that climatic changes (temperature, humidity, strong winds) can indirectly influence CoV ecology through affecting these vehicles. The PRCV appears to be airborne (Bourgueil et al., 1992; Wesley et al., 1990) traveling relatively long distances in aerosols; therefore relative humidity changes may significantly affect its survival as was shown for other viruses (Schoenbaum et al., 1990). The report of a severe outbreak of BCoV diarrheal disease in cattle in Italy during the warmer season (Decaro et al., 2008b) suggests that in the warmer seasons, birds and insects migrating between affected herds could promote BCoV spread. For TCoV in the US, the occurrence was highest in October and this seasonality was suggested to be associated with higher humidity and the fly population rise, contributing to mechanical spread of TCoV (Cavanagh et al., 2001). Collectively, CoV circulation and spread results from an interplay of multiple factors. Fomites and mechanical vehicles are also suspected to be important for aerosol spread of human CoVs (Dowell et al., 2004; Sizun et al., 2000); therefore studies of CoV survival on fomite surfaces and mechanical vehicles in different conditions are critical. Preliminary evidence suggests that various ecological influences including climate change may have played a role in recent host range and geographic expansions of avian pathogens (Fuller et al., 2012). Thus, the increasingly recognized diversity of CoVs in migratory birds suggests that climatic change may affect their migratory pathways and therefore CoV ecology.

Whether there is true seasonality or it is variable for different CoVs and geographical regions remains to be determined. Therefore, additional research to understand the association between climatic influences and CoV spread and circulation in natural reservoirs across diverse populations and geographical regions is needed. The accumulated epidemiological data can then be used to develop predictive strategies to avert outbreaks like SARS.

21.6 CONCLUSION

Coronaviruses are a vast and important group of large RNA viruses that possess unique molecular mechanisms of transcription and recombination, providing diverse models of pathogenesis and continually emerging new pathogens. The SARS pandemic and the discovery of the remarkable variety of animal and avian as well as novel human CoVs, revealed a very complex and sustainable ecology of these viruses. Coronaviruses can persist in diverse environments, infecting a variety of avian and mammalian species. Multiple viral genes contribute to CoV pathogenesis and result in different pathogenic phenotypes depending on interaction with host responses. The observations that coronavirus tissue tropism and host range variants already exist as quasispecies during replication in tissue culture or in avian/mammalian species indicate that CoVs may shift or extend their host range leading to the emergence of new CoV variants in humans. Although influenced by anthropogenic and possibly climatic factors, CoVs respond and evolve rapidly to adapt to new hosts/tissues/environments and may even manipulate these factors to maintain their persistence and abundance. More extensive molecular epidemiologic and genetic studies are needed to develop comprehensive models to predict CoV transmission, spread and disease in relation to probable climatic changes.

REFERENCES

Alekseev KP, Vlasova AN, Jung K, et al. 2008. Bovine-like coronaviruses isolated from four species of captive wild ruminants are homologous to bovine coronaviruses, based on complete genomic sequences. *J Virol* 82: 12422–31

An DJ, Jeoung HY, Jeong W, et al. 2011. Prevalence of Korean cats with natural feline coronavirus infections. *Virol J* 8: 455

Baric RS, Sullivan E, Hensley L, et al. 1999. Persistent infection promotes cross-species transmissibility of mouse hepatitis virus. *J Virol* 73: 638–49

Bastien N, Anderson K, Hart L, et al. 2005a. Human coronavirus NL63 infection in Canada. *J Infect Dis* 191: 503–6

Bastien N, Robinson JL, Tse A, et al. 2005b. Human coronavirus NL-63 infections in children: a 1-year study. *J Clin Microbiol* 43: 4567–73

Beaudette FR, Hudson, CB. 1937. Cultivation of the virus of infectious bronchitis. *J Am Vet Med Assoc* 90: 51–60

Benetka V, Kolodziejek J, Walk K, et al. 2006. M gene analysis of atypical strains of feline and canine coronavirus circulating in an Austrian animal shelter. *Vet Rec* 159: 170–4

Berberian G, Rosanova MT. 2012. Impact of climate change on infectious diseases. *Arch Argent Pediatr* 110: 39–45

Bourgueil E, Hutet E, Cariolet R, et al. 1992. Experimental infection of pigs with the porcine respiratory coronavirus (PRCV): measure of viral excretion. *Vet Microbiol* 31: 11–8

Brown MA, Troyer JL, Pecon-Slattery J, et al. 2009. Genetics and pathogenesis of feline infectious peritonitis virus. *Emerg Infect Dis* 15: 1445–52

Calisher CH, Childs JE, Field HE, et al. 2006. Bats: important reservoir hosts of emerging viruses. *Clin Microbiol Rev* 19: 531–45

Casanova L, Rutala WA, Weber DJ, et al. 2009. Survival of surrogate coronaviruses in water. *Water Res* 43: 1893–8

Casanova L, Rutala WA, Weber DJ, et al. 2010a. Coronavirus survival on healthcare personal protective equipment. *Infect Control Hosp Epidemiol* 31: 560–1

Casanova LM, Jeon S, Rutala WA, et al. 2010b. Effects of air temperature and relative humidity on coronavirus survival on surfaces. *Appl Environ Microbiol* 76: 2712–7

Cavanagh D. 1997. Nidovirales: a new order comprising Coronaviridae and Arteriviridae. *Arch Virol* 142: 629–33

Cavanagh D, Mawditt K, Sharma M, et al. 2001. Detection of a coronavirus from Turkey poults in Europe genetically related to infectious bronchitis virus of chickens. *Avian Pathol* 30: 355–68

Che XY, Di B, Zhao GP, et al. 2006. A patient with asymptomatic severe acute respiratory syndrome (SARS) and antigenemia from the 2003–2004 community outbreak of SARS in Guangzhou, China. *Clin Infect Dis* 43: e1–5

Cheever FS, Daniels JB, Pappenheimer AM, et al. 1949. A murine virus (JHM) causing disseminated encephalomyelitis with extensive destruction of myelin. *J Exp Med* 90: 181–210

Chen W, Yan M, Yang L, et al. 2005. SARS-associated coronavirus transmitted from human to pig. *Emerg Infect Dis* 11: 446–8

Chiu YC, Wu KL, Chou YP, et al. 2004. Diarrhea in medical care workers with severe acute respiratory syndrome. *J Clin Gastroenterol* 38: 880–2

Chu VC, McElroy LJ, Chu V, et al. 2006. The avian coronavirus infectious bronchitis virus undergoes direct low-pH-dependent fusion activation during entry into host cells. *J Virol* 80: 3180–8

Chu DK, Leung CY, Gilbert M, et al. 2011. Avian coronavirus in wild aquatic birds. *J Virol* 85: 12815–20

Cowley TJ, Weiss SR. 2010. Murine coronavirus neuropathogenesis: determinants of virulence. *J Neurovirol* 16: 427–34

Cowley TJ, Long SY, Weiss SR. 2010. The murine coronavirus nucleocapsid gene is a determinant of virulence. *J Virol* 84: 1752–63

Cui LJ, Zhang C, Zhang T, et al. 2011. Human coronaviruses HCoV-NL63 and HCoV-HKU1 in hospitalized children with acute respiratory infections in Beijing, *China. Adv Virol* 2011: 129134

Dea S, Verbeek AJ, Tijssen P. 1990. Antigenic and genomic relationships among turkey and bovine enteric coronaviruses. *J Virol* 64: 3112–8

Decaro N, Martella V, Elia G, et al. 2007. Molecular characterisation of the virulent canine coronavirus CB/05 strain. *Virus Res* 125: 54–60

Decaro N, Campolo M, Lorusso A, et al. 2008a. Experimental infection of dogs with a novel strain of canine coronavirus causing systemic disease and lymphopenia. *Vet Microbiol* 128: 253–60

Decaro N, Mari V, Desario C, et al. 2008b. Severe outbreak of bovine coronavirus infection in dairy cattle during the warmer season. *Vet Microbiol* 126: 30–9

Decaro N, Martella V, Elia G, et al. 2008c. Biological and genetic analysis of a bovine-like coronavirus isolated from water buffalo (*Bubalus bubalis*) calves. *Virology* 370: 213–22

Decaro N, Mari V, Campolo M, et al. 2009. Recombinant canine coronaviruses related to transmissible gastroenteritis virus of Swine are circulating in dogs. *J Virol* 83: 1532–7

Decaro N, Mari V, Elia G, et al. 2010. Recombinant canine coronaviruses in dogs, Europe. *Emerg Infect Dis* 16: 41–7

Dennehy JJ, Friedenberg NA, Holt RD, et al. 2006. Viral ecology and the maintenance of novel host use. *Am Nat* 167: 429–39

Ding Y, He L, Zhang Q, et al. 2004. Organ distribution of severe acute respiratory syndrome (SARS) associated coronavirus (SARS-CoV) in SARS patients: implications for pathogenesis and virus transmission pathways. *J Pathol* 203: 622–30

Dobson A. 2009. Climate variability, global change, immunity, and the dynamics of infectious diseases. *Ecology* 90: 920–7

Domingo E. 1998. Quasispecies and the implications for virus persistence and escape. *Clin Diagn Virol* 10: 97–101

Domingo E, Baranowski E, Ruiz-Jarabo CM, et al. 1998a. Quasispecies structure and persistence of RNA viruses. *Emerg Infect Dis* 4: 521–7

Domingo E, Escarmis C, Sevilla N, et al. 1998b. Population dynamics in the evolution of RNA viruses. *Adv Exp Med Biol* 440: 721–7

Domingo E, Martin V, Perales C, et al. 2006. Viruses as quasispecies: biological implications. *Curr Top Microbiol Immunol* 299: 51–82

Dominguez SR, O'Shea TJ, Oko LM, et al. 2007. Detection of group 1 coronaviruses in bats in North America. *Emerg Infect Dis* 13: 1295–300

Donaldson EF, Haskew AN, Gates JE, et al. 2010. Metagenomic analysis of the viromes of three North American bat species: viral diversity among different bat species that share a common habitat. *J Virol* 84: 13004–18

Dowell SF, Ho MS. 2004. Seasonality of infectious diseases and severe acute respiratory syndrome-what we don't know can hurt us. *Lancet Infect Dis* 4: 704–8

Dowell SF, Simmerman JM, Erdman DD, et al. 2004. Severe acute respiratory syndrome coronavirus on hospital surfaces. *Clin Infect Dis* 39: 652–7

Doyle LP, Hutchings LM. 1946. A transmissible gastroenteritis in pigs. *J Am Vet Med Assoc* 108: 257–9

Drosten C, Gunther S, Preiser W, et al. 2003a. Identification of a novel coronavirus in patients with severe acute respiratory syndrome. *N Engl J Med* 348: 1967–76

Drosten C, Preiser W, Gunther S, et al. 2003b. Severe acute respiratory syndrome: identification of the etiological agent. *Trends Mol Med* 9: 325–7

Erles K, Brownlie J. 2009. Sequence analysis of divergent canine coronavirus strains present in a UK dog population. *Virus Res* 141: 21–5

Fielding BC. 2011. Human coronavirus NL63: a clinically important virus? *Future Microbiol* 6: 153–9

Frieman M, Yount B, Agnihothram S, et al. 2012. Molecular determinants of severe acute respiratory syndrome coronavirus pathogenesis and virulence in young and aged mouse models of human disease. *J Virol* 86: 884–97

Fukutomi T, Tsunemitsu H, Akashi H. 1999. Detection of bovine coronaviruses from adult cows with epizootic diarrhea and their antigenic and biological diversities. *Arch Virol* 144: 997–1006

Fuller T, Bensch S, Muller I, et al. 2012. The ecology of emerging infectious diseases in migratory birds: an assessment of the role of climate change and priorities for future research. *Ecohealth* 9: 80–8

Gillim-Ross L, Taylor J, Scholl DR, et al. 2004. Discovery of novel human and animal cells infected by the severe acute respiratory syndrome coronavirus by replication-specific multiplex reverse transcription-PCR. *J Clin Microbiol* 42: 3196–206

Glass WG, Subbarao K, Murphy B, et al. 2004. Mechanisms of host defense following severe acute respiratory syndrome-coronavirus (SARS-CoV) pulmonary infection of mice. *J Immunol* 173: 4030–9

Gonzalez JM, Gomez-Puertas P, Cavanagh D, et al. 2003. A comparative sequence analysis to revise the current taxonomy of the family Coronaviridae. *Arch Virol* 148: 2207–35

Gorbalenya AE, Enjuanes L, Ziebuhr J, et al. 2006. Nidovirales: evolving the largest RNA virus genome. *Virus Res* 117: 17–37

Gorham JR, Evermann JF, Ward A, et al. 1990. Detection of coronavirus-like particles from mink with epizootic catarrhal gastroenteritis. *Can J Vet Res* 54: 383–4

Gough PM, Jorgenson RD. 1983. Identification of porcine transmissible gastroenteritis virus in house flies (*Musca domestica* Linneaus). *Am J Vet Res* 44: 2078–82

Gough RE, Drury SE, Culver F, et al. 2006. Isolation of a coronavirus from a green-cheeked Amazon parrot (*Amazon viridigenalis* Cassin). *Avian Pathol* 35: 122–6

Graham RL, Baric RS. 2010. Recombination, reservoirs, and the modular spike: mechanisms of coronavirus cross-species transmission. *J Virol* 84: 3134–46

Greer A, Ng V, Fisman D. 2008. Climate change and infectious diseases in North America: the road ahead. *Can Med Assoc J* 178: 715–22

de Groot RJ, Baker SC, Baric R, et al. 2011. Family coronaviridae. In: AMQ King, E Lefkowitz, MJ Adams, EB Carstens (eds), *Ninth Report of the International Committee on Taxonomy of Viruses*, pp. 806–28. Oxford: Elsevier

Guan Y, Zheng BJ, He YQ, et al. 2003. Isolation and characterization of viruses related to the SARS coronavirus from animals in southern China. *Science* 302: 276–8

Guan Y, Peiris JS, Zheng B, et al. 2004. Molecular epidemiology of the novel coronavirus that causes severe acute respiratory syndrome. *Lancet* 363: 99–104

Haelterman EO. 1962. *Epidemiological studies of transmissible gastroenteritis of swine*. Presented at the 66th Annual Meeting of the US Livestock Sanitary Association, Washington, DC

Haelterman EO, Hutchings LM. 1956. Epidemic diarrheal disease of viral origin in newborn swine. *Ann N Y Acad Sci* 66: 186–90

Hales S, de Wet N, Maindonald J, et al. 2002. Potential effect of population and climate changes on global distribution of dengue fever: an empirical model. *Lancet* 360: 830–4

Hamre D, Procknow JJ. 1966. A new virus isolated from the human respiratory tract. *Proc Soc Exp Biol Med* 121: 190–3

Han MG, Cheon DS, Zhang X, et al. 2006. Cross-protection against a human enteric coronavirus and a virulent bovine enteric coronavirus in gnotobiotic calves. *J Virol* 80: 12350–6

Harley D, Bi P, Hall G, et al. 2011. Climate change and infectious diseases in Australia: future prospects, adaptation options, and research priorities. *Asia Pac J Public Health* 23: 54S-66

Hasoksuz M, Alekseev K, Vlasova A, et al. 2007. Biologic, antigenic, and full-length genomic characterization of a bovine-like coronavirus isolated from a giraffe. *J Virol* 81: 4981–90

Heller LK, Gillim-Ross L, Olivieri ER, et al. 2006. Mustela vison ACE2 functions as a receptor for SARS-coronavirus. *Adv Exp Med Biol* 581: 507–10

Hensley LE, Holmes KV, Beauchemin N, et al. 1998. Virus-receptor interactions and interspecies transfer of a mouse hepatitis virus. *Adv Exp Med Biol* 440: 33–41

Herrewegh AA, Vennema H, Horzinek MC, et al. 1995. The molecular genetics of feline coronaviruses: comparative sequence analysis of the ORF7a/7b transcription unit of different biotypes. *Virology* 212: 622–31

Herrewegh AA, Smeenk I, Horzinek MC, et al. 1998. Feline coronavirus type II strains 79–1683 and 79–1146 originate from a double recombination between feline coronavirus type I and canine coronavirus. *J Virol* 72: 4508–14

Hogan RJ, Gao G, Rowe T, et al. 2004. Resolution of primary severe acute respiratory syndrome-associated coronavirus infection requires Stat1. *J Virol* 78: 11416–21

Holland J, Spindler K, Horodyski F, et al. 1982. Rapid evolution of RNA genomes. *Science* 215: 1577–85

Huang CY, Hsu YL, Chiang WL, et al. 2009. Elucidation of the stability and functional regions of the human coronavirus OC43 nucleocapsid protein. *Protein Sci* 18: 2209–18

Hughes LA, Savage C, Naylor C, et al. 2009. Genetically diverse coronaviruses in wild bird populations of northern England. *Emerg Infect Dis* 15: 1091–4

Ismail MM, Cho KO, Ward LA, et al. 2001. Experimental bovine coronavirus in Turkey poults and young chickens. *Avian Dis* 45: 157–63

Jin L, Cebra CK, Baker RJ, et al. 2007. Analysis of the genome sequence of an alpaca coronavirus. *Virology* 365: 198–203

Jung K, Alekseev KP, Zhang X, et al. 2007. Altered pathogenesis of porcine respiratory coronavirus in pigs due to immunosuppressive effects of dexamethasone: implications for corticosteroid use in treatment of severe acute respiratory syndrome coronavirus. *J Virol* 81: 13681–93

Jung K, Renukaradhya GJ, Alekseev KP, et al. 2009. Porcine reproductive and respiratory syndrome virus modifies innate immunity and alters disease outcome in pigs subsequently infected with porcine respiratory coronavirus: implications for respiratory viral co-infections. *J Gen Virol* 90: 2713–23

Kim SW, Ramakrishnan, M.A., Raynor, P.C., et al. 2007. Effects of humidity and other factors on the generation and sampling of a coronavirus aerosol. *Aerobiologia* 23: 239–48

King AMQ, Adams MJ, Carstens EB, et al. (eds) 2011. *Virus Taxonomy: Ninth Report of the International Committee on Taxonomy of Viruses*. Oxford: Elsevier

Ksiazek TG, Erdman D, Goldsmith CS, et al. 2003. A novel coronavirus associated with severe acute respiratory syndrome. *N Engl J Med* 348: 1953–66

Lafferty KD. 2009. The ecology of climate change and infectious diseases. *Ecology* 90: 888–900

Lai MM. 1992. RNA recombination in animal and plant viruses. *Microbiol Rev* 56: 61–79

Lai MMC, Holmes KV. 2001. Coronaviruses. In: BN Fields, DM Knipe, PM Howley, et al. (eds), *Fields Virology*, pp. 1163–85. Philadelphia: Lippincott Williams & Wilkins

Lai MM, Baric RS, Makino S, et al. 1985. Recombination between nonsegmented RNA genomes of murine coronaviruses. *J Virol* 56: 449–56

Lau SK, Woo PC, Li KS, et al. 2005. Severe acute respiratory syndrome coronavirus-like virus in Chinese horseshoe bats. *Proc Natl Acad Sci U S A* 102: 14040–5

Laude H, Van Reeth K, Pensaert M. 1993. Porcine respiratory coronavirus: molecular features and virus–host interactions. *Vet Res* 24: 125–50

Lee HK, Tso EY, Chau TN, et al. 2003. Asymptomatic severe acute respiratory syndrome-associated coronavirus infection. *Emerg Infect Dis* 9: 1491–2

Leung WK, To KF, Chan PK, et al. 2003. Enteric involvement of severe acute respiratory syndrome-associated coronavirus infection. *Gastroenterology* 125: 1011–7

Leung TF, Li CY, Lam WY, et al. 2009. Epidemiology and clinical presentations of human coronavirus NL63 infections in Hong Kong children. *J Clin Microbiol* 47: 3486–92

Li W, Shi Z, Yu M, et al. 2005a. Bats are natural reservoirs of SARS-like coronaviruses. *Science* 310: 676–9

Li W, Zhang C, Sui J, et al. 2005b. Receptor and viral determinants of SARS-coronavirus adaptation to human ACE2. *EMBO J* 24: 1634–43

Liu S, Chen J, Kong X, et al. 2005. Isolation of avian infectious bronchitis coronavirus from domestic peafowl (*Pavo cristatus*) and teal (Anas). *J Gen Virol* 86: 719–25

Lorusso A, Decaro N, Schellen P, et al. 2008. Gain, preservation, and loss of a group 1a coronavirus accessory glycoprotein. *J Virol* 82: 10312–7

Lorusso A, Desario C, Mari V, et al. 2009. Molecular characterization of a canine respiratory coronavirus strain detected in Italy. *Virus Res* 141: 96–100

Madu IG, Chu VC, Lee H, et al. 2007. Heparan sulfate is a selective attachment factor for the avian coronavirus infectious bronchitis virus Beaudette. *Avian Dis* 51: 45–51

Martina BE, Haagmans BL, Kuiken T, et al. 2003. Virology: SARS virus infection of cats and ferrets. *Nature* 425: 915

McArdle F, Bennett M, Gaskell RM, et al. 1992. Induction and enhancement of feline infectious peritonitis by canine coronavirus. *Am J Vet Res* 53: 1500–6

McIntosh K, Kapikian AZ, Turner HC, et al. 1970. Seroepidemiologic studies of coronavirus infection in adults and children. *Am J Epidemiol* 91: 585–92

Mihindukulasuriya KA, Wu G, St Leger J, et al. 2008. Identification of a novel coronavirus from a beluga whale by using a panviral microarray. *J Virol* 82: 5084–8

Morillas-Marquez F, Martin-Sanchez J, Diaz-Saez V, et al. 2010. Climate change and infectious diseases in Europe: leishmaniasis and its vectors in Spain. *Lancet Infect Dis* 10: 216–7

Mossel EC, Huang C, Narayanan K, et al. 2005. Exogenous ACE2 expression allows refractory cell lines to support severe acute respiratory syndrome coronavirus replication. *J Virol* 79: 3846–50

Mullis L, Saif LJ, Zhang Y, et al. 2012. Stability of bovine coronavirus on lettuce surfaces under household refrigeration conditions. *Food Microbiol* 30: 180–6

Muradrasoli S, Balint A, Wahlgren J, et al. 2010. Prevalence and phylogeny of coronaviruses in wild birds from the Bering Strait area (Beringia). *PLoS One* 5: e13640

Nagaraja KV, and Pomeroy BS. 1997. Coronaviral enteritis of turkeys (bluecomb disease). In: BW Calnek, HJ Barnes, CW Beard, et al. (eds), *Diseases of Poultry*, pp. 686–92. Ames: Iowa State University Press

Navas-Martin S, Hingley ST, Weiss SR. 2005. Murine coronavirus evolution in vivo: functional compensation of a detrimental amino acid substitution in the receptor binding domain of the spike glycoprotein. *J Virol* 79: 7629–40

Oh JS, Song DS, Park BK. 2003. Identification of a putative cellular receptor 150 kDa polypeptide for porcine epidemic diarrhea virus in porcine enterocytes. *J Vet Sci* 4: 269–75

Ostfeld RS. 2009. Climate change and the distribution and intensity of infectious diseases. *Ecology* 90: 903–5

Park SJ, Jeong C, Yoon SS, et al. 2006. Detection and characterization of bovine coronaviruses in fecal specimens of adult cattle with diarrhea during the warmer seasons. *J Clin Microbiol* 44: 3178–88

Park SJ, Kim GY, Choy HE, et al. 2007. Dual enteric and respiratory tropisms of winter dysentery bovine coronavirus in calves. *Arch Virol* 152: 1885–900

Patz JA, Githeko, AK, McCarty, JP, et al. 2003. Climate change and infectious diseases. In: AJ McMichael, DH Campbell-Lendrum, CF Corvalan (eds), *Climate Change and Human Health: Risks and Responses*, pp. 103–32. Geneva: World Health Organization

Pedersen NC. 1976. Morphologic and physical characteristics of feline infectious peritonitis virus and its growth in autochthonous peritoneal cell cultures. *Am J Vet Res* 37: 567–72

Pedersen NC. 2009. A review of feline infectious peritonitis virus infection: 1963–2008. *J Feline Med Surg* 11: 225–58

Pedersen NC, Boyle JF, Floyd K, et al. 1981. An enteric coronavirus infection of cats and its relationship to feline infectious peritonitis. *Am J Vet Res* 42: 368–77

Pedersen NC, Black JW, Boyle JF, et al. 1984a. Pathogenic differences between various feline coronavirus isolates. *Adv Exp Med Biol* 173: 365–80

Pedersen NC, Evermann JF, McKeirnan AJ, et al. 1984b. Pathogenicity studies of feline coronavirus isolates 79-1146 and 79-1683. *Am J Vet Res* 45: 2580–5

Pedersen NC, Allen CE, Lyons LA. 2008. Pathogenesis of feline enteric coronavirus infection. *J Feline Med Surg* 10: 529–41

Peiris JS, Guan Y, Yuen KY. 2004. Severe acute respiratory syndrome. *Nat Med* 10: S88–97

Peiris JS, Chu CM, Cheng VC, et al. 2003a. Clinical progression and viral load in a community outbreak of coronavirus-associated SARS pneumonia: a prospective study. *Lancet* 361: 1767–72

Peiris JS, Lai ST, Poon LL, et al. 2003b. Coronavirus as a possible cause of severe acute respiratory syndrome. *Lancet* 361: 1319–25

Pensaert M, Callebaut P, Vergote J. 1986. Isolation of a porcine respiratory, non-enteric coronavirus related to transmissible gastroenteritis. *Vet Q* 8: 257–61

Pensaert M, Cox E, van Deun K, et al. 1993. A sero-epizootiological study of porcine respiratory coronavirus in Belgian swine. *Vet Q* 15: 16–20

Perlman S, Dandekar AA. 2005. Immunopathogenesis of coronavirus infections: implications for SARS. *Nat Rev Immunol* 5: 917–27

Perlman S, Netland J. 2009. Coronaviruses post-SARS: update on replication and pathogenesis. *Nat Rev Microbiol* 7: 439–50

Petersen NC, Boyle JF. 1980. Immunologic phenomena in the effusive form of feline infectious peritonitis. *Am J Vet Res* 41: 868–76

Pfefferle S, Oppong S, Drexler JF, et al. 2009. Distant relatives of severe acute respiratory syndrome coronavirus and close relatives of human coronavirus 229E in bats, Ghana. *Emerg Infect Dis* 15: 1377–84

Pilchard EI. 1965. Experimental transmission of transmissible gastroenteritis virus by starlings. *Am J Vet Res* 26: 1177–9

Qu D, Zheng B, Yao X, et al. 2005. Intranasal immunization with inactivated SARS-CoV (SARS-associated coronavirus) induced local and serum antibodies in mice. *Vaccine* 23: 924–31

Roberts A, Vogel L, Guarner J, et al. 2005. Severe acute respiratory syndrome coronavirus infection of golden Syrian hamsters. *J Virol* 79: 503–11

Rosenthal J. 2009. Climate change and the geographic distribution of infectious diseases. *Ecohealth* 6: 489–95

Rota PA, Oberste MS, Monroe SS, et al. 2003. Characterization of a novel coronavirus associated with severe acute respiratory syndrome. *Science* 300: 1394–9

Rottier PJ, Nakamura K, Schellen P, et al. 2005. Acquisition of macrophage tropism during the pathogenesis of feline infectious peritonitis is determined by mutations in the feline coronavirus spike protein. *J Virol* 79: 14122–30

Saif LJ, Sestak, K. 2006. Transmissible gastroenteritis and porcine respiratory coronavirus. In: BE Straw, JJ Zimmerman, S D'Allaire, et al. (eds), *Diseases of Swine*, pp. 489–516. Ames: Blackwell

Saif LJ, Brock KV, Redman DR, et al. 1991. Winter dysentery in dairy herds: electron microscopic and serological evidence for an association with coronavirus infection. *Vet Rec* 128: 447–9

Sanchez CM, Izeta A, Sanchez-Morgado JM, et al. 1999. Targeted recombination demonstrates that the spike gene of transmissible gastroenteritis coronavirus is a determinant of its enteric tropism and virulence. *J Virol* 73: 7607–18

Schoenbaum MA, Zimmerman JJ, Beran GW, et al. 1990. Survival of pseudorabies virus in aerosol. *Am J Vet Res* 51: 331–3

Shi X, Gong E, Gao D, et al. 2005. Severe acute respiratory syndrome associated coronavirus is detected in intestinal tissues of fatal cases. *Am J Gastroenterol* 100: 169–76

Shuman EK. 2010. Global climate change and infectious diseases. *N Engl J Med* 362: 1061–3

Sims AC, Burkett SE, Yount B, et al. 2008. SARS-CoV replication and pathogenesis in an in vitro model of the human conducting airway epithelium. *Virus Res* 133: 33–44

Sizun J, Yu MW, Talbot PJ. 2000. Survival of human coronaviruses 229E and OC43 in suspension and after drying on surfaces: a possible source of hospital-acquired infections. *J Hosp Infect* 46: 55–60

Smits SL, de Lang A, van den Brand JM, et al. 2010. Exacerbated innate host response to SARS-CoV in aged non-human primates. *PLoS Pathog* 6: e1000756

Spaan W, Cavanagh D, Horzinek MC. 1988. Coronaviruses: structure and genome expression. *J Gen Virol* 69 (Pt 12): 2939–52

Subbarao K, McAuliffe J, Vogel L, et al. 2004. Prior infection and passive transfer of neutralizing antibody prevent replication of severe acute respiratory syndrome coronavirus in the respiratory tract of mice. *J Virol* 78: 3572–7

Tang XC, Zhang JX, Zhang SY, et al. 2006. Prevalence and genetic diversity of coronaviruses in bats from China. *J Virol* 80: 7481–90

Taylor LH, Latham SM, Woolhouse ME. 2001. Risk factors for human disease emergence. *Philos Trans R Soc Lond B: Biol Sci* 356: 983–9

Tong S, Li Y, Rivailler P, et al. 2012. A distinct lineage of influenza A virus from bats. *Proc Natl Acad Sci U S A* 109: 4269–74

Tsunemitsu H, el-Kanawati ZR, Smith DR, et al. 1995. Isolation of coronaviruses antigenically indistinguishable from bovine coronavirus from wild ruminants with diarrhea. *J Clin Microbiol* 33: 3264–9

Tu C, Crameri G, Kong X, et al. 2004. Antibodies to SARS coronavirus in civets. *Emerg Infect Dis* 10: 2244–8

Tyrrell DA, Bynoe ML. 1966. Cultivation of viruses from a high proportion of patients with colds. *Lancet* 1: 76–7

Tyrrell DA, Almeida JD, Cunningham CH, et al. 1975. Coronaviridae. *Intervirology* 5: 76–82

Vabret A, Mourez T, Dina J, et al. 2005. Human coronavirus NL63, France. *Emerg Infect Dis* 11: 1225–9

Vaughn EM, Halbur PG, Paul PS. 1995. Sequence comparison of porcine respiratory coronavirus isolates reveals heterogeneity in the S, 3, and 3–1 genes. *J Virol* 69: 3176–84

Vennema H, Poland A, Foley J, et al. 1998. Feline infectious peritonitis viruses arise by mutation from endemic feline enteric coronaviruses. *Virology* 243: 150–7

Vignuzzi M, Stone JK, Arnold JJ, et al. 2006. Quasispecies diversity determines pathogenesis through cooperative interactions in a viral population. *Nature* 439: 344–8

Vijaykrishna D, Smith GJ, Zhang JX, et al. 2007. Evolutionary insights into the ecology of coronaviruses. *J Virol* 81: 4012–20

Vijgen L, Keyaerts E, Moes E, et al. Complete genomic sequence of human coronavirus OC43: molecular clock analysis suggests a relatively recent zoonotic coronavirus transmission event. *J Virol* 79: 1595–604

Vijgen L, Keyaerts E, Lemey P, et al. Evolutionary history of the closely related group 2 coronaviruses: porcine hemagglutinating encephalomyelitis virus, bovine coronavirus, and human coronavirus OC43. *J Virol* 80: 7270–4

Vlasova AN, Halpin R, Wang S, et al. 2011. Molecular characterization of a new species in the genus Alphacoronavirus associated with mink epizootic catarrhal gastroenteritis. *J Gen Virol* 92: 1369–79

Vogel L, Van der Lubben M, te Lintelo EG, et al. 2010. Pathogenic characteristics of persistent feline enteric coronavirus infection in cats. *Vet Res* 41: 71

Walsh EE, Falsey AR, Hennessey PA. 1999. Respiratory syncytial and other virus infections in persons with chronic cardiopulmonary disease. *Am J Respir Crit Care Med* 160: 791–5

Watanabe S, Masangkay JS, Nagata N, et al. 2010. Bat coronaviruses and experimental infection of bats, the Philippines. *Emerg Infect Dis* 16: 1217–23

Weiss SR, Navas-Martin S. 2005. Coronavirus pathogenesis and the emerging pathogen severe acute respiratory syndrome coronavirus. *Microbiol Mol Biol Rev* 69: 635–64

Wesley RD. 1999. The S gene of canine coronavirus, strain UCD-1, is more closely related to the S gene of transmissible gastroenteritis virus than to that of feline infectious peritonitis virus. *Virus Res* 61: 145–52

Wesley RD, Woods RD, Hill HT, et al. 1990. Evidence for a porcine respiratory coronavirus, antigenically similar to transmissible gastroenteritis virus, in the United States. *J Vet Diagn Invest* 2: 312–7

Wesley RD, Woods RD, Cheung AK. 1991. Genetic analysis of porcine respiratory coronavirus, an attenuated variant of transmissible gastroenteritis virus. *J Virol* 65: 3369–73

Wilder-Smith A, Teleman MD, Heng BH, et al. 2005. Asymptomatic SARS coronavirus infection among healthcare workers, Singapore. *Emerg Infect Dis* 11: 1142–5

Wilson N, Slaney D, Baker MG, et al. 2011. Climate change and infectious diseases in New Zealand: a brief review and tentative research agenda. *Rev Environ Health* 26: 93–9

Wise AG, Kiupel M, Maes RK. 2006. Molecular characterization of a novel coronavirus associated with epizootic catarrhal enteritis (ECE) in ferrets. *Virology* 349: 164–74

Wise AG, Kiupel M, Garner MM, et al. 2010. Comparative sequence analysis of the distal one-third of the genomes of a systemic and an enteric ferret coronavirus. *Virus Res* 149: 42–50

Woo PC, Lau SK, Huang Y, et al. 2009. Coronavirus diversity, phylogeny and interspecies jumping. *Exp Biol Med (Maywood)* 234: 1117–27

Woo PC, Huang Y, Lau SK, et al. 2010. Coronavirus genomics and bioinformatics analysis. *Viruses* 2: 1804–20

Woo PC, Lau SK, Lam CS, et al. 2012. Discovery of seven novel mammalian and avian coronaviruses in Deltacoronavirus supports bat coronaviruses as the gene source of Alphacoronavirus and Betacoronavirus and avian coronaviruses as the gene source of Gammacoronavirus and Deltacoronavirus. *J Virol* 86: 3995–4008

Woods RD, Cheville NF, Gallagher JE. 1981. Lesions in the small intestine of newborn pigs inoculated with porcine, feline, and canine coronaviruses. *Am J Vet Res* 42: 1163–9

Wu D, Tu C, Xin C, et al. 2005. Civets are equally susceptible to experimental infection by two different severe acute respiratory syndrome coronavirus isolates. *J Virol* 79: 2620–5

Yeager CL, Ashmun RA, Williams RK, et al. 1992. Human aminopeptidase N is a receptor for human coronavirus 229E. *Nature* 357: 420–2

Yoshikawa T, Hill T, Li K, et al. 2009. Severe acute respiratory syndrome (SARS) coronavirus-induced lung epithelial cytokines exacerbate SARS pathogenesis by modulating intrinsic functions of monocyte-derived macrophages and dendritic cells. *J Virol* 83: 3039–48

Zhang XM, Herbst W, Kousoulas KG, et al. 1994. Biological and genetic characterization of a hemagglutinating coronavirus isolated from a diarrhoeic child. *J Med Virol* 44: 152–61

Zhang CY, Wei JF, He SH. 2006. Adaptive evolution of the spike gene of SARS coronavirus: changes in positively selected sites in different epidemic groups. *BMC Microbiol* 6: 88

Zhang X, Hasoksuz M, Spiro D, et al. 2007. Quasispecies of bovine enteric and respiratory coronaviruses based on complete genome sequences and genetic changes after tissue culture adaptation. *Virology* 363: 1–10

Zhang X, Alekseev K, Jung K, et al. 2008. Cytokine responses in porcine respiratory coronavirus-infected pigs treated with corticosteroids as a model for severe acute respiratory syndrome. *J Virol* 82: 4420–8

Zhong X, Guo Z, Yang H, et al. 2006. Amino terminus of the SARS coronavirus protein 3a elicits strong, potentially protective humoral responses in infected patients. *J Gen Virol* 87: 369–73

22

IMPACT OF ENVIRONMENTAL AND SOCIAL FACTORS ON ROSS RIVER VIRUS OUTBREAKS

Craig R. Williams

Sansom Institute for Health Research, University of South Australia, Adelaide, SA, Australia; National Centre for Epidemiology and Population Health, The Australian National University, Canberra, ACT, Australia

David O. Harley

National Centre for Epidemiology and Population Health, The Australian National University, Canberra, ACT, Australia

TABLE OF CONTENTS

Viral Infections and Global Change, First Edition. Edited by Sunit K. Singh.

22.1 INTRODUCTION

Ross River virus (RRV) is a zoonotic mosquito-borne arbovirus (arthropod-borne virus) belonging to the genus *Alphavirus*; it is endemic and enzootic in Australia and Papua New Guinea (Harley et al., 2001). Ross River virus was first isolated from *Aedes vigilax* trapped beside Ross River, Townsville, Queensland, in 1959 (Doherty et al., 1968), but is thought to have caused outbreaks in humans in Australia as early as 1886 (Kelly-Hope et al., 2004). The virus belongs, phylogenetically, among the Old World alphaviruses, a grouping that includes chikungunya, o'nyong-nyong, Barmah Forest, and Semliki Forest (Harley et al., 2001; Powers et al., 2001). The Old World alphaviruses generally cause rheumatic symptoms, in contrast to New World alphaviruses, which cause encephalitis (Harley et al., 2001; Powers et al., 2001). Interest in these viruses has increased with the spread of chikungunya virus into Europe (Rezza et al., 2007) and the occurrence, on the Indian Ocean Island of Reunion, of a large epidemic with severe disease and deaths (Economopoulou et al., 2009; Gerardin et al., 2008; Renault et al., 2007).

The clinical manifestations of RRV disease arise from immune responses directed towards persisting RRV antigens with the production of pro-inflammatory arthrogenic mediators (Harley and Suhrbier, 2013). RRV commonly causes arthralgia and may cause arthritis in multiple peripheral joints (polyarthritis), along with rash and systemic manifestations such as lethargy and myalgia (Harley et al., 2001, 2002; Mylonas et al., 2002). A significant minority of patients suffer postinfective fatigue for up to 12 months (Hickie et al., 2006). The major protective immune response is via virus-specific antibodies, probably neutralizing, and immunity is thought to be lifelong (Harley and Suhrbier, 2013).

In Australia, RRV is notified to the Commonwealth Department of Health and Ageing (National Notifiable Diseases Surveillance System) (Begg et al., 2008), most commonly on the basis of a single serological test positive for IgM (Harley et al., 2001). There were around 5000 notified cases per year in Australia in the period 1991–2000 (Harley et al., 2001). From July 1, 2006, to June 30, 2007, there were 3369 notifications in Australia; the mean for the 5 years (July 1, 2001, to June 30, 2006) before this was 3395 (Liu et al., 2008). Not all cases of disease are notified (Begg et al., 2008), and there is some evidence that not all those notified are truly cases of RRV disease, but are other infections including Barmah Forest virus disease (Harley et al., 2001; Lloyd et al., 2001; Mackenzie et al., 1993).

22.2 HISTORY OF MOSQUITO-BORNE EPIDEMIC POLYARTHRITIS OUTBREAKS IN AUSTRALIA AND THE PACIFIC

Outbreaks of seasonal "epidemic polyarthritis" in Australia were first recorded in the late nineteenth century and were only later associated with mosquito bites and pathogens such as RRV. The earliest known outbreak in Australia was thought to have been in 1886 in Victoria (although some authorities believe that this was a dengue outbreak) (Kelly-Hope et al., 2004; Lee et al., 1987). Further outbreaks have been documented throughout Australia since that time, with the number and size of recognized outbreaks increasing through time, presumably due in part to improved diagnostic tools and public health reporting services (Russell, 2002). Particularly large outbreaks have been recorded at times when large parts of Australia experienced well above-average rainfall (Kelly-Hope et al., 2004). A large outbreak of over 2000 notified cases occurred in Southeastern Australia, particularly in the states of Victoria and South Australia during 2011 (Australian Department of Health and Ageing, 2011).

Most recorded RRV disease cases occur in Australia. The virus is sometimes active in Papua New Guinea and the Solomon Islands (Russell, 2002; Scrimgeour et al., 1987), but there have been no recorded infections acquired in New Zealand (Kelly-Hope et al., 2002). This is despite the frequency of travel between Australia and New Zealand and the presence in the latter of at least one plausible vector species (the tree-hole and container-breeding *Aedes notoscriptus*). There has, however, been a significant outbreak in the Pacific Ocean island nations of Fiji, American Samoa, and the Cook Islands in 1979–1980 (Aaskov et al., 1981; Rosen et al., 1981; Tesh et al., 1981). This included approximately 500 000 infections in Fiji alone. On the basis of RRV disease in two tourists returning to Canada, it was speculated that the virus continues to circulate in Fiji (Klapsing et al., 2005).

22.3 RRV TRANSMISSION CYCLES HAVE A VARIETY OF ECOLOGIES

Ross River virus is primarily a zoonosis, with the virus maintained in reservoir hosts, mostly thought to be marsupial mammals, particularly macropodids (kangaroos and wallabies) (Russell and Dwyer, 2000). There is evidence that other marsupials such as brushtail possums (Boyd et al., 2001) and New Holland mice (Gard et al., 1973) and non-marsupials such as flying foxes (Harley et al., 2000; Ryan et al., 1997) may also serve as reservoirs in some circumstances.

For the most part, transmission to humans occurs when bitten by mosquitoes that have acquired infection from a reservoir host. Nonzoonotic (i.e., human–mosquito–human) transmission is possible and is thought to have occurred during the 1979–1980 outbreak in the South Pacific Islands (Aaskov et al., 1981; Kay and Aaskov, 1989) and at various times thereafter in the same region (Klapsing et al., 2005). It is plausible that non-zoonotic transmission occurs in large outbreaks in Australia as well (Harley et al., 2001).

Ross River virus transmission occurs over almost all of the Australian continent and offshore islands and hence occurs in a variety of ecological settings. RRV is maintained in diverse transmission cycles involving different combinations of vector and reservoir species (Kelly-Hope et al., 2004; Russell, 2002). Numerous mosquito species have been implicated in transmission.

For instance, RRV may be transmitted in coastal areas by mosquitoes breeding in brackish and saline marshland. Both the northern and southern saltmarsh mosquitoes (*Ae. vigilax* and *Aedes camptorhynchus*) are competent vectors of RRV and occur in high abundance in summer in coastal areas. In particular, *Ae. vigilax* is prolific in coastal samphire and mangrove marshland, breeding in saline and hypersaline pools created from tides. This is in contrast to *Ae. camptorhynchus*, which, although common in coastal areas, is better adapted to brackish water pools created from rainfall or groundwater. Further, the distribution of *Ae. camptorhynchus* is limited to Southern Australia and is sympatric in places with *Ae. vigilax*. *Ae. camptorhynchus* can also be found in inland areas affected by dryland salinity (Lindsay et al., 2007; Walker et al., 2009). So, even with ecologically similar saline-tolerant mosquito species, a variety of RRV ecologies occur. In areas with brackish wetlands in Eastern and Northern Australia, *Verrallina* species mosquitoes occur and are thought to play a role in early season amplification cycles in reservoir hosts—further evidence for the ecological complexity of RRV (Jeffery et al., 2006).

In inland areas, RRV is transmitted by a variety of mosquito species breeding in freshwater groundpools. Important vectors include *Culex sitiens* group species (in particular *Culex annulirostris*) breeding in floodwaters and irrigated areas (Russell, 1998). *Coquillettidia* species in more permanent water bodies may be important vectors. To add further complexity, *Aedes* species with desiccation-resistant eggs may undergo population explosions in response to rainfall in arid areas and also serve as potential cryptic sources of virus persistence in the absence of water and mosquito activity through transovarial virus transmission (Dhileepan et al., 1996).

Such diversity also extends to the abiotic environmental triggers, beyond mosquito abundance, that can lead to RRV epidemics. Despite the identification of above-average rainfall as a common association with RRV outbreaks (Kelly-Hope et al., 2004), numerous other additional triggers such as river height, tide height and/or frequency, temperature, reservoir host abundance, and age structure may be important in different locations.

22.4 TYPICAL ENVIRONMENTAL DETERMINANTS OF RRV ACTIVITY

Characterization of these varied "virus ecologies" has been done in order to develop management strategies that can be used to predict and manage RRV disease. Considerable research effort has gone toward the relation between RRV and abiotic variables including temperature, rainfall, and tides (Bi et al., 2009; Done et al., 2002; Gatton et al., 2005; Ryan et al., 1999; Tong and Hu, 2001, 2002; Tong et al., 2004, 2005; Williams et al., 2009a, b; Woodruff et al., 2002, 2006). These studies have been reviewed by Jacups et al. (2008) and Tong et al. (2008).

Rainfall is almost universally included in RRV predictive models (Jacups et al., 2008), with increased rainfall typically associated with increased disease rates. Positive associations between rainfall and RRV have been demonstrated Australia-wide (Jacups et al., 2008). Temperature increases are also commonly included in models as RRV predictors. Further, time-lagged variables of rainfall and temperature several months prior to RRV activity are included in several models, with plausible links to virus amplification cycles in nonhuman reservoir hosts (Jacups et al., 2008). For instance, higher rainfall totals in preceding months may enable strong early season mosquito breeding and virus amplification in reservoir hosts. This amplification may then heighten risk of RRV transmission to humans in subsequent months.

Tidal activity variables, associated with the abundance of coastal RRV vectors such as *Ae. vigilax*, have been included in some predictive models (e.g., Tong and Hu, 2002; Whelan et al., 1997). River height has also been included as a significant predictor of RRV activity for a regionally specific model for one section of the River Murray (Williams et al., 2009a, b).

Mosquito abundance is unsurprisingly a significant predictor of RRV activity (reviewed by Jacups et al., 2008) and has been shown to improve the accuracy of predictive models (Woodruff et al., 2006). The inclusion of mosquito abundance data provides a quantitative evaluation to support the intuitive link between the abundance of particular mosquito species and disease risk. However, there clearly is no simple link between mosquito abundance and RRV risk. Factors such as the extent of virus amplification in reservoir hosts in preceding months, daily survivorship of adult, host-seeking mosquitoes, and the extent of existing immunity in the human population are all important considerations.

Ecological models for small regions have greater accuracy and are more ecologically plausible than models for very large regions (Gatton et al., 2005; Jacups et al., 2008). Having to employ multiple models, each with its own particular input data requirements, may prove a barrier to the widespread application of such models for RRV forecasting. In one region of Australia (the River Murray Valley of South Australia), regular 3-month forecasts of RRV activity are made using regionally specific models, with forecasts then distributed to health authorities (e.g., Rau and Williams, 2012). Rainfall cut points for outbreaks have been identified for a number of Northern Territory centers (Jacups et al., 2011).

22.5 SOCIAL DETERMINANTS OF RRV DISEASE ACTIVITY

Important social predictors for Australian vector-borne diseases have rarely been identified. Nonetheless, it is well understood from incidence mapping by statistical local area (e.g., Liu et al., 2008) that RRV incidence is highest in rural and regional areas. The relatively lower incidence of RRV in large, urbanized centers is most likely related to the altered ecology of cities, in which the natural ecology of RRV is disturbed.

In general terms, there is no clear sex effect for RRV infection rates, with a variety of ratios reported (Harley et al., 2001; Liu et al., 2008). Sex ratios for RRV vary between locations and can be related to factors as varied as workforce composition in regions of high risk, to the deployment of serologically naïve foreign military personnel for exercises in Australia (Harley et al., 2001). Similarly, although RRV is most common in those aged 25–49 (Harley et al., 2001; Liu et al., 2008), the age class most at risk in a particular region will depend on other social or employment factors.

Lower levels of income and education are associated with reduced knowledge about disease transmission and protective measures against mosquito bites (Hu et al., 2007). Protective measures have been demonstrated as effective against RRV disease (Harley et al., 2005). Despite this, socioeconomic condition of a particular locality was not found to be predictive of RRV activity in Queensland, whereas ecological factors (such as rainfall) were (Hu et al., 2010). These findings were based on analyses at the population level and do not necessarily translate to predicting RRV risk for individuals.

Certain human behaviors such as the use of insect repellents are known to decrease the risk of RRV disease, while camping greatly increases risk (Harley et al., 2005). For any individual, behaviors and practices that permit increased exposure to biting mosquitoes in areas where likely vertebrate hosts (such as kangaroos and wallabies) are abundant will likely increase RRV risk.

22.6 A CONCEPTUAL FRAMEWORK FOR UNDERSTANDING THE INFLUENCE OF ENVIRONMENTAL AND SOCIAL FACTORS ON RRV DISEASE ACTIVITY

Because of the likely climate sensitivity of RRV (McMichael and Woodruff, 2008) and the volume of research on environmental and behavioral determinants for the disease (e.g., Bi et al., 2009; Gatton et al., 2005; Ryan et al., 1999; Tong and Hu, 2002; Tong et al., 2005; Weinstein, 1997; Woodruff et al., 2006), we propose a hierarchical conceptual framework

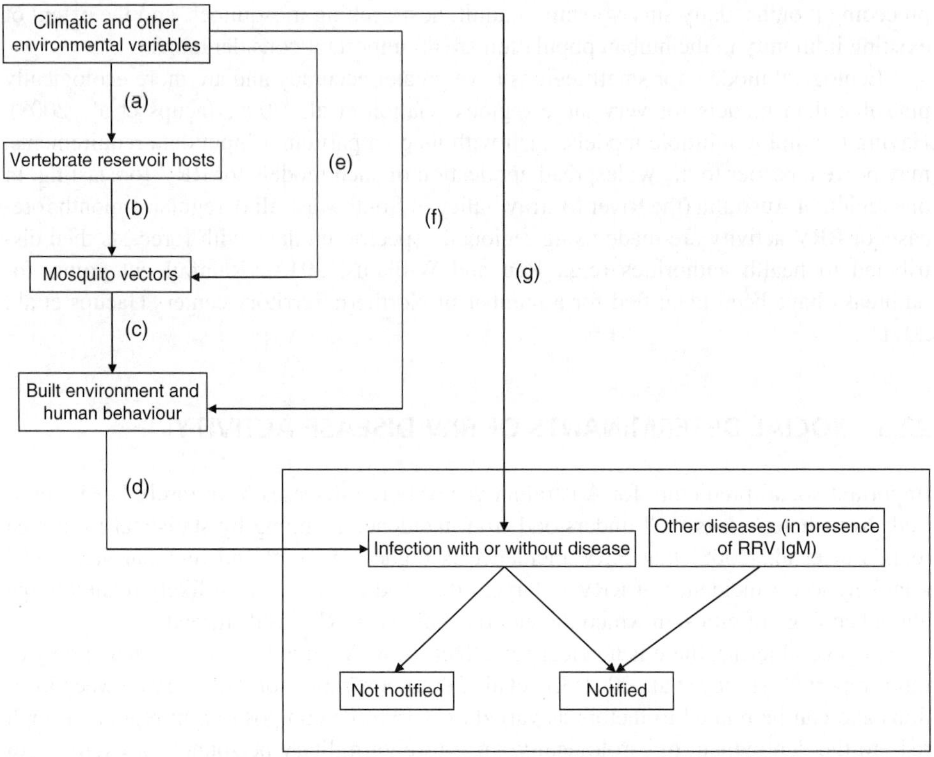

Figure 22.1. A conceptual model for understanding the determinants of Ross River virus infection and disease risk.

(Victora et al., 1997) for understanding the relation between distal (such as temperature) and more proximal (such as the presence of vector mosquitoes around dwellings and the use of repellents) determinants for RRV disease risk. Findings from two studies (Ryan et al., 1999; Woodruff et al., 2006) that include climatic variables, as well as vector data, as determinants for RRV disease risk are discussed in light of the model, and the implications for studies including only climate and disease data are discussed.

The following conceptual model provides a framework for understanding determinants for RRV infection and disease risk (Figure 22.1).

Risk and protective factors for disease are placed in four categories. From distal to proximal these are:

- Climatic and other environmental variables;
- Vertebrate reservoir hosts;
- Mosquito vectors;
- Built environment and human behavior.

The first of these (*climatic and other environmental variables*) includes tide heights, rainfall, temperature and humidity, environmental modification (e.g., runnelling (Hulsman et al., 1989), land clearing, and other changes in the landscape), and control measures such as larviciding and adulticiding. The second (*vertebrate reservoir hosts*) includes all aspects of vertebrate host biology and ecology that alter human risk for RRV infection such as host distribution, abundance, and prevalence of protective immunity. *Mosquito vectors* encompasses all aspects of vector ecology and biology that impact on the likelihood of transmission of RRV, including distribution, blood meal preference, and prevalence of RRV carriage in the vectors. *Built environment and human behavior* includes all aspects of behavior and the built environment that could influence the likelihood of RRV infection. The former includes clothing color, use of protective measures against mosquitoes, and leisure activities such as camping, all known to influence risk (Harley et al., 2005). The latter includes aspects of human environments such as office buildings and homes, including air-conditioning and insect screens.

The remaining elements of the model, enclosed by a box, provide a basis for understanding a significant, and thus far unrecognized, threat to the validity of studies of RRV risk. All studies linking RRV with climate have relied on routinely collected surveillance data. These data represent a subset of people infected with RRV and subsequently notified to health authorities, with a proportion of cases of other diseases misclassified as RRV due to the presence of RRV IgM, detected either incidentally or because a person has symptoms consistent with RRV disease, those caused by parvovirus B19 infection, for example (Harley et al., 2001; Mackenzie et al., 1993).

The most distal determinant for infection (*climatic and other environmental variables*) influences risk both via intermediates (pathways a–f) and directly (pathway g). The next two proximal determinants (*vertebrate reservoir hosts* and *mosquito vectors*) determine infection risk via those determinants proximal to them (via pathways b, c, and d).

22.6.1 Climatic and other variables: pathway a

Rainfall increases the abundance of grass, food for kangaroos and wallabies, and hence increases breeding with consequent greater entry of immunologically naïve animals to the population (McMichael and Woodruff, 2008). Presumably, other vertebrate hosts of importance are also affected by climate (Harley et al., 2001).

22.6.2 Vertebrate host reservoirs: pathway b

Vertebrate host factors influence the likelihood of mosquito infection and the ecology of mosquito vectors. In relation to the former, the determinants for importance as an arbovirus reservoir host are well described and include aspects of ecology and reproduction, attractiveness and accessibility to vectors, and prevalence of immunity to infection (Harley et al., 2001; Scott, 1988). Plausible evidence that at least some of these mechanisms determine human RRV disease risk exists; it is hypothesized that low spring rainfall reduces mosquito numbers and hence enzootic transmission; therefore, the prevalence of protective immunity in kangaroos decreases, leading to increased risk in subsequent years for transmission to humans (McMichael and Woodruff, 2008), and empirical data from Southeastern Australia support these hypothesized mechanisms (Woodruff et al., 2002).

In many mosquito species a blood meal, sometimes two, is required for ovarian development (Clements, 1992). The likelihood of a female mosquito obtaining one or more blood meals is determined by animal behavior and abundance. In Vietnam the abundance of *Culex gelidus*, from which RRV has been isolated in North Queensland (Harley et al., 2000), has been found to be influenced by host abundance (Hasegawa et al., 2008). *A priori* it is biologically plausible that vector numbers and fitness will be influenced by availability of vertebrates, and consequently of blood meals, and there is empirical evidence that vector numbers are influenced by vertebrate numbers.

22.6.3 Mosquito vectors: pathway c

Clearly, barriers exist between a vector mosquito with virus-infected salivary glands and transmission with subsequent infection of human host. These include protective behaviors, for example, wearing light-colored clothing and using mosquito protective measures (Harley, 2005). Features of housing design, air-conditioning, for example, are also included; while this is not demonstrated to be significantly protective for RRV (Harley et al., 2005), it is protective against American encephalitic arboviruses (Gahlinger et al., 1986). It is also biologically plausible that the use of air-conditioning should protect against RRV infection. There is some evidence, and it is logical conceptually, to envisage mediation of the risk determined by vector factors (including number, proportion RRV infected, and degree of anthropophagy) via determinants such as those outlined earlier.

22.6.4 Human behavior and the built environment: pathway d

Human behavior and the built environment are considered proximal to mosquito vectors, because they play a mediating role between an infected vector mosquito and the human host, in a way that vertebrate reservoir hosts, for example, do not. Other than nonmodifiable characters of the human host influencing attractiveness to vector mosquitoes, for example, factors included in this category are the last defense against the bite of an infected mosquito.

22.6.5 Climatic influences on mosquitoes: pathway e

Climate influences mosquito distribution and abundance through breeding sites (rainfall leading to pooling of water in grassy areas provides breeding sites for *Cx. annulirostris*, an important vector (Dale and Morris, 1996; Harley et al., 2001, for example)) and mortality effects (decreased humidity leads to increased mosquito mortality (Clements and Paterson, 1981)). Increased temperature may increase larval development rate for saltmarsh species such as the important RRV vector *Ae. vigilax*, permitting greater population sizes to emerge prior to the next tidal inundation of habitat (Kokkinn et al., 2009; Williams et al., 2009a, b). Increased temperature may decrease the extrinsic incubation period (EIP; the length of time from ingestion of an infected blood meal to infection of salivary glands by RRV) as for other arboviruses (Shope, 1991), although work with *Ae. vigilax*, an important RRV vector, has suggested that, at least from 18°–32°C, this may not be biologically important (Kay and Jennings, 2002).

22.6.6 Climatic influences on housing and human behavior: pathway f

Housing design is responsive to climate (O'Brien and Hes, 2008; Sadafi et al., 2008; Su and Aynsley, 2008). Aspects of climate influence human behavior; for example, rain and temperature will have a significant effect on the likelihood that people will go camping, a known risk factor for RRV disease (Harley et al., 2005).

22.6.7 Climatic influences on immune function: pathway g

Because climate and environment are known to influence human immune function, the possibility exists that the likelihood of a person infected with RRV developing symptoms (and consequently attending a doctor and being notified to health authorities) will be influenced by the environment in which they live and by future changes in the environment. As a specific example, chemical environmental contaminants may impair immune function (Crisp et al., 1998).

22.7 CLIMATE CHANGE AND RRV

Climate change is a major threat to human health (Butler and Harley, 2010; McMichael et al., 2006), particularly because of its potential to influence the incidence of infectious diseases (Harley and McMichael, 2008; McMichael and Woodruff, 2008). Many of the more dire predictions of the effects of climate change on health relate to vector-borne diseases, which is understandable given the strong links between weather, climate, and vector-borne disease ecology described earlier. For other vector-borne diseases, globally increased range has been predicted for dengue (Hales et al., 2002), and an increased duration of transmission season (Tanser et al., 2003) and increased global disease burden (Patz et al., 2005) have been predicted for malaria under climate change. The most convincing instance of climate change influencing the distribution of vector-borne disease relates to the veterinary arbovirus bluetongue (Purse et al., 2005). There is, however, considerable debate and some skepticism regarding the likely future impact of climate change on vector-borne diseases (Hay et al., 2002; Reiter, 1998, 2001; Reiter et al., 2004; Rogers and Randolph, 2000, 2006; Russell, 2009). In the case of RRV, it is likely that changes to climate will have an impact on transmission. However, one may infer from the diversity of ecological circumstances leading to RRV outbreaks (as described in this chapter) that the impact of climate change on transmission is unlikely to be uniform and thereby difficult to predict and generalize about. Nonetheless, alterations to the seasonality and distribution of high-risk regions for RRV are possible as the climate warms and rainfall patterns change (Harley et al., 2011).

Furthermore, it is possible that RRV activity may decrease in areas experiencing reduced rainfall and higher temperatures, as both factors will lead to decreased availability of mosquito breeding habitat and higher mortality rates for adult mosquitoes. These proposed negative impacts of climate change on mosquito populations may, however, be countered by increased transmission caused by shorter extrinsic incubation periods for RRV in mosquitoes, due to higher temperatures. In short, predictions about the impact of climate change on RRV activity need to make use of the many region-specific epidemiological models described earlier.

22.8 CONCLUSION

Ross River virus is a native Australian zoonotic arbovirus that causes disease in humans in diverse ecological circumstances. The many attempts to develop a better understanding of RRV ecology and epidemiology have led to the identification of above-average rainfall, together with high mosquito abundance, as powerful predictors for outbreaks in humans. However, this is an oversimplification that is not particularly useful in predicting RRV activity in particular regions with enough certainty to warrant public health interventions. The most accurate predictive models are those based on smaller regional scales. While the utility of such region-specific models is reduced by the data required to generate predictions, health authorities should when possible use such tools to plan RRV management.

ACKNOWLEDGEMENT

The South Australian Department of Health provided project funding for the development of RRV predictive tools. We are grateful to Adrian Sleigh and Vicky Ng for their conversation and comments on RRV ecology.

REFERENCES

Aaskov, J.G., Mataika, J.U., Lawrence, G.W., et al. (1981). An epidemic of Ross River virus infection in Fiji, 1979. *Am J Trop Med Hyg* 30: 1053–1059.

Australian Department of Health and Ageing. (2011). National Notifiable Diseases Surveillance System. Available from URL: http://www.health.gov.au/internet/main/publishing.nsf/content/cda-surveil-nndss-nndssintro.htm (accessed 20 December 2011).

Begg, K., Roche, P., Owen, R., et al. (2008). Australia's notifiable diseases status, 2006: annual report of the National Notifiable Diseases Surveillance System. *Commun Dis Intell* 32(2): 139–207.

Bi, P., Hiller, J.E., Cameron, A.S., et al. (2009). Climate variability and Ross River virus infections in Riverland, South Australia, 1992–2004. *Epidemiol. Infect.* 137: 1486–1493.

Boyd, A.M., Hall, R.A., Gemmell, R.T., et al. (2001). Experimental infection of Australian brushtail possums, *Trichosurus vulpecula* (Phalangeridae: Marsupialia), with Ross River and Barmah Forest viruses by use of a natural mosquito vector system. *Am J Trop Med Hyg* 65: 777–782.

Butler, C.D. and Harley, D. (2010). Primary, secondary and tertiary effects of eco-climatic change: the medical response. *Postgrad Med J* 86(1014): 230–234.

Clements, A.N. (1992). The Biology of Mosquitoes: Development, Nutrition and Reproduction. Chapman and Hall, London.

Clements, A.N. and Paterson, G.D. (1981). The analysis of mortality and survival rates in wild populations of mosquitoes. *J Appl Ecol* 18(2): 373–399.

Crisp, T.M., Clegg, E.D., Cooper, R.L., et al. (1998). Environmental endocrine disruption: an effects assessment and analysis. *Environ Health Perspect* 106(Suppl. 1): 11–56.

Dale, P.E.R. and Morris, C.D. (1996). *Culex annulirostris* breeding sites in urban areas: using remote sensing and digital image analysis to develop a rapid predictor of potential breeding areas. *J Am Mosq Control Assoc* 12(2): 316–320.

Dhileepan, K., Azuolas, J.K., and Gibson, C.A. (1996). Evidence of vertical transmission of Ross River and Sindbis viruses (Togaviridae: Alphavirus) by mosquitoes (Diptera: Culicidae) in southeastern Australia. *J Med Entomol* 33: 180–182.

Doherty, R.L., Standfast, H.A., Wetters, E.J., et al. (1968). Virus isolation and studies of arthropod-borne virus infections in a high rainfall area of North Queensland. *Trans R Soc Trop Med Hyg* 62: 862–867.

Done, S., Holbrook, N., and Beggs, P. (2002). The Quasi-Biennial Oscillation and Ross River virus incidence in Queensland, Australia. *Int J Biometeorol* 46(4): 202–207.

Economopoulou, A., Dominguez, M., Helynck, B., et al. (2009). Atypical Chikungunya virus infections: clinical manifestations, mortality and risk factors for severe disease during the 2005–2006 outbreak on Reunion. *Epidemiol Infect* 137(4): 534–541.

Gahlinger, P.M., Reeves, W.C., and Milby, M.M. (1986). Air conditioning and television as protective factors in arboviral encephalitis risk. *Am J Trop Med Hyg* 35(3): 601–610.

Gard, G., Marshall, I.D., and Woodroofe, G.M. (1973). Annually recurrent epidemic poly-arthritis and Ross River virus activity in a coastal area of New South Wales. II. Mosquitoes, viruses and wildlife. *Am J Trop Med Hyg* 22: 551–560.

Gatton, M., Kay, B., and Ryan, P. (2005). Environmental predictors of Ross River virus disease outbreaks in Queensland, Australia. *Am J Trop Med Hyg* 72(6): 792–799.

Gerardin, P., Barau, G., Michault, A., et al. (2008). Multidisciplinary prospective study of mother-to-child chikungunya virus infections on the island of La Reunion. *PLoS Med* 5(3): 413–422.

Hales, S., de Wet, N., Maindonald, J., et al. (2002). Potential effect of population and climate changes on global distribution of dengue fever: an empirical model. *Lancet* 360(9336): 830–834.

Harley, D., Ritchie, S., Bain, C., et al. (2005). Risks for Ross River virus disease in tropical Australia. *Int J of Epidemiol* 34(3): 548–555.

Harley, D. and McMichael, A. (2008). Global climate change and infectious diseases: paradigms, impacts and future challenges. *Infect Chemother* 40(Suppl. 2): S136–S143.

Harley, D. and Suhrbier, A. (2013) Ross River virus disease. In: Hunter's Tropical Medicine and Emerging Infectious Diseases, 9th Edition (eds. A.J. Magill, E.T. Ryan, T. Solomon, et al.). Elsevier, London.

Harley, D., Ritchie, S., Phillips, D., et al. (2000). Mosquito isolates of Ross River virus from Cairns, Queensland, Australia. *Am J Trop Med Hyg* 62(5): 561–565.

Harley, D., Sleigh, A., and Ritchie, S. (2001). Ross River virus transmission, infection and disease: a cross-disciplinary review. *Clin Microbiol Rev* 14(4): 909–932.

Harley, D., Bossingham, D., Purdie, D.M., et al. (2002). Ross River virus disease in tropical Queensland: evolution of rheumatic manifestations in an inception cohort followed for six months. *Med J Aust* 177 (7): 352–355.

Harley, D., Ritchie, S., Bain, C., et al. (2005). Risks for Ross River virus disease in tropical Australia. *Int J Epidemiol* 34: 548–555.

Harley, D., Bi, P., Hall, G., et al. (2011). Climate change and infectious diseases in Australia: future prospects, adaptation options and research priorities. *Asia Pac J Public Health* 23(Suppl. 2), 54S–66S.

Hasegawa, M., Tuno, N., Yen, N.T., et al. (2008). Influence of the distribution of host species on adult abundance of Japanese encephalitis vectors *Culex vishnui* subgroup and *Culex gelidus* in a rice-cultivating village in Northern Vietnam. *Am J Trop Med Hyg* 78(1): 159–168.

Hay, S.I., Cox, J., Rogers, D.J., et al. (2002). Climate change and the resurgence of malaria in the East African highlands. *Nature* 415: 905–909.

Hickie, I., Davenport, T., Wakefield, D., et al. (2006). Post-infective and chronic fatigue syndromes precipitated by viral and non-viral pathogens: prospective cohort study. *BMJ* 333(7568): 575.

Hu, W., Tong, S., Mengersen, K., et al. (2007). Exploratory spatial analysis of social and environmental factors associated with the incidence of Ross River virus in Brisbane, Australia. *Am J Trop Med Hyg* 76: 814–819.

Hu, W., Clements, A., Williams, G., et al. (2010). Bayesian spatiotemporal analysis of socio-ecologic drivers of Ross River virus transmission in Queensland, Australia. *Am J Trop Med Hyg* 83: 722–728.

Hulsman, K., Dale, P.E., and Kay, B.H. (1989). The runnelling method of habitat modification: an environment-focused tool for salt marsh mosquito management. *J Am Mosq Control Assoc* 5(2): 226–234.

Jacups, S.P., Whelan, P.I., and Currie, B.J. (2008). Ross River virus and Barmah Forest virus infections: a review of history, ecology, and predictive models, with implications for tropical northern Australia. *Vector Borne Zoonotic Dis* 8(2): 283–298.

Jacups, S.P., Whelan, P.I., and Harley, D.O. (2011) Arbovirus models to provide practical management tools for mosquito control and disease prevention in the Northern Territory, Australia. *J Med Entomol* 48(2): 453–460.

Jeffery, J.A.L., Kay, B.H., and Ryan, P.A. (2006). Role of *Verralina funerea* (Diptera: Culicidae) in transmission of Barmah Forest virus and Ross River virus in coastal areas of eastern Australia. *J Med Entomol* 43: 1239–1247.

Kay, B. H., and Aaskov, J. O. (1989). Ross River virus (epidemic poly arthritis). In: The Arboviruses: Epidemiology and Ecology(ed. T.P. Monath). Vol.4. Boca Raton, FL: CRC Press, Inc.

Kay, B.H. and Jennings, C.D. (2002). Enhancement or modulation of the vector competence of *Ochlerotatus vigilax* (Diptera: Culicidae) for Ross River virus by temperature. *J Med Entomol* 39(1): 99–105.

Kelly-Hope, L., Kay, B., and Purdie, D. (2002). The risk of Ross River and Barmah Forest virus disease in Queensland: implications for New Zealand. *Aust N Z J Public Health* 26: 69–77.

Kelly-Hope, L., Purdie, D., and Kay, B.H. (2004). Ross river virus disease in Australia, 1886–1998, with analysis of risk factors associated with outbreaks. *J Med Entomol* 41: 133–140.

Klapsing, P., MacLean, J.D., Glaze, S., et al. (2005). Ross River virus disease reemergence, Fiji, 2003–2004. *Emerg Infect Dis* 11: 613–615.

Kokkinn, M.J., Duval, D.J., and Williams, C.R. (2009). Modelling the ecology of the coastal mosquitoes *Aedes vigilax* and *Aedes camptorhynchus* at Port Pirie, South Australia. *Med Vet Entomol* 23(1): 85–91.

Lee, D.J., Hicks, M.M., Griffiths, M., et al. (1987). The Culicidae of the Australasian Region. Vol. 4, Entomology Monograph No. 2 (ed. M.L. Debenham). Australian Government Publishing Services, Canberra.

Lindsay, M.D.A., Jardine, A., Johansen, C.A., et al. (2007). Mosquito (Diptera: Culicidae) fauna in inland areas of south-west Western Australia. *Aust J Entomol* 46: 60–64.

Liu, C., Begg, K., Johansen, C., et al. (2008). Communicable Diseases Network Australia National Arbovirus and Malaria Advisory Committee annual report, 2006–07. *Commun Dis Intell* 32(1): 31–47.

Lloyd, A., Hickie, I., Vollmer-Conna, U., et al. (2001). Characterisation of prolonged ill health following Ross River virus infection: a prospective cohort study. *Arbo Res Aust* 8: 236.

Mackenzie, J.S., Broom, A.K., Calisher, C.H., et al. (1993). Diagnosis and reporting of arbovirus infections in Australia. *Arbo Res Aust* 6: 89–93.

McMichael, A.J. and Woodruff, R.E. (2008). Climate change and infectious diseases. In: The Social Ecology of Infectious Diseases (eds. K.H. Mayer and H.F. Pizer). Elsevier, Amsterdam.

McMichael, A.J., Woodruff, R.E., and Hales, S. (2006). Climate change and human health: present and future risks. *Lancet* 367(9513): 859–869.

Mylonas, A.D., Brown, A.M., Carthew, T.L., et al. (2002). Natural history of Ross River virus-induced epidemic polyarthritis. *Med J Aust* 177(7): 356–360.

O'Brien, D. and Hes, D. (2008). The third way: developing low environmental impact housing prototypes for hot/humid climates. *Int J Housing Sci Appl* 32(4): 311–322.

Patz, J.A., Campbell-Lendrum, D., Holloway, T., et al. (2005). Impact of regional climate change on human health. *Nature* 438(7066): 310–317.

Powers, A.M., Brault, A.C., Shirako, Y., et al. (2001). Evolutionary relationships and systematics of the alphaviruses. *J Virol* 75(21): 10118–10131.

Purse, B.V., Mellor, P.S., Rogers, D.J., et al. (2005). Climate change and the recent emergence of bluetongue in Europe. *Nat Rev Microbiol* 3(2): 171–181.

Rau, G. and Williams, C.R. (2012). South Australian Mosquito Intelligence Report No. 19. Available from URL: http://www.unisa.edu.au/sansominstitute/mosquito/ (accessed 27 April 2012).

Reiter, P. (1998). Global-warming and vector-borne disease in temperate regions and at high altitude. *Lancet* 351: 839–840.

Reiter, P. (2001). Climate change and mosquito-borne disease. *Environ Health Perspect* 109(Suppl. 1): 141–161.

Reiter, P., Thomas, C.J., Atkinson, P.M., et al. (2004). Global warming and malaria: a call for accuracy. *Lancet Infect Dis* 4(6): 323–324.

Renault, P., Solet, J.-L., Sissoko, D., et al. (2007). A major epidemic of chikungunya virus infection on Reunion Island, France, 2005–2006. *Am J Trop Med Hyg* 77(4): 727–731.

Rezza, G., Nicoletti, L., Angelini, R., et al. (2007). Infection with chikungunya virus in Italy: an outbreak in a temperate region. *Lancet* 370(9602): 1840–1846.

Rogers, D.J. and Randolph, S.E. (2000). The global spread of malaria in a future, warmer world. *Science* 289: 1763–1766.

Rogers, D.J. and Randolph, S.E. (2006). Climate change and vector-borne diseases. *Adv Parasitol* 62: 345–384.

Rosen, L., Gubler, D.J., and Bennett, P.H. (1981) Epidemic polyarthritis (Ross River) virus infection in the Cook Islands. *Am J Trop Med Hyg* 30: 1294–1302.

Russell, R.C. (1998). Mosquito-borne arboviruses in Australia: the current scene and implications of climate change for human health. *Int J Parasitol* 28: 955–969.

Russell, R.C. (2002). Ross River virus: ecology and distribution. *Ann Rev Entomol* 47: 1–31.

Russell, R.C. (2009). Mosquito-borne disease and climate change in Australia: time for a reality check. *Aust J Entomol* 48: 1–7.

Russell, R.C. and Dwyer, D.E. (2000). Arboviruses associated with human disease in Australia. *Microbes Infect* 2: 1693–1704.

Ryan, P.A., Martin, L., Mackenzie, J.S., et al. (1997). Investigation of gray-headed flying foxes, *Pteropus poliocephalus* (Megachiroptera: Pteropodidae) and mosquitoes in the ecology of Ross River virus in Australia. *Am J Trop Med Hyg* 57: 476–482.

Ryan, P.A., Do, K.A., and Kay, B.H. (1999). Spatial and temporal analysis of Ross River virus disease patterns at Maroochy Shire, Australia: association between human morbidity and mosquito (Diptera: Culicidae) abundance. *J Med Entomol* 36(4): 515–521.

Sadafi, N., Salleh, E., Haw, L.C., et al. (2008). Potential thermal impacts of internal courtyard in terrace houses: a case study in tropical climate. *J Appl Sci* 8(15): 2770–2775.

Scott, T.W. (1988). Vertebrate host ecology. In: The Arboviruses: Epidemiology and Ecology (ed. T.P. Monath). CRC Press, Boca Raton, FL.

Shope, R. (1991). Global climate change and infectious diseases. *Environ Health Perspect* 96: 171–174.

Su, B. and Aynsley, R. (2008). Roof thermal design for naturally ventilated houses in a hot humid climate. *Int J Ventil* 7(4): 369–378.

Tanser, F.C., Sharp, B., and le Sueur, D. (2003). Potential effect of climate change on malaria transmission in Africa. *Lancet* 362(9398): 1792–1798.

Tesh, R.B., McLean, R.G., Shroyer, D.A., et al. (1981). Ross River virus (Togaviridae: Alphavirus) infection (epidemic polyarthritis) in American Samoa. *Trans R Soc Trop Med Hyg* 75: 426–431.

Tong, S. and Hu, W. (2001). Climate variation and incidence of Ross River virus in Cairns, Australia: a time-series analysis. *Environ Health Perspect* 109(12): 1271–1273.

Tong, S. and Hu, W. (2002). Different responses of Ross River virus to climate variability between coastline and inland cities in Queensland, Australia. *Occup Environ Med* 59(11): 739–744.

Tong, S., Hu, W., and McMichael, A.J. (2004). Climate variability and Ross River virus transmission in Townsville Region, Australia, 1985–1996. *Trop Med Int Health* 9(2): 298–304.

Tong, S., Hu, W., Nicholls, N., et al. (2005). Climatic, high tide and vector variables and the transmission of Ross River virus. *Intern Med J* 35(11): 677–680.

Tong, S., Dale, P., Nicholls, N., et al. (2008). Climate variability, social and environmental factors, and Ross River virus transmission: research development and future research needs. *Environ Health Perspect* 116(12): 1591–1597.

Victora, C.G., Huttly, S.R., Fuchs, S.C., et al.(1997). The role of conceptual frameworks in epidemiological analysis: a hierarchical approach. *Int J Epidemiol* 26(1): 224–227.

Walker, K.F., Madden, C.P., Williams, C.R., et al. (2009). Freshwater invertebrates. In: Natural History of the Riverland and Murraylands (ed. J.T. Jennings). Royal Society of South Australia, Adelaide.

Weinstein, P. (1997). An ecological approach to public health intervention: Ross River virus in Australia. *Environ Health Perspect* 105(4): 364–366.

Whelan, P., Merianos, A., Hayes, G., et al. (1997). Ross River virus transmission in Darwin Northern Territory, Australia. *Arbo Res Aust* 7: 337–345.

Williams, C.R., Fricker, S.R., and Kokkinn, M.J. (2009a). Environmental and entomological factors determining Ross River virus activity in the River Murray Valley of South Australia. *Aust N Z J Public Health* 33(3): 284–288.

Williams, C.R., Williams, S.R., Nicholson, J., et al. (2009b). Diversity and seasonal succession of coastal mosquitoes (Diptera: Culicidae) in the northern Adelaide region of South Australia. *Aust J Entomol* 48: 107–112.

Woodruff, R., Guest, C., Garner, M., et al. (2002). Predicting Ross River virus epidemics from regional weather data. *Epidemiology* 13(4): 384–393.

Woodruff, R., Guest, C., Garner, M., et al. (2006). Early warning of Ross River virus epidemics: combining surveillance data on climate and mosquitoes. *Epidemiology* 17(5): 569–575.

INFECTION PATTERNS AND EMERGENCE OF O'NYONG-NYONG VIRUS

Ann M. Powers

Centers for Disease Control and Prevention, Fort Collins, CO, USA

TABLE OF CONTENTS

23.1	Introduction	433
23.2	History of outbreaks	434
23.3	Clinical manifestations	435
23.4	Epidemiology	435
23.5	Factors affecting emergence	437
	23.5.1 Etiologic agent: *viral genomics and antigens encoded*	438
	23.5.2 Transmission parameters	439
	23.5.3 Zoonotic maintenance	439
	23.5.4 Environmental influences	440
23.6	Conclusion	440
	References	441

23.1 INTRODUCTION

Emergence of a novel pathogen or reemergence of a known pathogen is dependent upon numerous factors ranging from specific characteristics of the agent to environmental conditions. Importantly, it is the interaction of these factors that determines whether an emergence event will occur or if the pathogen remains in an undetected condition causing no known human or animal infections. In the specific case of vector-borne pathogens, the

Viral Infections and Global Change, First Edition. Edited by Sunit K. Singh.
© 2014 John Wiley & Sons, Inc. Published 2014 by John Wiley & Sons, Inc.

cycle is further complicated by the presence of dual life cycles: one involving the vertebrate reservoir host and the second the infection of and transmission by an invertebrate vector. Due to multiple changes in the ecology and pathogen characteristics of many vector-borne zoonotic cycles, pathogens are emerging according to unexpected patterns. Identifying where and when these pathogens will emerge remains a significant scientific challenge.

23.2 HISTORY OF OUTBREAKS

One example of a vector-borne pathogen that has been responsible for outbreaks of tremendous scope is o'nyong-nyong virus (ONNV) (Table 23.1). The virus was first identified in 1959 when one of the largest arboviral epidemics ever documented began in the Acholi region of Uganda and resulted in over two million cases as it swept across central Africa east into Kenya and Tanzania, south to Mozambique, and west as far as Senegal. During the course of the outbreak, it was estimated that over 50% of the population in the affected areas was infected (Johnson, 1988; Williams et al., 1965b; Williams and Woodall, 1961). The single outbreak lasted over 3 years with little explanation for the termination of the epidemic (Bowen et al., 1973; Haddow et al., 1960; Johnson, 1988). It was speculated that the supply of susceptible individuals simply was depleted, but documentation of the cause of cessation does not exist.

TABLE 23.1. History of ONNV Activity

Year(s)	Location	Event	Magnitude	Reference
1959–1962	Uganda, Kenya, Tanzania, Mozambique	Human outbreak	Est. ~two million cases	Williams et al. (1965b)
~1960–1969	Kenya	Serosurvey (antibody detection)	~1500 seropositive individuals	Bowen et al. (1973); Marshall et al. (1982)
1966	Nigeria	Virus isolation (Igbo-Ora strain) from febrile human	Two individuals	Moore et al. (1975)
1967	Central African Republic	Serosurveys, febrile humans	Hundreds seropositive	Chippaux and Chippaux-Hyppolite (1968)
1969	Nigeria	Virus isolation (Igbo-Ora strain) from febrile human	Single individual	Moore et al. (1975)
1974–1975	Ghana, Nigeria (Sierra Leone)	Seropositive travelers	Four individuals	Woodruff et al. (1978)
1978	Kenya	Isolation from *An. funestus* mosquito pool	Single isolate	Johnson (1988)
1985	Cote d'Ivoire	Febrile humans	~33 cases	Lhuillier et al. (1988)
1996–1997	Uganda	Human outbreak	Unknown	Sanders et al. (1999)
2003	Cote d'Ivoire	Human outbreak	~30 cases	Posey et al. (2005)
2004	Chad	Virus isolation from febrile human	Single case	Bessaud et al. (2006)

Between 1962 and 1996, there was no significant epidemic ONNV activity reported. The only evidence that ONNV had not gone extinct was the occasional report of human infection, virus isolation from a mosquito in Kenya, and several serosurveys. Interestingly, some of these infections were determined to be caused by Igbo-Ora virus, a serological subtype of ONNV never associated with a large outbreak (Henderson et al., 1970; Lhuillier et al., 1988; Marshall et al., 1982; Moore et al., 1975; Olaleye et al., 1989, 1990; Rodhain et al., 1989; Woodruff et al., 1978).

In 1996, the second major documented ONNV epidemic occurred in the Rakai District of southern Uganda (Kiwanuka et al., 1999; Lanciotti et al., 1998; Rwaguma et al., 1997; Sanders et al., 1999). This outbreak was focused in the rural south-central regions of Uganda extending primarily into lake and swamp areas. While the number of cases documented during this outbreak was orders of magnitude lower than the 1959–1962 outbreak, it was still so extensive that the infection rate in affected areas was estimated to be nearly 70% (Sanders et al., 1999).

One other outbreak of febrile illness found to be caused by ONNV was reported in 2003 among Liberian refugees in western Cote d'Ivoire (Posey et al., 2005). The number of cases reported in this outbreak was quite small, but due to the unstable political environment, the actual scope of the outbreak could not be investigated. However, of the 8000 refuges awaiting resettlement, fewer than 50 were ill and able to be surveyed.

Curiously, the virus most closely related to ONNV, chikungunya virus (CHIKV), exhibits outbreak patterns that are quite distinct from that of ONNV. While there have been only a small handful of outbreaks of ONNV (with one of these being exceptionally large), CHIKV causes more frequent and geographically disbursed outbreaks. This likely reflects differences in the transmission patterns and will be addressed later.

23.3 CLINICAL MANIFESTATIONS

A classically reported triad of signs for ONNV infection includes fever, arthralgia, and rash (which is noted for being irritating). One other prominent indicator of ONNV disease includes posterior cervical lymphadenopathy, which is virtually never reported in infections due to the closely related CHIKV (Kiwanuka et al., 1999; Shore, 1961). The fever that is documented with ONNV infection also differs from CHIKV in that it usually ranges from 99 to 101 °F and lasts for only 2–3 days (in contrast to CHIKV fever, which is routinely greater than 100.5 °F and may last for up to a week). Furthermore, no biphasic fevers have been reported for ONNV. The overall clinical illness is otherwise remarkably similar to closely related viruses such as CHIKV and Ross River virus (RRV). Some additional symptoms that are frequently noted with ONNV infection include conjunctivitis, photophobia, lumbar back pain, and headache. The presence of the lymphadenopathy in combination with fever and polyarthralgia seems to provide the best combination of both specificity and sensitivity in clinical diagnosis (Kiwanuka et al., 1999).

23.4 EPIDEMIOLOGY

Geographically, ONNV is one of the most restricted alphaviruses with its range limited to select portions of Africa (Figure 23.1). The virus has only been identified in Uganda, Kenya, Tanzania, Malawi, Mozambique, Cameroon, Chad, Cote d'Ivoire, and the Central African Republic (Bessaud et al., 2006; Johnson, 1988; Kokernot et al., 1965a, b; Posey

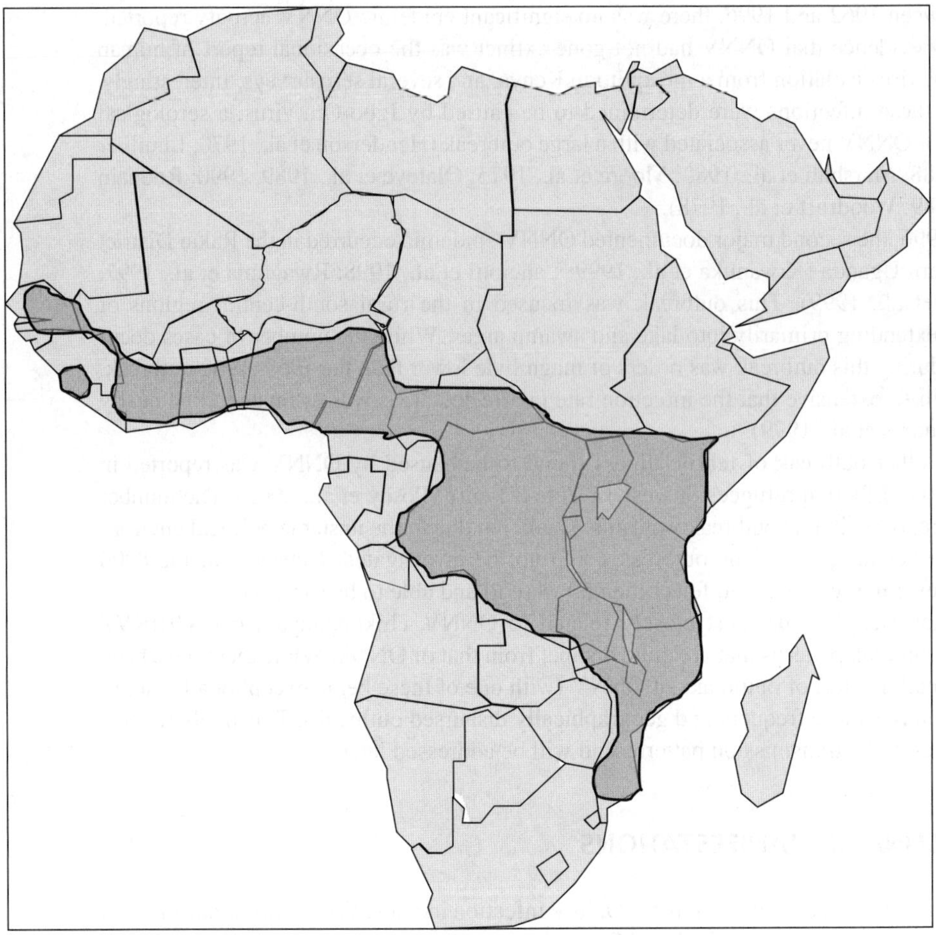

Figure 23.1. ONNV distribution map. Approximate known geographic distribution of ONNV. Positive areas determined by human febrile cases, human and/or animal serosurveys, and mosquito collections. For color detail, please see color plate section.

et al., 2005; Salim and Porterfield, 1973; Woodall et al., 1963). The antigenic subtype Igbo-Ora has additionally been identified in the Central African Republic and Nigeria extending the geographic range of ONNV to western Africa (Chippaux and Chippaus-Hyppolite, 1968; Johnson, 1988; Moore et al., 1975; Olaleye et al., 1988; Tomori et al., 1981). All of the genetic data on ONNV indicates that all strains are highly conserved (Powers et al., 2000). In contrast, the closely related CHIKV is distributed throughout Africa, Southeast Asia, and most of Oceania and exhibits much more genetic diversity than ONNV.

The two largest ONNV epidemics were both extensively characterized epidemiologically. Of major significance were the findings of the rapidity of movement through the affected areas as well as the sheer magnitude of the proportion of the population affected (Sanders et al., 1999; Williams et al., 1965b). For example, in 1959–1961, over two million people were infected during the 2-year outbreak and the virus spread across much of east and central portions of Africa. In the first year of the outbreak, the disease was estimated to be moving at the staggering rate of 1.7 miles per day. Similarly, during the 1996 outbreak, while the absolute number of affected individuals was not reported, infection rates were estimated to be as high as 68% with attack rates over 40% (Sanders et al., 1999).

While mortality has never been reported due to ONNV infection, the associated morbidity is so incapacitating that inactivity for several weeks is not uncommon. When this is combined with the accompanying high attack rate, significant drain on local health-care systems and economies (due to high absenteeism) is not uncommon. For example, in the initial outbreak, up to 10% of the local workforce was ill at any given time with 25% being incapacitated for 5 days or more (Williams et al., 1965b). In the areas where the virus has been detected, this pathogen remains a significant public health threat, and understanding how the virus is maintained is of particular importance.

Epidemiologically, a major distinction of ONNV is not only that it is transmitted by different species of mosquitoes from all other alphaviruses but that the ONNV vectors are in a completely different family. In fact, ONNV is unique among alphaviruses in that it is principally vectored by anopheline mosquitoes, while all the other alphaviruses utilize predominantly culicine vectors in their transmission cycles. To date, two mosquito species, *Anopheles gambiae* and *Anopheles funestus*, have been identified as the major vectors of ONNV (i.e., most abundant and/or harboring virus that was isolated) in both of the major recorded outbreaks. Both of these species are noted to be primarily endophilic, further supporting the idea that epidemic transmission of ONNV is most certainly between man and infected mosquitoes without the involvement of another vertebrate during outbreaks (Corbet et al., 1961; Johnson, 1988; Lutwama et al., 1999; Williams et al., 1965a). Interestingly, an isolate of ONNV was recovered from a pool of *Mansonia uniformis* during the 1996–1997 outbreak, and previous literature has suggested that *Mansonia* species mosquitoes may be ONNV vectors (Haddow et al., 1960; Lutwama et al., 1999). However, this isolate was the first ever recorded from a *Mansonia* species, and too few of this species were collected in 1997 to suggest a major role in the epidemic. Therefore, the epidemic and maintenance potential of mosquitoes in this genus remains to be determined. During a nonepidemic period (1978), an isolate of ONNV was obtained from *An. funestus*, suggesting that this mosquito may also be involved in the enzootic maintenance cycle (Johnson et al., 1981). To date, no other vector species have been implicated in transmission of ONNV.

While numerous field and laboratory studies have been conducted on the mosquito vectors involved in the transmission ONNV, there is little information available regarding vertebrate hosts involved in viral maintenance. Most of the speculation regarding potential vertebrate reservoirs has been derived from serological studies that have demonstrated the presence of antibodies against these viruses (Bedekar and Pavri, 1969; Johnson et al., 1977; Marshall et al., 1982; Mcintosh, 1961). Serosurveys have identified domestic livestock (camels, cattle, sheep, and goats) as well as several species of rodents including *Aethomys kaiseri*, *Dasymys imcomtus*, *Arvicanthis niloticus*, *Mastomys natalensis*, and *Pelomys isseli* as potential reservoirs of ONNV (Johnson et al., 1977; Olaleye et al., 1988). Additionally, laboratory studies have shown the development of antibody in monkeys (Binn et al., 1967; Paul and Singh, 1968). Interestingly, no evidence of antibodies in birds was found in >1200 birds from wild collections.

23.5 FACTORS AFFECTING EMERGENCE

O'nyong-nyong virus has shown itself to be an exceptionally unusual arbovirus both in its epidemic form and in the curious lack of information regarding its zoonotic maintenance. The paucity of knowledge regarding this virus makes it even more challenging to determine where and when it may reemerge in epidemic form. However, there are several factors that have been examined in some level of detail with ONNV, and understanding the combined roles of these elements may provide clues for the future of the virus. These elements include analysis of viral genetic elements, identification of potential invertebrate vectors and zoonotic reservoir hosts, and characterization of environmental factors that may influence reemergence. The role of each of these components and how they interact will be outlined in the succeeding text.

23.5.1 Etiologic agent: *viral genomics and antigens encoded*

The primary factors regulating transmission, maintenance, and emergence of a virus are the genetics and expressed elements of the virus itself. Like all alphaviruses, ONNV has a genome consisting of a linear, positive-sense, single-stranded RNA molecule of approximately 11.8 kb in size. Control of viral replication is encoded in the nonstructural proteins present in the 5′ two thirds of the genome. The structural genes are collinear with the 3′ one third and constitute the elements packaged in infectious virions. These structural proteins are produced by translation of an mRNA generated from an internal, subgenomic promoter immediately downstream of the nonstructural open reading frame. O'nyong-nyong virus has a 3′ noncoding region of ~425 nucleotides, and the repeat sequence elements present in this region that may participate in regulation of genetic replication are distinct from all other alphaviruses (Levinson et al., 1990; Pfeffer et al., 1998).

The nonstructural genes, designated nsP1–4, are essential for viral replication. Each has specific functions in this process: (i) nsP1 is required for initiation of synthesis of minus strands (Sawicki and Sawicki, 1998; Sawicki et al., 1981; Wang et al., 1991) and caps the viral RNAs during transcription (Mi and Stollar, 1991; Strauss and Strauss, 1994); (ii) nsP2 protein has RNA helicase and nonstructural polyprotein protease activity (Ding and Schlesinger, 1989; Hardy and Strauss, 1989); (iii) the functions of the nsP3 gene are not fully understood (Strauss et al., 1988), but it has been shown to tolerate numerous mutations, including large deletions, and still produce infectious virus (Davis et al., 1989; Li et al., 1990); and (iv) the final nonstructural protein, nsP4, is the viral RNA-dependent RNA polymerase due to the characteristic GDD motif it contains (Kamer and Argos, 1984).

The structural gene products include a capsid protein, two major envelope surface glycoproteins (E2 and E1), and two small peptides (E3 and 6K) (Strauss and Strauss, 1994). The highly conserved capsid protein associates positive-sense viral RNA to generate nucleocapsids. The cytoplasmic nucleocapsids move to the cell membrane where they associate with E1/E2 dimeric complexes (Anthony and Brown, 1991) before budding from the cell (Ekstrom et al., 1994; Garoff et al., 1998). The E2 protein has been found to be the primary determinant of antigenicity and cell receptor binding in both the vertebrate host and the insect vector for other alphaviruses. For example, a single mutation in the Venezuelan equine encephalitis virus (VEEV) E2 glycoprotein delayed replication by almost 2 days in mice (Davis et al., 1994), while a VEEV strain with a single mutation in the E2 was restricted in its ability to disseminate from the midgut of mosquitoes (Woodward et al., 1991). The closely related CHIKV was found to infect *Aedes albopictus* more easily with a single E2 mutation (Tsetsarkin et al., 2007).

Interestingly, it still remains to be determined if E2 plays a similar role in ONNV infection patterns. One study suggested that all of the structural proteins are necessary for anopheline vector specificity (Vanlandingham et al., 2006), while more recent work has shown that the nsP3 protein is the most important element for infection of *An. gambiae* (Saxton-Shaw et al., 2012). Additional studies demonstrate a role for a regulatory stop codon located between nsP3 and nsP4 in mosquito infectivity (Myles et al., 2006). These works suggest that ONNV may regulate its pattern of mosquito infection at the replication (regulatory) level rather than by binding or entry limitations. A similar postulation with another alphavirus, Sindbis virus, suggested that nsP3 mutations significantly reduced the replicative ability of the virus in mosquito cells (Lastarza et al., 1994). This suggested that this gene may aid in defining cell type and invertebrate vector specificity as was recently found with ONNV. This departure from mosquito infection patterns observed with most other alphaviruses may have implications for a lack of adaptation to alternate vectors as replication genes are frequently less prone to rapid mutation than are surface-exposed structural protein genes.

23.5.2 Transmission parameters

The identification of factors affecting transmission of virus by an invertebrate is often only examined during outbreaks or just after these events. With a virus such as ONNV where only two major outbreaks have occurred, this significantly limits the opportunities to determine mosquito parameters of transmission.

As noted previously, *An. gambiae* and *An. funestus* have been implicated as the outbreak vectors due to their prevalence during epidemics combined with the occasional finding of virus in these species. Since these species are so unexpected for alphavirus transmission, additional factors have been considered in determining vector status. For example, during the first outbreak, most people reported being bitten only inside their huts and at night. Nearly all mosquitoes collected exhibiting these traits were the two species noted. Additionally, these species were found extensively in the types of ecologies where there were clinical cases, but *An. funestus* were not found in areas without epidemic activity. Common vectors of the closely related CHIKV also were considered as options for vectoring ONNV, but there were no reports of *Aedes aegypti* either biting or present in outbreak areas. While the CHIKV vector *Aedes simpsoni* was present, it was not biting humans, indicating it was unlikely to be serving as an ONNV vector. Finally, as abundance is a major consideration in vector status, even unconventional vectors that were present in significant numbers have been considered as vectors of the virus. Bedbugs were found abundantly during the 1959–1962 outbreak but discounted as the virus was found to move too quickly to be spread only by slow-moving bedbugs. While they may have been involved in some transmission with a single hut or family complex, they could not have been responsible for such rapid movement between villages and across the continent. This rapid movement would require a vector with the capacity to move hundreds of yards daily reinforcing the finding of mosquitoes as the primary vectors. Interestingly, while bedbugs would seem an unexpected species to consider as an arbovirus vector, another bug species (*Oeciacus vicarius*) is known to vector the alphavirus Fort Morgan virus (Hayes et al., 1977). Overall, the important consideration in determining vector status is evaluating the combined body of evidence encompassing a variety of possible parameters. In the case of ONNV, all the known evidence implicates anophelines in spite of the unlikely nature of these as an alphavirus vector.

23.5.3 Zoonotic maintenance

In humans, ONNV is highly immunogenic and generates strong antibody responses. For this reason, there has been speculation that a similar phenomenon occurs in other vertebrates and the maintenance reservoir could be easily identified. As noted previously, the very few serosurveys performed have implicated several species of mice and possibly nonhuman primates as potential reservoirs. Recent field studies in East Africa further implicate several species of mice and rats as possible reservoirs along with domestic livestock including goats, sheep, and cattle (Powers et al., unpublished). Unfortunately, ONNV exhibits a close but unidirectional serological relationship with its nearest relative, CHIKV. Single-dose exposure antisera have indicated that CHIKV immune serum can neutralize both CHIKV and ONNV while ONNV antiserum neutralizes only homologous virus (Chanas et al., 1976; Karabatsos, 1975; Porterfield, 1961). Additional studies utilizing panels of monoclonal antibodies found that the majority of CHIKV-specific monoclones would react with ONNV but as few as 12% of the monoclonal antibodies generated against ONNV were reactive against CHIKV (Blackburn et al., 1995). These findings suggest that any serosurvey demonstrating the presence of ONNV could also mistakenly reflect a CHIKV infection making identification of a vertebrate reservoir more challenging. However, neutralizing antibody against ONNV in the absence of any neutralization against CHIKV should be a strong indication of ONNV-specific infection.

23.5.4 Environmental influences

While many studies of arbovirus movement focus on the virus, the vector, or the vertebrate reservoir, one final component that is important to consider is the environment. Considering ecological parameters may be even more relevant as global climate changes occur since these changes could impact all the other factors as well.

In considering the role of environment during the major epidemics, it is important to note the ecological habitat of the known vectors and/or reservoirs. *An. funestus* larvae are associated with clear water, moderate shade, and some growing vegetation, while *An. gambiae* larvae are found in small, fresh-formed rainwater pools devoid of vegetation or shade. Both are found in association with man and animals. Curiously, while water is important for both of these vectors, it was unusually dry during the first outbreak, a characteristic that would seem unexpected during an arboviral outbreak. Interestingly, this same trait was associated with a large CHIKV outbreak in Kenya in 2004 (Chretien et al., 2007). Here, the lack of water resulted in changing human behavior that resulted in increasing proximity of the mosquitoes to local residents, thus exacerbating the outbreak conditions.

To date, no specific environmental factors have been shown to affect the presence or outcome of an ONNV epidemic, but it is imperative to consider this aspect with ever changing environments due to human modification. As an example, the Kano plain on the border of Lake Victoria recently underwent environmental changes when land use policies allowed the change of this grassland ecology to irrigated rice-growing land (Surtees, 1970; Surtees et al., 1970). This new ecology is now a suitable habitat for anophelines and, thus, ONNV. Events like this, combined with natural changes in ecology, could drastically affect the known patterns of arboviral transmission.

23.6 CONCLUSION

O'nyong-nyong virus is one of the most interesting examples of an arbovirus reemerging periodically to cause human epidemics. A combination of factors including climate, ecology, host status, virology, genetics, and mosquito vectorial capacity has combined under precisely defined conditions at select times in history to result in the emergence of epidemic ONNV. In the intervening years, the virus is somehow maintained causing no known human or animal disease. The fact that ONNV utilizes unique mosquito vectors, has highly conserved genetics, and has a completely uncharacterized zoonotic cycle suggests that the differences it exhibits from close relatives may ultimately lead to discoveries as to how these zoonotic pathogens are both maintained and reemerge. Ultimately, the goal of such scientific discovery would be to eliminate the disease caused by viruses such as ONNV. What we can be certain of is that combating the emergence of arboviruses will require the integrated study of the viruses, the vectors, the hosts, and their combined ecology as we more fully recognize the importance of each of these components in viral infections.

REFERENCES

Anthony, R.P., Brown, D.T., 1991. Protein–protein interactions in an alphavirus membrane. *J Virol* 65, 1187–1194.

Bedekar, S.D., Pavri, K.M., 1969. Studies with chikungunya virus. I. Susceptibility of birds and small mammals. *Indian J Med Res* 57, 1181–1192.

Bessaud, M., Peyrefitte, C.N., Pastorino, B.A., et al., 2006. O'nyong-nyong virus, chad. *Emerg Infect Dis* 12, 1248–1250.

Binn, L.N., Harrison, V.R., Randall, R., 1967. Patterns of viremia and antibody observed in rhesus monkeys inoculated with chikungunya and other serologically related group A arboviruses. *Am J Trop Med Hyg* 16, 782–785.

Blackburn, N.K., Besselaar, T.G., Gibson, G., 1995. Antigenic relationship between chikungunya virus strains and o'nyong nyong virus using monoclonal antibodies. *Res Virol* 146, 69–73.

Bowen, E.T., Simpson, D.I., Platt, G.S., et al., 1973. Large scale irrigation and arbovirus epidemiology, Kano Plain, Kenya. II. Preliminary serological survey. *Trans R Soc Trop Med Hyg* 67, 702–709.

Chanas, A.C., Johnson, B.K., Simpson, D.I., 1976. Antigenic relationships of alphaviruses by a simple micro-culture cross-neutralization method. *J Gen Virol* 32, 295–300.

Chippaux, A., Chippaux-Hyppolite, C., 1968. Signs of the spread of O'Nyong-nyong virus in the Central African Republic. *Med Trop (Mars)* 28, 346–362.

Chretien, J.P., Anyamba, A., Bedno, S.A., et al., 2007. Drought-associated chikungunya emergence along coastal East Africa. *Am J Trop Med Hyg* 76, 405–407.

Corbet, P.S., Williams, M.C., Gillett, J.D., 1961. O'Nyong-nyong fever: an epidemic virus disease in East Africa. IV. Vector studies at epidemic sites. *Trans R Soc Trop Med Hyg* 55, 463–480.

Davis, N.L., Grieder, F.B., Smith, J.F., et al., 1994. A molecular genetic approach to the study of Venezuelan equine encephalitis virus pathogenesis. *Arch Virol* 9, 99–109.

Davis, N.L., Willis, L.V., Smith, J.F., et al., 1989. In vitro synthesis of infectious venezuelan equine encephalitis virus RNA from a cDNA clone: analysis of a viable deletion mutant. *Virology* 171, 189–204.

Ding, M.X., Schlesinger, M.J., 1989. Evidence that Sindbis virus NSP2 is an autoprotease which processes the virus nonstructural polyprotein. *Virology* 171, 280–284.

Ekstrom, M., Liljestrom, P., Garoff, H., 1994. Membrane protein lateral interactions control Semliki Forest virus budding. *EMBO J* 13, 1058–1064.

Garoff, H., Hewson, R., Opstelten, D.J.E., 1998. Virus maturation by budding. *Microbiol Mol Biol Rev* 62, 1171–1190.

Haddow, A.J., Davies, C.W., Walker, A.J., 1960. O'nyong-nyong fever: an epidemic virus disease in east Africa I. Introduction. *Trans R Soc Trop Med Hyg* 54, 517.

Hardy, W.R., Strauss, J.H., 1989. Processing the nonstructural proteins of sindbis virus: nonstructural proteinase is in the C-terminal half of nsP2 and functions both in cis and in trans. *J Virol* 63, 4653–4664.

Hayes, R.O., Francy, D.B., Lazuick, J.S., et al., 1977. Role of the cliff swallow bug (*Oeciacus vicarius*) in the natural cycle of a western equine encephalitis-related alphavirus. *J Med Entomol* 14, 257–262.

Henderson, B.E., Kirya, G.B., Hewitt, L.E., 1970. Serological survey for arboviruses in Uganda, 1967–69. *Bull World Health Organ* 42, 797–805.

Johnson, B.K., 1988. O'nyong-nyong virus disease. In: Monath, T.P. (Ed.), The Arboviruses: Epidemiology and Ecology, Vol. III. CRC Press, Boca Raton, FL, pp. 217–223.

Johnson, B.K., Chanas, A.C., Shockley, P., et al., 1977. Arbovirus isolations from, and serological studies on, wild and domestic vertebrates from Kano Plain, Kenya. *Trans R Soc Trop Med Hyg* 71, 512–517.

Johnson, B.K., Gichogo, A., Gitau, G., et al., 1981. Recovery of o'nyong-nyong virus from *Anopheles funestus* in Western Kenya. *Trans R Soc Trop Med Hyg* 75, 239–241.

Kamer, G., Argos, P., 1984. Primary structural comparison of RNA-dependent polymerases from plant, animal and bacterial viruses. *Nucleic Acids Res* 12, 7269–7282.

Karabatsos, N., 1975. Antigenic relationships of group A arboviruses by plaque reduction neutralization testing. *Am J Trop Med Hyg* 24, 527–532.

Kiwanuka, N., Sanders, E.J., Rwaguma, E.B., et al., 1999. O'nyong-nyong fever in south-central Uganda, 1996–1997: clinical features and validation of a clinical case definition for surveillance purposes. *Clin Infect Dis* 29, 1243–1250.

Kokernot, R.H., Casaca, V.M., Weinbren, M.P., et al., 1965a. Survey for antibodies against arthropod-borne viruses in the sera of indigenous residents of Angola. *Trans R Soc Trop Med Hyg* 59, 563–570.

Kokernot, R.H., Szlamp, E.L., Levitt, J., *et al.*, 1965b. Survey for antibodies against arthropod-borne viruses in the sera of indigenous residents of the Caprivi Strip and Bechuanaland Protectorate. *Trans R Soc Trop Med Hyg* 59, 553–562.

Lanciotti, R.S., Ludwig, M.L., Rwaguma, E.B., et al., 1998. Emergence of epidemic o'nyong-nyong fever in Uganda after a 35-year absence: genetic characterization of the virus. *Virology* 252, 258–268.

Lastarza, M.W., Grakoui, A., Rice, C.M., 1994. Deletion and duplication mutations in the C-terminal nonconserved region of Sindbis virus nsP3: effects on phosphorylation and on virus replication in vertebrate and invertebrate cells. *Virology* 202, 224–232.

Levinson, R.S., Strauss, J.H., Strauss, E.G., 1990. Complete sequence of the genomic RNA of O'nyong-nyong virus and its use in the construction of alphavirus phylogenetic trees. *Virology* 175, 110–123.

Lhuillier, M., Cunin, P., Mazzariol, M.J., et al., 1988. A rural epidemic of Igbo Ora virus (with inter-human transmission) in the Ivory Coast 1984–1985. *Bull Soc Pathol Exot Filiales* 81, 386–395.

Li, G.P., La Starza, M.W., Hardy, W.R., et al., 1990. Phosphorylation of sindbis virus nsP3 in vivo and in vitro. *Virology* 179, 416–427.

Lutwama, J.J., Kayondo, J., Savage, H.M., et al., 1999. Epidemic O'Nyong-nyong fever in southcentral Uganda, 1996–1997: entomologic studies in Bbaale village, Rakai District. *Am J Trop Med Hyg* 61, 158–162.

Marshall, T.F., Keenlyside, R.A., Johnson, B.K., et al., 1982. The epidemiology of O'nyong-nyong in the Kano Plain, Kenya. *Ann Trop Med Parasitol* 76, 153–158.

McIntosh, B.M., 1961. Susceptibility of some African wild rodents to infection with various arthropod-borne viruses. *Transfusion* 55, 63–68.

Mi, S., Stollar, V., 1991. Expression of sindbis virus nsP1 and methyltransferase activity in *Escherichia coli*. *Virology* 184, 423–427.

Moore, D.L., Causey, O.R., Carey, D.E., et al., 1975. Arthropod-borne viral infections of man in Nigeria, 1964–1970. *Ann Trop Med Parasitol* 69, 49–64.

Myles, K.M., Kelly, C.L., Ledermann, J.P., et al., 2006. Effects of an opal termination codon preceding the nsP4 gene sequence in the O'Nyong-nyong virus genome on *anopheles gambiae* infectivity. *J Virol* 80, 4992–4997.

Olaleye, O.D., Oladosu, L.A., Omilabu, S.A., et al., 1989. Complement fixing antibodies against arboviruses in horses at Lagos, Nigeria. *Rev Elev Med Vet Pays Trop* 42, 321–325.

Olaleye, O.D., Omilabu, S.A., Baba, S.S., 1990. Growth of Igbo-Ora virus in some tissue cultures. *Acta Virol* 34, 367–371.

Olaleye, O.D., Omilabu, S.A., Fagbami, A.H., 1988. Igbo-Ora virus (an alphavirus isolated in Nigeria): a serological survey for haemagglutination inhibiting antibody in humans and domestic animals. *Trans R Soc Trop Med Hyg* 82, 905–906.

Paul, S.D., Singh, K.R., 1968. Experimental infection of Macaca radiata with Chikungunya virus and transmission of virus by mosquitoes. *Indian Journal of Medical Research* 56, 802–811.

Pfeffer, M., Kinney, R.M., Kaaden, O.R., 1998. The alphavirus 3′-nontranslated region: size heterogeneity and arrangement of repeated sequence elements. *Virology* 240, 100–108.

Porterfield, J.S., 1961. Cross-neutralization studies with group A arthropod-borne viruses. *Bull World Health Organ* 24, 735.

Posey, D.L., O'Rourke, T., Roehrig, J.T., et al., 2005. O'Nyong-nyong fever in West Africa. *Am J Trop Med Hyg* 73, 32.

Powers, A.M., Brault, A.C., Tesh, R.B., et al., 2000. Re-emergence of chikungunya and o'nyong-nyong viruses: evidence for distinct geographical lineages and distant evolutionary relationships. *J Gen Virol* 81, 471–479.

Rodhain, F., Gonzalez, J.P., Mercier, E., et al., 1989. Arbovirus infections and viral haemorrhagic fevers in Uganda: a serological survey in Karamoja district, 1984. *Trans R Soc Trop Med Hyg* 83, 851–854.

Rwaguma, E.B., Lutwama, J.J., Sempala, S.D., et al., 1997. Emergence of epidemic O'nyong-nyong fever in southwestern Uganda, after an absence of 35 years. *Emerg Infect Dis* 3, 77.

Salim, A.R., Porterfield, J.S., 1973. A serological survey on arbovirus antibodies in the Sudan. *Trans R Soc Trop Med Hyg* 67, 206–210.

Sanders, E.J., Rwaguma, E.B., Kawamata, J., et al., 1999. O'nyong-nyong fever in south-central Uganda, 1996–1997: description of the epidemic and results of a household-based seroprevalence survey. *J Infect Dis* 180, 1436–1443.

Sawicki, D.L., Sawicki, S.G., 1998. Role of the nonstructural polyproteins in alphavirus RNA synthesis. *Adv Exp Med Biol* 440, 187–198.

Sawicki, D.L., Sawicki, S.G., Keranen, S., et al., 1981. Specific Sindbis virus-coded function for minus-strand RNA synthesis. *J Virol* 39, 348–358.

Saxton-Shaw, K.D., Ledermann, J.P., Borland, E.M., et al., 2012. O'nyong nyong virus molecular determinants of unique vector specificity reside in non-structural protein 3. *PLoS Negl Trop Dis* 7, e1931.

Shore, H., 1961. O'nyong-nyong fever: an epidemic virus disease in East Africa. III. Some clinical and epidemiological observations in the Northern Province of Uganda. *Trans R Soc Trop Med Hyg* 55, 361.

Strauss, E.G., Levinson, R., Rice, C.M., et al., 1988. Nonstructural proteins nsP3 and nsP4 of Ross River and O'nyong-nyong viruses: sequence and comparison with those of other alphaviruses. *Virology* 164, 265–274.

Strauss, J.H., Strauss, E.G., 1994. The alphaviruses: gene expression, replication, and evolution [published erratum appears in *Microbiol Rev* 1994 Dec;58(4):806]. [Review] [629 refs]. *Microbiol Rev* 58, 491–562.

Surtees, G., 1970. Large-scale irrigation and arbovirus epidemiology, Kano Plain, Kenya. I. Description of the area and preliminary studies on the mosquitoes. *J Med Entomol* 7, 509–517.

Surtees, G., Simpson, D.I., Bowen, E.T., et al., 1970. Ricefield development and arbovirus epidemiology, Kano Plain, Kenya. *Trans R Soc Trop Med Hyg* 64, 511–522.

Tomori, O., Monath, T.P., O'Connor, E.H., et al., 1981. Arbovirus infections among laboratory personnel in Ibadan, Nigeria. *Am J Trop Med Hyg* 30, 855–861.

Tsetsarkin, K.A., Vanlandingham, D.L., McGee, C.E., et al., 2007. A single mutation in chikungunya virus affects vector specificity and epidemic potential. *PLoS Pathog* 3, e201.

Vanlandingham, D.L., Tsetsarkin, K., Klingler, K.A., et al., 2006. Determinants of vector specificity of o'nyong nyong and chikungunya viruses in anopheles and aedes mosquitoes. *Am J Trop Med Hyg* 74, 663–669.

Wang, Y.F., Sawicki, S.G., Sawicki, D.L., 1991. Sindbis virus nsP1 functions in negative-strand RNA synthesis. *J Virol* 65, 985–988.

Williams, M.C., Woodall, J.P., 1961. O'nyong-nyong fever: an epidemic virus disease in East Africa. II. Isolation and some properties of the virus. *Trans R Soc Trop Med Hyg* 55, 135–141.

Williams, M.C., Woodall, J.P., Corbet, P.S., et al., 1965a. O'nyong-nyong fever: an epidemic virus disease in East Africa. VIII. Virus isolations from *Anopheles* mosquitoes. *Trans R Soc Trop Med Hyg* 59, 300–306.

Williams, M.C., Woodall, J.P., Gillett, J.D., 1965b. O'nyong-nyong fever: an epidemic virus diesease in East Africa. VII. Virus isolations from man and serological studies up to July 1961. *Trans R Soc Trop Med Hyg* 59, 186–197.

Woodall, J.P., Williams, M.C., Lule, M., 1963. O'nyong nyong, East African Virus Research Institute Report No. 13, Entebbe, p. 31.

Woodruff, A.W., Bowen, E.T., Platt, G.S., 1978. Viral infections in travellers from tropical Africa. *Br Med J* 1, 956–958.

Woodward, T.M., Miller, B.R., Beaty, B.J., et al., 1991. A single amino acid change in the E2 glycoprotein of Venezuelan equine encephalitis virus affects replication and dissemination in *Aedes aegypti* mosquitoes. *J Gen Virol* 72, 2431–2435.

24

ZOONOTIC HEPATITIS E: ANIMAL RESERVOIRS, EMERGING RISKS, AND IMPACT OF CLIMATE CHANGE

Nicole Pavio and Jérôme Bouquet

UMR 1161 Virology, ANSES, Laboratoire de Santé Animale, Maisons-Alfort, France
UMR 1161 Virology, INRA, Maisons-Alfort, France
UMR 1161 Virology, Ecole Nationale Vétérinaire d'Alfort, Maisons-Alfort, France

TABLE OF CONTENTS

Viral Infections and Global Change, First Edition. Edited by Sunit K. Singh.

24.1 INTRODUCTION

Hepatitis E virus (HEV) is responsible for enterically transmitted hepatitis and has become a major public health concern in many tropical and subtropical countries (Purcell and Emerson, 2008). The mortality rate associated with HEV infection (1–4%) is higher than that of hepatitis A (0.1–2%) (Peron et al., 2007) and can reach 20% during pregnancy in some endemic regions (Purcell and Emerson, 2008). Hepatitis E virus infection in developing countries is mostly a waterborne disease associated with widespread epidemics due to contamination of water and water supplies in poor sanitation conditions (Arankalle et al., 1995; Purcell and Emerson, 2008). In industrialized countries, however, including many European countries, the Unites States, and Japan, acute hepatitis E occurs sporadically and the contamination pathways are still poorly understood. So far, no case has been linked to contaminated water (Purcell and Emerson, 2008). The peculiarity of HEV is that, of all known major hepatitis viruses (A, B, C, and D), it is the only one which can infect animal species other than primates. In 1997, swine HEV was discovered in pigs in the United States (Meng et al., 1997). More recently, avian HEV (aHEV), rat HEV (rHEV), and rabbit HEV (rbHEV) strains were identified in their respective species in the United States, Germany, and China (Haqshenas et al., 2001; Johne et al., 2009; Zhao et al., 2009). Accumulating evidence indicates that hepatitis E is a zoonotic disease, and swine (and possibly other animal species) are reservoirs. Swine and human strains of HEV are mostly genetically closely related and, in some cases, even indistinguishable (Bouquet et al., 2011; Lu et al., 2006). Direct transmission through contact or the consumption of contaminated food products such as pork has been reported (Yazaki et al., 2003). Thus, zoonotic transmission of hepatitis E is an important public health issue with respect to food safety and zoonotic risk (Meng, 2009). Considering the high endemicity of HEV in humans and animals and its water- and foodborne transmission pathways, changes in environmental factors will, *a fortiori*, impact the risk of human exposure.

24.2 HEV BIOLOGY AND CLASSIFICATION

Hepatitis E virus was identified in the 1980s in India as a viral agent responsible for widespread waterborne epidemics distinct from hepatitis A virus (HAV). Hepatitis E virus was first observed by electron microscopy in a human fecal sample from an acute case. The HEV genome was amplified and sequenced in the 1990s. Hepatitis E virus has unique features in its genomic organization and amino acid sequence, so a new family was created in 2004. It is classified in the *Hepeviridae* family and genus *Hepevirus* (Emerson et al., 2004). The HEV is a small, nonenveloped virus of approximately 30 nm in diameter. The viral capsid has an icosahedral symmetry and is composed of 60 copies of a single protein: core or open reading frame-2 (ORF-2). The HEV genome is a single-stranded RNA molecule of positive polarity, capped and polyadenylated, of 7.2 Kb that encodes three ORFs (Figure 24.1). The first one, ORF-1, corresponds to nonstructural proteins and carries the consensus motif of several enzymatic functions such as methyltransferase, papain-like cysteine protease, helicase, and RNA-dependent RNA polymerase (RdRp). The second, ORF-2, encodes a capsid protein of 660 amino acids (~88 Kd) with three putative sites of N-glycosylation. The third one, ORF-3, encodes a small phosphoprotein of 123 amino acids associated with the cytoskeleton, but ORF-3's function is not known. However, ORF-3 is necessary for *in vivo* viral infection in cynomolgus monkeys (Graff et al., 2005).

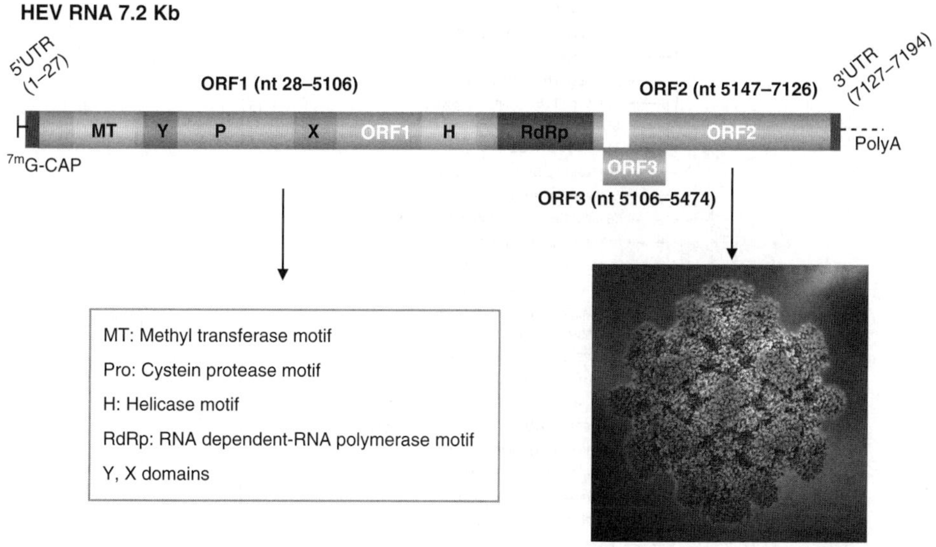

Figure 24.1. Hepatitis E virus genome organization and schematic structure. For color detail, please see color plate section.

Hepatitis E virus genomes show significant genetic variability and can be divided into at least four major mammalian genotypes (1–4) plus two distinct avian and rat HEVs (Figure 24.2). aHEV and rHEV share approximately 50–60% of nucleotide sequence identity with mammalian genotypes, so they may constitute a separate new genus distinct from the *Hepevirus* genus. At the moment, HEV classification is still under consideration by the International Committee on Taxonomy of Viruses (ICTV); a new "unassigned" genus was recently suggested for aHEV (http://ictvonline.org/virusTaxonomy.asp?version=2009). The four major mammalian HEVs share 73–74% identity: genotypes 1 and 2 are present exclusively in humans and mainly associated with widespread waterborne epidemics in tropical and subtropical regions (Lu et al., 2006); genotypes 3 and 4 are found both in humans and other mammalian reservoirs (swine, wild boars, deer, and mongooses) and are responsible for sporadic cases of hepatitis E. Genotype 3 is found mainly in Europe, the United States, and Japan, and genotype 4 has been described in Asia (China, Japan, and India) and occasionally in Europe (Belgium and France) (Hakze-van der Honing et al., 2011; Tesse et al., 2012). Different aHEV strains have been identified recently from China, Australia, and Europe (Hungary) with approximately 80% nucleotide sequence identity with the prototype aHEV from the United States (Bilic et al., 2009; Zhao et al., 2010). It suggests that aHEVs could be separated into three distinct genotypes according to their geographic distribution. A genetically distinct HEV was also recently isolated from rats in Germany and urban rats in Los Angeles, California (Figure 24.2). Sequence comparison with human and avian strains revealed only 59.9% and 49.9% sequence identity, respectively. Similarly, the deduced amino acid sequence for the complete capsid protein revealed 56.2% and 42.9% identity with human and avian strains, respectively (Johne et al., 2009; Purcell et al., 2011). Very recently, a novel strain of HEV was isolated from farm rabbits in China and the United States. The rbHEV is most closely related to genotype 3 HEV, with nearly 82% nucleotide sequence homology (Cossaboom et al., 2011; Zhao et al., 2009) (Figure 24.2). Further investigations into these new HEV sequences are warranted prior to definitive classification. The zoonotic potential of these other HEVs is discussed in the zoonotic hepatitis E (Section 24.5).

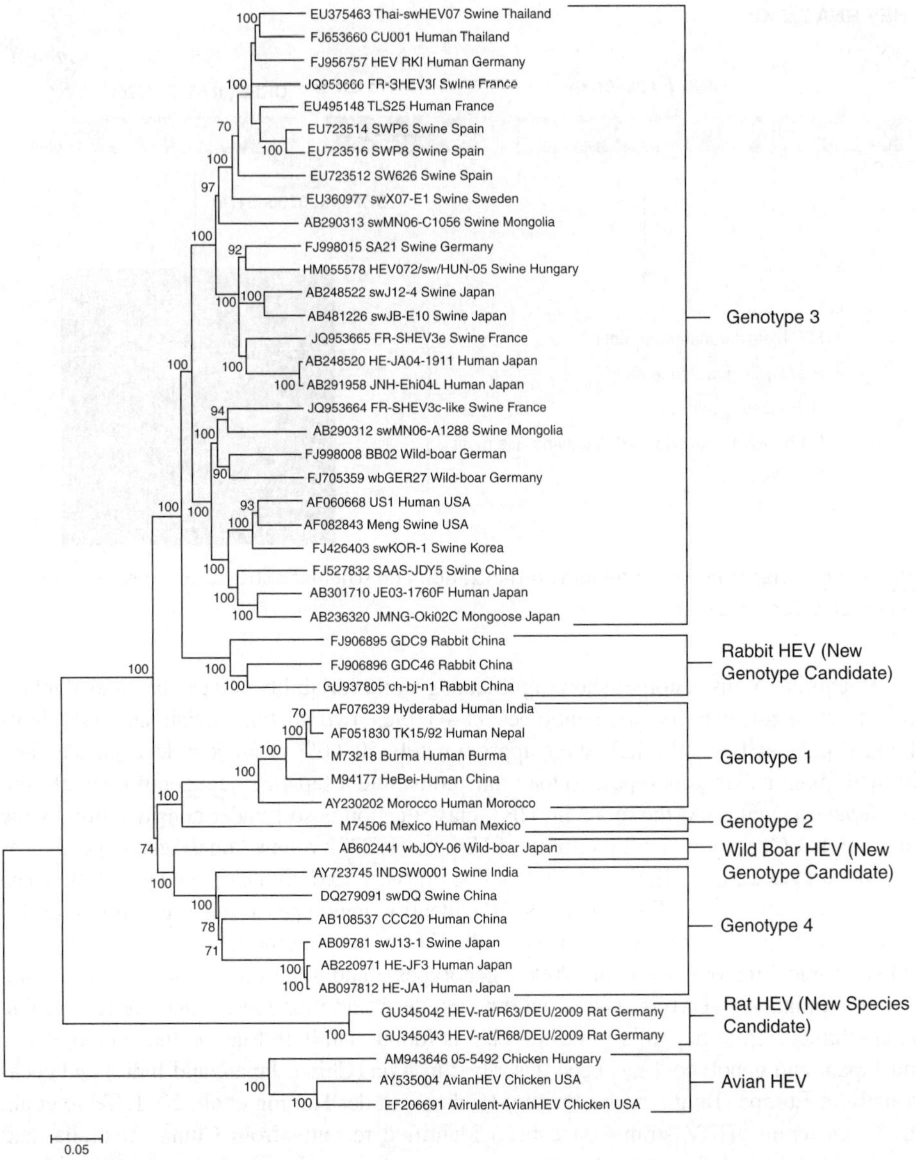

Figure 24.2. Phylogenetic tree of 48 full-length HEV sequences. Tree constructed by the neighbor-joining method, based on the ClustalW alignment of 48 HEV isolates retrieved from GenBank. Bootstrap values obtained from 1 000 replicates and over 70% are indicated at each node. Potential genotypes and branches are also indicated.

Like many other RNA viruses, HEV has a high intrinsic genetic variability due to the low fidelity of its RdRp and is present within an infected host as quasispecies (Bouquet et al., 2012; Grandadam et al., 2004). Furthermore, intragenotype recombination events have been described for HEV genotype 1 (van Cuyck et al., 2005). Genetic recombination between HEV RNA and cellular messengers has been also observed using an *in vitro* model of HEV replication (Shukla et al., 2011). Recombination confers more efficient replication to HEV in the model (Shukla et al., 2012). Thus, HEV has many ways of evolving genetically, and more virulent variants may emerge in the future.

Hepatitis E virus biology and multiplication cycle are still hypothetical since this virus does not grow efficiently *in vitro*. A few models have been developed in hepatoma cell lines (Emerson et al., 2010; Okamoto, 2011), but viral productions are low and require high initial viral titer.

24.3 PATHOGENESIS IN HUMANS

In tropical and subtropical areas, there are high rates of HEV seroprevalence, which range from 27% to 80% (Abe et al., 2006) in the general population. Waterborne HEV epidemics affect thousands of people, mostly young adults (Teo, 2010). In contrast, in nonendemic regions, seroprevalence can vary widely, from a few percent (2–7.8%) in countries in Europe and South America to several percent (18.2–52.2%) in France, the United States, Russia, the United Kingdom, Hong Kong, Korea, and China (Table 24.1). Men are more frequently

TABLE 24.1. Hepatitis E Prevalence in Nonendemic Regions

Country	Population	Seroprevalence (%)	References
Argentina	HIV	6.6	Fainboim et al. (1999)
Bolivia	Rural area	7.3	Bartoloni et al. (1999)
Brazil	BD	2.3	Bortoliero et al. (2006)
China	BD	32.6	Guo et al. (2010)
Cuba	Random	5.3	Quintana et al. (2005)
Denmark	BD	20.6	Christensen et al. (2008)
France	BD	3.2–52.2	Boutrouille et al. (2007) and Mansuy et al. (2011)
Germany	HC prof.	3.9	Nubling et al. (2002)
Greece	HD	4.8–9.7	Stefanidis et al. (2004)
	BD	0.23	Dalekos et al. (1998)
Hong Kong	Unknown	18.8	Wong et al. (2004)
Hungary	Acute hep.	9.6	Reuter et al. (2009)
Italy	BD	2.9	Vulcano et al. (2007)
	Swine W	33	
Japan	BD	8.9	Abe et al. (2006)
Korea	BD	18	Choi et al. (2003)
Moldavia	Swine W	51.1	Drobeniuc et al. (2001)
New Zealand	BD	4	Dalton et al. (2007)
The Netherlands	BD	2	Bouwknegt et al. (2008a)
	Swine W	11	
Portugal	BD	4	Macedo et al. (1998)
	HD	6.8	
Spain	BD	4.1	Galiana et al. (2008)
	Swine W	18.8	
Sweden	BD	9.3	Olsen et al. (2006)
	Swine W	13	
Switzerland	Sewage W	3.3	Jeggli et al. (2004)
Taiwan	BD	8.9	Lee et al. (2005)
	HD	31	
Turkey	Urban/rural	3.8	Cesur et al. (2002)
United Kingdom	BD	16–25	Dalton et al. (2008)
United States	BD	18–21	Kuniholm et al. (2009)
	Swine W	26	Meng et al. (2002)

BD, blood donors; HD, hemodialyzed patients; Swine W, professionals in swine-related occupations; HIV, patients with HIV; HC prof., health-care professionals; acute hep., acute hepatitis patients; Sewage W, professionals exposed to sewage.

exposed than women, and personnel with animal-related occupations, such as pig breeders, veterinarians, or slaughterhouse personnel, are statistically more frequently positive for anti-HEV antibodies (Bouwknegt et al., 2008a; Drobeniuc et al., 2001; Meng et al., 2002; Olsen et al., 2006). Hunters also appear to have a higher risk of exposure. According to a study performed in the south of France, contact with wildlife (hunters) increases the risk of HEV exposure (80% in hunters vs. 52.5% in the general population) (Mansuy et al., 2011).

Viral hepatitis E may occur in various clinical forms: acute, chronic, fulminant, or associated with neurologic symptoms.

24.3.1 Acute hepatitis

The clinical outcome in patients from industrialized countries is similar to that of those from endemic areas. Nevertheless, the mortality rate seems to be higher in developing areas, where it can reach 11% (Peron et al., 2007). Very severe forms are observed, particularly among individuals with an underlying liver condition or active alcohol abuse (Dalton et al., 2011).

The classical clinical HEV infection varies from subclinical forms to acute hepatitis. The mean incubation period is 40 days, after which symptoms appear. Symptoms can be nonspecific, including fever, pain, myalgia, vomiting, anorexia, and pruritus, or more evocative, such as jaundice and dark urine (Dalton et al., 2007; Mansuy et al., 2009). Hepatitis E virus viremia and viral excretion in feces start during the incubation period, 1–2 weeks before the onset of symptoms and transaminase increase. In some cases, this transient increase in serum liver enzymes may be absent (Nicand et al., 2001). HEV RNA is cleared from the blood from a few days to 2 weeks after the onset of clinical symptoms and from feces 2 weeks after onset.

24.3.2 Chronic hepatitis

In nonendemic areas, chronic hepatitis E has been observed in populations with compromised immune response such as transplant recipients following an immunosuppressive treatment (Haagsma et al., 2008; Kamar et al., 2008). Most of them had nonspecific symptomatic and chronic HEV infection diagnosed following an investigation of liver enzyme abnormalities. These patients frequently developed (60%) chronic liver disease (steatosis, fibrosis, cirrhosis) confirmed by histologic hepatitis abnormalities, persistent high levels of aminotransferase, and persistent presence of HEV RNA in serum. Reducing immunosuppressive therapy doses resulted in viral clearance in more than 30% of patients (Kamar et al., 2011b). Similarly, HEV has been described as persisting in HIV-infected patients in a developed area. In these cases, it was linked to a low CD4 cell count (Dalton et al., 2009). However, HIV-positive patients are not considered at higher risk of chronic hepatitis E than the general population in nonendemic regions (Renou et al., 2009). In tropical and subtropical areas, no study has investigated HIV and HEV coinfection.

24.3.3 Fulminant hepatitis

In some endemic areas, fulminant hepatic failure (FHF), with high mortality rates, has been described in pregnant women infected during the third term with HEV genotype 1. In contrast to areas such as Egypt or Southern India, pregnant women were not found to be at higher risk of developing FHF in these endemic areas (Navaneethan et al., 2008; Stoszek et al., 2006). A 20-year study reported an absence of correlation between HEV infection and FHF during pregnancy (Bhatia et al., 2008). Thus, other factors such as coinfections, viral load, or HEV strains (genotype) might play a role in the severity of symptoms (Renou et al., 2008). It is worth noting that cases of FHF during pregnancy were associated only with HEV genotype 1. Up to now, no case of fulminant hepatitis E has been reported in pregnant women infected with genotype 3 or 4 strains in nonendemic areas. A recent case of genotype 3 HEV infection in a pregnant woman failed to produce symptoms and was not transmitted to the fetus (Anty et al., 2012). It has been hypothesized that HEV genotypes 3 and 4 could be less virulent in humans (Purcell and Emerson, 2008). Nevertheless, FHF has been reported with genotype 3 or 4 in male patients with major underlying liver conditions (Dalton et al., 2007).

24.3.4 Neurologic disorders

Recently, several observations have reported on neurologic complications in acute or chronic cases of hepatitis E. In particular, a study including 126 patients with autochthonous genotype 3 infection from the United Kingdom and France reports seven (5.5%) neurologic disorders such as inflammatory polyradiculopathy, Guillain–Barre syndrome, bilateral brachial neuritis, encephalitis, and ataxia/proximal myopathy. Possible links between these neurologic disorders and extrahepatic manifestation of HEV infection are corroborated by the presence of HEV RNA in cerebrospinal fluid of patients with chronic HEV infection. Outcomes ranged from complete resolution to improvement with residual neurologic deficits to no improvement in one case (Kamar et al., 2011a).

24.4 ANIMAL RESERVOIRS

Domestic pigs are undoubtedly the main animal reservoirs of HEV worldwide (Pavio et al., 2010). However, anti-HEV antibodies have been detected in a large range of other animal species including wild boars, deer, rats, dogs, cats, mongooses, cows, sheep, goats, avian species, rabbits, and horses, suggesting that they are exposed to HEV or HEV-like agents. While viruses of genotype 3 and/or 4 have been identified in wild boars, deer, and mongooses, aHEV in poultry, rHEV in urban rats, and rbHEV in farm rabbits, the source of seropositivity in other animals remains unknown (Pavio et al., 2010).

24.4.1 HEV in pigs

Pigs become naturally infected at 8–10 weeks of age following a natural decrease in passive immunity gained through colostrum absorption (60% of piglets). The infection route is fecal–oral, through close contact between animals and natural coprophagia. Hepatitis E virus infection remains asymptomatic, with neither hyperthermia nor a loss in either appetite or weight and no change in normal behavior. Thus, HEV in swine herds causes no economic loss for breeders. As in humans, HEV replicates actively in the liver and is excreted in feces through the bile. Active viral shedding is observed at 13–16 weeks of age, and seroconversion appears between 15 and 18 weeks of age. Several studies on HEV prevalence in swine herds have been conducted in different countries from endemic and nonendemic regions, and all of them converge to a very widespread distribution of HEV in swine, with 50–100% of positive herds. Within each herd, seroprevalence rates vary (from a few to 100%) depending on the age of the animals tested, as the rate is usually higher in pigs over 4 months old (Fernandez-Barredo et al., 2006; Seminati et al., 2008; Takahashi et al.,

2005). By slaughter age, around 20 weeks, most animals have eliminated the virus. However, the molecular detection of HEV in pig livers collected at slaughter houses or in grocery stores has shown that from 2% to 11% of livers are positive for HEV RNA. Thus, HEV infection in pigs might be chronic in some cases, with a prolonged period of excretion, or breeding management may favor late infection of naïve animals.

Hepatitis E virus strains amplified from swine in both endemic and nonendemic regions are of genotype 3 or 4. In regions where sporadic cases of HEV are reported, the swine HEV sequences identified are genetically very close to local human HEV sequences (Bouquet et al., 2011), suggesting zoonotic transmission.

24.4.2 Prevalence of HEV in wild animals

Several studies have been conducted on the presence of HEV in wild boars and deer in Europe and Japan. The seroprevalence rates observed varied from 3% to 43% in wild boars and 2% to 35% in deer (Pavio et al., 2010). The prevalence of HEV RNA ranged from 3% to 34% in wild boars. In these studies, sample sizes were generally small and from various origins such as liver, sera, bile, and feces with little data on the age of the animals tested, making it difficult to estimate the infection level of these two species. The HEV strains amplified from wild boars and deer are of genotype 3 or 4 and are genetically very closely related to human and swine strains described in the same geographic regions (Reuter et al., 2009), suggesting possible interspecies transmission. However, an unrecognized genotype was isolated once in a wild boar from Japan (Takahashi et al., 2011) (Figure 24.2).

The presence of genotype 3 HEV in Japanese mongooses (8.3%, 7/84) has been reported (Li et al., 2006; Nakamura et al., 2006), although no human case associated with this reservoir has been described so far.

24.4.3 Prevalence of HEV in avian, rats, and rabbits

In contrast to infection of swine, which is mainly asymptomatic in animals, aHEV can cause poultry diseases such as big liver and spleen disease and hepatitis–splenomegaly syndrome, which resembles acute hepatitis E. Studies in Australia, North America, Europe, and China have shown that aHEV can be enzootic in chicken flocks (Marek et al., 2010; Sun et al., 2004a; Zhao et al., 2010). High seroprevalence (up to 80%) has been described in chickens over 40 weeks (Peralta et al., 2009). Genetically distinct aHEVs were identified in the aforementioned areas (Figure 24.3), suggesting an ancient ancestor (other genus) and an independent evolution into three different genotypes (Bilic et al., 2009). Apparently avirulent strains of aHEV isolated from healthy flocks are genetically related to virulent ones. The virulence determinants responsible for severity of symptoms have not yet been identified.

rHEV was first identified in Germany in *Rattus norvegicus* and then in urban rats from California, United States, and Vietnam (Johne et al., 2009; Li et al., 2011; Purcell et al., 2011). The seroprevalence of rHEV in rats from Vietnam or the United States was estimated to range from 20% to 80% in adult rats (over 200 g), respectively. The experimental infection of rats reveals poor transmissibility with mild hepatitis (rare liver lesions) and normal liver enzymes (Purcell et al., 2011). The rHEV sequences identified are genetically distinct from other HEV genotypes and aHEV and may constitute a new genus (Figure 24.2).

The natural infection of rex rabbits with a novel HEV (rbHEV) was first observed in China and later in farmed rabbits in the United States. The seroprevalence of rbHEV varies from 36% to 15–55% according to studies from the United States and China, respectively

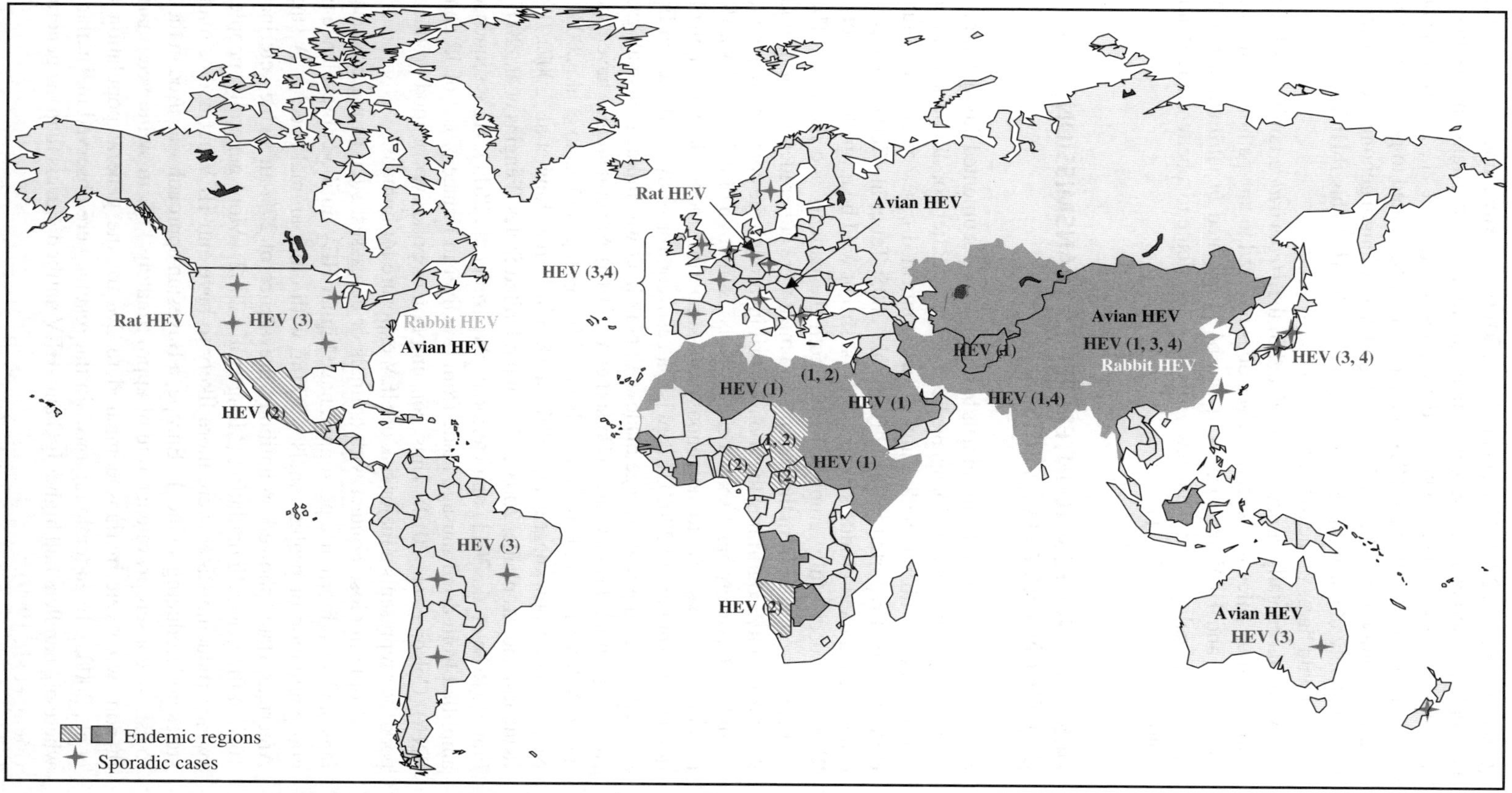

Figure 24.3. Geographic distribution of HEV and HEV-like viruses. For color detail, please see color plate section.

(Cossaboom et al., 2011; Geng et al., 2010, 2011b). The dynamics of natural infection have not yet been characterized. Experimental infection caused a transient rise in liver enzymes in some cases, associated with mild liver lesions (Ma et al., 2010). Depending on the genomic region compared, rbHEV sequences may be related to genotype 3 (>80%) or may be a new genotype. Further analyses using more full-length genomic sequences are required to definitively classify this HEV (Geng et al., 2011b). This rbHEV has not been investigated in wild hares or jackrabbits.

More anecdotally, a new virus related to the *Hepeviridae* family has been isolated in cutthroat trout. Infection with this virus does not lead to any pathology in fish, but, unlike other HEV and HEV-like viruses, it can be propagated *in vitro* and establish persistent infections of cell culture. This new virus may contribute to a better understanding of HEV biology (Batts et al., 2011).

24.5 ZOONOTIC AND INTERSPECIES TRANSMISSION OF HEV AND HEV-LIKE VIRUSES

Zoonotic transmissions from animals to humans are documented through direct and indirect evidence. Direct transmissions have been described after consumption of infected meat or direct contact with infected animals. Two confirmed cases of zoonotic transmission of HEV were first reported in Japan (Li et al., 2005; Tei et al., 2003) after consumption of contaminated meat products of sika deer (sushi) or wild boar (grilled). In both cases, HEV RNA was successfully amplified from both the patients and the meat. The HEV viral sequences recovered from the patients and meat were either identical or near identical with 99.95% identity, confirming the zoonotic origin through the consumption of meat products. A study in Germany has also associated the consumption of offal (41% vs. 19%) and wild boar meat (20% vs. 7%) to indigenous cases supporting the foodborne origin of HEV infection (Wichmann et al., 2008). Likewise, a case–control study conducted in France clearly showed that eating raw sausage made with pork liver was linked to a small group of indigenous hepatitis E human cases (Colson et al., 2010). Several studies have reported detecting HEV RNA in commercial pork livers: 1.9% in Japan (Yazaki et al., 2003), 6.5% in the Netherlands (Bouwknegt et al., 2007), 10.8% in Korea (Jung et al., 2007), 0.83% in India (Kulkarni and Arankalle, 2008), 11% in the United States (Feagins et al., 2007), and 4% in France (Rose et al., 2011) and Germany (Wenzel et al., 2011). HEV sequences were also amplified from dried sausage made from raw liver in France (7 out of 12 contained HEV RNA) (Colson et al., 2010). Thus, consumption of pork liver or food products containing pork liver represents a major risk of HEV exposure for humans.

People in close contact with animals or carcasses, such as pig breeders, veterinarians, and slaughterhouse personnel, have higher HEV antibody prevalence than the general population or professionals working with other animal species. A survey in eight American states showed an antibody prevalence of 27% in swine veterinarians versus 16% in the general population (Meng et al., 2002). More generally, people from major swine-producing states were more likely to have anti-HEV antibodies than those from states not producing swine. In Europe, a Bayesian approach was used in the Netherlands to estimate a seroprevalence rate of approximately 11% in swine veterinarians and 6% in non-swine veterinarians compared to 2% for the general population (Bouwknegt et al., 2008a). In endemic regions, similar results were observed in Thailand and China where pig handlers had higher IgG anti-HEV antibody levels than the general population (Meng et al., 1999).

Thus, HEV genotypes 3 and 4—found in the pig reservoir—seem to be easily transmitted to humans. This observation is also supported by experimental models of cross-species infection.

Inoculation of rhesus macaques (*Macaca mulatta*) with swine HEV of genotype 3 or 4 led to a productive infection (Arankalle et al., 2006; Meng et al., 1998b) with a limited increase in enzymes, fecal excretion, and seroconversion (Arankalle et al., 2006) (Table 24.2). In the same way, genotype 3 and 4 HEVs present in humans are infectious when used to inoculate pigs (Feagins et al., 2008b; Meng et al., 1998b). A comparison of pathological lesion profiles after infection of rhesus macaques with human or swine HEV genotype 3 strains did not reveal major differences, suggesting that genotype 3 or 4 HEV strains have no specific host restriction between swine and primates (Halbur et al., 2001; Meng et al., 1998b). To support the absence of a host barrier for HEV genotype 3, a molecular study using high-throughput sequencing was performed on HEV genetic consensus sequences and viral quasispecies during controlled interspecies transmission from human to pigs. This study revealed that there is no change in the consensus sequence and that major viral quasispecies are conserved, confirming the absence of host pressure or host barrier (Bouquet et al., 2012).

Swine are naturally infected by genotype 3 or 4 HEV but are not susceptible to infections by HEV genotype 1 (Pakistani strain Sar-55) nor genotype 2 (Mexican strain) (Meng et al., 1998a) (Table 24.2). The molecular mechanism underlying this host restriction remains unknown. Pig infection cannot therefore be used as a surrogate model to study

TABLE 24.2. Barrier Species Crossing of HEVs

Genotype	Natural host	Exp. models	Serology	Infection	Reference
1	Human	Macaque	+	+	Tsarev et al. (1993)
		Swine	−	−	Meng et al. (1998a)
		Rabbit	+	−	Ma et al. (2010)
2	Human	Macaque	+	+	Li et al. (2006)
		Swine	−	−	Meng et al. (1998a)
3	Human	Macaque	+	+	Erker (1999)
		Swine	+	+	Meng et al. (1998a)
	Swine	Macaque	+	+	Meng et al. (1998b)
		Swine	+	+	Meng et al. (1998b)
4	Human	Macaque	+	+	Ma et al. (2009)
		Swine	+	+	Feagins et al. (2008b)
		Rabbit	+	+	Ma et al. (2010)
	Swine	Macaque	+	+	Arankalle et al., (2006)
		Swine	+	+	
Rabbit	Rabbit	Rabbit	+	+	Ma et al. (2010)
		Swine	+	+	Cossaboom et al. (2012)
Rat	Rat	Rat	+	+	Purcell et al. (2011)
		Macaque	−	−	Purcell et al. (2011)
		Swine	−	−	Cossaboom et al. (2012)
Avian	Chicken	Chicken	+	+	Huang et al. (2004)
		Turkey	+	+	Sun et al. (2004b)
		Macaque	−	−	Huang et al. (2004)
		Swine	−	−	Pavio et al. (2010)

HEV infection in humans. Furthermore, experimental infection of pregnant gilts with swine HEV genotype 3 did not lead to fulminant hepatitis, abortion, or vertical transmission (Kasorndorkbua et al., 2003).

For the other HEV or HEV-like viruses present in rabbits, rats, or avian, there is no evidence of transmission to humans. Up to now, no case of direct zoonotic transmission has been observed with rbHEV, rHEV, or aHEV. This may be due to a lack of serologic or molecular tools to detect all HEV-related viruses in human cases. Although studies have shown that these viruses may have common epitopes and that antibodies against HEV (anti-HEV) can recognize recombinant antigen made from rbHEV, rHEV, or aHEV (Cossaboom et al., 2012; Guo et al., 2006; Haqshenas et al., 2002). aHEV and rHEV appear to be genetically distant from other HEV genotypes (1–4) or rbHEV, so molecular tools for their detection are not available for humans.

The host range of rbHEV has not yet been extensively studied. A molecular study based on identification of possible host range determinants in the rabbit strain suggests that rbHEV may cross the species barrier (Geng et al., 2011a). A recent study has shown the possible transmission of a rabbit strain to pigs (Cossaboom et al., 2012) (Table 24.2). On the other hand, the sensitivity of rabbits to genotypes 1 and 4 has been assessed. Infection and viral excretion were observed only in a few animals (2/9) inoculated with genotype 4 and none in those inoculated with genotype 1. Seroconversion was observed in both groups (Ma et al., 2010). The restricted transmission to rabbits of HEV genotypes present in humans (1 and 4) suggests that the rabbit reservoir is unlikely to host these HEVs and cannot therefore be considered to constitute a minor risk of exposure for humans.

The host range of aHEV appears limited. Although aHEV can infect turkeys (Sun et al., 2004b), attempts to infect pigs and rhesus monkeys with aHEV were unsuccessful (Huang et al., 2004; Pavio et al., 2010). In addition, a swine genotype 3 HEV failed to infect chickens under experimental conditions (Pavio et al., 2010), suggesting that interspecies transmission is not very probable.

The host range of rHEV has not yet been extensively studied. One study reported the absence of infection, viral excretion, or seroconversion after inoculation of rhesus monkeys with rHEV (Purcell et al., 2011). This result suggests that rHEV is not a source of human infection.

In conclusion, until further investigations clarify the possible transmission of rbHEV to nonhuman primates, only genotypes 3 and 4 identified in pigs, wild boars, deer, or mongooses are a source of zoonotic HEV transmission to humans.

24.6 HEV IN THE ENVIRONMENT

In tropical and subtropical countries, HEV has been detected in sewage, river water, and drinking water, confirming the waterborne origin of HEV (Hazam et al., 2010; Kitajima et al., 2009; Parashar et al., 2011). In Northern countries, besides contamination through consumption of infected food or direct contact with infected animals, HEV of animal origin can also be found in the environment. Large quantities of the virus can be excreted in animal feces (Kasorndorkbua et al., 2005) and persist in manure storage facilities (earthen lagoons or concrete pits) (Kasorndorkbua et al., 2005). Manure spreading might constitute a risk for contaminating field-grown fruit or vegetables. As shown very recently, HEV can be found in field-grown strawberries, suggesting that either the field or irrigation water is contaminated by HEV (Brassard et al., 2012). Regarding HEV contamination of surface water, a study in the Netherlands showed that 17% of samples collected were positive

(Rutjes et al., 2009). The possible HEV contamination of the environment is thus of concern and must be further studied. In contrast to HAV, which is frequently transmitted through shellfish such as mussels or oysters, there is no evidence of HEV transmission through such seafood products. In Japan, HEV RNA was found in one type of clam (*Corbicula japonica*), but there was no linked human case (Li et al., 2007). One report mentioned a possible seafood origin for several grouped cases of hepatitis E that occurred during a world ship cruise (Said et al., 2009). Little is known about HEV resistance in a high-salt environment.

24.7 CLIMATE CHANGE AND IMPACT ON HEV EXPOSURE

As for other waterborne diseases, climate change directly affects HEV propagation since the water cycle is drastically modified. The quantity and quality of water resources available to meet human and environmental requirements are subject to major disruption. Climate change can lead to both floods and drought. Global warming causes a rise in sea level and seriously affects coastal aquifers, a major source of urban and regional water supply systems. Global warming and higher water temperatures can also exacerbate many forms of water pollution. Water supply reliability, health, agriculture, energy, and aquatic ecosystems will all be impacted by these changes.

In some developing countries, drought and/or local conflicts have contributed to population movements. The limited source of water supplies in such situations has led to major hepatitis E epidemics (Guerrero-Latorre et al., 2011; Guthmann et al., 2006), with a high mortality rate among pregnant women (Boccia et al., 2006). On the other hand, it has been shown in the past that seasonal flooding can be responsible for contaminated water supplies by urban sewage, causing thousands of hepatitis E cases (Labrique et al., 2010; Mamun Al et al., 2009). Thus, in both schemes, droughts followed by floods are climatic factors contributing to increased exposure to HEV and widespread epidemics in developing countries. Increased climate change will raise the risk of HEV exposure through contaminated drinking water.

In industrialized countries, sanitation is not usually a problem, but floods might occasionally contribute to HEV exposure.

Furthermore, most of the world's coastal cities were established over the last few millennia, a period when global sea level was near constant. Since the mid-nineteenth century, the sea level has been rising. During the twentieth century, it rose about 15–20 cm (Climate Institute: http://www.climate.org/). An increase in sea level can have a dramatic impact on many coastal environments. Up to now, the survival of HEV in salty seawater has not been evaluated. Since HEV might be transmitted through consumption of shellfish, a rise in sea level may contribute to the spread of HEV in coastal environments.

HEV resistance to UV rays is not known, so any positive effects of higher UV light exposure on surface water or fields cannot be evaluated.

Climate change is also the most challenging event facing human and animal populations; it affects the population dynamics of wild animals, reproductive success, and population densities of some species (Singh et al., 2011). In some areas, wild fauna such as wild boars, known to be a reservoir of HEV, are in contact with domestic animals, fostering the transmission of pathogens. Larger populations of susceptible hosts—wild animals—contribute to higher interanimal transmission rates and consequently increased probability of contamination of domestic pigs, which excrete many pathogens with zoonotic potential, including HEV. Transmission of HEV by water and food may increase significantly.

In recent years, food habits have also evolved, and there is an emerging widespread tendency to eat raw food products. With global warming and increased local temperatures, people prefer barbecues to slowly simmered dishes. Several cases of HEV have been associated with raw or undercooked dishes. Exposure to HEV through contaminated food products consumed raw (Colson et al., 2010) or lightly cooked (Li et al., 2005) may increase the number of hepatitis E cases in the future.

24.8 PREVENTION

In tropical and subtropical countries, HEV prevention focuses mainly on sanitation, the goal being to limit contamination of drinking water. Particular information must be disseminated and surveillance increased for certain members of the population, such as pregnant women, who are at higher risk of fulminant hepatitis. In Northern countries, where zoonotic transmission seems to be the main route of hepatitis E infection, consumers of raw or undercooked meat should be advised of the potential risk. Since infectious HEV can be inactivated at high temperatures (>71 °C), cooking must be recommended for food product containing pork liver (Feagins et al., 2008a). Populations at risk of severe forms, such as pregnant women, transplant recipients, and individuals with underlying liver conditions, should be properly informed on the possible risk of consuming raw pork, deer, or wild boar meat or having contact with infected pigs. Similarly, slaughterhouse personnel, pig handlers, and veterinarians must all take hygienic measures after handling the meat or animals. In swine, since HEV seems very contagious between animals (Bouwknegt et al., 2008b), further investigations should determine whether farming procedures could limit HEV dissemination. HEV infection can also be prevented with an efficient immunization program (Purcell and Emerson, 2008). Since HEV has only one serotype, and natural infection leads to protective antibodies (Purcell and Emerson, 2008), HEV is a good candidate for the development of an effective vaccine. A widespread vaccination campaign in developing countries would reduce major waterborne epidemics affecting thousands of people. Travelers to endemic regions could also benefit from vaccination. In industrialized countries, populations at risk such as transplant recipients or persons with underlying liver conditions could also be a target for vaccination. The safety and efficacy of an HEV recombinant protein vaccine was evaluated in a phase 2 randomized, double-blind placebo-controlled trial (Shrestha et al., 2007). This vaccine—based on the HEV capsid protein—showed good efficacy (88.5% after one dose and 95.5% after three doses) in the prevention of hepatitis E in patients (Shrestha et al., 2007).

24.9 CONCLUSION

In the past 10 years, zoonotic HEV has been of increasing concern, mostly in countries with sporadic cases. Further studies should be conducted to identify the possible zoonotic origin of hepatitis E where human genotypes (1 and 2) are endemic. In previously named nonendemic regions, it appears that HEV is endemic in both the swine population and other species. It is now well known that many sporadic cases in those regions were of animal origin and that HEV strains were circulating freely between humans and animals. Direct transmission through contact or the consumption of contaminated food has been clearly identified. Other indirect transmission pathways are suspected, such as vegetables irrigated with contaminated water or shellfish grown in polluted water, and need to be investigated. Due to its transmission as a waterborne or foodborne pathogen, HEV infection is directly impacted by climate change and will be of increasing concern for the next decade. Since it is a rapidly evolving virus with a high mutation rate, its virulence must be placed under close surveillance. Prevention and preparedness measures for future HEV outbreaks arising from climate change should be taken to limit epidemics in both developing countries and Northern countries.

ACKNOWLEDGEMENT

JB was supported by a Ph.D. grant from ANSES. NP received funding from the European Union Seventh Framework Programme (FP7/2007-2013) under grant agreement no. 278433-PREDEMICS.

REFERENCES

Abe, K., Li, T. C., Ding, X., et al. (2006). International collaborative survey on epidemiology of hepatitis E virus in 11 countries. *Southeast Asian J Trop Med Public Health* **37**(1), 90–95.

Anty, R., Ollier, L., Peron, J. M., et al. (2012). First case report of an acute genotype 3 hepatitis E infected pregnant woman living in South-Eastern France. *J Clin Virol* **54**(1), 76–78.

Arankalle, V. A., Tsarev, S. A., Chadha, M. S., et al. (1995). Age-specific prevalence of antibodies to hepatitis A and E viruses in Pune, India, 1982 and 1992. *J Infect Dis* **171**(2), 447–450.

Arankalle, V. A., Chobe, L. P., and Chadha, M. S. (2006). Type-IV Indian swine HEV infects rhesus monkeys. *J Viral Hepat* **13**(11), 742–745.

Bartoloni, A., Bartalesi, F., Roselli, M., et al. (1999). Prevalence of antibodies against hepatitis A and E viruses among rural populations of the Chaco region, south-eastern Bolivia. *Trop Med Int Health* **4**(9), 596–601.

Batts, W., Yun, S., Hedrick, R., et al. (2011). A novel member of the family Hepeviridae from cutthroat trout (*Oncorhynchus clarkii*). *Virus Res* **158**(1–2), 116–123.

Bhatia, V., Singhal, A., Panda, S. K., et al. (2008). A 20-year single-center experience with acute liver failure during pregnancy: is the prognosis really worse? *Hepatology* **48**(5), 1577–1585.

Bilic, I., Jaskulska, B., Basic, A., et al. (2009). Sequence analysis and comparison of avian hepatitis E viruses from Australia and Europe indicate the existence of different genotypes. *J Gen Virol* **90**(Pt 4), 863–873.

Boccia, D., Guthmann, J. P., Klovstad, H., et al. (2006). High mortality associated with an outbreak of hepatitis E among displaced persons in Darfur, Sudan. *Clin Infect Dis* **42**(12), 1679–1684.

Bortoliero, A. L., Bonametti, A. M., Morimoto, H. K., et al. (2006). Seroprevalence for hepatitis E virus (HEV) infection among volunteer blood donors of the Regional Blood Bank of Londrina, State of Parana, Brazil. *Rev Inst Med Trop Sao Paulo* **48**(2), 87–92.

Bouquet, J., Tesse, S., Lunazzi, A., et al. (2011). Close similarity between sequences of hepatitis E virus recovered from humans and swine, France, 2008–2009. *Emerg Infect Dis* **17**(11), 2018–2025.

Bouquet, J., Cheval, J., Rogee, S., et al. (2012). Identical consensus sequence and conserved genomic polymorphism of hepatitis E virus during controlled interspecies transmission. *J Virol* **86**(11): 6238–6245.

Boutrouille, A., Bakkali-Kassimi, L., Cruciere, C., et al. (2007). Prevalence of anti-hepatitis E virus antibodies in French blood donors. *J Clin Microbiol* **45**(6), 2009–2010.

Bouwknegt, M., Engel, B., Herremans, M. M., et al. (2008a). Bayesian estimation of hepatitis E virus seroprevalence for populations with different exposure levels to swine in the Netherlands. *Epidemiol Infect* **136**(4), 567–576.

Bouwknegt, M., Frankena, K., Rutjes, S. A., et al. (2008b). Estimation of hepatitis E virus transmission among pigs due to contact-exposure. *Vet Res* **39**(5), 40.

Bouwknegt, M., Lodder-Verschoor, F., van der Poel, W. H., et al. (2007). Hepatitis E virus RNA in commercial porcine livers in the Netherlands. *J Food Prot* **70**(12), 2889–2895.

Brassard, J., Gagné, M.J., Généreux, M., et al. (2012). Detection of human food-borne and zoonotic viruses on irrigated, field-grown strawberries. *Appl Environ Microbiol* **78**(10), 3763–3766.

Cesur, S., Akin, K., Dogaroglu, I., et al. (2002). Hepatitis A and hepatitis E seroprevalence in adults in the Ankara area. *Mikrobiyol Bul* **36**(1), 79–83.

Choi, I. S., Kwon, H. J., Shin, N. R., et al. (2003). Identification of swine hepatitis E virus (HEV) and prevalence of anti-HEV antibodies in swine and human populations in Korea. *J Clin Microbiol* **41**(8), 3602–3608.

Christensen, P. B., Engle, R. E., Hjort, C., et al. (2008). Time trend of the prevalence of hepatitis E antibodies among farmers and blood donors: a potential zoonosis in Denmark. *Clin Infect Dis* **47**(8), 1026–1031.

Colson, P., Borentain, P., Queyriaux, B., et al. (2010). Pig liver sausage as a source of hepatitis E virus transmission to humans. *J Infect Dis* **202**(6), 825–834.

Cossaboom, C. M., Cordoba, L., Dryman, B. A., et al. (2011). Hepatitis E virus in rabbits, Virginia, USA. *Emerg Infect Dis* **17**(11), 2047–2049.

Cossaboom, C. M., Cordoba, L., Sanford, B. J., et al. (2012). Cross-species infection of pigs with a novel rabbit, but not rat, strain of hepatitis E virus isolated in the United States. *J Gen Virol* **93**(Pt 8), 1687–1695.

van Cuyck, H., Fan, J., Robertson, D. L., et al. (2005). Evidence of recombination between divergent hepatitis E viruses. *J Virol* **79**(14), 9306–9314.

Dalekos, G. N., Zervou, E., Elisaf, M., et al. (1998). Antibodies to hepatitis E virus among several populations in Greece: increased prevalence in an hemodialysis unit. *Transfusion* **38**(6), 589–595.

Dalton, H. R., Thurairajah, P. H., Fellows, H. J., et al. (2007). Autochthonous hepatitis E in southwest England. *J Viral Hepat* **14**(5), 304–309.

Dalton, H. R., Stableforth, W., Thurairajah, P., et al. (2008). Autochthonous hepatitis E in Southwest England: natural history, complications and seasonal variation, and hepatitis E virus IgG seroprevalence in blood donors, the elderly and patients with chronic liver disease. *Eur J Gastroenterol Hepatol* **20**(8), 784–790.

Dalton, H. R., Bendall, R. P., Keane, F. E., et al. (2009). Persistent carriage of hepatitis E virus in patients with HIV infection. *N Engl J Med* **361**(10), 1025–1027.

Dalton, H.R., Bendall, R.P., Rashid, M., et al. (2011). Host risk factors and autochthonous hepatitis E infection. *Eur J Gastroenterol Hepatol* **23**(12), 1200–1205.

Drobeniuc, J., Favorov, M. O., Shapiro, C. N., et al. (2001). Hepatitis E virus antibody prevalence among persons who work with swine. *J Infect Dis* **184**(12), 1594–1597.

Emerson, S. U., Anderson, D., Arankalle, A., et al. (2004). Hepevirus. *In* "Virus Taxonomy. Eighth Report of the International Committee on Taxonomy of Viruses" (M. A. M. C. M. Fauquet, J. Maniloff, U. Desselberger, et al., Eds.), pp. 851–855. Academic Press, London.

Emerson, S. U., Nguyen, H. T., Torian, U., et al. (2010). Release of genotype 1 hepatitis E virus from cultured hepatoma and polarized intestinal cells depends on open reading frame 3 protein and requires an intact PXXP motif. *J Virol* **84**(18), 9059–9069.

Erker, J. C., Desai, S.M., and Mushahwar I. K.(1999). Rapid detection of Hepatitis E virus RNA by reverse transcription-polymerase chain reaction using universal oligonucleotide primers. *J Virol Methods* **81**(1-2), 109–13.

Fainboim, H., Gonzalez, J., Fassio, E., et al. (1999). Prevalence of hepatitis viruses in an anti-human immunodeficiency virus-positive population from Argentina. A multicentre study. *J Viral Hepat* **6**(1), 53–57.

Feagins, A. R., Opriessnig, T., Guenette, D. K., et al. (2007). Detection and characterization of infectious Hepatitis E virus from commercial pig livers sold in local grocery stores in the USA. *J Gen Virol* **88**(3), 912–917.

Feagins, A. R., Opriessnig, T., Guenette, D. K., et al. (2008a). Inactivation of infectious hepatitis E virus present in commercial pig livers sold in local grocery stores in the United States. *Int J Food Microbiol* **123**(1–2), 32–37.

Feagins, A. R., Opriessnig, T., Huang, Y. W., et al. (2008b). Cross-species infection of specific-pathogen-free pigs by a genotype 4 strain of human hepatitis E virus. *J Med Virol* **80**(8), 1379–1386.

Fernandez-Barredo, S., Galiana, C., Garcia, A., et al. (2006). Detection of hepatitis E virus shedding in feces of pigs at different stages of production using reverse transcription-polymerase chain reaction. *J Vet Diagn Invest* **18**(5), 462–465.

Galiana, C., Fernandez-Barredo, S., Garcia, A., et al. (2008). Occupational exposure to hepatitis E virus (HEV) in swine workers. *Am J Trop Med Hyg* **78**(6), 1012–1015.

Geng, J., Wang, L., Wang, X., et al. (2010). Study on prevalence and genotype of hepatitis E virus isolated from Rex Rabbits in Beijing, China. *J Viral Hepat* **18**(9), 661–667.

Geng, J., Fu, H., Wang, L., et al. (2011a). Phylogenetic analysis of the full genome of rabbit hepatitis E virus (rbHEV) and molecular biologic study on the possibility of cross species transmission of rbHEV. *Infect Genet Evol* **11**(8), 2020–2025.

Geng, Y., Zhao, C., Song, A., et al. (2011b). The serological prevalence and genetic diversity of hepatitis E virus in farmed rabbits in China. *Infect Genet Evol* **11**(2), 476–482.

Graff, J., Nguyen, H., Yu, C., et al. (2005). The open reading frame 3 gene of hepatitis E virus contains a cis-reactive element and encodes a protein required for infection of macaques. *J Virol* **79**(11), 6680–6689.

Grandadam, M., Tebbal, S., Caron, M., et al. (2004). Evidence for hepatitis E virus quasispecies. *J Gen Virol* **85**(Pt 11), 3189–3194.

Guerrero-Latorre, L., Carratala, A., Rodriguez-Manzano, J., et al. (2011). Occurrence of water-borne enteric viruses in two settlements based in Eastern Chad: analysis of hepatitis E virus, hepatitis A virus and human adenovirus in water sources. *J Water Health* **9**(3), 515–524.

Guo, H., Zhou, E. M., Sun, Z. F., et al. (2006). Identification of B-cell epitopes in the capsid protein of avian hepatitis E virus (avian HEV) that are common to human and swine HEVs or unique to avian HEV. *J Gen Virol* **87**(Pt 1), 217–223.

Guo, Q. S., Yan, Q., Xiong, J. H., et al. (2010). Prevalence of hepatitis E virus in Chinese blood donors. *J Clin Microbiol* **48**(1), 317–318.

Guthmann, J. P., Klovstad, H., Boccia, D., et al. (2006). A large outbreak of hepatitis E among a displaced population in Darfur, Sudan, 2004: the role of water treatment methods. *Clin Infect Dis* **42**(12), 1685–1691.

Haagsma, E. B., van den Berg, A. P., Porte, R. J., et al. (2008). Chronic hepatitis E virus infection in liver transplant recipients. *Liver Transpl* **14**(4), 547–553.

Hakze-van der Honing, R. W., van Coillie, E., Antonis, A. F., et al. (2011). First isolation of hepatitis E virus genotype 4 in Europe through swine surveillance in the Netherlands and Belgium. *PLoS One* **6**(8), e22673.

Halbur, P. G., Kasorndorkbua, C., Gilbert, C., et al. (2001). Comparative pathogenesis of infection of pigs with hepatitis E viruses recovered from a pig and a human. *J Clin Microbiol* **39**(3), 918–923.

Haqshenas, G., Shivaprasad, H. L., Woolcock, P. R., et al. (2001). Genetic identification and characterization of a novel virus related to human hepatitis E virus from chickens with hepatitis-splenomegaly syndrome in the United States. *J Gen Virol* **82**(Pt 10), 2449–2462.

Haqshenas, G., Huang, F. F., Fenaux, M., et al. (2002). The putative capsid protein of the newly identified avian hepatitis E virus shares antigenic epitopes with that of swine and human hepatitis E viruses and chicken big liver and spleen disease virus. *J Gen Virol* **83**(Pt 9), 2201–2209.

Hazam, R. K., Singla, R., Kishore, J., et al. (2010). Surveillance of hepatitis E virus in sewage and drinking water in a resettlement colony of Delhi: what has been the experience? *Arch Virol* **155**(8), 1227–1233.

Huang, F. F., Sun, Z. F., Emerson, S. U., et al. (2004). Determination and analysis of the complete genomic sequence of avian hepatitis E virus (avian HEV) and attempts to infect rhesus monkeys with avian HEV. *J Gen Virol* **85**(Pt 6), 1609–1618.

Jeggli, S., Steiner, D., Joller, H., et al. (2004). Hepatitis E, *Helicobacter pylori*, and gastrointestinal symptoms in workers exposed to waste water. *Occup Environ Med* **61**(7), 622–627.

Johne, R., Plenge-Bonig, A., Hess, M., et al. (2010). Detection of a novel hepatitis E-like virus in faeces of wild rats using a nested broad-spectrum RT-PCR. *J Gen Virol* **91**(Pt 3):750–758.

Jung, K., Kang, B., Song, D. S., et al. (2007). Prevalence and genotyping of hepatitis E virus in swine population in Korea between 1995 and 2004: a retrospective study. *Vet J* **173**(3), 683–687.

Kamar, N., Selves, J., Mansuy, J. M., et al. (2008). Hepatitis E virus and chronic hepatitis in organ-transplant recipients. *N Engl J Med* **358**(8), 811–817.

Kamar, N., Bendall, R. P., Peron, J. M., et al. (2011a). Hepatitis E virus and neurologic disorders. *Emerg Infect Dis* **17**(2), 173–179.

Kamar, N., Garrouste, C., Haagsma, E. B., et al. (2011b). Factors associated with chronic hepatitis in patients with hepatitis E virus infection who have received solid organ transplants. *Gastroenterology* **140**(5), 1481–1489.

Kasorndorkbua, C., Thacker, B. J., Halbur, P. G., et al. (2003). Experimental infection of pregnant gilts with swine hepatitis E virus. *Can J Vet Res* **67**(4), 303–306.

Kasorndorkbua, C., Opriessnig, T., Huang, F. F., et al. (2005). Infectious swine hepatitis E virus is present in pig manure storage facilities on United States farms, but evidence of water contamination is lacking. *Appl Environ Microbiol* **71**(12), 7831–7837.

Kitajima, M., Matsubara, K., Sour, S., et al. (2009). First detection of genotype 3 hepatitis E virus RNA in river water in Cambodia. *Trans R Soc Trop Med Hyg* **103**(9), 955–957.

Kulkarni, M. A., and Arankalle, V. A. (2008). The detection and characterization of hepatitis E virus in pig livers from retail markets of India. *J Med Virol* **80**(8), 1387–1390.

Kuniholm, M. H., Purcell, R. H., McQuillan, G. M., et al. (2009). Epidemiology of hepatitis E virus in the United States: results from the Third National Health and Nutrition Examination Survey, 1988–1994. *J Infect Dis* **200**(1), 48–56.

Labrique, A. B., Zaman, K., Hossain, Z., et al. (2010). Epidemiology and risk factors of incident hepatitis E virus infections in rural Bangladesh. *Am J Epidemiol* **172**(8), 952–961.

Lee, C. C., Shih, Y. L., Laio, C. S., et al. (2005). Prevalence of antibody to hepatitis E virus among haemodialysis patients in Taiwan: possible infection by blood transfusion. *Nephron Clin Pract* **99**(4), c122–c127.

Li, T. C., Chijiwa, K., Sera, N., et al. (2005). Hepatitis E virus transmission from wild boar meat. *Emerg Infect Dis* **11**(12), 1958–1960.

Li, T. C., Saito, M., Ogura, G., et al. (2006). Serologic evidence for hepatitis E virus infection in mongoose. *Am J Trop Med Hyg* **74**(5), 932–936.

Li, T. C., Miyamura, T., and Takeda, N. (2007). Detection of hepatitis E virus RNA from the bivalve Yamato-Shijimi (*Corbicula japonica*) in Japan. *Am J Trop Med Hyg* **76**(1), 170–172.

Li, T. C., Yoshimatsu, K., Yasuda, S. P., et al. (2011). Characterization of self-assembled virus-like particles of rat hepatitis E virus generated by recombinant baculoviruses. *J Gen Virol* **92**(Pt 12), 2830–2837.

Li X., Kamili, S., and Krawczynski, K. (2006). Quantitative detection of hepatitis E virus RNA and dynamics of viral replication in experimental infection. *J Viral Hepat* **13**(12), 835–839.

Lu, L., Li, C., and Hagedorn, C. H. (2006). Phylogenetic analysis of global hepatitis E virus sequences: genetic diversity, subtypes and zoonosis. *Rev Med Virol* **16**(1), 5–36.

Ma, H., Song, X., Harrison, T.J., et al. (2009). Immunogenicity and efficacy of a bacterially expressed HEV ORF3 peptide, assessed by experimental infection of primates. *Arch Virol* **154**(10):1641–1648.

Ma, H., Zheng, L., Liu, Y., et al. (2010). Experimental infection of rabbits with rabbit and genotypes 1 and 4 hepatitis E viruses. *PLoS One* **5**(2), e9160.

Macedo, G., Pinto, T., Sarmento, J. A., et al. (1998). The first assessment of hepatitis E virus seroprevalence in northern Portugal. *Acta Med Port* **11**(12), 1065–1068.

Mamun Al, M., Rahman, S., Khan, M., et al. (2009). HEV infection as an aetiologic factor for acute hepatitis: experience from a tertiary hospital in Bangladesh. *J Health Popul Nutr* **27**(1), 14–19.

Mansuy, J. M., Abravanel, F., Miedouge, M., et al. (2009). Acute hepatitis E in south-west France over a 5-year period. *J Clin Virol* **44**(1), 74–77.

Mansuy, J. M., Bendall, R., Legrand-Abravanel, F., et al. (2011). Hepatitis E virus antibodies in blood donors, France. *Emerg Infect Dis* **17**(12), 2309–2312.

Marek, A., Bilic, I., Prokofieva, I., et al. (2010). Phylogenetic analysis of avian hepatitis E virus samples from European and Australian chicken flocks supports the existence of a different genus within the Hepeviridae comprising at least three different genotypes. *Vet Microbiol* **145**(1–2), 54–61.

Meng, X. J. (2010). Hepatitis E virus: animal reservoirs and zoonotic risk. *Vet Microbiol* **140**(3-4): 256–265.

Meng, X. J., Purcell, R. H., Halbur, P. G., et al. (1997). A novel virus in swine is closely related to the human hepatitis E virus. *Proc Natl Acad Sci USA* **94**(18), 9860–9865.

Meng, X. J., Halbur, P. G., Haynes, J. S., et al. (1998a). Experimental infection of pigs with the newly identified swine hepatitis E virus (swine HEV), but not with human strains of HEV. *Arch Virol* **143**(7), 1405–1415.

Meng, X. J., Halbur, P. G., Shapiro, M. S., et al. (1998b). Genetic and experimental evidence for cross-species infection by swine hepatitis E virus. *J Virol* **72**(12), 9714–9721.

Meng, X. J., Dea, S., Engle, R. E., et al. (1999). Prevalence of antibodies to the hepatitis E virus in pigs from countries where hepatitis E is common or is rare in the human population. *J Med Virol* **59**(3), 297–302.

Meng, X. J., Wiseman, B., Elvinger, F., et al. (2002). Prevalence of antibodies to hepatitis E virus in veterinarians working with swine and in normal blood donors in the United States and other countries. *J Clin Microbiol* **40**(1), 117–122.

Nakamura, M., Takahashi, K., Taira, K., et al. (2006). Hepatitis E virus infection in wild mongooses of Okinawa, Japan: demonstration of anti-HEV antibodies and a full-genome nucleotide sequence. *Hepatol Res* **34**(3), 137–140.

Navaneethan, U., Al Mohajer, M., and Shata, M. T. (2008). Hepatitis E and pregnancy: understanding the pathogenesis. *Liver Int* **28**(9), 1190–1199.

Nicand, E., Grandadam, M., Teyssou, R., et al.(2001). Viraemia and faecal shedding of HEV in symptom-free carriers. *Lancet* **357**(9249), 68–69.

Nubling, M., Hofmann, F., and Tiller, F. W. (2002). Occupational risk for hepatitis A and hepatitis E among health care professionals? *Infection* **30**(2), 94–97.

Okamoto, H. (2011). Hepatitis E virus cell culture models. *Virus Res* **161**(1), 65–77.

Olsen, B., Axelsson-Olsson, D., Thelin, A., et al. (2006). Unexpected high prevalence of IgG-antibodies to hepatitis E virus in Swedish pig farmers and controls. *Scand J Infect Dis* **38**(1), 55–58.

Parashar, D., Khalkar, P., and Arankalle, V. A. (2011). Survival of hepatitis A and E viruses in soil samples. *Clin Microbiol Infect* **17**(11), E1–E4.

Pavio, N., Meng, X. J., and Renou, C. (2010). Zoonotic hepatitis E: animal reservoirs and emerging risks. *Vet Res* **41**(6), 46.

Peralta, B., Biarnes, M., Ordonez, G., et al. (2009). Evidence of widespread infection of avian hepatitis E virus (avian HEV) in chickens from Spain. *Vet Microbiol* **137**(1–2), 31–36.

Peron, J. M., Bureau, C., Poirson, H., et al. (2007). Fulminant liver failure from acute autochthonous hepatitis E in France: description of seven patients with acute hepatitis E and encephalopathy. *J Viral Hepat* **14**(5), 298–303.

Purcell, R. H., and Emerson, S. U. (2008). Hepatitis E: an emerging awareness of an old disease. *J Hepatol* **48**(3), 494–503.

Purcell, R. H., Engle, R. E., Rood, M. P., et al. (2011). Hepatitis E virus in rats, Los Angeles, California, USA. *Emerg Infect Dis* **17**(12), 2216–2222.

Quintana, A., Sanchez, L., Larralde, O., et al. (2005). Prevalence of antibodies to hepatitis E virus in residents of a district in Havana, Cuba. *J Med Virol* **76**(1), 69–70.

Renou, C., Pariente, A., Nicand, E., et al. (2008). Pathogenesis of hepatitis E in pregnancy. *Liver Int* **28**(10), 1465.

Renou, C., Lafeuillade, A., Pavio, N., et al. (2010). Response to Madejon et al.: are HIV-infected patients at risk of HEV infection? *J Viral Hepat* **17**(5):380.

Reuter, G., Fodor, D., Forgach, P., et al. (2009). Characterization and zoonotic potential of endemic hepatitis E virus (HEV) strains in humans and animals in Hungary. *J Clin Virol* **44**(4), 277–281.

Rose, N., Lunazzi, A., Dorenlor, V., et al. (2011). High prevalence of hepatitis E virus in French domestic pigs. *Comp Immunol Microbiol Infect Dis* **34**(5), 419–427.

Rutjes, S. A., Lodder, W. J., Lodder-Verschoor, F., et al. (2009). Sources of hepatitis E virus genotype 3 in the Netherlands. *Emerg Infect Dis* **15**(3), 381–387.

Said, B., Ijaz, S., Kafatos, G., et al. (2009). Hepatitis E outbreak on cruise ship. *Emerg Infect Dis* **15**(11):1738–1744.

Seminati, C., Mateu, E., Peralta, B., et al. (2008). Distribution of hepatitis E virus infection and its prevalence in pigs on commercial farms in Spain. *Vet J* **175**(1), 130–132.

Shrestha, M. P., Scott, R. M., Joshi, D. M., et al. (2007). Safety and efficacy of a recombinant hepatitis E vaccine. *N Engl J Med* **356**(9), 895–903.

Shukla, P., Nguyen, H. T., Torian, U., et al. (2011). Cross-species infections of cultured cells by hepatitis E virus and discovery of an infectious virus–host recombinant. *Proc Natl Acad Sci USA* **108**(6), 2438–2443.

Shukla, P., Nguyen, H. T., Faulk, K., et al. (2012). Adaptation of a genotype 3 hepatitis E virus to efficient growth in cell culture depended on an inserted human gene segment acquired by recombination. *J Virol* **86**(10):5697–5707.

Singh, B. B., Sharma, R., Gill, J. P., et al. (2011). Climate change, zoonoses and India. *Rev Sci Tech* **30**(3), 779–788.

Stefanidis, I., Zervou, E. K., Rizos, C., et al. (2004). Hepatitis E virus antibodies in hemodialysis patients: an epidemiological survey in central Greece. *Int J Artif Organs* **27**(10), 842–847.

Stoszek, S. K., Engle, R. E., Abdel-Hamid, M., et al. (2006). Hepatitis E antibody seroconversion without disease in highly endemic rural Egyptian communities. *Trans R Soc Trop Med Hyg* **100**(2), 89–94.

Sun, Z. F., Larsen, C. T., Dunlop, A., et al. (2004a). Genetic identification of avian hepatitis E virus (HEV) from healthy chicken flocks and characterization of the capsid gene of 14 avian HEV isolates from chickens with hepatitis-splenomegaly syndrome in different geographical regions of the United States. *J Gen Virol* **85**(Pt 3), 693–700.

Sun, Z. F., Larsen, C. T., Huang, F. F., et al. (2004b). Generation and infectivity titration of an infectious stock of avian hepatitis E virus (HEV) in chickens and cross-species infection of turkeys with avian HEV. *J Clin Microbiol* **42**(6), 2658–2662.

Takahashi, M., Nishizawa, T., Tanaka, T., et al. (2005). Correlation between positivity for immunoglobulin A antibodies and viraemia of swine hepatitis E virus observed among farm pigs in Japan. *J Gen Virol* **86**(Pt 6), 1807–1813.

Takahashi, M., Nishizawa, T., Sato, H., et al. (2011). Analysis of the full-length genome of a hepatitis E virus isolate obtained from a wild boar in Japan that is classifiable into a novel genotype. *J Gen Virol* **92**(Pt 4), 902–908.

Tei, S., Kitajima, N., Takahashi, K., et al. (2003). Zoonotic transmission of hepatitis E virus from deer to human beings. *Lancet* **362**(9381), 371–373.

Teo, C. G. (2010). Much meat, much malady: changing perceptions of the epidemiology of hepatitis E. *Clin Microbiol Infect* **16**(1), 24–32.

Tesse, S., Lioure, B., Fornecker, L., et al. (2012). Circulation of genotype 4 hepatitis E virus in Europe: first autochthonous hepatitis E infection in France. *J Clin Virol* **54**(2):197–200.

Tsarev, S. A., Emerson, S. U., Tsareva, T. S., et al. (1993). Variation in course of hepatitis E in experimentally infected cynomolgus monkeys. *J Infect Dis* **167**(6), 1302–1306.

Vulcano, A., Angelucci, M., Candelori, E., et al. (2007). HEV prevalence in the general population and among workers at zoonotic risk in Latium Region. *Ann Ig* **19**(3), 181–186.

Wenzel, J. J., Preiss, J., Schemmerer, M., et al. (2011). Detection of hepatitis E virus (HEV) from porcine livers in Southeastern Germany and high sequence homology to human HEV isolates. *J Clin Virol* **52**(1), 50–54.

Wichmann, O., Schimanski, S., Koch, J., et al. (2008). Phylogenetic and case–control study on hepatitis E virus infection in Germany. *J Infect Dis* **198**(12), 1732–1741.

Wong, K. H., Liu, Y. M., Ng, P. S., et al. (2004). Epidemiology of hepatitis A and hepatitis E infection and their determinants in adult Chinese community in Hong Kong. *J Med Virol* **72**(4), 538–544.

Yazaki, Y., Mizuo, H., Takahashi, M., et al. (2003). Sporadic acute or fulminant hepatitis E in Hokkaido, Japan, may be food-borne, as suggested by the presence of hepatitis E virus in pig liver as food. *J Gen Virol* **84**(Pt 9), 2351–2357.

Zhao, C., Ma, Z., Harrison, T. J., et al. (2009). A novel genotype of hepatitis E virus prevalent among farmed rabbits in China. *J Med Virol* **81**(8), 1371–1379.

Zhao, Q., Zhou, E. M., Dong, S. W., et al. (2010). Analysis of avian hepatitis E virus from chickens, China. *Emerg Infect Dis* **16**(9), 1469–1472.

[illegible] (2012). Analysis of the full-length genome of a hepatitis E virus isolate [illegible] *Vet Microbiol* [illegible]

Tei, S., Kitajima, N., Takahashi, K., et al. (2003). Zoonotic transmission of hepatitis E virus from deer to human beings. *Lancet* 362(9381): 371–373.

Teo, C. G. (2010). Much meat, much malady: changing perceptions of the epidemiology of hepatitis E. *Clin Microbiol Infect* 16: 24–32.

[illegible] (2010). [illegible] hepatitis E [illegible] *J Clin Virol* [illegible]

[illegible] (2013). [illegible] *J Med Virol* [illegible]

[illegible] (2007). [illegible] *J Clin Microbiol* [illegible]

Wenzel, J.J., Preiss, J., Schemmerer, M., et al. (2011). Detection of hepatitis E virus (HEV) from porcine livers in Southeastern Germany and high sequence homology to human HEV isolates. *J Clin Virol* 52(1): 50–54.

Wichmann, O., Schimanski, S., Koch, J., et al. (2008). Phylogenetic and case-control study on hepatitis E virus infection in Germany. *J Infect Dis* 198: 1732–1741.

Wong, K.H., Liu, Y.M., Ng, P.S., et al. (2004). Epidemiology of hepatitis A and hepatitis E infection and their determinants in adult Chinese community in Hong Kong. *J Med Virol* 72(4): 538–544.

[illegible] (2014). [illegible] hepatitis E [illegible] [illegible]

Zhao, C., Ma, Z., Harrison, T.J., et al. (2009). A novel genotype of hepatitis E virus prevalent among farmed rabbits in China. *J Med Virol* 81: 1371–1379.

[illegible] (2010). [illegible]

IMPACT OF CLIMATE CHANGE ON OUTBREAKS OF ARENAVIRAL INFECTIONS

James Christopher Clegg

Les Mandinaux, Le Grand Madieu, France

TABLE OF CONTENTS

25.1	Introduction	467
25.2	Natural history of arenaviruses	468
25.3	Predicted climate changes	470
25.4	Arenaviral diseases and climate change	471
	References	473

25.1 INTRODUCTION

The *Arenaviridae* (Salvato et al., 2012) are a family of 23 enveloped viruses with bisegmented RNA genomes of around 10 kb in total length. Each of these RNA segments encodes two viral proteins in an ambisense arrangement. Their principal hosts are members of the mammalian order *Rodentia*. In terms of human health, they constitute an important family because six of these viruses infect humans and can cause serious and frequently fatal hemorrhagic fevers. With the exception of the type species *Lymphocytic choriomeningitis virus* (LCMV), which is widely distributed in Europe, Asia, and the Americas, each individual arenavirus species is found in a relatively localized area in Africa or in North or South America. The purposes of this review are to outline the general features of arenavirus diseases and to consider how currently predicted global climate changes might change the geographic distribution and human impact of these dangerous viral pathogens during the twenty-first century.

Viral Infections and Global Change, First Edition. Edited by Sunit K. Singh.
© 2014 John Wiley & Sons, Inc. Published 2014 by John Wiley & Sons, Inc.

25.2 NATURAL HISTORY OF ARENAVIRUSES

The arenaviruses are principally viruses of rodents of the family *Muridae*. Each virus is associated primarily with a single rodent species (the reservoir species), although in several cases infected animals of another species have been detected from time to time. Viruses are spread among populations through excretion in body fluids, and vertical transmission from mother to offspring also contributes to maintaining infection. Although we have insufficient knowledge of the details of the dynamics of virus–host interactions, the infected rodents show little or no overt disease, and their fitness appears not to be impaired to any significant extent. It is currently thought that the arenavirus–rodent host associations observed today are the results of coevolution of parasites and hosts. It has been a long-standing observation that the geographic range of a host rodent is usually more extensive than that of its associated arenavirus. However, it is likely that such apparent discrepancies may be clarified by recent molecular approaches to host rodent taxonomy (Coulibaly-N'Golo et al., 2011; Lecompte et al., 2006; Salazar-Bravo et al., 2002).

Some members of the arenavirus family are important causes of viral hemorrhagic fevers when humans become infected. These include *Lassa*, *Junín*, *Guanarito*, and *Machupo viruses*, which have caused quite large outbreaks, and *Sabiá* and *Chapare viruses*, which are known to have caused disease in a few cases (including laboratory workers). The geographic locations where these viruses have been found are indicated in Figure 25.1.

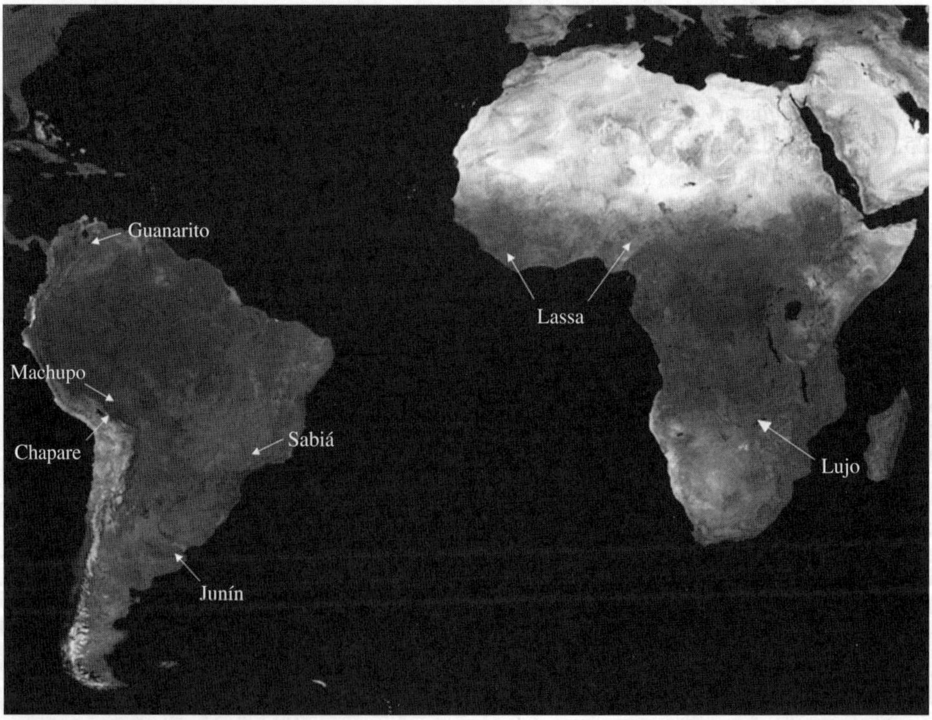

Figure 25.1. Geographic locations of arenaviruses associated with human hemorrhagic fevers. Map provided by NASA Visible Earth. Available at http://visibleearth.nasa.gov/. For color detail, please see color plate section.

The hemorrhagic fevers caused by these viruses are similar to each other, usually presenting as a nonspecific illness with symptoms including fever, headache, dizziness, asthenia, sore throat, pharyngitis, cough, retrosternal and abdominal pain, and vomiting. In severe cases, facial edema, hemorrhagic conjunctivitis, moderate bleeding (from nose, gums, vagina, etc.), and exanthema frequently occur. Neurological signs may develop and progress to confusion, convulsion, coma, and death. Case fatality rates range from 5% to 20% for hospitalized cases. In contrast, the type species LCMV causes aseptic meningitis or meningoencephalitis with an overall case fatality of less than 1%, but it has also been associated with hemorrhagic fever-like infections in organ transplant recipients (Fischer et al., 2006).

Lassa fever was first recognized in the 1960s and the causative arenavirus was isolated in 1969. It is now known to be present in large areas of both savannah and forest zones of sub-Saharan west Africa. The principal foci are in the west in the border regions of Guinea, Sierra Leone, and Liberia and in the east in Nigeria. The rodent host of Lassa virus is the multimammate rat *Mastomys natalensis*. This often-quoted relationship has recently been elegantly confirmed, by simultaneous sequence analysis of both infecting virus and infected rodent, in Guinea (Lecompte et al., 2006). Lassa virus was found only in *M. natalensis*, and not in another *Mastomys* species nor in 12 other rodent genera. *M. natalensis* is a peridomestic rodent that infests houses and food stores. Infection of humans can occur in the process of catching and preparing the animals for food, as well as by contact with animal excreta or contaminated materials. Lujo virus, an arenavirus recognized recently in a Zambian patient, is only distantly related to Lassa virus or to other African arenaviruses. It has caused a cluster of fatal human infections in South Africa (Briese et al., 2009).

The arenavirus Junín is the causative agent of Argentine hemorrhagic fever, first described in 1955. When first encountered, human cases were limited to an area of 16 000 km^2 in the humid pampas in the north of Buenos Aires province. However, the endemoepidemic area now extends to over 150 000 km^2, reaching north of Buenos Aires, south of Santa Fe, southeast of Cordoba, and northeast of La Pampa provinces. The human population at risk is estimated to be around five million. The virus is carried mainly by the vesper mouse *Calomys musculinus*, but other rodents (*Calomys laucha* and *Akodon azarae*) have also been implicated. These rodents mainly infest maize crops, and most human infections are seen in agricultural workers.

Venezuelan hemorrhagic fever is caused by Guanarito virus, which is carried by the cane mouse *Zygodontomys brevicauda*. Persons most affected are male agricultural workers around the town of Guanarito in Portuguesa State and adjacent parts of Barinas State in Venezuela. The virus was discovered in 1989 and the disease incidence has exhibited cyclical behavior with a period of 4–5 years.

Machupo virus is the cause of Bolivian hemorrhagic fever, which was first recognized in 1959 in the remote, sparsely populated savannah of Beni State, Bolivia. Ecological studies indicate that the rodent *Calomys callosus* is the principal animal reservoir. Agricultural workers are those most at risk, in the fields and in houses to which rodents have easy access. There were several local outbreaks of the disease in the 1960s, but the incidence fell markedly in the following decade following the institution of rodent control measures.

A small outbreak of viral hemorrhagic disease was caused in 2003–2004 in Cochabamba, Bolivia, by an arenavirus distinct from Machupo virus, the causative agent of the previously recognized Bolivian hemorrhagic fever (Delgado et al., 2008). It has been named Chapare virus. There is as yet no information about the extent of the public health threat from this virus, nor the identity of its normal rodent host species.

A single case of hemorrhagic fever caused naturally by the arenavirus Sabiá has been described. It occurred in Sabiá village, near São Paulo, Brazil. No natural rodent host has been identified. Two laboratory infections by the virus have also occurred.

It seems that arenavirus-caused disease in humans is an accidental product of their encounters with infected rodents and their excreta and body fluids. Infection of humans can occur through contact with rodent excreta or materials contaminated with them or ingestion of contaminated food. Direct contact of broken or abraded skin with rodent excreta is likely to be an important route, and inhalation of small droplets or particles containing rodent urine or saliva is also thought to be a significant source of infection. The nature of these incidental contacts evidently depends on the details of the living patterns and habits of both the rodent carriers and the human population. Where infected rodents prefer a field habitat, infection is primarily associated with agricultural workers. Where the rodents infest dwellings and other buildings, infection occurs in a domestic setting.

25.3 PREDICTED CLIMATE CHANGES

This discussion is mainly based on the findings of the Fourth Assessment Report of the Intergovernmental Panel on Climate Change (IPCC). This is the most recent published report, which appeared in 2007. The overwhelming scientific consensus is that anthropogenic greenhouse gas (GHG) emission is causing global warming at a rate quite without precedent in the Earth's climate history (Meehl et al., 2007). The current rate of increase in global mean surface temperature of about 0.2 °C per decade is projected to continue until around 2030, irrespective of whether GHG emissions continue at present rates or whether reductions can be achieved. This implies mean surface temperatures in the period 2011–2030 about 0.66 °C warmer than in the period 1980–1999. Further into the future, there is greater uncertainty because of increasing differences among the various scenarios modeled. These scenarios cover a range of possibilities for the mitigation (or lack of it) of GHG emissions. For a number of plausible scenarios, the best estimates of the IPCC for the increase in global mean surface temperature for the period 2090–2099 relative to 1980–1999 range from 1.8 to 4.0 °C. Surface temperature increases on land are predicted to be roughly twice this global mean, that is, in the range 3.6–8 °C by the end of this century. On the global scale, it is predicted that there will be more frequent and more extreme heat waves, fewer cold periods, and increased and more intense rainfall in regional tropical precipitation maxima. In subtropical and mid-latitudes, precipitation will decrease, but intense rainfall events interspersed with long periods of drought will become more common. Sea levels are expected to rise globally on the order of 0.5 m by the end of the century, but there is a great deal of uncertainty in making these estimates and in assessing their possible impact in particular geographic regions.

Disturbingly, more recent scientific and economic reviews of present GHG emission rates and their likely future trends and the lack of significant political progress in moving towards their reduction indicate that it is increasingly unlikely that any prospective global agreement can stabilize atmospheric GHGs at 450 ppm or even at 650 ppm CO_2 equivalent (Anderson and Bows, 2008; Anderson et al., 2008; Clark et al., 2008; Garnaut et al., 2008; Hansen et al., 2008). Hence, the IPCC scenarios are very likely to significantly underestimate the degree of climate change in the future.

For the purpose of this review, two specific geographic regions are of particular importance. These are the locations where Lassa fever is currently endemic (sub-Saharan west Africa) and the broader region where South American hemorrhagic fevers are found (South America). The East African Lujo virus will not be considered further, since very little is presently known about its natural history. It is worth pointing out though that the example of Lujo virus suggests that other pathogenic arenaviruses may well remain as yet undiscovered and that climate change-driven population movements or other factors may lead to human infections. Unfortunately, there are major difficulties in moving from global-scale climate change predictions towards more detailed descriptions of future outcomes at a regional level. These are particularly acute in west Africa, because of the relatively sparse data on past and current weather conditions, the complex nature of the terrain, and the influence of ocean basins. It is predicted (Christensen et al., 2007; Conway 2009) that Africa as a whole will warm more than the global annual mean throughout the year. Drier regions will warm more than the moister tropics. Changes in rainfall in the Sahel, the Guinean coast, and the southern Sahara in this century remain very difficult to predict because of shortcomings in the current models, which result in systematic errors, disagreements among different climate models, and inability to simulate correctly twentieth-century conditions. Key features such as the frequency and spatial distribution of tropical cyclones affecting Africa cannot be reliably assessed. Nonetheless, the frequency of extremely wet seasons is likely to rise markedly, as is also the case in East Africa. The west African coast and the Gulf of Guinea are thought to be at high risk of flooding due to sea-level rise (Boko et al., 2007).

In South America, the annual surface temperature increase is predicted to be similar to the global mean (Christensen et al., 2007). This represents an increase in the range of 3°–4 °C by the end of the century. Systematic differences among different models, together with large variations in predictions of changes in El Niño amplitude, and the height and sharpness of the Andes mountains make assessments at regional scale over much of Central and South America very unreliable. Rainfall changes show that regional differences are likely to occur; most models suggest a wetter climate around the Rio de la Plata but reduced precipitation in parts of northern South America. Extremes of weather and climate are likely to occur more frequently. Water stress will increase as a result of glacier retreat or disappearance in the Andes, leading to highly adverse effects on agriculture.

25.4 ARENAVIRAL DISEASES AND CLIMATE CHANGE

A well-recognized example of a change in the incidence of an infectious viral disease already exists. The spread of bluetongue virus disease in cattle from the Middle East and north Africa first to southern and then to northern Europe has been attributed to ongoing climate changes in these regions (Purse et al., 2008; Weaver and Reisen, 2009). The main factors that can affect the burden of infectious diseases in humans are (i) changes in abundance, virulence, or transmissibility of infectious agents; (ii) an increase in probability of exposure of humans; and (iii) an increase in the susceptibility of humans to infection and to the consequences of infection. A wide range of biological, physicochemical, behavioral, and social drivers can influence one or more of these factors (Wilson, 1995). In particular, alterations in the environment, brought about by currently predicted climate changes, clearly have the potential to affect, to a greater or lesser extent, all three of these factors. We need to consider the possible effects of climate changes within the currently known endemic areas of each arenavirus disease and also the extent to which such changes may influence transfer and persistence of arenavirus diseases to hitherto unaffected regions. However, it must be appreciated that the reliability of any such predictions remains very low, not only because of the relatively coarse scale of available climate change predictions but also because of our lack of reliable data on the current incidence of these diseases. This applies particularly to the prevalence of Lassa fever and other possible arenavirus diseases in Africa.

When we examine how these factors could affect arenavirus-caused disease, all three are likely to exert significant influence. In the first category, there are likely to be changes in the abundance of arenaviruses, in the sense that the reservoir host rodent populations are likely to be affected one way or another by changes in climate. Thus prolonged drought in a particular region may lead to reduction in population size, while increased seasonal rain may lead to a population explosion. Such events have been observed for other rodent-borne zoonoses, as documented in the IPCC Report (Confalonieri et al., 2007). In the case of another rodent-borne virus disease, hantavirus pulmonary syndrome (HPS), there is evidence that El Niño Southern Oscillation-induced increases in rainfall in the Four Corners region of the southwestern United States led to increases in the population of the rodent reservoir *Peromyscus maniculatus* and subsequent emergence of the disease in the human population (Glass et al., 2002). There may be a similar explanation for the emergence of HPS in Panama in 2000, following increases in the peri-domestic rodent population following heavy rainfall and flooding in the surrounding areas (Bayard et al., 2004). Further discussion of the influence of climatic factors on mammalian vector ecology and impact on the burden of zoonotic disease can be found in Mills et al. (2010). In Guinea, there is significant risk of infection by Lassa fever throughout the year, but there is an increased rate of infection during the dry season as a result of the greater number of rodents, some carrying the virus, infesting dwellings (Fichet-Calvet et al., 2007). In an analogous manner, the greater expected frequency of extreme weather events may lead to scenarios with increased rodent–human contact following heavy rain and subsequent flooding and consequent increases in the incidence of arenaviral diseases. More recently, analysis of geographic location of Lassa fever outbreaks in humans and incidence of infected rodents in relation to environmental variables across sub-Saharan west Africa has emphasized the important role of rainfall as a predictor of disease incidence (Fichet-Calvet and Rogers, 2009). Lassa fever outbreaks occurred in areas with annual precipitation in the range 1500–3000 mm. Areas with either less rainfall or more rainfall were not subject to Lassa fever outbreaks. It is thus very likely that changes in climate leading to altered rainfall intensity and/or distribution very likely alter the geographic and temporal distribution of Lassa fever disease outbreaks. However, detailed predictions must await the development of accurate local climate models.

Climate change in Venezuela, Bolivia, and Argentina may lead to changes in agricultural land use, with relocalization of crop-growing areas that are becoming unsuitable for agricultural use to others with more favorable climates. Where arenaviral diseases are carried by rodents infesting crops, as is the case with Venezuelan, Bolivian, and Argentine hemorrhagic fevers, there will be corresponding changes in the geographic location of rodents and thus disease.

It is unlikely that changes in virus virulence will result directly from climatic changes, but it is conceivable that virus transmissibility could be influenced. Arenaviruses are enveloped viruses that are not particularly robust when exposed to high temperatures or low humidity. Thus some climatic factors may be expected to influence the survival of the viruses in the environment either negatively or positively. However, it is difficult to estimate the importance of these effects compared with other climatic impacts on disease incidence.

It is very likely that in some environments, climate change will increase the probability of human exposure to arenavirus infections, whereas in others it will decrease the probability. Such effects are likely to be mediated through changes in the probability of encounters with reservoir rodents and contact with their excreta or contaminated materials. We can envisage direct effects of climate on the size and behavior of virus-carrying rodent populations, as discussed earlier, as well as on the human populations themselves, through changing land use (for instance, irrigation) triggered by increasing temperatures, fluctuating weather conditions, and the resultant disturbance of local landscapes. Climate change is likely to lead to mass migration and movement of populations, with consequent stresses associated with inadequate shelter and overcrowding. Such considerations are likely to be more significant in respect of Lassa fever compared with the South American arenaviral hemorrhagic fevers, because of the much larger human populations in the endemic areas. As well as possible changes in areas favorable for food production, flooding along the west African coast as a result of storm surges and sea-level rise could drive large-scale population movements in the area. It has been projected that the 500 km of this coast between Accra and the Niger delta will be a continuous urban megalopolis of some 50 million people by 2020 (Boko et al., 2007; Hewawasam, 2002). It has already been shown that Lassa fever can be a significant risk in refugee camps in Guinea (Bonner et al., 2007; Fair et al., 2007). There is increased risk in areas where there are higher numbers of infected rodents (Fair et al., 2007) and in poor quality housing and households with reduced levels of hygiene (Bonner et al., 2007). Thus populations driven onto higher ground by coastal flooding may be at increased risk from Lassa fever (among other diseases) unless sufficiently adequate housing and rodent control measures can be provided. It should be noted that rodent control is the key measure in any program to mitigate arenavirus disease in humans. However, authorities in the Lassa fever endemic regions, which include some of the most underdeveloped countries in the world, do not currently have the necessary resources and infrastructure to mount effective healthcare or disease prevention programs.

Finally, it is possible that predicted climate change may lead to more frequent transfer of arenavirus-infected patients to regions of the world without experience of these diseases. Although natural rodent vectors will be almost certainly absent, it is important that infected persons be swiftly recognized and diagnosed so that further transmission during patient care can be avoided. This can readily be achieved through careful barrier nursing techniques, but the fear engendered by viral hemorrhagic fevers, including those caused by arenaviruses, can place a heavy burden on hospital systems. It would be prudent if such considerations were included in healthcare planning to meet the challenges of global climate change.

REFERENCES

Anderson, K., Bows, A. 2008. Reframing the climate change challenge in light of post-2000 emission trends. *Philos. Trans. R. Soc. A* **366**:3863–3882.

Anderson, K., Bows, A., Mander, S. 2008. From long-term targets to cumulative emission pathways: reframing UK climate policy. *Energy Policy* **36**:3714–3722.

Bayard, V., Kitutani, P.T., Barria, E.O., et al. 2004. Outbreak of hantavirus pulmonary syndrome, Los Santos, Panama, 1999–2000. *Emerg. Infect. Dis.* **10**:1635–1642. Available at wwwnc.cdc.gov/eid/r/vol10no9/04-0143.htm (accessed July 8, 2013).

Boko, M., Niang, I., Nyong, A., et al. 2007. Africa. In: *Climate Change 2007: Impacts, Adaptation and Vulnerability. Contribution of Working Group II to the Fourth Assessment Report of the Intergovernmental Panel on Climate Change*, eds. M.L. Parry, O.F. Canziani, J.P. Palutikof, et al., 433–467. Cambridge, Cambridge University Press. Available at http://www.ipcc.ch/pdf/assessment-report/ar4/wg2/ar4-wg2-chapter9.pdf (accessed February 13, 2012).

Bonner, P.C., Schmidt, W.P., Belmain, S.R., et al. 2007. Poor housing quality increases risk of rodent infestation and Lassa fever in refugee camps of Sierra Leone. *Am. J. Trop. Med. Hyg.* **77**:169–175.

Briese, T., Paweska, J.T., McMullan, L.K., et al. 2009. Genetic detection and characterization of Lujo virus, a new hemorrhagic fever-associated arenavirus from southern Africa. *PLoS Pathog.* **5**(5):e1000455.

Christensen, J.H., Hewitson, B., Busuioc, A., et al. 2007. Regional climate projections. In: *Climate Change 2007: The Physical Science Basis. Contribution of Working Group I to the Fourth Assessment Report of the Intergovernmental Panel on Climate Change*, eds. S. Solomon, D. Qin, M. Manning, et al., 848–940. Cambridge, Cambridge University Press. Available at http://www.ipcc.ch/pdf/assessment-report/ar4/wg1/ar4-wg1-chapter11.pdf (accessed February 13, 2012).

Clark, P.U., Weaver, A.J., Brook, E., et al. 2008. Abrupt Climate Change. A Report by the U.S. Climate Change Science Program and the Subcommittee on Global Change Research. U.S. Geological Survey. Available at http://www.climatescience.gov/Library/sap/sap3-4/final-report/default.htm#finalreport (accessed June 5, 2013).

Confalonieri, U., Menne, B., Akhtar, R., et al. 2007. Human health. In: *Climate Change 2007: Impacts, Adaptation and Vulnerability. Contribution of Working Group II to the Fourth Assessment Report of the Intergovernmental Panel on Climate Change*, eds. M.L. Parry, O.F. Canziani, J.P. Palutikof, et al., 391–431. Cambridge, Cambridge University Press. Available at http://www.ipcc.ch/pdf/assessment-report/ar4/wg2/ar4-wg2-chapter8.pdf (accessed February 13, 2012).

Conway, G. 2009. The science of climate change in Africa: impacts and adaptation. Discussion Paper No. 1, Grantham Institute for Climate Change. Available at https://workspace.imperial.ac.uk/climatechange/public/pdfs/discussion_papers/Grantham_Institue_-_The_science_of_climate_change_in_Africa.pdf (accessed February 13, 2012).

Coulibaly-N'Golo, D., Allali, B., Kouassi, S.K., et al. 2011. Novel arenavirus sequences in *Hylomyscus* sp. and *Mus (Nannomys) setulosus* from Côte d'Ivoire: implications for evolution of arenaviruses in Africa. *PLoS One* **6**(6):e20893.

Delgado, S., Erickson, B.R., Agudo, R., et al. 2008. Chapare virus, a newly discovered arenavirus isolated from a fatal hemorrhagic fever case in Bolivia. *PLoS Pathog.* **4**(4):e1000047.

Fair, J., Jentes, E., Inapogui, A., et al. 2007. Lassa virus-infected rodents in refugee camps in Guinea: a looming threat to public health in a politically unstable region. *Vector Borne Zoonotic Dis.* **7**:167–171.

Fichet-Calvet, E., Rogers, D.J. 2009. Risk maps of Lassa fever in west Africa. *PLoS Negl. Trop. Dis.* **3**(3):e388.

Fischer, S.A., Graham, M.B., Kuehnert, M.J., et al. 2006. Transmission of lymphocytic choriomeningitis virus by organ transplantation. *N. Engl. J. Med.* **354**:2235–2249.

Fichet-Calvet, E., Lecompte, E., Koivogui, L., et al. 2007. Fluctuation of abundance and Lassa virus prevalence in *Mastomys natalensis* in Guinea, West Africa. *Vector Borne Zoonotic Dis.* **7**:119–128.

Garnaut, R., Howes, S., Jotzo, F., et al. 2008. Emissions in the platinum age: the implications of rapid development for climate-change mitigation. *Oxford Rev. Econ. Policy* **24**:377–401.

Glass, G.E., Yates, T.L., Fine J.B., et al. 2002. Satellite imagery characterizes local animal reservoir populations of Sin Nombre virus in the southwestern United States. *Proc. Natl. Acad. Sci. U.S.A.* **99**:16817–16822.

Hansen, J., Sato, M., Kharecha, P., et al. 2008. Target atmospheric CO_2: where should humanity aim? *Open Atmos. Sci. J.* **2**:217–231.

Hewawasam, I. 2002. *Managing the Marine and Coastal Environment of Sub-Saharan Africa: Strategic Directions for Sustainable Development*. World Bank, Washington, DC.

Lecompte, E., Fichet-Calvet, E., Daffis, S., et al. 2006. *Mastomys natalensis* and Lassa fever, west Africa. *Emerging Infect. Dis.* **12**:1971–1974.

Meehl, G.A., Stocker, T.F., Collins, W.D., et al. 2007. Global climate projections. In: *Climate Change 2007: The Physical Science Basis. Contribution of Working Group I to the Fourth Assessment Report of the Intergovernmental Panel on Climate Change*, ed. S. Solomon, D. Qin, M. Manning,

Z., et al., 747–845. Cambridge, Cambridge University Press. Available at http://www.ipcc.ch/pdf/assessment-report/ar4/wg1/ar4-wg1-chapter10.pdf (accessed February 13, 2012).

Mills, J.N., Gage, K.L., Khan, A.S. 2010. Potential influence of climate change on vector-borne and zoonotic diseases: a review and proposed research plan. *Environ. Health Perspect.* **118**:1507–1514. Available at http://www.ncbi.nlm.nih.gov/pmc/articles/PMC2974686/ (accessed July 8, 2012).

Purse, B.V., Brown, H.E., Harrup, L., et al. 2008. Invasion of bluetongue and other orbivirus infections into Europe: the role of biological and climatic processes. *Rev. Sci. Tech.* **27**:427–442.

Salazar-Bravo, J., Dragoo, J.W., Bowen, M.D., et al. 2002. Natural nidality in Bolivian hemorrhagic fever and the systematics of the reservoir species. *Infect. Genet. Evol.* **1**:191–199.

Salvato, M.S., Clegg, J.C.S., Buchmeier, M.J., et al. 2012. Arenaviridae. In: *Virus Taxonomy: Classification and Nomenclature of Viruses. 9th Report of the International Committee on Taxonomy of Viruses*, eds. A.M.Q. King, M.J. Adams, E.B. Carstens, et al., 715–723. London, Elsevier/Academic Press.

Weaver, S.C., Reisen, W.K. 2009. Present and future arboviral threats. *Antiviral Res.* **85**:328–345.

Wilson, M.E. 1995. Infectious diseases: an ecological perspective. *Br. Med. J.* **311**:1681–1684.

Z[illegible] 1982, 540. Cambridge: Cambridge University Press. Available at [illegible] (accessed February 15, 2012).

M[illegible] [illegible] 2010. [illegible] climate change [illegible] and [illegible]: a review and proposed research plan. [illegible] 2010. Available at [illegible] (accessed [illegible] 2012).

P[illegible] 2005. Invasion of [illegible] and other [illegible] [illegible]

Salazar, [illegible] 2012. [illegible]

[illegible], M., [illegible] Buchmann, M. [illegible] et al. 2012. [illegible] London: Elsevier, Academic Press.

W[illegible], G., [illegible] 2009. Present and future [illegible]

W[illegible] 1971. [illegible]

EMERGING AND REEMERGING HUMAN BUNYAVIRUS INFECTIONS AND CLIMATE CHANGE

Laura J. Sutherland

Case Western Reserve University, Cleveland, Ohio, USA

Assaf Anyamba

Universities Space Research Association and Biospheric Science Laboratory, NASA Goddard Space Flight Center, Greenbelt, MD, USA

A. Desiree LaBeaud

Children's Hospital Oakland Research Institute, Center for Immunobiology and Vaccine Development, Oakland, CA, USA

TABLE OF CONTENTS

26.1	Introduction	478
26.2	*Bunyaviridae* family	478
	26.2.1 *Hantavirus*	479
	26.2.2 *Nairovirus*	480
	26.2.3 *Orthobunyavirus*	481
	26.2.4 *Phlebovirus*	482
26.3	Climate change and *bunyaviridae*: Climatic influences on transmission cycles and subsequent risk for transmission of bunyaviruses	482
	26.3.1 Arboviral bunyaviruses	482
	26.3.2 Non-arboviral bunyaviruses	484
26.4	Disease spread due to growing geographic distribution of competent vectors	485
	26.4.1 Physical movement of vectors	485
	26.4.2 Expansion of suitable range	485

Viral Infections and Global Change, First Edition. Edited by Sunit K. Singh.
© 2014 John Wiley & Sons, Inc. Published 2014 by John Wiley & Sons, Inc.

26.1 INTRODUCTION

The *Bunyaviridae* family includes a growing number of viruses that have contributed to the burden of emerging and reemerging infectious diseases around the globe. Many of these viruses cause severe clinical outcomes in human and animal populations, the results of which can be detrimental to public health and the economies of affected communities. The threat to endemic and nonnative regions is particularly high, and national and international public health agencies are often on alert. Many of the bunyaviruses cause severe clinical disease including hemorrhage, organ failure, and death leading to their high-risk classification. Hantaviruses and Rift Valley fever virus (RVFV) (genus *Phlebovirus*) are National Institute of Allergy and Infectious Diseases Category A priority pathogens in the United States. Viral hemorrhagic fevers, a classification that includes many bunyaviruses, are immediately notifiable in the European Union. The emergence of new and reemerging bunyaviruses has resulted in numerous human and animal fatalities. Outbreaks of Rift Valley fever (RVF) in East Africa (1997/1998, 2006/2007), Sudan (2007), Southern Africa (2008–2010), Kenya (1997/1998, 2006/2007) (Anyamba et al., 2009, 2010; Breiman et al., 2010; Grobbelaar et al., 2011; Woods et al., 2002) and Saudi Arabia & Yemen (2000, 2010) (Food and Agriculture Organization, 2000; Hjelle and Glass, 2000; Madani et al., 2003) and the emergence of Sin Nombre virus (1993) (Hjelle and Glass, 2000) and most recently Schmallenberg virus (2011) (DEFRA, 2012) are prime examples of the devastating and worldwide toll bunyaviruses have on health and economies.

Climate variability (precipitation and temperature in particular) greatly influence the ecological conditions that drive arboviral disease outbreaks across the globe. Several human and animal disease outbreaks have been influenced by changes in climate associated with the El Niño Southern Oscillation (ENSO) phenomenon including the bunyaviruses RVFV and Sin Nombre (an etiologic agent of hantavirus pulmonary syndrome (HPS)), as well as Murray Valley encephalitis, chikungunya, and malaria to name but a few (Anyamba et al., 2009; Bouma and Dye, 1997; Chretien et al., 2007; Engelthaler et al., 1999; Kovats et al., 2003; Linthicum et al., 1999; Nicholls, 1986). Most bunyaviruses exhibit episodic outbreak patterns with seasonal or annual trends dependent upon climate conditions, vector abundance, and the proximity of a susceptible population. The implications for continued climate change are dire, especially with regard to vector-borne diseases, many of which can cause severe morbidity, sequelae, and death. Increased rainfall and widening endemicity as a result of climate change, compounded by the emergence of new viruses, poses a serious threat to a greater geographic range beyond the regions of endemicity.

26.2 *BUNYAVIRIDAE* FAMILY

Members of the *Bunyaviridae* family are single-stranded RNA viruses with a tripartite genome including a large (L), medium (M), and small (S) RNA segment. Most bunyaviruses are antisense, although some phleboviruses have an ambisense S segment (Baron and Shope, 1996). Virions are enveloped and as such are susceptible to environmental degradation and can be destroyed using lipid solvents, detergents, and low pH solutions. Most bunyaviruses, although not all, are transmitted by arthropods including mosquitoes, ticks,

TABLE 26.1. Bunyaviruses That are Known to Cause Significant Morbidity

Genus	Virus	Primary vector	Affected species	Principle geographic location
Hantavirus	Hantaan	Rodent	Human	Asia
	Puumala	Rodent	Human	Europe
	Seoul	Rodent	Human	Asia, Europe
	Sin Nombre	Rodent	Human	Americas
Nairovirus	CCHF	Tick	Human, bovine, caprine, ovine, cervids	Africa, Asia
	Nairobi sheep disease	Tick	Ovine, human	Africa
Orthobunyavirus	Akabane	Midge, mosquito	Bovine, caprine, ovine	Africa, Asia, Australia
	California encephalitis	Mosquito	Human	North America
	Jamestown Canyon	Mosquito	Human	North America
	La Crosse	Mosquito	Human	North America
	Schmallenberg	Midge	Ovine, bovine	Europe
	Snowshoe hare	Mosquito	Human	North America
Phlebovirus	Naples	Sand fly	Human	Europe, Africa
	Punta Toro	Sand fly	Human	Panama
	RFV	Mosquito	Human, ovine, bovine, caprine, primate, canine, feline, rodent	Africa

midges, and sand flies. Arboviral (arthropod-borne) outbreaks are intimately tied to vector competence, the ecology of which is often a dynamic correlation between climate factors (temperature, rainfall, duration of seasons), reservoir species, and agricultural fecundity. With a broad range of susceptible species, many bunyaviruses are zoonotic and can cause disease in humans as well as many animal species.

Bunyaviruses are especially malignant and pose a threat to numerous animal and human populations across many different continents. While there are many known human, animal, and plant bunyaviruses (a list that grows annually), our discussion herein will be restricted to those that pose a grave threat to human and/or animal life (Table 26.1). The *Bunyaviridae* family poses a serious global threat, one that will continue as new viruses emerge and endemicity expands with the changing climate and globalization.

26.2.1 *Hantavirus*

Hantaviruses are found in a diverse number of geographic locations including the Americas, Asia, and Europe, but no clinical cases (only antibody demonstrations) have been reported in Africa or the Middle East (Acha et al., 2003). Each viral species is transmitted by a specific or closely related endemic rodent or insectivore and, as such, is specific to a particular geographic region (Krüger et al., 2011). How the virus is maintained in the wild among susceptible rodents is currently unknown. Native rodents do not show clinical symptoms but do pass virus through excreta that can contaminate the environment and put humans at risk. Virus, while susceptible to dehydration, can remain viable outside the host for many days when protected by moist excreta (Spickler, 2008). Humans acquire infections directly by animal bite or contact with the saliva, feces, or urine of infected animals. Humans can also become infected if they inhale aerosolized virus from these rodent bodily fluids, usually when humans disturb rodent excreta and nests (Acha et al., 2003).

Hantavirus species can cause a wide range of clinical symptoms in humans. While some viral species such as Puumala cause a mild clinical disease, other species can have particularly severe clinical outcomes. Two life-threatening conditions are associated with the hantaviruses: HPS in the Western Hemisphere/New World and hemorrhagic fever with renal syndrome (HFRS) in Asia and Europe (Zhang, 2009).

HPS is a particularly severe manifestation of the many hantaviral diseases. In 1993, a new disease emerged in the Southwestern United States and was initially called "hantavirus pulmonary syndrome" (Acha et al., 2003; Schmaljohn and Hjelle, 1997). It was found to be caused by Sin Nombre virus and spread through the deer mouse (*Peromyscus maniculatus*). The HPS definition still encompasses a narrow clinical spectrum caused by a broad range of viruses within the Hantavirus genus. The initial clinical symptoms of HPS during the first 3–5 days include fever, myalgia, headache, nausea, chills, and dizziness. Patients present with cough and tachypnea progressing to pulmonary edema and hypoxia followed by cardiac abnormalities. With the onset of cardiopulmonary distress, patients decline rapidly requiring hospitalization and ventilation. Even with timely care, HPS has an estimated 40–60% case fatality rate (Acha et al., 2003; Schmaljohn and Hjelle, 1997).

HFRS occurs following infection with, although certainly not limited to, Hantaan virus in Asia and Puumala and Dobrava viruses in Europe. Clinical progression of disease follows five phases: febrile, hypotensive, oliguric, diuretic, and finally convalescent as the kidneys recover function (Acha et al., 2003; Krüger et al., 2011; Schmaljohn and Hjelle, 1997). HFRS patients have abrupt clinical symptoms including fever, chills, prostration, headache, and backache. Additional gastrointestinal symptoms include abdominal pain, nausea, and vomiting. There are also reports of visual impairment/light sensitivity and petechial rash of the trunk or palate. Following the prodromal phase (multiple days to 1 week), proteinuria is observed and renal function declines with subsequent hypotension and oliguria. Most fatalities, as a result of irreversible shock, occur during the hypotensive phase (Schmaljohn and Hjelle, 1997). Hemorrhagic signs include petechiae, hematuria, and melena especially in severe cases. The case fatality rate following HFRS is <0.4% for Puumala virus, 7–12% for Dobrava virus, and 10–15% for Hantaan virus (Acha et al., 2003; Spickler, 2008).

Emerging hantaviruses continue to be identified (Hjelle and Torres-Pérez, 2010). The Hantavirus genus is both genetically and geographically diverse. Numerous genotypes have been identified in both murid and cricetid rodents (the classic reservoir host), while newly identified genotypes with sometimes-unknown pathology have been identified in insectivores (shrews and moles, order *Soricomorpha*) (Kang et al., 2011; Klempa et al., 2007) and bats (order *Chiroptera*) (Hjelle and Torres-Pérez, 2010). There is still much to learn about hantaviruses, but it is clear that they will continue to cause disease in humans around the globe.

26.2.2 *Nairovirus*

The *Nairovirus* genus is endemic to diverse geographic ranges that overlap that of its tick vector including Asia, the Middle East, eastern Europe, and Africa (Sang et al., 2011; Zavitsanou et al., 2009). The *Nairovirus* genus is divided into seven serogroups that delineate 37 viruses. Hard ticks of the family *Ixodidae* primarily transmit the nairoviruses. Subsequent exposure to blood, excreta, and tissues of infected humans and animals, in certain serogroups, contributes to secondary infections particularly among caretakers and animal handlers. Many nairoviruses infect animals, but only three are known to cause disease in humans: Crimean–Congo hemorrhagic fever (CCHF), Dugbe, and Nairobi sheep viruses.

Crimean–Congo hemorrhagic fever virus (CCHFV) is naturally maintained in an enzootic cycle between asymptomatic animals and multiple tick species. The most common vectors of CCHFV are from the genus *Hyalomma* (Sang et al., 2011). Many mammalian species have been found to harbor CCHFV including hares and hedgehogs, cattle, sheep, goats, horses, and swine (Acha et al., 2003). Wild animals have been shown to maintain antibody levels against CCHFV including giraffes and rhinos (Spickler, 2009). Birds, while refractory to infection, may play a role in the distribution of infected ticks along migratory routes (Zavitsanou et al., 2009). Small vertebrates, including hares and rodents, amplify the virus and transmit to feeding tick larvae. These ticks maintain the virus throughout their entire life cycle from larvae to adult (transstadial), and in some instances, transovarial, and venereal transmission of virus has occurred; co-feeding transmission has also been implicated (EFSA, 2010). Infected adult ticks, feeding on larger domestic ruminants including cattle, sheep, and goats, pass the virus. Domestic animals remain viremic for 1 week following infection (Spickler, 2009). Humans in close contact with infected ticks, as well as with the infected body fluids expressed during hemorrhage (human), slaughter, and birth, are at greatest risk for disease.

Onset of CCHF is rapid with fever, myalgia, neck pain and stiffness, and photophobia. Gastrointestinal symptoms may be present including nausea, vomiting, and diarrhea. After a few days, patients may become confused or agitated with mood swings followed by a generalized lethargy and depression. CCHF is also accompanied by hemorrhagic conditions including petechiae, ecchymoses, melena, hematuria, and epistaxis. Severe cases end with hepatorenal and pulmonary failure; the case fatality rate is 30% (Acha et al., 2003).

26.2.3 *Orthobunyavirus*

The orthobunyaviruses represent a diverse genus of viruses that can result in severe health outcomes including neuroinvasive disease. This genus is divided into 18 distinct serogroups as defined by serologic testing. Of particular importance to humans is the California serogroup that includes La Crosse virus (LCV).

LCV is a zoonotic arboviral disease that circulates between small mammalian hosts (squirrels and chipmunks) and forest-dwelling mosquitoes, *Aedes triseriatus* (Acha et al., 2003). The virus is transmitted both transovarially and sexually in *Aedes*. LCV is most commonly reported in the northern, midwestern, and eastern states of the United States where deciduous forests harbor both host and vector species. Humans are dead-end hosts for LCV. The spectrum of clinical symptoms can be diverse and many individuals develop a subclinical infection. Others experience headache, nausea, vomiting, and fatigue. LCV can result in acute inflammation of the human brain. Severe disease, most often in the pediatric population, can progress to memory loss, seizures, coma, and death (<1%). Neurologic sequelae marked by intellectual deficit and recurring seizures can occur.

New orthobunyaviruses continue to emerge. In late 2011, an unprecedented number of stillbirths with fetal abnormalities were observed in cattle and sheep within Germany and the Netherlands. This new virus, thus far isolated in midges, has been named Schmallenberg virus (DEFRA, 2012; OIE, 2012). Adult cows initially show signs of fever, diarrhea, loss of appetite, and lastly abortion. Offspring of infected animals can be of live birth or stillborn with physical abnormalities including arthrogryposis, ankylosis, or cognitive deficit. The extent of fetal anomaly depends upon the stage of gestation at which infection occurs. Although not known to cause human illness, the economic and agricultural impact of such a virus has been quite serious and the virus continues to spread across the European continent posing problems for trade.

26.2.4 *Phlebovirus*

The *Phlebovirus* genus is comprised of 70 distinct serogroups, of which only a few have been researched in depth and fewer than ten have been linked with disease in humans. Phleboviruses cause a variety of clinical symptoms including self-limiting febrile illness, retinitis, encephalitis, and fatal hemorrhagic fever.

RVF is an economically devastating zoonotic arboviral disease caused by RVFV. Endemic to sub-Saharan Africa, repeated outbreaks have occurred in Kenya, Egypt, Madagascar, and South Africa (Anyamba et al., 2010; Chevalier et al., 2009; Favier et al., 2006; Mohamed et al., 2010; Nguku et al., 2010; Woods et al., 2002). RVFV outbreaks also occur in the Middle East (Balkhy and Memish, 2003; Gerdes, 2004). RVFV is primarily transmitted via mosquitoes, although human contact with infective animal products and aerosolized virus during slaughter and birth can also pose a significant transmission risk (Gerdes, 2004; King et al., 2010; LaBeaud et al., 2011a, b, c; Shope et al., 1982). Sustained RVF outbreaks require a multifaceted convergence of necessary factors: capable vectors, susceptible animals and humans in close proximity, and sufficiently heavy rains. It has been shown that RVFV circulates among endemic animals and humans during interepidemic periods, but only during high-rain seasons are large-scale outbreaks observed (Evans et al., 2008; LaBeaud et al., 2008, 2011a, b, c).

Pregnant domestic livestock are particularly susceptible to severe RVF disease. Human outbreaks of RVFV are often foreshadowed by large "abortion storms" among domestic livestock. The reported rates of abortion in pregnant ewes are 5% to almost 100% and 10–85% in cattle (Spickler, 2006). Young animals also succumb to infection with 90–100% of newborn lambs and 10–70% of calves dying from RVF. Adult animals also die from infection with reported mortality rates among sheep at 5–100% and less than 10% of adult cattle.

Human RVFV infection is typically asymptomatic or, at most, results in a mild to moderate flulike illness with fever, headache, joint pain, and weakness. In a small subset of the population, RVF causes lasting ocular disease (0.5–2%) that contributes to vision loss among affected individuals (Al-Hazmi et al., 2003; LaBeaud et al., 2008, 2011a, b, c). Less than 1% of RVF patients have meningoencephalitis or hemorrhagic fever. Those with meningoencephalitis who survive often suffer long-term neurologic and psychologic sequelae. The case fatality rate following hemorrhagic fever is 50%.

The effect of a RVFV outbreak on an economy is significant, and the impending threat to naïve but susceptible populations is a concern for many governments and producers. The United States alone exported $8.1 billion beef-related products in 2011. The World Organisation for Animal Health (OIE) imposes a 4-year trade ban on any country with confirmed RVFV transmission, a restriction that is only lifted after 6 months of being disease-free (Linthicum et al., 2007; Little, 2009; World Organization for Animal Health, 2011). Commercial veterinary vaccines, both inactivated and live-attenuated, are available against RVFV in endemic areas of sub-Saharan Africa. There is no commercially available human vaccine.

26.3 CLIMATE CHANGE AND *BUNYAVIRIDAE*: CLIMATIC INFLUENCES ON TRANSMISSION CYCLES AND SUBSEQUENT RISK FOR TRANSMISSION OF BUNYAVIRUSES

26.3.1 Arboviral bunyaviruses

26.3.1.1 *Mosquitoes.* Four of the five bunyavirus genera are arthropod-borne. Changes in climate affect disease vectors in a number of ways including mosquito survival;

susceptibility to viruses; mosquito population growth rate, distribution, and seasonality; replication and extrinsic incubation period of a virus in the mosquito; and transmission patterns and seasonality (Epstein, 2005; Gubler, 2001). The life cycles of many invertebrates are inextricably linked to temperature, rainfall, and daylight patterns, and any changes herein can have resounding affects on invertebrate vectors.

Mosquito life cycles are particularly susceptible to climatic factors. Taking blood meals is required for the development of eggs, and female mosquitoes will feed repeatedly until repletion (Beaty et al., 1983). High temperatures following complete blood meals quicken the development of eggs within the female mosquito. On average this process lasts between 3 and 4 days (ECDC, 2007; Estrada-Franco et al., 1995). Increases in temperature could both shorten the time necessary for egg development and increase the number of reproductive periods and, subsequently, the absolute number of competent vector offspring produced each season. Mosquito development rates also depend on the ambient temperature. Ideal ranges are between 25 °C and 30 °C, but with consistent temperatures above 25 °C, the developmental time between egg and adult can be as short as 1 week (ECDC, 2007; Estrada-Franco et al., 1995; Turell, 1989; Turell et al., 2001). With regional temperatures on the rise, more eggs can be laid and subsequently more larvae can hatch during each reproductive cycle, thereby creating a much larger population of adult mosquitoes to transmit diseases to susceptible hosts.

East Africa experiences dramatic seasonal rainfall patterns. Cyclic patterns of rainfall and drought contribute greatly to the emergence of arboviral bunyaviruses, and this phenomenon has been observed prior to epizootic RVFV outbreaks in Kenya. *Aedes* mosquitoes lay eggs *above* water lines, a trait that allows for periodic hatching of larvae only during appropriate (sustained wet) conditions. Floodwater *Aedes* species mosquitoes, the vectors of RVFV and LCV, naturally undergo a diapause state to allow survival against environmental extremes including drought and winter. This contributes to vector survival even during long drought periods and has been scientifically shown to aid in the spread of some bunyaviruses including RVFV and LCV (Acha et al., 2003). During short rains, shallow depressions in the landscape fill with water (called *dambos*) and, if sufficient depths are reached, invigorate dormant floodwater *Aedes* mosquito eggs and allow hatching. In the case of RVFV, *Aedes* eggs can be transovarially infected (vertical transmission), thereby replenishing the area with a population of viable and infected mosquitoes, a cycle that can repeat during periods of infrequent rainfall (Mondet et al., 2005). While many African *Aedes* mosquitoes feed on livestock animals, *Culex* mosquitoes primarily provide the bridge between these amplifying domestic livestock and nearby humans, resulting in the epidemic transmission of RVFV. When extreme weather patterns bring excessive rains, much larger bodies of water are formed and provide the ideal environment for reproducing *Aedes* and *Culex* mosquitoes (Figure 26.1).

The conflux of infective *Aedes*, amplifying cattle, and susceptible *Culex* provides the necessary bridge to bring RVFV to humans (Linthicum et al., 1985). RVFV epidemics occur in East Africa every 5–15 years. Epidemics and epizootics are closely linked to the excessive rainfall during ENSO events as sea surface temperatures (SST) warm, not only resulting in a bloom of capable vectors but also providing an ideal environment that places these vectors and susceptible humans in close proximity (Anyamba et al., 2009; Linthicum et al., 1999). Heavy rainfall is often a great relief in drought-ravaged regions, and the growing bodies of water that result cause a mass congregation of people and their animals.

Drought can also impact vector survival. Decreases in rainfall can reduce standing water levels, altering the location and number of small water sources, making ideal habitats for container-breeding mosquitoes.

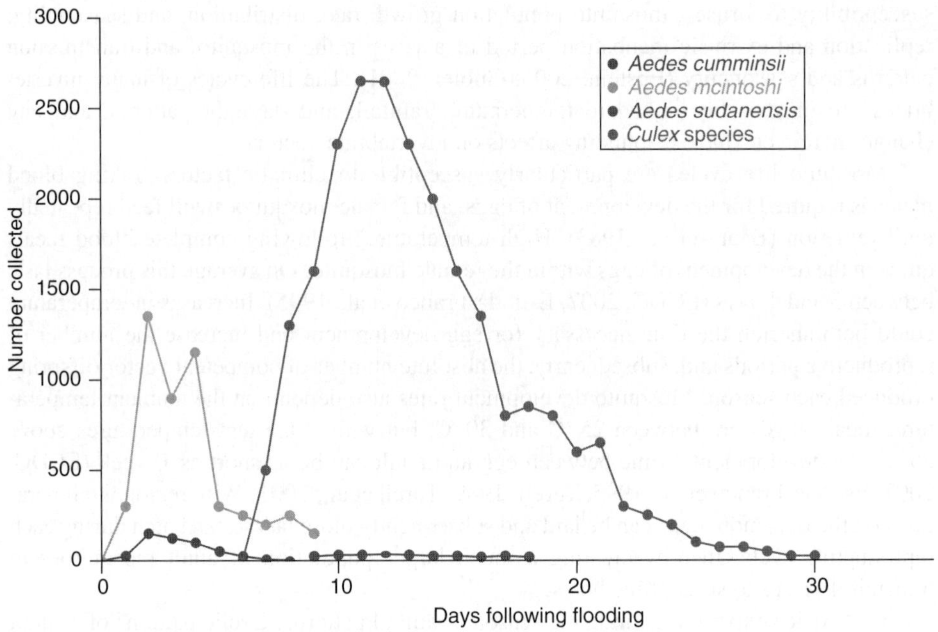

Figure 26.1. Mosquito population succession in a *dambo* habitat (reconstructed after Linthicum et al., 1983). For color detail, please see color plate section.

26.3.1.2 ***Ticks.*** Like mosquitoes, the successful propagation of ticks is influenced by appropriate weather patterns and the availability of a blood meal. Ticks thrive in warm, humid environments (Süss et al., 2008). Conversely, drought periods increase tick mortality (Githeko et al., 2000). The consequences of climate change are ideal for tick reproduction and success. Increases in global temperatures have the consequence of shortening colder seasons and extending warmer ones, essentially eliminating the seasonal interruption in viral transmission (World Health Organization, 2009). With a warmer winter, tick mortality decreases and the surviving adult ticks can maintain infective levels between seasons. Furthermore, increased numbers of mammalian hosts will survive during temperate winters, providing a plethora of blood meals to feed this increased number of ticks (Githeko et al., 2000).

26.3.2 Non-arboviral bunyaviruses

Diseases that do not require a vector for transmission, such as hantaviruses, can also be impacted by climate change as the ecological habitat of rodents is uniquely susceptible to changes in temperature and rainfall. Seasonal weather patterns are often implicated in increases of human hantaviruses (Engelthaler et al., 1999; Hjelle and Glass, 2000; Hjelle and Torres-Pérez, 2010; Jonsson et al., 2010; Schmaljohn and Hjelle, 1997). Outbreaks of Sin Nombre virus in the Southwestern United States are often witnessed when warm winters are followed by wet springs (Calisher et al., 2005a, b). The availability of food (vegetation) for rodents increases in years with the sustained rainfall associated with the ENSO, providing ample food supplies for rodent populations. Reproductive blooms occur during these fertile summer months resulting in ample offspring to replenish the reservoir host pool. Naïve second-generation young adults are susceptible to infection and able to transmit as they scavenge and fight for food and resources during the subsequent drought period and before the coming winter hibernation season (Calisher et al., 1999; Engelthaler et al., 1999).

While climatic changes do not directly increase viral acquisition or susceptibility of humans per se, these weather patterns do create a "perfect storm" whereby large populations of infected rodents and humans are in closer contact. Public health agencies, in the months and years following ENSO events, have witnessed increases in the number of cases of HFRS (Engelthaler et al., 1999; Hjelle and Glass, 2000; Zhang et al., 2010). Risky behaviors (handling rodent nests and excreta and rodent trappings) are frequently engaged in during the spring months, as people resume their outdoor activities and clean recreational structures (cabins, sheds, etc.) (Jonsson et al., 2010). Disturbing rodent nests and excreta aerosolizes virus that can be inhaled by humans and cause disease.

The linkages that result in increased HFRS cases are complex involving rainfall, vegetation and growth, and complex ecological interactions between reservoirs, prey and predators, and other mammalian species. The role of climate change plays a unique role in the dynamics of each of these parameters and, as such, in disease transmission.

26.4 DISEASE SPREAD DUE TO GROWING GEOGRAPHIC DISTRIBUTION OF COMPETENT VECTORS

26.4.1 Physical movement of vectors

The ENSO phenomenon affects multiple weather conditions including temperature, precipitation, and wind speed and direction. At a local level, these conditions can alter or impede migratory routes for many native bird species, some of which provide blood meals for vector species. Migratory birds can act as a vehicle for disease transmission as they transport infected vector species across and into new regions (Gray et al., 2009).

26.4.2 Expansion of suitable range

The globe is warming. Vertebrates thrive within a specified temperature range and, given slight extensions of temperature on either side of that spectrum, can easily adapt to new bioclimatic conditions. It is important to emphasize that temperature increases can shift the entire range for vector species to a more northerly direction. As global temperatures rise, geographic boundaries expand allowing the successful introduction and subsequent reproduction of nonnative vector species in new territories (Githeko et al., 2000; World Health Organization, 2009). Even after the first generation of vectors die, transovarially transmitted bunyaviruses can remain within the ecosystem. This is not a new phenomenon; many arboviral diseases have spread outside their endemic zones as vectors take up new residence and autochthonous transmission proceeds (Chretien and Linthicum, 2007; Gould and Higgs, 2009).

It has been postulated that the expansion of blue tongue virus (a disease of livestock) has been aided by a warming climate across southern Europe (Purse et al., 2005). Schmallenberg virus, likely transmitted by the same midge vectors, has benefited from the extensive range of these vector species. The broadening geographic range of Culicoides midges, as a consequence of this warmer weather, has aided the spread of vector-borne diseases across larger geographic regions (more northward) and for longer periods of time (through the unseasonably warm winter months) (Purse et al., 2005). Additionally, the placement of infective vectors within an area of immunologically naïve hosts increases the risk for human and animal morbidity and continued viral spread.

26.5 USING CLIMATE AS A MEANS FOR OUTBREAK PREDICTION

26.5.1 Climatic influences

Because bunyaviral outbreaks are linked to weather patterns, researchers can use climate modeling and risk monitoring to predict disease outbreaks, many months in advance, thereby affording public health agencies advanced notification. These methods have been employed to a greater extent in recent years. The benefits of disease prediction are innumerable: the ability for targeted and appropriate implementation of vector control methods, animal vaccinations and trade/movement restrictions, and public education to reduce both human and animal disease. Such an application has been most successful for RVF. Given that the impetus for RVFV outbreaks is often the unusually heavy rainfall associated with local and regional climate variability, the application of climate modeling to RVFV outbreak prediction will be discussed in detail.

Researchers have long observed the association between weather patterns and bunyavirus outbreaks. This is most easily observed during RVFV outbreaks. Davies et al. (1985) reported a composite statistic of surplus rainfall alongside the occurrence of RVFV outbreaks between 1950 and 1982 in Kenya. The four epizootics that occurred during that time were associated with periods of high surplus rainfall (Davies et al., 1985). Studies of outbreaks elsewhere, in South Africa (Swanepoel, 1976, 1981) and West Africa (Bicout and Sabatier, 2004), have additionally supported this relationship. The sustained high precipitation floods *dambos* that support the hatching and continued development of primary and secondary mosquito vectors, increase the likelihood of risky behaviors that place susceptible animals and humans in close proximity to those vectors, and facilitate transmission of RVFV.

The rainfall patterns associated with these outbreaks are inextricably linked to coincident effects of the warm phase of ENSO and warming events in the western equatorial Indian Ocean (increased SST) that result in above-normal and extended rainfall over East Africa (Anyamba and Tucker, 2002; Birkett and Murtugudde, 1999; Cane, 1983; Linthicum et al., 1999; Nicholson, 1986; Ropelewski and Halpert, 1987; Saji et al., 1999). ENSO events are marked by an oscillation between two weather phenomena that have varying effects on the global climate and weather patterns relevant to disease outbreaks. El Niño, the warming ENSO event, refers to a large-scale ocean–atmosphere climate phenomenon that is linked to periodic warming in SST across the Pacific. The opposite ENSO event is La Niña, the cold phase (Figure 26.2). The changes that occur across the sea surface of the Pacific have marked consequences on the temperature and precipitation of tropical climates, especially in the Horn of Africa. El Niño events, or Pacific warming, result in increased rainfall across East Africa, a phenomenon that is largely reversible during cycles of La Niña (Anyamba et al., 2012). More than 90% of RVFV outbreak events since 1950 have occurred during warm ENSO events (Linthicum et al., 1999). The interepidemic period of RVFV generally occurs during La Niña events (the cold phase of ENSO), which causes diminished rainfall and drought in East Africa (Anyamba and Tucker, 2002).

Additional ecological parameters conducive to outbreaks have been identified. The vigorous growth of vegetation and cooler temperatures during periods of extended and above-normal rainfall are ecologically appropriate for vector development and propagation and are an additional key to predicting epizootic/epidemic conditions. While *dambos* can be spotted aerially using aircraft-mounted radar and Landsat Thematic Mapper (TM) data, large-scale efforts require high-temporal resolution data so that appropriate pretreatment and vector control efforts can be implemented during high-risk periods (Pope et al., 1994). Such high-temporal resolution data is generated in the form of the normalized difference vegetation index (NDVI) from measurement made by the National Oceanic and Atmospheric Administration (NOAA) series of Polar Operational Environmental Satellites (POES). The

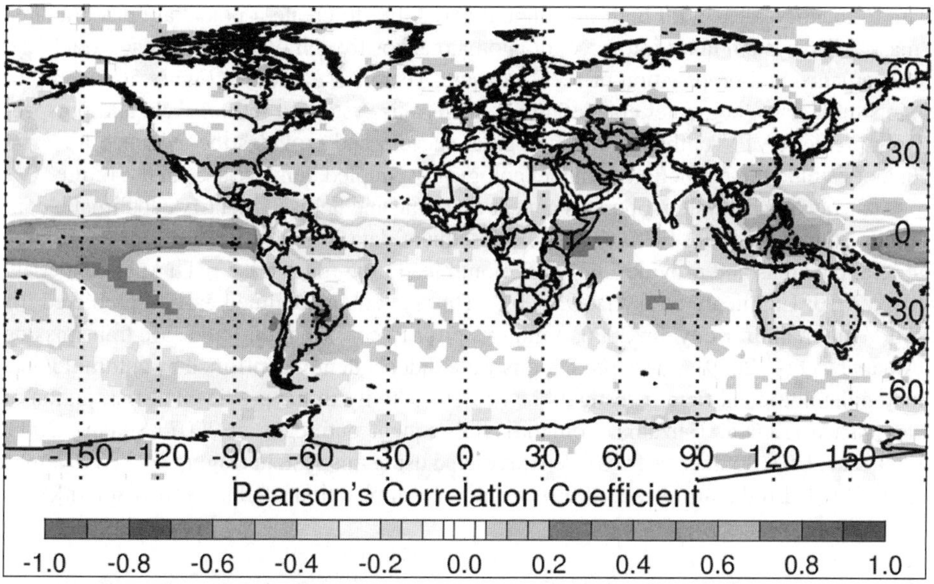

Figure 26.2. Correlation of SST and rainfall anomalies illustrates ENSO teleconnection patterns. There is a tendency for above (below)-normal rainfall during *El Niño* (*La Niña*) events over East Africa (Southern Africa, Southeast Asia). Similar differential anomaly patterns were observed for other regions, especially within the global tropics. These extremes (above or below) in rainfall influence regional ecology and consequently dynamics of mosquito disease vector populations and patterns of mosquito-borne disease outbreaks (Anyamba et al., 2012). For color detail, please see color plate section.

reliability (operational), spatial coverage (global), and temporal frequency (daily) of NOAA satellites has made such data a more cost-effective and successful means of monitoring vegetative habitats conducive to vector development. NDVI data when employed in a time series fashion indicates mosquito habitats (Linthicum et al., 1990, 1987, 1994). These anomalous NDVI changes signal ecological changes conducive to mosquito development and foreshadow the onset of increased RVF cases, allowing for improved prevention and control.

26.5.2 Risk mapping and predictions

A monitoring and risk mapping system has been developed that utilizes multiple satellite measurements including SST, outgoing longwave radiation, rainfall, and NDVI-derived landscape data, to assess and predict the climatic and ecological parameters conducive to RVFV outbreaks (Anyamba et al., 2002, 2010; Linthicum et al., 2007). Disease risk mapping and prediction systems take numerous factors into account and have been improved upon as we gain increasingly more sophisticated climatic and ecological data before, during, and after outbreaks. Current systems factor multiple variables including (i) monitoring phase and amplitude anomalies of global SST and western equatorial Indian Ocean SST anomalies within the Niño 3.4 region, (ii) monitoring patterns of outgoing longwave radiation anomalies to infer and detect large scale changes and shift in the major atmospheric centers of tropical convection as resulting from ENSO, and (iii) monitoring anomalous NDVI patterns over Africa as a proxy for ecological dynamics. Historic data also informs risk mapping; epizootic/epidemic regions are mapped and combined with long-term NDVI and rainfall data to identify areas with high interannual variability to create a potential epizootic area "mask." The system takes into account anomalous positive NDVI changes that would lead to emergence and propagation of RVF mosquito vectors as shown in Figure 26.1 and pinpoints these areas at greater epizootic risk. Using this system to retroactively and prospectively predict RVFV outbreaks has had great success and provides the evidence for valid epizootic/epidemic prediction of other ecologically coupled bunyaviruses in the future.

The developed system operates at near real time and helps to identify ecoclimatic conditions conducive to RVFV outbreaks. Subsequent data provided can be applied to large geographic areas and provides a 3- to 5-month lead time for disease risk (Anyamba et al., 2006, 2010; Linthicum et al., 1999, 2007). This system, as outlined earlier, predicted the RVFV outbreak in Kenya in 2006, 3 months prior to confirmation of disease transmission (Anyamba et al., 2009). These predictions were independently confirmed by entomological and epidemiological field investigations of virus activity in the predicted at-risk region of East Africa (Kenya, Tanzania, Somalia). Subsequent outbreaks in 2007 (Sudan), 2008, 2009, and 2010 (Southern Africa) were predicted using this system (Figure 26.3) (Anyamba et al., 2010; Linthicum et al., 1999). Continued application of this system provides an

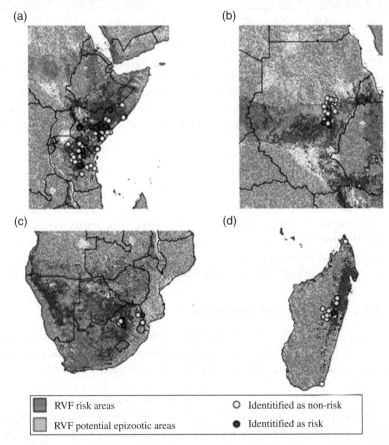

Figure 26.3. Summary RVF risk map, of Eastern Africa (September, 2006–May, 2007). Areas shown in green represent RVF potential epizootic areas, areas shown in red represent pixels that were mapped by the prediction system to be at risk for RVF activity during the respective time periods, blue dots indicate human cases identified to be in the RVF risk areas, and yellow dots represents human cases in areas not mapped to be at risk. For color detail, please see color plate section.

important tool for the early prevention and control of animal and human diseases at the local, national, and international level.

26.6 FUTURE PROBLEMS

The bunyaviruses are of great public health significance because of their geographic diversity and capacity to cause severe disease and death in multiple species. New bunyaviruses will continue to emerge, and known viruses will be found in new regions as favorable conditions expand into new territories. The expansion of endemic ranges threatens human and animal health. Competent vector and reservoir species in the vicinity of susceptible and immunologically naïve humans and animals can maintain newly introduced viruses and cause significant morbidity and mortality. Emerging viruses are identified annually and, as evidenced most recently by Schmallenberg virus, have the propensity to cause significant health and economic consequences. Public health agencies must remain vigilant by reporting unusual clusters of morbidity and mortality. Import and export agencies must also be on high alert and comply with regulations to halt disease spread between endemic and non-endemic regions.

As has been discussed, the geographic range, frequency, and outbreak patterns of bunyaviruses are likely to be influenced by changes in climate. These outbreaks can occur following dramatic and nuanced changes in rainfall, temperature, and wind patterns, the consequences of which have already been observed. Vector-borne diseases and non-arthropod-borne diseases, such as those in the Hantavirus genus, are highly susceptible to climatic variance in a number of ways. Both the proliferation and behaviors of mosquito and tick vectors and reservoir species are largely dependent on abiotic factors in the ecosystem. Changes in weather patterns can lengthen transmission seasons, increase risky behaviors, and shorten extrinsic incubation times. Warning systems used by researchers and government organizations can largely take advantage of this knowledge by using risk mapping and predictions based on known weather phenomena. Continued long-term studies of climate and atmospheric conditions will allow for more advanced predictions in more regions of the globe. With advanced warning, public health officials can unveil targeted vector control, vaccination, and educational campaigns to at-risk communities well in advance of outbreak conditions, thereby thwarting disease spread.

REFERENCES

Acha, P., Szyfres, B., and Pan American, H.O., 2003. *Zoonoses and Communicable Diseases Common to Man and Animals*. PAHO, Washington, DC.

Anyamba, A., and Tucker, C.J., 2002. From El Nino to La Nina: vegetation response patterns over East and Southern Africa during the 1997–2000 period. *J. Clim.* **15**, 3096–3103.

Anyamba, A., Linthicum, K.J., and Mahoney, R., 2002. Mapping potential risk of Rift Valley fever outbreaks in African savannas using vegetation index time series data. *Photogramm. Eng. Remote Sens.* **68**, 2, 137–145.

Al-Hazmi, M., Ayoola, E.A., Abdurahman, M., et al., 2003. Epidemic Rift Valley fever in Saudi Arabia: a clinical study of severe illness in humans. *Clin. Infect. Dis.* **36**, 3, 245–252.

Anyamba, A., Chretien, J., Small, J., et al., 2006. Developing global climate anomalies suggest potential disease risks for 2006–2007. *Int. J. Health Geograph.* **5**, 60–60.

Anyamba, A., Chretien, J.P., Small, J., et al., 2009. Prediction of a Rift Valley fever outbreak. *Proc. Natl. Acad. Sci. U.S.A.* **106**, 3, 955–959.

Anyamba, A., Linthicum, K.J., Small, J., et al., 2010. Prediction, assessment of the Rift Valley fever activity in East and Southern Africa 2006–2008 and possible vector control strategies. *Am. J. Trop. Med. Hyg.* **83**, 2, 43–51.

Anyamba, A., Linthicum, K.J., Small, J.L., et al., 2012. Climate teleconnections and recent patterns of human and animal disease outbreaks. *PLoS Negl. Trop. Dis.* **6**, 1, e1465.

Balkhy, H.H., and Memish, Z.A., 2003. Rift Valley fever: an uninvited zoonosis in the Arabian peninsula. *Int. J. Antimicrob. Agents* **21**, 2, 153–157.

Baron, S., and Shope, R.E., 1996. *Bunyaviruses.* University of Texas Medical Branch at Galveston, Galveston (TX).

Beaty, B.J., Bishop, D.H., Gay, M., et al., 1983. Interference between bunyaviruses in *Aedes triseriatus* mosquitoes. *Virology* **127**, 1, 83–90.

Bicout, D.J., and Sabatier, P., 2004. Mapping Rift Valley fever vectors and prevalence using rainfall variations. *Vector Borne Zoonotic Dis.* **4**, 1, 33–42.

Birkett, C., and Murtugudde, R., 1999. Indian Ocean climate event brings floods to East Africa's lakes and the Sudd Marsh. *Geophys. Res. Lett.* **26**, 8, 1031–1034.

Bouma, M.J., and Dye, C., 1997. Cycles of malaria associated with El Niño in Venezuela. *JAMA* **278**, 21, 1772–1774.

Breiman, R.F., Minjauw, B., Sharif, S.K., et al., 2010. Rift Valley fever: scientific pathways toward public health prevention and response. *Am. J. Trop. Med. Hyg.* **83**, 2, Suppl. 1–4.

Calisher, C.H., Sweeney, W., Mills, J.N., et al., 1999. Natural history of Sin Nombre virus in western Colorado. *Emerging Infect. Dis.* **5**, 1, 126–134.

Calisher, C.H., Mills, J.N., Sweeney, W.P., et al., 2005a. Population dynamics of a diverse rodent assemblage in mixed grass-shrub habitat, southeastern Colorado, 1995–2000. *J. Wildl. Dis.* **41**, 1, 12–28.

Calisher, C.H., Root, J.J., Mills, J.N., et al., 2005b. Epizootiology of Sin Nombre and El Moro Canyon hantaviruses, southeastern Colorado, 1995–2000. *J. Wildl. Dis.* **41**, 1, 1–11.

Cane, M.A., 1983. Oceanographic events during El Niño. *Science* **222**, 4629, 1189–1195.

Chevalier, V., Thiongane, Y., and Lancelot, R., 2009. Endemic transmission of Rift Valley fever in Senegal. *Transboundary Emerg. Dis.* **56**, 9–10, 372–374.

Chretien, J., and Linthicum, K.J., 2007. Chikungunya in Europe: what's next? *Lancet* **370**, 9602, 1805–1806.

Chretien, J., Anyamba, A., Bedno, S.A., et al., 2007. Drought-associated chikungunya emergence along coastal East Africa. *Am. J. Trop. Med. Hyg.* **76**, 3, 405–407.

Davies, F.G., Linthicum, K.J., and James, A.D., 1985. Rainfall and epizootic Rift Valley fever. *Bull. World Health Organ.* **63**, 5, 941–943.

DEFRA, Department for Environment Food and Rural Affairs, 2012. *Schmallenberg Virus*. Available: http://www.defra.gov.uk/animal-diseases/a-z/schmallenberg-virus/. Last updated: May 8, 2012 (May 8, 2012).

European Centre for Disease Prevention Control (ECDC), 2007. *Consultation on Vector-Related Risk for Chikungunya Virus Transmission in Europe*. ECDC, Paris.

EFSA Panel on Animal and Welfare (AHAW), 2010. Scientific opinion on the role of tick vectors in the epidemiology of Crimean Congo hemorrhagic fever and African swine fever in Eurasia. *EFSA J.* **8**, 8, 1703 [156 pp.]. Available: www.efsa.europa.eu/efsajournal.htm (July 8, 2013).

Engelthaler, D.M., Mosley, D.G., Cheek, J.E., et al., 1999. Climatic and environmental patterns associated with hantavirus pulmonary syndrome, Four Corners region, United States. *Emerging Infect. Dis.* **5**, 1, 87–94.

Epstein, P.R., 2005. Climate change and human health. *N. Engl. J. Med.* **353**, 14, 1433–1436.

Estrada-Franco, J., Craig, G.B., and Pan American, H.O., 1995. *Biology, Disease Relationships, and Control of Aedes albopictus*. Pan American Health Organization, Pan American Sanitary Bureau, Regional Office of the World Health Organization (1995), Washington DC.

Evans, A., Gakuya, F., Paweska, J.T., et al., 2008. Prevalence of antibodies against Rift Valley fever virus in Kenyan wildlife. *Epidemiol. Infect.* **136**, 9, 1261–1269.

Favier, C., Chalvet-Monfray, K., Sabatier, P., et al., 2006. Rift Valley fever in West Africa: the role of space in endemicity. *Trop. Med. Int. Health* **11**, 12, 1878–1888.

Food and Agriculture Organization, 2000. Update on Rift Vally Fever outbreaks in Saudi Arabia and Yemen. EMPRES Transboundary Animal Diseases Bulletin No. 15-2000, pp. 3–4.

Gerdes, G.H., 2004. Rift Valley fever. *Rev.-Off. Int. Epizoot.* **23**, 2, 613–623.

Githeko, A.K., Lindsay, S.W., Confalonieri, U.E., et al., 2000. Climate change and vector-borne diseases: a regional analysis. *Bull. World Health Organ.* **78**, 9, 1136–1147.

Gould, E.A., and Higgs, S., 2009. Impact of climate change and other factors on emerging arbovirus diseases. *Trans. R. Soc. Trop. Med. Hyg.* **103**, 2, 109–121.

Gray, J.S., Dautel, H., Estrada-Peña, A., et al., 2009. Effects of climate change on ticks and tick-borne diseases in Europe. *Interdiscip. Perspect. Infect. Dis.* **2009**, 1–12.

Grobbelaar, A.A., Weyer, J., Leman, P.A., et al., 2011. Molecular epidemiology of Rift Valley fever virus. *Emerg. Infect. Dis.* **17**, 12, 2270–2276.

Gubler, D.J., 2001. Human arbovirus infections worldwide. *Ann. N. Y. Acad. Sci.* **951**, 13–24.

Hjelle, B., and Glass, G.E., 2000. Outbreak of hantavirus infection in the Four Corners region of the United States in the wake of the 1997–1998 El Niño -southern oscillation. *J. Infect. Dis.* **181**, 5, 1569–1573.

Hjelle, B., and Torres-Pérez, F., 2010. Hantaviruses in the Americas and their role as emerging pathogens. *Viruses* **2**, 12, 2559–2586.

Jonsson, C.B., Figueiredo, L.T.M., and Vapalahti, O., 2010. A global perspective on hantavirus ecology, epidemiology, and disease. *Clin. Microbiol. Rev.* **23**, 2, 412–441.

Kang, H.J., Kadjo, B., Dubey, S., et al., 2011. Molecular evolution of Azagny virus, a newfound hantavirus harbored by the West African pygmy shrew (*Crocidura obscurior*) in CÃ´te d'Ivoire. *Virol. J.* **8**, 373.

King, C.H., Kahlon, S.S., Muiruri, S., et al., 2010. Facets of the Rift Valley fever outbreak in Northeastern Province, Kenya, 2006–2007. *Am. J. Trop. Med. Hyg.* **82**, 3, 363–363.

Klempa, B., Fichet-Calvet, E., Lecompte, E., et al., 2007. Novel hantavirus sequences in Shrew, Guinea. *Emerging Infect. Dis.* **13**, 3, 520–522.

Kovats, R.S., Bouma, M.J., Hajat, S., et al., 2003. El Niño and health. *Lancet* **362**, 9394, 1481–1489.

Krüger, D., Schonrich, G., and Klempa, B., 2011. Human pathogenic hantaviruses and prevention of infection. *Hum. Vaccin.* **7**, 6, 685–693.

LaBeaud, A.D., Muchiri, E.M., Ndzovu, M., et al., 2008. Interepidemic Rift Valley fever virus seropositivity, northeastern Kenya. *Emerging Infect. Dis.* **14**, 8, 1240–1246.

LaBeaud, A.D., Bashir, F., and King, C.H., 2011a. Measuring the burden of arboviral diseases: the spectrum of morbidity and mortality from four prevalent infections. *Popul. Health Metrics* **9**, 1, 1.

LaBeaud, A.D., Muiruri, S., Sutherland, L.J., et al., 2011b. Postepidemic analysis of Rift Valley fever virus transmission in northeastern Kenya: a village cohort study. *PLoS Negl. Trop. Dis.* **5**, 8, e1265.

LaBeaud, A.D., Sutherland, L.J., Muiruri, S., et al., 2011c. Arbovirus prevalence in mosquitoes, Kenya. *Emerging Infect. Dis.* **17**, 2, 233–241.

Linthicum, K.J., Anyamba, A., Tucker, C.J., et al., 1999. Climate and satellite indicators to forecast Rift Valley fever epidemics in Kenya. *Science* **285**, 5426, 397–400.

Linthicum, K.J., Anyamba, A., Britch, S.C., et al., 2007. A Rift Valley fever risk surveillance system for Africa using remotely sensed data: potential for use on other continents. *Vet. Ital.* **43**, 3, 663–674.

Linthicum, K.J., Bailey, C.L., Tucker, C.J., et al., 1990. Application of polar-orbiting, meteorological satellite data to detect flooding of Rift Valley fever virus vector mosquito habitats in Kenya. *Med. Vet. Entomol.* **4**, 4, 433–438.

Linthicum, K.J., Bailey, C.L., and Davies, F.G., 1987. Detection of Rift Valley fever viral activity in Kenya by satellite remote sensing imagery. *Science* **235**, 4796, 1656–1659.

Linthicum, K.J., Bailey, C.L., Tucker, C.J., et al., 1994. Observations with NOAA and SPOT satellites on the effect of man-made alterations in the ecology of the Senegal River basin in Mauritania on Rift Valley fever virus transmission. *Sistema Terra* **3**, 44–47.

Linthicum, K.J., Davies, F.G., Bailey, C.L., et al., 1983. Mosquito species succession in a dambo in an East African forest. *Mosq. News* **43**, 464–470.

Linthicum, K.J., Kaburia, H., and Davies, F.G., 1985. A blood meal analysis of engorged mosquitoes found in Rift Valley fever epizootics areas in Kenya. *J. Am. Mosq. Control Assoc.* **1**, 1, 93–95.

Little, P.D., 2012. Hidden value on the hoof: cross-border livestock trade in Eastern Africa. *Common Market for Eastern and Southern Africa Comprehensive African Agriculture Development Program, Policy Brief Number 2*. Available: http://www.nepad-caadp.net/pdf/COMESA CAADP Policy Brief 2 Cross Border Livestock Trade (2).pdf. Last updated: February 2009 (May 15, 2012).

Madani, T.A., Al-Mazrou, Y., Al-Jeffri, M., et al., 2003. Rift Valley fever epidemic in Saudi Arabia: epidemiological, clinical, and laboratory characteristics. *Clin. Infect. Dis.* **37**, 8, 1084–1092.

Mohamed, M., Mosha, F., Mghamba, J., et al., 2010. Epidemiologic and clinical aspects of a Rift Valley fever outbreak in humans in Tanzania, 2007. *Am. J. Trop. Med. Hyg.* **83**, 2, 22–27.

Mondet, B., Diaïté, A., Ndione, J., et al., 2005. Rainfall patterns and population dynamics of *Aedes* (Aedimorphus) *vexans* arabiensis, Patton 1905 (Diptera: Culicidae), a potential vector of Rift Valley fever virus in Senegal. *J. Vector Ecol.* **30**, 1, 102–106.

Nguku, P.M., Sharif, S.K., Mutonga, D., et al., 2010. An investigation of a major outbreak of Rift Valley fever in Kenya: 2006–2007. *Am. J. Trop. Med. Hyg.* **83**, 2, 05–13.

Nicholls, N., 1986. A method for predicting murray valley encephalitis in Southeast Australia using the southern oscillation. *Aust. J. Exp. Biol. Med.* **64**, 587–594.

Nicholson, S.E., 1986. The quasi-periodic behavior of rainfall variability in Africa and its relationship to the southern oscillation. *Arch. Meteorol. Geophys. Bioclimatol. Ser. A* **34**, 3–4, 311–348.

Pope, K.O., Rejmankova, E., Savage, H.M., et al., 1994. Remote sensing of tropical wetlands for malaria control in Chiapas, Mexico. *Ecol. Appl.* **4**, 1, 81–90.

Purse, B.V., Mellor, P.S., Rogers, D.J., et al., 2005. Climate change and the recent emergence of bluetongue in Europe. *Nat. Rev. Microbiol.* **3**, 2, 171–181.

Ropelewski, C.F., and Halpert, M.S., 1987. Global and regional scale precipitation patterns associated with the El Niño/Southern Oscillation. *Mon. Wea. Rev.* **115**, 1606–1626.

Saji, N.H., Goswami, B.N., Vinayachandran, P.N., et al., 1999. A dipole mode in the tropical Indian Ocean. *Nature* **401**, 6751, 360–363.

Sang, R., Lutomiah, J., Koka, H., et al., 2011. Crimean-Congo hemorrhagic fever virus in Hyalommid ticks, northeastern Kenya. *Emerging Infect. Dis.* **17**, 8, 1502–1505.

Schmaljohn, C., and Hjelle, B., 1997. Hantaviruses: a global disease problem. *Emerging Infect. Dis.* **3**, 2, 95–104.

Shope, R.E., Peters, C.J., and Davies, F.G., 1982. The spread of Rift Valley fever and approaches to its control. *Bull. World Health Organ.* **60**, 3, 299–304.

Spickler, A.R., 2006. *Rift Valley Fever*. Available: http://www.cfsph.iastate.edu/Factsheets/pdfs/rift_valley_fever.pdf. Last updated: November 8, 2006 (May 1, 2012).

Spickler, A.R., 2008. *Hantavirus*. Available: http://www.cfsph.iastate.edu/Factsheets/pdfs/hantavirus.pdf. Last updated: September 20, 2008 (May 1, 2012).

Spickler, A.R., 2009. *Crimean-Congo Hemorrhagic Fever*. Available: http://www.cfsph.iastate.edu/Factsheets/pdfs/crimean_congo_hemorrhagic_fever.pdf. Last updated: August 20, 2009 (May 1, 2012).

Süss, J., Klaus, C., Gerstengarbe, F., et al., 2008. What makes ticks tick? Climate change, ticks, and tick-borne diseases. *J. Travel Med.* **15**, 1, 39–45.

Swanepoel, R., 1976. Studies on the epidemiology of Rift Valley fever. *J. S. Afr. Vet. Assoc.* **47**, 2, 93–94.

Swanepoel, R., 1981. Observations on Rift Valley fever in Zimbabwe. *Contrib. Epidemiol. Biostat.* **3**, 83–91.

Turell, M.J., 1989. Effect of environmental temperature on the vector competence of *Aedes fowleri* for Rift Valley fever virus. *Res. Virol.* **140**, 2, 147–154.

Turell, M.J., O'Guinn, M.L., Dohm, D.J., et al., 2001. Vector competence of North American mosquitoes (Diptera: Culicidae) for West Nile virus. *J. Med. Entomol.* **38**, 2, 130–134.

Woods, C.W., Karpati, A.M., Grein, T., et al., 2002. An outbreak of Rift Valley fever in Northeastern Kenya, 1997–98. *Emerging Infect. Dis.* **8**, 2, 138–144.

World Health Organization, 2009. *Protecting Health from Climate Change Connecting Science, Policy, and People*. World Health Organization Regional Office for Europe, Geneva.

World Organization for Animal Health, 2011. *Terrestrial Animal Health Code*, 20th edn. OIE, Paris.

Zavitsanou, A., Babatsikou, F., and Koutis, C., 2009. Crimean Congo hemorrhagic fever: an emerging tick-borne disease. *Health Sci. J.* 3, 1, 10–18.

Zhang, W., Guo, W., Fang, L., et al., 2010. Climate variability and hemorrhagic fever with renal syndrome transmission in Northeastern China. *Environ. Health Perspect.* **118**, 7, 915–920.

Zhang, Y., 2009. Hantaviruses in rodents and humans, Inner Mongolia Autonomous Region, China. *Emerging Infect. Dis.* **15**, 6, 885–891.

27

EMERGING TREND OF ASTROVIRUSES, ENTERIC ADENOVIRUSES, AND ROTAVIRUSES IN HUMAN VIRAL GASTROENTERITIS

Daniel Cowley

Enteric Virus Group, Murdoch Childrens Research Institute, Royal Children's Hospital, Parkville, VIC, Australia

Celeste Donato and Carl D. Kirkwood

Enteric Virus Group, Murdoch Childrens Research Institute, Royal Children's Hospital, Parkville, VIC, Australia
Department of Microbiology, La Trobe University, Bundoora, VIC, Australia

TABLE OF CONTENTS

Viral Infections and Global Change, First Edition. Edited by Sunit K. Singh.

27.1 INTRODUCTION

Acute gastroenteritis is a leading cause of global morbidity and mortality, with an estimated two billion cases of disease each year (WHO, 2011). Disease presentation occurs with a variety of enteric symptoms including nausea, vomiting, abdominal pain, fever, and diarrhea. Estimates in 2011 indicate that globally diarrhea is responsible for 15% of deaths among children aged <5 years, with developing countries in Africa, the Eastern Mediterranean, and Southeast Asia having the highest disease burden (WHO, 2011). In developed countries acute gastroenteritis is often perceived as a minor and self-limiting illness, causing significant morbidity but low mortality. However, the associated high level of healthcare usage and absenteeism results in a substantial disruption to society and the economy. The economic impact estimates vary due to study population and methodology; in Australia community gastroenteritis costs the economy AUD$343 million (Hellard et al., 2003), and in the Netherlands estimates are €345 million (van den Brandhof et al., 2004). In the United States, foodborne infection is estimated to cost US$77.7 billion through direct and indirect economic costs; pathogens that cause acute gastroenteritis form a significant proportion of this estimate (Scharff, 2012).

The etiology of acute gastroenteritis includes a variety of parasitic, bacterial, and viral pathogens. The relative prevalence of different pathogens is dependent upon the community setting, age, season, and testing methodology. In 1991, the World Health Organization (WHO) reported a case–control study of the etiology of gastroenteritis in children (<35 months old) from five countries: China, India, Mexico, Myanmar, and Pakistan. The pathogens most strongly associated with disease were rotavirus, *Shigella* species, and enterotoxigenic *Escherichia coli* (Huilan et al., 1991). Rotaviruses remain the single most important cause of life-threatening acute diarrhea in children <5 years old globally (Tate et al., 2012). A recent prospective study performed in the United States in adults presenting to emergency departments tested for a comprehensive panel of viral, bacterial, and parasitic pathogens. The most commonly detected pathogens were norovirus (26%), rotavirus (18%), and *Salmonella* species (5.3%) (Bresee et al., 2012). A large prospective community cohort study of adults and children in the United Kingdom identified that norovirus and *Campylobacter* species were the most common cause of gastroenteritis (Tam et al., 2012). Similarly, prospective and retrospective studies performed in a variety of settings including the Netherlands and England have reported comparable findings (Amar et al., 2007; de Wit et al., 2001). These studies demonstrate that globally viruses play a major role in the etiology of gastroenteritis in both children and adults.

Enteric viruses represent a wide spectrum of viral genera that invade and replicate in the mucosa of the intestinal track. Enteric viruses are capable of causing localized inflammation at any level of the intestinal track, predominantly in small intestinal mucosa, resulting in acute gastroenteritis (reviewed in Bishop and Kirkwood, 2008). Prominent examples of enteric viruses causing acute gastroenteritis include rotaviruses (Bishop et al., 1973), adenoviruses (Flewett et al., 1974a), and astroviruses (Madeley and Cosgrove, 1975b). These viruses may be refractory to efforts designed to reduce the incidence of infection and represent significant public health challenges. This chapter will focus on the emerging trends of rotaviruses, enteric adenoviruses, and astroviruses. Emphasis will be made to key virological concepts and public health considerations of these pathogens.

27.2 EMERGING TRENDS IN ROTAVIRUSES

27.2.1 Rotavirus classification

Rotaviruses were first characterized in 1963 from fecal samples obtained from monkeys and mice using electron microscopy (Adams and Kraft, 1963; Malherbe and Harwin, 1963). Human rotavirus was subsequently reported in 1973 by Ruth Bishop and colleagues in Australia using electron microscopy of intestinal biopsies collected from infants with gastroenteritis (Bishop et al., 1974).

Rotavirus has a distinct wheel-like appearance when viewed under an electron microscope (EM) (*rota* is Latin for wheel), and the mature, infectious rotavirus virion (virus particle) is approximately 75 nm in diameter (Estes and Kapikian, 2007; Flewett et al., 1974b). Rotavirus virions are nonenveloped, possess icosahedral symmetry, and are comprised of three concentric protein layers (core, inner capsid, and outer capsid) encasing the 11 segments of double-stranded RNA (Estes and Kapikian, 2007). The 11 genome segments encode for six structural proteins (VP1, VP2, VP3, VP4, VP6, and VP7) and six nonstructural proteins (NSP1, NSP2, NSP3, NSP4, NSP5/NSP6). Ten of the genes are monocistronic and encode a single protein; genome segment 11 encodes for two proteins NSP5 and NSP6 (Estes and Kapikian, 2007). Each gene has various roles in virus replication, viral morphogenesis, maturation, and virulence (Estes and Kapikian, 2007).

The *Rotavirus* genus is classified into the *Reoviridae* family, which also includes the genera *Aquareovirus*, *Coltivirus*, *Cypovirus*, *Fijivirus*, *Idnoreovirus*, *Mycoreovirus*, *Orbivirus*, *Orthoreovirus*, *Oryzavirus*, *Phytoreovirus*, and *Seadornavirus* (Attoui et al., 2012). Within the *Rotavirus* genus there are seven rotavirus groups (A–G) that are defined by distinct antigenic and genetic differences in the *VP6* gene (Estes and Kapikian, 2007). Group A rotaviruses are the predominant cause of disease in humans as well as a wide range of mammalian and some avian species (Estes and Kapikian, 2007). Group B strains represent a small group of viruses that cause severe gastroenteritis that have been predominantly detected in adults in China, Bangladesh, and India (Sanekata et al., 2003). Group C strains cause sporadic outbreaks predominantly in children in numerous countries worldwide including Australia, Brazil, Argentina, and the United States (Bridger et al., 1986; Castello et al., 2000; Jiang et al., 1995). Group D, E, F, and G strains are only detected in animals (Estes and Kapikian, 2007).

Group A rotavirus strains can be classified into subgroups (SG I, II, I + II, or non-I, non-II) based on the antigenic specificity of the VP6 protein. Genetic analysis allows classification into genogroups I (previously subgroup I) and II (previously subgroup II, I + II, or non-I, non-II) (Iturriza Gomara et al., 2002). A binary classification system for rotavirus based on the outer capsid proteins VP7 and VP4 was established in 1989 and initially used the antigenic reactivity of the two outer capsid proteins to denote P serotypes for VP4 (protease sensitive) and G serotypes for VP7 (glycoprotein). Antigenic classification has been replaced by classification based on genetic sequence of the genes encoding the VP7 and VP4 proteins, resulting in the designation of G and P genotypes (Graham and Estes, 1985). To date, 27 G genotypes and 35 P genotypes have been identified (Matthijnssens et al., 2011). Recently, a whole-genome classification system based on nucleotide sequence of gene coding regions has been described. The nomenclature for the complete genome assignment of VP7-VP4-VP6-VP1-VP2-VP3-NSP1-NSP2-NSP3-NSP4-NSP5/NSP6 is Gx-P[x]-Ix-Rx-Cx-Mx-Ax-Nx-Tx-Ex-Hx, respectively. To date, 27 G, 35 P, 16 I, 9 R, 9 C, 8 M, 16 A, 9 N, 12 T, 14 E, and 11 H genotypes have been identified. A nomenclature for strain names has also been proposed: RV group/species of origin/country of identification/common name/year of identification/G and P types (Matthijnssens et al., 2011).

27.2.2 Epidemiology of human rotaviruses

A considerable burden of disease can be attributed to rotavirus in both developing and developed countries. Rotavirus causes 114 million episodes of diarrhea annually worldwide, resulting in 24 million clinic visits and 2.4 million hospitalizations in children under 5 years of age (Glass et al., 2006). The mortality rates associated with rotavirus disease are unevenly distributed; of the estimated 453 000 annual deaths, the overwhelming majority occur in developing nations in Asia and sub-Saharan Africa (Tate et al., 2012). Group A rotavirus strains cause disease in the elderly, sporadic disease in adults, and outbreaks in hospital wards and institutions. Rotavirus infection can result in prolonged symptomatic infections in immunocompromised pediatric and adult patients.

In countries with a temperate climate, the majority of rotavirus disease occurs in the winter months. However, in regions with a tropical climate, rotavirus infection occurs year-round with higher rates of detection observed in cooler, drier months (Cook et al., 1990). An 11-year study in Australia revealed higher weekly temperatures and humidity correlated with a decrease in rotavirus admissions in the following week (D'Souza et al., 2008).

27.2.3 Clinical symptoms and pathogenesis

Rotavirus is transmitted by the fecal–oral route; a small infectious dose (<100 virus particles) can initiate an infection. Transmission of the virus occurs through close personal contact and contact with contaminated surfaces and objects (fomites). Rotavirus virions are resistant to physical inactivation and can survive at ambient temperatures for long periods of time (Estes and Kapikian, 2007). The incubation period is 1–4 days and typically is less than 48 h. The clinical symptoms usually last 4–7 days and include profuse, watery, nonbloody diarrhea, vomiting, and fever that can lead to dehydration (Estes and Kapikian, 2007). Severe complications of rotavirus infections rarely occur but may involve the spread of rotavirus beyond the intestine. Small quantities of rotavirus antigens, RNA, and infectious virus have been detected in serum, cerebrospinal fluid, and several organs (liver, heart, lung, and kidneys) (Grimwood and Lambert, 2009). Rotavirus is routinely shed in the stool for 7–10 days, and a third of infants often continue to shed virus for up to 21 days (Richardson et al., 1998). Rotavirus replicates in the mature enterocytes in the small intestine, and rotavirus-associated diarrhea is attributed to several mechanisms associated with virus replication including malabsorption due to the destruction of enterocytes, villus ischemia, effects of the virus-encoded toxin (NSP4), and the activation of the enteric nervous system (Estes and Kapikian, 2007).

27.2.4 Genomic diversity of rotaviruses

Rotavirus strains are able to evolve rapidly, employing a several of mechanisms to generate diversity in the wild-type strain population. The generation of spontaneous sequential point mutations (genetic drift) occurs due to the error-prone nature of the rotavirus RNA-dependent RNA polymerase (RdRp) (Taniguchi and Urasawa, 1995). Genetic analysis has shown that sequential point mutations accumulate during a single epidemic season and the accumulation of point mutations can result in antibody escape mutants (Taniguchi and Urasawa, 1995).

Genomic rearrangement involves alterations within a single genome segment in the form of deletions, duplications, or insertions of sequence (Taniguchi and Urasawa, 1995). Rearrangement is most commonly observed in rotavirus strains excreted by chronically infected immunocompromised children and animals and has also been observed *in vitro* after the serial passage of strains at high multiplicity in cell culture systems (Taniguchi and Urasawa, 1995). Rearrangement is more commonly detected in *NSP1*, *NSP3*, and *NSP5* genes (Matthijnssens et al., 2006). Most rearranged rotavirus strains are not defective and can replace non-rearranged RNA segments structurally and functionally as most instances of rearrangement involve head-to-tail duplication that occurs downstream of the open reading frame (Estes and Kapikian, 2007).

Genomic reassortment (genetic shift) occurs when single cells are infected with two rotavirus strains. The segmented nature of the rotavirus genome facilitates the exchange of genes between strains to produce new gene combinations, which often leads to novel viral strains (Ramig and Ward, 1991). Reassortment occurs more frequently and provides greater diversity than genetic drift. It has been shown that all 11 genes can undergo reassortment. However, some genome constellations are more favorable than others and more commonly isolated (Chaimongkol et al., 2012; Ward and Knowlton, 1989). In settings where vaccines have been introduced, reassortment between wild-type strains and vaccine strains has been detected. In Nicaragua, two G1P[8] strains have been isolated with *NSP2* genes of RotaTeq vaccine origin (Bucardo et al., 2012). Reassortment has also been identified between two of the parental RotaTeq vaccine strains; studies have characterized symptomatic infections caused by G1P[8] vaccine strains derived from the reassortment between the parental strains G1P[5] and G6P[8] (Donato et al., 2012).

The introduction of animal rotavirus strains into the human population (zoonotic transmission) also increases the genetic diversity within the strains causing human infection (Taniguchi and Urasawa, 1995). Few human infections occur where strains possess all 11 genes of animal strain origins, suggesting interspecies transmissions of complete animal strains are uncommon (Gentsch et al., 2005; Santos and Hoshino, 2005). Reassortment represents a more common mechanism for the introduction of single or multiple animal rotavirus genes to enter the human strain population (Santos and Hoshino, 2005). Zoonotic transmission tends to occur more often in countries where contact with domesticated animals is higher or from animals that reside in close contact with humans such as cats and dogs (Tsugawa and Hoshino, 2008).

27.2.5 Rotavirus vaccines and the impact of vaccine introduction on the burden of rotavirus disease

The development of a rotavirus vaccine has been a WHO priority, aimed to reduce the global burden of disease caused by rotavirus infection. The basis for rotavirus vaccine development with live oral viruses was based on the observation that a primary natural rotavirus infection produced protection against severe disease on subsequent reinfections (Bishop et al., 1983).

The first commercially available rotavirus vaccine, RotaShield (Wyeth Lederle Vaccines, Marietta, PA, USA), was an oral, tetravalent rhesus–human reassortant vaccine administered in a three-dose schedule at 2, 4, and 6 months of age. RotaShield was licensed in the United States in 1998 and subsequently withdrawn from the market in late 1999 due to an association with intussusception. An estimated one excess case per 10 000–15 000 infants vaccinated was identified (CDC, 1999).

Two live oral vaccines Rotarix (GlaxoSmithKline Biologicals, Rixensart, Belgium) and RotaTeq (Merck & Co., PA, USA) are licensed and available for use globally. RotaTeq is a live-attenuated pentavalent vaccine that contains five genetically distinct human–bovine reassortant virus strains (Vesikari et al., 2006). Each reassortant strain contains a gene encoding one of the outer capsid proteins of human strains (G1, G2, G3, G4, or P[8]), within a bovine rotavirus backbone (G6P[5]). RotaTeq is administered as a three-dose schedule, recommended at 2, 4, and 6 months of age. Rotarix is a monovalent vaccine containing a single strain (RIX4414) of the most prevalent human genotype worldwide (G1P[8]) and is administered as a 2-dose schedule at 2 and 4 months of age (Vesikari et al., 2007).

There are several other vaccines in various stages of development and licensure. In China, a lamb strain of rotavirus has been licensed since 2000, although no clinical trial results or vaccine effectiveness data have been reported. Asymptomatic neonatal rotavirus strains isolated in India and Australia and a human–bovine reassortant strain combination are undergoing clinical trials. Other vaccine candidates such as synthetic viral proteins, empty viruslike particles, synthetic peptides, and DNA vaccines are also in early stages of development (Estes and Kapikian, 2007).

The introduction of rotavirus vaccines into the routine vaccination programs of numerous countries has resulted in a significant reduction in rotavirus-associated hospitalizations, emergency room visits, and episodes of gastroenteritis in many countries including Brazil, Belgium, the United States, Australia, Nicaragua, Austria, and Mexico (Buttery et al., 2011; Curns et al., 2010; Patel et al., 2009; Paulke-Korinek et al., 2011; Richardson et al., 2010; Zeller et al., 2010). Evidence of herd immunity has also been observed in numerous vaccine settings with deceases in rotavirus disease observed in infants too young and children and adults too old to be vaccinated (Buttery et al., 2011; Paulke-Korinek et al., 2011; Richardson et al., 2010; Zeller et al., 2010). This is thought to be due to decreased transmission of rotavirus strains in the community.

Changes in the rotavirus season have been detected in some countries since vaccine introduction. In the United States prior to vaccine introduction, there were a predictable rotavirus seasonality and geographic pattern of spread across the country. Since vaccine introduction, there has been a shift in the onset of the season by 1–2 months; the season is considerably shorter and the geographic pattern of spread no longer identified (Curns et al., 2010; Dennehy, 2012; Tate et al., 2009). This delay in the onset of the season and shorter duration of season has also been described in Australia and Belgium since the introduction of rotavirus vaccines (Buttery et al., 2011; Zeller et al., 2010).

27.2.6 Globally emerging rotavirus genotypes

There have been 27 G genotypes and 35 P genotypes described to date; of these 12 G and 15 P genotypes are known to infect humans (Matthijnssens et al., 2011; Patel et al., 2011). Genotype G1P[8], G2P[4], G3P[8], G4P[8], and G9P[8] strains cause over 90% of rotavirus disease worldwide. In North America, Europe, and Australia, they represent over 90% of characterized isolates, but in South America and Africa, they represent 83% and 55% of isolates, respectively (Santos and Hoshino, 2005). The emergence of novel genotypes has been observed in developing and developed nations, and these novel or rare genotypes include G5, G6, G8, G9, G10, and G12 and highlight the worldwide diversity of rotavirus strains circulating in the community. In some countries unusual strains can be detected in high frequencies such as G8P[6] in Malawi, G5P[8] in Brazil, and G10P[11] in India (Gentsch et al., 2005). There is a concern that Rotarix and RotaTeq may prove to be less efficient in regions of the world where a higher prevalence of unusual genotypes exists. RotaTeq and Rotarix may provide protection against some emerging strains, including G5P[8] and G9P[8], because of shared P genotype antigens. Other strains, such as G9P[6] and G8P[6], that do not share either genotype antigen may prove to challenge the efficacy of these vaccines.

27.3 EMERGING TRENDS IN ENTERIC ADENOVIRUSES

27.3.1 Adenovirus classification

Adenoviruses were first described in 1953 following culture of tonsillar and adenoidal tissues from children (Rowe et al., 1953). They are nonenveloped icosahedral viruses 70–90 nm in diameter and composed of a nucleocapsid that surrounds a double-stranded linear DNA genome. The adenovirus genome is between 26 and 48 kbp in length and complex, encoding approximately 40 different polypeptides through alternative splicing mechanisms (Harrach et al., 2011). The icosahedral capsid consists of 240 non-vertex capsomers (hexons) that are 8–10 nm in diameter and 12 vertex capsomers (penton bases), each with a fiber that protrudes from the virion surface (Liu et al., 2010; Reddy et al., 2010; Stewart et al., 1993). The fiber is involved in attachment to cell surface receptors (Bergelson et al., 1997; Bewley et al., 1999) and may demonstrate variation in length (Kidd et al., 1993). The fiber, penton, and hexon are major antigenic determinates of the virus (Harrach et al., 2011).

The family *Adenoviridae* is composed of five genera identified in a variety of animal hosts including *Atadenovirus* (birds, sheep, and cows), *Aviadenovirus* (birds), *Ichtadenovirus* (fish), *Mastadenovirus* (mammals), and *Siadenovirus* (birds and frogs). There are seven species of human adenoviruses (HAdVs) (A–G) within the genus *Mastadenovirus* that cause infections to the respiratory, gastrointestinal, urinary, and ocular surface mucosas (Harrach et al., 2011). Individual HAdV types were originally differentiated based on serological methods. More recently, genomic and bioinformatic approaches have been employed to characterize adenoviruses. Serological typing, which relies on epitopes encoded by a relatively small region of the viral genome, may generate discordant results when compared to whole-genome analysis (Singh et al., 2012) and led to a debate in the assignment of novel HAdV types (de Jong et al., 2008; Jones et al., 2007). Efforts have been made to develop an integrated adenovirus designation system that utilizes full genome, phylogenetics, and serological data (Aoki et al., 2011; Seto et al., 2011), although several matters remain under consideration, including defining intertypic recombinants and the extent of genome sequence required to assign new HAdV types. Currently, more than 50 HAdV types have been recognized and classified into the seven species. Certain HAdV types are predominantly associated with specific pathology; these include adenoidal–pharyngeal conjunctivitis (3, 4, 7, 14), acute respiratory outbreaks (4, 7, 14, 21), epidemic keratoconjunctivitis (8, 19, 37, 53, 54), and acute gastroenteritis (40, 41) (Harrach et al., 2011).

27.3.2 Epidemiology of enteric adenoviruses HAdV-F40 and HAdV-F41

The first identification of adenovirus particles in human feces was performed in 1974 by Flewett and colleagues through the electron microscopy of negatively stained extracts of diarrheal samples (Flewett et al., 1974a). Adenoviruses associated with diarrhea were subsequently isolated and characterized as HAdV-F40 and HAdV-F41 within species F (de Jong et al., 1983; Wigand et al., 1983) and referred to as enteric adenoviruses. Enteric adenoviruses have been detected in stool samples from children with gastroenteritis in multiple countries. The incidence of adenovirus-related gastroenteritis varies, ranging from 1.5% to 11% (reviewed in Wold and Horwitz, 2007). HAdV-F40 and HAdV-F41 are associated with acute gastroenteritis that requires hospitalization in approximately 5% of young children (Barnes et al., 1998; Jin et al., 2009). Furthermore, HAdV-F40 and HAdV-F41 are reported as the cause of gastroenteritis in children presenting to general practice clinics (Tam et al., 2012).

Enteric adenoviruses are significant pathogens associated with sporadic cases as well as outbreaks of gastroenteritis in settings such as kindergartens, schools, and hospitals (Akihara et al., 2005; Chiba et al., 1983; Van et al., 1992). Infections occur at unpredictable intervals year-round with no seasonal prevalence (Barnes et al., 1998; de Jong et al., 1993). Asymptomatic infection with HAdV-F40 and HAdV-F41 has been reported in epidemics of gastroenteritis (Van et al., 1992). Transmission is by a variety of routes, with person-to-person spread playing a major role. Surfaces and objects contaminated with enteric adenoviruses can also be a source of transmission (Boone and Gerba, 2007). Enteric adenoviruses can survive on porous and nonporous contaminated surfaces albeit for periods shorter than for other enteric viruses (Abad et al., 1994). Enteric adenoviruses also demonstrate unique physiochemical properties that enable them to be a successful enteric pathogen, including resistance to acid and proteolytic treatment (Favier et al., 2004). These attributes contribute to environmental survival and facilitate the spread of infection to susceptible individuals.

27.3.3 Clinical symptoms and pathogenesis

Diarrhea due to enteric adenovirus infection is more common in infants <12 months of age when compared to older children (Brandt et al., 1985). The duration of clinical symptoms varies with a mean of 7–10 days, although prolonged illness is also noted (Grimwood et al., 1995; Krajden et al., 1990). Adenovirus-associated diarrhea can occur in immunocompromised hosts such as those with HIV-1 (Cunningham et al., 1988; Grohmann et al., 1993). However, the significance of diarrhea in this context is unclear (Durepaire et al., 1995; Kaljot et al., 1989) due to the coinfection with other pathogens and frequent shedding of a diverse range of adenovirus species in stool (Curlin et al., 2010). The natural history of infection and the development of immunity are unclear. Seroepidemiological studies have demonstrated that antibodies against HAdV-F40 and HAdV-F41 are common in children and adults (Kidd et al., 1983; Shinozaki et al., 1987) suggesting that exposure is widespread. No vaccines are currently available against HAdV-F40 and HAdV-F41.

Enteric adenoviruses replicate within the epithelial cells of the small intestine (Whitelaw et al., 1977). The mechanisms that cause diarrhea are unclear, but destruction of infected epithelial cells has a role. The limited capacity to cultivate enteric adenoviruses *in vitro* has reduced the ability to obtain insights into the molecular basis of viral replication and disease. HAdV-F40 and HAdV-F41 demonstrate fastidious growth characteristics making cultivation *in vitro* difficult; multiple blocks to productive infection have been demonstrated including sensitivity to type I interferon (Sherwood et al., 2007), deficiencies in viral gene transactivation (Takiff and Straus, 1982), and blocks to the release of progeny virus (Brown et al., 1992). The most reliable growth has been achieved in human embryonic kidney cells (293 cells) immortalized by transfection with regions of HAdV-5 (Takiff et al., 1981). Recent reports also demonstrate that the culture of HAdV-F40 and HAdV-F41 in cell lines that express viral transactivator proteins (Kim et al., 2010) or suppression of type I interferon production (Sherwood et al., 2007, 2012) supports productive infection.

27.3.4 Diversity and evolution of enteric adenoviruses

Comparative sequence analysis of hexon genes of the entire spectrum of HAdVs has revealed a high degree of genetic diversity both within and between adenovirus species (Ebner et al., 2005). At a population level, longitudinal analysis of enteric adenovirus infections has identified changes in the dominance and frequency of HAdV-F40 and HAdV-F41 infections (de Jong et al., 1993; Grimwood et al., 1995). However, limited information is available regarding the longitudinal genetic and antigenic changes in circulating enteric adenoviruses. Early reports using restriction endonuclease digestion of extracted viral DNA demonstrate the circulation of HAdV-F40 and HAdV-F41 genetic variants in different geographic regions (Kidd, 1984; Kidd et al., 1984). This technique has also demonstrated successive changes within the pattern of circulating enteric adenovirus types within populations (Scott-Taylor and Hammond, 1995). Multiple circulating HAdV-F41 genome types forming two distinct genome-type clusters with antigenic differences have been detected in Japan, Vietnam, and Korea (Fukuda et al., 2006; Li et al., 2004). Distinct HAdV-F41 variants have also been reported in India, with sequence analysis of hypervariable regions within the hexon gene and the fiber shaft region also demonstrating circulation of unique antigenic variants (Dey et al., 2011). These data suggest existence of a diverse population of enteric adenoviruses and that immunological pressure on circulating strains may lead to antigenic changes. With the increasing analysis of circulating strains, particularly the availability of full genomes, it is likely that greater understanding of the genetic diversity and evolution of enteric adenoviruses will be obtained.

27.3.5 Emerging human enteric adenovirus species

The increasing use of molecular epidemiology has identified other adenovirus species in addition to the "classic" enteric adenoviruses that are associated with gastroenteritis. A high prevalence of species D adenoviruses has been identified in fecal specimens from urban Kenyans with diarrhea (Magwalivha et al., 2010). In this study, the predominant species and genotypes identified from rural settings were the common enteric HAdV species: HAdV-F40, HAdV-F41, and HAdV-A31. In contrast in older children from urban settings, HAdV-D types were identified and associated with diarrhea. In a large cohort of fecal specimens (>900) collected in Bangladesh, predominantly HAdV-F40 was found, followed by species D types HAdV-D9 and HAdV-D10; interestingly an absence of HAdV-F41 was noted (Dey et al., 2009). The emergence of species G adenovirus (HAdV-G52) phylogenetically related to a simian adenovirus has been identified in California and associated with an outbreak of gastroenteritis in a convalescent home (Jones et al., 2007). It has been suggested that this novel HAdV-G52 may be a cause of gastroenteritis of unknown etiology, although a prospective study of 209 diarrheic stools and 45 effluent sewage samples undertaken in a different geographic region failed to detect it (Banyai et al., 2009).

Recombination is an important mechanism of adenovirus evolution and may lead to novel types that are associated with disease. Recently, a novel human recombinant species labeled HAdV-A31 MZ has been reported to be associated with acute gastroenteritis in Japan. Analysis of the genome of this virus indicated that its hexon sequence is the result of recombination between HAdV-A31 and HAdV-A12 (Matsushima et al., 2011). Another recombinant adenovirus, HAdV-D58, that contains a novel hexon and recombinant fiber gene has been reported in an AIDS patient with chronic diarrhea (Liu et al., 2011). Similar novel recombinant adenoviruses within species D (HAdV-D65) have been detected in the feces of four children in Bangladesh with acute gastroenteritis. Sequence analysis indicated hexon genes were closely related to HAdV-10, penton base genes of HAdV-37 and HAdV-58, and a fiber gene of HAdV-9 (Matsushima et al., 2012). Whether these recombinant adenovirus types represent isolated cases in cohorts with reduced immunocompetence or are an emerging public health threat (Matsushima et al., 2012) remains to be determined.

27.4 EMERGING TRENDS IN ASTROVIRUSES

27.4.1 Astrovirus classification

Astroviruses were first described in 1975 during outbreaks of acute nonbacterial gastroenteritis in infants (Appleton and Higgins, 1975; Madeley and Cosgrove, 1975b). Astroviruses are small nonenveloped, icosahedral virions typically 28–30 nm in diameter; their name derives from the characteristic star-shaped appearance in viral particles when viewed using EM (Madeley and Cosgrove, 1975a). The viral genome comprises a single-stranded positive-sense RNA of between 6.4 and 7.9 kb that includes a 5′ untranslated region (UTR), followed by three open reading frames, ORF1a, ORF1b, and ORF2, a 3′ UTR, and a polyadenylated trail (Jiang et al., 1993). ORF1a encodes a polypeptide that is cleaved into the viral serine protease (Speroni et al., 2009) and other nonstructural proteins (NSP) (Geigenmuller et al., 2002; Kiang and Matsui, 2002); limited information is available regarding the cleavage pattern and functions of these proteins. ORF1b encodes the viral RdRp, expressed through a frameshift RNA stem-loop structure between ORF1a and ORF1b (Jiang et al., 1993; Lewis et al., 1994). ORF2 is expressed from a subgenomic RNA and encodes for the viral capsid protein (Willcocks and Carter, 1993).

The family *Astroviridae* is composed of two genera, *Avastrovirus* that infects birds and *Mamastrovirus* that infects mammals. Mamastroviruses are ubiquitous, detected in a diverse range of mammalian hosts including dogs (Williams, 1980), cats (Hoshino et al., 1981), bats (Chu et al., 2008), cheetahs (Atkins et al., 2009), marine mammals (Rivera et al., 2010), and rats (Chu et al., 2010). The predominant feature of infection with mamastroviruses is gastroenteritis. The current classification of *Mamastrovirus* species is based on the host from which they were isolated and does not correspond to genetic differences between viruses. Recent efforts have been made to redefine the classification of mamastroviruses using the complete amino acid sequence of the capsid region. This classification defines two main genogroups within *Mamastrovirus*; each genogroup includes astroviruses infecting different host species and can be further subdivided based on both genetic and host species criteria (Bosch et al., 2011). There are currently eight species of human astroviruses (HAstVs) that were initially identified based on reactivity to polyclonal antisera. Molecular techniques have largely supplanted serology in the identification of HAstVs, with a correlation between serotype and sequence data (Noel et al., 1995).

27.4.2 Epidemiology of HAstVs

HAstVs are endemic pathogens and recognized worldwide as a common cause of diarrhea in young children, with epidemiological studies indicating that HAstVs are responsible for 0.6–10% of cases of acute gastroenteritis (Bon et al., 1999; Jeong et al., 2011; Mustafa et al., 2000; Phan et al., 2004; Utagawa et al., 1994). A longitudinal study in Mayan children in Mexico has provided insights into the natural history of HAstV infection. This study found a high prevalence (61%) of HAstV infection in a birth cohort of 271 children followed for 3 years. Infection occurred primarily in infants <12 months old and showed a high rate of asymptomatic infection and prolonged shedding (2–17 weeks) in many infants (Maldonado et al., 1998). Seroepidemiological studies demonstrate that antibodies against HAstV are frequently detected, increasing with age to reach 90% by 5–10 years (Kriston et al., 1996; Kurtz and Lee, 1978; Mitchell et al., 1999). Coinfection with other enteric viruses such as rotavirus is also noted (Herrmann et al., 1991). These studies demonstrate that infection with HAstV is common and frequently occurs during early childhood. Infection in the elderly (Gray et al., 1987), immunocompromised (Grohmann et al., 1993; Wunderli et al., 2011), and healthy, immunocompetent adults has also been reported (Belliot et al., 1997).

HAstV infections may demonstrate peaks during winter and rainy seasons in temperate and tropical climates, respectively, although infections are noted year-round (Herrmann et al., 1991; Mustafa et al., 2000). Outbreaks have been reported in hospitals, day-care centers, schools, and nursing homes. Infection with HAstV frequently occurs via the fecal–oral route via person-to-person spread. Significantly, astroviruses can persist for prolonged periods on contaminated surfaces (Abad et al., 2001) and have been detected in environmental water samples from an area where a concurrent gastroenteritis outbreak was reported (Pinto et al., 1996). These studies suggest that fomites and environmental persistence may facilitate the spread of HAstV.

27.4.3 Clinical symptoms and pathogenesis

The clinical symptoms of gastroenteritis due to HAstV is characterized by abdominal pain, diarrhea, vomiting, nausea, fever, and malaise (Herrmann et al., 1991). In the immunocompetent host, disease is self-limited lasting for 5–6 days on average. The duration and severity of illness are regarded as being shorter than that caused by rotavirus (Dennehy et al., 2001). Atypical clinical presentation such as prolonged disease, viral shedding, and disseminated infection has been described in immunocompromised hosts (Cubitt et al., 1999; Sebire et al., 2004). HAstV-4 has recently been reported to cause protracted disease and disseminated infection in severely immunocompromised infants (Wunderli et al., 2011). Similarly, astrovirus has been detected in the brain tissue of an immunocompromised patient with encephalitis (Quan et al., 2010).

There is limited literature that describes the histopathology and pathogenesis of HAstV infections, attributed to the lack of a small animal model to understand astrovirus disease. Sebire et al. (2004) reported the histopathological findings of astrovirus infection in a bone marrow transplant recipient with severe diarrhea. Extensive evidence of astrovirus replication was noted with progressively more astrovirus antigen in the duodenal and jejunal sections of the small intestine, localized to enterocytes at the villus tips. However, there was limited evidence of inflammation or morphological abnormalities. Similarly, experimental models of astrovirus enteritis in turkeys, calves, and gnotobiotic lambs show mild histopathological changes in the intestine with limited cell death, despite enteric infection and diarrhea (Koci et al., 2003; Snodgrass et al., 1979; Woode et al., 1984). These data suggest that astrovirus-induced diarrhea is due to a mechanism other than inflammation and destruction of the intestinal epithelium. *In vitro* studies undertaken using a Caco-2 tissue culture model demonstrated that astrovirus infection increases epithelial barrier permeability independent of viral replication, likely dependent on the viral capsid (Moser et al., 2007). In addition, the turkey animal model of infection suggests that astrovirus infection induced sodium malabsorption of epithelium cells through redistribution of specific sodium transporters (Nighot et al., 2010). These studies partially explain the mechanisms of the osmotic diarrhea that characterizes astrovirus infections.

Early reports demonstrate that astroviruses enter cells by receptor-mediated endocytosis (Donelli et al., 1992). The crystal structure of the astrovirus capsid spike has recently been reported, with homology modeling suggesting a putative receptor binding site with amino acid compositions characteristic for polysaccharide recognition (Dong et al., 2011); however, the precise cellular receptor remains to be identified. The viral replication cycle and the cellular factors that regulate astrovirus infection are poorly understood, with much information inferred from the large body of literature on other positive-strand RNA viruses. A recent review by De Benedictis et al. provides a comprehensive summary on the replication and molecular biology of astroviruses (De Benedictis et al., 2011).

27.4.4 Diversity and evolution of HAstVs

There are currently eight species (or genotypes) of HAstVs (HAstV-1 to HAstV-8) that cause gastroenteritis in humans (Bosch et al., 2011). Serotype and molecular epidemiological studies have been conducted in numerous countries and demonstrate that HAstV-1 is the predominant circulating genotype, followed by HAstV-2, HAstV-3, HAstV-4, HAstV-5, and occasionally HAstV-8. HAstV-6 and HAstV-7 are rarely identified (Espul et al., 2004; Gabbay et al., 2007; Guix et al., 2002; Liu et al., 2007; Mustafa et al., 2000; Palombo and Bishop, 1996). Differences may exist in particular geographic regions. For example, in Mexico the prevalence of HAstV-1 strains was low compared to other types, with HAstV-2 the predominant genotype, followed by HAstV-4, HAstV-3, and HAstV-7 (Guerrero et al., 1998; Walter et al., 2001b). In Ghana an absence of HAstV-1 was noted, with the predominant genotype HAstV-8, followed by HAstV-2, HAstV-5, and HAstV-6 (Silva et al., 2008). Recent reports also demonstrate HAstV-8 circulating in several geographic regions including Saudi Arabia (Tayeb et al., 2010) and Madagascar (Papaventsis et al., 2008). Together, these studies demonstrate that multiple HAstV genotypes may circulate within populations. It is unknown whether there is an association between genotype and disease potential, although one study has suggested that higher fecal viral loads and persistent disease are associated with HAstV-3 when compared to other genotypes (Caballero et al., 2003).

Sequence diversity within the HAstV genome may occur through the accumulation of nucleotide mutations and RNA recombination. Recent analysis of complete HAstV genomes demonstrated a high evolutionary rate comparable to other RNA viruses and the identification of several potential recombination points (Babkin et al., 2012). This study, while limited by analysis of available complete genomes, identified the potential for significant sequence diversity to exist in circulating HAstV strains. Several studies have attempted to assess the degree of sequence diversity within HAstV genotypes. This has led to the identification of multiple lineages within each genotype. HAstV-1 has been divided into at least six lineages HAstV-1a–HAstV-1f (Gabbay et al., 2007), HAstV-2 into at least three lineages HAstV-2a–HAstV-2c (De Grazia et al., 2011), HAstV-3 into two lineages

(Liu et al., 2008), and HAstV-4 into two lineages, HAstV-4a and HAstV-4b (Espul et al., 2004). Coinfection by two different genotypes in a single person also provides an opportunity for recombination to occur. Significantly, recombinant HAstV genomes have been reported in diarrhea specimens (Verma et al., 2010; Walter et al., 2001a; Wolfaardt et al., 2011), demonstrating that this mechanism of HAstV evolution takes place between circulating genotypes. Whether sequence diversity and the evolution of HAstV influence immune escape and disease potential remains unclear. Longitudinal studies identifying variation in capsid nucleotide sequences have demonstrated a lack of corresponding amino acid changes (Mustafa et al., 2000; Palombo and Bishop, 1996; Schnagl et al., 2002), although these studies are limited by analysis of a relatively conserved part of the capsid protein. Additional sequence data, particularly the full-genome analysis of circulating strains, are required to characterize the evolutionary history of HAstVs.

27.4.5 Emerging HAstV species

The application of random Sanger sequencing and next-generation pyrosequencing in pathogen discovery from diarrhea specimens has led to the characterization of previously unrecognized astroviruses. Using these techniques, several highly divergent astroviruses have been recently discovered in human diarrhea specimens: HAstV-MLB1, HAstV-MLB2 (Finkbeiner et al., 2008a, b, 2009a), HMOAstV-A/HAstV-VA2 (Finkbeiner et al., 2009a; Kapoor et al., 2009), HMOAstV-B (Kapoor et al., 2009), HMOAstV-C/HAstV-VA1 (Finkbeiner et al., 2009c; Kapoor et al., 2009), and HAstV-VA3 (Finkbeiner et al., 2009a). The classification and nomenclature of several novel astrovirus species identified independently but genetically similar remain to be resolved (Bosch et al., 2011).

HAstV-MLB1 was first identified in 2008 from the stool of a 3-year-old boy in Australia (Finkbeiner et al., 2008a, b) and represents the most well characterized of the novel HAstVs. The entire genome of this virus has been sequenced and characterized, demonstrating that while most closely related to HAstVs, it is highly divergent from them (Finkbeiner et al., 2008b). A subsequent phylogenetic comparison of HAstV-MLB1 ORF1b with a newly identified astrovirus from urban brown rats (*Rattus norvegicus*) demonstrates that they are related. Molecular evolutionary clock methods demonstrate that they may share a most common recent ancestor virus estimated to date from AD1054 (Chu et al., 2010). These data suggest that HAstV-MLB1 may be of animal origin. A prevalence study demonstrated that HAstV-MLB1 was circulating in diarrhea fecal specimens from St. Louis, USA (Finkbeiner et al., 2009b); similarly it has been detected in specimens from Mexico (Banyai et al., 2010), Nigeria (Kapoor et al., 2009), India (Finkbeiner et al., 2009a), and Hong Kong (Chu et al., 2010). These studies demonstrated that HAstV-MLB1 is likely to be circulating globally. The significance of HAstV-MLB1 as a cause of gastroenteritis remains debatable. A recent case–control study undertaken in a longitudinal birth cohort in Vellore, India, failed to demonstrate that HAstV-MLB1 was associated with diarrhea (Holtz et al., 2011), although the authors conceded that additional studies were required to definitely exclude HAstV-MLB1 as a cause of diarrhea in infants.

Two independent reports identified novel HAstVs genetically similar to mink and ovine astroviruses (Finkbeiner et al., 2009c; Kapoor et al., 2009). A novel astrovirus, HAstV-VA1 was identified in an outbreak of gastroenteritis of unknown etiology in a child-care setting in Virginia. This astrovirus is highly divergent but most related to ovine and mink astroviruses (Finkbeiner et al., 2009c). A separate study during the same period also identified three divergent HAstVs that were phylogenetically related to each other, with the closest relatives being mink and ovine astroviruses (Kapoor et al., 2009). One astrovirus characterized in this study, termed HMOAstV-C, identified in a fecal sample in Nepal was highly similar to HAstV-VA1. HAstV-VA1 has also been detected in the stool of a patient in the Netherlands (Smits et al., 2010). These reports suggest that these divergent HAstVs are common infectious agents circulating in the human population. A recent report demonstrating that antibodies to HMOAstV-C are widespread in both adults and children (Burbelo et al., 2011) supports this conclusion. The identification of bat astroviruses phylogenetically related to mink, ovine, and the recently discovered divergent human astroviruses provides evidence that these viruses may undergo cross-species transmission (Xiao et al., 2011).

REFERENCES

Abad, F. X., Pinto, R. M., and Bosch, A. (1994). Survival of enteric viruses on environmental fomites. *Appl Environ Microbiol* **60**(10), 3704–10.

Abad, F. X., Villena, C., Guix, S., et al. (2001). Potential role of fomites in the vehicular transmission of human astroviruses. *Appl Environ Microbiol* **67**(9), 3904–7.

Adams, W. R., and Kraft, L. M. (1963). Epizootic diarrhea of infant mice: identification of the etiologic agent. *Science* **141**(3578), 359–60.

Akihara, S., Phan, T. G., Nguyen, T. A., et al. (2005). Existence of multiple outbreaks of viral gastroenteritis among infants in a day care center in Japan. *Arch Virol* **150**(10), 2061–75.

Amar, C. F., East, C. L., Gray, J., et al. (2007). Detection by PCR of eight groups of enteric pathogens in 4,627 faecal samples: re-examination of the English case-control Infectious Intestinal Disease Study (1993–1996). *Eur J Clin Microbiol Infect Dis* **26**(5), 311–23.

Aoki, K., Benko, M., Davison, A. J., et al. (2011). Toward an integrated human adenovirus designation system that utilizes molecular and serological data and serves both clinical and fundamental virology. *J Virol* **85**(11), 5703–4.

Appleton, H., and Higgins, P. G. (1975). Letter: Viruses and gastroenteritis in infants. *Lancet* **1**(7919), 1297.

Atkins, A., Wellehan, J. F., Jr., Childress, A. L., et al. (2009). Characterization of an outbreak of astroviral diarrhea in a group of cheetahs (*Acinonyx jubatus*). *Vet Microbiol* **136**(1–2), 160–5.

Attoui, H., Mertens, P. P. C., Becnel, J., et al. (2012). Family—Reoviridae. *In* "Virus Taxonomy" (M. Q. K. Andrew, L. Elliot, J. A. Michael, et al., Eds.), pp. 541–637. Elsevier, San Diego.

Babkin, I. V., Tikunov, A. Y., Zhirakovskaia, E. V., et al. (2012). High evolutionary rate of human astrovirus. *Infect Genet Evol* **12**(2), 435–42.

Banyai, K., Martella, V., Meleg, E., et al. (2009). Searching for HAdV-52, the putative gastroenteritis-associated human adenovirus serotype in Southern Hungary. *New Microbiol* **32**(2), 185–8.

Banyai, K., Meleg, E., Moschidou, P., et al. (2010). Detection of newly described astrovirus MLB1 in stool samples from children. *Emerg Infect Dis* **16**(1), 169; author reply 169–70.

Barnes, G. L., Uren, E., Stevens, K. B., et al. (1998). Etiology of acute gastroenteritis in hospitalized children in Melbourne, Australia, from April 1980 to March 1993. *J Clin Microbiol* **36**(1), 133–8.

Belliot, G., Laveran, H., and Monroe, S. S. (1997). Outbreak of gastroenteritis in military recruits associated with serotype 3 astrovirus infection. *J Med Virol* **51**(2), 101–6.

Bergelson, J. M., Cunningham, J. A., Droguett, G., et al. (1997). Isolation of a common receptor for Coxsackie B viruses and adenoviruses 2 and 5. *Science* **275**(5304), 1320–3.

Bewley, M. C., Springer, K., Zhang, Y. B., et al. (1999). Structural analysis of the mechanism of adenovirus binding to its human cellular receptor, CAR. *Science* **286**(5444), 1579–83.

Bishop, R., and Kirkwood, C. (2008). Enteric viruses, 3rd ed. *In* "Encyclopedia of Virology" (B. W. J. Mahy, and M. H. V. Van Regenmortel, Eds.), pp. 116–123, 5 vols. Academic Press, Amsterdam; Boston.

Bishop, R. F., Davidson, G. P., Holmes, I. H., et al. (1973). Letter: Evidence for viral gastroenteritis. *N Engl J Med* **289**(20), 1096–7.

Bishop, R. F., Davidson, G. P., Holmes, I. H., et al. (1974). Detection of a new virus by electron microscopy of faecal extracts from children with acute gastroenteritis. *Lancet* **1**(7849), 149–51.

Bishop, R. F., Barnes, G. L., Cipriani, E., et al. (1983). Clinical immunity after neonatal rotavirus infection. A prospective longitudinal study in young children. *N Engl J Med* **309**(2), 72–6.

Bon, F., Fascia, P., Dauvergne, M., et al. (1999). Prevalence of group A rotavirus, human calicivirus, astrovirus, and adenovirus type 40 and 41 infections among children with acute gastroenteritis in Dijon, France. *J Clin Microbiol* **37**(9), 3055–8.

Boone, S. A., and Gerba, C. P. (2007). Significance of fomites in the spread of respiratory and enteric viral disease. *Appl Environ Microbiol* **73**(6), 1687–96.

Bosch, A., Guix, S., Krishna, N. K., et al. (2011). Family Astroviridae. *In* "Virus Taxonomy: Classification and Nomenclature of Viruses: Ninth Report of the International Committee on Taxonomy of Viruses" (A. M. Q. King, E. Lefkowitz, M. J. Adams, et al., Eds.), pp. 953–959. Academic Press, London; Waltham.

Brandt, C. D., Kim, H. W., Rodriguez, W. J., et al. (1985). Adenoviruses and pediatric gastroenteritis. *J Infect Dis* **151**(3), 437–43.

Bresee, J. S., Marcus, R., Venezia, R. A., et al. (2012). The etiology of severe acute gastroenteritis among adults visiting emergency departments in the United States. *J Infect Dis* **205**(9), 1374–81.

Bridger, J. C., Pedley, S., and McCrae, M. A. (1986). Group C rotaviruses in humans. *J Clin Microbiol* **23**(4), 760–3.

Brown, M., Wilson-Friesen, H. L., and Doane, F. (1992). A block in release of progeny virus and a high particle-to-infectious unit ratio contribute to poor growth of enteric adenovirus types 40 and 41 in cell culture. *J Virol* **66**(5), 3198–205.

Bucardo, F., Rippinger, C. M., Svensson, L., et al. (2012). Vaccine-derived NSP2 segment in rotaviruses from vaccinated children with gastroenteritis in Nicaragua. *Infect Genet Evol* **12**(6), 1282–94.

Burbelo, P. D., Ching, K. H., Esper, F., et al. (2011). Serological studies confirm the novel astrovirus HMOAstV-C as a highly prevalent human infectious agent. *PLoS One* **6**(8), e22576.

Buttery, J. P., Lambert, S. B., Grimwood, K., et al. (2011). Reduction in rotavirus-associated acute gastroenteritis following introduction of rotavirus vaccine into Australia's National Childhood vaccine schedule. *Pediatr Infect Dis J* **30**(1 Suppl.), S25–9.

Caballero, S., Guix, S., El-Senousy, W. M., et al. (2003). Persistent gastroenteritis in children infected with astrovirus: association with serotype-3 strains. *J Med Virol* **71**(2), 245–50.

Castello, A. A., Arguelles, M. H., Villegas, G. A., et al. (2000). Characterization of human group C rotavirus in Argentina. *J Med Virol* **62**(2), 199–207.

CDC (1999). From the Centers for Disease Control and Prevention. Withdrawal of rotavirus vaccine recommendation. *JAMA* **282**(22), 2113–4.

Chaimongkol, N., Khamrin, P., Malasao, R., et al. (2012). Genotypic linkages of gene segments of rotaviruses circulating in pediatric patients with acute gastroenteritis in Thailand. *Infect Genet Evol* **12**(7), 1381–91.

Chiba, S., Nakata, S., Nakamura, I., et al. (1983). Outbreak of infantile gastroenteritis due to type 40 adenovirus. *Lancet* **2**(8356), 954–7.

Chu, D. K., Poon, L. L., Guan, Y., et al. (2008). Novel astroviruses in insectivorous bats. *J Virol* **82**(18), 9107–14.

Chu, D. K., Chin, A. W., Smith, G. J., et al. (2010). Detection of novel astroviruses in urban brown rats and previously known astroviruses in humans. *J Gen Virol* **91**(Pt 10), 2457–62.

Cook, S. M., Glass, R. I., LeBaron, C. W., et al. (1990). Global seasonality of rotavirus infections. *Bull World Health Org* **68**(2), 171–7.

Cubitt, W. D., Mitchell, D. K., Carter, M. J., et al. (1999). Application of electronmicroscopy, enzyme immunoassay, and RT-PCR to monitor an outbreak of astrovirus type 1 in a paediatric bone marrow transplant unit. *J Med Virol* **57**(3), 313–21.

Cunningham, A. L., Grohman, G. S., Harkness, J., et al. (1988). Gastrointestinal viral infections in homosexual men who were symptomatic and seropositive for human immunodeficiency virus. *J Infect Dis* **158**(2), 386–91.

Curlin, M. E., Huang, M. L., Lu, X., et al. (2010). Frequent detection of human adenovirus from the lower gastrointestinal tract in men who have sex with men. *PLoS One* **5**(6), e11321.

Curns, A. T., Steiner, C. A., Barrett, M., et al. (2010). Reduction in acute gastroenteritis hospitalizations among US children after introduction of rotavirus vaccine: analysis of hospital discharge data from 18 US states. *J Infect Dis* **201**(11), 1617–24.

D'Souza, R. M., Hall, G., and Becker, N. G. (2008). Climatic factors associated with hospitalizations for rotavirus diarrhoea in children under 5 years of age. *Epidemiol Infect* **136**(1), 56–64.

De Benedictis, P., Schultz-Cherry, S., Burnham, A., et al. (2011). Astrovirus infections in humans and animals—molecular biology, genetic diversity, and interspecies transmissions. *Infect Genet Evol* **11**(7), 1529–44.

De Grazia, S., Platia, M. A., Rotolo, V., et al. (2011). Surveillance of human astrovirus circulation in Italy 2002–2005: emergence of lineage 2c strains. *Clin Microbiol Infect* **17**(1), 97–101.

de Jong, J. C., Wigand, R., Kidd, A. H., et al. (1983). Candidate adenoviruses 40 and 41: fastidious adenoviruses from human infant stool. *J Med Virol* **11**(3), 215–31.

de Jong, J. C., Bijlsma, K., Wermenbol, A. G., et al. (1993). Detection, typing, and subtyping of enteric adenoviruses 40 and 41 from fecal samples and observation of changing incidences of infections with these types and subtypes. *J Clin Microbiol* **31**(6), 1562–9.

de Wit, M. A., Koopmans, M. P., Kortbeek, L. M., et al. (2001). Etiology of gastroenteritis in sentinel general practices in the Netherlands. *Clin Infect Dis* **33**(3), 280–8.

de Jong, J. C., Osterhaus, A. D., Jones, M. S., et al. (2008). Human adenovirus type 52: a type 41 in disguise? *J Virol* **82**(7), 3809; author reply 3809–10.

Dennehy, P. H. (2012). Effects of vaccine on rotavirus disease in the pediatric population. *Curr Opin Pediatr* **24**(1), 76–84.

Dennehy, P. H., Nelson, S. M., Spangenberger, S., et al. (2001). A prospective case-control study of the role of astrovirus in acute diarrhea among hospitalized young children. *J Infect Dis* **184**(1), 10–5.

Dey, S. K., Shimizu, H., Phan, T. G., et al. (2009). Molecular epidemiology of adenovirus infection among infants and children with acute gastroenteritis in Dhaka City, Bangladesh. *Infect Genet Evol* **9**(4), 518–22.

Dey, R. S., Ghosh, S., Chawla-Sarkar, M., et al. (2011). Circulation of a novel pattern of infections by enteric adenovirus serotype 41 among children below 5 years of age in Kolkata, India. *J Clin Microbiol* **49**(2), 500–5.

Donato, C. M., Ch'ng, L. S., Boniface, K. F., et al. (2012). Identification of strains of RotaTeq rotavirus vaccine in infants with gastroenteritis following routine vaccination. *J Infect Dis* **206**(3), 377–83.

Donelli, G., Superti, F., Tinari, A., et al. (1992). Mechanism of astrovirus entry into Graham 293 cells. *J Med Virol* **38**(4), 271–7.

Dong, J., Dong, L., Mendez, E., et al. (2011). Crystal structure of the human astrovirus capsid spike. *Proc Natl Acad Sci USA* **108**(31), 12681–6.

Durepaire, N., Ranger-Rogez, S., Gandji, J. A., et al. (1995). Enteric prevalence of adenovirus in human immunodeficiency virus seropositive patients. *J Med Virol* **45**(1), 56–60.

Ebner, K., Pinsker, W., and Lion, T. (2005). Comparative sequence analysis of the hexon gene in the entire spectrum of human adenovirus serotypes: phylogenetic, taxonomic, and clinical implications. *J Virol* **79**(20), 12635–42.

Espul, C., Martinez, N., Noel, J. S., et al. (2004). Prevalence and characterization of astroviruses in Argentinean children with acute gastroenteritis. *J Med Virol* **72**(1), 75–82.

Estes, M. K., and Kapikian, A. Z. (2007). Rotaviruses, 5th ed. *In* "Fields Virology" (B. N. Fields, D. M. Knipe, and P. M. Howley, Eds.), pp. 1917–1974, 2 vols. Wolters Kluwer Health/Lippincott Williams & Wilkins, Philadelphia.

Favier, A. L., Burmeister, W. P., and Chroboczek, J. (2004). Unique physicochemical properties of human enteric Ad41 responsible for its survival and replication in the gastrointestinal tract. *Virology* **322**(1), 93–104.

Finkbeiner, S. R., Allred, A. F., Tarr, P. I., et al. (2008a). Metagenomic analysis of human diarrhea: viral detection and discovery. *PLoS Pathog* **4**(2), e1000011.

Finkbeiner, S. R., Kirkwood, C. D., and Wang, D. (2008b). Complete genome sequence of a highly divergent astrovirus isolated from a child with acute diarrhea. *Virol J* **5**, 117.

Finkbeiner, S. R., Holtz, L. R., Jiang, Y., et al. (2009a). Human stool contains a previously unrecognized diversity of novel astroviruses. *Virol J* **6**, 161.

Finkbeiner, S. R., Le, B. M., Holtz, L. R., et al. (2009b). Detection of newly described astrovirus MLB1 in stool samples from children. *Emerg Infect Dis* **15**(3), 441–4.

Finkbeiner, S. R., Li, Y., Ruone, S., et al. (2009c). Identification of a novel astrovirus (astrovirus VA1) associated with an outbreak of acute gastroenteritis. *J Virol* **83**(20), 10836–9.

Flewett, T. H., Bryden, A. S., and Davies, H. (1974a). Diagnostic electron microscopy of faeces. I. The viral flora of the faeces as seen by electron microscopy. *J Clin Pathol* **27**(8), 603–8.

Flewett, T. H., Bryden, A. S., Davies, H., et al. (1974b). Relation between viruses from acute gastroenteritis of children and newborn calves. *Lancet* **2**(7872), 61–3.

Fukuda, S., Kuwayama, M., Takao, S., et al. (2006). Molecular epidemiology of subgenus F adenoviruses associated with pediatric gastroenteritis during eight years in Hiroshima Prefecture as a limited area. *Arch Virol* **151**(12), 2511–7.

Gabbay, Y. B., Leite, J. P., Oliveira, D. S., et al. (2007). Molecular epidemiology of astrovirus type 1 in Belem, Brazil, as an agent of infantile gastroenteritis, over a period of 18 years (1982–2000): identification of two possible new lineages. *Virus Res* **129**(2), 166–74.

Geigenmuller, U., Chew, T., Ginzton, N., et al. (2002). Processing of nonstructural protein 1a of human astrovirus. *J Virol* **76**(4), 2003–8.

Gentsch, J. R., Laird, A. R., Bielfelt, B., et al. (2005). Serotype diversity and reassortment between human and animal rotavirus strains: implications for rotavirus vaccine programs. *J Infect Dis* **192**(Suppl. 1), S146–59.

Glass, R. I., Parashar, U. D., Bresee, J. S., et al. (2006). Rotavirus vaccines: current prospects and future challenges. *Lancet* **368**(9532), 323–32.

Graham, D. Y., and Estes, M. K. (1985). Proposed working serologic classification system for rotaviruses. *Annales de l'Institut Pasteur/Virologie* **136**, 5–12.

Gray, J. J., Wreghitt, T. G., Cubitt, W. D., et al. (1987). An outbreak of gastroenteritis in a home for the elderly associated with astrovirus type 1 and human calicivirus. *J Med Virol* **23**(4), 377–81.

Grimwood, K., Carzino, R., Barnes, G. L., et al. (1995). Patients with enteric adenovirus gastroenteritis admitted to an Australian pediatric teaching hospital from 1981 to 1992. *J Clin Microbiol* **33**(1), 131–6.

Grimwood, K., and Lambert, S. B. (2009). Rotavirus vaccines: opportunities and challenges. *Hum Vaccin* **5**(2), 57–69.

Grohmann, G. S., Glass, R. I., Pereira, H. G., et al. (1993). Enteric viruses and diarrhea in HIV-infected patients. Enteric Opportunistic Infections Working Group. *N Engl J Med* **329**(1), 14–20.

Guerrero, M. L., Noel, J. S., Mitchell, D. K., et al. (1998). A prospective study of astrovirus diarrhea of infancy in Mexico City. *Pediatr Infect Dis J* **17**(8), 723–7.

Guix, S., Caballero, S., Villena, C., et al. (2002). Molecular epidemiology of astrovirus infection in Barcelona, Spain. *J Clin Microbiol* **40**(1), 133–9.

Harrach, B., Benko, M., Both, G. W., et al. (2011). Family Adenoviridae. *In* "Virus Taxonomy: Classification and Nomenclature of Viruses: Ninth Report of the International Committee on Taxonomy of Viruses" (A. M. Q. King, E. Lefkowitz, M. J. Adams, et al., Eds.), pp. 125–141. Academic Press, London; Waltham.

Hellard, M. E., Sinclair, M. I., Harris, A. H., et al. (2003). Cost of community gastroenteritis. *J Gastroenterol Hepatol* **18**(3), 322–8.

Herrmann, J. E., Taylor, D. N., Echeverria, P., et al. (1991). Astroviruses as a cause of gastroenteritis in children. *N Engl J Med* **324**(25), 1757–60.

Holtz, L. R., Bauer, I. K., Rajendran, P., et al. (2011). Astrovirus MLB1 is not associated with diarrhea in a cohort of Indian children. *PLoS One* **6**(12), e28647.

Hoshino, Y., Zimmer, J. F., Moise, N. S., et al. (1981). Detection of astroviruses in feces of a cat with diarrhea. *Brief report. Arch Virol* **70**(4), 373–6.

Huilan, S., Zhen, L. G., Mathan, M. M., et al. (1991). Etiology of acute diarrhoea among children in developing countries: a multicentre study in five countries. *Bull World Health Org* **69**(5), 549–55.

Iturriza Gomara, M., Wong, C., Blome, S., et al. (2002). Molecular characterization of VP6 genes of human rotavirus isolates: correlation of genogroups with subgroups and evidence of independent segregation. *J Virol* **76**(13), 6596–601.

Jeong, A. Y., Jeong, H. S., Jo, M. Y., et al. (2011). Molecular epidemiology and genetic diversity of human astrovirus in South Korea from 2002 to 2007. *Clin Microbiol Infect* **17**(3), 404–8.

Jiang, B., Dennehy, P. H., Spangenberger, S., et al. (1995). First detection of group C rotavirus in fecal specimens of children with diarrhea in the United States. *J Infect Dis* **172**(1), 45–50.

Jiang, B., Monroe, S. S., Koonin, E. V., et al. (1993). RNA sequence of astrovirus: distinctive genomic organization and a putative retrovirus-like ribosomal frameshifting signal that directs the viral replicase synthesis. *Proc Natl Acad Sci USA* **90**(22), 10539–43.

Jin, Y., Cheng, W. X., Yang, X. M., et al. (2009). Viral agents associated with acute gastroenteritis in children hospitalized with diarrhea in Lanzhou, China. *J Clin Virol* **44**(3), 238–41.

Jones, M. S., 2nd, Harrach, B., Ganac, R. D., et al. (2007). New adenovirus species found in a patient presenting with gastroenteritis. *J Virol* **81**(11), 5978–84.

Kaljot, K. T., Ling, J. P., Gold, J. W., et al. (1989). Prevalence of acute enteric viral pathogens in acquired immunodeficiency syndrome patients with diarrhea. *Gastroenterology* **97**(4), 1031–2.

Kapoor, A., Li, L., Victoria, J., et al. (2009). Multiple novel astrovirus species in human stool. *J Gen Virol* **90**(Pt 12), 2965–72.

Kiang, D., and Matsui, S. M. (2002). Proteolytic processing of a human astrovirus nonstructural protein. *J Gen Virol* **83**(Pt 1), 25–34.

Kidd, A. H. (1984). Genome variants of adenovirus 41 (subgroup G) from children with diarrhoea in South Africa. *J Med Virol* **14**(1), 49–59.

Kidd, A. H., Banatvala, J. E., and de Jong, J. C. (1983). Antibodies to fastidious faecal adenoviruses (species 40 and 41) in sera from children. *J Med Virol* **11**(4), 333–41.

Kidd, A. H., Berkowitz, F. E., Blaskovic, P. J., et al. (1984). Genome variants of human adenovirus 40 (subgroup F). *J Med Virol* **14**(3), 235–46.

Kidd, A. H., Chroboczek, J., Cusack, S., et al. (1993). Adenovirus type 40 virions contain two distinct fibers. *Virology* **192**(1), 73–84.

Kim, M., Lim, M. Y., and Ko, G. (2010). Enhancement of enteric adenovirus cultivation by viral transactivator proteins. *Appl Environ Microbiol* **76**(8), 2509–16.

Koci, M. D., Moser, L. A., Kelley, L. A., et al. (2003). Astrovirus induces diarrhea in the absence of inflammation and cell death. *J Virol* **77**(21), 11798–808.

Krajden, M., Brown, M., Petrasek, A., et al. (1990). Clinical features of adenovirus enteritis: a review of 127 cases. *Pediatr Infect Dis J* **9**(9), 636–41.

Kriston, S., Willcocks, M. M., Carter, M. J., et al. (1996). Seroprevalence of astrovirus types 1 and 6 in London, determined using recombinant virus antigen. *Epidemiol Infect* **117**(1), 159–64.

Kurtz, J., and Lee, T. (1978). Astrovirus gastroenteritis age distribution of antibody. *Med Microbiol Immunol* **166**(1–4), 227–30.

Lewis, T. L., Greenberg, H. B., Herrmann, J. E., et al. (1994). Analysis of astrovirus serotype 1 RNA, identification of the viral RNA-dependent RNA polymerase motif, and expression of a viral structural protein. *J Virol* **68**(1), 77–83.

Li, L., Shimizu, H., Doan, L. T., et al. (2004). Characterizations of adenovirus type 41 isolates from children with acute gastroenteritis in Japan, Vietnam, and Korea. *J Clin Microbiol* **42**(9), 4032–9.

Liu, M. Q., Yang, B. F., Peng, J. S., et al. (2007). Molecular epidemiology of astrovirus infection in infants in Wuhan, China. *J Clin Microbiol* **45**(4), 1308–9.

Liu, M. Q., Peng, J. S., Tang, L., et al. (2008). Identification of new subtype of astrovirus type 3 from an infant with diarrhea in Wuhan, China. *Virology* **375**(1), 301–6.

Liu, H., Jin, L., Koh, S. B., et al. (2010). Atomic structure of human adenovirus by cryo-EM reveals interactions among protein networks. *Science* **329**(5995), 1038–43.

Liu, E. B., Ferreyra, L., Fischer, S. L., et al. (2011). Genetic analysis of a novel human adenovirus with a serologically unique hexon and a recombinant fiber gene. *PLoS One* **6**(9), e24491.

Madeley, C. R., and Cosgrove, B. P. (1975a). Letter: 28 nm particles in faeces in infantile gastroenteritis. *Lancet* **2**(7932), 451–2.

Madeley, C. R., and Cosgrove, B. P. (1975b). Letter: Viruses in infantile gastroenteritis. *Lancet* **2**(7925), 124.

Magwalivha, M., Wolfaardt, M., Kiulia, N. M., et al. (2010). High prevalence of species D human adenoviruses in fecal specimens from Urban Kenyan children with diarrhea. *J Med Virol* **82**(1), 77–84.

Maldonado, Y., Cantwell, M., Old, M., et al. (1998). Population-based prevalence of symptomatic and asymptomatic astrovirus infection in rural Mayan infants. *J Infect Dis* **178**(2), 334–9.

Malherbe, H., and Harwin, R. (1963). The cytopathic effects of vervet monkey viruses. *S Afr Med J* **37**, 407–11.

Matsushima, Y., Shimizu, H., Phan, T. G., et al. (2011). Genomic characterization of a novel human adenovirus type 31 recombinant in the hexon gene. *J Gen Virol* **92**(Pt 12), 2770–5.

Matsushima, Y., Shimizu, H., Kano, A., et al. (2012). Novel human adenovirus strain, Bangladesh. *Emerg Infect Dis* **18**(5), 846–8.

Matthijnssens, J., Ciarlet, M., McDonald, S. M., et al. (2011). Uniformity of rotavirus strain nomenclature proposed by the Rotavirus Classification Working Group (RCWG). *Arch Virol* **156**(8), 1397–413.

Matthijnssens, J., Rahman, M., and Van Ranst, M. (2006). Loop model: mechanism to explain partial gene duplications in segmented dsRNA viruses. *Biochem Biophys Res Commun* **340**(1), 140–4.

Mitchell, D. K., Matson, D. O., Cubitt, W. D., et al. (1999). Prevalence of antibodies to astrovirus types 1 and 3 in children and adolescents in Norfolk, Virginia. *Pediatr Infect Dis J* **18**(3), 249–54.

Moser, L. A., Carter, M., and Schultz-Cherry, S. (2007). Astrovirus increases epithelial barrier permeability independently of viral replication. *J Virol* **81**(21), 11937–45.

Mustafa, H., Palombo, E. A., and Bishop, R. F. (2000). Epidemiology of astrovirus infection in young children hospitalized with acute gastroenteritis in Melbourne, Australia, over a period of four consecutive years, 1995 to 1998. *J Clin Microbiol* **38**(3), 1058–62.

Nighot, P. K., Moeser, A., Ali, R. A., et al. (2010). Astrovirus infection induces sodium malabsorption and redistributes sodium hydrogen exchanger expression. *Virology* **401**(2), 146–54.

Noel, J. S., Lee, T. W., Kurtz, J. B., et al. (1995). Typing of human astroviruses from clinical isolates by enzyme immunoassay and nucleotide sequencing. *J Clin Microbiol* **33**(4), 797–801.

Palombo, E. A., and Bishop, R. F. (1996). Annual incidence, serotype distribution, and genetic diversity of human astrovirus isolates from hospitalized children in Melbourne, Australia. *J Clin Microbiol* **34**(7), 1750–3.

Papaventsis, D. C., Dove, W., Cunliffe, N. A., et al. (2008). Human astrovirus gastroenteritis in children, Madagascar, 2004–2005. *Emerg Infect Dis* **14**(5), 844–6.

Patel, M., Pedreira, C., De Oliveira, L. H., et al. (2009). Association between pentavalent rotavirus vaccine and severe rotavirus diarrhea among children in Nicaragua. *JAMA* **301**(21), 2243–51.

Patel, M. M., Steele, D., Gentsch, J. R., et al. (2011). Real-world impact of rotavirus vaccination. *Pediatr Infect Dis J* **30**(1 Suppl.), S1–5.

Paulke-Korinek, M., Kundi, M., Rendi-Wagner, P., et al. (2011). Herd immunity after two years of the universal mass vaccination program against rotavirus gastroenteritis in Austria. *Vaccine* **29**(15), 2791–6.

Phan, T. G., Okame, M., Nguyen, T. A., et al. (2004). Human astrovirus, norovirus (GI, GII), and sapovirus infections in Pakistani children with diarrhea. *J Med Virol* **73**(2), 256–61.

Pinto, R. M., Abad, F. X., Gajardo, R., et al. (1996). Detection of infectious astroviruses in water. *Appl Environ Microbiol* **62**(8), 3073.

Quan, P. L., Wagner, T. A., Briese, T., et al. (2010). Astrovirus encephalitis in boy with X-linked agammaglobulinemia. *Emerg Infect Dis* **16**(6), 918–25.

Ramig, R. F., and Ward, R. L. (1991). Genomic segment reassortment in rotavirus and other reoviridae. *Adv Virus Res* **39**, 163–207.

Reddy, V. S., Natchiar, S. K., Stewart, P. L., et al. (2010). Crystal structure of human adenovirus at 3.5 A resolution. *Science* **329**(5995), 1071–5.

Richardson, S., Grimwood, K., Gorrell, R., et al. (1998). Extended excretion of rotavirus after severe diarrhoea in young children. *Lancet* **351**(9119), 1844–8.

Richardson, V., Hernandez-Pichardo, J., Quintanar-Solares, M., et al. (2010). Effect of rotavirus vaccination on death from childhood diarrhea in Mexico. *N Engl J Med* **362**(4), 299–305.

Rivera, R., Nollens, H. H., Venn-Watson, S., et al. (2010). Characterization of phylogenetically diverse astroviruses of marine mammals. *J Gen Virol* **91**(Pt 1), 166–73.

Rowe, W. P., Huebner, R. J., Gilmore, L. K., et al. (1953). Isolation of a cytopathogenic agent from human adenoids undergoing spontaneous degeneration in tissue culture. *Proc Soc Exp Biol Med* **84**(3), 570–3.

Sanekata, T., Ahmed, M. U., Kader, A., et al. (2003). Human group B rotavirus infections cause severe diarrhea in children and adults in Bangladesh. *J Clin Microbiol* **41**(5), 2187–90.

Santos, N., and Hoshino, Y. (2005). Global distribution of rotavirus serotypes/genotypes and its implication for the development and implementation of an effective rotavirus vaccine. *Rev Med Virol* **15**(1), 29–56.

Scharff, R. L. (2012). Economic burden from health losses due to foodborne illness in the United States. *J Food Prot* **75**(1), 123–31.

Schnagl, R. D., Belfrage, K., Farrington, R., et al. (2002). Incidence of human astrovirus in central Australia (1995 to 1998) and comparison of deduced serotypes detected from 1981 to 1998. *J Clin Microbiol* **40**(11), 4114–20.

Scott-Taylor, T. H., and Hammond, G. W. (1995). Local succession of adenovirus strains in pediatric gastroenteritis. *J Med Virol* **45**(3), 331–8.

Sebire, N. J., Malone, M., Shah, N., et al. (2004). Pathology of astrovirus associated diarrhoea in a paediatric bone marrow transplant recipient. *J Clin Pathol* **57**(9), 1001–3.

Seto, D., Chodosh, J., Brister, J. R., et al. (2011). Using the whole-genome sequence to characterize and name human adenoviruses. *J Virol* **85**(11), 5701–2.

Sherwood, V., Burgert, H. G., Chen, Y. H., et al. (2007). Improved growth of enteric adenovirus type 40 in a modified cell line that can no longer respond to interferon stimulation. *J Gen Virol* **88** (Pt 1), 71–6.

Sherwood, V., King, E., Totemeyer, S., et al. (2012). Interferon treatment suppresses enteric adenovirus infection in a model gastrointestinal cell-culture system. *J Gen Virol* **93**(Pt 3), 618–23.

Shinozaki, T., Araki, K., Ushijima, H., et al. (1987). Antibody response to enteric adenovirus types 40 and 41 in sera from people in various age groups. *J Clin Microbiol* **25**(9), 1679–82.

Silva, P. A., Stark, K., Mockenhaupt, F. P., et al. (2008). Molecular characterization of enteric viral agents from children in northern region of Ghana. *J Med Virol* **80**(10), 1790–8.

Singh, G., Robinson, C. M., Dehghan, S., et al. (2012). Overreliance on the hexon gene, leading to misclassification of human adenoviruses. *J Virol* **86**(8), 4693–5.

Smits, S. L., van Leeuwen, M., van der Eijk, A. A., et al. (2010). Human astrovirus infection in a patient with new-onset celiac disease. *J Clin Microbiol* **48**(9), 3416–8.

Snodgrass, D. R., Angus, K. W., Gray, E. W., et al. (1979). Pathogenesis of diarrhoea caused by astrovirus infections in lambs. *Arch Virol* **60**(3–4), 217–26.

Speroni, S., Rohayem, J., Nenci, S., et al. (2009). Structural and biochemical analysis of human pathogenic astrovirus serine protease at 2.0 A resolution. *J Mol Biol* **387**(5), 1137–52.

Stewart, P. L., Fuller, S. D., and Burnett, R. M. (1993). Difference imaging of adenovirus: bridging the resolution gap between X-ray crystallography and electron microscopy. *EMBO J* **12**(7), 2589–99.

Takiff, H. E., and Straus, S. E. (1982). Early replicative block prevents the efficient growth of fastidious diarrhea-associated adenoviruses in cell culture. *J Med Virol* **9**(2), 93–100.

Takiff, H. E., Straus, S. E., and Garon, C. F. (1981). Propagation and in vitro studies of previously non-cultivable enteral adenoviruses in 293 cells. *Lancet* **2**(8251), 832–4.

Tam, C. C., Rodrigues, L. C., Viviani, L., et al. (2012). Longitudinal study of infectious intestinal disease in the UK (IID2 study): incidence in the community and presenting to general practice. *Gut* **61**(1), 69–77.

Taniguchi, K., and Urasawa, S. (1995). Diversity in rotavirus genomes. *Semin Virol* **6**, 123–31.

Tate, J. E., Panozzo, C. A., Payne, D. C., et al. (2009). Decline and change in seasonality of US rotavirus activity after the introduction of rotavirus vaccine. *Pediatrics* **124**(2), 465–71.

Tate, J. E., Burton, A. H., Boschi-Pinto, C., et al. (2012). 2008 estimate of worldwide rotavirus-associated mortality in children younger than 5 years before the introduction of universal rotavirus vaccination programmes: a systematic review and meta-analysis. *Lancet Infect Dis* **12**(2), 136–41.

Tayeb, H. T., Al-Ahdal, M. N., Cartear, M. J., et al. (2010). Molecular epidemiology of human astrovirus infections in Saudi Arabia pediatric patients. *J Med Virol* **82**(12), 2038–42.

Tsugawa, T., and Hoshino, Y. (2008). Whole genome sequence and phylogenetic analyses reveal human rotavirus G3P[3] strains Ro1845 and HCR3A are examples of direct virion transmission of canine/feline rotaviruses to humans. *Virology* **380**(2), 344–53.

Utagawa, E. T., Nishizawa, S., Sekine, S., et al. (1994). Astrovirus as a cause of gastroenteritis in Japan. *J Clin Microbiol* **32**(8), 1841–5.

van den Brandhof, W. E., De Wit, G. A., de Wit, M. A., et al. (2004). Costs of gastroenteritis in The Netherlands. *Epidemiol Infect* **132**(2), 211–21.

Van, R., Wun, C. C., O'Ryan, M. L., et al. (1992). Outbreaks of human enteric adenovirus types 40 and 41 in Houston day care centers. *J Pediatr* **120**(4 Pt 1), 516–21.

Verma, H., Chitambar, S. D., and Gopalkrishna, V. (2010). Astrovirus associated acute gastroenteritis in western India: predominance of dual serotype strains. *Infect Genet Evol* **10**(4), 575–9.

Vesikari, T., Karvonen, A., Prymula, R., et al. (2007). Efficacy of human rotavirus vaccine against rotavirus gastroenteritis during the first 2 years of life in European infants: randomised, double-blind controlled study. *Lancet* **370**(9601), 1757–63.

Vesikari, T., Matson, D. O., Dennehy, P., et al. (2006). Safety and efficacy of a pentavalent human-bovine (WC3) reassortant rotavirus vaccine. *N Engl J Med* **354**(1), 23–33.

Walter, J. E., Briggs, J., Guerrero, M. L., et al. (2001a). Molecular characterization of a novel recombinant strain of human astrovirus associated with gastroenteritis in children. *Arch Virol* **146**(12), 2357–67.

Walter, J. E., Mitchell, D. K., Guerrero, M. L., et al. (2001b). Molecular epidemiology of human astrovirus diarrhea among children from a periurban community of Mexico City. *J Infect Dis* **183**(5), 681–6.

Ward, R. L., and Knowlton, D. R. (1989). Genotypic selection following coinfection of cultured cells with subgroup 1 and subgroup 2 human rotaviruses. *J Gen Virol* **70**(Pt 7), 1691–9.

Whitelaw, A., Davies, H., and Parry, J. (1977). Electron microscopy of fatal adenovirus gastroenteritis. *Lancet* **1**(8007), 361.

WHO (2011). World Health Statistics 2011. World Health Organization, Geneva.

Wigand, R., Baumeister, H. G., Maass, G., et al. (1983). Isolation and identification of enteric adenoviruses. *J Med Virol* **11**(3), 233–40.

Willcocks, M. M., and Carter, M. J. (1993). Identification and sequence determination of the capsid protein gene of human astrovirus serotype 1. *FEMS Microbiol Lett* **114**(1), 1–7.

Williams, F. P., Jr. (1980). Astrovirus-like, coronavirus-like, and parvovirus-like particles detected in the diarrheal stools of beagle pups. *Arch Virol* **66**(3), 215–26.

Wold, W. S. M., and Horwitz, M. S. (2007). Adenoviruses, 5th ed. *In* "Fields Virology" (B. N. Fields, D. M. Knipe, and P. M. Howley, Eds.), pp. 2396–2436, 2 vols. Wolters Kluwer Health/Lippincott Williams & Wilkins, Philadelphia.

Wolfaardt, M., Kiulia, N. M., Mwenda, J. M., et al. (2011). Evidence of a recombinant wild-type human astrovirus strain from a Kenyan child with gastroenteritis. *J Clin Microbiol* **49**(2), 728–31.

Woode, G. N., Pohlenz, J. F., Gourley, N. E., et al. (1984). Astrovirus and Breda virus infections of dome cell epithelium of bovine ileum. *J Clin Microbiol* **19**(5), 623–30.

Wunderli, W., Meerbach, A., Gungor, T., et al. (2011). Astrovirus infection in hospitalized infants with severe combined immunodeficiency after allogeneic hematopoietic stem cell transplantation. *PLoS One* **6**(11), e27483.

Xiao, J., Li, J., Hu, G., et al. (2011). Isolation and phylogenetic characterization of bat astroviruses in southern China. *Arch Virol* **156**(8), 1415–23.

Zeller, M., Rahman, M., Heylen, E., et al. (2010). Rotavirus incidence and genotype distribution before and after national rotavirus vaccine introduction in Belgium. *Vaccine* **28**(47), 7507–13.

28

EMERGING HUMAN NOROVIRUS INFECTIONS

Melissa K. Jones, Shu Zhu, and Stephanie M. Karst

Department of Molecular Genetics and Microbiology, University of Florida College of Medicine, Gainesville, FL, USA

TABLE OF CONTENTS

28.1 INTRODUCTION

Human noroviruses (HuNoVs), non-enveloped plus-stranded RNA viruses 7.4–7.7 kb in length, are the leading cause of gastroenteritis outbreaks across the globe (Koo et al., 2010), causing over 20 million cases of acute diarrheal disease, 70000 hospitalizations, and 800 deaths annually in the United States (CDC, 2012a). Moreover, HuNoVs are responsible for a large majority of foodborne illness compared to other enteric pathogens in the United States

Viral Infections and Global Change, First Edition. Edited by Sunit K. Singh.

(Scallan et al., 2011). While healthy adults typically develop an acute gastroenteritis that resolves in 1–3 days upon HuNoV infection, several risk groups—infants, young children, the elderly, and immunocompromised and transplant patients—are susceptible to more severe, prolonged, and even life-threatening disease (Karst, 2010). Highlighting this point, it has been estimated that HuNoVs cause nearly 1 million clinic visits and 200 000 deaths of children <5 years old in developing nations each year (Patel, 2008); malnutrition undoubtedly contributes to the severity of HuNoV infections in this setting. Although HuNoVs are highly genetically variable and segregate into three genogroups within the *Norovirus* (NoV) genus, these being further segregated into over 30 genotypes, a specific subcluster of HuNoV strains (genogroup II, genotype 4 (GII.4)) has caused 70–80% of all outbreaks in recent years. In fact, GII.4 HuNoVs have been responsible for at least five pandemics since 2002 (Updated Norovirus Outbreak Management and Disease Prevention Guidelines, 2012). Although efforts are underway to develop effective HuNoV vaccines (Atmar et al., 2011), several significant challenges remain—the high degree of genetic and potentially antigenic variability among virus strains and the regular emergence of new pandemic strains will likely require a frequently modified vaccine formulation, and the lack of lasting protective immunity elicited by at least some HuNoV strains during natural infections (Johnson et al., 1990; Parrino et al., 1977) may translate to inefficient long-term vaccine efficacy as well. Because of these hurdles, it is especially important to consider the impact that human behavior and environmental changes have on HuNoV epidemiology to pinpoint effective measures to reduce the spread of HuNoVs to populations at risk for severe disease. In this chapter we will summarize the current understanding of NoV pathogenesis and epidemiology, and we will discuss potential effects of globalization of the food market, increased population size of immunocompromised individuals, and climate change on HuNoV epidemiology.

28.2 NOROVIRUS EPIDEMIOLOGY

NoVs are a genetically diverse family of plus-stranded RNA viruses. They segregate into five genogroups (GI–GV), three of which contain primarily human pathogens (Fankhauser et al., 1998; Zheng et al., 2006). These genogroups are further subdivided into distinct clusters or genotypes based on genetic similarity, with intra-genogroup members sharing at least 40% identity in their capsid sequences and intra-genotype members sharing at least 56% identity in their capsid sequences (Zheng et al., 2006). This high degree of genetic variability represents another complication in HuNoV diagnostic strategies, in addition to their uncultivability. While members of both GI and GII genogroups commonly cause human disease throughout the world, the GII.4 HuNoV cluster has been solely responsible for at least five pandemics since 2002 (Karst, 2010). The GII.4 HuNoV variants display an epochal pattern of evolution similar to influenza virus, with a novel pandemic strain arising every 2–4 years to replace the previously circulating strain (Siebenga et al., 2007). It is unclear why this particular genotype has become so globally dominant, but possibilities include increased virulence and/or infectivity (Desai et al., 2012; Friesema et al., 2009; Huhti et al., 2011), higher titers of shed virus contributing to increased transmission (Bucardo et al., 2008), more efficient person-to-person transmission (Kroneman et al., 2008; Verhoef, 2009), and more rapid evolution (Bull et al., 2010). As reviewed recently by Bull et al., possible factors contributing to the more rapid evolution of GII.4 HuNoVs compared to other genotypes include (i) a larger pool of susceptible hosts due to broader receptor usage, (ii) herd immunity (most likely short term) driving the emergence of antigenic variants, (iii) limited sequence space that restricts the number of residues able to undergo positive selection, and

(iv) lower fidelity of the GII.4 RNA-dependent RNA polymerase (RdRp) (Bull and White, 2011). Collectively, these observations are suggestive that GII.4 HuNoVs are globally dominant due to efficient antigenic escape, altered receptor usage, and potentially increased virulence properties. However, multiple confounding issues must be considered. First, comparisons of pathogenic properties of HuNoV genotypes all relied on retrospective analyses of clinical data sets that were not collected in a controlled fashion. A major limitation in HuNoV studies of this type is the inability to assess individuals' previous HuNoV exposure histories. Since nearly 100% of adults have been exposed to at least one HuNoV (Donaldson et al., 2010), this likely has a significant influence on patient susceptibility profiles to specific genotypes. Moreover, these studies were not designed to control for host genetic, nutritional, or microbial differences or environmental variability, all of which can affect viral evolution. Second, the lack of a small animal model or cell culture system for HuNoVs prevents true virulence and antigenicity assessments of viral variants.

Although GII.4 HuNoVs are clearly a major cause of HuNoV outbreaks and are associated with childhood diarrheal disease, GII.3 HuNoVs also frequently cause childhood diarrhea (Barreira et al., 2010; Bok et al., 2009; Boon et al., 2011; Dey et al., 2010; Dove et al., 2005; Phan et al., 2005). In fact, GII.3 NoVs are often the dominant genotype identified in pediatric diarrheal patients in developing nations.

28.3 FEATURES OF NOROVIRUS OUTBREAKS

HuNoVs are responsible for a majority of gastroenteritis outbreaks worldwide. Recently, they have been recognized as the leading cause of foodborne illness (CDC, 2011) and hospital ward closures associated with nosocomial outbreaks (Hansen et al., 2007). Outbreaks generally occur in semi-closed communities. For example, HuNoV outbreaks reported in the United States in 2010–2011 occurred in long-term care facilities (59%), restaurants (8%), parties and events (7%), hospitals (4%), schools (4%), and cruise ships (4%) (CDC, 2012a) (Figure 28.1). One reason for the explosive nature of HuNoV outbreaks in these confined settings is that they are infectious through multiple transmission routes—they spread efficiently via fecal–oral contamination of a food source, water, or fomites and person-to-person (Figure 28.2): Food products can become contaminated during cultivation, harvesting, or processing. For example, shellfish grown in fecally contaminated harvest waters are a common source of HuNoV outbreaks; produce (e.g., leafy greens and soft red fruits) are also common sources likely due to exposure to fecally contaminated irrigation waters (Jones and Karst, 2012). Another primary point of HuNoV entry into the food chain is food handling by an infected individual, highlighted by numerous outbreaks linked to symptomatic food handlers (Anderson et al., 2001; Becker et al., 2000; Daniels et al., 2000; Friedman et al., 2005; Grotto et al., 2004; Herwaldt et al., 1994; Lo et al., 1994; Parashar et al., 1998; Patterson et al., 1997; Vivancos et al., 2009). This is particularly of concern for produce and ready-to-eat foods. NoVs can also survive for long periods of time on fomites such as food preparation surfaces (Cannon et al., 2006; Doultree et al., 1999; Fallahi and Mattison, 2011) and are resistant to commonly used industrial sanitizers (D'Souza and Su, 2010). In addition to fecal–oral transmission, HuNoVs can spread efficiently person-to-person. Supporting this mode of transmission, a significant risk factor for becoming HuNoV infected is contact with an infected individual (e.g., Cáceres et al., 1998; Kaplan et al., 1982; Kappus et al., 1982; Marks et al., 2000). The high secondary attack rate of HuNoVs among close contacts, ranging from 14% to 33% in representative outbreaks (Alfano-Sobsey et al., 2012; Baron et al., 1982; Kaplan et al., 1982; Kappus et al., 1982), also supports person-to-person spread (v).

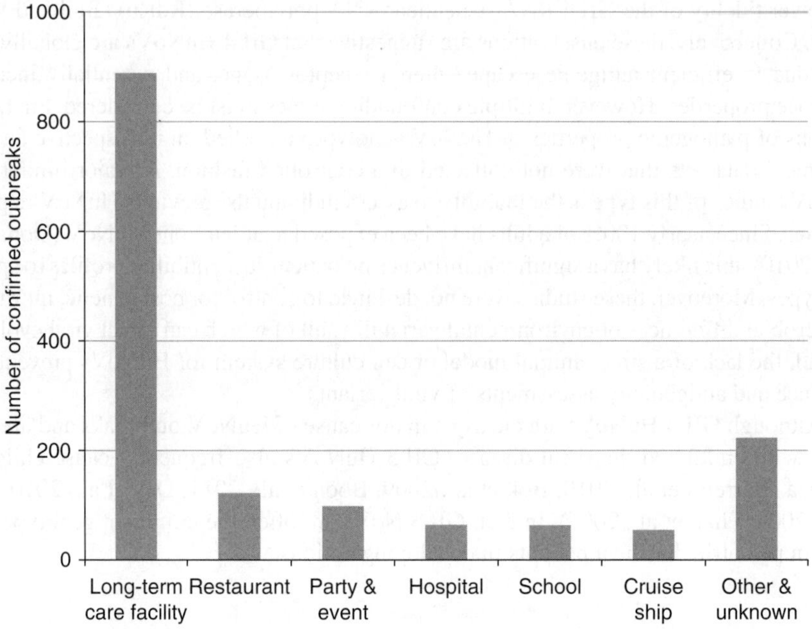

Figure 28.1. Setting of 1518 confirmed norovirus outbreaks in the United States from 2010–2011. Data are reproduced from http://www.cdc.gov/features/dsnorovirus/figure2.html. For color detail, please see color plate section.

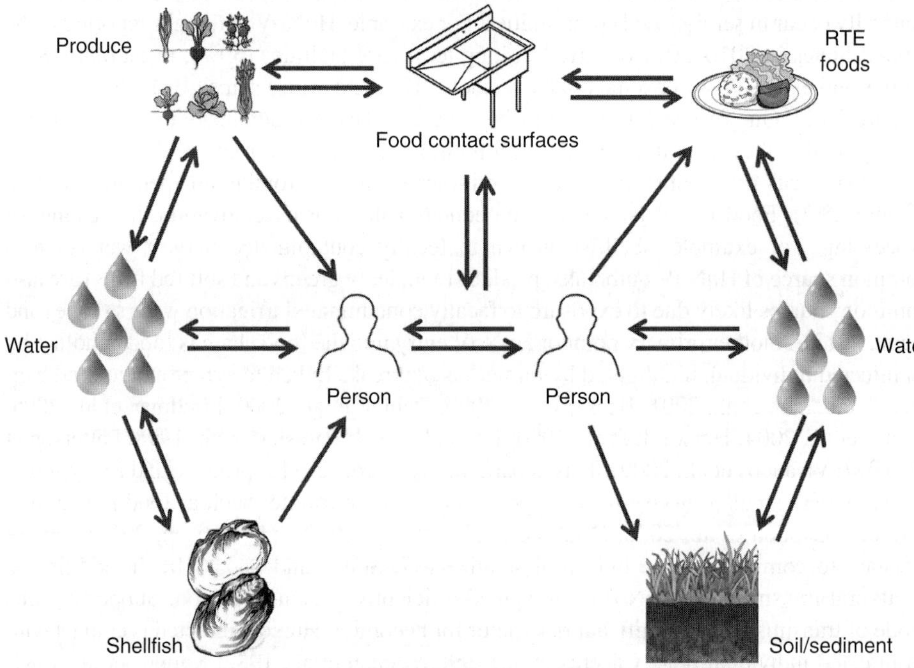

Figure 28.2. Routes of norovirus transmission. Human noroviruses spread efficiently person-to-person and through exposure to contaminated foods (e.g., produce, ready-to-eat [RTE] foods, and shellfish), water, and fomites. They are extremely stable in water and soils, thus these environmental sources as well as infected people can all act as norovirus reservoirs.

Finally, HuNoVs can disseminate via aerosolized vomitus particles (Chadwick and McCann, 1994; Lopman, 2011; Marks et al., 2000; O'Neill and Marks, 2005).

Additional factors contribute to the widespread nature of HuNoV outbreaks. For example, HuNoVs are highly infectious—as evidenced by their high secondary attack rate—even at low doses (Teunis et al., 2008). Furthermore, prolonged shedding from infected persons well after symptom resolution (Atmar et al., 2008; Patterson et al., 1993; Rockx et al., 2002) and extremely high levels of virus shed in feces (e.g., as high as 10^{10} HuNoV genome copies per gram of feces have been detected by RT-PCR (Atmar et al., 2008; Chan et al., 2006)) also hinder outbreak control.

28.4 CLINICAL FEATURES OF NOROVIRUS INFECTION

HuNoV infection in healthy adults follows a rapid course, with symptoms arising 24–28 h following exposure and lasting 12–72 h (Estes et al., 2006). Symptoms include nausea, vomiting, and diarrhea and less frequently abdominal cramps, low-grade fever, and malaise. Risk groups are susceptible to more severe and prolonged HuNoV infections. For example, severe infections lasting up to 6 weeks have been documented in infants and young children (Kirkwood and Streitberg, 2008; Murata et al., 2007; Sakai et al., 2001; Zintz et al., 2005). The elderly are prone to severe and even fatal HuNoV infections (Lopman et al., 2003, 2004; MMWR, 2007). In transplant patients and other immunocompromised individuals, HuNoV infections can become chronic leading to significant weight loss and death (Gallimore et al., 2004, 2006; Kaufman et al., 2005; Morotti et al., 2004; Roddie et al., 2009; Saif et al., 2011; Westhoff et al., 2009). The potential implications of chronically infected hosts are discussed in more detail in the succeeding text.

Although there is no direct proof that HuNoV infections are more severe in the malnourished host, clinical features of infections in developed versus developing countries provide strong circumstantial evidence: While HuNoV infections cause self-limited gastroenteritis in children in developed countries that rarely requires treatment, they are responsible for 3–20% of severe pediatric diarrheal cases that require hospitalization in developing countries (Al-Ali et al., 2011; Jakab et al., 2010; Kaplan et al., 2011; Kirby et al., 2011; Moyo et al., 2011; Nataraju et al., 2010; O'Ryan et al., 2010; Ozkul et al., 2011; Rahouma et al., 2011; Tamura et al., 2010; Thongprachum et al., 2010; Zhang et al., 2011). In fact, it has been estimated that HuNoVs cause over 1 million hospitalizations and 200 000 deaths in primarily children <5 years of age in developing nations annually, presumably a large proportion of which are malnourished (Patel, 2008). Thus, the malnourished host should be considered an additional HuNoV risk group. The epidemiological lag in recognizing the importance of HuNoV infections in severe childhood diarrheal disease can be explained by their refractory cultivability (Duizer et al., 2004; Lay et al., 2010) and their extreme genetic variability (Karst, 2010; described in more detail in the succeeding text), the combination of which historically hampered HuNoV diagnosis. The advent of RT-PCR for NoV detection and the identification of a conserved region within the NoV genome (Bull et al., 2005) have provided a solution to these challenges. In fact, there is emerging evidence that HuNoVs are second only to rotaviruses in causing severe childhood gastroenteritis (Barreira et al., 2010; Glass et al., 2009). With these new insights into the clinical importance of HuNoV infections in causing childhood morbidity, research efforts to fully characterize the HuNoV disease burden in developing parts of the world are absolutely essential.

Improved HuNoV diagnostics has also revealed their association with traveler's diarrhea (Ahn et al., 2011; Chapin et al., 2005; Ko et al., 2005). Moreover, there have been numerous anecdotal reports of HuNoV associations with clinical outcomes other than gastroenteritis, including infantile necrotizing enterocolitis (Bagci et al., 2010; Stuart et al., 2010; Turcios-Ruiz et al., 2008; Tzialla et al., 2011), encephalopathy (Ito et al., 2006; Obinata et al., 2010), benign pediatric seizures (Bartolini et al., 2011; Chan et al., 2011; Chen et al., 2009; Kawano et al., 2007), postinfectious irritable bowel syndrome (Marshall et al., 2007), and exacerbation of inflammatory bowel disease (Khan et al., 2009). Particularly concerning based on the ability of related animal viruses to cause fatal hemorrhagic-like and systemic diseases (Hurley et al., 2004; Parra and Prieto, 1990; Pedersen et al., 2000), a number of individuals during a 2002 HuNoV outbreak presented with headaches, neck stiffness, light sensitivity, diminished alertness, and disseminated intravascular coagulation in addition to gastroenteritis (MMWR, 2002). While these reports clearly do not prove causation, they do underscore the importance of considering HuNoVs as potential etiologies of diverse pathologies.

28.5 HOST SUSCEPTIBILITY

Many HuNoVs have been demonstrated to bind histo-blood group antigens (HBGAs), neutral carbohydrates expressed on the surface of intestinal epithelial cells and secreted into the gut lumen (Hutson et al., 2002; Tan and Jiang, 2005). Two human volunteer studies have clearly demonstrated an association between expression of HBGAs on mucosal surfaces and susceptibility to HuNoV infection—~20% of the population have an inactive fucosyltransferase 2 (FUT2) gene and thus do not express mucosal HBGAs; these individuals, referred to as nonsecretors, are resistant to both a GI.1 and a GII.4 HuNoV strain, whereas a majority of secretors with an active FUT2 gene are susceptible (Frenck et al., 2012;Hutson et al., 2005; Lindesmith et al., 2003). It is unclear what role HBGAs play in HuNoV replication since they do not facilitate virus entry into nonpermissive cells (Duizer et al., 2004; Guix et al., 2007). It is possible that they mediate attachment to the intestinal wall, facilitating virion interaction with an entry receptor on the same or neighboring cells. Interestingly, the genetically diverse HuNoVs display distinct binding patterns to the polymorphic HBGAs, with some evidence for a shared intra-genogroup binding interface for the GII HuNoVs (recently reviewed in Tan and Jiang, 2011). There is evidence that sequential pandemic GII.4 HuNoV strains display variable HBGA binding patterns, providing emerging virus strains the potential advantage of infecting a naïve subset of the population that expresses a distinct type of HBGA (Lindesmith et al., 2008; Siebenga et al., 2009). There is extensive and promising research now focused on identifying prophylactic compounds (Feng and Jiang, 2007; Hansman et al., 2012; Rademacher et al., 2011) or antibodies (Lindesmith et al., 2012; Parra et al., 2012) that block HuNoV–HBGA interactions.

28.6 EFFECT OF INCREASED SIZE OF IMMUNOCOMPROMISED POPULATION

One HuNoV risk group that has dramatically increased in size over recent years is immunocompromised/transplant patients. Multiple factors are responsible for this rise including advances in modern medicine in terms of transplantation success and immunosuppressive but effective cancer drugs and HIV infections leading to AIDS. In fact, it has been estimated that 3.6% of the population in the United States, or ca. 10 million people, fall into one of these categories (Kemper et al., 2002). An extensive literature describes severe and prolonged HuNoV infections of immunocompromised individuals, with symptoms and fecal shedding lasting many months and even years (e.g., Frange et al., 2012; Gallimore et al., 2004, 2006; Kaufman et al., 2003, 2005; Morotti et al., 2004; Roddie et al., 2009; Saif et al., 2011; Westhoff et al., 2009). Moreover, it has been widely speculated that chronically infected immunocompromised patients are reservoirs of emergent HuNoVs based on the high degree of viral evolution in these hosts (Bull et al., 2011; Carlsson et al., 2009; Hoffmann et al., 2012; Nilsson et al., 2003; Schorn et al., 2010; Siebenga et al., 2008). For example, viral diversity increased over time in one immunocompromised individual with over 50 co-circulating viral variants detected at each of three time points; in contrast, only limited diversity was observed in a healthy host (Bull et al., 2011). An increase in the number of chronically infected immunocompromised hosts that support a high degree of HuNoV evolution may result in more frequent emergence of phenotypically distinct virus strains. Novel viral variants may be antigenically distinct, may acquire enhanced virulence or transmission properties, or may display distinct HBGA binding profiles compared to previously circulating strains. Chronically infected immunodeficient hosts also represent potential pools of carriers that could reseed the world with previously circulating strains or recombinant HuNoVs if infected with multiple strains. There is precedent for this idea—immunodeficient patients can be long-term carriers of vaccine-derived poliovirus and are considered a major barrier to poliovirus eradication (Minor, 2009).

28.7 EFFECT OF GLOBALIZATION OF THE FOOD MARKET ON NOROVIRUS SPREAD

Growing international trade in conjunction with increasing consumer demand for exotic and non-regional foods has led to globalization of our food market. Unfortunately, this has also led to the worldwide dissemination of foodborne disease. In fact, disease outbreaks in the United States resulting from the consumption of imported foods are on the rise (CDC, 2012b). For HuNoVs, characteristics such as frequent transmission through foods, environmental stability, and a low infectious dose make their global spread increasingly common. The two foods most often associated with HuNoV outbreaks are shellfish and produce; as much as 85% and 60%, respectively, of these foods consumed in the United States are imported (Surveillance for Norovirus Outbreaks, 2012).

Shellfish have been responsible for numerous outbreaks of HuNoV disease across the globe. Notable outbreaks over the last decade related to imported shellfish include several concurrent outbreaks across Australia (Webby et al., 2007) and Singapore (Ng et al., 2005), one in England (Gallimore et al., 2005), and another cropping of concurrent outbreaks in the United States (David et al., 2007) and New Zealand (Simmons et al., 2007) and additionally in France and Italy (Guyader et al., 2006). Most recently, in 2011 internationally shipped oysters were responsible for outbreaks of HuNoV disease in numerous countries, including the United States (MMWR, 2012; Taiwan Department of Health, 2012). In addition to shellfish, imported produce has also been responsible for numerous HuNoV outbreaks. Frozen raspberries were identified as the causative agent in outbreaks in Sweden in 2004 and 2006 (Hjertqvist et al., 2006; Le Guyader et al., 2004), multiple outbreaks in Denmark and France in 2005 (Cotterelle et al., 2005; Falkenhorst et al., 2005), and multiple large outbreaks in Finland in 2009 (Sarvikivi et al., 2012). Furthermore, imported lettuce from France was associated with a large, widespread HuNoV outbreak in Denmark and Norway (Ethelberg et al., 2010).

Numerous factors have contributed to the global spread of HuNoVs via food products. Some of these factors include varying sanitation standards among countries and inadequate monitoring of growing areas and water supplies. In many countries, the use of partially treated sewage as a source of irrigation water is prohibited; however, in other locales this type of effluent is commonly used to irrigate farms that produce fruits and vegetables (Amahmid et al., 1999; Friedel et al., 2000). This practice has directly led to outbreaks of enteric pathogens from produce (Steele and Odumeru, 2004). In the case of shellfish, exposure of harvest waters to raw or minimally treated sewage is a primary contributor to HuNoV contamination of these foods. Because of this, most industrialized nations monitor their coastal waters for the presence of fecal contamination. Inadequate monitoring has resulted in cases of HuNoV disease (Kingsley et al., 2002). In fact, the most recent global outbreak of HuNoV from oysters resulted from contamination so extensive that even canned and cooked products were recalled due to concern that processing treatments would not be sufficient in lowering such high concentrations of virus (http://www.fda.gov/Safety/Recalls/ucm306377.htm). However, this preventative measure alone is not adequate to prevent disease since HuNoV outbreaks have occurred when both the water and food product met all bacteriological standards (Gallimore et al., 2005). Furthermore, survey studies revealed the presence of HuNoVs in oysters harvested from countries all around the world and highlighted that even oysters harvested by routinely monitored water supplies may still be contaminated (Boxman et al., 2006; Cheng et al., 2005). Another compounding factor associated with HuNoV disease and the global distribution of foods is the common (and often required) practice of refrigeration and/or freezing. Cold temperatures are known to enhance the stability of NoVs, so cold transport and storage may facilitate the international spread of this particular pathogen (Butot et al., 2008).

In response to the rise of foodborne infections and the issues related to the global spread of foodborne disease, many nations and organizations have developed surveillance systems to monitor, investigate, and aid in early outbreak identification. The European Union operates Enter-net, which focuses on international surveillance of human gastrointestinal illness. In the United States, the CDC operates a surveillance system called FoodNet to track all foodborne disease outbreaks. In 2009, the CDC also unveiled CaliciNet, a database of NoV genomes aimed at linking outbreaks caused by a common source. Similarly, Australia's OzFoodNet uses foodborne disease surveillance to aid in identification, investigation, and prevention of foodborne disease in their country. The WHO, in conjunction with the Food and Agricultural Organization (FAO) of the United Nations, has also established the International Food Safety Authorities Network. This surveillance network is designed to aid in the exchange of information on food safety issues of global interest and promote collaboration between nations. This organization also specifically monitors for potential food safety events that would have an international impact and aims to identify nations that appear to repeatedly export products associated with disease in order to prompt investigation of a persistent source of contamination or systemic problem.

In addition to international monitoring for disease, an internationally recognized set of food safety standards has been developed to normalize measures of food quality among countries. The International Standards for food quality (ISO) has developed microbial standards, among other things, for the safety and quality of foods at all levels of the food chain. Currently, there are 160 member nations that adopt ISO standards on some level. Large food-producing corporations have also developed a collaborative partnership called the Global Food Safety Initiative, which seeks to collectively increase the safety and quality of food. Integrated systems of disease tracking and food quality standards like these will undoubtedly aid in public health efforts to control the spread of HuNoV infections.

28.8 EFFECT OF CLIMATE CHANGE

While the effect of climate change on HuNoV epidemiology is currently unclear, HuNoVs do display seasonality. Evidence for seasonal transmission of HuNoV infections can be traced back over 80 years to 1929 when Zahorsky (1929) coined the term "winter vomiting disease" to describe a highly infectious disease with the predominant clinical symptom of vomiting, sometimes associated with abdominal cramps and diarrhea. However, it was not until 1972 that Kapikian et al. (1972) finally provided definitive proof of a viral etiology for one such outbreak occurring in Norwalk, Ohio, using immune electron microscopy of filtered stool samples. Since then, HuNoVs have been recognized as the leading cause of acute gastroenteritis in all age groups. It is now well documented that HuNoVs display a seasonal pattern of infection at least in the northern hemisphere, with peaks in the number of outbreaks in February–March (e.g., Greer et al., 2009; Oldak et al., 2012; Phillips et al., 2010; Verhoef et al., 2008). Possible explanations include temperature and humidity effects on virus stability. Indeed, NoVs are more stable at colder temperatures in mineral and tap waters (Ngazoa et al., 2008), on food preparation surfaces (Cannon et al., 2006; Fallahi and Mattison, 2011), and in food products (Horm and D'Souza, 2011). Furthermore, both low temperatures and dry conditions enhance NoV stability in soil (Fallahi and Mattison, 2011; Meschke and Sobsey, 2003). Underscoring the extreme stability of NoVs in water, a recent study by Seitz et al. (2011) reported that a HuNoV remains detectable in groundwater for 3 years and maintains its infectivity for at least 2 months at room temperature. Considering possible effects of climate change on HuNoV epidemiology, clearly changes in global temperature patterns could affect the seasonality of outbreaks. Moreover, shifts in rain and wind patterns may have significant implications. For example, Bruggink et al. reported a correlation between the monthly incidence of HuNoV-associated gastroenteritis outbreaks and average monthly rainfall in Australia from 2002 to 2007 (Bruggink and Marshall, 2010).

REFERENCES

Ahn, J.Y., Chung, J.-W., Chang, K.-J., et al., 2011. Clinical characteristics and etiology of travelers' diarrhea among Korean travelers visiting South-East Asia. *Journal of Korean Medical Science* **26**, 196.

Al-Ali, R.M., Chehadeh, W., Hamze, M., et al., 2011. First description of gastroenteritis viruses in Lebanese children: A pilot study. *Journal of Infection and Public Health* **4**, 59–64.

Alfano-Sobsey, E., Sweat, D., Hall, A., et al., 2012. Norovirus outbreak associated with undercooked oysters and secondary household transmission. *Epidemiology and Infection* **140**, 276–282.

Amahmid, O., Asmama, S., Bouhoum, K., 1999. The effect of waste water reuse in irrigation on the contamination level of food crops by Giardia cysts and Ascaris eggs. *International Journal of Food Microbiology* **49**, 19–26.

Anderson, A.D., Garrett, V.D., Sobel, J., et al., 2001. Multistate outbreak of Norwalk-like virus gastroenteritis associated with a common caterer. *American Journal of Epidemiology* **154**, 1013–1019.

Atmar, R.L., Opekun, A.R., Gilger, M.A., et al., 2008. Norwalk virus shedding after experimental human infection. *Emerging Infectious Diseases* **14**, 1553–1557.

Atmar, R.L., Bernstein, D.I., Harro, C.D., et al., 2011. Norovirus vaccine against experimental human Norwalk virus illness. *The New England Journal of Medicine* **365**, 2178–2187.

Bagci, S., Eis-Hübinger, A.M., Yassin, A.F., et al., 2010. Clinical characteristics of viral intestinal infection in preterm and term neonates. *European Journal of Clinical Microbiology and Infectious Diseases* **29**, 1079–1084.

Baron, R.C., Murphy, F.D., Greenberg, H.B., et al., 1982. Norwalk gastrointestinal illness: An outbreak associated with swimming in a recreational lake and secondary person-to-person transmission. *American Journal of Epidemiology* **115**, 163–172.

Barreira, D.M.P.G., Ferreira, M.S.R., Fumian, T.M., et al., 2010. Viral load and genotypes of noroviruses in symptomatic and asymptomatic children in Southeastern Brazil. *Journal of Clinical Virology* **47**, 60–64.

Bartolini, L., Mardari, R., Toldo, I., et al., 2011. Norovirus gastroenteritis and seizures: An atypical case with neuroradiological abnormalities. *Neuropediatrics* **42**, 167–169.

Becker, K.M., Moe, C.L., Southwick, K.L., et al., 2000. Transmission of Norwalk virus during football game. *New England Journal of Medicine* **343**, 1223–1227.

Bok, K., Abente, E.J., Realpe-Quintero, M., et al., 2009. Evolutionary dynamics of GII.4 noroviruses over a 34-year period. *Journal of Virology* **83**, 11890–11901.

Boon, D., Mahar, J.E., Abente, E.J., et al., 2011. Comparative evolution of GII.3 and GII.4 norovirus over a 31-year period. *Journal of Virology* **85**, 8656–8666.

Boxman, I.L.A., Tilburg, J.J.H.C., te Loeke, N.A.J.M., et al., 2006. Detection of noroviruses in shellfish in the Netherlands. *International Journal of Food Microbiology* **108**, 391–396.

Bruggink, L.D., Marshall, J.A., 2010. The incidence of norovirus-associated gastroenteritis outbreaks in Victoria, Australia (2002–2007) and their relationship with rainfall. *International Journal of Environmental Research and Public Health* **7**, 2822–2827.

Bucardo, F., Nordgren, J., Carlsson, B., et al., 2008. Pediatric norovirus diarrhea in nicaragua. *Journal of Clinical Microbiology* **46**, 2573–2580.

Bull, R.A., White, P.A., 2011. Mechanisms of GII.4 norovirus evolution. *Trends in Microbiology* **19**, 233–240.

Bull, R.A., Hansman, G.S., Clancy, L.E., et al., 2005. Norovirus recombination in ORF1/ORF2 overlap. *Emerging Infectious Diseases* **11**, 1079–1085.

Bull, R.A., Eden, J.-S., Rawlinson, W.D., et al., 2010. Rapid evolution of pandemic noroviruses of the GII.4 lineage. *PLoS Pathogens* 6, e1000831.

Bull, R.A., Eden, J.-S., Luciani, F., et al., 2011. Contribution of intra- and inter-host dynamics to norovirus evolution. *Journal of Virology* **86**, 3219–3229.

Butot, S., Putallaz, T., Sánchez, G., 2008. Effects of sanitation, freezing and frozen storage on enteric viruses in berries and herbs. *International Journal of Food Microbiology* **126**, 30–35.

Cáceres, V.M., Kim, D.K., Bresee, J.S., et al., 1998. A viral gastroenteritis outbreak associated with person-to-person spread among hospital staff. *Infection Control and Hospital Epidemiology* **19**, 162–167.

Cannon, J.L., Papafragkou, E., Park, G.W., et al., 2006. Surrogates for the study of norovirus stability and inactivation in the environment: A comparison of murine norovirus and feline calicivirus. *Journal of Food Protection* **69**, 2761–2765.

Carlsson, B., Lindberg, A.M., Rodriguez-Díaz, J., et al., 2009. Quasispecies dynamics and molecular evolution of human norovirus capsid P region during chronic infection. *Journal of General Virology* **90**, 432–441.

CDC—2011 Estimates of Foodborne Illness, 2011. URL http://www.cdc.gov/foodborneburden/2011-foodborne-estimates.html. Accessed on July 08, 2013.

CDC—Norovirus—Trends and Outbreaks, 2012a. URL http://www.cdc.gov/norovirus/trends-outbreaks.html. Accessed on June 06, 2013.

CDC research shows outbreaks linked to imported foods increasing, 2012b. CDC Online Newsroom. URL http://www.cdc.gov/media/releases/2012/p0314_foodborne.html. Accessed on June 06, 2013.

Chadwick, P.R., McCann, R., 1994. Transmission of a small round structured virus by vomiting during a hospital outbreak of gastroenteritis. *Journal of Hospital Infection* **26**, 251–259.

Chan, M.C.W., Sung, J.J.Y., Lam, R.K.Y., et al., 2006. Fecal viral load and norovirus-associated gastroenteritis. *Emerging Infectious Diseases* **12**, 1278–1280.

Chan, C.V., Chan, C.D., Ma, C., et al., 2011. Norovirus as cause of benign convulsion associated with gastro-enteritis. *Journal of Paediatrics and Child Health* **47**, 373–377.

Chapin, A.R., Carpenter, C.M., Dudley, W.C., et al., 2005. Prevalence of norovirus among visitors from the United States to Mexico and Guatemala who experience traveler's diarrhea. *Journal of Clinical Microbiology* **43**, 1112–1117.

Chen, S.-Y., Tsai, C.-N., Lai, M.-W., et al., 2009. Norovirus infection as a cause of diarrhea-associated benign infantile seizures. *Clinical Infectious Diseases* **48**, 849–855.

Cheng, P.K.C., Wong, D.K.K., Chung, T.W.H., et al., 2005. Norovirus contamination found in oysters worldwide. *Journal of Medical Virology* **76**, 593–597.

Cotterelle, B., Drougard, C., Rolland, J., et al., 2005. Outbreak of norovirus infection associated with the consumption of frozen raspberries, France, March 2005. *Euro Surveillance* **10**, E050428.1.

D'Souza, D.H., Su, X., 2010. Efficacy of chemical treatments against murine norovirus, feline calicivirus, and MS2 bacteriophage. *Foodborne Pathogens and Disease* **7**, 319–326.

Daniels, N.A., Bergmire-Sweat, D.A., Schwab, K.J., et al., 2000. A foodborne outbreak of gastroenteritis associated with Norwalk-like viruses: First molecular traceback to deli sandwiches contaminated during preparation. *Journal of Infectious Diseases* **181**, 1467–1470.

David, S.T., McIntyre, L., MacDougall, L., et al., 2007. An outbreak of norovirus caused by consumption of oysters from geographically dispersed harvest sites, British Columbia, Canada, 2004. *Foodborne Pathogens and Disease* **4**, 349–358.

Desai, R., Hembree, C.D., Handel, A., et al., 2012. Severe outcomes are associated with genogroup 2 genotype 4 norovirus outbreaks: A systematic literature review. *Clinical Infectious Diseases* **55**, 189–193.

Dey, S.K., Phathammavong, O., Okitsu, S., et al., 2010. Seasonal pattern and genotype distribution of norovirus infection in Japan. *The Pediatric Infectious Disease Journal* **29**, e32–e34.

Donaldson, E.F., Lindesmith, L.C., LoBue, A.D., et al., 2010. Viral shape-shifting: Norovirus evasion of the human immune system. *Nature Reviews Microbiology* **8**, 231–241.

Doultree, J.C., Druce, J.D., Birch, C.J., et al., 1999. Inactivation of feline calicivirus, a Norwalk virus surrogate. *Journal of Hospital Infection* **41**, 51–57.

Dove, W., Cunliffe, N.A., Gondwe, J.S., et al., 2005. Detection and characterization of human caliciviruses in hospitalized children with acute gastroenteritis in Blantyre, Malawi. *Journal of Medical Virology* **77**, 522–527.

Duizer, E., Schwab, K.J., Neill, F.H., et al., 2004. Laboratory efforts to cultivate noroviruses. *Journal of General Virology* **85**, 79–87.

Estes, M.K., Prasad, B.V., Atmar, R.L., 2006. Noroviruses everywhere: Has something changed? *Current Opinion in Infectious Diseases* **19**, 467–474.

Ethelberg, S., Lisby, M., Bottiger, B., et al., 2010. Outbreaks of gastroenteritis linked to lettuce, Denmark, January 2010. *Euro Surveillance* **15**.

Falkenhorst, G., Krusell, L., Lisby, M., et al., 2005. Imported frozen raspberries cause a series of norovirus outbreaks in Denmark, 2005. *Euro Surveillance* **10**, E050922.2.

Fallahi, S., Mattison, K., 2011. Evaluation of murine norovirus persistence in environments relevant to food production and processing. *Journal of Food Protection®* 74, 1847–1851.

Fankhauser, R.L., Noel, J.S., Monroe, S.S., et al., 1998. Molecular epidemiology of "Norwalk-like viruses" in outbreaks of gastroenteritis in the United States. *Journal of Infectious Diseases* **178**, 1571–1578.

Feng, X., Jiang, X., 2007. Library screen for inhibitors targeting norovirus binding to histo-blood group antigen receptors. *Antimicrobial Agents and Chemotherapy* **51**, 324–331.

Frange, P., Touzot, F., Debré, M., et al., 2012. Prevalence and clinical impact of norovirus fecal shedding in children with inherited immune deficiencies. *Journal of Infectious Diseases* **206**, 1269–1274.

Frenck, R., Bernstein, D.I., Xia, M., et al., 2012. Predicting susceptibility to norovirus GII.4 using a human challenge model. *Journal of Infectious Diseases* **206**, 1386–1393.

Friedel, J.K., Langer, T., Siebe, C., et al., 2000. Effects of long-term waste water irrigation on soil organic matter, soil microbial biomass and its activities in central Mexico. *Biology and Fertility of Soils* **31**, 414–421.

Friedman, D.S., Heisey-Grove, D., Argyros, F., et al., 2005. An outbreak of norovirus gastroenteritis associated with wedding cakes. *Epidemiology and Infection* **133**, 1057–1063.

Friesema, I., Vennema, H., Heijne, J., et al., 2009. Differences in clinical presentation between norovirus genotypes in nursing homes. *Journal of Clinical Virology* **46**, 341–344.

Gallimore, C.I., Cheesbrough, J.S., Lamden, K., et al., 2005. Multiple norovirus genotypes characterised from an oyster-associated outbreak of gastroenteritis. *International Journal of Food Microbiology* **103**, 323–330.

Gallimore, C.I., Lewis, D., Taylor, C., et al., 2004. Chronic excretion of a norovirus in a child with cartilage hair hypoplasia (CHH). *Journal of Clinical Virology* **30**, 196–204.

Gallimore, C.I., Taylor, C., Gennery, A.R., et al., 2006. Environmental monitoring for gastroenteric viruses in a pediatric primary immunodeficiency unit. *Journal of Clinical Microbiology* **44**, 395–399.

Glass, R.I., Parashar, U.D., Estes, M.K., 2009. Norovirus gastroenteritis. *New England Journal of Medicine* **361**, 1776–1785.

Greer, A.L., Drews, S.J., Fisman, D.N., 2009. Why "winter" vomiting disease? Seasonality, hydrology, and norovirus epidemiology in Toronto, Canada. *Ecohealth* **6**, 192–199.

Grotto, I., Huerta, M., Balicer, R.D., et al., 2004. An outbreak of norovirus gastroenteritis on an Israeli military base. *Infection* **32**, 339–343.

Guix, S., Asanaka, M., Katayama, K., et al., 2007. Norwalk virus RNA is infectious in mammalian cells. *Journal of Virology* **81**, 12238–12248.

Guyader, F.S.L., Bon, F., DeMedici, D., et al., 2006. Detection of multiple noroviruses associated with an international gastroenteritis outbreak linked to oyster consumption. *Journal of Clinical Microbiology* **44**, 3878–3882.

Hansen, S., Stamm-Balderjahn, S., Zuschneid, I., et al., 2007. Closure of medical departments during nosocomial outbreaks: Data from a systematic analysis of the literature. *Journal of Hospital Infection* **65**, 348–353.

Hansman, G.S., Shahzad-ul-Hussan, S., McLellan, J.S., et al., 2012. Structural basis for norovirus inhibition and fucose mimicry by citrate. *Journal of Virology* **86**, 284–292.

Herwaldt, B.L., Lew, J.F., Moe, C.L., et al., 1994. Characterization of a variant strain of Norwalk virus from a food-borne outbreak of gastroenteritis on a cruise ship in Hawaii. *Journal of Clinical Microbiology* **32**, 861–866.

Hjertqvist, M., Johansson, A., Svensson, N., et al., 2006. Four outbreaks of norovirus gastroenteritis after consuming raspberries, Sweden, June–August 2006. *Eurosurveillance*. URL http://www.eurosurveillance.org/ViewArticle.aspx?ArticleId=3038. Accessed on June 06, 2013.

Hoffmann, D., Hutzenthaler, M., Seebach, J., et al., 2012. Norovirus GII.4 and GII.7 capsid sequences undergo positive selection in chronically infected patients. *Infection, Genetics and Evolution* **12**, 461–466.

Horm, K.M., D'Souza, D.H., 2011. Survival of human norovirus surrogates in milk, orange, and pomegranate juice, and juice blends at refrigeration (4 °C). *Food Microbiology* **28**, 1054–1061.

Huhti, L., Szakal, E.D., Puustinen, L., et al., 2011. Norovirus GII-4 causes a more severe gastroenteritis than other noroviruses in young children. *Journal of Infectious Diseases* **203**, 1442–1444.

Hurley, K.E., Pesavento, P.A., Pedersen, N.C., et al., 2004. An outbreak of virulent systemic feline calicivirus disease. *Journal of the American Veterinary Medical Association* **224**, 241–249.

Hutson, A.M., Atmar, R.L., Graham, D.Y., et al., 2002. Norwalk virus infection and disease is associated with ABO histo-blood group type. *Journal of Infectious Diseases* **185**, 1335–1337.

Hutson, A.M., Airaud, F., LePendu, J., et al., 2005. Norwalk virus infection associates with secretor status genotyped from sera. *Journal of Medical Virology* **77**, 116–120.

Ito, S., Takeshita, S., Nezu, A., et al., 2006. Norovirus-associated encephalopathy. *Pediatric Infectious Disease Journal* **25**, 651–652.

Jakab, F., Németh, V., Oldal, M., et al., 2010. Epidemiological and clinical characterization of norovirus infections among hospitalized children in Baranya County, Hungary. *Journal of Clinical Virology* **49**, 75–76.

Johnson, P.C., Mathewson, J.J., DuPont, H.L., et al., 1990. Multiple-challenge study of host susceptibility to Norwalk gastroenteritis in US adults. *Journal of Infectious Diseases* **161**, 18–21.

Jones, M.K., Karst, S.M., 2012. Noroviruses, in: *Foodborne Infections and Intoxications*. Elsevier, Amsterdam.

Kapikian, A.Z., Wyatt, R.G., Dolin, R., et al., 1972. Visualization by immune electron microscopy of a 27-nm particle associated with acute infectious nonbacterial gastroenteritis. *Journal of Virology* **10**, 1075–1081.

Kaplan, J.E., Schonberger, L.B., Varano, G., et al., 1982. An outbreak of acute nonbacterial gastroenteritis in a nursing home. Demonstration of person-to-person transmission by temporal clustering of cases. *American Journal of Epidemiology* **116**, 940–948.

Kaplan, N.M., Kirby, A., Abd-Eldayem, S.A., et al., 2011. Detection and molecular characterisation of rotavirus and norovirus infections in Jordanian children with acute gastroenteritis. *Archives of Virology* **156**, 1477–1480.

Kappus, K.D., Marks, J.S., Holman, R.C., et al., 1982. An outbreak of Norwalk gastroenteritis associated with swimming in a pool and secondary person-to-person transmission. *American Journal of Epidemiology* **116**, 834–839.

Karst, S.M., 2010. Pathogenesis of noroviruses, emerging RNA viruses. *Viruses* **2**, 748–781.

Kaufman, S.S., Chatterjee, N.K., Fuschino, M.E., et al., 2003. Calicivirus enteritis in an intestinal transplant recipient. *American Journal of Transplantation* **3**, 764–768.

Kaufman, S.S., Chatterjee, N.K., Fuschino, M.E., et al., 2005. Characteristics of human calicivirus enteritis in intestinal transplant recipients. *Journal of Pediatric Gastroenterology and Nutrition* **40**, 328–333.

Kawsano, G., Oshige, K., Syutou, S., et al., 2007. Benign infantile convulsions associated with mild gastroenteritis: A retrospective study of 39 cases including virological tests and efficacy of anticonvulsants. *Brain and Development* **29**, 617–622.

Kemper, A.R., Davis, M.M., Freed, G.L., 2002. Expected adverse events in a mass smallpox Vaccination Campaign. Effective Clinical Practice. URL http://www.acponline.org/clinical_information/journals_publications/ecp/marapr02/kemper.htm. Accessed on June 06, 2013.

Khan, R.R., Lawson, A.D., Minnich, L.L., et al., 2009. Gastrointestinal norovirus infection associated with exacerbation of inflammatory bowel disease. *Journal of Pediatric Gastroenterology and Nutrition* **48**, 328–333.

Kingsley, D.H., Meade, G.K., Richards, G.P., 2002. Detection of both hepatitis A virus and Norwalk-like virus in imported clams associated with food-borne illness. *Applied and Environmental Microbiology* **68**, 3914–3918.

Kirby, A., Al-Eryani, A., Al-Sonboli, N., et al., 2011. Rotavirus and norovirus infections in children in Sana'a, Yemen. *Tropical Medicine and International Health* **16**, 680–684.

Kirkwood, C.D., Streitberg, R., 2008. Calicivirus shedding in children after recovery from diarrhoeal disease. *Journal of Clinical Virology* **43**, 346–348.

Ko, G., Garcia, C., Jiang, Z.-D., et al., 2005. Noroviruses as a cause of traveler's diarrhea among students from the United States visiting Mexico. *Journal of Clinical Microbiology* **43**, 6126–6129.

Koo, H.L., Ajami, N., Atmar, R.L., et al., 2010. Noroviruses: The principal cause of foodborne disease worldwide. *Discovery Medicine* **10**, 61–70.

Kroneman, A., Verhoef, L., Harris, J., et al., 2008. Analysis of integrated virological and epidemiological reports of norovirus outbreaks collected within the foodborne viruses in Europe network from 1 July 2001 to 30 June 2006. *Journal of Clinical Microbiology* **46**, 2959–2965.

Lay, M.K., Atmar, R.L., Guix, S., et al., 2010. Norwalk virus does not replicate in human macrophages or dendritic cells derived from the peripheral blood of susceptible humans. *Virology* **406**, 1–11.

Le Guyader, F.S., Mittelholzer, C., Haugarreau, L., et al., 2004. Detection of noroviruses in raspberries associated with a gastroenteritis outbreak. *International Journal of Food Microbiology* **97**, 179–186.

Lindesmith, L., Moe, C., Marionneau, S., et al., 2003. Human susceptibility and resistance to Norwalk virus infection. *Nature Medicine* **9**, 548–553.

Lindesmith, L.C., Donaldson, E.F., LoBue, A.D., et al., 2008. Mechanisms of GII.4 norovirus persistence in human populations. *PLoS Medicine* 5, e31.

Lindesmith, L.C., Beltramello, M., Donaldson, E.F., et al., 2012. Immunogenetic mechanisms driving norovirus GII.4 antigenic variation. *PLoS Pathogens* 8, e1002705.

Lo, S.V., Connolly, A.M., Palmer, S.R., et al., 1994. The role of the pre-symptomatic food handler in a common source outbreak of food-borne SRSV gastroenteritis in a group of hospitals. *Epidemiology and Infection* **113**, 513–521.

Lopman, B., 2011. Air sickness: Vomiting and environmental transmission of norovirus on aircraft. *Clinical Infectious Diseases* **53**, 521–522.

Lopman, B.A., Adak, G.K., Reacher, M.H., et al., 2003. Two epidemiologic patterns of norovirus outbreaks: Surveillance in England and Wales, 1992–2000. *Emerging Infectious Diseases* **9**, 71–77.

Lopman, B.A., Reacher, M.H., Vipond, I.B., et al., 2004. Clinical manifestation of norovirus gastroenteritis in health care settings. *Clinical Infectious Diseases* **39**, 318–324.

Marks, P.J., Vipond, I.B., Carlisle, D., et al., 2000. Evidence for airborne transmission of Norwalk-like virus (NLV) in a hotel restaurant. *Epidemiology and Infection* **124**, 481–487.

Marshall, J.K., Thabane, M., Borgaonkar, M.R., et al., 2007. Postinfectious irritable bowel syndrome after a food-borne outbreak of acute gastroenteritis attributed to a viral pathogen. *Clinical Gastroenterology and Hepatology* **5**, 457–460.

Meschke, J.S., Sobsey, M.D., 2003. Comparative reduction of Norwalk virus, poliovirus type 1, F+ RNA coliphage MS2 and *Escherichia coli* in miniature soil columns. *Water Science and Technology* **47**, 85–90.

Minor, P., 2009. Vaccine-derived poliovirus (VDPV): Impact on poliomyelitis eradication. *Vaccine* **27**, 2649–2652.

Morotti, R.A., Kaufman, S.S., Fishbein, T.M., et al., 2004. Calicivirus infection in pediatric small intestine transplant recipients: Pathological considerations. *Human Pathology* **35**, 1236–1240.

Moyo, S., Gro, N., Matee, M., et al., 2011. Age specific aetiological agents of diarrhoea in hospitalized children aged less than five years in Dar es Salaam, Tanzania. *BMC Pediatrics* **11**, 19.

Murata, T., Katsushima, N., Mizuta, K., et al., 2007. Prolonged norovirus shedding in infants under 6 months of age with gastroenteritis. *Pediatric Infectious Disease Journal* **26**, 46–49.

Nataraju, S.M., Ganesh, B., Das, S., et al., 2010. Emergence of noroviruses homologous to strains reported from Djibouti (Horn of Africa), Brazil, Italy, Japan and USA among children in Kolkata, India. *European Review for Medical and Pharmacological Sciences* **14**, 789–794.

Ng, T.L., Chan, P.P., Phua, T.H., et al., 2005. Oyster-associated outbreaks of norovirus gastroenteritis in Singapore. *Journal of Infection* **51**, 413–418.

Ngazoa, E.S., Fliss, I., Jean, J., 2008. Quantitative study of persistence of human norovirus genome in water using TaqMan real-time RT-PCR. *Journal of Applied Microbiology* **104**, 707–715.

Nilsson, M., Hedlund, K.-O., Thorhagen, M., et al., 2003. Evolution of human calicivirus RNA in vivo: Accumulation of mutations in the protruding P2 domain of the capsid leads to structural changes and possibly a new phenotype. *Journal of Virology* **77**, 13117–13124.

Norovirus activity—United States, 2006–2007, 2007. *Morbidity and Mortality Weekly Report (MMWR)* **56**, 842–846.

Notes from the field: Norovirus infections associated with frozen raw oysters – Washington, 2011, 2012. *Morbidity and Mortality Weekly Report (MMWR)* **61**, 110.

O'Neill, P.D., Marks, P.J., 2005. Bayesian model choice and infection route modelling in an outbreak of norovirus. *Statistics in Medicine* **24**, 2011–2024.

Obinata, K., Okumura, A., Nakazawa, T., et al., 2010. Norovirus encephalopathy in a previously healthy child. *Pediatric Infectious Disease Journal* **29**, 1057–1059.

Oldak, E., Sulik, A., Rozkiewicz, D., et al., 2012. Norovirus infections in children under 5 years of age hospitalized due to the acute viral gastroenteritis in northeastern Poland. *European Journal of Clinical Microbiology and Infectious Diseases* **31**, 417–422.

O'Ryan, M.L., Peña, A., Vergara, R., et al., 2010. Prospective characterization of norovirus compared with rotavirus acute diarrhea episodes in Chilean children. *The Pediatric Infectious Disease Journal* **29**, 855–859.

Outbreak of acute gastroenteritis associated with Norwalk-like viruses among British military personnel—Afghanistan, May 2002, 2002. *Morbidity and Mortality Weekly Report (MMWR)* **51**, 477–479.

Ozkul, A.A., Kocazeybek, B.S., Turan, N., et al., 2011. Frequency and phylogeny of norovirus in diarrheic children in Istanbul, Turkey. *Journal of Clinical Virology* **51**, 160–164.

Parashar, U.D., Dow, L., Fankhauser, R.L., et al., 1998. An outbreak of viral gastroenteritis associated with consumption of sandwiches: Implications for the control of transmission by food handlers. *Epidemiology and Infection* **121**, 615–621.

Parra, F., Prieto, M., 1990. Purification and characterization of a calicivirus as the causative agent of a lethal hemorrhagic disease in rabbits. *Journal of Virology* **64**, 4013–4015.

Parra, G.I., Abente, E.J., Sandoval-Jaime, C., et al., 2012. Multiple antigenic sites are involved in blocking the interaction of GII.4 norovirus capsid with ABH histo-blood group antigens. *Journal of Virology* **86**, 7414–7426.

Parrino, T.A., Schreiber, D.S., Trier, J.S., et al., 1977. Clinical immunity in acute gastroenteritis caused by Norwalk agent. *New England Journal of Medicine* **297**, 86–89.

Patel, M.M., 2008. Systematic literature review of role of noroviruses in sporadic gastroenteritis. *Emerging Infectious Diseases* **14**, 1224–1231.

Patterson, T., Hutchings, P., Palmer, S., 1993. Outbreak of SRSV gastroenteritis at an international conference traced to food handled by a post-symptomatic caterer. *Epidemiology and Infection* **111**, 157–162.

Patterson, W., Haswell, P., Fryers, P.T., et al., 1997. Outbreak of small round structured virus gastroenteritis arose after kitchen assistant vomited. *Communicable Disease Report – CDR Review* **7**, R101–R103.

Pedersen, N.C., Elliott, J.B., Glasgow, A., et al., 2000. An isolated epizootic of hemorrhagic-like fever in cats caused by a novel and highly virulent strain of feline calicivirus. *Veterinary Microbiology* **73**, 281–300.

Phan, T.G., Nguyen, T.A., Nishimura, S., et al., 2005. Etiologic agents of acute gastroenteritis among Japanese infants and children: Virus diversity and genetic analysis of sapovirus. *Archives of Virology* **150**, 1415–1424.

Phillips, G., Tam, C.C., Rodrigues, L.C., et al., 2010. Prevalence and characteristics of asymptomatic norovirus infection in the community in England. *Epidemiology and Infection* **138**, 1454–1458.

Rademacher, C., Guiard, J., Kitov, P.I., et al., 2011. Targeting norovirus infection—Multivalent entry inhibitor design based on NMR experiments. *Chemistry* **17**, 7442–7453.

Rahouma, A., Klena, J.D., Krema, Z., et al., 2011. Enteric pathogens associated with childhood diarrhea in tripoli-libya. *American Journal of Tropical Medicine and Hygiene* **84**, 886–891.

Rockx, B., de Wit, M., Vennema, H., et al., 2002. Natural history of human calicivirus infection: A prospective cohort study. *Clinical Infectious Diseases* **35**, 246–253.

Roddie, C., Paul, J.P.V., Benjamin, R., et al., 2009. Allogeneic hematopoietic stem cell transplantation and norovirus gastroenteritis: A previously unrecognized cause of morbidity. *Clinical Infectious Diseases* **49**, 1061–1068.

Saif, M.A., Bonney, D.K., Bigger, B., et al., 2011. Chronic norovirus infection in pediatric hematopoietic stem cell transplant recipients: A cause of prolonged intestinal failure requiring intensive nutritional support. *Pediatric Transplantation* **15**, 505–509.

Sakai, Y., Nakata, S., Honma, S., et al., 2001. Clinical severity of Norwalk virus and Sapporo virus gastroenteritis in children in Hokkaido, Japan. *Pediatric Infectious Disease Journal* **20**, 849–853.

Sarvikivi, E., Roivainen, M., Maunula, L., et al., 2012. Multiple norovirus outbreaks linked to imported frozen raspberries. *Epidemiology and Infection* **140**, 260–267.

Scallan, E., Hoekstra, R.M., Angulo, F.J., et al., 2011. Foodborne illness acquired in the United States—Major pathogens. *Emerging Infectious Diseases* **17**, 7–15.

Schorn, R., Höhne, M., Meerbach, A., et al., 2010. Chronic norovirus infection after kidney transplantation: Molecular evidence for immune-driven viral evolution. *Clinical Infectious Diseases* **51**, 307–314.

Seitz, S.R., Leon, J.S., Schwab, K.J., et al., 2011. Norovirus infectivity in humans and persistence in water. *Applied and Environmental Microbiology* **77**, 6884–6888.

Siebenga, J.J., Vennema, H., Renckens, B., et al., 2007. Epochal evolution of GGII.4 norovirus capsid proteins from 1995 to 2006. *Journal of Virology* **81**, 9932–9941.

Siebenga, J.J., Beersma, M.F.C., Vennema, H., et al., 2008. High prevalence of prolonged norovirus shedding and illness among hospitalized patients: A model for in vivo molecular evolution. *Journal of Infectious Diseases* **198**, 994–1001.

Siebenga, J.J., Vennema, H., Zheng, D., et al., 2009. Norovirus illness is a global problem: Emergence and spread of norovirus GII.4 variants, 2001–2007. *Journal of Infectious Diseases* **200**, 802–812.

Simmons, G., Garbutt, C., Hewitt, J., et al., 2007. A New Zealand outbreak of norovirus gastroenteritis linked to the consumption of imported raw Korean oysters. *New Zealand Medical Journal* **120**, U2773.

Steele, M., Odumeru, J., 2004. Irrigation water as source of foodborne pathogens on fruit and vegetables. *Journal of Food Protection* **67**, 2839–2849.

Stuart, R.L., Tan, K., Mahar, J.E., et al., 2010. An outbreak of necrotizing enterocolitis associated with norovirus genotype GII.3. *Pediatric Infectious Disease Journal* **29**, 644–647.

Surveillance for Norovirus Outbreaks, 2012. CDC Features. URL http://www.cdc.gov/features/dsnorovirus/index.html. Accessed on June 06, 2013.

Taiwan Department of Health, 2012. Taiwan Health News. Department of Health, Taiwan, ROC. URL http://www.doh.gov.tw/EN2006/DM/DM2_p01.aspx?class_no=387&now_fod_list_no=9073&level_no=1&doc_no=85510. Accessed on June 06, 2013.

Tamura, T., Nishikawa, M., Anh, D.D., et al., 2010. Molecular epidemiological study of rotavirus and norovirus infections among children with acute gastroenteritis in Nha Trang, Vietnam, December 2005–June 2006. *Japanese Journal of Infectious Diseases* **63**, 405–411.

Tan, M., Jiang, X., 2005. Norovirus and its histo-blood group antigen receptors: An answer to a historical puzzle. *Trends in Microbiology* **13**, 285–293.

Tan, M., Jiang, X., 2011. Norovirus–host interaction: Multi-selections by human histo-blood group antigens. *Trends in Microbiology* **19**, 382–388.

Teunis, P.F.M., Moe, C.L., Liu, P., et al., 2008. Norwalk virus: How infectious is it? *Journal of Medical Virology* **80**, 1468–1476.

Thongprachum, A., Khamrin, P., Chaimongkol, N., et al., 2010. Evaluation of an immunochromatography method for rapid detection of noroviruses in clinical specimens in Thailand. *Journal of Medical Virology* **82**, 2106–2109.

Turcios-Ruiz, R.M., Axelrod, P., St. John, K., et al., 2008. Outbreak of necrotizing enterocolitis caused by norovirus in a neonatal intensive care unit. *Journal of Pediatrics* **153**, 339–344.

Tzialla, C., Civardi, E., Borghesi, A., et al., 2011. Emerging viral infections in neonatal intensive care unit. *Journal of Maternal-Fetal and Neonatal Medicine*. **24**, 156–158, Informa Healthcare.

Updated Norovirus Outbreak Management and Disease Prevention Guidelines, 2012. http://www.cdc.gov/mmwr/preview/mmwrhtml/rr6003a1.htm. Accessed on June 06, 2013.

Verhoef, L.P.B., 2009. Selection tool for foodborne norovirus outbreaks. *Emerging Infectious Diseases* **15**, 31–38.

Verhoef, L., Depoortere, E., Boxman, I., et al., 2008. Emergence of new norovirus variants on spring cruise ships and prediction of winter epidemics. *Emerging Infectious Diseases* **14**, 238–243.

Vivancos, R., Shroufi, A., Sillis, M., et al., 2009. Food-related norovirus outbreak among people attending two barbeques: Epidemiological, virological, and environmental investigation. *International Journal of Infectious Diseases* **13**, 629–635.

Webby, R.J., Carville, K.S., Kirk, M.D., et al., 2007. Internationally distributed frozen oyster meat causing multiple outbreaks of norovirus infection in Australia. *Clinical Infectious Diseases* **44**, 1026–1031.

Westhoff, T.H., Vergoulidou, M., Loddenkemper, C., et al., 2009. Chronic norovirus infection in renal transplant recipients. *Nephrology, Dialysis, Transplantation* **24**, 1051–1053.

Zahorsky, J., 1929. Hyperemesis hiemis or the winter vomiting disease. *Archives of Paediatrics* **46**, 391–395.

Zhang, S., Chen, T., Wang, J., et al., 2011. Symptomatic and asymptomatic infections of rotavirus, norovirus, and adenovirus among hospitalized children in Xi'an, China. *Journal of Medical Virology* **83**, 1476–1484.

Zheng, D.-P., Ando, T., Fankhauser, R.L., et al., 2006. Norovirus classification and proposed strain nomenclature. *Virology* **346**, 312–323.

Zintz, C., Bok, K., Parada, E., et al., 2005. Prevalence and genetic characterization of caliciviruses among children hospitalized for acute gastroenteritis in the United States. *Infection, Genetics and Evolution* **5**, 281–290.

Tenaillon, [illegible] 2011. Nonsynonymous substitution. Multi-[illegible] within the sampled group [illegible] *Biochem. Mol. Biol.* 19:[illegible]

Thomas, P. M., [illegible] 2008. [illegible] Mole infections [illegible] *Biochem. Wildl.* 20: 1069–1276.

Thomas, [illegible] Chapin, P., [illegible] N., et al. 2010. [illegible] to an unique [illegible] apply [illegible] for rapid detection of [illegible] by [illegible] mass spectrometry. The [illegible] *Journal* [illegible] 2196–2199.

Tafrate-Klein, R. M., Axford, [illegible] Plane, [illegible] et al. 2008. [illegible] of [illegible] patients [illegible] in a pediatric intensive care unit. *Journal of Pediatrics* 152: 689–694.

Tebelli, C., [illegible] A., et al. 2011. Interspinal [illegible] of [illegible] *Journal of [illegible] and [illegible]* 24: 1554–1556. [illegible]

UK Health Protection Agency. Management and [illegible] of [illegible] Guidelines. 2010. http://www. [illegible] (accessed on [illegible] 2013).

Voight, L. R. 2008. Selection [illegible] anthrax [illegible] *Emerging Infectious Diseases* 14: [illegible]

Vlasov, E., Lepojeva, [illegible] et al. 2008. Enterococci [illegible] [illegible] *Journal of Clinical Microbiology* 46: [illegible]

Wanda, [illegible] S., [illegible] M., et al. 2009. Food-related [illegible] [illegible] *Infectious Diseases* 18: 228–235.

Wells, [illegible] 2008. [illegible] outbreaks of [illegible] 24: [illegible]

Whitley, [illegible] M., [illegible] et al. 2007. [illegible] *Journal* 24: [illegible]

Zhao, [illegible] *Applied* [illegible] 45: [illegible]

Zhang, [illegible] 2011. [illegible] Molecular [illegible] [illegible]

Zhang, D., [illegible] et al. 2008. [illegible] *Nature* 496: [illegible]

Zhu, C., [illegible] 2012. [illegible] [illegible] *Journal of Clinical [illegible]* 50: [illegible]

29

EMERGENCE OF NOVEL VIRUSES (TOSCANA, USUTU) IN POPULATION AND CLIMATE CHANGE

Mari Paz Sánchez-Seco Fariñas and Ana Vazquez

Laboratory of Arbovirus and Imported Viral Diseases, Virology Department, National Center of Microbiology, Institute of Health "Carlos III", Ctra Pozuelo-Majadahonda, Madrid, Spain

TABLE OF CONTENTS

Viral Infections and Global Change, First Edition. Edited by Sunit K. Singh.

29.1 INTRODUCTION

Emerging viral infections have been reported with increased frequency in recent years, like Chikungunya virus in northern Italy, West Nile virus (WNV) in North America, bluetongue virus in northern Europe, and Usutu virus (USUV) and Schmallenberg virus in Europe. Besides new Toscana virus (TOSV) infections have been documented in endemic areas or in others where it had never been detected before. Emerging infectious diseases are diseases that have newly appeared in a population or that have been known for some time but with a rapid increase in their incidence or their geographical ranges (Brown, 2004). The reasons for disease emergence are multiple, but there are three main factors: global trends, climate change, and some biological viral characteristics (ability to change and adaptation).

Since the 1980s, the emergence of many previously unrecognized infectious diseases and the reemergence of known infectious diseases that were thought to be under control have been observed. This trend has continued until the present time, and many infectious pathogens, predominantly viruses, have been newly identified in recent years (Hoffmann et al., 2012; Kuiken et al., 2003). Most of the emerging viruses are zoonotic, which means they can infect both animals and humans (Jones et al., 2008) humans being dead-end hosts in most cases. Among emerging viruses, arboviruses play a major role.

Arboviruses are viruses that survive in nature by transmission from infected to susceptible hosts (vertebrates) by certain species of arthropods (mosquitoes, ticks, sandflies, midges, etc.). The appearance of arboviruses is strongly influenced by climatic changes because these affect the arthropods (some characteristics of their life cycle and distribution) and equally affect the arboviruses (dispersal patterns, evolution, and the efficiency to be transmitted from arthropods to vertebrate hosts) (Gould and Higgs, 2009). However, even when linkages between disease dynamics and climate change are relatively strong, there are other factors that may influence (Reiter, 2001). These other factors are extreme climatic events (natural disasters), or changes associated with global trends like changes in human behavior (leading to higher risk of human/virus exposure), or anthropogenic effects on the environment (urbanization, deforestation, changes in farming and animal husbandry, use of pesticides, increase of international trade and travel, use and development of irrigation, increase in the emissions of greenhouse gases, availability of potable water, risk of flooding, etc.), among others (Pherez, 2007). Global trends have resulted in a global reemergence of epidemic vector-borne diseases affecting both humans and animals over the past 30 years (Gubler, 2009).

The rising global travel and trade activities facilitate the introduction, into novel environments, of mosquito-borne viruses and competent mosquito vectors, some of which may become of public health or veterinary concern. Moreover, the effects of global warming, and the virus ability to evolve, will increase the efficiency of viral replication of these viruses and their permanent establishment in less competent vector species and in unexpected host species (Weissenböck et al., 2010).

This chapter describes the spread and emergence of two important arboviral diseases (TOSV and USUV) as well as the possible role of climatic change in these variations, bearing in mind some biological viral characteristics like the ability of change and adaptation.

29.2 TOSV

29.2.1 Virus properties and classification

TOSV belongs to the phlebotomus (or sandfly) antigenic group in the *Phlebovirus* genus (*Bunyaviridae* family). It is antigenically related to other members of the family showing various degrees of cross-reactivity with them in serological tests but mainly with the rest of the members of the sandfly Naples group (Naples, Karimabad and Tehran viruses).

Like other members of the family Bunyaviridae, the MW of virions is 300×10^6–400×10^6; they have a S_{20W} of 350–500S, are spherical or pleomorphic, 80–120 nm in diameter, and display surface glycoprotein projections of 5–10 nm, which are embedded in a lipid bilayered envelope. The diameter of viral ribonucleoproteins is 2–2.5 nm, and they display helical symmetry. The virions are sensitive to heat, lipid solvents, detergents, and formaldehyde.

The viral genome comprises three molecules of negative or ambisense ssRNA, named L (large, 6400 nt approximately), M (medium, 4200 nt approximately), and S (small, 1800 nt approximately), which total 11–12 kb. The terminal nt sequences of genome segments are conserved at their ends. The S RNA exhibits an ambisense coding strategy: it is transcribed by the virions RNA polymerase to a subgenomic virus-complementary sense mRNA that encodes the N protein and, from a full-length antigenome S RNA, to a subgenomic virus-sense mRNA that encodes for a nonstructural protein (NSm). Viral RNAs are not polyadenylated and are truncated relative to the genomic RNAs at the 3′; mRNAs have 5′-methylated caps and 10–18 non-templated nt at the 5′ end, which are derived from host-cell mRNAs. So, there are four structural proteins, two external glycoproteins (Gn and Gc, 50–72 and 55–75 aa, respectively), a nucleocapsid protein (N, 48–54 aa), and a large (L, 238–241 aa) transcriptase protein. Nonstructural protein is expressed by the S segment and by the M segment.

N protein is the most abundant as it is its mRNA (its amount is ten times that of mRNA of M gene (Cusi et al., 2001). It is highly immunogenic, is the most conserved, and is the target of IgG, IgM, and IgA (Collao et al., 2009; Schwarz et al., 1969a). The complex made by the interaction with RNA serves as the template for transcription being N protein and the polymerase, L protein, the minimal protein requirement for an active transcription complex (Accardi et al., 2001).

Neutralizing and hemagglutinating antibodies have been detected in Gn and Gc of other members of the genus (Pifat and Smith, 1987; Saluzzo et al., 1989). All stages of replication occur in the cytoplasm. After attachment mediated by glycoproteins and an as yet unidentified host receptor, the virus enters and uncoats by endocytosis of virions and fusion of viral membranes with endosomal ones. The primary transcription starts with the synthesis of mRNA complementary to viral RNA by the virion-associated polymerase using host-derived capped primers. Following transcription, free ribosomes start translation of L and S mRNAs, and then membrane-bound ribosomes act on M segment mRNAs. Primary glycosylation of nascent envelope proteins and co-translational cleavage of a precursor to yield Gn and Gc and NSm takes place. Then antigenome RNA synthesis and encapsidation occurs, serving as templates for genomic RNA. Finally, amplified synthesis of the mRNA species and ambisense transcription precedes morphogenesis, which includes accumulation of Gn and Gc in the Golgi, terminal glycosylation, and acquisition of modified host membranes. Fusion of cytoplasmic vesicles with the plasma membrane and release of mature virions take place to end the cycle.

Persistently TOSV-infected cultures showed a selective graded resistance to the replication of other phleboviruses being more permissive to infection with members of other serogroups (sandfly fever Sicilian virus (SFSV)) than to superinfection with more related viruses such as sandfly fever Naples virus (SFNV) or TOSV itself (Verani et al., 1984).

29.2.2 Clinical picture and geographical distribution

A substantial proportion of infection likely results in asymptomatic or paucisymptomatic cases (Braito et al., 1997; Hemmersbach-Miller et al., 2004). After an incubation period ranging from a few days to 2 weeks, the disease appears as meningitis with headache, fever, nausea, and vomiting (Schwarz et al, 1995a). Neck rigidity and Kernig signs are also usually present. Clinical abnormalities usually resolve in a few days, and patients recover spontaneously within 1–2 weeks; in a few cases, severe and persistent headache is the only recorded sequela and a case of deafness has been described (Calisher et al., 1987; Ehrnst et al., 1985; Martinez-Garcia et al., 2008). A small number of severe cases including severe meningoencephalitis or encephalitis have also been described (Dionisio et al., 2001).

After the isolation from its vector in Italy in 1971, many reports indicate that it is endemic in Italy. The virus has been detected in nonimmune population acquiring infection in endemic countries. So the first description of TOSV infection cases was in Cyprus, Spain, Turkey, Portugal, Greece, and France affecting soldiers or tourists (Dobler et al., 1997; Ehrnst et al., 1985; Eitrem et al., 1990, 1991; Schwarz et al., 1995b, c 1996b).

Most cases have been reported in residents or travelers in central Italy or Spain and sporadically from other Mediterranean countries, such as Portugal, Cyprus, France, and Greece (Calisher et al., 1987; Charrel et al., 2005; Echevarria et al., 2003; Eitrem et al., 1991; Hemmersbach-Miller et al., 2004; Navarro et al., 2004; Nicoletti et al., 1991; Schwarz et al., 1995a).

The classical distribution of TOSV reflects the range of its main vector, *Phlebotomus perniciosus*, between 20° and 45° north latitude, mainly surrounding the Mediterranean Sea.

The situation described shows the nowadays scenery of TOSV infections; however, due to climate change, this could vary in near future. In fact, in 2001 *P. perniciosus* was found for the first time in Germany (where some autochthonous cases of leishmaniasis, also transmitted by these vectors, have been described since then), and other species of sandflies have been detected in northern latitudes, reaching Cochem (50°) (Naucke et al., 2008). This means that apparition of autochthonous TOSV infections is probable in northern latitudes, and, in fact, it seems that the first autochthonous TOSV infections have been detected in the Upper Rhine Valley in southern Germany (Ursula Meyer-König, personal communication).

Another important aspect in the consideration of TOSV as an emerging virus is that in recent years many more infections have been described in endemic areas or in others where it has never been detected before. This is due to the expansion of the vector and the virus but also to better diagnostic tools and a higher clinical suspicion.

29.2.3 Phylogenetic studies: distribution of genotypes

For TOSV, two lineages with different geographical distribution have been described (Figure 29.1a) (Collao et al., 2009). The TOSV A genotype circulates in Italy, France, and Portugal, and the TOSV B genotype circulates in Spain, Portugal, and France. Both genotypes have been reported in France (Charrel et al., 2007); however, the co-circulation of both genotypes in Portugal is confusing (Liu et al., 2003; Venturi et al., 2007). Geographical differences in genotype distribution may relate to differences in vector distribution. Two different subtypes of *P. perniciosus* are described within the Mediterranean Basin; one has been found in northeast Africa, Malta, Italy, and France, while the second is present in northwest Africa and the Iberian Peninsula (Esseghir et al., 1997). The differentiation of virus genotypes may have been facilitated by sandflies flying in short hops, the possible genetic incompatibility of sandfly populations, and vector displacement. Interestingly, in France where co-circulation of both TOSV genotypes had been demonstrated, sequences of TOSV were also detected in *Sergentomyia minuta* (Charrel et al., 2006), suggesting that the virus can be present in other still undetermined species.

Variation within genotypes has also been described, and four different clusters were observed studying strains belonging to genotype A (Venturi et al., 2007). The precise biological significance of these facts is still not well understood; however, the low nucleotide substitution rate found in the M gene suggests a high degree of purifying selection that is in agreement with the lack of a strong selective pressure from the immune system of a vertebrate host (Collao et al., 2009; Venturi et al., 2007).

29.2.4 Ecology

TOSV is transmitted by *P. perniciosus* and *Phlebotomus perfiliewi*. Its occurrence is seasonal, and the highest incidence takes place during temperate seasons, late spring, and summer months, depending on temperatures and the rainy season.

The sandfly seeks a blood meal in the early evening and is small enough to penetrate mosquito netting. The extrinsic incubation period in the vector is about 7–10 days.

Despite demonstration of infection in some animals (Navarro-Mari et al., 2011), a vertebrate reservoir has not been identified. TOSV was isolated from the brain of *Pipistrellus kuhlii* that may play a role in the biological cycle of this agent (Verani et al., 1988).

(a)

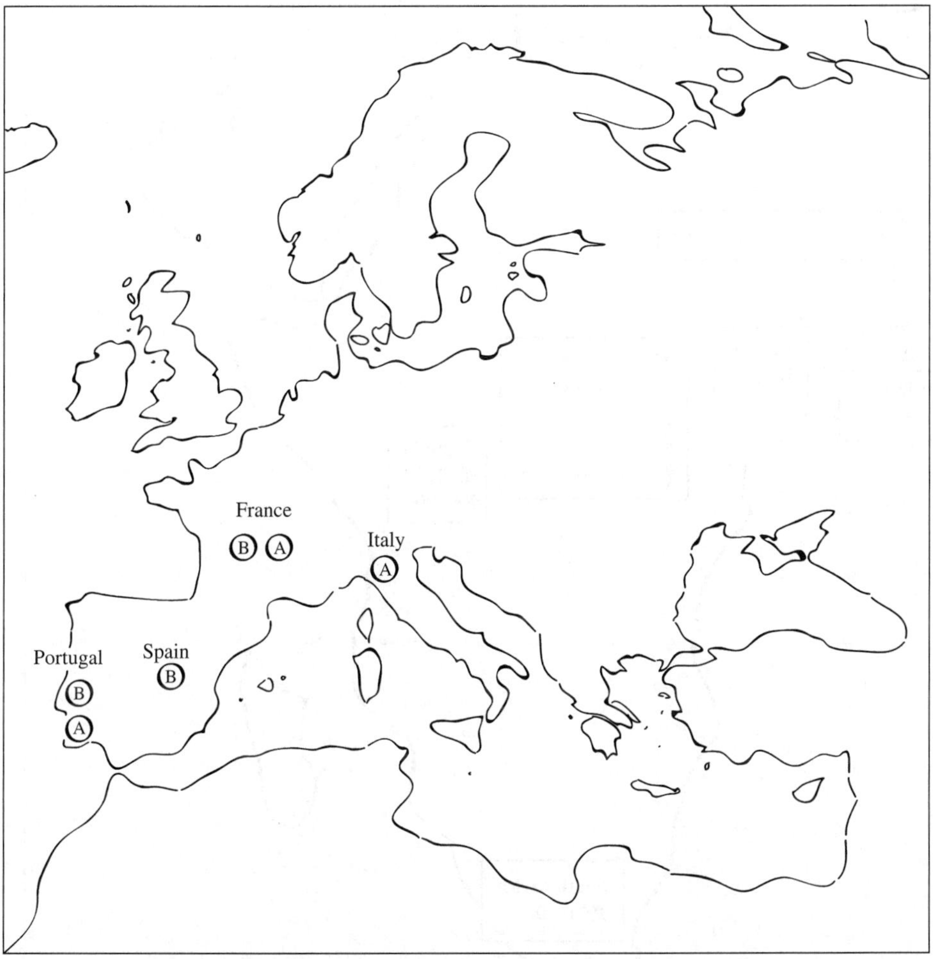

Figure 29.1. (a) Geographical distribution of TOSV genotypes.

(b)

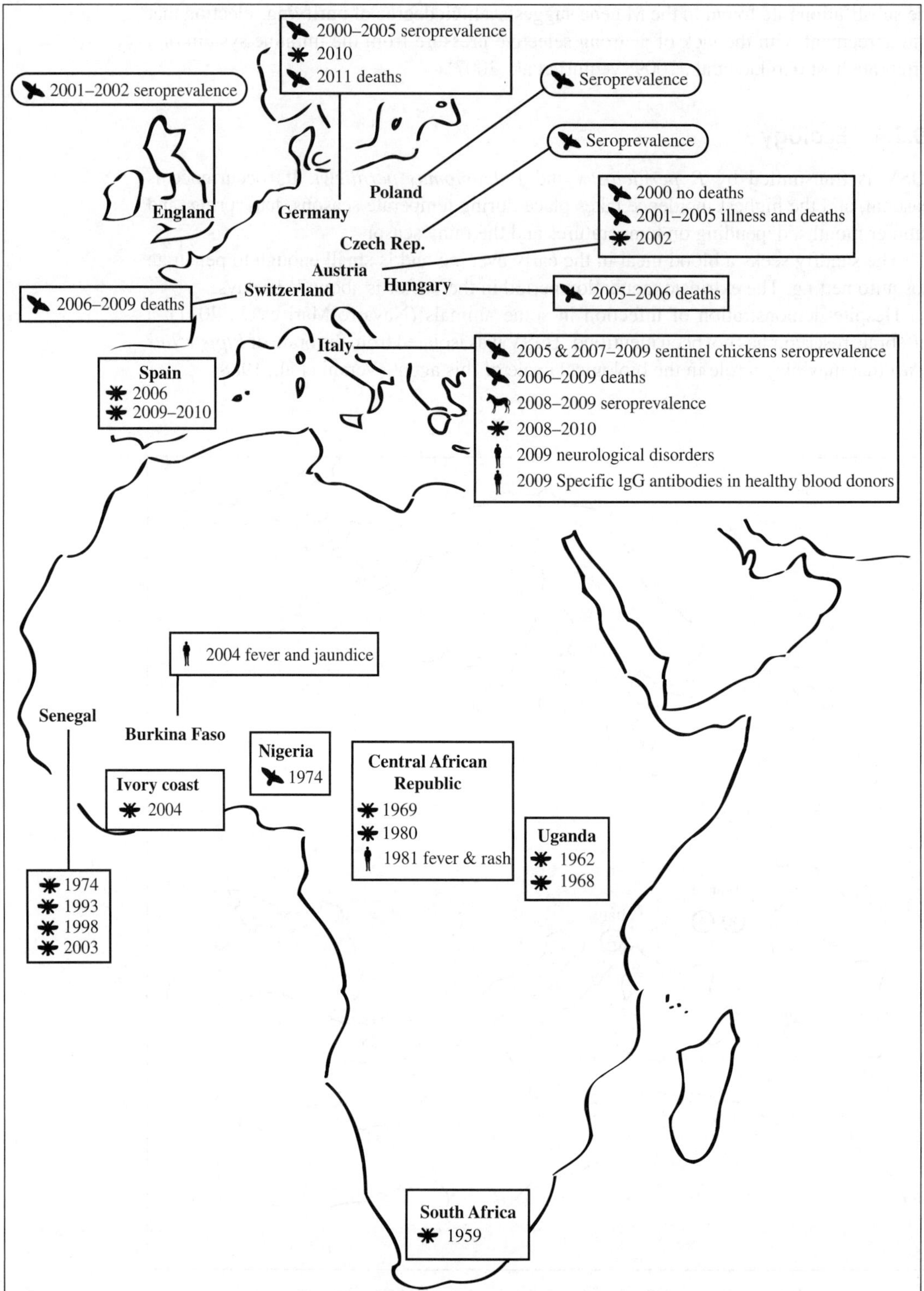

Figure 29.1. (b) Geographical distribution of USUV.

The fly will take a blood meal from a variety of species (humans, cattle, canines, equines, birds, etc.), and in many cases host choice is related to its availability as suggested for *P. perniciosus* and *P. perfiliewi* (Bongiorno et al., 2003).

Transovarial and venereal transmission were observed after experimental infection of *P. perniciosus* (Ciufolini et al., 1989; Tesh and Duboise, 1987; Tesh et al., 1992). Of course, viruses can overwinter via transovarial transmission and also by diapause in the fourth larval stage.

29.2.5 Laboratory diagnosis

Two critical factors for achieving successful laboratory diagnosis are the time elapsed from the illness's onset to recovery of samples and adequate conservation of those samples. Suitable techniques must also be applied to optimize diagnostic accuracy. Given the relatively nonspecific nature of the clinical illness, TOSV must be suspected when patients in an endemic area have the described clinical symptoms.

Direct diagnosis of acute viral neurological infections is known to be difficult, due to the rapid viremic phase and the presence of a low viral load at the time of clinical symptoms, often corresponding to the moment of hospitalization and clinical sampling. When techniques based on serology are used, cross-reactions can occur.

29.2.5.1 ***Virus Isolation.*** Virus isolation may be tried in the first 2–4 days of illness. Several animals are suitable, but cell culture, mainly Vero cells, is more frequently used. Other cell types such as BHK-21, CV-1, SW-13, or LLC-MK2 can be utilized. Viral isolation in cell culture is time consuming and less efficient than molecular techniques since TOSV was isolated in only 14% of the PCR-positive CSF specimens (Charrel et al., 2005). The main advantage of this method is its wide spectrum of detection since most members of this group can grow in the mentioned cell types.

29.2.5.2 ***Molecular Detection.*** Molecular detection based on RT nested PCRs has been extensively applied in diagnosis for TOSV detection, and differential diagnosis of TOSV and enteroviral infections is also possible using a multiplex RT nested PCR (Schwarz et al., 1995b, c; Valassina et al., 1996, 2002). Recently, methods based on real-time polymerase chain reaction (RT-PCR) were developed for the detection of TOSV, SFSV, and Rift Valley fever virus (RVFV) (Perez-Ruiz et al., 2007; Weidmann et al., 2008). PCR methods for RNA viruses, due to their high variability, need to be designed after an in-depth study of as many sequences as possible in order to avoid false-negative results from lack of hybridization of the primers and the target. Molecular detection of different members of the genus, TOSV, and other known phleboviruses, as well as new ones, was achieved using consensus PCRs designed in the N or L gene (Charrel et al., 2009; Collao et al., 2009; Lambert and Lanciotti, 2009; Sanbonmatsu-Gamez et al., 2005; Sánchez-Seco et al., 2003; Zhioua et al., 2010). A combination of these consensus primers with degenerated primers able to detect alphaviruses or flaviviruses has also been described and used for viral screening in arthropod samples, and a method in which a mix of consensus primers for *Phlebovirus* and *Nairovirus* genera is used has also recently been described (Lambert and Lanciotti, 2009; Sánchez-Seco et al., 2009).

29.2.5.3 ***Serological Diagnosis.*** Serological diagnosis can be achieved by detecting a fourfold rise in neutralization, or enzyme-linked immunosorbent assay (ELISA), titers between acute and convalescent sera or by demonstration of IgM antibodies acutely. For TOSV, indirect diagnosis is very useful by assaying for the specific IgM as a marker of acute infection. Immunoenzymatic techniques have been widely used in preparing tests where the recombinant nucleoprotein expressed in *Escherichia coli* is used as the specific antigen so as to exclude cross-reaction with SFNV or SFSV (Ciufolini et al., 1999; Schwarz et al., 1998; Soldateschi et al., 1999; Valassina et al., 1998). N protein is the major antigen responsible for the humoral response as measured using viral-protein-specific immunoassays (Magurano and Nicoletti, 1999), and antibodies against it have partial neutralizing activity (Cusi et al., 2001), but only the combination of N and Gc proteins induces 100% protection against challenge with TOSV in mice. Detection of anti-Gn and anti-Gc yields irregular results, perhaps due to lack of conformational maintenance of the epitopes involved in the humoral response (Di Bonito et al., 2002; Magurano and Nicoletti, 1999). The fact that N protein is the most conserved structural protein in this genus and glycoproteins the ones with the highest diversity could affect serological results depending on the antigen used (Collao et al., 2009).

A few immunological assays are available commercially, and some of them are widely used. ELISA and immunofluorescence are very useful and easy to perform. The hemagglutinating activity of the envelope glycoproteins is revealed with goose red blood cells by the hemagglutination test, and inhibition of hemagglutination provides highly specific results. Complement fixation or, especially, neutralizing assays are more complex to carry out but provide us with specific results because cross-reactivity exists between members of the genus *Phlebovirus* and specifically between TOSV and other serotypes of SFNV.

29.2.6 Prevention of transmission and treatment

The mild nature of phlebotomus fever has not encouraged studies on the possibility of treating the disease with specific antiviral compounds. In fact, the pharmacological treatment of phlebotomus fever patients is symptomatic. At present, the only method effective in controlling TOSV infection is to reduce human contact with the vector. Insecticides are extremely effective in the control of peridomestic sandfly species but are of little value for sylvan species. Mechanical means (e.g., protective clothing, bed nets, screening of windows) and use of repellents such as diethyltoluamide can also be used to prevent human–vector contact.

29.3 USUV

29.3.1 Virus: properties and classification

USUV is an African mosquito-borne virus that belongs to the *Flavivirus* genus (*Flaviviridae* family) (Kuno et al., 1998).

Flaviviruses particles are small (~50 nm in diameter), spherical, and surrounded by a lipid envelope. Virions are composed of a single, positive-strand RNA genome of about 11 kb in length. The genome has one open reading frame encoding a single polyprotein, flanked by 5′ and 3′ terminal noncoding regions that form specific secondary stem-loop structures required for genome replication (Rice, 1996). The genome encodes three structural proteins (capsid, membrane, and envelope) that constitute the virus particle and seven nonstructural proteins (NS1, NS2A, NS2B, NS3, NS4A, NS4B, and NS5) that are essential for viral replication (Chambers et al., 1990).

USUV is classified within the Japanese encephalitis virus (JEV) antigenic group in the mosquito-borne cluster, together with Cacipacore virus (CPCV), Koutango virus (KOUV), JEV, Murray Valley encephalitis virus (MVEV) and its subtype Alfuy virus (ALFV),

St. Louis encephalitis virus (SLEV), WNV and its subtype Kunjin virus (KUNV), and Yaounde virus (YAOV) (Calisher and Gould, 2003; Gould, 2002). The majority of viruses that belong to this antigenic group are important human or animal pathogens and are transmitted mainly by *Culex* mosquitoes (Gould et al., 2003; Poidinger et al., 1996).

29.3.2 History and geographical distribution

USUV was originally isolated from a mosquito (*Culex neavei*) in 1959 in South Africa (Williams et al., 1964). Further USUV strains were detected from different bird and mosquito species in Africa in subsequent years, but human disease (rash, fever, and jaundice) has only been reported in two cases, in the Central African Republic (1981) and in Burkina Faso (2004). In the summer of 2001, USUV was introduced to Austria and was spread through Europe, causing disease in birds and humans. In Austria, the virus caused fatalities in birds, between 2001 and 2005 (Chvala et al., 2004; Meister et al., 2008; Weissenböck et al., 2002, 2003). In the following years, the virus spread to Hungary (2005) (Bakonyi et al., 2007), Italy (2006) (Manarolla et al., 2010), Switzerland (2006) (Steinmetz et al., 2011), and Germany (2011) (Becker et al., 2012), causing outbreaks in free and captive wild bird populations. USUV infection has also been demonstrated serologically in wild bird hosts in England (2001–2002) (Buckley et al., 2006), Germany (2007) (Linke et al., 2007), Italy (2007) (Lelli et al., 2008), Morocco (2008) (Figuerola et al., 2009), Poland, and Czech Republic (Hubálek et al., 2008b). Moreover, USUV has been detected in *Culex* mosquitoes in Spain (2006 and 2009) (Busquets et al., 2008; Vázquez et al., 2011a,b), Italy (2009) (Calzolari et al., 2010; Tamba et al., 2010), and Germany (2010) (Jöst et al., 2011). In the summer of 2009, the first cases of USUV-related neurological diseases were reported in patients in Italy (Cavrini et al., 2009; Pecorari et al., 2009) (Figure 29.1b).

Until its spread to Europe, USUV was not considered as a potential pathogen for humans, because the virus had never been associated with severe neither fatal diseases in animals or humans before (Nikolay et al., 2011).

29.3.3 Ecology: vector, host, and incidental host

The life cycle of USUV seems to be similar to WNV. The virus is maintained through a bird–mosquito life cycle, being several avian species (mainly orders Passeriformes and Strigiformes) the principal host and ornithophilic mosquitoes (mainly *Culex* mosquitoes) the principal vectors. Other mammals, like humans and horses, are incidental hosts. To date, transmission cycles of USUV are unclear, as the available information about the primary vectors, vertebrate hosts, and the ecology of the virus are scarce.

29.3.3.1 ***Vectors.*** Different mosquito species, belonging to the genera *Culex*, *Coquillettidia*, *Mansonia*, and *Culiseta*, have been found infected with USUV in Africa (Adam and Digouette, 2005; Nikolay et al., 2011). In Europe, USUV has been detected in Italy (*Culex pipiens* and *Aedes albopictus*), Spain (*Cx. pipiens* and *Culex perexiguus*) and Germany (*Cx. pipiens*), showing that the virus is detected mainly from *Culex* mosquitoes (especially *Cx. pipiens*) in European countries.

The host-feeding habits form a very important factor to identify potential vertebrate hosts and then to gain further insight into its transmission cycle in nature. In Europe, the *Cx. pipiens* complex contains two distinct biotypes: pipiens and molestus, which are morphologically indistinguishable but differ in physiology and behavior. *Culex pipiens pipiens* is predominantly ornithophilic, but in contrast, *Cx. pipiens molestus* is mainly anthropophilic (Medlock et al., 2005). In addition, hybrids are common in areas of overlap (Reusken et al., 2010).

Recent studies about the host-feeding preference in Europe revealed that mosquitoes of *Cx. pipiens* complex fed on a high diversity of avian and mammalian hosts (Muñoz et al., 2011; Roiz et al., 2012), and these vectors have been demonstrated to serve as both a cycle enzootic and a bridge vector for humans (Farajollahi et al., 2011; Hamer et al., 2008).

29.3.3.2 ***Host.*** Wild birds are the principal reservoir for USUV, and migratory birds could play a key role in the introduction of USUV into new areas (Weissenböck et al., 2002).

In Africa, USUV has been isolated from nonmigratory bird species, but in the spread of USUV to Europe, the virus has been found in numerous migratory wild birds (Nikolay et al., 2011). In the Austrian, Italian, Hungarian, and German outbreaks, blackbirds (*Turdus merula*) were the species more affected by USUV infection, but the house sparrows (*Passer domesticus*) were the species most affected in the outbreak that occurred in Switzerland (Steinmetz et al., 2011). Other bird species that have been infected in Europe by USUV are barn swallow, great tit, blue tit, great grey owl, and common starling, between others (Becker et al., 2012; Steinmetz et al., 2011). Serological evidence of infection, seroconversion, and neutralizing antibodies has been detected in sentinel healthy chickens and resident birds in United Kingdom (Buckley et al., 2006), Italy (Lelli et al., 2008), Germany (Linke et al., 2007), Austria (Chvala et al., 2007), Morocco (Figuerola et al., 2009), Czech Republic, and Poland (Hubálek et al., 2008a). Therefore, it seems that USUV is not necessarily pathogenic for all avian species infected and that the herd immunity developed in these birds is a possible explanation for the significant decrease in USUV-associated bird mortalities after the outbreaks in several European countries (Meister et al., 2008). Moreover, antibody-positive results in juvenile birds have been observed in UK, which implies that this virus is being actively transmitted to the indigenous UK bird population (Buckley et al., 2003, 2006). The sentinel and domestic chickens (Chvala et al., 2005; Lelli et al., 2008; Rizzoli et al., 2007) and domestic geese (Chvala et al., 2006) seem to be resistant to USUV infection.

29.3.3.3 ***Incidental Host.*** USUV infection has been described also in mammals, such as humans and horses. USUV can play a role as a human pathogen, causing a benign asymptomatic infection, mild illness, and/or severe neurological disease. In humans, the first two cases of USUV infections occurred in Africa, with symptoms such as fever, rash, and jaundice (Nikolay et al., 2011). But recently, the first human cases of USUV neuroinvasive illness were reported in Italy. These cases affected two immunocompromised patients (Cavrini et al., 2009; Pecorari et al., 2009). The two infections could be consistent with local transmission either directly through a mosquito bite or indirectly through an infected donor, because both patients had received blood transfusions in the same period of time. In this sense, the presence of USUV antibodies has been demonstrated in serum samples from healthy volunteer blood donors with no history of other flavivirus infection (Gaibani et al., 2012). Moreover, USUV genome has been identified in immunocompetent patients with acute meningoencephalitis (Cavrini et al., 2011). USUV infection in healthy horses has been described for first time in Italy (Savini et al., 2011). In 2008, the incidence of USUV in sentinel horses was high and decreased markedly in the following year. So, the presence of active virus circulation was demonstrated. It seems also that the virus is able to infect horses without developing illness and horses appear to be good sentinel animals to monitor USUV circulation.

29.3.4 Pathology

29.3.4.1 ***Birds.*** Pathogenicity of USUV in avian species varies. Affected birds with USUV showed some unspecific clinical signs like apathy, ruffled plumage, increased water intake, neurological disturbances suspicious of encephalitis, and deaths. The disease has been characterized by encephalitis, myocardial degeneration, and necrosis in the liver and spleen. Immunohistologically, viral protein was abundantly found in the cytoplasm of cerebral neurons and glial cells but also in cardiac myocytes, endothelial cells, and white blood cells of various organs. Comparisons of pathological alterations revealed similar lesions in birds infected in the Austrian, Hungarian, Italian, German, and Swiss USUV outbreaks; however, not all of the lesions mentioned earlier are present (Bakonyi et al., 2007; Becker et al., 2012; Chvala et al., 2007; Steinmetz et al., 2011; Weissenböck et al., 2002, 2003).

29.3.4.2 ***Humans.*** The clinical picture in human infections varies from asymptomatic infection to neurological disease. The asymptomatic infections have been described in healthy volunteer blood donors by the detection of USUV antibodies in serum samples (Gaibani et al., 2012). Mild illness is characterized by unspecific symptoms, like fever, rash, and jaundice (Nikolay et al., 2011). And finally, neurological disease has been associated to acute meningoencephalitis in immunocompetent patients (Cavrini et al., 2011) and in two immunocompromised ones. One of the immunocompromised patients suffered a diffuse large B cell lymphoma and the other received an orthotopic liver transplant (Cavrini et al., 2009; Pecorari et al., 2009). Both received blood transfusions. The common clinical symptoms were persistent fever of 39.5°C, headache, and neurological disease (impaired neurological functions) resembling the related WNV neuroinvasive disease. One of them developed a fulminant hepatitis, a pathology that had been described previously in rare cases of WNV infection. Whether this new tropism was associated with new characteristics of the infecting viruses, with a possible inoculation route through transfusion, and/or to the underlying diseases of the patients still remains unclear. Until now, the number of cases of human USUV infections is rather limited, and more clinical work on infected patients is necessary to define its real pathogenicity.

29.3.5 Laboratory diagnosis

Clinical suspicion of USUV infection in human and birds requires laboratory confirmation, as often the clinical symptoms and pathological lesions are unspecific. Laboratory diagnoses are based on direct methods (detection of the virus or viral antigen) and also indirect methods (detection of the antibody response to the infection) (Vazquez et al., 2011a,b).

By analogy to the current knowledge about the pathogenesis of WNV-related illness in humans, it is assumed that USUV incubation period will be 2–14 days. The virus will be detectable in CSF and serum in the acute stage of the disease, and the IgM antibodies will appear 5 days after the onset of fever. Antibodies may persist in serum for many months after infection (Solomón, 2004). Diagnosis of USUV is not easy, particularly in areas where circulation along with other cross-reacting flaviviruses occurs. In fact, a false-positive result of a WNV RT-PCR was reported in a patient with viremia caused by USUV in Italy (Gaibani et al., 2010). Consequently, it is necessary to develop molecular and serological USUV-specific diagnostic tools for human and animal infection.

29.3.5.1 ***Direct Methods.*** In acute samples, it is necessary to use direct methods for USUV diagnosis.

29.3.5.1.1 Genomic Amplification. USUV is present at very low levels in human or equine samples; therefore, an amplification of the genomic material is used to enhance the detection rate of USUV infections. There is a high variety of generic molecular methods to detect flavivirus, but in most of them it is necessary to sequence the produced amplicons to identify specifically the virus responsible of the infection (Johnson et al., 2010; Moureau et al., 2007). New specific methods are being designed to identify and distinguish USUV infection from other arboviruses, particularly from members of the JEV group that co-circulate with USUV

(Hubálek, 2008; Nikolay et al., 2011). Recently, a novel specific RT-PCR assay was developed to detect the USUV genome in human plasma, serum, and CSF samples (Cavrini et al., 2011).

29.3.5.1.2 Cell Culture. The classical method of flavivirus isolation is intracerebral inoculation of suckling mice or inoculation of embryonated eggs. But a variety of primary cells and established cell lines have been used for flavivirus isolation and propagation in routine diagnostic applications. The appearance on cytopathic effects (CPEs) and plaque formation varies depending of the viruses and host cell types. USUV can infect cell cultures of various tissue types derived from a wide variety of animal species like human (HeLa), green monkey (Vero), equine (ED), bovine (MDBK), porcine (PK-15), rabbit (RK-13), canine (MDCK, DK), feline (CR), hamster (BHK-21, BF), rat (C6), turtle (TH1), horse kidney (EqK), chicken embryo fibroblast (CEF), and goose embryo fibroblast (GEF). Viral multiplication has been detected in all tested mammalian cell types, but only Vero, PK-15, and GEF cells developed CPE after USUV infection. By this reason, these three cell lines and cell culture are the most appropriate for diagnostic (Bakonyi et al., 2005, 2007).

29.3.5.1.3 Other Direct Methods. Tissues and other organs can be examined by several methods, like histopathology, immunohistochemistry (IHC), and *in situ* hybridization (ISH). Histopathology shows macroscopical lesions in the affected organs. IHC assay by the avidin–biotin complex technique and with rabbit USUV antiserum is used to detect viral antigen in tissue samples (Weissenböck et al., 2004). And ISH assays with a USUV-specific oligonucleotide probe (shown not to cross-react with WNV) are excellent confirmatory tools to demonstrate viral nucleic acid within tissue sections (Chvala et al., 2004).

29.3.5.2 *Indirect Methods.* Serological diagnosis is needed and important to identify infection after the viremic stage. The serological approaches for screening are IgM and/or IgG antibody detection (ELISA) and hemagglutination inhibition (HI). Sera reacting to both WNV and USUV were detected in some studies that used tests with low specificity such as HI (Meister et al., 2008) or ELISA (Lelli et al., 2008). In this way, a novel ELISA for detection of specific anti-USUV in humans has been developed and assessed in serum samples collected from healthy blood donors (Gaibani et al., 2012). Due to the strong cross-reaction between members of the JEV serogroup, the plaque reduction neutralization test (PRNT) or microneutralization assays (mNTA) are the standard assays that are used to discriminate immune responses against them, but none of them is routinely applicable for large screenings, and these techniques can be performed only in specialized laboratories that can handle hazardous viruses (Pierro et al., 2011). As an already available alternative, acute and convalescent sera should be tested for seroconversion of IgG antibodies using in-house or commercial ELISA tests based on WNV antigens.

29.3.6 Phylogenetic studies

Phylogenetic analysis of the USUV sequences available in GenBank shows mainly three different clusters. One of them corresponds to the strain detected in mosquitoes from Africa, the other one corresponds to the strain detected in mosquitoes from Spain, and the third one corresponds to a group of sequences detected in mosquitoes, humans, and birds from several central European countries (Austria, Hungary, Switzerland, Italy, and Germany) (Figure 29.2).

The percent of homology detected in the sequences available of USUV Spanish strain in the *NS5* gene shows that it is more related to African than to central European strains (97–98% and 95–96% homology, respectively). And the analysis realized between all central European USUV strains detected in birds, mosquitoes, and humans, in partial nucleotide sequences from the envelope and the *NS5* genes, revealed 99–100% identity (Table 29.1).

29.3 USUV

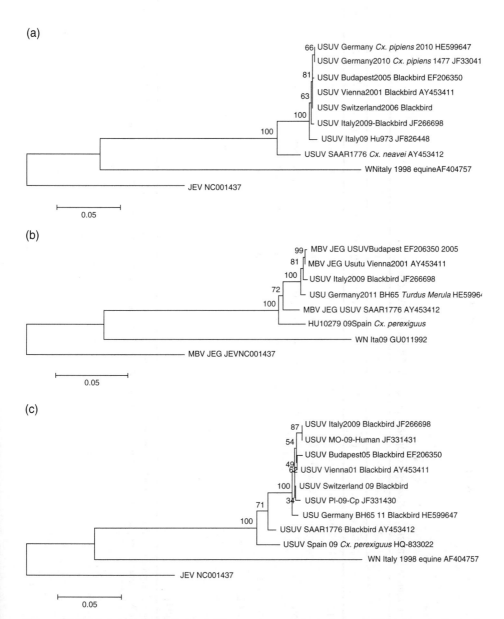

Figure 29.2. Phylogenetic analysis of different genome fragments of USUV. The phylogenetic trees are based on nucleotides sequences of USUV strains available in GenBank: (a) length of 451 nucleotides of NS5 gene; (b) length of 1691 nucleotides of NS5 gene; and (c) length of 800 nucleotides of envelope gene. Trees were built with the neighbor-joining (NJ) method and distance-p model, which calculated confidence values of 1000 bootstrapping trials. For each sequence, the GenBank accession number, strain designation and strain origin are provided. The trees are rooted with the same sequence of JEV.

The frequent recurrence of outbreaks and detections of phylogenetically homogeneous USUV in central European countries since 2001 is consistent with a single introduction followed by viral persistence in endemic foci in the area, rather than resulting from independent introductions from exogenous endemic foci. In Germany, after the initial

TABLE 29.1. Percentage of Nucleotidic Sequence Similarity Between European and South African USUV strains

Strains	2			3			4			5			6			7			8			9			10
	A	B	C	A	B	C	A	B	C	A	B	C	A	B	C	A	B	C	A	B	C	A	B	C	C
1. USUV_SAAR1776_Blackbird_AY453412	97	97	96	97	97	96	97	97	95	97	97	96	98	97		97	96		97	95	97	97			96
2. USUV_Vienna01_Blackbird_AY453411	–	–	–	100	100	100	100	100	99	99	99	100	95	96		100	100		100	99	99	99			100
3. USUV_Budapest05_Blackbird_EF206350				–	–	–	99	99	99	99	99	100	95	96		99	100		99	99	99	99			100
4. USUV_Italy2009Blackbird_JF266698							–	–	–	99	99	99	95	96		99	100		99	99	99	99			99
5. USUV_Germany_BH65_11_Blackbird_HE599647										–	–	–	95	96		99	100		99	99	99	99			100
6. USUV_Spain_09_Cx. perexiguus_HQ833022													–	–		95	–		–	95	95				
7. USUV_Switzweland_09_Blackbird																–	–		99	99	100				100
8. USUV_MO-09-Human_JF331434																			–	–	–	99			99
9. USUV_PI-09-Cp_JF331430																						–	–	–	
10. USUV_Germany_Cx. pipiens2010_JF330418																									–

The percentage of similarity has been calculated using the program MEGA5, neighbor-joining method, and distance-p model. A, Fragment analyzed of 451 nucleotides in length of NS5 gene; B, fragment analyzed of 1691 nucleotides in length of NS5 gene; C, fragment analyzed of 800 nucleotides in length of envelope gene. For each sequence, the GenBank accession number, strain designation, and strain origin are provided.

detection of USUV in *Cx. pipiens* mosquitoes in 2010 (Jöst et al., 2011), the same virus spread in 2011 and caused epizootics among wild and captive birds in southwest Germany (Becker et al., 2012). These data intensify the hypothesis of persistence of USUV in the affected areas and also an increased risk of possible USUV infections in humans. Retrospective analysis of environmental factors during the birds USUV outbreaks in Austria and Germany disclosed an influence on USUV dynamics of climatic factors affecting the mosquito population (Rubel et al., 2008). The higher average temperatures compared to the ones of previous years can led to an increase in the number of *Culex* generations and may have contributed to the rapid spread of USUV. If the high bird mortality in Europe is associated to a particular susceptibility of the host species (being completely naïve in the new endemic areas) or, alternatively, to the high pathogenicity of this USUV strain for certain bird species because of genetic mutations, need to be studied. When the complete sequences available of African and central European strains are compared, significant amino acid substitutions were distributed throughout the entire genome, which may influence in the apparent differences in virulence for birds of the two strains. In birds from Austria, Hungary, Switzerland, and Germany, no serological evidence for USUV circulation was found in recent years preceding the outbreaks, and due to this lack of immunity in the birds, there could be more susceptibility to USUV infection. Furthermore, in these European countries, in subsequent years, a less severe USUV activity, herd immunity, and antibodies against USUV have been detected in horses, sentinel animals, humans, and/or wild birds (Meister et al., 2008). Human disease seems to be very rare, and patients with neurological disease have been reported only in Italy. The host-related variables such as susceptibility, age, immunity, and general health status might affect the USUV pathogenicity for humans and need to be studied. Very scarce information is available on the distribution and pathogenicity of USUV in Africa and Spain, but they are apparently nonpathogenic for local bird populations. This could be due to these strains being nonvirulent viruses, by the lack of detection of disease, or due to the immunity and/or genetic resistance arising in the bird population as the result of exposure to USUV and other antigenically related flaviviruses over a long period of time.

These data may indicate that there were two different introductions of USUV from Africa to Europe (detected in Vienna and Spain), evolving the corresponding viruses independently. In Africa, USUV has been isolated several times in different geographical regions, but only one sequence is known. Several strains of USUV could be circulating in Africa, and different strains may have been introduced in Europe (Spanish and central European strains).

Complete sequence analysis of the viral genomes has proved to be a powerful tool for evaluating relatedness and reconstructing the evolutionary history and phylogeography of the viruses. More information about complete sequences and studies in animal models to compare the pathogenicity of more isolates of African, Spanish, and European USUV is necessary. These results would be very useful to understand in deep the origin and dispersal of USUV within Africa and Europe, its pathogenicity, host range, and the evolutionary origins and dispersal.

29.3.7 Treatment, prevention, and surveillance

Treatments for USUV are not available. Insecticides are extremely effective in the vector control, and the most effective method is to reduce human contact with the vector. Mechanical means (e.g., protective clothing, bed nets, screening of windows) and use of repellents are of importance to prevent human–vector contact. As treatments for USUV are not available, it is very important to establish a surveillance system (in birds, horses, mosquitoes, and human samples) to detect USUV activity and to assess the risk for public health (Vazquez et al., 2011a,b).

29.4 CONCLUSIONS

We have seen some characteristics of two viruses that may be present in new areas in the next years being the cause of neurological illness in nonimmune population. The adaptation of arthropods to human-altered environments led to their global distribution through dispersal via humans and, combined with their mixed feeding patterns on birds and mammals (including humans), increased the transmission of several animal pathogens to humans. As discussed previously, the most effective measure to prevent their expansion would be the control of their arthropod vectors. These viruses have been able to spread and establish in new areas, and this had been possible thanks to the environmental conditions, which allowed them to adapt to new hosts and vectors and establish an enzootic life cycle.

ACKNOWLEDGEMENT

We would like to thank Drs. N. Nowotny and T. Bakonyi for providing the NS5 and envelope USUV sequences from Swiss birds. In addition, we would like to acknowledge the scientific review of the manuscript to Dr. A. Tenorio and the graphic support to Gustavo Russo.

REFERENCES

Accardi, L., C. Prehaud, P. Di Bonito, et al. (2001). "Activity of Toscana and Rift Valley fever virus transcription complexes on heterologous templates." *J Gen Virol* **82**(Pt 4): 781–785.

Adam, F. and J. P. Digouette (2005). "Virus d'Afrique (base de données). Centre collaborateur OMS de référence et de recherche pour les arbovirus et les virus de fièvres hémorrhagiques (CRORA); Institut Pasteur de Dakar". Available from http://www.pasteur.fr/recherche/banques/CRORA [Accessed on June 06, 2013].

Bakonyi, T., H. Lussy, H. Weissenböck, et al. (2005). "In vitro host-cell susceptibility to Usutu virus." *Emerg Infect Dis* **11**(2): 298–301.

Bakonyi, T., K. Erdélyi, K. Ursu, et al. (2007). "Emergence of Usutu virus in Hungary." *J Clin Microbiol* **45**(12): 3870–3874.

Becker, N., H. Jöst, U. Ziegler, et al. (2012). "Epizootic emergence of usutu virus in wild and captive birds in Germany." *PLoS One* **7**(2): e32604.

Bongiorno, G., A. Habluetzel, C. Khoury, et al. (2003). "Host preferences of phlebotomine sand flies at a hypoendemic focus of canine leishmaniasis in central Italy." *Acta Trop* **88**(2): 109–116.

Braito, A., R. Corbisiero, S. Corradini, et al. (1997). "Evidence of Toscana virus infections without central nervous system involvement: a serological study." *Eur J Epidemiol* **13**(7): 761–764.

Brown, C. (2004). "Emerging zoonoses and pathogens of public health significance—an overview." *Rev Sci Tech* **23**(2): 435–442.

Buckley, A., A. Dawson, and E. A. Gould (2006). "Detection of seroconversion to West Nile virus, Usutu virus and Sindbis virus in UK sentinel chickens." *Virol J* **4**(3): 71.

Buckley, A., A. Dawson, S. R. Moss, et al. (2003). "Serological evidence of West Nile virus, Usutu virus and Sindbis virus infection of birds in the UK." *J Gen Virol* **84**: 2807–2817.

Busquets, N., A. Alba, A. Allepuz, et al. (2008). "Usutu virus sequences in *Culex pipiens* (Diptera: Culicidae), Spain." *Emer Infect Dis* **14**(5): 861–862.

Calisher, C. H. and E. A. Gould (2003). "Taxonomy of the virus family Flaviviridae." *Adv Virus Res* **59**: 1–19.

Calisher, C. H., A. N. Weinberg, D. J. Muth, et al. (1987). "Toscana virus infection in United States citizen returning from Italy." *Lancet* **1**(8525): 165–166.

Calzolari, M., P. Bonilauri, R. Bellini, et al. (2010). "Evidence of simultaneous circulation of west nile and usutu viruses in mosquitoes sampled in emilia-romagna region (Italy) in 2009." *PLoS One* **5**(12): e14324.

Cavrini, F., P. Gaibani, G. Longo., et al. (2009). "Usutu virus infection in a patient who underwent orthotropic liver transplantation, Italy, August–September 2009." *Euro Surveill* **14**(50): pii: 19448.

Cavrini, F., M. E. Della Pepa, P. Gaibani, et al. (2011). "A rapid and specific real-time RT-PCR assay to identify Usutu virus in human plasma, serum, and cerebrospinal fluid." *J Clin Virol* **50**(3): 221–223.

Chambers, T. J., C. S. Hahn, R. Galler, et al. (1990). "Flavivirus genome organization, expression and replication." *Ann Rev Microbiol* **44**: 649–688.

Charrel, R. N., P. Gallian, J. M. Navarro-Mari, et al. (2005). "Emergence of Toscana virus in Europe." *Emerg Infect Dis* **11**(11): 1657–1663.

Charrel, R. N., A. Izri, S. Temmam, et al. (2006). "Toscana virus RNA in *Sergentomyia minuta* files." *Emerg Infect Dis* **12**(8): 1299–1300.

Charrel, R. N., A. Izri, S. Temmam, et al. (2007). "Cocirculation of 2 genotypes of Toscana virus, southeastern France." *Emerg Infect Dis* **13**(3): 465–468.

Charrel, R. N., G. Moureau, S. Temmam, et al. (2009). "Massilia virus, a novel Phlebovirus (Bunyaviridae) isolated from sandflies in the Mediterranean." *Vector Borne Zoonotic Dis* **9**(5): 519–530.

Chvala, S., J. Kolodziejek, N. Nowotny, et al. (2004). "Pathology and viral distribution in fatal Usutu virus infections of birds from the 2001 and 2002 outbreaks in Austria." *J Comp Pathol* **131**(2–3): 176–185.

Chvala, S., T. Bakonyi, R. Hackl, et al. (2005). "Limited pathogenicity of Usutu virus for the domestic chicken (*Gallus domesticus*)." *Avian Pathol* **34**(5): 392–395.

Chvala, S., T. Bakonyi, R. Hackl, et al. (2006). "Limited pathogenicity of usutu virus for the domestic goose (Anser anser f. domestica) following experimental inoculation." *J Vet Med B Infect Dis Vet Public Health* **53**(4): 171–175.

Chvala, S., T. Bakonyi, C. Bukovsky, et al. (2007). "Monitoring of Usutu virus activity and spread by using dead bird surveillance in Austria, 2003–2005." *Vet Microbiol* **122**(3–4): 237–245.

Ciufolini, M. G., M. Maroli, E. Guandalini, et al. (1989). "Experimental studies on the maintenance of Toscana and Arbia viruses (Bunyaviridae: Phlebovirus)." *Am J Trop Med Hyg* **40**(6): 669–675.

Ciufolini, M. G., C. Fiorentini, P. di Bonito, et al. (1999). "Detection of Toscana virus-specific immunoglobulins G and M by an enzyme-linked immunosorbent assay based on recombinant viral nucleoprotein." *J Clin Microbiol* **37**(6): 2010–2012.

Collao, X., G. Palacios, S. Sanbonmatsu-Gámez, et al. (2009). "Genetic diversity of Toscana virus." *Emerg Infect Dis* **15**(4): 574–577.

Cusi, M. G., P. E. Valensin, M. Donati, et al. (2001). "Neutralization of Toscana virus is partially mediated by antibodies to the nucleocapsid protein." *J Med Virol* **63**(1): 72–75.

Di Bonito, P., S. Bosco, S. Mochi, et al. (2002). "Human antibody response to Toscana virus glycoproteins expressed by recombinant baculovirus." *J Med Virol* **68**(4): 615–619.

Dionisio, D., M. Valassina, M. G. Ciufolini, et al. (2001). "Encephalitis without meningitis due to sandfly fever virus serotype toscana." *Clin Infect Dis* **32**(8): 1241–1243.

Dobler, G., J. Treib, A. Haass, et al. (1997). "Toscana virus infection in German travellers returning from the Mediterranean." *Infection* **25**(5): 325.

Echevarria, J. M., F. de Ory, M. E. Guisasola, et al. (2003). "Acute meningitis due to Toscana virus infection among patients from both the Spanish Mediterranean region and the region of Madrid." *J Clin Virol* **26**(1): 79–84.

Ehrnst, A., C. J. Peters, B. Niklasson, et al. (1985). "Neurovirulent Toscana virus (a sandfly fever virus) in Swedish man after visit to Portugal." *Lancet* **1**(8439): 1212–1213.

Eitrem, R., S. Vene, and B. Niklasson (1990). "Incidence of sand fly fever among Swedish United Nations soldiers on Cyprus during 1985." *Am J Trop Med Hyg* **43**(2): 207–211.

Eitrem, R., B. Niklasson, and O. Weiland (1991). "Sandfly fever among Swedish tourists." *Scand J Infect Dis* **23**(4): 451–457.

Esseghir, S., P. D. Ready, R. Killick-Kendrick, et al. (1997). "Mitochondrial haplotypes and phylogeography of Phlebotomus vectors of Leishmania major." *Insect Mol Biol* **6**(3): 211–225.

Farajollahi, A., D. M. Fonseca, L. D. Kramer, et al. (2011). "'Bird biting' mosquitoes and human disease: a review of the role of *Culex pipiens* complex mosquitoes in epidemiology." *Infect Genet Evol* **11**(7): 1577–1585.

Figuerola, J., R. E. Baouab, R. Soriguer, et al. (2009). "West Nile virus antibodies in wild birds, Morocco, 2008." *Emerg Infect Dis* **15**(10): 1651–1653.

Gaibani, P., A. M. Pierro, F. Cavrini, et al. (2010). "False-positive transcription-mediated amplification assay detection of West Nile virus in blood from a patient with viremia caused by an Usutu virus infection." *J Clin Microbiol* **48**(9): 3338–3339.

Gaibani, P., A. Pierro, R. Alicino, et al. (2012). "Detection of Usutu-virus-specific IgG in blood donors from Northern Italy." *Vector Borne Zoonotic Dis.* **12**(5): 431–433.

Gould, E. A. (2002). "Evolution of the Japanese encephalitis serocomplex viruses." *Curr Top Microbiol Immunol* **267**: 391–404.

Gould, E. A. and S. Higgs (2009). "Impact of climate change and other factors on emerging arbovirus diseases." *Trans R Soc Trop Med Hyg* **103**(2): 109–121.

Gould, E. A., X. de Lamballerie, P. M. de A. Zanotto, et al. (2003). "Origins, evolution, and vector/host coadaptations within the genus Flavivirus." *Adv Virus Res* **59**: 277–314.

Gubler, D. J. (2009). "Vector-borne diseases." *Rev Sci Tech* **28**(2): 583–588.

Hamer, G. L., U. D. Kitron, J. D. Brawn, et al. (2008). "*Culex pipiens* (Diptera: Culicidae): a bridge vector of West Nile virus to humans." *J Med Entomol* **45**(1): 125–128.

Hemmersbach-Miller, M., P. Parola, R. N. Charrel, et al. (2004). "Sandfly fever due to Toscana virus: an emerging infection in southern France." *Eur J Intern Med* **15**(5): 316–317.

Hoffmann, B., M. Scheuch, D. Höper, et al. (2012). "Novel orthobunyavirus in cattle, europe, 2011." *Emerg Infect Dis* **18**(3): 469–472.

Hubálek, Z. (2008). "Mosquito-borne viruses in Europe." *Parasitol Res* **103**(Suppl. 1): 29–43.

Hubálek, Z., J. Halouzka, Z. Juricová, et al. (2008a). "Serologic survey of birds for West Nile flavivirus in southern Moravia (Czech Republic)." *Vector Borne Zoonotic Dis* **8**(5): 659–666.

Hubálek, Z., E. Wegner, P. Tryjanowski, et al. (2008b). "Serologic survey of potential vertebrate hosts for West Nile virus in Poland." *Viral Immunol* **21**(2): 247–253.

Johnson, N., P. R. Wakeley, K. L. Mansfield, et al. (2010). "Assessment of a novel real-time pan-flavivirus RT-polymerase chain reaction." *Vector Borne Zoonotic Dis* **10**(7): 665–671.

Jones, K. E., N. G. Patel, M. A. Levy, et al. (2008). "Global trends in emerging infectious diseases." *Nature* **451**(7181): 990–993.

Jöst, H., A. Bialonski, D. Maus, et al. (2011). "Isolation of usutu virus in Germany." *Am J Trop Med Hyg* **85**(3): 551–553.

Kuiken, T., R. Fouchier, G. Rimmelzwaan, et al. (2003). "Emerging viral infections in a rapidly changing world." *Curr Opin Biotechnol* **14**(6): 641–646.

Kuno, G., G. J. Chang, K. R. Tsuchiya, et al. (1998). "Phylogeny of the genus Flavivirus." *J Virol* **72**(1): 73–83.

Lambert, A. J. and R. S. Lanciotti (2009). "Consensus amplification and novel multiplex sequencing method for S segment species identification of 47 viruses of the Orthobunyavirus, Phlebovirus, and Nairovirus genera of the family Bunyaviridae." *J Clin Microbiol* **47**(8): 2398–2404.

Lelli, R., G. Savini, L. Teodori, et al. (2008). "Serological evidence of USUTU virus occurrence in north-eastern Italy." *Zoonoses Public Hlth* **55**(7): 361–367.

Linke, S., M. Niedrig, A. Kaiser, et al. (2007). "Serologic evidence of West Nile virus infections in wild birds captured in Germany." *Am J Trop Med Hyg* **77**(2): 358–364.

Liu, D. Y., R. B. Tesh, A. P. Travassos Da Rosa, et al. (2003). "Phylogenetic relationships among members of the genus Phlebovirus (Bunyaviridae) based on partial M segment sequence analyses." *J Gen Virol* **84**(Pt 2): 465–473.

Magurano, F. and L. Nicoletti (1999). "Humoral response in Toscana virus acute neurologic disease investigated by viral-protein-specific immunoassays." *Clin Diagn Lab Immunol* **6**(1): 55–60.

Manarolla, G., T. Bakonyi, D. Gallazzi, et al. (2010). "Usutu virus in wild birds in northern Italy." *Vet Microbiol* **141**(1–2): 159–163.

Martinez-Garcia, F. A., A. Moreno-Docon, M. Segovia-Hernández, et al. (2008). "Deafness as a sequela of Toscana virus meningitis." *Med Clin (Barc)* **130**(16): 639.

Medlock, J. M., K. R. Snow, and S. Leach (2005). "Potential transmission of West Nile virus in the British Isles: an ecological review of candidate mosquito bridge vectors." *Med Vet Entomol* **19**(1): 2–21.

Meister, T., H. Lussy, T. Bakonyi, et al. (2008). "Serological evidence of continuing high Usutu virus (Flaviviridae) activity and establishment of herd immunity in wild birds in Austria." *Vet Microbiol* **127**(3–4): 237–248.

Moureau, G., S. Temmam, J. P. Gonzalez, et al. (2007). "A real-time RT-PCR method for the universal detection and identification of flaviviruses." *Vector Borne Zoonotic Dis* **7**(4): 467–477.

Muñoz, J., R. Eritja, M. Alcaide, et al. (2011). "Host-feeding patterns of native *Culex pipiens* and invasive *Aedes albopictus* mosquitoes (Diptera: Culicidae) in urban zones from Barcelona, Spain." *J Med Entomol.* **48**(4): 956–960.

Naucke, T. J., B. Menn, D. Massberg, et al. (2008). "Sandflies and leishmaniasis in Germany." *Parasitol Res* **103**(Suppl. 1): S65–S68.

Navarro, J. M., C. Fernandez-Roldan, M. Pérez-Ruiz, et al. (2004). "Meningitis by Toscana virus in Spain: description of 17 cases." *Med Clin (Barc)* **122**(11): 420–422.

Navarro-Mari, J. M., B. Palop-Borras, M. Pérez-Ruiz, et al. (2011). "Serosurvey study of Toscana virus in domestic animals, Granada, Spain." *Vector Borne Zoonotic Dis* **11**(5): 583–587.

Nicoletti, L., P. Verani, S. Caciolli, et al. (1991). "Central nervous system involvement during infection by Phlebovirus toscana of residents in natural foci in central Italy (1977–1988)." *Am J Trop Med Hyg* **45**(4): 429–434.

Nikolay, B., M. Diallo, C. S. Boye, et al. (2011). "Usutu virus in Africa." *Vector Borne Zoonotic Dis* **11**(11): 1417–1423.

Pecorari, M., G. Longo, W. Gennaril, et al. (2009). "First human case of Usutu virus neuroinvasive infection, Italy, August–September 2009." *Euro Surveill* **14**(14): pii: 19446.

Perez-Ruiz, M., X. Collao, J. M. Navarro-Marí, et al. (2007). "Reverse transcription, real-time PCR assay for detection of Toscana virus." *J Clin Virol* **39**(4): 276–281.

Pherez, F. M. (2007). "Factors affecting the emergence and prevalence of vector borne infections (VBI) and the role of vertical transmission (VT)." *J Vector Borne Dis* **44**(3): 157–163.

Pierro, A., P. Gaibani, C. Manisera, et al. (2011). "Seroprevalence of West Nile virus-specific antibodies in a cohort of blood donors in northeastern Italy." *Vector Borne Zoonotic Dis* **11**(12): 1605–1607.

Pifat, D. Y. and J. F. Smith (1987). "Punta Toro virus infection of C57BL/6J mice: a model for phlebovirus-induced disease." *Microb Pathog* **3**(6): 409–422.

Poidinger, M., R. A. Hall, and J. S. Mackenzie (1996). "Molecular characterisation of the Japanese encephalitis serocomplex of the Flavivirus genus." *Virology.* **218**: 417–421.

Reiter, P. (2001). "Climate change and mosquito-borne disease." *Environ Health Perspect* **109** (1): 141–161.

Reusken, C. B., A. de Vries, J. Buijs, et al. (2010). "First evidence for presence of *Culex pipiens* biotype molestus in the Netherlands, and of hybrid biotype pipiens and molestus in northern Europe." *J Vector Ecol* **35**(1): 210–212.

Rice, C. M. (1996). "Flaviviridae: the viruses and their replication." *Virology* 931–960.

Rizzoli, A., R. Rosà, F. Rosso, et al. (2007). "West Nile virus circulation detected in northern Italy in sentinel chickens." *Vector Borne Zoonotic Dis* **7**(3): 411–417.

Roiz, D., A. Vázquez, R. Rosà, et al. (2012). "Blood meal analysis, flavivirus screening, and influence of meteorological variables on the dynamics of potential mosquito vectors of West Nile virus in northern Italy." *J Vector Ecol* **37**: 1–9.

Rubel, F., K. Brugger, M. Hantel, et al. (2008). "Explaining Usutu virus dynamics in Austria: model development and calibration." *Prev Vet Med* **85**(3–4): 166–186.

Saluzzo, J. F., G. W. Anderson, Jr., J. F. Smith, et al. (1989). "Biological and antigenic relationship between Rift Valley fever virus strains isolated in Egypt and Madagascar." *Trans R Soc Trop Med Hyg* **83**(5): 701.

Sanbonmatsu-Gamez, S., M. Perez-Ruiz, X. Collao, et al. (2005). "Toscana virus in Spain." *Emerg Infect Dis* **11**(11): 1701–1707.

Sánchez-Seco, M. P., J. M. Echevarria, L. Hernández, et al. (2003). "Detection and identification of Toscana and other phleboviruses by RT-nested-PCR assays with degenerated primers." *J Med Virol* **71**(1): 140–149.

Sánchez-Seco, M. P., A. Vázquez, X. Collao, et al. (2009). "Surveillance of arboviruses in Spanish wetlands: detection of new flavi- and phleboviruses." *Vector Borne Zoonotic Dis.* **10**(2): 203–206

Savini, G., F. Monaco, C. Terregino, et al. (2011). "Usutu virus in Italy: an emergence or a silent infection?" *Vet Microbiol* **151**(3–4): 264–274.

Schwarz, T. F., S. Gilch, and G. Jager (1995a). "Aseptic meningitis caused by sandfly fever virus, serotype Toscana." *Clin Infect Dis* **21**(3): 669–671.

Schwarz, T. F., G. Jager, S. Gilch, et al. (1995b). "Nested RT-PCR for detection of sandfly fever virus, serotype Toscana, in clinical specimens, with confirmation by nucleotide sequence analysis." *Res Virol* **146**(5): 355–362.

Schwarz, T. F., G. Jager, S. Gilch, et al. (1995c). "Serosurvey and laboratory diagnosis of imported sandfly fever virus, serotype Toscana, infection in Germany." *Epidemiol Infect* **114**(3): 501–510.

Schwarz, T. F., S. Gilch, C. Pauli, et al. (1996a). "Immunoblot detection of antibodies to Toscana virus." *J Med Virol* **49**(2): 83–86.

Schwarz, T. F., G. Jager, S. Gilch, et al. (1996b). "Travel-related vector-borne virus infections in Germany." *Arch Virol Suppl* **11**: 57–65.

Schwarz, T. F., S. Gilch, and H. M. Schätzl (1998). "A recombinant Toscana virus nucleoprotein in a diagnostic immunoblot test system." *Res Virol* **149**(6): 413–418.

Soldateschi, D., G. M. dal Maso, M. Valassina, et al. (1999). "Laboratory diagnosis of Toscana virus infection by enzyme immunoassay with recombinant viral nucleoprotein." *J Clin Microbiol* **37**(3): 649–652.

Solomón, T. (2004). "Flavivirus encephalitis." *N Engl J Med* **351**: 370–378.

Steinmetz, H. W., T. Bakonyi, H. Weissenböck, et al. (2011). "Emergence and establishment of Usutu virus infection in wild and captive avian species in and around Zurich, Switzerland—genomic and pathologic comparison to other central European outbreaks." *Vet Microbiol* **148**(2–4): 207–212.

Tamba, M., P. Bonilauri, R. Bellini, et al. (2010). "Detection of Usutu virus within a West Nile virus surveillance program in Northern Italy." *Vector Borne Zoonotic Dis*. September 17.

Tesh, R. B. and S. M. Duboise (1987). "Viremia and immune response with sequential phlebovirus infections." *Am J Trop Med Hyg* **36**(3): 662–668.

Tesh, R. B., J. Lubroth, H. Guzman, et al. (1992). "Simulation of arbovirus overwintering: survival of Toscana virus (Bunyaviridae: Phlebovirus) in its natural sand fly vector *Phlebotomus perniciosus*." *Am J Trop Med Hyg* **47**(5): 574–581.

Valassina, M., M. G. Cusi, P. E. Valensin, et al. (1996). "Rapid identification of Toscana virus by nested PCR during an outbreak in the Siena area of Italy." *J Clin Microbiol* **34**(10): 2500–2502.

Valassina, M., D. Soldateschi, G. M. D. Maso, et al. (1998). "Diagnostic potential of Toscana virus N protein expressed in *Escherichia coli*." *J Clin Microbiol* **36**(11): 3170–3172.

Valassina, M., M. Valentini, P. E. Valensin, et al. (2002). "Fast duplex one-step RT-PCR for rapid differential diagnosis of entero- or toscana virus meningitis." *Diagn Microbiol Infect Dis* **43**(3): 201–205.

Vazquez, A., M. Jimenez-Clavero, L. Franco, et al. (2011a). "Usutu virus—potential risk of human disease in Europe." *Euro Surveill* **16**(31): pii: 19935.

Vázquez, A., S. Ruiz, L. Herrero, et al. (2011b). "West Nile and Usutu viruses in mosquitoes in Spain, 2008–2009." *Am J Trop Med Hyg* **85**(1): 178–181.

Venturi, G., G. Madeddu, G. Rezza, et al. (2007). "Detection of Toscana virus central nervous system infections in Sardinia Island, Italy." *J Clin Virol* **40**(1): 90–91.

Verani, P., L. Nicoletti, and A. Marchi (1984). "Establishment and maintenance of persistent infection by the Phlebovirus Toscana in Vero cells." *J Gen Virol* **65**(Pt 2): 367–375.

Verani, P., M. G. Ciufolini, S. Caciolli, et al. (1988). "Ecology of viruses isolated from sand flies in Italy and characterized of a new Phlebovirus (Arabia virus)." *Am J Trop Med Hyg* **38**(2): 433–439.

Weidmann, M., M. P. Sánchez-Seco, A.A. Sall, et al. (2008). "Rapid detection of important human pathogenic phleboviruses." *J Clin Virol* **41**(2): 138–142.

Weissenböck, H., J. Kolodziejek, A. Url, et al. (2002). "Emergence of Usutu virus, an African mosquitoborne flavivirus of the Japanese encephalitis virus group, central Europe." *Emerg Infect Dis* **8**: 652–656.

Weissenböck, H., J. Kolodziejek, K. Fragner, et al. (2003). "Usutu virus activity in Austria, 2001–2002." *Microbes Infect* **5**: 1132–1136.

Weissenböck, H., T. Bakonyi, S. Chvala, et al. (2004). "Experimental Usutu virus infection of suckling mice causes neuronal and glial cell apoptosis and demyelination." *Acta Neuropathol* **108**(5): 453–460.

Weissenböck, H., Z. Hubálek, T. Bakonyi, et al. (2010). "Zoonotic mosquito-borne flaviviruses: worldwide presence of agents with proven pathogenicity and potential candidates of future emerging diseases." *Vet Microbiol* **140**(3–4): 271–280.

Williams, M. C., D. I. Simpson, A. J. Haddow, et al. (1964). "The isolation of West Nile Virus from man and of Usutu virus from the bird-biting mosquito Mansonia aurites (Theobald) in the Entebbe area of Uganda." *Ann Trop Med Parasitol* **58**: 367–374.

Zhioua, E., G. Moureau, I. Chelbi, et al. (2010). "Punique virus, a novel phlebovirus, related to sandfly fever Naples virus, isolated from sandflies collected in Tunisia." *J Gen Virol* **91**(Pt 5): 1275–1283.

Tamba, M., P. Bonilauri, R. Bellini, et al. (2010). "Detection of Usutu virus within a West Nile virus surveillance program in Northern Italy." *Vector Borne Zoonotic Dis.* September 17.

Tesh, R. B. and S. M. Duboise (1987). "Viremia and immune response with sequential phlebovirus infections." *Am J Trop Med Hyg* 36: 662–668.

Tesh, R. B., J. Lubroth, H. Guzman, et al. (1992). "Simulation of arbovirus overwintering: survival of Toscana virus (Bunyaviridae: Phlebovirus) in its natural sand fly vector *Phlebotomus perniciosus*." *Am J Trop Med Hyg* 47(5): 574–581.

Valassina, M., M. G. Cusi, P. E. Valensin, et al. (1998). "Rapid identification of Toscana virus by nested PCR during an outbreak in the Siena area of Italy." *J Clin Microbiol* 36(10): 2503–2504.

Valassina, M., D. Soldateschi, F. Dal Maso, et al. (1998). "Diagnostic potential of Toscana virus N protein expressed in *Escherichia coli*." *J Clin Microbiol* 36(12): 3670–3672.

Valassina, M., M. Valentini, A. Valensin, et al. (2002). "Fast duplex one-step RT-PCR for rapid differential diagnosis of entero- or Toscana virus meningitis." *Diagn Microbiol Infect Dis* 43(3): 201–205.

Vazquez, A., M. Jimenez-Clavero, L. Franco, et al. (2011). "Usutu virus – potential risk of human disease in Europe." *Euro Surveill* 16(31): pii: 19935.

Vazquez, A., S. Ruiz, L. Herrero, et al. (2011). "West Nile and Usutu viruses in mosquitoes in Spain, 2008–2009." *Am J Trop Med Hyg* 85(1): 178–181.

Venturi, G., C. Madeddu, G. Rezza, et al. (2007). "Detection of Toscana virus central nervous system infections in Sardinia Island, Italy." *J Clin Virol* 40(1): 90–91.

Verani, P., L. Nicoletti, and A. Marchi (1984). "Establishment and maintenance of persistent infection by the Phlebovirus Toscana in Vero cells." *Arch Virol* 65(2): 367–375.

Verani, P., M. G. Ciufolini, S. Caciolli, et al. (1988). "Ecology of viruses isolated from sand flies in Italy and characterization of a new Phlebovirus (Arbia virus)." *Am J Trop Med Hyg* 38(2): 433–439.

Weidmann, M., M. P. Sanchez-Seco, A. A. Sall, et al. (2008). "Rapid detection of important human pathogenic phleboviruses." *J Clin Virol* 41(2): 138–142.

Weissenböck, H., J. Kolodziejek, A. Url, et al. (2002). "Emergence of Usutu virus, an African mosquito-borne flavivirus of the Japanese encephalitis virus group, central Europe." *Emerg Infect Dis* 8: 652–656.

Weissenböck, H., J. Kolodziejek, K. Fragner, et al. (2003). "Usutu virus activity in Austria, 2001–2002." [illegible]

Weissenböck, H., T. Bakonyi, S. Chvala, et al. (2004). "Experimental Usutu virus infection of suckling mice causes neuronal and glial cell apoptosis and demyelination." *Acta Neuropathol* 108(5): 453–460.

Weissenböck, H., Z. Hubálek, T. Bakonyi, et al. (2010). "Zoonotic mosquito-borne flaviviruses: worldwide presence of agents with proven pathogenicity and potential candidates of future emerging diseases." *Vet Microbiol* 140(3–4): 271–280.

Williams, M. C., J. P. Woodall, A. J. Haddow, et al. (1964). "The isolation of West Nile virus from man and of Usutu virus from the bird-biting mosquito *Mansonia aurites* (Theobald) in the Entebbe area of Uganda." *Ann Trop Med Parasitol* 58: 367–374.

Xiao, S. Y., [illegible]

BORNA DISEASE VIRUS AND THE SEARCH FOR HUMAN INFECTION

Kathryn M. Carbone

Division of Intramural Research, NIDCR/NIH, Bethesda, MD, USA

Juan Carlos de laTorre

IMM-6 The Scripps Research Institute La Jolla, CA, USA

TABLE OF CONTENTS

30.1	Introduction	558
30.2	Long-standing controversy around BDV as a human pathogen	559
30.3	A negative is impossible to prove, but do we have enough evidence to stop looking?	560
30.4	Recent improvements in testing for evidence of BDV in human samples	562
	30.4.1 Serology	562
	30.4.2 Nucleic acid tests	563
30.5	The possibilities for clinical expression of human BDV infection are myriad and almost impossible to predict	563
30.6	Epidemiology: the "new" frontier of human BDV studies?	565
30.7	Where do we go from here?	566
	Acknowledgement	568
	References	568

Viral Infections and Global Change, First Edition. Edited by Sunit K. Singh.
© 2014 John Wiley & Sons, Inc. Published 2014 by John Wiley & Sons, Inc.

30.1 INTRODUCTION

Over a century before the specific agent of disease was identified (de la Torre et al., 1990; Lipkin, 1990), "classic" Borna disease (BD) was well known as an endemic and sometimes epidemic encephalitis and killer of horses and sheep in Central Europe (as reviewed in Staeheli et al., 2000). In the two decades since the discovery that the etiological agent of BD was an enveloped virus with a non-segmented, negative-strand (NNS) RNA genome, Borna disease virus (BDV) has been detected in a large variety of species, including the horse, sheep, cattle, bird, mouse, rat, rabbit, llama, ostrich, fox, dog, and monkey, on most continents of the world (as reviewed by Lipkin et al., 2011).

BDV is extraordinarily neurotropic and persistently infects the neurons, astrocytes, and oligodendrocytes in the central and peripheral nervous systems, as well as a small number of thymus-resident and blood-circulating fibroblastic stromal cells (Carbone et al., 1987; Carbone et al., 1989; Narayan et al., 1983a; Rubin et al., 1995; Sierra-Honigmann et al., 1993). BDV infection is associated with a wide range of disease manifestations ranging from asymptomatic infection to behavioral disease to fatal mononuclear encephalitis (Carbone et al., 2001; Narayan et al., 1983b; Stitz et al., 1995). BDV has also been reported to be the etiological agent of avian proventricular dilatation disease (PDD) (Staeheli et al., 2010). BDV has also served as an intriguing model system for studying the pathogenesis of many diseases, perhaps the most interesting of which is autism, where the virus produces behavioral (e.g., cognitive, social), neuropathological (e.g., cerebellar), and neurochemical (e.g., serotonin) changes reminiscent of those reported in children with autism (Pletnikov et al., 1999; Pletnikov et al., 2002). This wide range of disease manifestations reflects the fact that BDV both can cause immune-mediated neurological disease, that is, resulting largely from the direct and indirect damage due to the massive mononuclear inflammatory response generated in response to the virus, and can cause more subtle behavioral diseases, for example, when replicating in neural tissues without a major inflammatory response. Consistent with these findings, in tissue culture, BDV replicates in cells from a wide range of species, with predilection for cells of the nervous system, and often causes a persistent and non-lytic infection (Herzog and Rott, 1980).

BDV is an enveloped virus with a non-segmented, negative-strand (NNS) RNA genome (de la Torre, 1994; Schneemann et al., 1995). BDV virions have a spherical morphology with a diameter ranging from 70 to 130 nm, containing an internal electron-dense core (50–60 nm) and a limiting outer membrane envelope covered with spikes approximately 7.0 nm long (Kohno et al., 1999). BDV has evolved a non-cytolytic replication strategy where production of cell-free virus is conspicuously absent and cell-associated infectivity is often extremely low. The BDV genome (c. 8.9 kb), the smallest among known animal NNS RNA viruses, has a gene organization that is characteristic of the order Mononegavirales (MNV) (de la Torre, 1994; Schneemann et al., 1995). As with other MNV genomes, untranslated regions (UTRs) known as the leader (Le) and trailer (Tr) sequences are found at the 3' and 5' ends of the BDV genome, respectively. As with other MNV genomes, BDV exhibits limited terminal complementarity that reflects the presence of cis-acting sequences at the genome termini that play key roles in transcription and RNA replication, as well as genome packaging into virus particles (Conzelmann, 2004). The BDV genome contains six major open reading frames (ORFs) (de la Torre, 1994; Schneemann, et al., 1995) that code for polypeptides with predicted M_r of 40 kDa (p40), 23 kDa (p23), 10 kDa (p10), 16 kDa (p16), 57 kDa (p57), and 180 kDa (p190). Based on their positions in the viral genome (3′-N-p10/P-M-G-L-5'), together with their biochemical sequence and functional features, these polypeptides correspond to the nucleoprotein (N), phosphoprotein transcriptional activator (P), matrix (M), surface glycoprotein (G), and RNA-dependent RNA polymerase (L) polypeptides found in other NNS RNA viruses. The p10 ORF, also called X, encodes a polypeptide of 10 kDa that is readily detected in BDV-infected cells (Wehner et al., 1997), but does not appear to be present in virus particles (Schwardt et al., 2005). The synthesis of BDV p10 starts within the same mRNA transcription unit as P but 49 nt upstream, and it overlaps, in a different frame, with the 71 N-terminal amino acids of P. The G ORF overlaps, in a different frame, with the C-terminus of M. The G gene directs the synthesis of the G protein precursor GPC, which has a predicted M_r of c. 56 kDa but, due to its extensive glycosylation, migrates with a M_r of 84–94 kDa. GPC is posttranslationally cleaved by the cellular protease furin into GP-1 (GP_N) and GP-2 (GP_C) corresponding to the N- and C-terminal regions, respectively, of G (Gonzalez-Dunia et al., 1997; Richt et al., 1998). Antibodies to BDV G readily detect both GPC and GP-2 (c. 43 kDa) in BDV-infected cells, whereas the detection of GP-1 is complicated by the high content of N-glycans that shield antigenic sites (Kiermayer, 2002). The BDV L ORF contains conserved domains and motifs characteristic of MNV L proteins (de la Torre, 1994; Poch et al., 1990; Schneemann, et al., 1995). BDV has the property, unique among known animal NNS RNA viruses, of a nuclear site for the replication and transcription of its genome and the use of RNA splicing for the regulation of its genome expression. Likewise, recent evidence indicates that BDV uses a unique mechanism of terminal realignment and elongation to remove 5'-triphosphate nucleotides without compromising genome integrity (Martin et al., 2011). This process likely contributes to BDV persistence as 5′-triphosphate groups are potent pathogen-associated molecular patterns recognized by the host–pathogen recognition receptors responsible for triggering innate immune defense mechanisms. Based on its unique genetic and biological features, BDV has been determined to be the prototypic member of a new virus family, Bornaviridae, within the order MNV.

30.2 LONG-STANDING CONTROVERSY AROUND BDV AS A HUMAN PATHOGEN

The potential for BDV to infect and cause disease in humans has been extremely controversial since the original publication documenting serological evidence of human infection in 1985 (Rott et al., 1985). Thus, many studies on humans and BDV infection have been published, citing both positive and negative findings, and this topic has been the subject of a number of review articles (Carbone, 2001; Chalmers et al., 2005; Dürrwald et al., 2007; Schwemmle, 2001). Initial studies examining BDV infection in humans relied on subjective serological assays to detect human serum antibodies to BDV using differential signals on infected and normal cell substrates; this likely contributed to the inconsistency of findings among different laboratories. Subsequently, the identification and molecular characterization of the BDV genome in the early 1990s allowed the development of highly specific and sensitive genome-based diagnostic tests, including RT-PCR (Sierra-Honigmann et al., 1993).

Since the 1990s, using a variety of tests, humans with a large range of neurological and psychiatric disease diagnoses from a number of geographic regions have been tested for BDV infection with variable results. Several excellent reviews on the subject have discussed the limitations of these studies, including unvalidated testing methodologies and the absence of rigorous clinical study design, making it difficult to interpret both claims of human infection with BDV and those disease states found to be associated with BDV (Carbone, 2001; Chalmers et al., 2005; Dürrwald et al., 2007; Schwemmle, 2001). The various weaknesses affecting the testing methodology and clinical study design have been thoroughly documented in reviews cited in the chapter references and will not be rediscussed extensively here. These issues and others can be illustrated in just a few recent human studies cited in Table 30.1 and Table 30.2 as examples of "positive" and "negative" studies, respectively.

TABLE 30.1. Selected Negative Clinical Studies

Patient groups[a]	Location	Method	Genome	Results	References
Depression Schizophrenia Bipolar	Korea	RT-PCR (IFA)	P, N	0/198	Na et al. 2009
Normal (jockeys)	Korea	RT-PCR (IFA)	P, N	0/48	Song et al. 2011
Schizophrenia Bipolar disorder Major depression disorder	The United States	RT-PCR, ELISA, WB (IFA)	P, N	0/198	^Hornig et al. 2012

[a]Studies with (^) include patient/control clinical characterization

TABLE 30.2. Selected Positive Clinical Studies

Patient[a]	Location	Method	Genome	Results	References
Agricultural employees	The United Kingdom	ELISA (IFA)	N		Thomas et al. 2005
1994				12/525	
1996				15/489	
1999				11/422	
Schizophrenic/ schizoaffective	Brazil	RT-PCR	P	12/27	^Nunes et al. 2008
First-degree relatives—healthy				10/20	
First-degree relatives—mood disorder				9/24	
Healthy controls				4/27	
Viral encephalitis of unknown etiology	China, Japan	RT-PCR, WB	N, P	6/40	Li et al 2009
#Alcohol and drug dependence	Czech Republic	CIC ELISA	N, P	15/41 and 12/28	Rackova et al. 2010
Controls (blood donors)				47/126	

[a]Studies with (^) include patient/control clinical characterization
#D0 and D56 of hospitalization, respectively

Initial excitement in the possibility of the discovery of BDV-infected humans has waned, and many scientists now believe that BDV does not target humans at all. What seems to be currently underappreciated, however, are data from scientific disciplines that focus on the environmental setting, population characteristics, and socioeconomic properties surrounding the infection, rather than solely on the infected individual, that provide information that may be the key to resolve some of the controversy surrounding BDV infections of humans.

30.3 A NEGATIVE IS IMPOSSIBLE TO PROVE, BUT DO WE HAVE ENOUGH EVIDENCE TO STOP LOOKING?

Given the contrast between the scientific community's general agreement that natural animal infection occurs in a wide range of species and in many, if not all, of the world's continents and the controversy about the existence of human BDV infection, one might assume that the failure to prove human BDV infection means it does not exist. If not, then why does the latter possibility continue to be researched and debated if the failure so far to identify clear, incontrovertible evidence of humans infected with BDV may lead most to conclude that humans are not susceptible to BDV infection and that there is no human disease associated with BDV infection? In fact, many scientists who initially published evidence in support of human BDV infection are now firmly in the "nay" camp. So why continue to search?

The history of the search for viruses and viral diseases provides some clues to consider for improving the search for human BDV infection. The history of infectious diseases shows that many human pathogens are not indentified until the circumstances (e.g., society, behavior, environment) create the right opportunity for viruses to make themselves clearly known. Examples include polio paralysis from poliovirus (associated with improving hygiene and exposure to the virus at older ages) (Nathanson and Martin, 1979), influenza virus (the discovery of infectious "unfilterable" agents) (Smith et al., 1933), human immunodeficiency virus (HIV) (changes in human sexual behavior) (Barré-Sinoussi et al., 1983), and the SARS coronavirus (exposures to infected animals in live-animal markets) (Holmes, 2003). An untargeted search for HIV would have been unsuccessful in the late 1960s, even when scientists who had already identified similar animal retroviruses in mouse, goat, sheep, and horses predicted the high likelihood of the existence of a human retrovirus. Without the proper reagents and knowledge of the "at risk" populations or even the clinical syndrome associated with human disease, the search for the first human immunodeficiency retrovirus would have been guaranteed futile. Once the social and environmental setting supported the rapid spread of the virus, the long latency of the acquired immunodeficiency syndrome (AIDS) epidemic meant that the virus was widespread before it was firmly identified as a human pathogen. When the epidemiological awareness of AIDS and the testing methodology advances made the existence of this human retrovirus infection indisputable and scientists knew where and how to look for the virus, early and isolated cases were in fact identified decades before the 1980s AIDS epidemic occurred (Zhu et al., 1998). Consider, too, that although the AIDS epidemic started and was already widespread in the underdeveloped countries of Africa, the first cases to come to the attention of many developed countries were of individuals returning from Africa to their native European countries, where clinical medical access and diagnostic tests that were readily available uncovered the unusual syndrome of AIDS as a novel syndrome, that is, diseases may exist for a long time in areas where medical care is limited and go largely unrecognized for years.

How do these historical lessons apply to the search for human BDV infection? Since 1985 most searches for BDV in humans have occurred in developed countries and in humans with various neuropsychiatric diseases or exposure to domesticated large animal known to be targets of BD. But do we really know which people are most at risk for BDV infection? Are we looking in the right place and at the correct populations?

Since direct evidence of human BDV infection has not been proven beyond scientific doubt, is there indirect evidence that supports the likelihood that BDV infects humans and, thereby, might justify continuing to search for human BDV infection? *In vitro* studies have shown that human cells are remarkably easily infected and replicate BDV as well as cells from any natural host. Nonhuman primates can be experimentally infected with BDV and develop encephalitic and retinal disease (Stitz et al., 1981). Certainly there exist a number of cases of idiopathic encephalitis in humans as well as many psychiatric diseases without clear etiologies.

Recent publications noted the presence of ancient BDV genomic sequences within the germline human genome. These endogenous viruslike DNA sequences appear to be LINE element-facilitated integration derived from viral mRNA, including parts of the N ORF, suggesting that at some time in history, humans were infected by BDV or BDV-like viruses (Belyi et al., 2010; Horie et al., 2010). Some hypothesize that species that carry these expressed BDV sequences are *less* likely to be susceptible to BDV-associated disease due to the possibility that infection could produce double-stranded RNA, thus stimulating endogenous virus-killing interferon, or that defective assembly from the abnormal protein expressed by the endogenous ORF would interfere with the virus life cycle (Belyi et al, 2010). This hypothesis, however, seems difficult to reconcile with the fact that many human-derived cell lines are highly susceptible to BDV.

In sum, there is indirect evidence that the search for human BDV infection is not futile. However, this hypothesis begs the question, however: what experimental design changes are needed to improve the likelihood that human BDV research will come to some reasonably well-accepted conclusion, be that conclusion "yeah" or "nay"?

30.4 RECENT IMPROVEMENTS IN TESTING FOR EVIDENCE OF BDV IN HUMAN SAMPLES

Previous and extensive reviews have concluded that problems in BDV detection methodology and clinical study methodology have seriously limited the ability to draw conclusions from studies on BDV infection in humans. Testing methodologies that include serology and highly sensitive nucleic acid tests (Dürrwald et al., 2007) have concerns regarding the specificity and/or sensitivity of the assays, for example, contamination with laboratory virus. A brief summary of the testing methodology is presented here.

30.4.1 Serology

In 1985, the first report of anti-BDV antibodies in human sera was published using the indirect fluorescent antibody test (IFA) (Rott et al., 1985). This method detects the presence of anti-BDV antibodies by comparison of the patterns of binding of human serum to infected and uninfected cells. The IFA is highly subjective and reader dependent, and normalizing specificity and sensitivity in a *reproducible* fashion between laboratories is extraordinarily difficult. The Western blot (WB) provided an additional level of specificity as a serological test because the results could be reasonably compared to control lanes stained with infected animal serum and identity of specific BDV antigens in each human serum could be reasonably ascertained via the detection of specific-sized bands of proteins by serum antibodies (Waltrip et al., 1995). BDV is highly membrane bound, and purification of virus proteins from contaminating cellular components was difficult; however, the isolation of the BDV genome facilitated the *in vitro* production of large amounts of pure recombinant viral antigens and allowed the development of sensitive, quantitative serological assays such as ELISA. Since the recombinant BDV antigens are mainly produced in bacteria, the lack of specific posttranslational modifications may result in the loss of some epitopes. Nevertheless, the use of several different virus antigens to probe the polyclonal response in human sera minimized this concern. Notably, conformational epitope information may be provided by RIA tests that have been proposed to be more sensitive to the detection of conformational epitopes (Matsunaga et al., 2005). Measuring BDV circulating immune complexes (CIC) has been also proposed as a highly specific and sensitive assay to detect BDV infection in humans (Bode et al., 2001), but this assay has never been widely used and no evidence has been presented that the reported CIC contain bona fide BDV antigens. The overall meaning from the results of serological testing is further limited by having no agreement as to what signal constitutes a positive response (e.g., antibodies to which BDV proteins, or combination of proteins, constitute a positive response; no agreed-upon signal threshold of positivity; and lack of a positive human control serum), thus further complicating the analysis of serological results.

30.4.2 Nucleic acid tests

The presence of BDV nucleic acid in biological samples has been approached via several techniques suitable for RNA detection such as Northern blot, the more sensitive amplification techniques such as RT-PCR, and related techniques including real-time quantitative RT-PCR (RTqPCR). In naturally or experimentally infected animals, BDV may replicate to low titers, and even in known infected animals, the detection of BDV RNA may require the highly sensitive RT-PCR-based assays (Sierra-Honigmann et al., 1993). Because the high-sensitivity assays are prone to contaminations with laboratory-derived sources of viral RNA, to be reliable they must be properly controlled with implementation of strict quality control procedures. Further, all results from RT-PCR assays should be complemented with subsequent sequencing of the amplicons to confirm the BDV sequence. The analysis of amplicon-derived sequences also provides very valuable information to rule out, or identify, potential contamination problems (Dürrwald et al., 2007). More recently, the use of metagenomic analysis of unbiased high-throughput sequencing data obtained using RNA isolated from biological samples of interest has revolutionized the ability to identify candidate etiological agents in infectious diseases of unknown etiology (Mokili et al., 2012; Relman, 2011; Tang and Chiu, 2010). Future developments are likely to diminish the cost of the use of this technology, thus making its incorporation into the field of molecular epidemiology possible.

A summary of recent human BDV studies is summarized in Table 30.1 and Table 30.2, consisting of negative study outcomes and positive study outcomes, respectively. These studies highlight the dilemma in understanding the results of human BDV research, for example, studies of a wide range of types of human disease, testing methodologies, and geographic areas. The number of subjects studied is relatively small, and each study, in essence, is a "one-off" with little that can be done to summarize data between laboratories or combine results over time to see a specific trend. The good news, however, is that increasingly studies now utilize multiple testing methods on the same subjects, some are testing subjects repeatedly over time, and some studies are providing enough clinical study design information to allow more confidence in a meaningful comparison of subjects and controls. Still, even with these improvements, recognition of a "trend" that lends toward resolution of the "yeah" and "nay" human BDV-infection camps is not obviously forthcoming.

30.5 THE POSSIBILITIES FOR CLINICAL EXPRESSION OF HUMAN BDV INFECTION ARE MYRIAD AND ALMOST IMPOSSIBLE TO PREDICT

To date, most scientists have assumed that BDV might be an etiology of human psychiatric disease, and clinical studies have tended to focus on patients with a variety of psychiatric diseases. A wide range of subjects with other illnesses have also been tested. However, predicting the clinical expression of BDV infection in humans (if, indeed, it occurs) may be extremely difficult. While viral infections may contribute to a number of neurobiological diseases, our understanding of the specific pathogenesis and mechanisms of viral diseases of the nervous system is limited (as reviewed in van den Pol, 2009). The expression of viral-induced neurobehavioral disease depends on both viral and host factors, as well as route of infection and size of the initial virus inoculums, for example, disease outcome may vary by how the virus reaches the nervous system and route of entry (e.g., poliovirus, herpes simplex virus). Viruses may damage the nervous system or even kill the host via direct lytic damage to key neurons or, like rabies, through mechanisms where critical neuronal functions such as neurotransmitter release are affected to the degree that the neurons cease to perform their specialized roles, but the neurons themselves do not die (de la Torre et al., 1991). Indirect damage by the inflammatory response to virus infection can also be a significant outcome of viral infections. The brain is essentially a topographically organized organ with specific brain regions assigned specific functions; thus, the expression of a single type of disease can result from infection with different viruses that damage the same area of the brain. Similarly, many different diseases can be caused by a single virus depending upon differences in the major site of infection and pathological damage, as seen in herpes simplex encephalitis of the temporal lobe versus the common herpetic "cold sore" (Smith, 2012). Viral infections of the nervous system can even be subclinical, as the "normal viral flora of the brain" (e.g., JC virus) may lie dormant and unobserved until a trigger is introduced, such as when drug-induced immunosuppression results in JC virus-induced progressive multifocal leukoencephalopathy (Berger and Khalili, 2011). Age of the host at time of infection can significantly affect disease expression, for example, herpes simplex virus infections are often fatal in newborns and result in minor "cold sores" when infected as an adult. Other host genetic contributions to virus pathogenicity are seen, as in HIV infection where the genetically determined sequences of virus co-receptors on the surface of the cell (e.g., CCR3, CCR5) may affect the host's risk of infection and/or disease progression from HIV (He et al., 1997). The tight bond between differences in interactions of the host and the virus and variability in expression of viral diseases can make the demonstration of causal associations difficult, especially if the incubation period is long or the virus causes neurological disease in a minority of infected hosts.

BDV infection results in a myriad of disease states in the naturally or experimentally infected animal because this virus embodies many of the features that lead to variability in the neurobehavioral disease due to nervous system infection. For example, genetics of the host affects BD expression (e.g., classic fatal encephalitis of horses and sheep vs. birds with PDD vs. behavioral disease in mice). Within the same species, variability in BD outcome has been shown to be associated with the age of the host when infected, often due to differences in the susceptibility of the nervous system (e.g., neurodevelopmental damage) or immune state of the animal (e.g., immune tolerance vs. encephalitic response) at time of infection. In the experimental animal, changes in early BD outcomes are also seen depending upon the route of infection. Therefore, potential human disease caused by BDV infection could range from fatal encephalitis to focal damage of the eyes or gut to developmental brain damage to purely behavioral illness. Thus, the narrow and focused selection of subjects with specific diseases to study for evidence of BDV infection has given inconsistent results and is unlikely to identify human BDV infection, much less document association with a specific human disease.

On a molecular level, the mechanisms by which BDV damages the nervous system are also multifaceted. The analysis of experimental BDV infection of primary neurons demonstrated selective interference with proteins important in neurotransmission, neurogenesis, cytoskeleton dynamics, regulation of gene expression, and chromatin remodeling, as well as the adaptive response of neurons (Suberbielle et al., 2008). BDV protein-associated changes have been demonstrated in neurosynaptic adaptation and function or neurotransmitter levels in experimental settings (Peng et al., 2008). But not all of the BDV-associated disease is due to neuronal infection, as other studies using transgenic mice suggest that BDV proteins expressed in glial cells may also lead to a number of BDV-associated abnormalities (e.g., hyperactivity, spatial memory, and learning abnormalities) (Kamitani, 2003). *In vitro* studies demonstrate that neurogenesis of human cells can be altered by BDV infection (Brnic et al., 2012); these data suggest the possibility of variability in BD related to age at time of infection of humans. All these data suggest that focusing on a small number of subjects in a short list of human diseases whose manifestations may be compatible with BDV is somewhat akin to playing a lottery and bound to fail as we have little information to tell us the specific human diseases, if any, that are more or less likely to be associated with BDV infection in the human brain.

30.6 EPIDEMIOLOGY: THE "NEW" FRONTIER OF HUMAN BDV STUDIES?

Most of the information about BD is based on disease incidence reporting in domesticated livestock in endemic areas; relatively few formal epidemiological studies have been published. Based on these observations, classic BD in horses and sheep (Dürrwald et al., 2006) occurs in the form of recurrent epidemics; one of the first reported was an epidemic in cavalry horses in 1885. With improved sanitation and industrialization in Europe, a decreasing frequency of epidemics since the 1800s is associated. Variability in frequency is seen year to year, but when it does appear, sheep and horses tend to trend in the same direction together. Observers noted seasonal peaks in spring and early summer, thus drawing an association with release of livestock to graze on pasture, rodent mating/birthing activity, and soil work for planting. Suggestions have been made of a linking of increased frequency with density of livestock housed, traditional farm practices, the keeping of many different species together, and poorer hygiene.

Although many of the conclusions of trends are based on observations rather than formal scientific study, most of this information suggests a reservoir/vector exists for the virus outside of the domesticated livestock; as with many viruses, rodents or rodent-like species have been the main vectors considered and studied.

Laboratory studies of BDV-infected rodents have been extensive. There is substantial experimental evidence for horizontal and vertical transmission of BDV between laboratory rodents (Okamoto et al., 2003; Sauder and Staeheli, 2003). Rodent-borne viruses tend to cause increased incidence in late summer and autumn, as the rodents move inside the houses and barns in preparation for the winter. Experimental studies in rodents have commonly shown the possibility of persistent infection, mild disease, and secretion of live virus following neonatal or in utero infection, so that late spring/early summer epidemics in livestock may coincide with mating activity and births of small, persistently and neonatally infected animals in and around livestock.

Much of the testing of wild rodents occurred before nucleic acid tests for BDV were developed, and house mice and brown and black rats did not carry the virus even in endemic areas; even house mice of sheep farm with high incidence were found to be negative (Vahlenkamp et al., 2002). However, using nucleic acid testing methods in a study of wild moles, shrews, and mice in an area endemic for BD of horses and sheep, only the tested white-toothed shrews (*Crocidura leucodon*) were found to be persistently infected without disease (Hilbe et al., 2006; Puorger et al., 2010). Interestingly, these ground-dwelling insectivores are often found in the same area as endemic BD, and they may forage in the same pastures in spring/early summer as the farm animals. Unfortunately, although experimental horizontal transmission of BDV has been purported to occur by urine contact in neonatally infected rats, no shrew urine was examined in the initial study; a subsequent study did identify infected cells in the respiratory and urogenital tracts. Recently, experimental infection of wild, neonatal bank voles (*Myodes glareolus*) led to animals with widespread infectious virus, including secretion in the feces and urine (Kinnunen, 2011). Also of great interest, the authors reported finding BDV N and P DNA in some of the voles, suggesting the possibility of genomic integration. In addition to rodents, other wild animal vectors have been suggested, such as birds. Avian BDV was detected in half of a wild goose flock tested in North America; sequence data suggests an independent cluster with some identity with traditional non-avian BDV (Payne et al., 2011).

Clearly, BDV is a documented infection of several species of animals in the natural setting, and evidence is mounting that suggests BDV may be a vector-borne infection of domesticated large animals with a more complicated life cycle than previously suspected. It is reasonable to assume, therefore, that if human BDV occurs, it may be a zoonotic infection (i.e., spread from animal hosts to humans) rather than spread between humans. However, unlike previously suspected, it is possible that neonatally infected wild rodents or other small animal hosts, rather than large domesticated livestock, serve as the vector for humans.

30.7 WHERE DO WE GO FROM HERE?

The assessment of BDV as a suspected but unproven Level 2 zoonosis (Palmer et al., 2005) was based on poor specificity of human testing methodologies with a recommendation for enhanced surveillance. Because the serological response to BDV and viral spread within the infected animal is quite variable between species and can be variable even within a species, the most conservative approach to BDV diagnosis in humans lies in using high-quality, modern tests with validated sensitivity and specificity (e.g., RTqPCR in a setting of exquisitely careful avoidance of contamination and with sequence information, combined with ELISA and WB testing using at least two different BDV antigens) and, if possible, using positive controls using nonhuman primate samples. Using stringent criteria, at least two carefully controlled studies have failed to confirm evidence of BDV infection in specific human patient populations with psychiatric disease (see Table 30.1). Basic science advances have taken the field as far as it can go until new molecular testing methods are developed and until there is a human positive control. With identification of an infected human, isolation and sequencing of a virus strain from human samples can be utilized as a positive control to develop and validate new testing methods.

For decades, clinical research methodologies have been less than ideal in many published BDV human studies, based on a review of the reported subject and control definitions and clinical data with little detailed information on methods of diagnosis, stage of disease, and treatment status. In addition, sex, sociodemographic, or economic matching is rarely performed with control subjects. Without this information, any conclusions of association of BDV infection (even if proven) and human disease will remain unproven. Over time, some studies, however, have provided more information on well-screened patient and well-matched control subjects; however, this clinical design improvement has still not resolved the controversy of human BDV infection.

30.7 WHERE DO WE GO FROM HERE?

The recent negative human BDV studies that utilize higher standards of diagnostic testing and clinical design and execution leave it tempting to speculate that human BDV infection does not exist. Before recommending dropping further investigation, however, more consideration needs to be given to environmental and epidemiological data that should be taken into account when evaluating clinical study design.

Generically, habitat changes in the environment have led to an increase in the rate of emerging and reemerging infectious diseases, and many are zoonoses and RNA viruses that are believed to be highly adaptable to changes in host and environment (Vandegrift et al., 2011). What sort of environment and environmental changes might predict increased risk for human BDV infection, if it were a zoonotic infection of small wild rodent or rodent-like species? Movement of these presumed vector species to the budding pasture lands during the spring/summer in search of food and breeding could predict the observed increase in intranasal exposure of low-grazing herbivores (e.g., horses and sheep) to urine and other bodily secretions that may contain infectious virus. Infections of the young vector mammals born in the spring/early summer, which evidence widespread virus replication in nonneural tissues and low disease pathogenicity, would lead to increased deposition of those infected secretions and potentially enhance the intranasal infection of low-grazing animals.

Since humans, like large farm animals, might be expected to be exposed to BDV from the small vector animals, exposure to soil in endemic areas (e.g., planting, gardening, mowing), rather than exposure to large farm animals directly (e.g., animal husbandry, veterinarians, jockeys), may put humans at higher risk for BDV infection. Thus, activities like plowing fields in the process of planting crops are more likely to lead to aerosolization of the infected small mammal secretions and inhalation of the subsequent dust/debris than exposure to an infected horse or sheep. Further, asymptomatic infection with BDV, particularly in some species, appears common, suggesting that classical signs of encephalitic BD may only be the "tip of the iceberg"; long latency and subclinical infections may make BDV exposure and infection risk harder to identify without careful and *formal* epidemiological studies.

In domesticated large animals, unsanitary and crowded conditions appear to be associated with increased risk for BD; thus, there may be a higher risk of humans contracting BDV when humans are housed in close, unsanitary quarters. Most BDV studies in humans are from developed countries where presumably most of the individuals live in a minimally sanitary setting and without significant exposure to rodent excretions. If increased rodent exposure is linked to BDV infection in humans, it may be found more often in economically disadvantaged rural settings and crowded, unsanitary urban areas. So it is possible that most studies to date, even those that are well designed and executed, may be looking for human BDV infection in "all the wrong places" as it were. Few of the well-designed studies using multiple tests of BDV infectivity are focused on the socioeconomically disadvantaged rural and urban areas that might be expected to lead to increased exposure to rodent secretions. It is tempting to speculate that for decades BDV scientists have been looking under the wrong "lamppost" for evidence of BDV infection and need to focus on areas more likely to lead to greater exposures to BDV-infected small animals. If this is true, more studies of human BDV infection need to take place in undeveloped countries and socioeconomically blighted urban areas of developed countries in order to more completely evaluate evidence of human infection with BDV.

Diagnosing BDV infection in humans does not automatically, however, clearly point to the specific diseases that BDV may cause. A bit of a conundrum develops because in areas of socioeconomic deprivation, with limited medical care for even life-threatening medical disease, one may have more difficulty linking a specific, perhaps even somewhat subtle, medical condition with evidence of BDV infection even *IF* infection was conclusively demonstrated. Perhaps most efforts need to be focused on formal surveillance and epidemiological studies that can better direct scientists to simply demonstrating whether BDV clearly exists in the human host in the natural setting before we concern ourselves with what disease states BDV may cause in humans.

So, the question remains: should the investigations of human infection with BDV be pursued at all? Certainly, the presence of partial BDV genomes in human DNA suggests that at some time the virus or a related virus infected humans. While it has been postulated that these partial ORFs or the proteins they produce may provide protection from BDV through a natural "interference" of BDV replication at the protein or even RNA level and that the species with BDV sequences in the genome are less likely to suffer from infection and/or disease caused by BDV, macaques can be experimentally infected, and this species, too, contains BDV genome sequences in their DNA.

The most significant and perhaps most often overlooked gap in BDV human studies is in the application of solid epidemiological science to the question. In the past 10 years, more attention has been paid to this area, with several interesting studies and hypotheses generated. Taken together, these recent studies suggest that we may have been overlooking perhaps a more appropriate human target population at risk for BDV exposure.

ACKNOWLEDGEMENT

This research was supported in part by the Intramural Research Program of the NIH, NIDCR.

REFERENCES

Barré-Sinoussi, F, Chermann, JC, Rey, F, et al. 1983. Isolation of a T-lymphotropic retrovirus from a patient at risk for acquired immune deficiency syndrome (AIDS). *Science.* **220**:868–871.

Belyi, VA, Levin, AJ, Skalka, AM. 2010. Unexpected inheritance: Multiple integrations of ancient Bornavirus and Ebolavirus/Marburgvirus sequences in vertebrate genomes. *PLoS Pathog.* **6**(7): e1001030.

Berger, JR, Khalili, K. 2011. The pathogenesis of progressive multifocal leukoencephalopathy. *Discov Med.* **12**:495–503.

Bode, L, Reckwald, P, Severus, WE, et al. 2001. Borna disease virus-specific circulating immune complexes, antigenemia, and free-antibodies—the key marker triplet determining infection and prevailing in severe mood disorders. *Mol Psychiatry.* **6**:481–491.

Brnic, D, Stevanovic, V, Cochet, M, et al. 2012. Borna disease virus infects human neural progenitor cells and impairs neurogenesis. *J Virol.* **86**:2512–2522.

Carbone, KM. 2001. Borna disease virus and human disease. *Clin Microbiol Rev.* **14**:513–527.

Carbone, KM, Duchala, CS, Griffin, JW, et al. 1987. Pathogenesis of Borna disease in rats: Evidence that intra-axonal spread is the major route for virus dissemination and the determinant for disease incubation. *J Virol.* **61**:3431–3440.

Carbone, KM, Trapp, BD, Griffin, JW, et al. 1989. Astrocytes and Schwann cells are virus-host cells in the nervous system of rats with Borna disease. *J Neuropathol Exp Neurol.* **48**:631–644.

Carbone, KM, Rubin, SA, Nishino, Y, et al. 2001. Borna disease: Virus-induced neurobehavioral disease pathogenesis. *Curr Opin Microbiol.* **4**:467–475.

Chalmers, RM, Thomas, DRh, Salmon, RL. 2005. Borna disease virus and the evidence for human pathogenicity: A systematic review. *Q J Med.* **98**:255–274.

Conzelmann, KK. 2004. Reverse genetics of mononegavirales. *Curr Top Microbiol Immunol* **283**:1–41.

Dürrwald, R, Kolodziejek, J, Muluneh, A, et al. 2006. Epidemiological pattern of classical Borna disease and regional genetic clustering of Borna disease viruses points towards the existence of to-date unknown endemic reservoir host populations. *Microbes Infect.* **8**:917–929.

Dürrwald, R, Kolodziejek, J, Herzog, S, et al. 2007. *Rev Med Virol.* **17**:181–203.

Gonzalez-Dunia, D, Cubitt, B, Grasser, FA, et al. 1997. Characterization of Borna disease virus p56 protein, a surface glycoprotein involved in virus entry. *J Virol.* **71**:3208–3218.

He, J, Chen, Y, Farzan, M, et al. 1997. CCR3 and CCR5 are co-receptors for HIV-1 infection of microglia. *Nature.* **385**: 645–649.

Herzog, S, Rott, R. 1980. Replication of Borna disease virus in cell cultures. *Med Microbiol Immunol.* **168**:153–158.

Hilbe, M, Herrsche, R, Kolodziejek, J, et al. 2006. Shrews as reservoir hosts of Borna disease virus. *Emerging Infect Dis.* **12**:675–677.

Holmes, KV. 2003. SARS coronavirus: A new challenge for prevention and therapy. *J Clin Invest.* **111**:1605–1609.

Horie, M, Honda, T, Suzuki, Y, et al. 2010. Endogenous non-retroviral RNA virus elements in mammalian genomes. *Nature.* **463**:84–87.

Hornig, M, Briese, T, Licinio, J, et al. 2012. Absence of evidence for bornavirus infection in schizophrenia, bipolar disorder and major depressive disorder. *Mol Psychiatry.* **17**(5):486–493.

Kallio, ER, Heikkila, HP, Koskela, E, et al. 2011. Intracerebral Borna disease virus infection of bank voles leading to peripheral spread and reverse transcription of viral RNA. *PLoS One.* **6**(8):323622.

Kamitani, W, Ono, E, Yoshino, S, et al. 2003. Glial expression of Borna disease virus phosphoprotein induces behavioral and neurological abnormalities in transgenic mice. *Proc Natl Acad Sci USA.* **100**(15):8969–8974.

Kinnunen, PM, Inkeroinen, H, Ilander, M, et al. 1990. Isolation and characterization of the Borna disease agent cDNA clones. *Proc Natl Acad Sci USA.* **87**:4184–4188.

Kohno, T, Goto, T, Takasaki, T, et al. 1999. Fine structure and morphogenesis of Borna disease virus *J Virol.* **71**(1):76–6.

Kiermayer, S, Kraus, I, Richt, JA, et al. 2002. Identification of the amino terminal subunit of the glycoprotein of Borna disease virus. *FEBS Lett.* **531**(2):255–258.

Li, Q, Want, Z, Zhu, D, et al. 2009. Detection and analysis of Borna disease virus in Chinese patients with neurological disorders. *Eur J Neurol.* **16**(3):399–403.

Lipkin, WI, Travis, GH, Carbone, KM, et al. 1990. Isolation and characterization of Borna disease agent cDNA clones. *Proc Natl Acad Sci USA.* **87**(11):4184–4188.

Lipkin, WI, Briese, T, Hornig, M. 2011. Borna disease virus-fact and fantasy. *Virus Res.* **162**:162–172.

Martin, A, Hoefs, N, Tadewaldt, J, et al. 2011. Genomic RNAs of Borna disease virus are elongated on internal template motifs after realignment of 3′ termini. *Proc Natl Acad Sci USA.* **108**:7206–7211.

Matsunaga, H, Tanaka, S, Sasao, F, et al. 2005. Detection by radioligand assay of antibodies against Borna disease virus in patients with various psychiatric disorders. *Clin Diagn Lab Immunol.* **12**:671–676.

Mokili, JL, Rohwer, F, Dutilh, BE. 2012. Metagenomics and future perspectives in virus discovery. *Curr Opin Virol.* **2**:63–77.

Na, KS, Tae, SH, Song, JW, et al. 2009. Failure to detect borna disease virus antibody and RNA from peripheral blood mononuclear cells of psychiatric patients. *Psychiatry Investig.* **6**(4):306–312.

Narayan, O, Herzog, S, Frese, K, et al. 1983a. Pathogenesis of Borna disease in rats: Immune-mediated viral ophthalmoencephalopathy causing blindness and behavioral abnormalities. *J Infect Dis.* **148**:305–315.

Narayan, O, Herzog, S, Frese, K, et al. 1983b. Behavioral disease in rats caused by immunopathological responses to persistent borna virus in the brain. *Science.* **220**:1401–1403.

Nathanson, N, Martin, JR. 1979. The epidemiology of poliomyelitis: Enigmas surrounding its appearance, epidemicity and disappearance. *Am J Epidemiol.* **110**:672–692.

Nunes, SO, Itano, EN, Amarante, MK, et al. 2008. RNA from Borna disease virus in patients with schizophrenia, schizoaffective patients, and their biological relatives. *J Clin Lab Anal.* **22**(4):314–320.

Okamoto, M, Hagiwara, K, Kamitani, W, et al. 2003. Experimental vertical transmission of Borna disease virus in the mouse. *Arch Virol.* **148**:1557–1568.

Palmer, S, Brown, D, Morgan, D. 2005. Early quantitative risk assessment of emerging zoonotic potential of animal diseases. *BMJ.* **331**:1256–1260.

Payne, S, Covaleda, L, Jianhua, G, et al. 2011. Detection and characterization of a distinct Borna virus lineage from healthy Canada geese (*Branta canadensis*). *J Virol.* **85**:12053–12056.

Peng, G, Yan, Y, Zhu, C, et al. 2008. Borna disease virus P protein affects neural transmission through interactions with gamma-aminobutyric acid receptor-associated protein. *J Virol.* **84**:12487–12497.

Pletnikov, MV, Rubin, SA, Vasudevan, K, et al. 1999. Persistent neonatal Borna disease virus (BDV) infection of the brain causes chronic emotional abnormalities in adult rats. *Behav Brain Res.* **100**:43–50.

Pletnikov, MV, Moran, TH, Carbone, KM. 2002. Borna disease virus infection of the neonatal rat: Developmental brain injury model of autism spectrum disorders. *Front Biosci.* **7**:d593–d607.

Poch, O, Blumberg, BM, Bouqueleret, L, et al. 1990. Sequence comparison of five polymerases (L proteins) of unsegmented negative-strand RNA viruses: Theoretical assignment of functional domains. *J Gen Virol.* **71**:1153–1162.

Puorger, ME, Hilbe, M, Muller, JP, et al. 2010. Distribution of Borna disease virus antigen and RNA in tissues of naturally infected bicolored white-toothed shrews, *Crocidura leucodon*, supporting their role as reservoir host species. *Vet Pathol.* **42**:236–244.

Rackova, S, Janu, L, Kabickova, H. 2010. Borna disease virus (BDV) circulating immunocomplex positivity in addicted patients in the Czech Republic: A prospective cohort analysis. *BMC Psychiatry.* **10**:70.

Relman, DA. 2011 Microbial genomics and infectious diseases. *N Eng J Med.* **365**:347–357.

Richt, JA, Fürbringer, T, Koch, A, et al. 1998. Processing of Borna disease virus glycoprotein gp94 by the subtilisin-like endoprotease furin. *J Virol.* **72**:4528–4533.

Rott, R, Herzog, S, Fleischer, B, et al. 1985. Detection of serum antibodies to Borna disease virus in patients with psychiatric disorders. *Science.* **228**:755–756.

Rubin, SA, Sierra-Honigmann, AM, Lederman, HM, et al. 1995. Hematologic consequences of Borna disease virus infection of rat bone marrow and thymus stromal cells. *Blood.* **85**:2762–2769.

Sauder, C, Staeheli, P. 2003. Rat model of Borna disease virus transmission: Epidemiological implications. *J Virol.* **77**:12886–12890.

Schneemann, A, Schneider, PA, Lamb, RA, et al. 1995. The remarkable coding strategy of Borna disease virus: A new member of the nonsegmented negative strand RNA viruses. *Virology.* **210**:1–8.

Schwardt, M, Mayer, D, Frank, R, et al. 2005. The negative regulator of Borna disease virus polymerase is a non-structural protein. *J Gen Virol.* **86**:3163–3169.

Schwemmle, M. 2001. Borna disease virus infection in psychiatric patients: Are we on the right track? *Lancet Inf Dis.* **1**:46–52.

Sierra-Honigmann, AM, Rubin, SA, Estafanous, MG, et al. 1993. Borna disease virus in peripheral blood mononuclear and bone marrow cells of neonatally and chronically infected rats. *J Neuroimmunol.* **45**:31–36.

Smith, G. 2012. Herpesvirus transport to the nervous system and back again. *Ann Rev Microbiol.* **66**:53–76.

Smith, W, Andrewes, CH, Laidlaw, PP. 1933. A virus obtained from influenza patients. *Lancet.* **2**:66–68.

Song, JW, Na, KS, Tae, SH, et al. 2011. Borna disease virus antibody and RNA from peripheral blood nmononuclear cells of race horses and jockeys in Korea. *Psychiatry Investig.* **8**(1):58–60.

Staeheli, P, Sauder, C, Hausmann, J, et al. 2000. Epidemiology of Borna disease virus. *J Gen Virol.* **81**:2123–2135.

Staeheli, P, Rinder, M, Kaspers, B. 2010. Avian Borna virus associated with fatal disease in Psittacine birds. *J Virol.* **84**:6269–6275.

Stitz, L, Krey, H, Ludwig, H. 1981. Borna disease in rhesus monkeys as a model for uveo-cerebral symptoms. *J Med Virol.* **6**:333–340.

Stitz, L, Dietzschold, B, Carbone, KM. 1995. Immunopathogenesis of Borna disease. *Curr Top Microbiol Immunol.* **190**:75–92.

Suberbielle, E, Stella, A, Pont, F, et al. 2008. Proteomic analysis reveals selective impediment of neuronal remodeling upon Borna disease virus infection. *J Virol.* **82**:12265–12279.

Tang, P, Chiu, C. 2010. Metagenomics for the discovery of novel human viruses. *Future Microbiol.* **5**:117–189.

Thomas, DR, Chalmers, RM, Crook, B, et al. 2005. Borna disease virus and mental health: A cross-sectional study. *QJM.* **98**(4):247–254.

de la Torre, JC. 1994. Molecular biology of Borna disease virus: Prototype of a new group of animal viruses. *J Virol.* **68**:7669–7675.

de la Torre , JC, Carbone , KM , Lipkin , WI. 1990 . Molecular characterization of the Borna disease agent. *Virology* **179**:853–856.

de la Torre , JC, Borrow, P, Oldstone, MBA. 1991. Viral persistence and disease: Cytopathology in the absence of cytolysis. *Br. Med Bull.* **47**: 838–851.

Vahlenkamp, TW, Konrath, A, Weber, M., et al. 2002. Persistence of Borna disease virus in naturally infected sheep. *J Virol.* **76**:9735–9743.

Van den Pol, AN. 2009. Viral infection leading to brain dysfunction: More prevalent than appreciated? *Neuron.* **64**:17–20.

Vandegrift, KJ, Wale, N, Epstein, JH. 2011. An ecological and conservation perspective on advances in the applied virology of zoonoses. *Viruses.* **3**:379–397.

Waltrip, RWII, Buchanan, RW, Summerfelt, A, et al. 1995 Borna disease virus and schizophrenia. *Psychiatry Res.* **56**:33–44.

Wehner, T, Ruppert, A, Herden, C, et al. 1997. Detection of a novel Borna disease virus-encoded 10 kDa protein in infected cells and tissues. *J Gen Virol.* **78**:2459–2466.

Zhu, T, Korber, BT, Nahmias, AJ, et al. 1998. An African HIV-1 sequence from 1959 and implications for the origin of the epidemic. *Nature.* **391**:594–597.

31

TICK-TRANSMITTED VIRUSES AND CLIMATE CHANGE

Agustín Estrada-Peña

Department of Parasitology, Faculty of Veterinary Medicine, Zaragoza, Spain

Zdenek Hubálek and Ivo Rudolf

Institute of Vertebrate Biology, v.v.i., Academy of Sciences of the Czech Republic, and Masaryk University, Faculty of Science, Department of Experimental Biology, Brno, Czech Republic

TABLE OF CONTENTS

Viral Infections and Global Change, First Edition. Edited by Sunit K. Singh.

31.1 INTRODUCTION

Ticks are obligate hematophagous ectoparasites of wild and domestic animals and humans that are distributed from Arctic to tropical regions of the world. Globally, the recognized number of distinct and epidemiologically important diseases transmitted by ticks has increased considerably during the last 30 years. For example, more than 10 newly recognized spotted fever rickettsioses have been identified since 1984 (Paddock et al., 2008; Parola et al., 2005). In the United States, the list of national notifiable diseases included six tick-borne diseases, namely, Lyme disease (*Borrelia burgdorferi* s.l. infection), human granulocytic anaplasmosis (HGA, *Anaplasma phagocytophilum* infection), human babesiosis (*Babesia* spp.), human monocytic ehrlichiosis (*Ehrlichia chaffeensis* infection), Rocky Mountains spotted fever and Powassan disease, most of which have increased steadily in average annual incidence (Bacon et al., 2008). Although advances in molecular technology have contributed to the identification of these pathogens, rapidly expanding pathogen diagnosis and increasing incidence have raised concerns about the accuracy of case counts and epidemiology reports (Mantke et al., 2008). The problem of analyzing the incidence of tick-borne pathogens in humans is the concurrency of factors affecting the whole system such as climate, driving the life cycle of the ticks, the availability, occurrence and seasonal patterns of competent reservoirs, and social habits, leading the contact with tick-infested areas need to be considered. The systems of tick-borne pathogens are very complex in nature, and we should regard them as layers of information, each one increasing the complexity of the previous one until the whole system is covered.

While not specifically mentioned in the preceding text, the virus transmitted by tick bites remains a health problem in many parts of the world. Several events that occurred during the final decades of the twentieth century and the beginning of the twenty-first century suggest a rise of tick-borne viral infections worldwide. These events include recent national and regional epidemics of known diseases such as tick-borne encephalitis (TBE) in Central and Eastern Europe, Kyasanur forest disease (KFD) in Karnataka state in India, and Crimean–Congo hemorrhagic fever (CCHF) in northern Turkey and the southwestern regions of the Russian Federation (Maltezou et al., 2010; Pattnaik, 2006; Randolph, 2008a). Some of them, like TBE, may be also transmitted by milk intake, while others, like CCHF, may be transmitted to humans at abattoirs.

We want to review here some findings relating climate and the behavior in nature of some important tick-transmitted viruses, like the etiologic agents of TBE and CCHF. Our specific point in this review is that climate may be probably behind some of the recent (re) emergence of the reported active foci of the disease, driving the dynamics and the abundance patterns of ticks. However, a note of caution is issued about the lack of suitable data on the dynamics of the hosts and about the changes that climate may operate in social habits, which are difficult to quantify (but see Sumilo et al., 2007; Zeman et al., 2010). The effects of climate on tick-transmitted viruses are indirect and difficult to quantify. A simple approach might not be enough to capture the many levels at which climate operates driving these infections. Of course this is not the first time the topic has been reviewed. The interested readers will find a text on comparative tick bionomics and viruses (Sonenshine, 1974) and a comprehensive review by Randolph (2008b) focused not only on tick-transmitted viruses but on general tick-borne disease systems, as influenced by climate and other factors. We will also summarize some findings related to other tick-transmitted viruses as associated with human disease. We will use the abbreviations TOT, for transovarial transmission (in arthropods), and TST, for transstadial transmission (in arthropods).

31.2 TICKS IN NATURE

Ticks spend most of their life cycle in the environment, and all tick life cycle stages are dependent on a complex combination of climate variables for development and survival. In summary, ticks must develop from one stage to the next in the life cycle, following a sum of degree-days. While ticks are molting, there is a resulting mortality because of relatively unsuitable climate conditions. After adequate cuticle hardening, ticks quest for a host. Such an activity period results in further mortality because of water losses. Ticks are sensitive to changes in several limiting abiotic factors, including temperature, which affects the timing and speed of development, and atmospheric water deficit, which affects mortality. Changes in these variables shape the probabilities of a tick population to persist. Although surveillance and reporting of changes in the distribution of tick populations are generally inadequate, some well-documented reports support the slow but apparently continuous expansion of the historical frontiers of some tick species into areas where they were previously absent (reviewed by Gray et al., 2009). With this rationale, warmer temperatures have been suggested, together with host movements, as the main driver of some tick geographical range changes (Danielová et al., 2006; Lindgren et al., 2000; Ogden et al., 2004). However, the potential influence of changing rainfall patterns has largely been ignored although this may have a greater effect than temperature on the ability of tick populations to establish in new areas. Finally, there is little doubt that human-induced changes in abiotic (climate, land cover, habitat structure) and biotic (distribution and abundance of tick hosts) conditions have occurred over the past few decades, and there is equally indisputable evidence for the increase in recorded human cases of some tick-borne diseases (Randolph, 2009).

Host availability may modulate the dynamics of tick populations. Though many animal species can serve as tick hosts, there are several determinants of host suitability, and the specificity of tick–reservoir host–pathogen relationships is key to our understanding of the processes conditioning the transmission of pathogens by ticks (Randolph, 2009). Shelter and protection from environmental conditions are critical to tick survival, because questing and diapausing ticks are vulnerable to extreme temperature and humidity. The concerns about climate change added fuel to a debate about how predicted climate changes may alter tick–host–pathogen relationships and particularly tick potential for invasion of new areas and pathogen transmission. However, our efforts to disentangle such complex systems have so far scratched only the surface and are far from providing a complete answer to the many questions about the epidemiology of these processes. Invasive events (the transportation of an exotic tick species into an area far from its native range) are also well documented and seem to be related to unrestricted domestic animal movements or overabundance of certain wild hosts. The spread of ticks is a controversial issue because of a lack of empirical data and its importance in managing the further spread of prominent pathogens affecting human and animal health (Wilson, 2009).

31.3 FAMILY *FLAVIVIRIDAE*

31.3.1 Tick-borne encephalitis virus

Over the past decades, TBE has become a growing public health concern in Europe and Asia and is the most important viral tick-borne disease in Europe. It is also important in the Far East and in other parts of Asia. Protective vaccination is indicated for persons inhabiting or visiting natural foci of TBE. For this purpose, it is necessary to know where TBE virus (TBEV) occurs, where vectors are a potential hazard, and where as a consequence autochthonous TBE cases have been registered. Unlike Lyme borreliosis-endemic regions, TBE risk areas are distributed in a patchwork pattern, sometimes the situation remains stable, and sometimes changes occur due to altered climatic conditions or other factors. Adequate reviews about the topic exist (i.e., Süss, 2003; 2011) as well as comprehensive analyses of the human incidence rates in several zones of Eastern Europe (Sumilo et al., 2007; 2008).

There are three recognized TBE subtypes (Calisher, 1988; Calisher et al., 1989; Clarke, 1964; Dobler, 2010; Gritsun et al., 2003; Lindquist and Vapalahti, 2008; Rubin and Chumakov, 1980; Votyakov et al., 1978): (i) Western or European subtype (TBEV-W), also called Central European encephalitis virus (CEEV—topotype strains are Hypr and Neudoerfl) or sometimes "ricinus" subtype, whose varieties include Spanish sheep encephalitis (SSE), Turkish sheep encephalitis (TSE), and Greek goat encephalitis ("Vergina") viruses (Hubálek et al., 1995); (ii) (Ural-)Siberian subtype (TBEV-S: the prototype strains are Aina and Vasilchenko), sometimes called "persulcatus" subtype, causing Russian spring–summer encephalitis (RSSE:); and (iii) Far Eastern subtype (TBEV-FE with prototype strain Sofyin, isolated from human brain in Khabarovsk, 1937). A taxonomic and nomenclatural confusion around TBEV has repeatedly been emphasized (Calisher, 1988; Clarke, 1964; Holzmann et al., 1992; Stephenson, 1989). In addition, TBEV is very closely related to louping-ill virus (LIV), which should be regarded in fact as the fourth (or, historically, the first?) subtype of TBEV (see following text). According to Ecker et al. (1999), variation in amino acids within a subtype is up to 2% and between subtypes 5–6% (Lindquist and Vapalahti, 2008).

TBEV (its RSSEV subtype) was first isolated in 1937 (Chumakov and Zeitlenok, 1939), and CEEV (strain "256") from *I. ricinus* ticks was collected near Minsk, Belarus, in 1940 (Levkovich and Karpovich, 1962; Votyakov et al., 1978). Further isolations of CEEV were reported in Czechland from human patients and *I. ricinus* ticks in 1948–1949 (Gallia et al., 1949; Krejčí, 1949; Rampas and Gallia, 1949). Principal arthropod vectors are ticks of the genus *Ixodes*: *I. ricinus* for CEEV (TST, TOT: Benda, 1958b; Řeháček, 1962) and *I. gibbosus* (a marginal vector in the Mediterranean). Mean prevalence rate of CEEV in ticks in natural foci can reach 1%, but it is usually much lower, at about 0.1%. Occasional vectors are other tick species such as *Ixodes hexagonus*, while only sporadically metastriate tick species *Haemaphysalis inermis*, *H. concinna*, *H. punctata*, *Dermacentor marginatus*, *D. reticulatus* (Kožuch and Nosek, 1971; Křivanec et al., 1988; Naumov et al., 1980; Riedl et al., 1971), and *Hyalomma marginatum* (Crimea). The main vector for RSSEV is *I. persulcatus* (infection prevalence rates can reach frequently >2%; TST, TOT: Chunikhin, 1990), less often *Ixodes ovatus*, *Dermacentor silvarum*, *D. reticulatus*, *D. marginatus*, *H. concinna* (TOT), *Haemaphysalis longicornis*, and *H. japonica* (Naumov et al., 1980).

Competent vertebrate hosts of TBEV are small forest mammals—especially rodents and insectivores (*Apodemus flavicollis*, *A. sylvaticus*, *Myodes glareolus*, *M. rufocanus*, *Microtus agrestis*, *Sciurus vulgaris*, *Talpa europaea*, *Sorex araneus*, *Erinaceus concolor*), further goat, sheep, and rarely cattle. The role of some forest passerines and other birds as hosts of TBEV has not yet been fully elucidated; the virus was isolated occasionally from a number of other vertebrate species. Experimental viremia has been demonstrated in many mammalian, avian, amphibian, and reptilian species. Encephalitis with ataxia, jumping, tremor, and convulsions can affect lambs, kids, or, exceptionally, dogs. CEEV infection is usually subclinical in adult ruminants and pig; goats, sheep, and cows excrete virus in the milk (Benda, 1958a; Grešíková, 1958a,b; Smorodintsev et al., 1953; van Tongeren, 1955). TBEV (especially TBE-S and TBE-FE virus subtypes) occasionally kills birds of some species, for example, *Carduelis flammea*, *Passer domesticus*, and *Fulica atra*, and amphibians *Rana temporaria* and *Bufo bufo*.

Natural foci of TBE (and other tick-borne diseases) have been classified (Rosický, 1959) as "theriodic" (situated in deciduous and mixed forest ecosystems, often game preserves, where the main hosts of adult female vector ticks are deer and other wild mammals), "boskematic" (pastoral, where the main vectors of adult female vector ticks are grazed domestic ruminants), mixed "theriodic–boskematic," or "mountain" (Rosický and Bárdoš, 1966). Urban foci of CEE have also been described in Eurasia. In general, most natural foci of TBE are situated in forest (less pastoral) ecosystems.

There are two basic modes of human infection with TBE—by the bite of an infective tick or by consumption of infected raw (unpasteurized) goat (less often sheep or cow) milk or unpasteurized dairy products (Grešíková, 1972; Smorodintsev et al., 1953). Whereas the tick-transmitted cases are sporadic, the milk-borne infections usually affect whole families or population groups in outbreaks. For instance, a large milk-borne TBE epidemic occurred in Rožňava, East Slovakia, in 1951, when 660 persons were infected and 274 of them had to be hospitalized (Blaškovič, 1954). As much as 76% of human infections have been alimentary in Belarus (Ivanova, 1984). The virus may resist in milk at 60 °C for more than 10 m and partially even the pasteurization at 62 °C for 20 m, and it is not inactivated at pH 2.8 within 24 h/4°C. In addition, many laboratory infections (usually by infectious aerosol) have been reported in unvaccinated personnel.

TBEV circulates in a series of interactions between virus, vector ticks, and tick hosts and is able to persist in a given habitat over long periods of time (Nuttall, 1999). The occurrence of vector ticks and suitable vertebrates on which ticks can become infected is crucial for virus existence in a given area. The following mechanisms of virus transmission between ticks occur: (i) feeding/cofeeding (Alekseev and Chunikhin 1990; Labuda et al., 1993a, b), (ii) TOT, and perhaps (iii) sexual transmission. Cofeeding transmission is especially effective, and the virus can be transmitted through this mechanism from a feeding vector tick to cofeeding ticks even on immune hosts, while TOT is considerably less efficient. Studies have shown that tick saliva contains factors that modulate host inflammatory, coagulation, and immune response to improve tick blood feeding and pathogen transmission (Alekseev et al., 1991; Jones et al., 1989; Labuda et al., 1993a; Randolph, 2009). This so-called saliva-assisted transmission (SAT) was reviewed by Nuttall et al. (2008). Inoculation of salivary glands extracts and TBEV into laboratory animal hosts resulted in enhanced transmission from hosts to nymphal ticks when compared with pathogen inoculation alone (Alekseev et al., 1991; Labuda et al., 1993b). SAT helped to explain the mechanism behind the equally novel observation of TBEV transmission between cofeeding ticks in the absence of a systemic infection (Alekseev and Chunikhin, 1990; Labuda et al., 1993a,b; Randolph, 2009).

The non viremic (cofeeding) transmission imposes constraints because it requires cofeeding by at least two tick stages in synchrony in their seasonal activity (Randolph et al., 2000). The long and slow life cycle typical of temperate tick species, caused by low temperature-dependent developmental rates and overwinter diapause, slows the pace of pathogen transmission. As tick phenology is reset each year by winter conditions (Randolph et al., 2002), the critical stages (larvae and nymphs for TBEV) may emerge from diapause in more or less synchrony in the spring, depending on whether temperatures rise sufficiently rapidly to cross the threshold for larval activity (c. 10 °C mean daily maximum) soon after the threshold for nymphal activity (c. 7 °C mean daily maximum) (Randolph and Sumilo, 2007). The variability of thermal conditions associated with seasonal synchrony between tick stages has been identified as the key determinant of the focal distribution of TBEV across Europe, allowing the predicted risk of TBE to be mapped (Randolph et al., 2000).

Altogether, this information suggests that climate exerts an extreme control of the natural cycles of TBEV and delineates both their intensity (in terms of field tick prevalence rates) and their geographical distribution. According to the prevalent hypothesis outlined before, the climate at the beginning of the spring exerts a regulatory action on the synchrony of the active immature ticks, conditioning the necessary coexistence of nymphs and larvae on the same hosts. Because of the short time of feeding for both larvae and nymphs, small changes in the temperature in that period may promote a lack of synchronicity of a few days, enough to prevent the "backward" transmission of the virus. Therefore, the extreme lability of the TBEV foci would be primarily driven by very small changes in spring temperatures. The system is thus very local in its nature. These events have not yet been captured by a process-driven model, which could be a welcomed addition to our array of epidemiological tools, necessary to understand the TBEV epidemiology and design intervention for its prevention.

At a continental scale, it has been reported that *I. ricinus* and TBEV reach higher altitudes in the Czech mountains in a consistent pattern after the year 2000, higher than reported for the years 1970–1980 (Daniel et al., 2008; Danielová et al., 2006; Materna et al., 2008). The tick has been reported to spread north in Sweden (Eisen, 2008; Lindgren et al., 2000; Lindgren and Jaenson, 2006), Norway (Skarpaas et al., 2006), Finland (Jääskeläinen et al., 2006), as well as Germany (Hemmer et al., 2005; Süss et al., 2008) and west in Austria (Holzmann et al., 2009). Nearly all these data were collected along the fringes of tick distribution and do not apply to the core areas. It must be however realized that the series of cases in humans are not a direct mirror of the "activity" of TBE foci and that the mechanisms regulating the later are far more complex. This is why it has been proposed to check the active foci by direct examination of the ticks collected in the field by a PCR system (Gaumann et al., 2010).

A picture of the number of cases of TBE in European countries has been provided by Süss (2008). It has been speculated that host abundance, changes in social habits, economic fluctuations, environmental changes, and to a lesser extent climate changes have increased the incidence of TBE (Lindgren and Gustafson, 2001; Sumilo et al., 2006, 2007, 2008; Zeman and Benes, 2004). It is anyway difficult to correlate series of human clinical cases against basic climate features because climate has several collateral effects, not only affecting tick life cycle but also hosts and, most importantly, social habits. This has been demonstrated in a series of data for TBE cases in the countries of the Baltic Sea (Sumilo et al., 2007) and the Czech Republic (Zeman and Benes, 2004). Even if we regard the epidemiology of TBE from 1976 to 2007 in general, most questions remain to be answered (Süss, 2008). Thus, the political turnaround and the resulting socioeconomic changes and changes in the behavioral pattern of the exposed population in certain countries of the former Eastern Bloc at the beginning of the 1990s certainly are a significant influence factor. However, this does not explain the increase in the number of TBE cases since the 1990s in Sweden, Italy, Hungary, Czechland, Finland, and Germany. As a result, the TBE incidence in the German risk areas shows the same trend as in the Baltic States; the political turnaround, however, only took place in the eastern part of the country, where TBE incidence is very low compared to southern Germany and the influence on the total number of registered cases consequently is very low (Süss, 2008).

As mentioned earlier, it seems that a kind of chaotic system emanates from the several layers of complexity emanating from the epidemiology of TBEV. The first one is the basic layer of the impacts of climate on the tick populations, which may be extremely local as mentioned before. A process-driven model (i.e., Dobson et al., 2011) may be an excellent starting point to handle such impacts. Analysis of the long-term trends of climate and its impact on the suitability for *I. ricinus* has been already presented (Estrada-Peña and Venzal, 2006) based in a long series of climate data. Figure 31.1 shows a different kind of analysis, based on the relationships between the sites where the tick *I. ricinus* has been collected and reported and the long-term climate found at these sites. It thus represents the "mean" expected climate suitability for the tick in its distribution range. This index is not correlated with tick abundance, since it depends in local factors. It only provides with an estimation of how suitable climate factors have been in the last 30 years. An analysis of the trends in climate for the period 2000–2010 shows that climate has become clearly more favorable in wide regions of Northern Europe (Figure 31.1). Therefore, using such a basic and primary approach, climate has obvious effects on tick available range to be colonized. These figures, however, are not aimed to provide answers about the seasonal patterns of the tick stages.

A second layer of complexity has also prominent role on the local epidemiological patterns. Such a layer is related to the abundance of populations of reservoir host, their seasonality, and the abundance of hosts that feed large numbers of adult ticks (like ungulates). Both factors would affect the system by providing respectively a larger nonsystemic (cofeeding) transmission and an increased abundance of engorged females that produce more eggs and more ticks for the next year population. The TOT from tick to eggs is known anyway to contribute to the system (Matser et al., 2009). Finally, the peculiar human habits as operating in each country and the impact of climate on those would certainly manifest the third layer of complexity, distorting the previous, natural ones and surfacing into the reported pattern of human incidence rates. We should not discard that every layer of the system may respond to the impact of climate in different ways affected by local constraints.

31.3.2 Louping ill virus

LIV is very closely related to TBEV, in fact indistinguishable from it by conventional serological and cross protection tests (Calisher, 1988; Calisher et al., 1989; Clarke, 1962, 1964; Holzmann et al., 1992; Hubálek et al., 1995; Kopecký et al., 1991; Madrid and Porterfield, 1974; Rubin and Chumakov 1980; Shamanin et al., 1990; Shiu et al., 1991; Stephenson, 1989; Tsekhanovskaya et al., 1993; Venugopal et al., 1992) but also by nucleotide sequence homology of the E gene (Gao et al., 1993; Gould et al., 2003; Grard et al., 2007; Jääskeläinen et al., 2010; Venugopal et al., 1994). Several authors therefore suggested arrangement of LIV as another subtype of TBEV, while not as a separate virus (Grard et al., 2007; Hubálek et al., 1995).

The geographical distribution of LIV involves United Kingdom, Ireland, and Norway (the only country of continental Europe where a typical LIV strain was isolated: Gao et al., 1993). Natural foci there represent most often pastoral heather habitat ("tick–sheep cycle"). LIV does not occur outside Europe. The "louping-ill (LI)" disease of sheep has long been recognized in Scotland. The virus was first isolated from sheep brain in Selkirkshire, Scotland, in 1929 (prototype strain Moredun L1–31: Pool et al., 1930), and it is, in fact, the very first arthropod-borne virus isolated in Europe. The principal vector of LIV is the tick *I. ricinus*; LI is also transmissible by goat and sheep milk similarly as the other TBE subtypes. Vertebrate hosts are rodents (*A. sylvaticus*), insectivores (*S. araneus*), mountain hare (*Lepus timidus*), sheep, and red grouse (*Lagopus lagopus scoticus*: Reid, 1990).

Natural foci of LI are "boskematic" (pastoral: Rosický, 1959)—rough, poorly drained hill pastures and heather moorlands with bracken and moor grass—principally a sheep–tick or sheep–tick–grouse cycle (Reid, 1990; Smith and Varma, 1981).

The human illness is usually biphasic; the febrile phase, after a short period of improvement, is followed by high fever and symptoms of meningoencephalitis, headache,

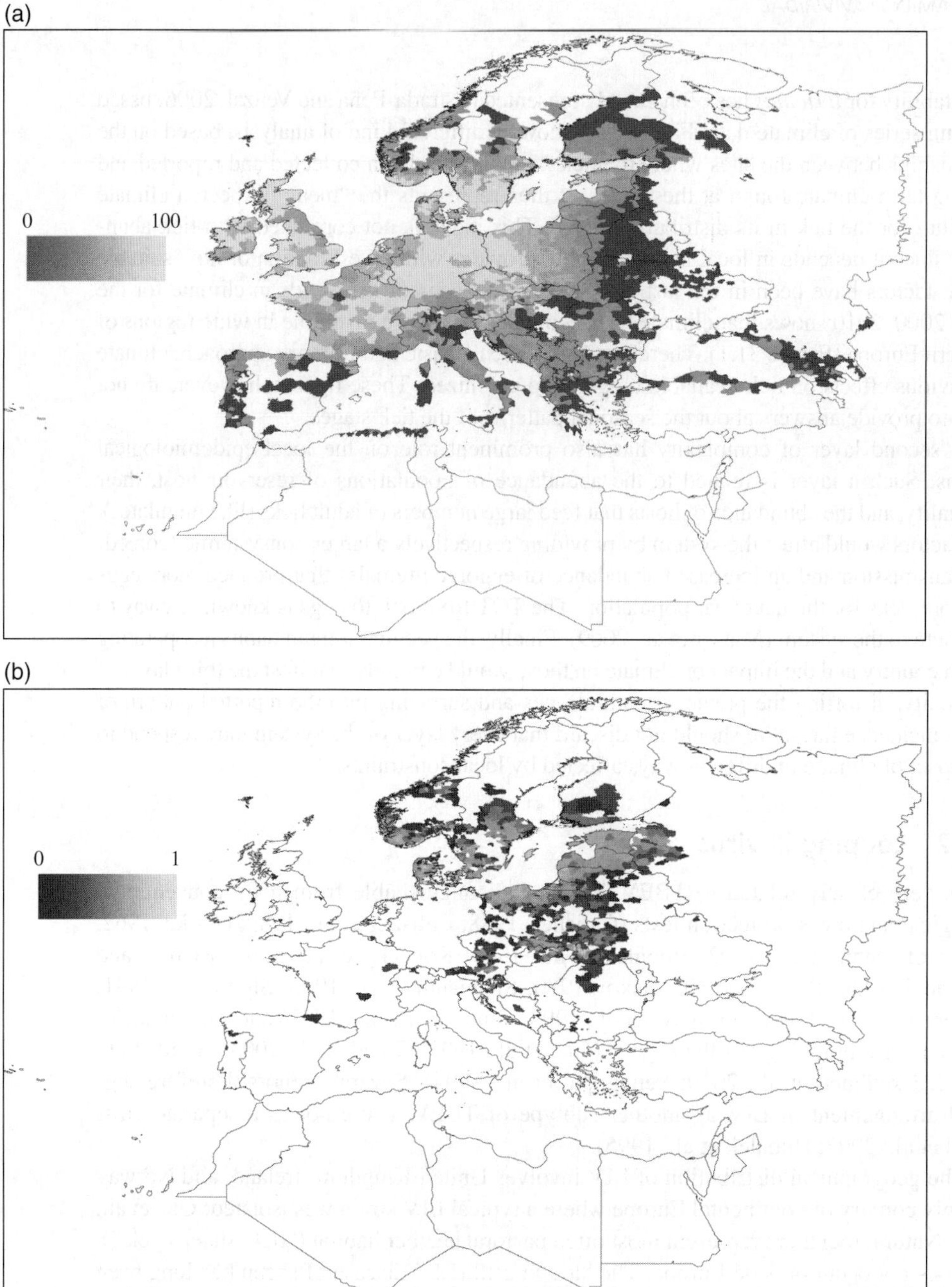

Figure 31.1. Predicted climate suitability for the tick *I. ricinus* in the western Paleartic. (**a**) Predicted climate suitability (0–100) was evaluated by a model trained with more than 4000 occurrence tick points and using Maxent as modeling software. The map is based on previous developments by Estrada-Peña et al. (2006). The ramp of colors shows the probability to find permanent populations of the tick, as driven only by climate conditions, including a set of remotely sensed monthly average temperature and monthly average vegetation stress (NDVI, a proxy for tick water stress) between the years 2000 and 2010. (**b**) Changes in climate suitability for *I. ricinus* in the period 2000–2010 (from 0, the minimum, to 1, the maximum) based in the same model. It is based on the modeling of climate suitability separately for each year, then evaluating the trend of such an index along the period 2000–2010. Both maps (**a** and **b**) are not a depiction of tick abundance but of the appropriateness of the climate for the development of the tick (**a**) and how such a factor evolved in time (**b**). Figures reproduced from Estrada-Peña et al. (2012). For color detail, please see color plate section.

weakness, stiff neck, conjunctivitis, retrobulbar pain, photophobia, myalgia, arthralgia, dysarthria, excessive sweating, nausea, vomiting, insomnia, drowsiness, confusion, tremors, nystagmus, and ataxia. Symptoms of LI in humans are usually milder than in TBE. Nineteen naturally acquired human cases and 26 laboratory infections with LIV have been described in Great Britain between 1934 and 1990 (Davidson et al., 1991), including one fatal encephalitis. LIV transmission to man is obviously infrequent in the United Kingdom because the vector ticks only occasionally bite people in endemic areas (similarly as with Lyme borreliosis). It is primarily an occupational disease, affecting shepherds, crofters, veterinary personnel, forestry workers, butchers, and laboratory personnel. However, human cases of LI with a milder symptomatology might remain underreported. Meningoencephalitis was demonstrated histologically in the deer (Reid et al., 1982), and LIV was isolated from a roe deer (Reid et al., 1976). LIV occasionally affects also cattle, pig (piglets), goat (kids), horse, dog, hare, and red grouse (with a mortality rate of 70–80% especially in juvenile birds: Reid et al., 1978; 1980). Typical course of LI in sheep is biphasic, with fever and weakness, followed by meningoencephalitis with ataxia, generalized tremor, jumping (to "loup" means to leap in vernacular Scottish), vigorous kicking, salivation, and champing of jaws, progressing to paralysis, coma, and death (lethality 40–60%). Concurrent tick-borne fever (*A. phagocytophilum* infection) and external stress enhance the disease course (Reid, 1990).

31.3.3 Powassan virus

This virus (and its variant genetic lineage "deer tick virus" (DTV) is a member of the TBE complex. The complete nucleotide sequence of the genome was determined (a total of 10 839 nucleotides). Powassan virus (POWV) was first isolated from the brain of a child who had died from encephalitis in Powassan, Ontario (Canada), in 1958, and the next year, the virus was also isolated from *Dermacentor andersoni* tick (Theiler and Downs, 1973). Its geographical distribution is North America (northern United States, Canada) and Russian Far East.

Vectors are the ixodid ticks *Ixodes cookei, I. marxi, I. scapularis* (DTV), and *D. andersoni*, while in Asian Russia, the tick *I. persulcatus* is the major vector, and probably a minor role is played by the species *D. silvarum, H. concinna, and H. neumanni* (Gritsun et al., 2003). Because POWV is secreted in milk of experimentally infected goats, it can also be transmitted by drinking raw milk and eating raw milk products.

Vertebrate hosts of POWV/DTV are small- and medium-sized mainly forest mammals, especially rodents such as woodchucks *Marmota monax, Peromyscus leucopus* (DTV), and *Tamiasciurus hudsonicus*, also skunk *Mephitis mephitis*, raccoon *Procyon lotor*, fox, and, in the Far Eastern Russia, for example, *Apodemus peninsulae*, *A. agrarius*, and *M. rufocanus*. Animal disease has usually an inapparent course. However, experimental inoculation of adult laboratory mice and *Macaca mulatta* monkeys and horse (Little et al., 1985) causes their encephalitis and death.

Powassan is an encephalitis with fever, headache, prostration, meningitis, spastic pareses, and rarely paralyses and sometimes results in death (fatality rate about 10%); neurological sequelae often persist. In general, it is an infrequent disease in North America. For instance, only 27 cases (without fatalities) were reported in the United States between 1958 (first human case) and 1998, but since the late 1990s, the incidence of human disease seems to be increasing (Hinten et al., 2008; Pesko et al., 2010); anyway, "the disease is probably underrecognized" (Hinten et al., 2008). In Far Eastern Russia, POWV co-circulates with TBEV, and 14 cases of POW disease were described between 1973 and 1988 (Gritsun et al., 2003).

31.3.4 Omsk hemorrhagic fever virus

The virus belongs to the so-called TBE antigenic complex, being related to Kyasanur forest disease virus (KFDV) and readily distinguishable from TBEV. Omsk hemorrhagic fever virus (OHFV) was first isolated in 1947 from human blood and *D. marginatus* ticks (Gritsun et al., 2003). Primary vectors are *D. reticulatus* (TST, TOT) and *Ixodes apronophorus*. Alimentary transmission (consumption of raw milk of goats and sheep or drinking contaminated water) has also been described, as well as direct contact—for example, in muskrat trappers (Theiler and Downs, 1973). Its geographical distribution are steppe ecosystem with lakes in southern and western Siberia—specifically the regions Omsk, Novosibirsk, Kurgan, and Tyumen.

The vertebrate hosts are rodents (muskrat *Ondatra zibethica*, imported to Siberia from Canada in 1928 and 1935–1939; *Arvicola terrestris*; *Microtus gregalis*), possibly also frogs and some birds (Růžek et al., 2010). The animal disease produces occasional epizootics (e.g., mass dying of muskrats in Siberia in the period 1946–1970).

The Omsk hemorrhagic fever is characterized by high fever (accompanied by chills, sometimes biphasic), headache, severe myalgia, cough, nausea, nasal bleeding, pharyngitis, conjunctivitis, hyperemia of the face, petechial rash, hemorrhages, and encephalitis (occasionally) with pareses, with a case fatality rate of 1–3% and long convalescence. Between 1946 and 1958, 972 human OHF cases were reported; thereafter the incidence declined remarkably. During 1988–1998, a total of 172 cases were reported from western Siberia (Gritsun et al., 2003). However, mild cases might have been misdiagnosed or not reported. Seasonal peaks of OHF occur in September and October.

31.3.5 Kyasanur Forest disease virus

This virus also belongs to the TBE antigenic complex. Very closely related to KFDV (in fact, its variants or subtypes) are the viruses Alkhumra virus (ALKV; its overall genomic homology with KFDV is 89%) and Nanjianyin virus (occurring in China). KFD was first recognized as a new hemorrhagic zoonotic disease in Shimoga district of Karnataka (then Mysore) State, India, in 1957 (Theiler and Downs, 1973). ALKV was first isolated from the blood of a butcher in Jeddah, Alkhumra district of Saudi Arabia, in 1995 (Madani, 2005; Pattnaik, 2006). KFD occurs in forest ecosystems in India and West China (province Yunnan, *Nanjianyin* virus) while ALKV in semidesert habitats in Saudi Arabia.

The major hematophagous vector (and reservoir) of KFDV is the tick *Haemaphysalis spinigera*, less important seem to be *H. turturis*, other *Haemaphysalis* spp., and possibly *Ornithodoros savignyi* in ALKV. Direct contact with infected sheep and goats and drinking raw milk seem to be important modes in the transmission of ALKV (Madani et al., 2011; Pattnaik, 2006), and mosquito bites were reported as the only risk factor in one-fifth of 78 Alkhumra hemorrhagic fever (AHF) patients examined, while only 3% of them reported history of tick bites (Madani et al., 2011). There is an occupational risk with ALKV infection (e.g., at slaughtering sheep).

The vertebrate hosts are monkeys, rat *Rattus blanfordi*, striped forest squirrel *Funambulus tristriatus*, insectivores (*Suncus murinus*), and bats (*Rhinolophus rouxii*) in KFD and probably sheep and goat with ALKV in Saudi Arabia. In animals, occasional epizootics like mass dying of primates (KFDV) may be observed. For instance, high mortality due to KFD, significantly reducing population density of local monkeys, was observed in the black-faced langur (*Semnopithecus entellus*) and the red-faced bonnet monkey (*Macaca radiata*) in the Kyasanur Forest in 1957 and later: 1965–1966 and 1969–1975 (Theiler and Downs, 1973, Pattnaik, 2006).

KFD and AHF present fever (often biphasic course), headaches (severe headache initiates the 2nd phase), malaise, myalgia, arthralgia, anorexia, backache, nausea and vomiting, diarrhea, abdominal pain, erythema on face, conjunctivitis, retro-orbital pain, bradycardia, pharyngitis, meningoencephalitis (in about 20% of cases), neck stiffness, impaired sleep, mental disturbance, hepatitis, hemorrhagic manifestations (nasal and gastrointestinal bleeding), leukopenia and thrombocytopenia, and elevated liver enzyme levels. Fatality rate is 2–15% (but in AHF can be as high as 25%). Convalescence is long—up to 4 weeks (Madani, 2005; Madani et al., 2011). Big outbreaks of KFD occurred in the Indian state Karnataka (then Mysore) in 1957 (hundreds of human cases) and later on in 1986 (213 cases with 14 fatalities). The KFD foci activated in the 1990s, and hundreds of human cases have been reported annually since 2001, with a spike of 915 cases in the year 2003 (Pattnaik, 2006). In Saudi Arabia, about 60 human cases of AHF in Jeddah and Makkah provinces occurred until 2003 (Pattnaik, 2006; the fatality rate was 25%). Additional c. 90 human cases were reported in Makkah province and a number of patients also in Najran, in the south border of Saudi Arabia, with Yemen during 2003–2009 (Madani et al., 2011).

31.4 FAMILY *BUNYAVIRIDAE*

31.4.1 Crimean-Congo hemorrhagic fever virus

CCHF is a serious human disease mainly transmitted by ticks of the genus *Hyalomma*. Since the first outbreak of CCHF described in Europe in 1945, several subsequent outbreaks have been reported worldwide in both newly discovered foci and foci at which the virus was known to be present. Interest in the disease increased after the recent epidemic in Turkey and new viral records reported in areas near Turkey such as the Balkans and Russia (Ergonul and Whitehouse, 2007). Studies have focused in outlining the probable routes for virus introduction into Western Europe from the original foci of the disease in Eastern Europe and Turkey (Gale et al., 2010). The finding of Crimean–Congo hemorrhagic fever virus (CCHFV) in Western Europe (Estrada-Peña et al., 2012a) encouraged the studies aimed to assess the endemic potential of the virus in Europe. These results demonstrated that the virus is not restricted to Eastern Europe, as obviously known, and that a viral strain circulates in southwestern Mediterranean. This increased the concerns about the spread of the virus into northern latitudes (Estrada-Peña et al., 2012b). The virus has the largest known distribution of any other tick-transmitted virus (Ergonul and Whitehouse, 2007). The virus is transmitted to reservoir mammals and humans through the bite of hard ticks, mainly of the genus *Hyalomma* (Hoogstraal, 1979). Humans may also become infected through direct contact with the blood or tissues of infected humans or livestock (Hoogstraal, 1979). Some other tick species from the genera *Dermacentor*, *Amblyomma*, *Rhipicephalus*, and *Haemaphysalis* have been found to harbor the virus in the field or have been artificially infected in the laboratory, but there is little evidence of their involvement in natural transmission or maintenance of foci (Watts et al., 1988). All natural reports linking the transmission of the virus by way of an infected vector have involved ticks of the genus *Hyalomma* (Watts et al., 1988). It would appear that additionally *Hyalomma* ticks are also necessary for the maintenance of active foci of the virus in the field, even within periods of silent activity. The principal species implicated in transmitting CCHFV in Eurasia are *H. marginatum*, *H. turanicum*, *H. anatolicum*, and *H. scupense* (including the former *H. detritum*, now considered a synonym of *H. scupense*, Guglielmone et al., 2009).

The tick genus *Hyalomma* is widespread in different ecological areas of the Palearctic and Afrotropical regions. The tick vector has three active stages. The immatures (larvae and nymphs) commonly feed on the same hosts, which are many species of small mammals and birds. It is thus a two-host tick, although it may behave as a three-host tick under some conditions (Hoogstraal, 1979). Large ungulates serve as hosts for the adults. Tick females contribute to the infection by TOT of the virus to the eggs. Feeding on infected reservoir hosts or through the nonsystemic (cofeeding) transmission of the virus might also infect ticks. The nonsystemic transmission may occur when uninfected ticks feed in near proximity to infected ones, which pass the virus with the saliva without host systematic infection (Gordon et al., 1993). It is well established that the immature tick stages (and not the adult) of *H. marginatum* infest birds and medium-sized mammals, while adults feed on large ungulates (Hoogstraal, 1979). As for any other species, certain conditions of temperature and humidity are needed for molting of immature stages of *H. marginatum* to adults (Estrada-Peña et al., 2011). Some species, like *H. scupense* (one- or two-host biology) and *H. anatolicum* (two- or three-host biology), prefer to feed on the same large ungulates (mostly cattle) during all developmental stages and then adopt a nidiculous life cycle, protecting them from extreme environmental conditions.

The virus has been reported to survive throughout the life of the tick and passes transstadially and transovarially. The long survival of the virus in ticks is important from the epidemiological point of view. However, there is still a dearth of knowledge regarding host exposure rates and host immune responses particularly in populations of short-lived birds, insectivores, and lagomorphs. Such animals have a high population turnover shown to be important in other tick-borne pathogens (i.e., TBEV) where such hosts develop antibodies to exposure in the nest during their first few days of life. The epidemiological potential, relating climate, ticks, and reservoirs of the active foci, is a very important part of the enzootic ecology of CCHFV. Recent reports of an increased incidence of CCHF stimulated speculation about the presumed effects of climate on the historical geographical range of *H. marginatum* ticks in the Palearctic region (Karti et al., 2004; Maltezou et al., 2010) and the probable spread of the pathogen. The tick is presumed to be the most prominent vector of the virus to humans in a large region extending from the Balkans in Europe to Pakistan and Afghanistan in the Middle East (Ergonul and Whitehouse, 2007). In an expert consultation organized by the European Center for Disease Control in 2008, a short-term priority was recognized to be "endemic regions in countries with CCHF in southeastern Europe should be further mapped on national and international levels, and the degree of CCHF risk in all countries should be estimated."

We ignore basic epidemiological parameters for the transmission of CCHFV and how changes in transmission rates among ticks and competent reservoir hosts affect virus circulation and geographical range. It is known that the tick larvae molt into nymphs while attached to the bird, lengthening the duration of host attachment (12–26 days) and so enabling the passive transport of the immature *Hyalomma* ticks by migrating birds over long distances (Hoogstraal et al., 1961). As an example, an adult male *Hyalomma rufipes* tick was identified on a horse in the Netherlands during a survey of ticks (Nijhof et al., 2007). As that horse was not imported, Nijhof et al. (2007) speculated that the tick was introduced as a nymph by a migratory bird from Africa. *H. rufipes* is endemic in many regions of Africa and has been recorded on migratory birds in spring in Europe (Molin et al., 2011). However, the species is not known to have permanent populations in Europe because it is an Afrotropical tick, which needs high temperatures for adequate molting (Estrada-Peña and Venzal, 2007). Every year, literally millions of passerine birds reach the European continent, parasitized by ticks coming from the northwestern coast of Africa and which serve as vectors of CCHFV, a pathogen that is known to exist in the area where the birds rest before the entry into southwestern Europe. How the climate could affect the flight of the migratory birds, how the ticks attached may enter at higher rates, and how many infected ticks may spread over the continent each year are key variables that have not yet been evaluated.

Ticks can disperse large distances only while on their hosts (Randolph, 1998). Therefore, the capacity for a population to spread depends on the availability and invading abilities of the potential hosts in combination with other factors that deeply affect the behavior of the host, such as habitat fragmentation and physical barriers to migration. The potential effects of the climate trend on the geographical range of arthropods are commonly evaluated by climate-matching models, a set of methods based on the recorded distribution that assess the potentially available range for a species according to its preferences for a group of explanatory variables (an example is provided in Figure 31.1). Process-driven models focus on each part of the life cycle and are regarded as an essential tool for research on tick-borne pathogen transmission rates (Randolph and Rogers, 2000). Efforts to build process-driven models have been focused on *I. ricinus* (Dobson et al., 2011), but until recently a process-driven model of the life cycle of *H. marginatum* was unavailable (Estrada-Peña et al., 2011).

It is now known that clinical cases of CCHF are not reported everywhere the tick vector exists, making evident that a complement of epidemiological factors are necessary to fire up a new focus or for reemergence of former ones. Studies have been carried out in South Africa (Swanepoel et al., 1983), Tanzania, and African countries from Senegal in the west to Kenya in the east (Hoogstraal, 1979). The field investigations that followed recognition of the disease included antibody sera collected from humans and livestock and a survey of the prevalence of the virus in questing and feeding ticks. Further studies were carried out in west Africa, mainly in Senegal and Mauritania (Chapman et al., 1991; Wilson, 2009; Zeller et al., 1994a, b, 1997). These studies highlighted the clear correlation between antibodies to the disease in livestock and humans and the distribution of ticks of the genus *Hyalomma* (Wilson et al., 2009). Humidity in Senegal varies from 200 mm in the Sahelian zone in the north to more than 1400 mm in the sub-Guinean zone in the south, and this is reflected in changing composition of the tick species across the country. Bioclimatic zones differed in the intensity with which CCHFV was transmitted. Evidence of infection in sheep was greatest in the northern, arid, sparsely vegetated zone of Senegal and decreased consistently toward the southern, moister, forested zone. The specific identity of the tick vectors that maintain CCHFV transmission in Senegal is unknown (Wilson et al., 2009) although their results indicated that *Hyalomma* species are important in the maintenance of local or regional foci of the disease. Further studies (Sylla et al., 2008) focused also on the effect of climate variables along a north–south gradient in Senegal as a marker for the dominant tick species, and in turn the serological prevalence of CCHFV in humans.

Such a kind of climate transition affecting the main vectors of CCHFV is harder to outline for other areas in Africa, because of the wide variety of habitats and species with well-varied climate preferences. Figure 31.2 includes the reported distribution of several species in the genus *Hyalomma* in both western Palearctic and Africa. Records in the Mediterranean basin correspond to *H. marginatum*, the main vector of the virus in the area. The other records correspond to several species of *Hyalomma* as reported in the Afrotropical region. Such a kind of detailed distribution is missing for other areas where *Hyalomma* ticks are known to be present. It has been however reported that warmer scenarios would favor the distribution of *Hyalomma* in South Africa (Estrada-Peña, 2003). In the western

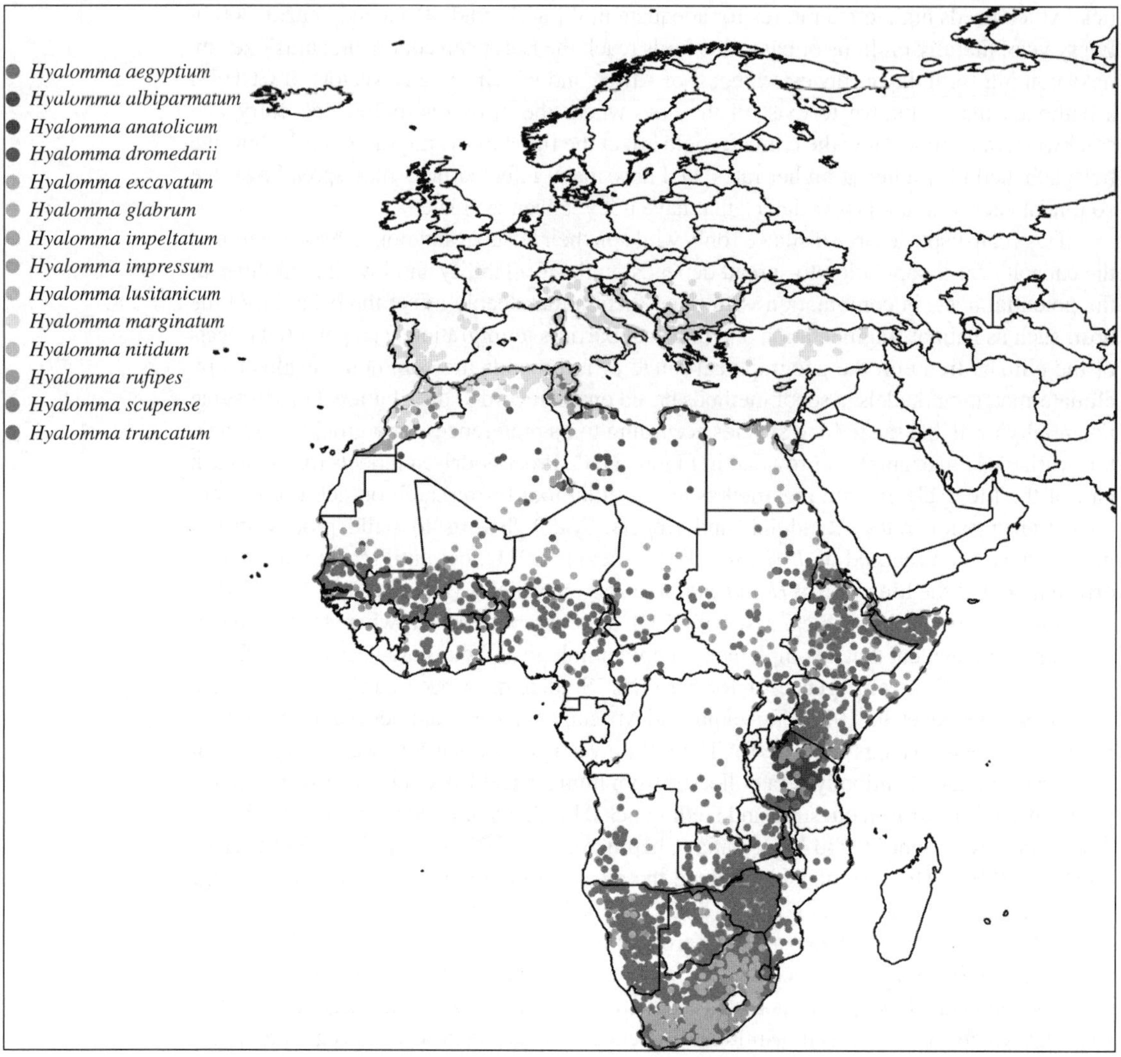

Figure 31.2. The distribution of the most prominent species of the genus *Hyalomma* in Africa. Only those records with accurate georeferences where included, that is, with a pair of coordinates or with an unambiguous name of a locality. The compilation is not intended to be exhaustive, and it provides only general information on the distribution of those species. For color detail, please see color plate section.

Palearctic, studies suggest that scenarios of warmer climate would increase the northern distribution limit of the tick (Figure 31.3) because it would improve the colder conditions in winter and would rise the number of days with temperature above the minimum threshold necessary for completion of development by the tick (Estrada-Peña et al., 2012b).

The recent epidemic of CCHF in Turkey began with some isolated cases in Tokat Province (Gozalan et al., 2004). The human health authorities soon realized that more clinical cases were being reported from neighboring sites and then later over a large territory in the country (Yilmaz et al., 2009), largely coinciding with the expected distribution of the tick *H. marginatum* in an early paper about the dynamics of the infection in that country (Estrada-Peña et al., 2007). The very focal nature of CCHF in Turkey

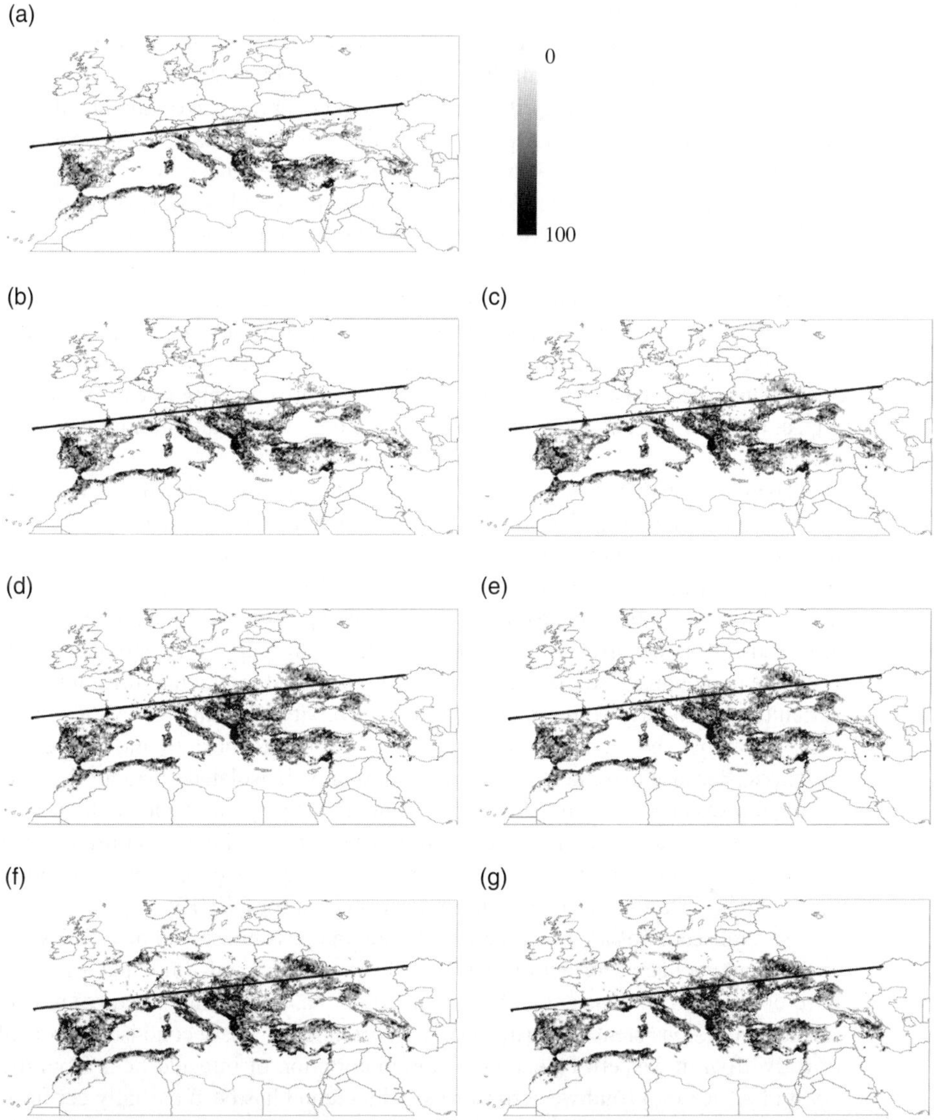

Figure 31.3. Changes in the computed probability of occupancy for *H. marginatum* in the western Palearctic. The measure is unitless, combining the net growth rate of the tick in the site and the connectivity of vegetation patches due to host movements. Data on probability of occupancy range from 0 to 100 (with data and methods reported by Estrada-Peña et al. 2012c). **(a)** Results for the current climate (1961–2000). **(b–g)** Recruitment was computed for climate projections for the years 2020 (**b, c**), 2050 (**d, e**), and 2080 (**f, g**) and scenarios **a** (**b, d, f**; hard growth, high emissions) and **b** (**c, e, g**; low growth, fewer emissions). The black line in every figure marks the approximate latitudinal level of the northern limit of the ticks under current climate conditions.

exhibited a strong correlation of the highest incidence in human cases with the presence of particular land use categories, significantly small and highly mosaic agricultural localities. This was consistent with the most common transmission mechanism reported, the bite of an infected tick, and with the increased densities of ticks in fragmented sites, which in turn provide an environment for higher exposure in humans.

Similar to other tick-borne diseases, climate trends have been commonly linked to outbreaks of clinical cases. Ticks are highly sensitive to small changes in climate, and long-term, sustained, and small differences in key climate variables may drive a serious change. An assessment of the effects of climate on the presence of human clinical cases of the disease in Turkey was carried out (Estrada-Peña et al., 2010). The analysis included monthly values of several climate variables and concluded that climate was not different in sites with active foci of the disease as compared with sites where *H. marginatum* is common but human cases are not reported. They concluded that it is not possible to predict, based solely on climate grounds, where new cases could appear in a reasonably near future. Climate is not the sole factor driving the virus amplification, even if climate in autumn/winter may have a strong regulating role on the survival of tick populations (Hoogstraal, 1979). It is noteworthy that *Hyalomma* endemic areas with mildest autumns and winters in the Mediterranean basin have historically been and are currently free of the disease, so the impact of a warmer climate seems not to be the only factor determining the occurrence of the disease.

There is however evidence that a fragmented landscape, with many small patches existing within a matrix of unsuitable tick habitat, may lead to isolated populations of both ticks and hosts, producing an amplification cycle whereby ticks feed on infected hosts (Estrada-Peña et al., 2010). For CCHFV eco-epidemiology, the degree of habitat patchiness contributes to the increased contact rate among reservoir hosts, humans, and ticks. It also leads to the occurrence of isolated amplification foci, with consequently higher tick exposure to humans (Estrada-Peña et al., 2009). Isolated fragments surrounded by a matrix agricultural land have commonly the poorest diversity of mammals. Although *Hyalomma* ticks can move horizontally, they require a vertebrate host to move over significant distances. Because these host populations are relatively isolated, there are few local movements of hosts and therefore new "naive" animals carrying uninfected ticks are not diluting the prevalence rates in the isolated patch; however, the immune response of such isolated populations against the ticks and the pathogen may seriously decrease the population of infected ticks. These aspects have never been studied for CCHFV.

Several scenarios are of concern regarding the possible spread of CCHFV to new foci or the reemergence of former ones. One is the impact that a warmer and drier climate may have on the distribution range of the ticks in Africa, where they already are occupying a climate niche consistent of warm and dry environments. The second is the probable spread of active foci in western Palearctic or the importation of infected ticks from the western coast of Africa into southwestern Europe. The second has been partially confirmed by the finding of CCHFV in *H. lusitanicum* ticks in southern Spain (Estrada-Peña et al., 2012b). The details around the findings suggest that the virus strain has been circulating in the area since a long time ago, because it has been detected in a tick that does not use migratory hosts (i.e., it is resident and restricted to some areas in Southern Europe). However, no clinical cases have been reported in such area in Spain nor in the near Portugal. All these extremes must to be confirmed before confirming the hypotheses of an old existence of the virus in Western Europe.

31.4.2 Henan virus

Henan fever bunyavirus (HNFV) is a new, emerging bunyavirus, not yet assigned to a genus (Xu et al., 2011; Yu et al., 2011). It is also called Huaiyangshan virus (HYSV: Chen et al., 2012) or severe fever with thrombocytopenia syndrome virus (SFTSV: Yu et al., 2011). The virus is distantly related to tick-borne Uukuniemi bunyavirus

(Xu et al., 2011; Yu et al., 2011). It extends in China (mainly northeast and central provinces); the natural foci are situated in woody and hilly areas.

The main tick vector is *Haemaphysalis longicornis*. However, the virus is also transmitted from person to person by contacting patient's blood (Bao et al., 2011; Liu et al., 2012). The vertebrate hosts are unknown, possibly rodents. Newborn mice are highly susceptible at experimental inoculation with HNFV: the mice that died developed extensive necrotic areas in liver while no obvious pathologic changes were seen in other organs. HNFV antigen and RNA were present in almost all organs, indicating a systemic infection (Chen et al., 2012).

Human disease consists of severe fever with thrombocytopenia syndrome (SFTS) or fever with thrombocytopenia and leukopenia syndrome (FTLS)—hemorrhagic fever-like disease. The key clinical features include fever, fatigue, diarrhea, abdominal pain, lymphocytopenia, and thrombocytopenia. In a clinical study, 8 of 49 patients (16.3%) with hemorrhagic fever caused by HYSV died; and the fatal outcome was associated with high viral RNA load in blood at admission, higher liver transaminase levels, and pronounced coagulation disturbances (Zhang et al., 2012). Other sources report a 21% case fatality rate among 171 patients by September 2010 (Dr. Wang Yu, personal communication). Human FTLS cases have been observed in China since 2006, and up to 2010, about 300 patients with this syndrome were confirmed (Liu et al., 2012; Xu et al., 2011).

31.4.3 Bhanja virus

This virus (synonym Palma virus, Filipe et al. 1994) is, together with two other African tick-borne viruses Kismayo and Forécariah, a member of Bhanja group that has not yet been assigned to a recognized genus of the family *Bunyaviridae*. Bhanja bunyavirus (BHAV) was isolated first from *Haemaphysalis intermedia* (syn. *Haemaphysalis parva*) ticks that had been collected from a paralyzed goat in Bhanjanagar (district Ganjam, Orissa State, India) in 1954 (Shah and Work, 1969). In Europe, the first isolation was from adult *H. punctata* collected in Italy in 1967 (Verani et al., 1970). The geographical distribution is Southern and partly Central Europe (Italy, Croatia, Bulgaria, Romania, Slovakia, Portugal). Outside Europe it has been reported from India, Kirghizia, Kazakhstan, Azerbaijan, Armenia, Senegal, Guinea, Nigeria, Cameroon, Central Africa, Kenya, and Somalia. Antibodies were detected in vertebrates of many additional countries (Spain, Moldova, Sri Lanka, Pakistan, Iran, Turkmenia, Uzbekistan, Tajikistan, Uganda, Tanzania, Egypt, and Tunisia).

The virus is transmitted by metastriate ixodid ticks: *Haemaphysalis intermedia, H. punctata, H. sulcata, D. marginatus, Rhipicephalus decoloratus, R. annulatus, R. geigyi, Amblyomma variegatum, Hyalomma marginatum, H. detritum, H. dromedarii, H. truncatum, H. asiaticum* (TOT), *Rhipicephalus bursa*, and *R. appendiculatus*. Probable vertebrate hosts for BHAV are sheep, goat, and cattle; in Africa, BHAV was also isolated from the four-toed hedgehog (*Atelerix albiventris*) and striped ground squirrel (*Xerus erythropus*). The virus does not usually cause apparent infection in adult animals but is pathogenic for young ruminants (lamb, kid, calf), causing fever and meningoencephalitis (Camicas et al. 1981; Mádr et al., 1984; Semashko et al., 1976; Theiler and Downs, 1973). Experimental encephalitis was produced in rhesus monkey (Balducci et al., 1970).

Natural foci of BHAV are boskematic (pastoral steppe or forest) steppe ecosystems in xerothermic areas or in karst habitats at more northern latitudes. Based on a comparison of several known natural foci of BHAV infection, their common and typical features were extracted and bio-indicator species (plants, animals) were selected that can be used for prediction of potential presence of BHAV in other geographical areas within Europe (Hubálek, 2009).

In humans, BHAV can cause febrile illness with headache, conjunctivitis, or sometimes meningoencephalitis with photophobia, vomiting, and pareses. About 10 natural and/or laboratory infections with BHAV have been described in humans, one of them serious, with quadriparesis (Calisher and Goodpasture 1975; Punda et al., 1980; Vesenjak-Hirjan et al., 1980). There is some occupational risk for shepherds and veterinary personnel. Probably, this has been an underdiagnosed disease in some countries.

31.4.4 Keterah virus

This virus has not yet been assigned to a genus. Keterah bunyavirus (KETV) was first isolated from larval *Argas pusillus* infesting *Scotophilus* bats in Malaysia, 1966 (Karabatsos, 1985), while an identical virus (Issyk-Kul virus) later from bats and their ticks *Argas vespertilionis* in Kirghizia (Lvov et al., 1973). Its geographical distribution involves Malaysia and Central Asia (Kirghizia, Tajikistan, Uzbekistan).

The vectors are soft ticks (*Argasidae*), possibly also biting midges (*Culicoides schultzei*) and mosquitoes. The vertebrate hosts are bats. The animal disease is asymptomatic in bats; in green monkeys it causes damage to visceral organs but without overt clinical symptoms. A human outbreak with more than 60 cases was described in southern Tajikistan, 1982 (Lvov et al., 1984).

31.5 FAMILY *REOVIRIDAE*

31.5.1 Colorado tick fever virus

The Colorado tick fever virus (CTFV) is transmitted by ixodid ticks (principal vector is *Dermacentor andersoni*, but CTFV has also been isolated from *D. occidentalis, D. parumapertus, D. albipictus*). It may be transmitted by blood transfusion, because the virus causes in humans persistent viremia up to 120 days, being localized in erythrocytes. This virus extends in North America, with natural foci occurring in the Rocky Mountains of the United States and Canada, most often at altitudes of 1200–3000 m above sea level.

The main vertebrate hosts are rodents (reservoirs: mainly *Spermophilus lateralis, Tamias minimus, T. amoenus, Tamiasciurus richardsoni, Erethizon dorsatum, Neotoma cinerea, Peromyscus maniculatus*). The disease in animals seems to be inapparent but teratogenic in mice. Colorado tick fever (CTF) is usually a biphasic fever disease in humans, with headache, myalgia and arthralgia, conjunctivitis, photophobia, sometimes orchitis, and affection of the CNS (mainly in children); temporary rash occurs less often (5–10% of patients) than in RMSF, and occasionally myopericarditis, pneumonia, and hepatitis occur. Rare complications with this disease have included aseptic meningitis, encephalitis, and hemorrhagic fever. Laboratory findings include leukopenia, thrombocytopenia, and mildly elevated liver enzyme levels. Mortality is low, but the convalescence long (fatigue, lethargy).

31.5.2 Kemerovo virus

Kemerovo virus (KEMV) was first isolated from ixodid ticks and a patient during an expedition to study RSSE in Siberia in 1962 (Chumakov et al., 1963). It is transmitted by *I. persulcatus*. Migratory birds have been implicated in the dispersal of KEMV over vast distances. For instance, KEMV was isolated from a migrating redstart *Phoenicurus phoenicurus* in Egypt in 1961 (Schmidt and Shope, 1971). Its vertebrate hosts are birds and rodents. The animal disease has an inapparent course but meningoencephalitis was observed in experimentally inoculated monkeys (Grešíková et al., 1966).

31.5.3 Tribeč virus

Tribeč virus (TRBV) is a member of Kemerovo antigenic group and the Kemerovo subgroup (Belhouchet et al., 2010), closely related to the Siberian KEMV by complement-fixation test but distinguishable by virus neutralization test (Libíková and Buckley, 1971) or RNA–RNA hybridization (Brown et al., 1988). Kemerovo group and other orbiviruses have a great reassortment potential (because of the segmented dsRNA), resulting in biological variability (Brown et al., 1988; Gorman, 1983). The reported synonyms and subtypes are Lipovník, Koliba, Cvilín, Brezová, Mircha, and Kharagysh virus. It is known from Slovakia, Czechland, Ukraine, Belarus, Russia, southern Norway, Italy, and exceptionally northern Africa. Natural foci of TRBV infections are both boskematic and theriodic (pastoral and mixed woodland ecosystems).

The first strains of TRBV were isolated from *I. ricinus* in three regions of Slovakia in 1963 (Grešíková et al., 1965; Libíková et al., 1964, 1965). TRBV is transmitted by ticks *I. ricinus* (TST), occasionally by *H. punctata* (Topciu et al., 1968). The vertebrate hosts of TRBV are rodents, for example, *Myodes glareolus* and *Microtus subterraneus*; hare *Lepus europaeus*; goat; European starling *Sturnus vulgaris*; and chaffinch *Fringilla coelebs* (Dobler et al., 2006; Grešíková et al., 1965; Skofertsa et al., 1974). Animal disease is unknown.

TRBV causes febrile illness or aseptic meningitis in humans occasionally; for example, about 15 patients with the CNS infection (meningitis) revealed seroconversion against TRBV in Czechland (Fraňková, 1981; Hubálek et al., 1987; Málková et al., 1986). The disease caused by TRBV is probably underdiagnosed. Additional studies are necessary to evaluate the public health importance of TRBV.

31.6 FAMILY *ORTHOMYXOVIRIDAE*

31.6.1 Thogoto virus

Thogoto virus (THOV) was first isolated from a mixed pool of *Rhipicephalus decoloratus* and *Rhipicephalus* spp. ticks collected on cattle in Thogoto Forest near Nairobi, Kenya, in 1960 (Haig et al., 1965). In Europe, it was first isolated from ticks collected on ruminants in Sicily in 1969 (Albanese et al., 1972) and then in Portugal in 1978 (Filipe and Calisher, 1984). Arthropod vectors of THOV are metastriate ticks only—*R. annulatus*, *A. variegatum*, *R. appendiculatus*, *R. sanguineus*, *R. bursa*, *R. evertsi*, other *Rhipicephalus* spp., *H. truncatum*, and *H. anatolicum*. THOV extends in areas of Kenya, Uganda, Ethiopia, Nigeria, Cameroon, Central Africa, Egypt, Iran, Portugal, and Sicily. Tick-infested domestic animals (e.g., camels) and migratory birds could disseminate the virus over a wide geographical range (Calisher et al., 1987). Natural foci are boskematic—pastoral xerothermic ecosystems.

Vertebrate hosts for THOV are camel and horse. Antibodies were also detected in sheep and goat. The infection course is usually inapparent in animals, but THOV can cause leukopenia in cattle and fever and abortion in sheep (Davies et al., 1984; Theiler and Downs, 1973).

Only two cases of human disease have been described, one with bilateral optic neuritis and another as a fatal meningoencephalitis with hepatitis although complicated by a sickle-cell disease (Moore et al., 1975; Theiler and Downs, 1973). THOV is probably contagious from man to man.

31.6.2 Dhori virus

Dhori virus (DHOV) was first isolated from *H. dromedarii* ticks collected on camels in Dhori, Gujarat State, India, in 1961 (Anderson and Casals, 1973). In Europe it has been recovered several times from *H. marginatum* and twice from *H. scupense* collected at Astrakhan, South Russia, since 1969 (as "Astra" virus (Bannova et al., 1974; Butenko et al., 1971, 1987)) and in Crimea (one strain—"Batken"); additional two strains were obtained from *H. scupense* near Astrakhan (Smirnova et al., 1988) and another one in southern Portugal in 1971 (Filipe and Casals, 1979). The reported Batken virus (Lvov et al., 1974) is a synonym of DHOV. It is known from areas in Portugal, Crimea, Astrakhan (southern Russia), Armenia, Azerbaijan, Kirghizia, Uzbekistan, India, and Egypt, while only antibodies were detected in vertebrates from Pakistan. Natural foci are boskematic, typically pastoral xerothermic ecosystems.

Principal arthropod vectors are metastriate ticks *Hyalomma dromedarii*, *H. marginatum*, *H. scupense*, and *D. marginatus*. Occasional isolations of DHOV were reported from *Ornithodoros lahorensis*. Vertebrate hosts are cattle, camel, horse, and bats (Kirghizia), but animal disease is asymptomatic. Natural foci of DHOV are boskematic (pastoral xerothermic and semidesert ecosystems).

DHOV produces an acute illness with severe fever, headache, general weakness, and retrobulbar pain, with encephalitis in c. 40% of patients and a long 2-month convalescence period. Five cases of severe laboratory infection (due to aerosol) have been described (Butenko et al., 1987). The virus could be contagious from man to man.

31.7 OTHER TICK-TRANSMITTED VIRUSES

Nine additional tick-borne viruses, occasionally causing clinical disease in humans, are only briefly presented in Table 31.1. No details exist about the impact of the climate on the spread of these viruses since they have been poorly investigated.

31.8 CONCLUSIONS

Talking in general terms, we are not yet able to evaluate the fine effects of climate trends on the epidemiology of the most prominent tick-borne viruses. We should keep in mind that these changes do not affect only the dynamics of tick vectors but also the abundance of hosts for immature stages of the tick or their migratory timings in the case of birds (which may be hosts for the immatures of the ticks) or even how climate may affect the densities of hosts in natural conditions. While available models might greatly contribute to understand the behavior of the ticks under variable climate conditions, we need yet to build upon those models to reach the necessary level of complexity. Local processes are not captured yet by these models, adding "noise" to the general background picture of the fine-scale distribution of ticks, their vectors, and the pathogens they transmit. There is an implicit need of further research at both local and regional scales. The first must be to finely capture the molecular details of the tick–pathogen relationships (Randolph, 2009), and the second should focus to describe such relationships under a generalist framework (Estrada-Peña et al., 2012).

TABLE 31.1. Tick-Transmitted Viruses Causing Disease in Humans only occasionally (Charrel et al. 2004; Karabatsos 1985; Labuda and Nuttall 2004; Theiler and Downs 1973)

Virus	Genus	Tick Vectors	Vertebrate Hosts	Geographical Distribution	Human Infection
Flaviviridae					
Tyuleniy	*Flavivirus*	*Ixodes uriae*	Seabirds (*Uria aalge*, *Eudyptula minor*), suslik (*Citellus undulatus*)	Coastal N. Russia (Murmansk), Norway (Lofoten), Asian Russia (Far East), western United States (Oregon), Canada	Three cases (malaise, laryngitis, lymphadenopathy, arthralgia, and skin petechiae) in biologists visiting seabird colonies
Bunyaviridae					
Soldado	*Nairovirus*	*Ornithodoros maritimus*, *O. capensis*	Seabirds (*Sterna fuscata*, *Larus argentatus*, *Rissa tridactyla*)	Trinidad, Hawaii, Texas, Ethiopia, Senegal, Seychelles, South Africa, Morocco, United Kingdom, Ireland, France, Iceland	Several cases (fever, pruritus, rhinopharyngitis)
Zirqa	*Nairovirus*	*Ornithodoros muesebecki*	Seabirds	Persian Gulf	Several cases (fever, headache, pruritus, erythema)
Punta Salinas	*Nairovirus*	*Ornithodoros amblus*, *Argas arboreus*	Colonial birds	Peru, Tanzania	Several cases (fever, headache, pruritus, erythema)
Dugbe	*Nairovirus*	*A. variegatum*, *R. decoloratus*, *H. truncatum*	Rodents	Tropical Africa	Two cases (encephalitis)
Nairobi sheep disease (syn. Ganjam)	*Nairovirus*	*R. appendiculatus*, *Haemaphysalis* spp.	Sheep, goat, *Arvicanthis abyssinicus*	Kenya, Uganda, South Africa, India	Six cases (fever, arthritis)
Avalon (syn. Paramushir)	*Nairovirus*	*I. uriae*, *Ixodes signatus*	Seabirds (*L. argentatus*)	France (Brittany), Asian Russia (Far East), Canada	Three cases (cervical adenopathy)
Wanovrie	Unassigned	*Hyalomma* spp.	?	India	One case (hemorrhagic fever)
Orthomyxoviridae					
Quaranfil	Unassigned	*A. arboreus*, *Argas reflexus*	Colonial birds (*Ardeola ibis*, pigeon)	Egypt, Yemen, Kuwait, Iran, Iraq, Afghanistan, Nigeria, South Africa	Two cases (fever)

? means not known.

ACKNOWLEDGEMENTS

We want to express our gratitude to our colleagues Jochen Süss (Friedrich-Loeffler-Institute, National Reference Laboratory for Tick-borne Diseases, Jena, Germany) and Hervé Zeller (European Center for Disease Prevention and Control, Stockholm, Sweden) for their comments and help in preparing the manuscript. We also thank the Operational Programme Education for Competiveness project CEB (CZ.1.07/2.3.00/20.0183).

REFERENCES

Albanese, N., Bruno-Smiraglia, C., Di Cuonzo, G., et al. 1972. Isolation of Thogoto virus from *Rhipicephalus bursa* ticks in western Sicily. *Acta Virologica* **16**:267.

Alekseev, A.N., Chunikhin, S.P. 1990. Transmission of tick-borne encephalitis virus between ixodid ticks co-feeding on animals without detectable viremia (in Russian). *Meditsinskaya Parazitologia* **2**:48–50.

Alekseev, A.N., Chunikhin, S.P., Rukhkyan, M.Y., et al. 1991. Possible role of Ixodidae salivary gland substrate as an adjuvant enhancing arbovirus transmission. *Meditsinskaya Parazitologiya i Parazitarnye Bolezni* **1**:28–31.

Anderson, C.R., Casals, J. 1973. Dhori virus, a new agent isolated from *Hyalomma dromedarii* in India. *Indian Journal of Medical Research* **61**:1416–1420.

Bacon, R.M., Kugler, K.J., Mead, P.S. 2008. Surveillance for Lyme disease United States, 1992–2006. *Morbidity and Mortality Weekly Report. Surveillance Summary* **57**:1–9.

Balducci, M., Verani, P., Lopes, M.C., et al. 1970. Experimental pathogenicity of Bhanja virus for white mice and Macaca mulatta monkeys. *Acta Virologica* **14**:237–243.

Bannova, G.G., Sarmanova, E.S., Karavanov, A.S., et al. 1974. Isolation of Dhori-Astra virus from *Hyalomma marginatum* ticks collected on cattle in Krasnodar region, USSR (in Russian). *Meditsinskaya Virusologija* (Moskva) **22**:162–164.

Bao, C.J., Guo, X.L., Qi, X., et al. 2011. A family cluster of infections by a newly recognized bunyavirus in eastern China, 2007: further evidence of person-to-person transmission. *Clinical Infectious Diseases* **53**:1208–1214.

Belhouchet, M., Jaafar, F.M., Tesh, R., et al. 2010. Complete sequence of Great Island virus and comparison with the T2 and outer-capsid proteins of Kemerovo, Lipovnik and Tribec viruses (genus *Orbivirus*, family Reoviridae). *Journal of General Virology* **91**:2985–2993.

Benda, R, 1958a. Experimental transmission of Czech tick-borne encephalitis to goats by infected *Ixodes ricinus* female ticks (in Czech). *Ceskoslovenska Epidemiologie Mikrobiologie a Imunologie* **7**:1–8.

Benda, R. 1958b. The common tick *Ixodes ricinus* L. as a reservoir and vector of tick-borne encephalitis virus, I, II. *Journal of Hygiene Epidemiology Microbiology and Immunology* **2**:314–344.

Blaškovič, D. (ed) 1954. An Outbreak of Encephalitis in the Rožňava Natural Focus of Infections (in Slovak). Vydav SAV, Bratislava, 314 pp.

Brown, S.E., Morrison, H.G., Buckley, S.M., et al. Genetic relatedness of the Kemerovo serogroup viruses: I. RNA–RNA blot hybridization and gene reassortment in vitro of the Kemerovo serocomplex. *Acta Virologica* **32**:369–378.

Butenko, A.M., Chumakov, M.P. 1971. Isolation of "Astra" arbovirus, new to the USSR, from *Hyalomma marginatum* ticks and *Anopheles hyrcanus* mosquitoes in Astrakhan region (in Russian). *Voprosy Meditsinskoi Virusologii* (Moskva) **2**:111–112.

Butenko, A.M., Leshchinskaya, E.V., Semashko, I.V., et al. 1987. Dhori virus—the causative agent of human disease. Five cases of laboratory infection (in Russian). *Vozprosy Virusologii* **32**:724–729.

Calisher, C.H. 1988. Antigenic classification and taxonomy of *flaviviruses* (family Flaviviridae) emphasizing a universal system for the taxonomy of viruses causing tick-borne encephalitis. *Acta Virologica* **32**:469–478.

Calisher, C.H., Goodpasture, H.C. 1975. Human infection with Bhanja virus. *American Journal of Tropical Medicine and Hygiene* **24**:1040–1042.

Calisher, C.H., Karabatsos, N., Filipe, A.R. 1987. Antigenic uniformity of topotype strains of Thogoto virus from Africa, Europe, and Asia. *American Journal of Tropical Medicine and Hygiene* **37**:670–673.

Calisher, C.H., Karabatsos, N., Dalrymple, J.M., et al. 1989. Antigenic relationships between flaviviruses as determined by cross-neutralization tests with polyclonal sera. *Journal of General Virology* **70**:37–43.

Camicas, J.L., Deubel, V., Heme, G., et al. 1981. Étude écologique et nosologique des arbovirus transmis par les tiques au Sénégal. II. Étude expérimentale du pouvoir pathogène du virus Bhanja pour les petits ruminants domestiques. *Revue de Élevage et Médicine Veterinaire des Pays Tropicau* **34**:257–261.

Chapman, L.E., Wilson, M.L., Hall, D.B., et al. 1991. Risk factors for Crimean-Congo hemorrhagic fever in rural northern Senegal. *Journal of Infectious Diseases* **164**:686–692.

Chen, X.P., Cong, M.L., Li, M.H., et al. 2012. Infection and pathogenesis of Huaiyangshan virus (a novel tick-borne bunyavirus) in laboratory rodents. *Journal of General Virology* **93**:1288–1293.

Chumakov, M.P., Zeitlenok, N.A. 1939. Tick-borne spring-summer encephalitis in the wider Ural area (in Russian). *Archiv Biologicheskikh Nauk* **56**:11–17.

Chumakov, M.P., Karpovich, L.G., Sarmanova, E.S., et al. Report on the isolation from *Ixodes persulcatus* ticks and from patients in Western Siberia of a virus differing from the agent of tick-borne encephalitis. *Acta Virologica* **7**:82–83.

Chunikhin, S.P. 1990. Experimental investigation on the ecology of TBE virus (in Russian). *Voprosy Virusologii* **35**:183–188.

Clarke, D.H. 1962. Antigenic relationships among viruses of the tick-borne encephalitis complex as studied by antibody absorption and agar gel precipitin techniques. In: Libíková H (ed) Biology of Viruses of the Tick-Borne Encephalitis Complex. Publ House Czech Acad Sci, Prague, pp 67–75.

Clarke, D.H. 1964. Further studies on antigenic relationships among the viruses of the group B tick-borne complex. *Bulletin World Health Organization* **31**:45–56.

Daniel, M., Danielová, V., Kříž, B., et al. 2004. An attempt to elucidate the increased incidence of tick-borne encephalitis and its spread to higher altitudes in the Czech Republic. *International Journal of Medical Microbiology* **293**, Suppl 37:55–62.

Daniel, M., Kříž, B., Danielová, V., et al. 2008. Sudden increase in tick-borne encephalitis cases in the Czech Republic, 2006. *International Journal of Medical Microbiology* **298** (S1): 81–87.

Danielová, V., Rudenko, N., Daniel, M., et al. 2006. Extension of the *Ixodes ricinus* ticks and agents of tick-borne diseases to mountain areas in the Czech Republic. *International Journal of Medical Microbiology* **296** (S1):48–53.

Davidson, M.M., Williams, H., MacLeod, J.A.J. 1991. Louping-ill in man: a forgotten disease. *Journal of Infection* **23**:241–249.

Davies, F.G., Soi, R.K., Wariru, B.N. 1984. Abortion in sheep caused by Thogoto virus. *Veterinary Record* **115**:654.

Dobler, G. 2010. Zoonotic tick-borne flaviviruses. *Veterinary Microbiology* **140**:221–228.

Dobler, G., Wölfel, R., Schmuser, H., et al. 2006. Seroprevalence of tick-borne and mosquito-borne arboviruses in European brown hares in Northern and Western Germany. *International Journal of Medical Microbiology* **296**:80–83.

Dobson, A.D.M., Finnie, T.J.R., Randolph, S.E. 2011. A modified matrix model to describe the seasonal population ecology of the European tick *Ixodes ricinus*. *Journal of Applied Ecology* **28**:1017–1028.

Ecker, M., Allison, S.L., Meixner, T., et al. 1999. Sequence analysis and genetic classification of tick-borne encephalitis viruses from Europe and Asia. *Journal of General Virology* **80**:179–185.

Eisen, L. 2008. Climate change and tick-borne diseases: a research field in need of long-term empirical field studies. *International Journal of Medical Microbiology* **298** (S1):12–18.

Ergonul, O., Whitehouse, C.A. 2007. Introduction. In: Ergonul and Whitehouse (eds) Crimean-Congo Hemorrhagic Fever. A Global Perspective. Springer, The Netherlands, pp: 3–13.

Estrada-Peña, A. 2003. Climate change decreases habitat suitability for some tick species (Acari: Ixodidae) in South Africa. *The Onderstepoort Journal of Veterinary Research* **70**:79–93.

Estrada-Peña, A. 2009. Tick-borne pathogens, transmission rates and climate change. *Frontiers Bioscience* **14**:2674–2687.

Estrada-Peña, A., Venzal, J.M. 2006. Changes in habitat suitability for the tick *Ixodes ricinus* (Acari: Ixodidae) in Europe (1900–1999). *EcoHealth* **3**:154–162.

Estrada-Peña, A., Venzal, J.M. 2007. Climate niches of tick species in the Mediterranean region: modeling of occurrence data, distributional constraints, and impact of climate change. *Journal of Medical Entomology* **44**:1130–1138.

Estrada-Peña, A., Vatansever, Z., Gargili, A., et al. 2010. The trend towards habitat fragmentation is the key factor driving the spread of Crimean-Congo haemorrhagic fever. *Epidemiology and Infection* **138**:1194–1203.

Estrada-Peña, A., Avilés, M., Martínez-Reoyo, M. J. 2011. A population model to describe the distribution and seasonal dynamics of the tick *Hyalomma marginatum* in the Mediterranean basin. *Transboundary & Emerging Diseases* **58**:213–223.

Estrada-Peña, A., Jameson, L., Medlock, J., et al. 2012a. Unraveling the ecological complexities of tick-associated crimean-congo hemorrhagic Fever Virus transmission: a gap analysis for the Western palearctic. *Vector-Borne and Zoonotic Diseases* **12**:743–752.

Estrada-Peña, A., Palomar, A.M., Santibáñez, P., et al. 2012b. Crimean-Congo hemorrhagic fever virus in ticks, southwestern Europe, 2010. *Emerging Infectious Diseases* **18**:179–180.

Filipe, A.R., Casals, J. 1979. Isolation of Dhori virus from *Hyalomma marginatum* ticks in Portugal. *Intervirology* **11**:124–127.

Filipe, A.R., Calisher, C.H. 1984. Isolation of Thogoto virus from ticks in Portugal. *Acta Virologica* **28**:152–155.

Filipe, A.R., Alves, M.J., Karabatsos, N., et al. 1994. Palma virus, a new Bunyaviridae isolated from ticks in Portugal. *Intervirology* **37**:348–351.

Franková, V. 1981. Meningoencephalitis caused by orbivirus infections in Czechoslovakia (in Czech). *Sbornik Lékařský* **83**:234–235.

Gale, P., Estrada-Peña, A., Martinez, M., et al. 2010. The feasibility of developing a risk assessment for the impact of climate change on the emergence of Crimean-Congo haemorrhagic fever in livestock in Europe: a Review. *Journal of Applied Microbiology* **108**:1859–1870.

Gallia, F., Rampas, J., Hollender, L. 1949. A laboratory infection with encephalitis virus (in Czech). *Časopis Lékařů českých* **88**:224–229.

Gao, G.F., Jiang, W.R., Hussain, M.H., et al. 1993. Sequencing and antigenic studies of a Norwegian virus isolated from encephalomyelitic sheep confirm the existence of louping-ill virus outside Great Britain and Ireland. *Journal of General Virology* **74**:109–114.

Gaumann, R., Mühlemann, K., Strasser, M., et al. 2010. High-throughput procedure for tick surveys of Tick-Borne Encephalitis virus and its application in a national surveillance study in Switzerland. *Applied and Environmental Microbiology* **76**:4241–4249.

Gordon, S.W., Linthicum, K.J., Moulton, J.R. 1993. Transmission of Crimean-Congo hemorrhagic fever virus in two species of *Hyalomma* ticks from infected adults to cofeeding immature forms. *American Journal of Tropical Medicine & Hygiene* **48**:576–580.

Gorman, B.M. 1983. On the evolution in orbiviruses. *Intervirology* **20**:169–180.

Gould, E.A., De Lamballerie, X., Zanotto, P.M.A., et al. 2003. Origins, evolution, and vector/host coadaptations within the genus *Flavivirus*. *Advances in Virus Research* **59**:277–314.

Gozalan, A., Akin, L., Rolain, J.M., et al. 2004. Epidemiological evaluation of a possible outbreak in and nearby Tokat province. *Mikrobiyol Bulletin* **38**:33–44.

Grard, G., Moureau, G., Charrel, R., et al. 2007. Genetic characterization of tick-borne flaviviruses: new insights into evolution, pathogenetic determinants and taxonomy. *Virology* **361**:80–92.

Gray, J.S., Dautel, H., Estrada-Pena, A., et al. 2009. Effects of climate change on ticks and tick-borne diseases in Europe. *Interdisciplinary Perspectives of Infectious Diseases*: article ID 593232.

Grešíková, M. 1958a. Recovery of the tick-borne encephalitis virus from the blood and milk of subcutaneously infected sheep. *Acta Virologica* **2**:113–119.

Grešíková, M. 1958b. Excretion of the tick-borne encephalitis virus in the milk of subcutaneously infected cows. *Acta Virologica* **2**:188–192.

Grešíková, M. 1972. Studies on tick-borne arboviruses isolated in Central Europe. *Biologické Prace* (Bratislava) **18**:1–116.

Grešíková, M., Nosek, J., Kožuch, O., et al. 1965. Study of the ecology of Tribeč virus. *Acta Virologica* **9**:83–88.

Grešíková, M., Rajčáni, R., Hrúzik, J. 1966. Pathogenicity of Tribeč virus for Macaca rhesus monkeys and white mice. *Acta Virologica* **10**:420.

Gritsun, T.S., Nuttall, P.A., Gould, E.A. 2003. Tick-borne flaviviruses. *Advances in Virus Research* **61**:317–371.

Guglielmone, A.A., Robbins, R.G., Apanaskevich, D.A., et al. 2010. The Argasidae, Ixodidae and Nuttalliellidae (Acari: Ixodida) of the world: a list of valid species names. *Zootaxa* **2528**:1–28.

Haig, D.A., Woodall, J.P., Danskin, D. 1965. Thogoto virus, a hitherto undescribed agent isolated from ticks in Kenya. *Journal of General Microbiology* **38**:389–394.

Hemmer, C., Littmann, M., Löbermann, M., et al. 2005. Tickborne meningoencephalitis, first case after 19 years in northeastern Germany. *Emerging Infectious Diseases* **11**:633–634.

Hinten, S.R., Beckett, G.A., Gensheimer, K.F., et al. 2008. Increased recognition of Powassan encephalitis in the United States, 1999–2005. *Vector-Borne Zoonotic Diseases* **8**:733–740.

Holzmann, H., Vorobyova, M.S., Ladyzhenskaya, I.P., et al. 1992. Molecular epidemiology of tick-borne encephalitis virus: cross-protection between European and Far Eastern subtypes. *Vaccine* **10**:345–349.

Holzmann, H., Aberle, S.W., Stiasny, K., et al. 2009. Tick-borne encephalitis from eating goat cheese in a mountain region of Austria. *Emerging Infectious Diseases* **15**:1671–1673.

Hoogstraal, H. 1979. The epidemiology of tick-borne Crimean-Congo hemorrhagic fever in Asia, Europe, and Africa. *Journal of Medical Entomology* **15**:307–417.

Hoogstraal, H., Kaiser, M. N., Traylor, M. A., et al. 1961. Ticks (Ixodidae) on birds migrating from Africa to Europe and Asia. *Bulletin of World Health Organization* **24**:197–212.

Hubálek, Z. 2009. Biogeography of tick-borne Bhanja virus (Bunyaviridae) in Europe. *Interdisciplinary Perspectives in Infectious Diseases* 11 pp. Available at http://wwwhindawicom/journals/ipid/2009/372691html.

Hubálek, Z., Havelka, I., Bárdoš, I., et al. 1987. Arbovirus infections in the Znojmo district (in Czech). *Ceskoslovenska Epidemiologie Mikrobiologie a Imunologie* **36**:337–344.

Hubálek, Z., Pow, I., Reid, H.W., et al. 1995. Antigenic similarity of Central European encephalitis and louping-ill viruses. *Acta Virologica* **39**:251–256.

Ivanova, L.M. 1984. Contemporary epidemiology of infections with natural focality in Russia (in Russian). *Meditsinskaya Parazitologija* **62**(2):17–21.

Jääskeläinen, A.E., Tikkakoski, T., Uzcategui, N.Y., et al. 2006. Siberian subtype tickborne encephalitis virus, Finland. *Emerging Infectious Diseases* **12**:1568–1571.

Jääskeläinen, A.E., Sironen, T., Murueva, G.B., et al. 2010. Tick-borne encephalitis virus in ticks in Finland, Russian Karelia and Buryatia. *Journal of General Virology* **91**:2706–2712.

Jones, L.D., Hodgson, E., Nuttall, P.A. 1989. Enhancement of virus transmission by tick salivary glands. *Journal of General Virology* **70**:1895–1898.

Karabatsos, N. (ed) 1985. International catalogue of arboviruses, including certain other viruses of vertebrates. 3rd ed American Society of Tropical Medicine and Hygiene, San Antonio, TX. With the 1986–1995. Supplements to the International catalogue. Ft Collins: CDC Div Vector-Borne Infect Dis.

Karti, S.S., Odabasi, Z., Korten, V., et al. 2004. Crimean-Congo hemorrhagic fever in Turkey. *Emerging Infectious Diseases* **10**:1379–1384.

Kopecký, J., Tomková, E., Grubhoffer, L., et al. 1991. Monoclonal antibodies to tick-borne encephalitis (TBE) virus: their use for differentiation of the TBE complex viruses. *Acta Virologica* **35**:365–372.

Kožuch, O., Nosek, J. 1971. Transmission of tick-borne encephalitis (TBE) virus by *Dermacentor marginatus* and *D. reticulatus* ticks. *Acta Virologica* **15**:334.

Krejčí, J. 1949. An epidemic of virus meningoencephalitis in the Vyškov area (Moravia) (in Czech). *Lékařské Listy* (Brno) **4**:73–75; 112–116; 132–134.

Křivanec, K., Kopecký, J., Tomková, E., et al. 1988. Isolation of TBE virus from the tick *Ixodes hexagonus*. *Folia Parasitologica* **35**:273–276.

Labuda, M., Jones, L.D., Williams, T., et al. 1993a. Enhancement of tick-borne encephalitis virus transmission by tick salivary gland extracts. *Medical & Veterinary Entomology* **7**:193–196.

Labuda, M., Jones, L.D., Williams, T., et al. 1993b. Efficient transmission of tick-borne encephalitis virus between cofeeding ticks. *Journal of Medical Entomology* **30**:295–299.

Levkovich, E.N., Karpovich, L.G. 1962. Study on biological properties of viruses of the tick-borne encephalitis complex in tissue culture. In: Libíková H (ed) Biology of Viruses of the Tick-Borne Encephalitis Complex. Publ House Czech Acad Sci, Prague, pp 161–165.

Libíková, H., Buckley, S.M. 1971. Serological characterization of Eurasian Kemerovo group viruses. II. Cross plaque neutralization tests. *Acta Virologica* **15**:79–86.

Libíková, H., Řeháček, J., Grešíková, M., et al. 1964. Cytopathic viruses isolated from *Ixodes ricinus* ticks in Czechoslovakia. *Acta Virologica* **8**:96.

Libíková, H., Řeháček, J., Somogyiová, J. 1965. Viruses related to the Kemerovo virus in Ixodes ricinus ticks in Czechoslovakia. *Acta Virologica* **9**:76–82.

Lindgren, E., Gustafson, R. 2001. Tick-borne encephalitis in Sweden and climate change. *The Lancet* **358**:16–18.

Lindgren, E., Jaenson, T.G.T. 2006. Lyme borreliosis in Europe: influences of climate and climate change, epidemiology, ecology and adaptation measures. Publications of the Regional WHO Office for Europe, EUR/04/5046250.

Lindgren, E., Tälleklint, L., Polfeldt, T. 2000. Impact of climatic change on the northern latitude limit and population density of the disease-transmitting European tick *Ixodes ricinus*. *Environmental Health Perspectives* **108**:119–123.

Lindquist, L., Vapalahti, O. 2008. Tick-borne encephalitis. *The Lancet* **371**:1861–1871.

Little, P.B., Thorsen, J., Moore, W., et al. 1985. Powassan viral encephalitis: a review and experimental studies in the horse and rabbit. *Veterinary Pathology* **22**:500–507.

Liu, Y., Li, Q., Hu, W., et al. 2012. Person-to-person transmission of severe fever with thrombocytopenia syndrome virus. *Vector Borne Zoonotic Diseases* **12**:156–160.

Lvov, D.K., Karas, F.R., Timofeev, E.M., et al. 1973. "Issyk-Kul" virus, a new arbovirus isolated from bats and *Argas (Carios) vespertilionis* (Latr., 1802) in the Kirghiz S.S.R. *Archiv fur die gesamte Virusforschung* **42**:207–209.

Lvov, D.K., Karas, F.R., Tsyrkin, Y.M., et al. 1974. Batken virus, a new arbovirus isolated from ticks and mosquitoes in Kirghiz SSR. *Archiv fur die gesamte Virusforschung* **44**:70–73.

Lvov, D.K., Kostyukov, M.A., Daniyarov, O.A., et al. 1984. Outbreak of arbovirus infection in the Tadzhik SSR due to the Issyk-Kul virus (Issyk-Kul fever) (in Russian). *Voprosy Virusologii* **29**:89–92.

Madani, T.A. 2005. Alkhumra virus infection, a new viral hemorrhagic fever in Saudi Arabia. *Journal of Infection* **51**:91–97.

Madani, T.A., Azhar, E.I., El-Abuelzein, T.M., et al. 2011. Alkhumra (Alkhurma) virus outbreak in Najran, Saudi Arabia: epidemiological, clinical, and laboratory characteristics. *Journal of Infection* **62**:67–76.

Mádr, V., Hubálek, Z., Zendulková, D. 1984. Experimental infection of sheep with Bhanja virus. *Folia Parasitologica* **31**:79–84.

Madrid, A.T. de, Porterfield, J.S. 1974 The flaviviruses (group B arboviruses): a cross-neutralization study. *Journal of General Virology* **23**:91–96.

Málková, D., Danielová, V., Holubová, J., et al. 1986. Less known arboviruses of Central Europe., *Rozpravy ČSAV (Praha), matematické a přírodní vědy* **96**:1–75.

Maltezou, H. C., Andonova, L., Andraghetti, R., et al. 2010. Crimean-Congo hemorrhagic fever in Europe: current situation calls for preparedness. *Eurosurveillance* **15**, pii = 19504.

Mantke, O.D., Schädler, R., Niedrig, M. 2008. A survey on cases of tick-borne encephalitis in European countries. *Eurosurveillance* **13**, pii = 18848.

Materna, J., Daniel, M., Metelka, L., et al. 2008. The vertical distribution, density and the development of the tick *Ixodes ricinus* in mountain areas influenced by climate change (The Krkonose Mts., Czech Republic). *International Journal of Medical Microbiology* **298** (S1):25–37.

Matser, A., Hartemink, N., Heesterbeek, H., et al. 2009. Elasticity analysis in epidemiology: an application to tick-borne infections. *Ecology Letters* **12**:1298–1305.

Molin, Y., Lindeborg, M., Nystrom, F., et al. 2011. Migratory birds, ticks, and *Bartonella*. *Infection Ecology Epidemiology* **1**:5997.

Moore, D.L., Causey, O.R., Carey, D.E., et al. 1975. Arthropod-borne viral infections of man in Nigeria, 1964–1970. *Annals of Tropical Medicine and Parasitology* **69**:49–64.

Naumov, R.L., Gutova, V.P., Chunikhin, S.P. 1980. Ixodid ticks and TBE virus. 1. Interaction of the virus with ticks of the genus Ixodes. 2. The genera *Dermacentor* and *Haemaphysalis* (in Russian) *Meditsindkaya Parazitologija* **49**(2):17–23; (3):66–69.

Nijhof, A.M., Bodaan, C., Postigo, M., et al. 2007. Ticks and associated pathogens collected from domestic animals in the Netherlands. *Vector Borne Zoonotic Diseases* **7**:585–595.

Nuttall, P.A., 1999. Pathogen-tick-host interactions: *Borrelia burgdorferi* and TBE virus. *Zentralblatt* fur *Bakteriologie, Mikrobiologie und Hygiene* **289**:492–505.

Nuttall, P. A., Labuda, M., Bowman, A. S. 2008. Saliva-assisted transmission of tick-borne pathogens. *Ticks: biology, disease and control*, 205–219.

Ogden, N.H., Lindsay, L.R., Beauchamp, G., et al. 2004. Investigation of the relationships between temperature and development rates of the tick *Ixodes scapularis* (Acari: Ixodidae) in the laboratory and field. *Journal of Medical Entomology* **41**:622–633.

Paddock, C.D., Finley, R.W., Wright, C.S., et al. 2008. *Rickettsia parkeri* rickettsiosis and its clinical distinction from Rocky Mountain spotted fever. *Clinical Infectious Diseases* **47**:1188–1196.

Parola, P., Paddock, C.D., Raoult, D. 2005. Tick-borne rickettsioses around the world: emerging diseases challenging old concepts. *Clinical Microbiology Review* **18**:719–756.

Pattnaik, P. 2006. Kyasanur forest disease: an epidemiological view in India. *Review of Medical Virology* **16**:151–165.

Pesko, K.N., Torres-Perez, F., Hjelle, B.L., et al. 2010. Molecular epidemiology of Powassan virus in North America. *Journal of General Virology* **91**:2698–2705.

Pool, W.A., Brownlee, A., Wilson, D.R. 1930. The etiology of "louping-ill." *Journal of Comparative Pathology and Therapy* **43**:253–290.

Punda, V., Beus, I., Calisher, C.H., et al. 1980. Laboratory infections with Bhanja virus. *Zentralblatt fur Bakteriologie, Mikrobiologie und Hygiene* Suppl **9**:273–275.

Rampas, J., Gallia, F. 1949 Isolation of encephalitis virus from *Ixodes ricinus* ticks (in Czech). *Časopis Lékařů českých* **88**:1179–1180.

Randolph, S.E. 1998. Ticks are not insects: consequences of contrasting vector biology for transmission potential. *Parasitology Today* **14**:186–192.

Randolph, S.E. 2008a. Tick-borne encephalitis in Central and Eastern Europe: consequences of political transition. *Microbes and Infection* **10**:209–216.

Randolph, S.E. 2008b. Tick-borne disease systems. *Revue scientifique et technique de Office international de Epizoties* **27**:1–15.

Randolph, S.E. 2009. Tick-borne diseases systems emerge from the shadows: the beauty lies in molecular details, the message in epidemiology. *Parasitology* **136**:1403–1413.

Randolph, S.E., Rogers, D.J. 2000. Fragile transmission cycles of tick-borne encephalitis virus may be disrupted by predicted climate change. *Proceedings of the Royal Society of London, Series B: Biological Sciences* **267**:1741–1744.

Randolph, S.E., Sumilo, D. 2007. Tick-borne encephalitis in Europe: dynamics of changing risk. In: Takken W, and Knols BGJ (eds) Emerging Pests and Vector-Borne Disease in Europe. Wageningen Academic Publishers, The Netherlands, pp. 187–206.

Randolph, S.E., Green, R.M., Peacey, M.F., et al. 2000. Seasonal synchrony: the key to tick-borne encephalitis foci identified by satellite data. *Parasitology* **121**:15–23.

Randolph, S.E., Green, R.M., Hoodless, A.N., et al. 2002. An empirical quantitative framework for the seasonal population dynamics of the tick *Ixodes ricinus*. *International Journal of Parasitology* **32**:979–989.

Řeháček, J. 1962. Transovarial transmission of tick-borne encephalitis virus by ticks. *Acta Virologica* **6**:220–226.

Reid, H.W. 1990. Louping-ill virus. In: Dinter Z, Morein B (eds) Virus infections in ruminants. Elsevier, Amsterdam, pp 279–289.

Reid, H.W., Barlow, R.M., Boyce, J.B., et al. 1976. Isolation of louping-ill virus from a roe deer *(Capreolus capreolus)*. *Veterinary Record* **98**:116.

Reid, H.W., Duncan, J.S., Phillips, J.D.P., et al. 1978. Studies on louping-ill virus (Flavivirus group) in wild red grouse (*Lagopus lagopus scoticus*). *Journal of Hygiene* **81**:321–329.

Reid, H.W., Moss, R., Pow, I., et al. 1980. The response of three grouse species (*Tetrao urogallus, Lagopus mutus, Lagopus lagopus*) to louping-ill virus. *Journal of Comparative Pathology* **90**:257–263.

Riedl, H., Kožuch, O., Sixl, W., et al. 1971. Isolierung des Zeckenencephalitisvirus aus der Zecke *Haemaphysalis concinna* Koch. *Archiv fur Hygiene und Bakteriologie* **154**:610–611.

Rosický, B. 1959. Notes on the classification of natural foci of tick-borne encephalitis in central and south-east Europe. *Journal of Hygiene, Epidemiology, Microbiology and Immunology* **3**:431–443.

Rosický, B., Bárdoš, V. 1966. A natural focus of tick-borne encephalitis outside the main distribution area of *Ixodes ricinus*. *Folia Parasitologica* **13**:103–112.

Rubin, S.G., Chumakov, M.P. 1980. New data on the antigenic types of tick-borne encephalitis (TBE) virus. *Zentralblatt Archive Hygiene Bakteriol fur Bakteriologie, Mikrobiologie und Hygiene* Suppl **9**:232–236.

Růžek, D., Yakimenko, V.V., Karan, L.S., et al. 2010. Omsk haemorrhagic fever. *The Lancet* **376**:2104–2113.

Schmidt, J.R., Shope, R.E. 1971. Kemerovo virus from a migrating common redstart of Eurasia. *Acta Virologica* **15**:112.

Semashko, I.V., Chumakov, M.P., Tsyakin, L.B., et al. 1976. Experimental pathogenicity of Bhanja virus for lambs at different infection routes (in Russian). *Ekologia Virusov* (Baku), pp. 184–18.

Shah, K.V., Work, T.H. 1969. Bhanja virus: a new arbovirus from ticks *Haemaphysalis intermedia* Warburton and Nuttall, 1909, in Orissa, India. *Indian Journal of Medical Research* **57**:793–798.

Shamanin, V.A., Pletnev, A.G., Rubin, S.G., et al. 1990. Differentiation of strains of tick-borne encephalitis virus by means of RNA–DNA hybridization. *Journal of General Virology* **71**:1505–1515.

Shiu, S.Y.W., Ayres, M.D., Gould, E.A. 1991. Genomic sequence of the structural proteins of louping-ill virus: comparative analysis with tick-borne encephalitis virus. *Virology* **180**:411–415.

Skarpaas, T., Golovljova, I., Vene, S., et al. 2006. Tick-borne encephalitis virus, Norway and Denmark. *Emerging Infectious Diseases* **12**:1136–1138.

Skofertsa, P.G., Korchmar, N.D., Yarovoi, P.I., et al. 1974. Isolation of Kharagysh virus of the Kemerovo group from *Sturnus vulgaris* in the Moldavian SSR (in Russian). In: Gaidamovich SYa (ed) Arbovirusy. Institut virusologii im D I Ivanovskogo, Moskva, pp 100–103.

Smirnova, S.E., Skvortsova, T.M., Sedova, A.G., et al. 1988. On the newly isolated strains of Batken virus (in Russian). *Voprosy Virusologii* **33**:360–362.

Smith, C.E.G., Varma, M.G.R. 1981. Louping-ill. In: Beran GW (ed) Viral Zoonoses, Vol **I**. CRC Press, Boca Raton, FL, pp. 191–200.

Smorodintsev, A.A., Alekseev, B.P., Gulamova, V.P., et al. 1953. Epidemiological features of biphasic virus meningoencephalitis (in Russian). *Zhurnal Mikrobiologii, Epidemiologii i Immunobiologii* (**5**):54–59.

Sonenshine, D.E. 1974. Vector population dynamics in relation to tick-borne arboviruses: a review. *Phytophatology* **64**:1060–1071.

Stephenson, J.R. 1989. Classification of tick-borne flaviviruses. *Acta Virologica* **33**:494.

Sumilo, D., Bormane, A., Asokliene, L., et al. 2006. Tick-borne encephalitis in the Baltic States: identifying risk factors in space and time. *International Journal of Medical Microbiology* **296**:76–79.

Sumilo, D., Asokliene, L., Bormane, A., et al. 2007. Climate change cannot explain the upsurge of tick-borne Encephalitis in the Baltics. *PLoS One* **2**, e500.

Sumilo, D., Bormane, A., Asokliene, L., et al. 2008. Socio-economic factors in the differential upsurge of tick-borne encephalitis in central and eastern Europe. *Review in Medical Virology* **18**:81–95.

Süss, J. 2003. Epidemiology and ecology of TBE relevant to the production of effective vaccines. *Vaccine* **21**(S1):19–35.

Süss, J. 2008. Tick-borne encephalitis in Europe and beyond—the epidemiological situation as of 2007. *Eurosurveillance* **13**:1–8.

Süss, J. 2011. Tick-borne encephalitis 2010: epidemiology, risk areas, and virus strains in Europe and Asia—an overview. *Ticks and Tick-Borne Diseases* **2**:2–15.

Süss, J., Klaus, C., Gerstengarbe, F.-W., et al. 2008. What makes ticks tick? Climate change, ticks and tick-borne diseases. *Journal of Travel Medicine* **15**:39–45.

Swanepoel, R., Struthers, J.K., Shepherd, A.J., et al. 1983. Crimean-Congo hemorrhagic fever in South Africa. *American Journal of Tropical Medicine and Hygiene* **32**:1407–1415.

Sylla, M., Molez, J.F., Cornet, J.P., et al. 2008. Variabilité climatique et repartition de la Fièvre Hémorragique de crime-Congo et de la Cowdriosis, maladies à tiques au Sénégal. *Acarologia* **48**:155–161.

Theiler, M., Downs, W.G. 1973. The Arthropod-Borne Viruses of Vertebrates. Yale University Press: New Haven, CT, 578 pp.

van Tongeren, H.A.E. 1955 Encephalitis in Austria. IV. Excretion of virus by milk of the experimentally infected goats. *Archiv fur die gesamte Virusforschung* **6**:158–162.

Topciu, V., Rosiu, N., Georgescu, L., et al. 1968. Isolation of a cytopathic agent from the tick *Haemaphysalis punctata*. *Acta Virologica* **12**:287.

Tsekhanovskaya, N.A., Matveev, L.E., Rubin, S.G., et al. 1993. Epitope analysis of tick-borne encephalitis (TBE) complex viruses using monoclonal antibodies to envelope glycoprotein of TBE virus (*persulcatus* subtype). *Virus Research* **30**:1–16.

Venugopal, K., Buckley, A., Reid, H.W., et al. 1992. Nucleotide sequence of the envelope glycoprotein of Negishi virus shows very close homology to louping-ill virus. *Virology* **190**:515–521.

Venugopal, K., Gritsun, T., Lashkevich, V.A., et al. 1994. Analysis of the structural protein gene sequence shows Kyasanur Forest disease virus as a distinct member in the tick-borne encephalitis virus serocomplex. *Journal of General Virology* **75**:227–232.

Verani, P., Balducci, M., Lopes, M.C., et al. 1970. Isolation of Bhanja virus from *Haemaphysalis* ticks in Italy. *American Journal of Tropical Medicine and Hygiene* **19**:103–105.

Votyakov, V.I., Protas, I.I., Zhdanov, V.M. 1978. Western Tick-Borne Encephalitis (in Russian). Nauka Belarus, Minsk, 256 pp.

Watts, D.M., Ksiasek, T.G., Linthicum, K.J., et al. 1988. Crimean-Congo hemorrhagic fever. In: Monath TP (ed.) The Arboviruses: Epidemiology and Ecology. CRC Press, Boca Raton, FL.

Wilson, M.L., Le Guenno, B., Guillaud, M., et al. 2009. Distribution of Crimean-Congo hemorrhagic fever viral antibody in Senegal: environmental and vectorial correlates. *American Journal of Tropical Medicine and Hygiene* **43**:557–566.

Xu, B., Liu, L., Huang, X., et al. 2011. Metagenomic analysis of fever, thrombocytopenia and leukopenia syndrome (FTLS) in Henan Province, China: discovery of a new bunyavirus. *PLoS Pathogens* **7**:e1002369.

Yilmaz, G.R., Buzgan, T., Irmak, H., et al. 2009. The epidemiology of Crimean-Congo hemorrhagic fever in Turkey, 2002–2007. *International Journal of Infectious Diseases* **13**:380–386.

Yu, X.J., Liang, M.F., Zhang, S.Y., et al. 2011. Fever with thrombocytopenia associated with a novel bunyavirus in China. *New England Journal of Medicine* **364**:1523–1532.

Zeller, H.G., Cornet, J.P., Camicas, J.L. 1994a. Experimental transmission of Crimean-Congo hemorrhagic fever virus by West African wild ground-feeding birds to *Hyalomma marginatum rufipes* ticks. *American Journal of Tropical Medicine and Hygiene* **50**:676–681.

Zeller, H.G., Cornet, J.P., Camicas J.L. 1994b. Crimean-Congo haemorrhagic fever virus infection in birds: field investigations in Senegal. *Research in Virology* **145**:105–109.

Zeller, H.G., Cornet, J.P., Diop, A., et al. 1997. Crimean-Congo haemorrhagic fever in ticks (Acari: Ixodidae) and ruminants: field observations of an epizootic in Bandia, Senegal (1989–1992). *Journal of Medical Entomology* **34**:511–516.

Zeman, P., Benes, C. 2004. A tick-borne encephalitis ceiling in Central Europe has moved upwards during the last 30 years: possible impact of global warming? *International Journal of Medical Microbiology* 293, Suppl **37**:48–54.

Zeman, P., Pazdiora, P., Benes, C. 2010. Spatio-temporal variation of tick-borne encephalitis (TBE) incidence in the Czech Republic: is the current explanation of the disease's rise satisfactory? *Ticks and Tick-Borne Diseases* **1**:129–140.

Zhang, Y.Z., He, Y.W., Dai, Y.A., et al. 2012. Hemorrhagic fever caused by a novel bunyavirus in China: pathogenesis and correlates of fatal outcome. *Clinical Infectious Diseases* **54**:527–533.

32

THE TICK–VIRUS INTERFACE

Kristin L. McNally and Marshall E. Bloom

Laboratory of Virology, Division of Intramural Research, National Institute of Allergy and Infectious Diseases, National Institutes of Health, Rocky Mountain Laboratories, Hamilton, MT, USA

TABLE OF CONTENTS

Viral Infections and Global Change, First Edition. Edited by Sunit K. Singh.

32.1 INTRODUCTION

Ticks transmit a variety of pathogens to vertebrates including the bacteria that cause Lyme disease and anaplasmosis, parasites that cause babesiosis, and viruses that cause encephalitis and hemorrhagic fevers (Burgdorfer et al., 1982; Pancholi et al., 1995; Spielman and Clifford, 1979; Work, 1958; Zilber and Clifford, 1946). These diseases cause significant morbidity and mortality worldwide, making ticks medically relevant disease vectors. In addition, the impact of tick-transmitted diseases to agriculturally important and domestic animals is an economic burden. Several characteristics of ticks make them efficient vectors of disease including their blood-feeding preferences, their heterophagous digestion, and their remarkable longevity. Interestingly, it appears that tick numbers are increasing in many parts of the world and expanding into geographical areas not traditionally considered within their range (Jaenson et al., 2012; Tokarevich et al., 2011). Global climate change is thought to be a major factor influencing ticks. However, changes in tick populations are the result of complex interactions between biotic and abiotic factors that are not fully understood. Regardless, increased tick populations will likely lead to greater transmission of tick-borne viral infections to humans and animals. Thus, it is important to understand the impact of climate change on ticks to better understand and control the diseases they transmit.

Ticks belong to the phylum Arthropoda, class Arachnida, subclass Acari, order Parasitiformes, and suborder Ixodida. Greater than 900 species of ticks are classified into two major families, Ixodidae and Argasidae, and a third monotypic family, Nuttalliellidae (Guglielmone et al., 2010). The ixodid, or hard-bodied ticks, are characterized by a sclerotized scutum and the apical position of their mouthparts. The ixodid life cycle contains four developmental stages: egg, larvae, nymph, and adult. Larval and nymphal stages are required to take a blood meal in order to molt to the next stage. Adults also take one blood meal and mating occurs on the host. Adult females feed to repletion and drop off of the host for oviposition. Ixodid tick feeding can last for days to weeks, and feeding ticks can ingest up to 100 times their weight. Excess water is secreted back into the host via the salivary glands to concentrate the nutrients in the blood meal.

The argasid ticks are characterized by a soft, leathery cuticle and the anterior ventral location of their mouthparts. The life cycle of soft ticks also comprises four developmental stages, but unlike ixodid ticks, argasid ticks have multiple nymphal stages, each requiring a blood meal. Mating occurs away from the host. In contrast to the prolonged feeding period characteristic of hard ticks, soft ticks feed rapidly to engorgement within minutes to hours and secrete excess fluid back into the host via the coxal glands.

Viruses can persist in ticks for the duration of the tick lifespan, contributing to the maintenance of viruses in nature (Davies et al., 1986). As a consequence, ticks play dual roles as virus reservoirs as well as virus vectors. Certain viruses are thought to spend the great majority of their life cycle in the tick vector. For example, it is estimated that greater than 95% of the life cycle of the flavivirus tick-borne encephalitis virus (TBEV) is spent in the tick vector (Nuttall and Labuda, 2003; Nuttall et al., 1991). The tick, therefore, exerts selective pressures over time that influence virus genotype and phenotype (Dzhivanian et al., 1988; Labuda et al., 1994). Despite the importance of understanding virus infection in the tick vector, studies are lacking.

Viruses that replicate in and are transmitted by ticks are arboviruses. All known tick-borne viruses have an RNA genome, with the exception of African swine fever virus (ASFV). Tick-borne arboviruses fall into several virus families: Asfarviridae, Flaviviridae, Reoviridae, Bunyaviridae, and Rhabdoviridae. Specific viruses within these families have been extensively reviewed (Labuda and Nuttall, 2003, 2004; Nuttall, 2009).

For some arboviruses, such as the tick-borne flaviviruses, transmission to mammals is required for virus maintenance in nature. Thus, arboviruses have evolved to successfully replicate in two very disparate systems: an arthropod vector and a mammalian host. These viruses are transmitted during tick feeding, creating a unique and complex tick–host–virus interface. The tick–virus interaction is the least well-characterized and thus the least well-understood part of this interface. The goal of this report is to summarize what is currently known about the interactions of viruses in the tick vector and to discuss the most important aims for future research.

32.2 VIRUSES WITHIN THE TICK VECTOR

32.2.1 Impact of virus infection on ticks

To achieve persistence, arbovirus infections must not have overt negative impacts on the tick vector. However, there are several examples of viruses that have measurable effects on their vectors. ASFV, in the family Asfarviridae, caused increased mortality in *Ornithodoros marocanus* larval and nymphal ticks compared with mock-infected ticks (Endris et al., 1992). ASFV-infected adult *Ornithodoros moubata* ticks also demonstrated increased mortality (Hess et al., 1989; Rennie et al., 2000). The mechanisms underlying these interesting observations were not characterized. Infection of *Ixodes persulcatus* with TBEV altered tick questing behavior (Alekseev et al., 1988). Infection of *Rhipicephalus appendiculatus* with the orthomyxovirus Thogoto virus inhibited salivary fluid secretion by ~25% *in vitro* compared to control ticks (Kaufman et al., 2002). Thus, a detrimental effect of virus infection was observed despite these ticks being the natural vectors. A more recent study demonstrated by microarray analysis that infection of *Ixodes scapularis* nymphs with Langat virus, a member of the TBEV serocomplex, altered the salivary gland transcript expression profile during feeding when compared to uninfected ticks (McNally et al., 2012). Together, these data demonstrate that while viral infections in ticks are generally not cytopathic in nature, they are also not entirely silent and can impact the survival, behavior, and gene expression of these important disease vectors. It would be interesting to determine how these measurable changes affect virus transmissibility. It is likely that if examined, impacts of other virus infections on tick vectors will be observed, potentially providing useful information for controlling tick populations.

32.2.2 Impact of the tick vector on viruses

Virus infection of ticks appears to be quite specific to the species of tick and the virus involved. Long-term virus persistence suggests that ticks have an enormous impact on virus evolution and perhaps even pathogenesis in mammalian hosts. Viruses are taken into the tick with the blood meal during feeding in the form of virions or infected host cells. The blood meal is imbibed into the midgut, where it is concentrated and eventually digested. Unlike insects, blood meal digestion in ticks is heterophagous, occurring intracellularly. The nature of this digestion initially protects virus from potentially hostile conditions in the midgut lumen. However, conditions in the midgut can be harsh due to the presence of reactive oxygen by-products as a result of blood digestion. How different viruses persist in this hostile environment is not well understood.

The ability of a virus to enter and replicate in midgut cells (i.e., midgut barrier) is an important factor that determines vector competence for a particular tick species (McKelvey et al., 1981; Nuttall et al., 1994) (Figure 32.1). It is assumed that recognition of and binding

Figure 32.1. Barriers to virus infection of a tick vector. Virus enters the tick with the blood meal during feeding and is imbibed into the midgut (light virions). Entry into midgut cells represents the first barrier to infection of the tick (arrow 1) and likely defines vector competence. Specific interactions between virus and midgut cell receptors have not been identified. Virus must then replicate in midgut cells and overcome the second barrier to infection, exiting the midgut (arrow 2), for spread throughout the tick. Virus must then enter the salivary glands (arrow 3) and replicate. Finally, virus must exit the salivary glands (arrow 4) via secretion into the saliva for transmission to a host (darkest particles). Green, salivary glands; blue, midgut; virions range in color from light pink to dark red; arrows represent barriers to infection. For color detail, please see color plate section.

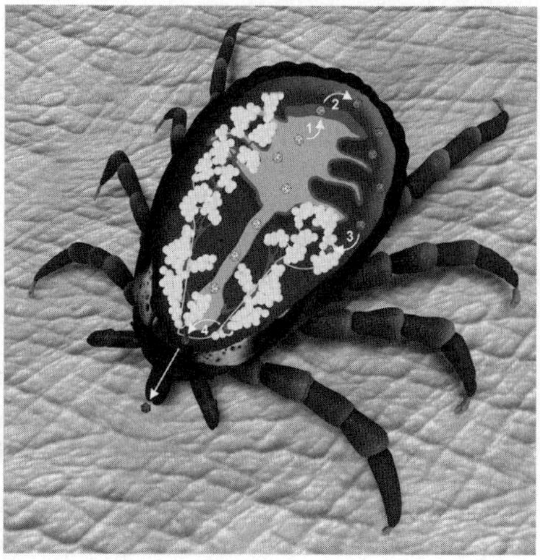

to receptors on tick midgut cells by virus is the first step in a complex series of events that leads to infection of the midgut and spread throughout the tick. Although this initial step is critical, no specific virus–midgut interactions have been identified. The generation of midgut cell lines from different species of ticks would be a valuable tool to examine these types of interactions.

Ultimately, the virus must enter the salivary glands and be secreted into the saliva for transmission to a host. The mechanisms by which viruses exit infected midgut cells, travel to the salivary glands, and are secreted into the saliva are still unclear. Each of these steps—entry into, replication in, and exit from salivary glands—represents significant barriers to infection that are not well understood (Figure 32.1). One study determined that transmission of the flavivirus deer tick virus from infected ticks to a rodent host occurred within 15 min of attachment, suggesting that the virus was present in the salivary glands before feeding commenced (Ebel and Kramer, 2004). How this compares with other tick-borne viruses and whether feeding induces some viruses to migrate to the salivary glands warrant further investigation. These steps are critically important to understand since secretion into saliva during feeding is the primary route of virus transmission to hosts (Chernasky and McLean, 1969; Nosek et al., 1972).

All tick-borne viruses except ASFV have an RNA genome. RNA viruses typically exhibit high mutation rates during replication, which often leads to heterogenous populations containing quasispecies. Interestingly, arboviruses appear to have relatively slow rates of evolution compared to other RNA viruses (Jenkins et al., 2002; Zanotto et al., 1996). This may be due to the requirement of these viruses to replicate in two very disparate systems: vertebrate and invertebrate (Steele and Nuttall, 1989). Quasispecies generation appears to be exacerbated in mosquitos compared to the vertebrate host, suggesting that mosquito vector strongly impacts virus evolution (Brackney et al., 2009). Whether ticks drive virus evolution through the generation of quasispecies similar to the mosquito vector is an interesting question.

In general, virus replication in ticks tends to be non-cytopathic and persistent, whereas in vertebrate hosts infection is acute and cytopathic. These conflicting replication strategies may involve genetic trade-off, where adaptation to replication in the tick may lead to reduced fitness in vertebrates, and vice versa (Jenkins et al., 2002). The relative contribution of each system to virus evolution likely varies depending on the infecting virus, vector species, and vertebrate hosts. This phenomenon can be modeled in cell culture where tick cells show little to no cytopathic effect (CPE) after virus infection, whereas susceptible mammalian cells demonstrate notable CPE (Senigl et al., 2006). While the complexity of virus replication and evolution is not exactly recapitulated in cell culture, these types of analyses allow for the dissection of the relative contributions of different systems on virus selection. For example, repeated passaging of Langat virus in either tick or murine cells produced virus populations that preferentially replicated in the cells they were passaged in (Mitzel et al., 2008). Adaptation of these viruses was associated with amino acid changes. When inoculated intraperitoneally into mice, the tick cell-adapted virus was attenuated as demonstrated by increased survival of the mice. Even more interesting, when virus was isolated from the *surviving* mice, additional changes were observed near the sites of the original tick cell-adapted changes. The authors speculated that the additional changes were compensatory to the initial mutations that occurred during passage in the tick cells. Studies aimed at determining the relative impacts of invertebrate and vertebrate systems on tick-borne viruses may identify viral proteins important for adaptation and thus virulence. Furthermore, elucidating the mechanisms underlying persistence in the tick vector may provide insight into the rare persistent flavivirus infections in humans (Gritsun et al., 2003; Murray et al., 2010).

32.2.3 Tick immunity

Like other invertebrates, ticks have no adaptive immune response. Thus, they must rely on an innate immune system, which provides an immediate and relatively nonspecific response to infection. The tick innate immune system consists of cellular responses such as phagocytosis, encapsulation, and nodulation as well as humoral factors secreted into the hemolymph. Together, this response serves to decrease pathogen infection to a tolerable level. To date, little is known about tick innate immunity against pathogen infection and even less is understood about how viruses are detected and mitigated in the tick vector.

Antimicrobial peptides (AMPs) are humoral factors secreted from hemocytes, midgut, and fat body tissues. Tick AMPs are classified as defensins, lectins, lysozymes, proteases, and protease inhibitors. Lectins are carbohydrate-binding proteins found in a broad range of organisms. Tick lectins function in hemagglutination and opsonization of microorganisms, and their expression is often tissue specific (Grubhoffer et al., 2004). The best-characterized tick lectin, Dorin M, from the hemolymph of *Ornithodoros tartakovskyi*, was shown to have hemagglutinating activity (Kovar et al., 2000). Dorin M was subsequently shown to be a fibrinogen-related protein, leading the authors to hypothesize that Dorin M may be involved in recognition of nonself molecules as well as pathogen transmission (Rego et al., 2006). Proteins with homology to lectins have been identified in several tick species including *Ixodes ricinus* (Kuhn et al., 1996) and *R. appendiculatus* (Kamwendo et al., 1993). The upregulation of tick lectins after blood feeding or pathogen infection suggests an important role in the immune response of the tick.

Defensins are naturally occurring antimicrobial proteins found in a wide variety of organisms that are characterized to function against bacteria, fungi, and viruses (Ganz, 2003). Tick defensins are upregulated in response to blood feeding and bacterial infection (Nakajima et al., 2001). Most tick defensins have preferential activity against gram-positive bacteria. However, defensin-related transcripts were downregulated in response to Langat virus infection in *I. scapularis* nymphs (McNally et al., 2012). The significance of this observation is unknown, but the differential regulation of defensin-like transcripts in response to virus infection is intriguing. One study reported that two defensin isoforms from *I. ricinus* showed no virucidal activity against TBEV or West Nile virus (WNV) *in vitro* (Chrudimská et al., 2011). In contrast to tick defensins, mammalian defensins have been shown to have antiviral activity against a variety of viruses (Daher et al., 1986), suggesting that further analysis may identify tick defensins with antiviral activity.

Lysozymes, proteases, and protease inhibitors are humoral factors hypothesized to function during the innate immune response in ticks. Cysteine, serine, and Kunitz-type protease inhibitors function during blood digestion and hemostasis. A cysteine protease inhibitor from *Haemaphysalis longicornis* has been shown to reduce *Babesia bovis* growth *in vitro* (Zhou et al., 2006), and a Kunitz-type protease inhibitor from *Dermacentor variabilis* has been shown to reduce rickettsia colonization in murine cell culture (Ceraul et al., 2008). However, these molecules have not yet been shown to play a direct role in the innate immune response in the tick vector.

The tick AMPs described to date appear to have preferential activity against bacteria, although antiviral activity has not been rigorously investigated. In fact, the differential regulation of defensins and proteases in response to Langat virus infection strongly suggests a role for these proteins in the immune response of the tick vector (McNally et al., 2012). It is likely that future work will reveal tick AMPs that function to recognize and mitigate virus infection.

32.2.4 Other mediators of immunity

Digestion of a blood meal is accompanied by proteolytic processing of ingested host hemoglobin, which leads to high levels of oxidative stress (Graca-Souza et al., 2006). Host hemoglobin fragments isolated from the midgut tissue of *O. moubata* or synthetic hemoglobin fragments whose sequences were isolated from the midgut of *Boophilus microplus* demonstrated antimicrobial activity against gram-positive bacteria (Fogaca et al., 1999; Nakajima et al., 2003). These results strongly suggest that by-products from hemoglobin digestion may function as a midgut defense mechanism in ticks. Whether this activity would control viruses has not yet been determined.

RNA interference (RNAi) may act as a natural defense mechanism against virus infection in a broad range of organisms. In *Aedes aegypti*, siRNA specific to viral genomes has been demonstrated in mosquitos infected with dengue virus or Sindbis virus (Cirimotich et al., 2009; Myles et al., 2008; Sánchez-Vargas et al., 2009). The phenomenon of RNAi has been shown to be active in several tick species including *Amblyomma americanum*, *I. scapularis*, *H. longicornis*, *Rhipicephalus sanguineus*, *D. variabilis*, and *B. microplus* (Aljamali et al., 2003; de la Fuente et al., 2006; Karim et al., 2004; Miyoshi et al., 2004; Narasimhan et al., 2004; Nijhof et al., 2007; Pedra et al., 2006). These studies utilized RNAi as a molecular tool to knock down specific genes to determine their function (de la Fuente et al., 2007). Although it has not yet been directly demonstrated, RNAi will likely play an important role in the innate immune response by controlling virus infection in ticks, similar to that shown for mosquitos.

Ticks are often infected with more than one pathogen (coinfections). In addition, ticks are colonized by endosymbionts. The presence of endosymbionts or multiple pathogens very likely impacts the immune response of the tick vector. Interaction between pathogens, competition between strains of the same pathogen, and interactions between pathogens and endosymbionts may impact pathogenesis as well as the survival and behavior of the tick vector. For example, *Borrelia* infection has been reported to suppress TBEV infection in *I. persulcatus* (Alekseev et al., 1996). In contrast, no evidence of interference between *Borrelia burgdorferi* and TBEV in *I. persulcatus* was found in field-collected ticks over a five-year study (Korenberg et al., 1999). These contradicting reports emphasize the need for future studies aimed at determining the impact of natural coinfections on the survival and transmission of each pathogen under defined experimental conditions.

Understanding the dynamics of coinfections in the tick vector may provide alternative solutions for controlling tick-borne pathogens. For example, *Ae. aegypti* mosquitos transinfected with *Wolbachia* have significantly reduced lifespans (McMeniman et al., 2009). Because many viruses must develop in the mosquito before they can be transmitted, called the extrinsic incubation period, older mosquitos are responsible for the majority of virus transmission. Thus, reducing the mosquito lifespan has the potential to reduce transmission of pathogens carried by this vector. *Wolbachia* infection of *Ae. aegypti* mosquitos also limited dengue virus or chikungunya virus from establishing a productive infection (Moreira et al., 2009). Another study demonstrated significantly decreased transmission of dengue virus from *Ae. aegypti* mosquitos that were infected with *Wolbachia* compared to control mosquitos (Bian et al., 2010). In addition, *Wolbachia* infection of *Culex quinquefasciatus* led to decreased WNV titers and reduced transmission rates (Glaser and Meola, 2010). These data suggest that infection of arthropods with *Wolbachia* bacteria may interfere with the replication and transmission of arthropod-borne viruses. While similar studies have not been pursued in ticks, this novel approach could lead to the control of tick-borne viruses as well.

In addition to the aforementioned mediators, a search of the *I. scapularis* genome reveals putative interferon-regulated genes, toll genes, tripartite motif-containing proteins, TNFα-associated factors, lectins, and many other putative immune-related genes, strongly suggesting that further examination of these proteins will reveal innate immune molecules that target viral infections (Lawson et al., 2009; Pagel Van Zee et al., 2007). The potential for hemostatic proteases to have a dual role in innate immunity is intriguing and highlights the interplay between hemostasis and immunity. Thus, molecules already characterized to have important functions during blood feeding or pathogen transmission could also be examined to determine if they are involved in the innate immune response against pathogens in the tick vector.

32.3 SALIVA-ASSISTED TRANSMISSION

The components of tick saliva play an important role in the survival and transmission of tick-borne viruses in the host as well as in the tick vector. The enhancement of pathogen transmission by saliva is called saliva-assisted transmission (SAT) (Nuttall and Jones, 1991). The first direct evidence for a SAT effect on virus infection was described in studies where guinea pigs were inoculated with Thogoto virus alone or a mixture of Thogoto virus and salivary gland extract (SGE) from partially fed *R. appendiculatus* ticks (Jones et al., 1989). The number of infected nymphs was significantly greater when fed on guinea pigs that received virus and SGE compared to the animals that received virus alone (Jones et al., 1989). SGE has also been shown to have an enhancing effect on tick-borne flavivirus transmission (Labuda et al., 1993).

Indirect evidence for SAT of other viruses has been demonstrated by transmission between infected and uninfected ticks feeding on the same host, termed non-viremic transmission. These studies demonstrate that uninfected ticks become infected with virus simply by feeding near ticks that are infected, and this occurs in the absence of host viremia. Evidence for non-viremic transmission has been reported for Crimean–Congo hemorrhagic fever virus and *Hyalomma marginatum*, WNV and *O. moubata*, and Bhanja virus and

Dermacentor marginatus and *R. appendiculatus* (Gordon et al., 1993; Labuda et al., 1997; Lawrie et al., 2004). Together, these studies imply that specific factors in tick saliva affect virus transmission, although none have been identified.

SAT also occurs for other tick-borne pathogens such as *B. burgdorferi*, where specific proteins responsible for increased survival or transmission have been identified. Salp15, a salivary gland protein from *I. scapularis*, is upregulated in response to *Borrelia* infection (Ramamoorthi et al., 2005). Salp15 directly binds to OspC of *Borrelia*, which protects the spirochete from antibody-mediated killing *in vitro* (Ramamoorthi et al., 2005). Mice immunized with Salp15 were significantly protected from tick-transmitted *Borrelia* infection (Dai et al., 2009). The Salp15 SAT effect was specific to *Borrelia* because *Anaplasma phagocytophilum* infection did not lead to upregulation of Salp15 and immunization of mice with Salp15 did not affect *A. phagocytophilum* transmission (Dai et al., 2009). This specificity of SAT factors for a specific pathogen suggests coevolution of the tick and pathogen. However, because ticks are frequently coinfected with multiple pathogens and SAT molecules may function through their immunomodulatory effects on the host, SAT factors should be analyzed for activity against multiple pathogens. For example, Salp15 is a multifunctional protein that binds to a C-type lectin on dendritic cells (Hovius et al., 2008). Because dendritic cells are thought to be a primary target of tick-borne flavivirus infection, it would be interesting to determine if Salp15 affects transmission of deer tick virus, a flavivirus for which the natural vector is *I. scapularis*. Another salivary protein from *I. scapularis*, Salp25D, was shown to have peroxidase activity, which offered a survival advantage to *Borrelia* by preventing ROS-mediated killing (Das et al., 2001; Narasimhan et al., 2007). Immunization of mice with Salp25D led to reduced acquisition of *Borrelia* spirochetes by ticks feeding on the immunized mice (Narasimhan et al., 2007). Salp15 and Salp25D represent tick proteins that function to enhance pathogen transmission either to the host (Salp15) or to the tick vector (Salp25D). Taken together, the results of these studies demonstrate the importance of understanding SAT factors in order to target them for the development of anti-tick and anti-pathogen vaccines.

One caveat to the studies involving saliva or SGE is that these materials are generally collected from adult ticks. However, nymphs are believed to be the most important developmental stage for pathogen transmission (Nosek et al., 1972; Nuttall et al., 1994). *I. scapularis* nymphs showed several differentially regulated salivary gland transcripts in response to Langat virus infection (McNally et al., 2012). The original hypothesis was that upregulated transcripts may be beneficial to Langat virus and downregulated transcripts may be inhibitory to virus replication. However, this idea may be an oversimplification. It would be accurate in the case of SAT molecules Salp15 and Salp16, which were upregulated in response to *Borrelia* or *Anaplasma*, respectively, and enhance infection (Ramamoorthi et al., 2005; Sukumaran et al., 2006). In contrast, innate immune-related molecules like defensins that counteract spirochete infections were also upregulated in response to infection. These examples underscore the complexity of tick proteins and the interplay between systems. Nevertheless, future studies aimed at identifying SAT factors in nymphs are warranted.

Evidence for the enhancement of virus transmission by salivary factors thus far suggests that they promote SAT through activity on the host. Although no direct interactions have been identified to date, it cannot be ruled out that some SAT factors may directly interact with viral proteins. Identification of viral SAT factors and characterization of the mechanism of their activity will lead to a better understanding of viral infections in the tick vector and will provide interesting targets for the development of vaccines. An anti-vector vaccine derived from a *R. appendiculatus* cement protein was shown to protect mice against lethal infection with tick-transmitted TBEV, demonstrating the feasibility of targeting vector proteins to reduce pathogen transmission (Labuda et al., 2006; Trimnell et al., 2005).

32.4 SUMMARY AND FUTURE DIRECTIONS

The tick–host–virus interface is a complex environment resulting from millions of years of coevolution. Studies dissecting tick–host interactions have provided important information about the host response to tick infestation and the basic mechanisms involved in blood feeding by ticks. Virus–host interactions are widely studied and provide critical information regarding viral pathogenesis and the host response to virus infection. However, these studies often exclude the role of the tick vector and tick saliva during infection. The tick–virus interface is the least studied and the least understood area of this triad, partly due to the very specific interactions between a particular virus and the species of tick. Although working with ticks can be logistically difficult and time consuming and present problems for biocontainment, future studies using tick vectors of viral infections should be expanded, particularly to the following areas of study.

32.4.1 Generation of tick cell lines

The development of tissue-specific cell lines from tick midguts or salivary glands would help to identify the potential interactions between viral proteins and receptors on these cell types important for entry, replication, or release. Elucidating these barriers to infection would provide critical insight into vector competence.

32.4.2 The role of endosymbionts and coinfections

Ticks often carry multiple pathogens and are colonized with endosymbionts. How endosymbionts and coinfections affect the immune response of the tick vector or the transmission of pathogens to a host is not well understood. Studies aimed at determining the impact of coinfections on virus infection and transmission would provide valuable information about the ecology and transmission of these pathogens. In addition, studies aimed at identifying endosymbionts in ticks and determining their impact on pathogen infection and transmission may provide insight for controlling tick populations and the diseases they transmit.

32.4.3 Tick innate immunity

The factors involved in tick innate immunity against bacterial pathogens are being elucidated. However, innate immunity against virus infection in ticks is largely uncharacterized. Studies aimed at identifying immune molecules that function to modulate virus infections in the tick vector are necessary. Identification of novel proteins that function to control tick-borne virus infections could lead to the development of therapeutics.

32.4.4 Identification and characterization of viral SAT factors

The phenomenon of RNAi provides a molecular tool to identify and characterize viral SAT factors. SAT factors may be specific to each virus–tick interaction and may be regulated by feeding or developmental stage, making this goal a difficult undertaking. Nevertheless, identification of viral SAT factors will provide targets for the development of novel anti-tick and antivirus vaccines. Advances in the *I. scapularis* genome project should aid these studies.

32.4.5 Viral persistence in tick vectors

Viral infection in ticks is largely a non-cytopathic, persistent infection. However, the mechanisms underlying persistence are not well understood. Future studies focused on understanding persistent infections in the tick vector and mammalian reservoirs may provide valuable insight into the rare persistent tick-borne viral infections in humans.

32.4.6 The impact of climate change on tick vectors and tick-borne diseases

The complex relationship between global climate change and tick-borne diseases is not well understood. Ticks are directly impacted by environmental factors such as temperature, precipitation, and humidity. These environmental factors also affect the availability of vegetation for questing and hosts to feed on. It appears that tick populations are increasing and their geographical distribution is expanding. In addition, the prevalence of tick-borne diseases in humans and animals appears to be increasing (Bacon et al., 2008; Hinten et al., 2008). Due to the sensitivity of ticks to environmental conditions, climate change is thought to have a major impact on tick populations and transmission of the pathogens they harbor. However, this concept remains controversial because the factors affecting ticks are complex. In fact, climate change has been proposed to be only a minor contributing factor to the increase in tick-borne diseases (Randolph, 2008). Thus, it is necessary to understand how global climate change impacts tick populations, as this information is critically important to understanding transmission of tick-borne viruses to humans and animals.

ACKNOWLEDGEMENTS

The authors thank Dr. H. Feldmann and Dr. S. Best for critical review of the manuscript and A. Mora and H. Murphy for graphical assistance. This work was supported by the Intramural Research Program of the National Institute of Allergy and Infectious Diseases, NIH.

REFERENCES

Alekseev AN, Burenkova LA, Chunikhin SP. 1988. Behavioral characteristics of *Ixodes persulcatus* P. Sch. Ticks infected with the tick-borne encephalitis virus. *Medical Parazitology* (Mosk) 2: 71–75.

Alekseev AN, Burenkova LA, Vasil'eva IS, et al. 1996. The functioning of foci of mixed tick-borne infections on Russian territory. *Medical Parazitology* (Mosk) 4: 9–16.

Aljamali MN, Bior AD, Sauer JR, et al. 2003. RNA interference in ticks: a study using histamine binding protein dsRNA in the female tick *Amblyomma americanum*. *Insect Molecular Biology* 12(3): 299–305.

Bacon RM, Kugeler KJ, Mead PS. 2008. Surveillance for Lyme disease United States, 1992–2008. *Morbidity and Mortality Weekly Report. Surveillance Summaries* 57: 1–9.

Bian G, Xu Y, Lu P, et al. 2010. The endosymbiotic bacterium *Wolbachia* induces resistance to dengue virus in *Aedes aegypti*. *PLoS Pathogens* 6(4): e1000833.

Brackney DE, Beane JE, Ebel GD. 2009. RNAi targeting of West Nile virus in mosquito midguts promotes virus diversification. *PLoS Pathogens* 5(7): e10000502.

Burgdorfer W, Barbour AG, Hayes SF, et al. 1982. Lyme disease – a tick-borne spirochetosis? *Science* 216: 1317–1319.

Ceraul SM, Dreher-Lesnick SM, Mulenga A, et al. 2008. Functional characterization and novel rickettsiostatic effects of a Kunitz-type serine protease inhibitor from the tick *Dermacentor variabilis*. *Infection and Immunity* 76(11): 5429–5435.

Chernasky MA, McLean DM. 1969. Localization of Powassan virus in *Dermacentor andersoni* ticks by immunofluorescence. *Canadian Journal of Microbiology* 15: 1399–1408.

Chrudimska T, Slaninova J, Rudenko N, et al. 2011. Functional characterization of two defensin isoforms of the hard tick *Ixodes ricinus*. *Parasites and Vectors* 4: 63–71.

Cirimotich CM, Scott JC, Phillips AT, et al. 2009. Suppression of RNA interference increases alphavirus replication and virus-associated mortality in *Aedes aegypti* mosquitoes. *BMC Microbiology* 9: 49–61.

Daher KA, Selsted ME, Lehrer RI. 1986. Direct inactivation of viruses by human granulocyte defensins. *Journal of Virology* 60(3): 1068–1074.

Dai J, Wang P, Adusumilli S, et al. 2009. Antibodies against a tick protein, Salp15, protect mice from the Lyme disease agent. *Cell Host & Microbe* 6(5): 482–492.

Das S, Banerjee G, DePonte K, et al. 2001. Salp25D, an *Ixodes scapularis* antioxidant, is 1 of 14 immunodominant antigens in engorged tick salivary glands. *Journal of Infectious Diseases* 184(8): 1056–1064.

Davies CR, Jones LD, Nuttall PA. 1986. Experimental studies on the transmission cycle of Thogoto virus, a candidate orthomyxovirus, in *Rhipicephalus appendiculatus* ticks. *American Journal of Tropical Medicine and Hygiene* 35: 1256–1262.

de la Fuente J, Almazán C, Blouin EF, et al. 2006. Reduction of tick infections with *Anaplasma marginale* and *A. phagocytophilum* by targeting the tick protective antigen subolesin. *Parasitology Research* 100(1): 85–91.

de la Fuente J, Kocan KM, Almazán C, et al. 2007. RNA interference for the study and genetic manipulation of ticks. *Trends in Parasitology* 23(9): 427–433.

Dzhivanian TL, Korolev MB, Karganova GG, et al. 1988. Changes in the host-dependent characteristics of the tick-borne encephalitis virus during its adaptation to ticks and its readaptation to white mice. *Voprosy Virusologii* 33: 589–595.

Ebel GD, Kramer LD. 2004. Short report: duration of tick attachment required for transmission of Powassan virus by deer ticks. *American Journal of Tropical Medicine and Hygiene* 71(3): 268–271.

Endris RG, Hess WR, Caiado JM. 1992. African swine fever virus infection in the Iberian soft tick, Ornithodoros (Pavlovskyella) marocanus (Acari: Argasidae). *Journal of Medical Entomology* 29(5): 874–878.

Fogaca AC, da Silva PI Jr, Miranda MT, et al. 1999. Antimicrobial activity of a bovine hemoglobin fragment in the tick *Boophilus microplus*. *Journal of Biological Chemistry* 274(36): 25330–25334.

Ganz, T. 2003. Defensins: antimicrobial peptides of innate immunity. *Nature Reviews Immunology* 3(9): 710–720.

Glaser RL, Meola MA. 2010. The native *Wolbachia* endosymbionts of *Drosophila melanogaster* and *Culex quinquefasciatus* increase host resistance to West Nile virus infection. *PLoS One* 5(8): e11977.

Gordon SW, Linthicum KJ, Moulton JR. 1993. Transmission of Crimean-Congo hemorrhagic fever virus in two species of *Hyalomma* ticks from infected adults to cofeeding immature forms. *American Journal of Tropical Medicine and Hygiene* 48: 576–580.

Graça-Souza AV, Maya-Monteiro C, Paiva-Silva GO, et al. 2006. Adaptations against heme toxicity in blood-feeding arthropods. *Insect Biochemistry and Molecular Biology* 36(4): 322–335.

Gritsun TS, Lashkevich VA, Gould EA. 2003. Tick-borne encephalitis. *Antiviral Research* 57(1–2): 129–146.

Grubhoffer L, Kovár V, Rudenko N. 2004. Tick lectins: structural and functional properties. *Parasitology* 129 Suppl: S113–S125.

Guglielmone AA, Robbins RG, Apanaskevich DA, et al. 2010. The Argasidae, Ixodidae and Nuttalliellidae (Acari: Ixodida) of the world: a list of valid species names. *Zootaxa* 2525: 1–28.

Hess WR, Endris RG, Lousa A, et al. 1989. Clearance of African swine fever virus from infected tick (Acari) colonies. *Journal of Medical Entomology* 26(4): 314–317.

Hinten SR, Beckett GA, Gensheimer KF, et al. 2008. Increased recognition of Powassan encephalitis in the United States, 1999–2005. *Vector Borne and Zoonotic Diseases* 8(6): 733–740.

Hovius JW, de Jong MA, den Dunnen J, et al. 2008. Salp15 binding to DC-SIGN inhibits cytokine expression by impairing both nucleosome remodeling and mRNA stabilization. *PLoS Pathogens* 4(2): e31.

Jaenson TG, Jaenson DG, Eisen L, et al. 2012. Changes in the geographical distribution and abundance of the tick *Ixodes ricinus* during the past 30 years in Sweden. *Parasites and Vectors* 5: 8–22.

Jenkins GM, Rambaut A, Pybus OG, et al. 2002. Rates of molecular evolution in RNA viruses: a quantitative phylogenetic analysis. *Journal of Molecular Evolution* 54(2): 156–165.

Jones LD, Hodgson E, Nuttall PA. 1989. Enhancement of virus transmission by tick salivary glands. *Journal of General Virology* 70: 1895–1898.

Kamwendo SP, Ingram GA, Musisi FL, et al. 1993. Characteristics of tick, *Rhipicephalus appendiculatus*, glands distinguished by surface lectin binding. *Annals of Tropical Medicine and Parasitology* 87(5): 525–535.

Karim S, Ramakrishnan VG, Tucker JS, et al. 2004. *Amblyomma americanum* salivary glands double-stranded RNA-mediated gene silencing of synaptobrevin homologue and inhibition of PGE2 stimulated proteins secretion. *Insect Biochemistry and Molecular Biology* 34(4): 407–413.

Kaufman WR, Bowman AS, Nuttall PA. 2002. Salivary fluid secretion in the Ixodid tick *Rhipicephalus appendiculatus* is inhibited by Thogoto virus infection. *Experimental and Applied Acarology* 25: 661–674.

Korenberg EI, Kovalevskii YV, Karavanov AS, et al. 1999. Mixed infection by tick-borne encephalitis virus and Borrelia in ticks. *Medical and Veterinary Entomology* 13(2): 204–208.

Kovar V, Kopacek P, Grubhoffer L. 2000. Isolation and characterization of Dorin M, a lectin from plasma of the soft tick *Ornithodoros moubata*. *Insect Biochemistry and Molecular Biology* 30(3): 195–205.

Kuhn KH, Uhlir J, Grubhoffer L. 1996. Ultrastructural localization of a sialic acid-specific hemolymph lectin in the hemocytes and other tissues of the hard tick *Ixodes ricinus* (Acari; Chelicerata). *Parasitology Research* 82(3): 215–221.

Labuda M, Nuttall PA. 2003. Viruses transmitted by ticks. In Ticks, Biology, Disease and Control. eds. Bowman AS, Nuttall PA. Cambridge, UK: Cambridge University Press.

Labuda M, Nuttall PA. 2004. Tick-borne viruses. *Parasitology* 129 Suppl: S221–S245.

Labuda M, Jones LD, Williams T, et al. 1993. Enhancement of tick-borne encephalitis virus transmission by tick salivary gland extracts. *Medical and Veterinary Entomology* 7: 193–196.

Labuda M, Jiang WR, Laluzova M, et al. 1994. Change in phenotype of tick-borne encephalitis virus following passage in *Ixodes ricinus* ticks and associated amino acid substitution in the envelope protein. *Virus Research* 31: 305–315.

Labuda M, Alves JG, Eleckova E, et al. 1997. Transmission of tick-borne bunyaviruses by cofeeding ixodid ticks. *Acta Virologica* 41: 325–328.

Labuda M, Trimnell AR, Licková M, et al. 2006. An antivector vaccine protects against a lethal vector-borne pathogen. *PLoS Pathogens* 2(4): e27.

Lawrie CH, Uzcátegui NY, Gould EA, et al. 2004. Ixodid and argasid tick species and West Nile virus. *Emerging Infectious Diseases* 10(4): 653–657.

Lawson D, Arensburger P, Atkinson P, et al. 2009. VectorBase: a data resource for invertebrate vector genomics. *Nucleic Acids Research* 37(Database issue): D583–D587.

McKelvey JJ Jr, Eldridge BF, Maramorosch K. 1981. Vectors of Disease Agents. Praeger, New York.

McMeniman CJ, Lane RV, Cass BN, et al. 2009. Stable introduction of a life-shortening *Wolbachia* infection into the mosquito *Aedes aegypti*. *Science* 323(5910): 141–144.

McNally KL, Mitzel DN, Anderson JM, et al. 2012. Differential salivary gland transcript expression profile in *Ixodes scapularis* nymphs upon feeding or flavivirus infection. *Ticks and Tick Borne Diseases* 3(1): 18–26.

Mitzel DN, Best SM, Masnick MF, et al. 2008. Identification of genetic determinants of a tick-borne flavivirus associated with host-specific adaptation and pathogenicity. *Virology* 381(2): 268–276.

Miyoshi T, Tsuji N, Islam MK, et al. 2004. Gene silencing of a cubilin-related serine proteinase from the hard tick *Haemaphysalis longicornis* by RNA interference. *Journal of Veterinary and Medical Science* 66(11): 1471–1473.

Moreira LA, Iturbe-Ormaetxe I, Jeffery JA, et al. 2009. A *Wolbachia* symbiont in *Aedes aegypti* limits infection with dengue, Chikungunya, and Plasmodium. *Cell* 139(7): 1268–1278.

Murray K, Walker C, Herrington E, et al. 2010. Persistent infection with West Nile virus years after initial infection. *Journal of Infectious Diseases* 201(1): 2–4.

Myles KM, Wiley MR, Morazzani EM, et al. 2008. Alphavirus-derived small RNAs modulate pathogenesis in disease vector mosquitos. *Proceedings of the National Academy of Sciences USA* 105(50): 19938–19943.

Nakajima Y, van der Goes van Naters-Yasui A, Taylor D, et al. 2001. Two isoforms of a member of the arthropod defensin family from the soft tick, *Ornithodoros moubata* (Acari: Argasidae). *Insect Biochemistry and Molecular Biology* 31(8): 747–751.

Nakajima Y, Ogihara K, Taylor D, et al. 2003. Antibacterial hemoglobin fragments from the midgut of the soft tick, *Ornithodoros moubata* (Acari: Argasidae). *Journal of Medical Entomology* 40(1): 78–81.

Narasimhan S, Montgomery RR, DePonte K, et al. 2004. Disruption of *Ixodes scapularis* anticoagulation by using RNA interference. *Proceedings of the National Academy of Sciences USA* 101(5): 1141–1146.

Narasimhan S, Sukumaran B, Bozdogan U, et al. 2007. A tick antioxidant facilitates the Lyme disease agent's successful migration from the mammalian host to the arthropod vector. *Cell Host & Microbe* 12(2): 7–18.

Nijhof AM, Taoufik A, de la Fuente J, et al. 2007. Gene silencing of the tick protective antigens, Bm86, Bm91 and subolesin, in the one-host tick *Boophilus microplus* by RNA interference. *International Journal for Parasitology* 37(6): 653–662.

Nosek J, Ciampor F, Kozuch O, et al. 1972. Localization of tick-borne encephalitis virus in alveolar cells of salivary glands of *Dermacentor marginatus* and *Haemaphysalis inermis* ticks. *Acta Virologica* 16: 493–497.

Nuttall PA. 2009. Molecular characterization of tick-virus interactions. *Frontiers in Bioscience* 14: 2466–2483.

Nuttall PA, Jones LD. 1991. Non-viraemic tick-borne virus transmission: mechanism and significance. In Modern Acarology, eds Dusbabek, F and Bukva, Vols 3–6. The Hague, The Netherlands: SPB Academic.

Nuttall PA, Labuda M. 2003. Dynamics of infection in tick vectors and at the tick–host interface. *Advances in Virus Research* 60: 233–272.

Nuttall PA, Jones LD, Davies CR. 1991. The role of arthropod vectors in arbovirus evolution. *Advances in Disease Vector Research* 7: 15–45.

Nuttall PA, Jones LD, Labuda M, et al. 1994. Adaptation of arboviruses to ticks. *Journal of Medical Entomology* 31: 1–9.

Pagel Van Zee J, Geraci NS, Guerrero FD, et al. 2007. Tick genomics: the Ixodes genome project and beyond. *International Journal for Parasitology* 37(12): 1297–1305.

Pancholi P, Kolbert CP, Mitchell PD, et al. 1995. Ixodes dammini as a potential vector of human granulocytic ehrlichiosis. *Journal of Infectious Diseases* 172: 1007–1012.

Pedra JH, Narasimhan S, Deponte K, et al. 2006. Disruption of the salivary protein 14 in *Ixodes scapularis* nymphs and impact on pathogen acquisition. *American Journal of Tropical Medicine and Hygiene* 75(4): 677–682.

Ramamoorthi N, Narasimhan S, Pal U, et al. 2005. The Lyme disease agent exploits a tick protein to infect the mammalian host. *Nature* 436(7050): 573–577.

Randolph SE. 2008. Dynamics of tick-borne disease systems: minor role of recent climate change. *Revue Scientifique et Technique* 27(2): 367–381.

Rego RO, Kovár V, Kopácek P, et al. 2006. The tick plasma lectin, Dorin M, is a fibrinogen-related molecule. *Insect Biochemistry and Molecular Biology* 36(4): 291–299.

Rennie L, Wilkinson PJ, Mellor PS. 2000. Effects of infection of the tick *Ornithodoros moubata* with African swine fever virus. *Medical and Veterinary Entomology* 14: 335–360.

Sánchez-Vargas I, Scott JC, Poole-Smith BK, et al. 2009. Dengue virus type 2 infections of *Aedes aegypti* are modulated by the mosquito's RNA interference pathway. *PLoS Pathogens* 5: e1000299.

Senigl F, Grubhoffer L, Kopecky J. 2006. Differences in the maturation of tick-borne encephalitis virus in mammalian and tick cell line. *Intervirology* 49(4): 239–248.

Spielman A, Clifford C. 1979. Human babesiosis on Nantucket Island, USA: description of the vector, Ixodes (Ixodes) dammini, n. sp. (Acarina: Ixodidae). *The Journal of Experimental Medicine* 15: 218–234.

Steele GM, Nuttall PA. 1989. Differences in vector competence of two species of sympatric ticks, *Amblyomma variegatum* and *Rhipicephalus appendiculatus*, for Dugbe virus (Nairovirus, Bunyaviridae). *Virus Research* 14: 73–84.

Sukumaran B, Narasimhan S, Anderson JF, et al. 2006. An *Ixodes scapularis* protein required for survival of *Anaplasma phagocytophilum* in tick salivary glands. *The Journal of Experimental Medicine* 203(6): 1507–1517.

Tokarevich NK, Tronin SS, Blinova OV, et al. 2011. The impact of climate change on the expansion of *Ixodes persulcatus* habitat and the incidence of tick-borne encephalitis in the north of European Russia. *Global Health Action* 4:8448.

Trimnell AR, Davies GM, Lissina O, et al. 2005. A cross-reactive tick cement antigen is a candidate broad-spectrum tick vaccine. *Vaccine* 23(34): 4329–4341.

Work TH. 1958. Russian spring-summer encephalitis virus in India. Kyasanur Forest disease. *Progress in Medical Virology* 1: 248–279.

Zanotto PM, Gould EA, Gao GF, et al. 1996. Population dynamics of flaviviruses revealed by molecular phylogenies. *Proceedings of the National Academy of Sciences USA* 93(2): 548–553.

Zhou J, Ueda M, Umemiya R, et al. 2006. A secreted cystatin from the tick *Haemaphysalis longicornis* and its distinct expression patterns in relation to innate immunity. *Insect Biochemistry and Molecular Biology* 36(7): 527–535.

Zilber A, Clifford C. 1946. Far eastern tick-borne spring-summer encephalitis. *American Reviews of Soviet Medicine* 80 (special supplement).

INDEX

Viral Infections and Global Change, First Edition. Edited by Sunit K. Singh.

INDEX

myriad of disease, 564
nervous system damage, 564
neurobiological diseases, 564
neuronal infection, 565
detection methodology, 562
environmental changes influence, 567
epidemiology, 565–6
as human pathogen, 559–62
human testing methodologies, 566
negative human BDV studies, 567
nucleic acid tests, 563
proventricular dilatation disease (PDD), 558
serology, 562–3
bunyaviridae family
arboviral bunyas
mosquitoes, 482–4
ticks, 484
Bhanja *bunyavirus*, 589–90
characteristics, 478–9
Crimean-Congo hemorrhagic fever virus, 584–8
disease spread, geographic distribution
physical movement, vectors, 485
suitable range expansion, 485
hantavirus, 479–80
Henan fever bunyavirus, 588–9
Keterah bunyavirus, 590
nairovirus, 480–481
non-arboviral bunyas, 484–5
orthobunyaviruses, 481
outbreak prediction, climate
climatic influences, 486–7
risk mapping and predictions, 487–9
phlebovirus, 482

chikungunya virus (CHIKV)
arboviral host range, 70–71
diagnostic virology, 257
VVD, emerging and reemerging, 11–12
cholera, predictive modeling, 245–6
chronic hepatitis, 450 *see also* hepatitis E virus (HEV)
CO_2 concentration elevation, mosquito development
lignin concentrations, 45
negative effects, 45
organic matter decomposition, 45
water, chemical properties, 44
coinfections, tick vector, 608, 611
Colorado tick fever virus (CTFV), 590
coronavirus (CoV)
animal migration, viral spread risk, 162
bat CoV, 402–3

bats, reservoir hosts, 201–3
characteristics, 393
classification, 397
cross-species transmission
alpha-type, 404–5
beta-type, 405–6
gamma type, 407
ecology in diverse hosts, 403
emerging diseases, 399, 402–3
feline infectious peritonitis, 394
genetic diversity, 394
genome organization *vs.* different genera, 395–6
infectious bronchitis virus, 394
murine hepatitis virus, 394
natural reservoirs, 402
outbreaks, anthropogenic factors and climate influence, 407–10
pathogenesis, 397, 399
pathotype, 400–401
phylogenetic tree, 398
severe acute respiratory syndrome, 394
cowpox
characteristics, 380–381
clinical features, 381
epidemiology, 381
as first basis of vaccination, 380
geographical distribution, 382
natural reservoirs, 381
Crimean-Congo hemorrhagic fever virus (CCHFV) *see also* nairovirus
climate transition, 585
diagnostic virology, 257
distribution, ticks, 585, 586
geographical distribution, 583–4
H. marginatum in western Palearctic, 586–7
Hyalomma, causative agent, 583
nairovirus, 481
spread, 588
stages, vector, 584
transmission, 584
Culex pipiens complex, arboviral host range, 66, 67
Culex tarsalis
behavior, 37
life history, 46–7

deforestation and fragmentation, viral zoonotic transfer, 80–81
dengue virus (DENV)
Aedes aegypti, 6
areas at risk, 8